Materials and Technology

VOLUME VI

Volume 1 Air, water, inorganic chemicals and nucleonics

2 Non-metallic ores, silicate industries and solid mineral fuels

3 Metals and ores

4 Petroleum and organic chemicals

5 Natural organic materials and related synthetic products

6 Wood, paper, textiles, plastics and photographic materials

7 Vegetable food products and luxuries

8 Edible oils and fats and animal food products

Index

Editors

L. W. Codd, M. A.

K. Dijkhoff, Chem. Drs.

J. H. Fearon, B. Sc., C. Eng., M.I.E.E.

C. J. van Oss, Ph. D.

H. G. Roebersen, Chem. Drs. †

E. G. Stanford, M. Sc., Ph. D., C. Eng., M.I.E.E., F. Inst. P.

Materials and Technology

A systematic encyclopedia of the technology of materials used in industry and commerce, including foodstuffs and fuels

Based upon a work originally devised by the late Dr. J. F. van Oss

VOLUME VI

Wood, paper, textiles, plastics and photographic materials

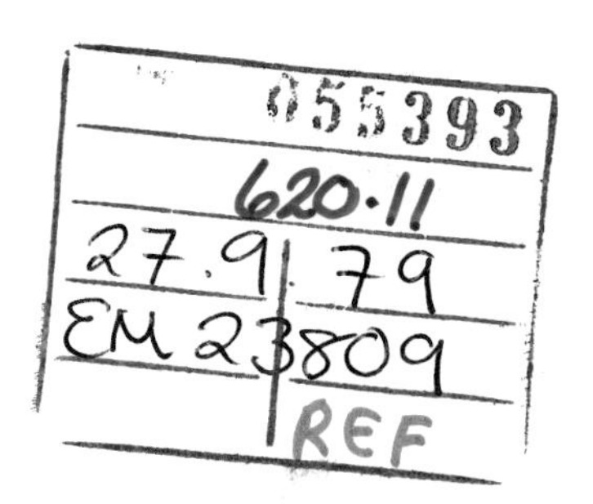

LONGMAN
J. H. DE BUSSY

Longman Group Ltd
London
Associated companies, branches and representatives throughout the world

J. H. de Bussy
172, Herengracht, Amsterdam 1001

First published 1973

Set in 9 on 11 point Times
and printed in the Netherlands
by De Bussy Ellerman Harms N.V., Amsterdam

ISBN 0 582 46206 1

Published in the United States and the Philippines by Barnes & Noble Books under the title

CHEMICAL TECHNOLGY: AN ENCYCLOPEDIC TREATMENT

Contents

Preface

This volume of *Materials and Technology*, like all the other volumes of the encyclopedia is written for people who want to know about materials, including those who have no advanced training in science or technology. Anyone who has studied physics and chemistry to sixth form level in a secondary school will have little difficulty in following the text.

Each chapter has been written by a specialist of group of specialists so as to present a comprehensive and up-to-date account of the subject, which will be of use not only to the general reader but also to those who work with materials in their professional capacity. Thus the encyclopedia wil serve the engineer who requires authoritative guidance on materials with which he is not familiar, the business man in commerce in industry, the craftsman who wants to acquire a sound knowledge of the materials used in his trade and, of course, the materials scientist for whom it will provide a work of reference covering the whole field of his subject in a convenient and manageable form.

It is also intended to serve a useful function in education as a link between formal textbooks and the practical applications of academic learning. University and college lecturers and school teachers should find it a valuable work of reference, as also will their students and pupils, in its description of the industrial uses of the materials and processes covered in their courses.

In order to give the reader a measure of the scale of activity in the various fields, quantitative information is given. Every effort has been made to include the most recent figures available but, as there is considerable variation in the speed of publication of such figures from country to country, and from industry to industry, the figures quoted can only be illustrative of trends. Because the encyclopedia can only be revised infrequently, guidance is given wherever possible to the sources consulted so that the most recent information may easily be sought out.

The reader must be warned that some of the information presented is protected by patents and before using it for commercial purposes careful enquiries should be made. Reference in this book to any particular product does not imply approval or recommandation of that product and the use of trade names in reference to any product does not imply that they are generic terms as they may or may not be protected by registered trademarks.

The publishers have made arrangements whereby they hope to be able to answer any enquiries arising out of the use of the encyclopedia. These should be sent to Longman Group Limited, Longman House, Burnt Mill, Harlow, Essex. They should be marked specifically 'For the attention of the Editor, Materials and Technology Encyclopedia'.

The Editors have set out to make *Materials and Technology* both authoritative and approachable: they believe it is unique in presenting so broad a field of technology in so manageable and understandable a form. They hope it will find a welcome with technologists seeking an insight into specialisations other than their own with which they come into contact professionally; with the practical men who need to know about the materials they must use; with businessmen requiring some knowledge of those technological processes with which their work is ultimately involved; with teachers and lecturers attempting to broaden the interests of their pupils as well as referring them to a uniquely useful source book; and particularly with the students and pupils themselves seeking to discover the practical uses to which science is put.

THE EDITORS

Authors

Dr. R. H. FARMER, Formerly of Forest Products Research Laboratory	Ch. 1 Wood
A. C. WARREN, Assistant Secretary, The National Association of Paper Merchants	Ch. 2 Paper
Dr. D. STARKIE, Formerly Director of Hosiery and Allied Trade Research Association and Research Director, English Rose Ltd.	Ch. 3 Natural Fibres
Dr. D. STARKIE	Ch. 4 Man-made Fibres
Dr. D. STARKIE	Ch. 5 Production of Yarn
Dr. D. STARKIE	Ch. 6 Production of Fabrics
Dr. D. STARKIE	Ch. 7 After-Treatment of Yarns and Fabrics
Dr. D. STARKIE (Editor) and others	Ch. 8 Synthetic Resins and Plastics

Acknowledgements

We are grateful to the following for their kind permission to reproduce the figures listed below:

Figs. 1.1, 1.2, 1.5, 1.6, 1.7, 1.8, 1.9, 1.10, 1.16, 1.20, 1.24, 1.25, 1.26, 1.27, 1.29, 1.33, 1.36, 1.37, 1.38, 1.39, 1.40, 1.41, 1.42, 1.43, 1.44, 1.47: Crown copyright reproduced by courtesy of the Controller of Her Majesty's Stationery Office.
Figs. 1.11, 1.12, 1.13, 1.14: by courtesy of Bowater Paper Corporation Ltd.
Fig. 1.19: reproduced from *British Commonwealth Forest Terminology,* Part 2, Fig. 3, page 225, 1957, published by the Commonwealth Forestry Association.
Figs 1.21, 1.23: by courtesy of Wadkin Ltd.
Fig. 1.22: by courtesy of Stenner of Tiverton Ltd.
Figs. 1.45, 1.46: by courtesy of Hickson's Timber Products Ltd.
Figs. 2.7, 2.12, 2.18, 2.22, 2.23: by courtesy of Bowater Paper Corporation Ltd.
Fig. 2.8: by courtesy of Wiggins Teape Ltd.
Fig. 2.13: by courtesy of Reed International Ltd.
Fig. 2.14: by courtesy of W. Green Son& Waite Ltd.
Figs. 3.1, 3.3, 3.32, 3.33: by courtesy of the Shirley Institute.
Fig. 3.2: plate 16 from *An Introduction to the Botany of Tropical Plants* by Leslie S. Corby, published by Longman, by courtesy of The Public Relations Officer, Sudan Government.
Figs. 3.5, 3.6, 3.35: by courtesy of Lambeg Industrial Research Association.
Fig. 3.8: from *An Introduction to the Botany of Tropical Plants* by Leslie S. Corby, plate 19, published by Longman, by courtesy of Dr. D. Chatterjee, Systematic Botanist of the Indian Agricultural Research Institute.
Fig. 3.10: plate 20 from *An Introduction to the Botany of Tropical Plants* by Leslie S. Corby, published by Longman, by courtesy of Mr. H. Lyon, Sudan Government, Ministry of Agriculture.
Fig. 3.11: plate 17 from *An Introduction to the Botany of Tropical Plants* by Leslie S. Corby, published by Longman, by courtesy of Mr. H. Lyon, Sudan Government, Ministry of Agriculture.

Fig. 3.12: plate 22 from *An Introduction to the Botany of Tropical Plants* by Leslie S. Corby, published by Longman, by courtesy of The Kenya Information Office.
Fig. 3.13: by courtesy of The Tropical Products Institute.
Fig. 3.14: plate 23 from *An Introduction to the Botany of Tropical Plants* by Leslie S. Corby, published by Longman, by courtesy of Mr. G. W. Smith of the Imperial College of Tropical Agriculture.
Figs. 3.15, 3.18, 3.36: by courtesy of the International Wool Secretariat.
Figs. 3.16(a), 3.16(b): by courtesy of the British Wool Marketing Board.
Fig. 3.34: by courtesy of Courtaulds Limited.
Figs. 4.3, 4.6, 4.8, 4.11. 4.12, 4.13, 4.16, 4.17, 4.18, 4.20, 4.21, 4.23, 4.25: by courtesy of Courtaulds Limited.
Fig. 4.29: by courtesy of Imperial Chemical Industries Ltd., Fibres Division.
Fig. 5.4: from the West Yorkshire Folk Museum.
Figs. 5.8, 5.11, 5.15, 5.17, 5.21, 5.22: by courtesy of Courtaulds Limited.
Figs. 5.18, 5.24: by courtesy of Platt International Limited.
Figs. 5.33, 5.34: by kind permission of Heathcoat Yarns and Fibres Limited.
Figs. 6.1, 6.31, 6.32, 6.40, 6.41: by courtesy of Courtaulds Limited.
Figs. 6.2, 6.4: by kind permission of Savio (Great Britain) Limited.
Figs. 6.7, 6.10, 6.29: by kind permission of the Science Museum, London, Crown Copyright.
Fig. 6.13: lent to the Science Museum, London by N. Corah, Leicester.
Fig. 6.19: by courtesy of The Bentley Engineering Group Limited.
Fig. 6.21: by courtesy of Platt International Limited.
Fig. 6.24: by kind permission from the Victoria and Albert Museum, Crown Copyright.
Fig. 6.25: by kind permission of the Victoria and Albert Museum, Crown Copyright.
Fig. 6.33: an International Wool Secretariat picture.
Figs. 6.34. 6.36: by kind permission of Hall's Barton Ropery Company Limited.
Figs. 7.1, 7.2, 7.6, 7.8, 7.9, 7.10, 7.11, 7.12, 7.13, 7.14: by courtesy of Courtaulds Limited.
Fig. 7.4: by courtesy of the International Wool Secretariat.
Fig. 7.5: by courtesy of Sir James Farmer Norton and Company Limited.
Fig. 8.5: by kind permission of B.P. Chemicals (UK) Limited.
Figs. 8.6, 8.8, 8.9, 8.15, 8.22, 8.24: by courtesy of British Industrial Plastics Limited.
Figs. 8.10, 8.14, 8.17: by kind permission of Imperial Chemical Industries Limited.
Fig. 8.18: by kind permission of Shell (U.K.).
Fig. 8.19: by courtesy of British Petroleum Chemicals International Limited.
Fig. 8.20: by courtesy of Bakelite Xylonite Limited.
Figs. 8.21, 8.23: by kind permission of Imperial Chemical Industries, Plastics Division.
Fig. 9.20: by courtesy of Graphic Arts Technical Foundation from *Encyclopedia of Chemical Technology*, Second Edition, Volume 16, p. 504 by Kirk and Othmer. (Interscience, New York).
Fig. 9.21: by courtesy of Philips N.V. Eindhoven.
Fig. 9.23: by courtesy of D. van Nostrand Co. Inc., Princeton, from *Photography Its Materials and Processes*, page 406, by C. B. Neblette.

Symbols and Abbreviations

Symbol	*Name of unit*
*	radioactive
a	are (= 100 m^2) (= 119.599 yd^2)
A	ampere
Å	angström (= 10^{-8} m) (= 0.003 937 01μ in)
a.c.	alternating current
at	technical atmosphere
atm	standard atmosphere (= 101.325 kN/m^2) (= 14.6959 lbf/in^2)
b	bar (10^5 N/m^2) (= 14.5038 lbf/in^2)
Bé	Beaumé's scale
BP	boiling point
Btu	British thermal unit (= 1.05506 kJ)
bu	bushel (= 36.368 7 dm^3)
c	centi (= 10^{-2})
C	coulomb
°C	degree Celsius (= 5/9 (°F – 32)) (temperature value)
cal	calorie (International table)
cd	candela
cg	centigram
Ci	Curie
cm	centimetre (= 0.393 701 in)
c/s	cycles per second
cwt	hundred weight (= 50.8023 kg) (= 112 lb)
d	deci (= 10^{-1})
da	deca (= 10^1)
dag	decagram
d.c.	direct current
degC	degree Celsius (temperature interval)
degF	degree Fahrenheit (temperature interval)
dg	decigram
dm	decimetre
dyn	dyne (= 10^{-5}N) (= 0.224829 x 10^{-5} lbf)
ε	molar extinction coefficient
erg	erg (= 10^{-7}J) (= 0.737 562 x 10^{-7} ft lbf) electron-volt (= 1.60209 x 10^{-19}J)
F	Farad
°F	degree Fahrenheit (= $\frac{9}{5}$ °C + 32) (temperature value)
fl oz	Fluid ounce (= 28.4131 cm^3)
ft	foot (= 0.3048 m) (= 12 in)
ft H_2O	foot water (= 2989.07 N/m^2)
g	gram
G	giga (= 10^9)
gal	UK gallon (= 4.596 litres) (= 4.546 09 dm^3) (cf. US gallon= 3.78541 dm^3)
gr	grain (= 64.798 9 mg)
h	hecto (= 10^2)
H	henry

Symbol	*Name of unit*
ha	hectare (= 10000 m²) (= 2.471 05 acres)
hp	horsepower (= 745.700 W)
Hz	hertz (= 1cycle/sec.)
in	inch (= 2.54 cm)
in Hg	conventional inch of mercury (= 3386.39 N/m²) (= 33.8639 mb)
in H_2O	conventional inch of water (= 249.089 N/m²)
J	joule (= 0.737562 ft lbf)
k	kilo (= 10^3)
K	degree kelvin (= °C + 273)
kcal	kilocalorie
kg	kilogram
kgf	kilogram force (= 9.806 65 N) (= 2.204 62 lbf)
kJ	kilojoule
km	kilometre
kp	kilopond (= kgf)
kW	kilowatt
l	litre (= approx 1 dm³) (= 0.220 0 gal) (= 0.24642 US gal)
lb	pound (= 0.45359237 kg)
lbf	pound force (= 4.44822 N)
lm	lumen
lx	lux (= 1 lm per m²)
m	metre (= 1.09361 yd)
m	milli (= 10^{-3})
M	mega (= 10^6)
mb	millibar (= 100 N/m²)
m.g.d.	million gallons per day
mile	mile (= 1.60934 km)
ml	millilitre
mm	millimetre
mmHg	conventional millimetre of mercury (= 133.322 N/m²) (= 0.0393701 in Hg)
mm H_2O	conventional millimetre of water (= 9.80665 N/m²)
money	£ (= UK pound unless stated to the contrary)

Symbol	*Name of unit*
$ (= US dollar unless stated to the contrary)	
M.P.	melting point
μ	micro (= 10^{-6})
μb	microbar (= 0.1 N/m²)
μin	microinch (= 0.0254 μm) (= 0.000001 in)
μm	micrometre (micron) (39.3701 m in)
μmHg	micron of mercury (= 0.133322 N/m²)
n	nano (= 10^{-9})
N	newton (= 0.224809 lbf)
n mile	international nautical mile (= 1852 m) (*cf.* UK nautical mile = 1853.18 m)
oz	ounce (= 28.3495 g)
Oz apoth	apothecaries' ounce (= 31.1035 g) (= oz tr)
ozf	ounce force (= 0.278014 N)
oz tr	troy ounce (= 31.1035 g) (= oz apoth)
Ω	ohm
p	pico (= 10^{-12})
P	poise (= 0.1 N s/m²) (= 2.08854 x 10^{-3} lbf s/ft²)
Pl	poiseville (= N m²/s)
pH value	measure of acidity/alkalinity
pK_A	measure of strength of acid
pK_B	measure of strength of alkali
p.p.m.	parts per million
p.s.i.	poundweight per square inch (= 6894.76 N/m²) (= 68.9476 mb)
pt	pint (= 0.568261 dm³)
pz	pieze (= 10^3 N/m²)
PVC	polyvinylchloride
q	quintal (= 100 kg)
qt	Imperial quart (1.13652 dm³)
°R	degree Rankine (°F + 459.67)
rad	radian
s	second

Symbol	*Name of unit*
St	stokes (= 10^{-4} m^2/s) (= 558.001 in^2/h)
t	metric ton (= tonne) (= 1000 kg) (= 0.984207 tons) (= 2204.6 lb)
T	tera (= 10^{12})
ton	Imperial ton (= 1016.05 kg) (= 2240 lb) (= long ton) (*cf.* US ton = 2000 lb = short ton)
tonf	ton force (= 9964.02 N)
V	volt
W	Watt (= J/s)
Wb	Weber
wt	weight
w/w	weight for weight
yd	yard (= 0.9144 m)

Table of Chemical Elements

*These elements, like the transuranic elements (see below), have been produced by artificial means and do not occur naturally (at least, not in any appreciable amount).

Element	*Symbol*	*Atomic Number*	*Atomic Weight*
Actinium	Ac	89	227
Aluminium	Al	13	26.98
Antimony	Sb	51	121.76
Argon	A	18	39.944
Arsenic	As	33	74.91
*Astatine	At	85	(210)
Barium	Ba	56	137.36
Beryllium (Glucinium)	Be (Gl)	4	9.013
Bismuth	Bi	83	209
Boron	B	5	10.82
Bromine	Br	35	79.916
Cadmium	Cd	48	112.41
Caesium	Cs	55	132.91
Calcium	Ca	20	40.08
Carbon	C	6	12.01
Cerium	Ce	58	140.13
Chlorine	Cl	17	35.457
Chromium	Cr	24	52.01
Cobalt	Co	27	58.94
Copper	Cu	29	63.54
Dysprosium	Dy	66	162.46
Erbium	Er	68	167.3
Europium	Eu	63	152
Fluorine	F	9	19
Francium	Fr	87	223
Gadolinium	Gd	64	156.9
Gallium	Ga	31	69.72
Germanium	Ge	32	72.60
Gold	Au	79	197
Hafnium	Hf	72	178.5
Helium	He	2	4.003
Holmium	Ho	67	164.94
Hydrogen	H	1	1.008
Indium	In	49	114.76
Iodine	I	53	126.91
Iridium	Ir	77	192.2
Iron	Fe	26	55.85
Krypton	Kr	36	83.80
Lanthanum	La	57	138.92
Lead	Pb	82	207.21
Lithium	Li	3	6.940
Lutetium (Cassiopeium)	Lu (Cp)	71	174.99
Magnesium	Mg	12	24.32
Manganese	Mn	25	54.94
Mercury	Hg	80	200.61
Molybdenum	Mo	42	95.95
Neodymium	Nd	60	144.27
Neon	Ne	10	20.183
Nickel	Ni	28	58.7
Niobium (Columbium)	Nb (Cb)	41	92.91
Nitrogen	N	7	14.008

Element	*Symbol*	*Atomic Number*	*Atomic Weight*
Osmium	Os	76	190.2
Oxygen	O	8	16
Palladium	Pd	46	106.7
Phosphorus	P	15	30.974
Platinum	Pt	78	195.23
Polonium (Radium F)	Po	84	210
Potassium	K	19	39.100
Praseodymium	Pr	59	140.92
*Prometheum	Pm	61	(145)
Protactinium	Pa	91	231
Radium	Ra	88	226.05
Radon (Niton)	Rn (Nt)	86	222
Rhenium	Re	75	186.31
Rhodium	Rh	45	102.91
Rubidium	Rb	37	85.48
Ruthenium	Ru	44	101.1
Samarium	Sm	62	150.43
Scandium	Sc	21	44.96
Selenium	Se	34	78.96
Silicon	Si	14	28.09
Silver	Ag	47	107.873
Sodium	Na	11	22.997
Strontium	Sr	38	87.63
Sulphur	S	16	32.066
Tantalum	Ta	73	180.88
*Technetium	Tc	43	(98.91)
Tellurium	Te	52	127.61
Terbium	Tb	65	158.9
Thallium	Tl	81	204.39
Thorium	Th	90	232.1
Thulium	Tm	69	168.94
Tin	Sn	50	118.70
Titanium	Ti	22	47.90
Tungsten (Wolfram)	W	74	183.92
Uranium	U	92	238.07
Vanadium	V	23	50.95
Xenon	Xe	54	131.3
Ytterbium	Yb	70	173
Yttrium	Y	39	89
Zinc	Zn	30	65.38
Zirconium	Zr	40	91.22

TRANSURANIC ELEMENTS

Atomic Number	*Element*	*Symbol*
93	Neptunium	Np
94	Plutonium	Pu
95	Americium	Am
96	Curium	Cm
97	Berkelium	Bk
98	Californium	Cf
99	Einsteinium	E
100	Fermium	Fm
101	Mendelevium	Mv
102	Nobellium	No

Units Conversion Table

This table has been compiled with reference to units and magnitudes appropriate to the technical and commercial operations with which these volumes are concerned, and in the light of the present transitional situation. It therefore includes many units which, though widely employed in imperial or metric usage, fall outside the strict Système International des Unités.

IMPERIAL/US AND METRIC/SI

Length

1 mile	=	1.6093	km	1 km	=	0.6214	mile
1 yd	=	0.9144	m	1 m	=	1.0936	yd
1 ft	=	0.3048	m	1 cm	=	0.3938	in
1 in	=	2.54	cm	1 mm	=	39.37	'thou'
1 mil or 'thou' (1/1000 in)	=	0.0254	mm				

Area

1 mile^2	=	2.590	km^2	1 ha	=	2.471	acres
	or	259	ha		*or*	0.386	mile^2
1 acre	=	4047	m^2	1 km^2	=	247.1	acres
	or	0.4047	ha	1 m^2	=	1.196	yd^2
1 yd^2	=	0.8361	m^2	1 cm^2	=	0.1550	in^2
1 ft^2	=	0.0930	m^2				
1 in^2	=	645.2	mm^2				

Volume, Capacity

1 yd^3	=	0.7646	m^3	1 m^3	=	1.3079	yd^3
1 ft^3	=	0.02832	m^3		*or*	35.315	ft^3
1 in^3	=	16.387	cm^3	1 dm^3	=	0.0353	ft^3
1 gal	=	4.546	l	1 cm^3	=	0.0610	in^3
1 US gal	=	3.785	l		*or*	0.0351	fluid oz
1 pint	=	0.5682	l	1 l	=	0.220	gal
1 fluid oz	=	28.413	cm^3		*or*	1.760	pints
					or	0.2642	US gal

Velocity

1 mile/h	=	1.6093	km/h	1 km/h	=	0.6214	mile/h
1 ft/min	=	0.00508	m/s	1 m/s	=	3.2808	ft/s
1 ft/s	=	0.3048	m/s	1 mm/s	=	0.0394	in/s
1 in/s	=	25.40	mm/s				

IMPERIAL/US AND METRIC/SI

Mass

1 ton	=	1016	kg	1 tonne	=	2204.7	lb
(2240 lb)				(1000 kg)	*or*	0.9842	ton
1 short ton	=	907.19	kg		*or*	1.1023	short ton
(2000 lb)				1 quintal	=	220.47	lb
1 cwt	=	50.802	kg	(100 kg)			
(112 lb)				1 kg	=	2.2047	lb
1 stone	=	6.350	kg	1 g	=	0.0353	oz
(14 lb)							
1 lb	=	0.4536	kg				
1 oz	=	28.349	g				

Mass per Unit Length

1 ton/mile	=	631.3	kg/km	1 tonne/km	=	1.584	ton/mile
1 lb/yd	=	0.4961	kg/m		*or*	1.774	short ton/mile
1 lb/ft	=	1.488	kg/m	1 kg/m	=	2.016	lb/yd
				1 tonne/m	=	0.900	ton/yd

Length per Unit Mass

1 yd/lb	=	2.016	m/kg	1 m/kg	=	0.490	yd/lb
1 in/oz	=	0.7005	cm/g	1 cm/g	=	0.0721	in/oz

Mass per Unit Area

1 ton/mile^2	=	392.3	kg/km^2	1 tonne/ha	=	892.2	lb/acre
1 ton/acre	=	0.2511	kg/m^2				
1 lb/ft^2	=	4.882	kg/m^2	1 kg/m^2	=	0.2048	lb/ft^2
1 lb/in^2	=	70.31	g/cm^2	1 kg/cm^2	=	14.22	lb/in^2

Area per Unit Mass (Specific Surface)

1 mile^2/ton	=	2549	m^2/kg	1 ha/tonne	=	2.511	acre/ton
1 yd^2/ton	=	0.823	m^2/tonne	1 m^2/kg	=	0.542	yd^2/lb
1 ft^2/lb	=	0.205	m^2/kg				

Volume per Unit Mass (Specific Volume)

1 ft^3/ton	=	0.0279	l/kg	1 m^3/tonne	=	1.332	yd^3/ton
1 ft^3/lb	=	62.428	l/kg				
1 gal/lb	=	10.022	l/kg	1 l/kg	=	0.995	gal/lb
1 in^3/lb	=	36.127	cm^3/kg				

Mass Rate of Flow

1 ton/h	=	1016	kg/h	1 tonne/h	=	0.984	ton/h
1 lb/h	=	0.454	kg/h				
1 lb/s	=	0.454	kg/s	1 kg/s	=	2.2047	lb/s

Volume Rate of Flow

1 ft^3/s	=	28.32	l/s	1 l/s	=	0.353	ft^3/s
(1 cusec)	*or*	1019	m^3/h				
1 gal/min	=	272.76	l/h	1 l/h	=	0.220	gal/h
1 US gal/min	=	227.10	l/h		*or*	0.264	US gal/h

Density

1 ton/yd^3	=	1.329 tonnes/m^3	1 tonne/m^3	=	0.7525 ton/yd^3
1 lb/ft^3	=	16.018 kg/m^3	1 kg/m^3	=	0.0624 lb/ft^3
1 lb/in^3	=	27.680 g/cm^3	1 g/cm^3	=	0.0361 lb/in^3
1 lb/gal	=	0.10 g/cm^3			
1 oz/gal	=	6.236 g/l	1 g/l	=	0.1600 oz/gal

Force

1 pound force (lbf)	=	0.454 kgf	1 kgf (kp)	=	2.205 lbf
	or	4.449 N			
1 poundal (pdl)	=	0.1383 N	1 N	=	0.225 lbf
				or	7.233 pdl
1 ton force (tonf)	=	1.016 tf	1 tonne force (tf)	=	0.984 tonf
1 tonf	=	9.964 kN	1 kN	=	0.1004 tonf

Pressure

1 lbf/in^2	=	0.0703 kgf/cm^2	1 kgf/cm^2	=	14.22 lbf/in^2
	or	6.8947 kN/m^2	1 kN/m^2	=	0.145 lbf/in^2
1 lbf/ft^2	=	47.880 N/m^2			
1 tonf/ft^2	=	10.936 tf/m^2	1 tf/m^2	=	0.914 tonf/ft^2
	or	0.1072 N/mm^2			
1 atm (760 mm Hg)	=	101.325 kN/m^2	1 mb	=	2.088 lbf/ft^2
1 in Hg	=	3386.4 N/m^2			0.019 lbf/in^2
1 in H_2O	=	249.1 N/m^2			

Energy, Work, Heat

1 Btu (0.252 kcal)	=	1055 J	1 kcal	=	3.9683 Btu
1 therm (100000 Btu)	=	105.51 x 10^6 J	1 thermie (10^{6}cal$_{15}$)	=	0.309 therm
1 ft lbf	=	1.3558 J	1 J	=	0.7375 ft lbf
	or	3.766 x 10^{-7} kWh	1 kWh	=	2.6552 ft lbf
1 ft pdl	=	0.0421 J			
1 hph	=	2.6845 x 10^6 J	1 J	=	37.25 x 10$^-$ hph

Power

1 hp	=	1.0139 met hp	1 met hp	=	0.9863 hp
1 hp	=	0.7457 kW (1 cv)	1 kW	=	1.3410 hp
1 ft lbf/s	=	1.3558 W	1 kW	=	737.56 ft lbf/s

Various Heat Factors

1 Btu/h	=	0.293 W	1 W	=	3.4121 Btu/h
1 Btu/lb	=	2326 J/kg	1 J/kg	=	0.4299 x 10^{-3} Btu/lb
1 therm/gal	=	2.321 x 10^4 J/cm^3	1 J/cm^3	=	4.3089 therm/gal
1 Btu/ft^3	=	0.0372 J/cm^3	1 J/cm^3	=	26.839 Btu/ft^3
1 Btu/lb deg F	=	4.187 x 10^3 J/kg deg C	1 J/kg deg C	=	0.239 x 10^3 Btu/lb deg F
1 Btu/ft^{2}h	=	3.1546 x 10^{-4} W/cm^2	1 W/cm^2	=	3160 Btu/ft^2 h
1 Btu/ft^2 h °F	=	5.6783 x 10^{-4} W/cm^2 °C	1 W/cm^2 °C	=	1760 Btu/ft^2 h °F
1 Btu/ft h °F	=	1.7307 x 10^{-2} W/n °C	1 W/cm °C	=	57.78 Btu/ft h °F

General references

Although a list of references is included in most chapters, additional information may be found in the following major reference works.

L. F. and M. FIESER. *Advanced organic chemistry*
New York, Reinhold, 1961

L. I. FINAR. *Organic Chemistry*, Vol. 1
London, Longman, 1973 (6th edition)

R. E. KIRK and D. F. OTHMER. *Encyclopedia of Chemical Technology*
New York, Wiley, 1963-69 (2nd edition)

E. H. RODD (Ed.). *Chemistry of carbon compounds. A modern comprehensive treatise*, 5 vol. in 10 parts
Amsterdam, Elsevier, 1951-62

N. I. SAX. *Dangerous properties of industrial materials*
New York, Reinhold, 1968 (3rd edition)

Most of statistical data in this work has been taken from the following references. The most recent editions of these references should be consulted for more up to date statistics. In addition in most countries there is a government bureau of statistics which publishes annual volumes containing data on production, consumption and trade.

Periodicals (annual) of international production and trade statistics:

Industrie and Handwerk Fachserie D.
Reihe 8 Industrie des Auslandes
II Verarbeitende Industrie
Stuttgart, W. Kohlhammer.

Statistical Yearbook of the United Nations.
New York.

Commodity Yearbook.
New York, Commodity research bureau Inc.

Periodicals (weekly and monthly) including international production and trade statistics.

Chemical Age (Incorporating *Chemical Trade Journal*).
London, Benn Brothers.

Chemische Industrie.
Düsseldorf, Verlag Handelsblatt G.m.b.H.

Periodicals (annual) of production and trade statistics:

United Kingdom:

Annual statement of the trade of the United Kingdom. Vol II and III.
Imports and exports by commodity.
London, Her Majesty's Stationery Office.

Census of production of the Board of trade of the U.K. (a four year periodical).
London, H.M.S.O.

Annual Abstracts of Statistics.
London, H.M.S.O.

Accounts relating to Trade and Navigation of the U.K. (December number).
London, H.M.S.O.

U.S.A.:

Statistical Abstracts of the United States.
U.S. imports of merchandise for consumption.
U.S. exports of domestic merchandise.
U.S. Department of commerce.
Washington, Bureau of the census.

Chemical and Engineering News (the first September number). An American Chemical Society publication.
New York, Reinhold Corp.

CHAPTER 1

Wood

1.1 WOOD IN GENERAL

In any discussion of wood as a material it is natural that a good deal of emphasis is placed on its less favourable properties – its variability, the presence of defects such as knots, its susceptibility to decay and insect attack, its liability to swell and shrink, etc. – because it is only through adequate knowledge of these characteristics that proper allowance can be made for them and wood can be used to the best advantage. However, in spite of these limitations wood is a very valuable and versatile material with an extremely wide range of uses in building, furniture, packaging, boat building, transport vehicles and many other spheres. It is therefore worth while at the outset to list some of the favourable features of wood which enable it to hold its place in competition with other materials. The following are among its more important assets:

Attractive and varied appearance.
High strength/weight ratio compared with most materials.
Ease of machining, working with hand tools, and jointing with nails, screws, metal connectors, or glues.
Durability, either natural or enhanced by treatment with preservatives.
Resistance to corrosive influences and effects of weather.
Good thermal insulation and acoustic properties.
Compatibility with paints and other finishes.
Wide range of properties available by selection of suitable species of tree.
Low cost.

Two other points of a general nature should be emphasized at this stage. The first is that, unlike our mineral resources, some of which threaten to become exhausted within the foreseeable future, wood is a renewable material. As supplies are used up they are constantly being replaced by the continued growth of trees and replanting of felled areas. It is beyond the scope of the present chapter to consider problems of forest management, and it will be sufficient to state that by implementation of sound management policies – that is, by proper control of felling and planting and by care of the growing crop – a forest area will continue to yield a sustained output of timber for an indefinitely long time. Not all the world's forests are yet under proper management plans, and it is the objective of forest departments in the various territories to control forestry operations so that the reserves are not depleted. In this way timber supplies on a world-wide basis will be assured in perpetuity.

The second point is that wood differs from most other materials in that it is an organic material produced as a result of living processes. This fact is of fundamental importance in the consideration of its structure and properties, which are determined by its biological function and can only be modified to a limited extent. The fine structure of wood is much more complex than that of man-made materials and, being a natural product, its properties are subject to greater variability than are those of materials manufactured under controlled conditions. This variability is additional to the variation between timbers of different species. Allowance for natural variability, for example in strength properties, must be made in the design of timber structures and components.

Another consequence of its natural origin is that timber is only available in limited shapes and sizes. Furthermore, its properties are highly dependent upon the direction in which they are measured in relation to the longitudinal axis of the tree (the grain direction of the wood). These limitations are overcome to a large extent in the production of the various wood-based materials described in section 1.5. Some of these materials have, in addition, certain other improved properties.

Finally, wood, in common with most other materials, is subject to deterioration when used in unfavourable conditions. In the case of wood the principal causes of deterioration are biological in nature (fungal decay and attack by wood-boring insects), but it will be seen in section 1.6 that such deterioration can often be avoided by control of the environment or it can be effectively eliminated by treatment of the wood with preservatives. Wood is more resistant than many other materials to attack by chemicals and corrosive atmospheres. The existence of wooden beams and framing and interior woodwork in buildings that are many hundreds of years old provides ample evidence of its long life and resistance to deterioration under conditions favourable to its use.

1.1.1 History

Wood has been used by man from the very earliest times because it is a readily available and easily worked natural material. Its more important early uses included the building of huts for shelter, construction of boats for fishing and travel, primitive ploughs for tilling the soil, wagons, and weapons of war (spears, bows and arrows), while in terms of quantity probably its major use was for fuel for heating and cooking.

All the great maritime races of the past, from the Phoenicians and Vikings to the later European nations, have depended upon plentiful supplies of timber for the construction of their boats and ships. The great battle fleets of Drake, Nelson and others and those of their enemies, and also the ships of the explorers of the fifteenth and sixteenth centuries such as Columbus, Vasco da Gama, Magellan and Cabot were built of wood. In this sense the availability of suitable timber in the countries concerned and the skill of their people in building ships may be said to have played a part in influencing the course of events in this period. These voyages of exploration and conquest would not have been possible unless the ships were available in which to make them.

Another very important use of wood in the Middle Ages and in later periods was in the building of houses, churches and other buildings. The roofs of churches and other large buildings were commonly of timber construction and timber-framed houses were very widely built, many of them being still in use. Wood is seen at its best in the beautiful Gothic carving found in so many cathedrals and churches, as well as in the domestic furniture associated with the names of Sheraton, Chippendale and others.

The development of iron and steel production in the 18th and 19th centuries had two important consequences for the timber-producing countries and timber-using industries. Timber was replaced to an increasing extent by iron and steel in some of its major uses, particularly in constructional work and ship-building, but in spite of this there was no reduction in the demand for timber in the industrialized countries. With expanding populations and new industrial uses, e.g. for railway sleepers, mining timber, etc., their timber requirements continued to increase. In Great Britain the timber needs were far too great to be met from home production and importation on a large scale became necessary.

Secondly, the development of the iron and steel industries led to a very large demand for charcoal which was used for iron smelting. This was the cause of serious devastation of forests in a number of countries in the period up to about the end of the 18th century, when smelting with coke was introduced and the need for charcoal diminished. The wholesale destruction of forests which resulted from the need for charcoal for smelting was further intensified by a growing demand for other wood products. In particular potash was obtained from wood ashes and was used in the glass and textile industries and in manufacture of soap, and pitch (derived from wood distillation) was used as a sealer and preservative in shipbuilding. The woodlands of Britain and of some other European countries have never recovered from the overcutting that took place to satisfy the needs for these products as well as to meet timber requirements.

At the present time the use of wood in its natural state appears to be declining slowly in Britain and in some other countries. In some fields it is being replaced by alternative materials while in others its use continues to expand. However, there is a marked increase in output of products in which wood is mechanically or chemically modified, especially board materials and pulp and paper products. Such wood-based products are expected to constitute the major outlets of wood utilization in the future.

1.1.2 Structure

For an understanding of the properties of wood as a material it is necessary to consider first the manner of its formation in the tree.

A tree grows as a result of two separate processes. First, elongation of the stem and branches occurs as a result of growth at their ends. This is called primary or apical growth and the ultimate form of the mature tree, which varies from species to species, is determined by the increase in length and the extent to which branching takes place at the growing points.

In addition to the primary growth, increase in thickness of the stem and branches (secondary growth) also takes place throughout the life of the tree and is of greater importance than the primary growth from the point of view of the features of the wood that is formed. Secondary growth results from the activity of a thin growing layer (the cambium), one cell in thickness, which surrounds the wood of the stem and lies between it and the bark. Each cambial cell is capable of dividing with formation of a pair of cells, one of which remains as a cambial cell, while the other eventually forms part of the xylem or wood or, less frequently, the phloem or bark. This process of cell division is repeated many times during the growing season, so that each year a fresh band of wood is built up around the stem, beneath the bark, and the stem and branches continually increase in thickness during the whole life of the tree. The amount of bark formed is comparatively small in relation to the wood and, furthermore, the phloem cells are compressed by bark pressure and portions of the outer bark may be lost from time to time. For these reasons the increase in thickness of the bark is relatively small and the bark only makes up a small fraction of the total volume of the stem.

a. Softwoods and hardwoods. Commercial timbers fall into two main classes, generally known as softwoods and hardwoods. These terms, however, are somewhat misleading because they imply a division based on a physical property of the timber, whereas in fact it rests purely on a botanical distinction. The trees that produce these two different classes of timber are themselves quite distinct. The softwoods are derived from coniferous or cone-bearing trees which usually have needle-shaped leaves and seeds that are not enclosed in a seed-case (gymnosperms). The hardwoods are derived from the broad-leaved trees having seeds that are enclosed in a seed-case (angiosperms), and are found in great variety in many parts of the world. There are important differences in structure between softwoods and hardwoods which affect the properties of the timbers, but it should be noted that the physical softness or hardness of a timber is not a reliable guide to its classification. There are some timbers belonging to the class of hardwoods, e.g. lime, willow, poplar, obeche, balsa, which are softer and lighter in weight than many of the softwoods, and conversely a few of the softwoods are relatively hard and heavy.

b. Anatomical features of wood. In cross-section the stem of a tree is seen to consist of concentric growth rings, or annual rings, which result from the production of a fresh layer of woody tissue during each year's growth. Wood cells are tubular in shape and the cells that are laid down in the early part of the growing season are thin-walled and have a wide lumen, or central cavity, so the 'spring wood' or 'early wood' has a relatively low density. Later in the season the increase in thickness of the stem is slower and the cell walls are thicker. The 'summer wood' or 'late wood' is therefore higher in density than the spring wood, and is also often darker in colour. The prominence of the annual rings depends upon the contrast in colour and density between spring wood and summer wood.

The annual rings, although always present, are not equally distinct in the wood of all trees. The conifers generally show clearly marked annual rings, but the latter are much less easily visible in many of the hardwoods, and may be very difficult to distinguish in the wood of trees grown in tropical climates where the seasonal differences are small.

Another feature which is clearly visible in cross-section is the distinction between sapwood and heartwood. The wood of the outermost growth rings of the stem of a tree (sapwood) is light in colour and performs certain physiological functions, of which the most important is sap conduction. After a few years' growth, however, there is a more or less abrupt change in colour of the wood of many species of tree, the wood in the centre of the stem (heartwood) being darker. The heartwood no longer participates in sap conduction or in the storage of reserve food, and may be regarded as dead tissue. Its dark colour in many species is due to the deposition of extraneous materials such as resins, tannins and colouring matters. In some trees (e.g. beech, ash, spruce) there is no clear, visible distinction between sapwood and heartwood. Sapwood differs from heartwood in some of its technical properties, e.g. moisture content, resistance to decay and insect attack, permeability and colour, but there is not usually any appreciable difference between the two in their mechanical properties.

Fig. 1.1 Section of stem of a coniferous tree.

c. Fine structure. The fine structure of wood, which has been explored in detail with the aid of the optical microscope and the electron microscope, is highly complex and cannot be described in detail here. The following brief outline will serve to indicate the nature of the material and the way in which differences in structure which exist between different species of wood influence their individual properties.

Wood, like all biological materials, is composed of cells. About 90 to 95% of these are highly elongated and lie more or less parallel to the axis of the stem of the tree, i.e. in the direction of the grain. They are cemented together by the intercellular material (mainly lignin) which imparts rigidity to the structure. In softwoods, these cells are called *tracheids* (though they are often referred to as fibres) and are up to 3 to 4 mm long. They are responsible for the strength properties of the material as well as serving as conducting tissue for the transport of fluids. In hardwoods, they are termed *fibres* and are generally about 1 mm long, and they again form the main supporting tissue, but conduction takes place through a parallel system of *vessels* or pores. In both softwoods and hardwoods, *parenchyma* cells, which provide food storage, occur, mainly in the form of *rays*, which are horizontal strands of cells running across the grain in a radial direction. The structure and arrangement of the various types of cell differ widely between different wood species and greatly affect the texture and the strength. They are also of value for identifying wood species. The thickness of the cell walls, particularly of fibres and tracheids, differs markedly between species and is largely responsible for

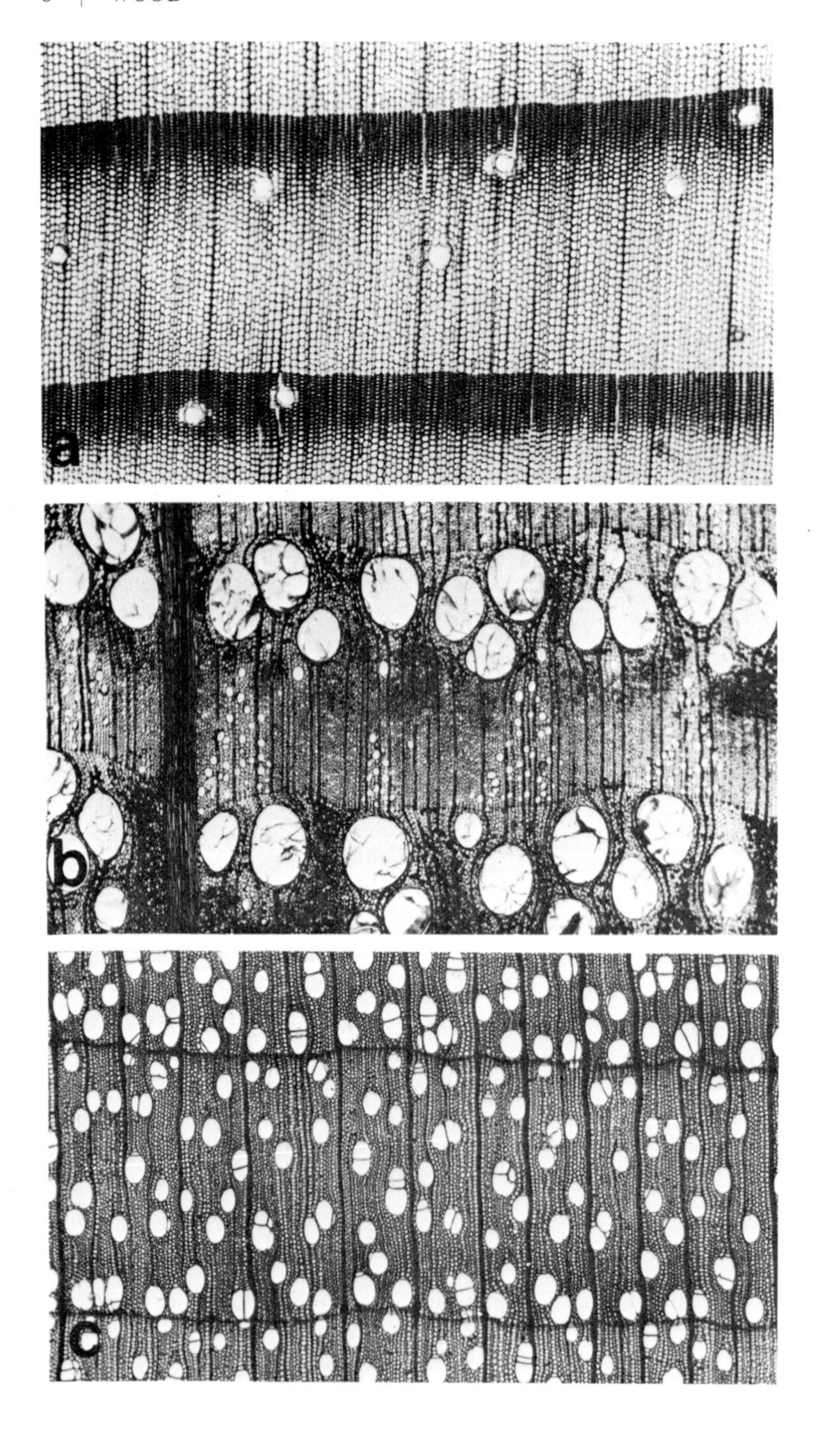

Fig. 1.2 Photomicrographs of cross-sections showing the structure, of (a) a softwood (Scots pine), (b) a ring-porous hardwood (oak), (c) a diffuse-porous hardwood (birch). (Magnification × 25 approx.)

the differences in density between timbers, those with thick cell walls being higher in density than the thinner-walled species.

The presence of vessels in hardwoods and their absence in softwoods is an important structural difference between the two types. In some hardwoods, e.g. beech and birch, the vessels are of uniform size across the growth rings and these timbers are known as *diffuse-porous* woods. In contrast, in certain timbers the pores in the spring wood are considerably larger than those in the summer wood zone. Such timbers, examples of which are oak, elm and teak, are known as *ring-porous* woods. These different structures are clearly visible in photomicrographs of cross-sections.

1.1.3 Chemical composition

Wood substance is an organic material, composed mainly of carbon, hydrogen and oxygen, together with small amounts of nitrogen, and of mineral elements (principally calcium, potassium and magnesium) which are found in the ash when wood is burned. The elementary composition of the dry wood of all species is fairly constant and approximates to C 49 to 50%, H 6.1%, O 44 to 45%. The principal organic components of wood are cellulose, hemicelluloses and lignin. All of these form large, complex molecules which are very closely associated with one another, and may even be chemically combined in the wood.

As explained in the preceding section, the wood of any species of tree consists essentially of a large number of fibres (together with a small proportion of cells of other types), which are bonded together to form a solid and rigid structure. These fibres are hollow, cylindrical structures, about 1 to 4 mm in length, the walls of which are composed mainly of cellulose and its associated polysaccharides (hemicelluloses), while the bonding material which unites them into a solid mass consists essentially of lignin. The ultra-fine structure of the fibre walls is complex – they are composed of

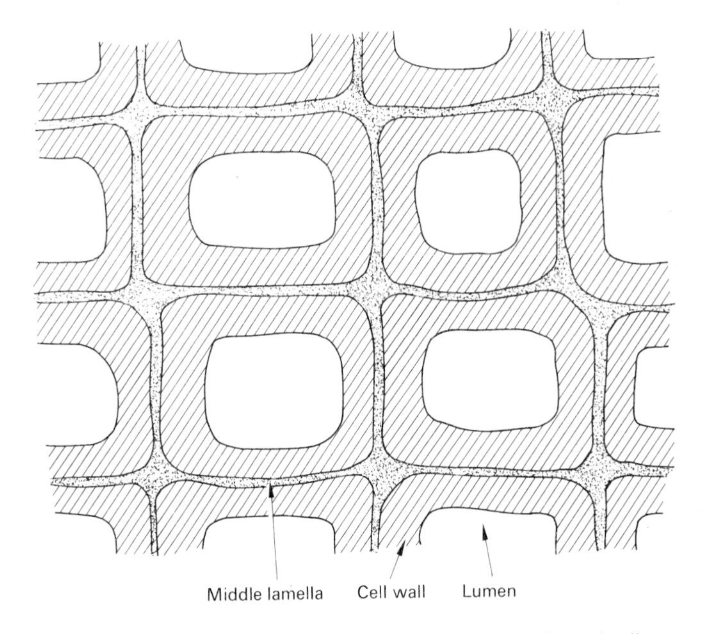

Fig. 1.3 Diagram of enlarged cross-section of a group of wood cells.

smaller units, called microfibrils, the orientation of which differs in different parts of the cell wall – but the important point to note is that the cell wall material consists largely of cellulose and other polysaccharides (though there is some penetration of lignin into them), while the intercellular substance is mainly lignin. These two constituents are well fitted by their chemical nature for their respective roles in the structure of woody tissue.

Cellulose is a long-chain linear polymer, built up of glucose units linked together in such a way as to form relatively straight chains. The number of units in the chain molecule of native cellulose is probably in the region of 10000. These chains lie substantially parallel to one another in a regular arrangement, forming bundles from which the microfibrils of the cell wall are built up. It is apparent that cellulose is well suited, by virtue of its molecular structure, to build up parallel bundles of chains, and thus to fulfil its function as a fibre-forming material, and it is, in fact, the basic fibre-forming substance in all the important vegetable fibres (cotton, flax, hemp, jute, sisal, etc.).

The hemicelluloses, which are closely associated with cellulose, are similar to it in nature but form shorter and less regular molecular chains which may be built up from a number of different sugar units. Lignin, on the other hand, does not possess a linear molecule, but is an amorphous material having a three-dimensional polymeric structure. It thus has the properties necessary for it to perform its function as a bonding or cementing material between the fibres. It will be seen from the above that the tensile strength of wood along the direction of the grain is a result of the strength properties of cellulosic fibres of which it is largely made, while its rigidity is due to the bonding together of the fibres by the cementing material, lignin, and the stiffening effect of the lignin present in the cell wall.

The proportions in which the three basic components of wood substance, cellulose, hemicelluloses and lignin, are present do not vary very much from species to species, though there is a broad difference in composition between softwoods and hardwoods, the softwoods containing rather more lignin and less hemicellulose than the hardwoods. In general, dry wood substance contains about 40 to 50% of cellulose, 15 to 30% of hemicelluloses and 20 to 35% of lignin. It should be noted that many tropical hardwoods differ from the temperate hardwoods in having higher lignin contents, which may be as high as, or even higher than, those of the softwoods. There are also differences between softwoods and hardwoods in the composition of the hemicelluloses and of the lignin present in them.

The analysis of wood presents difficulties because, owing to the very close association between the components, it is not generally possible to isolate them quantitatively, and indirect methods of analysis often have to be used. The analytical data arrived at depend to some extent on the procedures by which they are obtained. The gravimetric

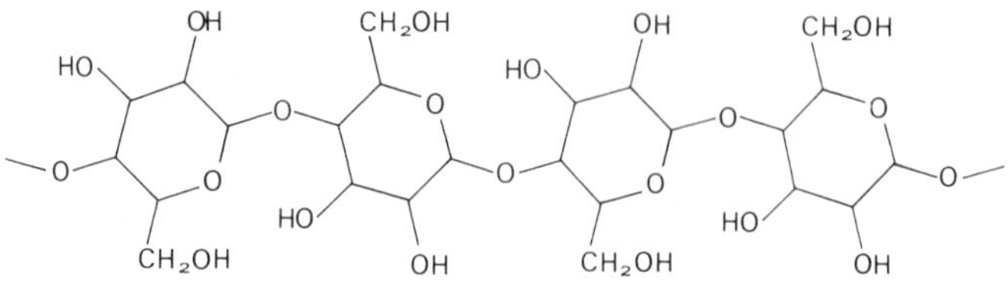

Fig. 1.4 Spatial arrangement of portion of a cellulose molecule (after Preston).

methods formerly used have been largely superseded, at least in investigations on the polysaccharides of wood, by chromatographic procedures.

It may be noted that the polysaccharides in wood, and to a lesser extent the lignin, have in their molecules large numbers of hydroxyl groups. This is a point of some importance because these groups are responsible for the affinity of wood for moisture. The wood-moisture relationships will be discussed in greater detail later (see section 1.1.6).

Extractives. The principal components of the wood cell wall – polysaccharides and lignin – are essentially similar in all species of wood, although certain differences exist in the detailed chemical structure of these components and in the proportions in which they are present. In addition to these structural components, however, all woods also contain smaller amounts of minor or extraneous components, which are much more diverse in their chemical nature. These extraneous components do not form part of the cell wall structure but are probably present as cell contents or deposited within the cell walls. They may be extracted to a large extent by means of inert solvents without destroying the structure of the wood, and for this reason they are often termed extractives. They vary in amount from less than 1% to 20 to 30% or more by weight of the wood, and they include many different classes of organic compounds ranging in complexity from relatively simple molecules, such as sugars and phenols, to highly complex colouring matters, tannins, resins, alkaloids, terpenes, etc. Few generalizations concerning them can therefore be made, but these components are of considerable interest because of the influence they have on some of the properties (e.g. colour, natural durability, toxic or irritant properties) of the woods in which they occur. A few of them, notably the tannins, and turpentine and rosin, also form the basis of important industries.

In most species of tree the extractives are found principally in the heartwood and are only present in much smaller amounts in the sapwood. The sapwood is therefore generally deficient in those properties for which extractives are responsible. For example, in many timbers the heartwood is distinguished from the sapwood by its darker colour and its greater resistance to decay.

The nitrogen content of most woods is very small, being generally less than 0.1% in species grown in temperate climates, and often slightly higher in tropical species. The nitrogen is present largely as protein, which is an essential constituent of all living matter, together with free amino acids which are found mainly in the sapwood. There is some evidence that the wood-boring insects which attack some species of wood are dependent on the nitrogen compounds in the wood for their nutritional requirements, and that the amount of nitrogen present may be a controlling factor which determines whether or not a wood will support the growth of these insects.

The mineral content of woods grown in temperate regions is generally less than 1% by weight, but may be greater in some tropical timbers. The components most commonly present are calcium, potassium and magnesium, combined in the form of carbonates, phosphates, silicates and sulphates and as salts of organic acids, but less common elements are found in certain species. Wood ash has a high content of potash and was at one time used in the manufacture of glass and soap, and as a fertilizer.

Most woods contain only very small amounts of silica, but this substance occurs in larger amounts in a few tropical hardwoods in which it is present as discrete grains, visible under the microscope. It is of some technical importance because of its abrasive action on tools; it is found that when a wood contains more than about 0.5% of silica it

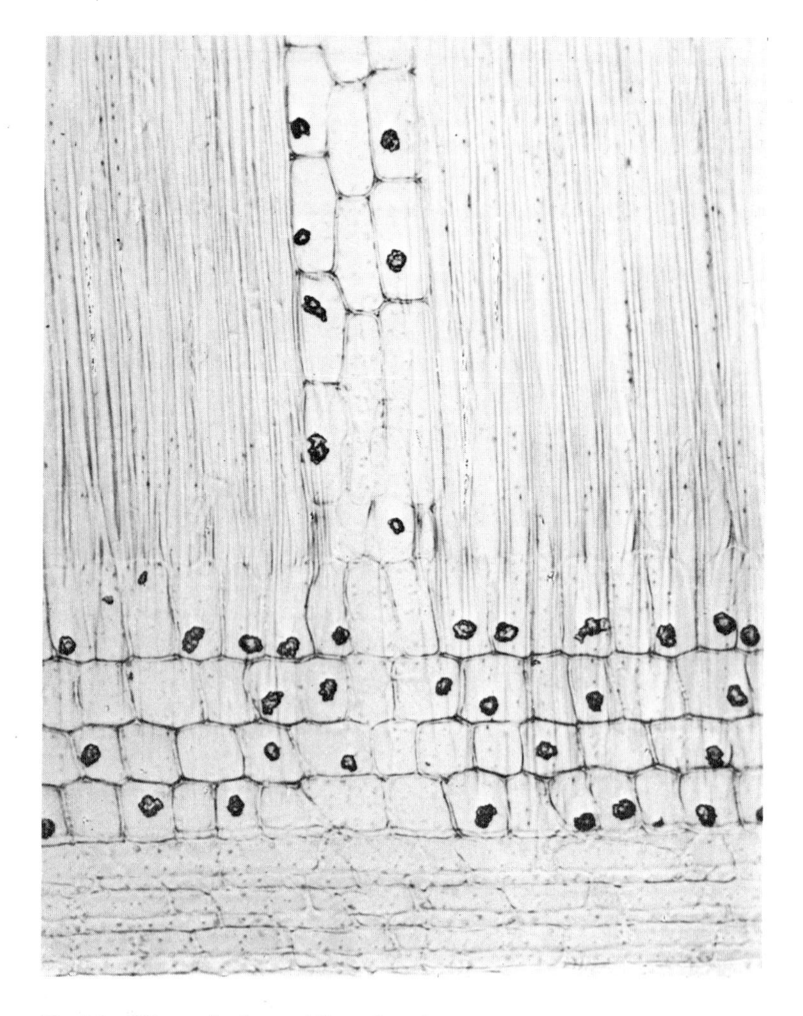

Fig. 1.5 Silica grains in wood (basralocus).

causes unduly rapid blunting of cutting edges. Special measures have to be taken when sawing siliceous timbers.

1.1.4 Physical properties

Owing to the anisotropic character of wood, resulting from its method of growth and anatomical structure, the values of some of its physical properties depend upon the direction in which they are measured relative to the axis of the stem. The three principal planes in which wood is examined are the radial, tangential and transverse plane.

a. Density. The density of a timber is one of its most important single characteristics because many of its other properties, particularly strength, are highly dependent upon it.

The density of a piece of wood varies with the amount of water that it contains. For this reason it is important, when quoting the density of a timber, to state the moisture

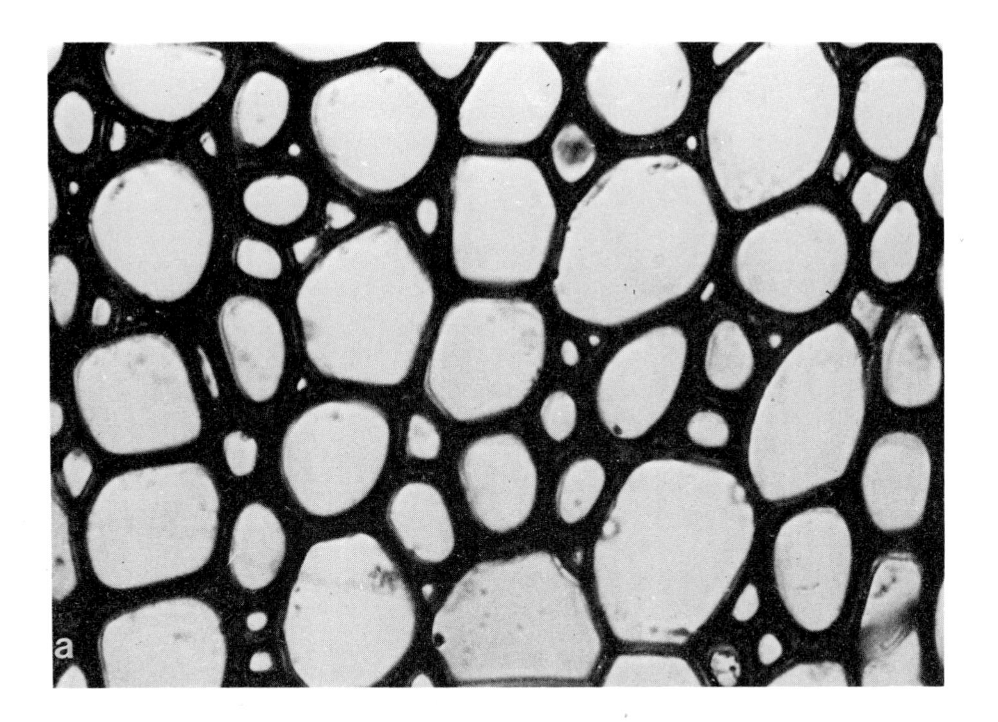

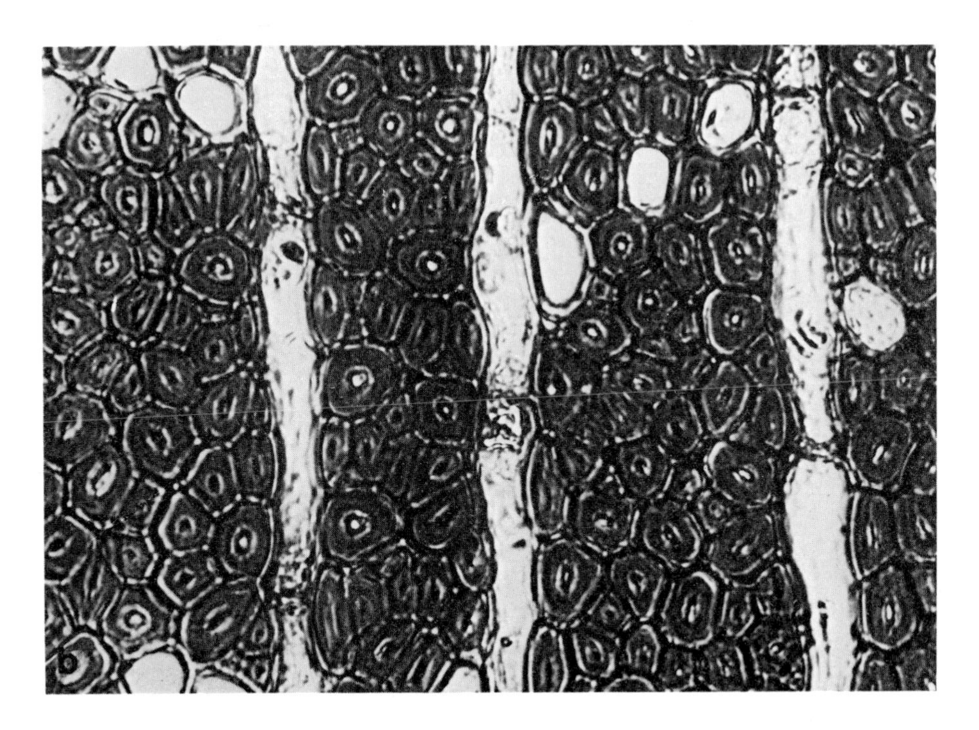

Fig. 1.6 Photomicrographs of cross-sections of (a) a low density wood (balsa), and (b) a high density wood (lignum vitae) (Magnification × 200).

content to which it refers. When comparing one timber with another it is usual to quote their densities at 12% (or sometimes 15%) moisture content. The density of a timber at any other moisture content within the range from about 5 to 25% can be estimated with fair accuracy by adding or subtracting 0.5% of the given density for each 1% moisture content above or below 12%.

A piece of dry wood consists of the solid material of the cell walls, and the cell cavities which are mainly hollow but may contain small quantities of gums, resins, etc. The specific gravity of the solid cell wall material is similar in all species of wood and is approximately 1.5. The differences in density which exist between timbers and between different specimens of the same timber are due, in the main, to differences in the ratio of cell wall volume to void space and are largely governed by the thickness of the cell walls. Thus the cell walls of a light weight timber such as balsa (average density about 10 lb/ft^3, 160 kg/m^3) are very thin, while in a heavy timber (e.g. lignum vitae, density about 77 lb/ft^3 (1,233 kg/m^3) they are much thicker and the void space is correspondingly reduced. In the descriptions of the common commercial timbers given in section 1.2 the average densities of the timbers are quoted and these figures give an indication of the range of density that exists in timbers in common use.

In addition to the wide range of density occurring in timbers of different species, there is also an appreciable variation between different samples of the same species. When figures for density of a timber are quoted it must be realized that these are average values and that some variation on either side of the average must be expected.

Variations in density within a species are due to variations within a tree and to differences between trees. One of the principal factors influencing the density of wood is the rate of growth of the tree, but the effect is not the same in all species. In the temperate softwoods, slowly grown wood has narrow growth rings having a relatively high proportion of the thicker walled, and consequently higher density, summer wood, and

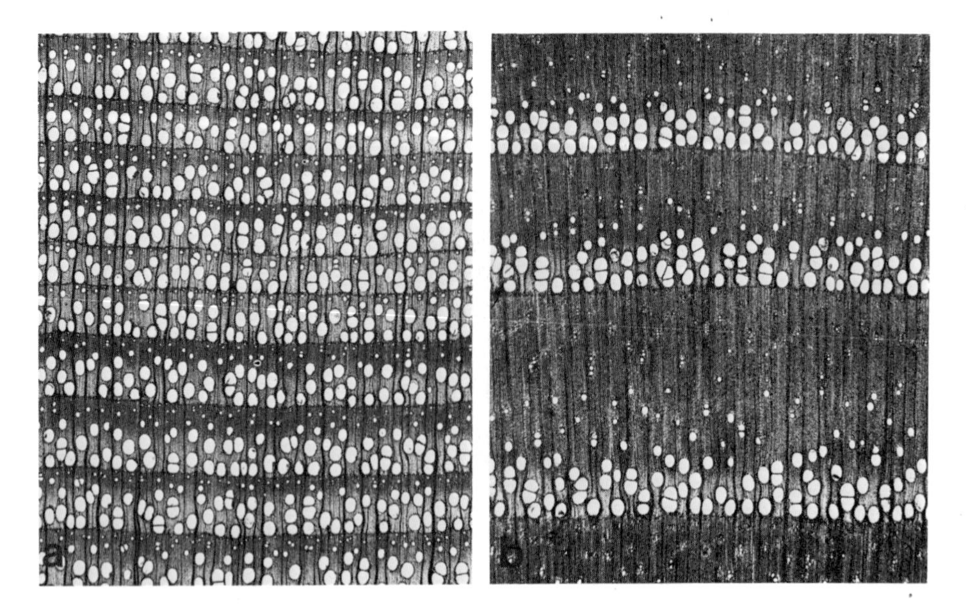

Fig. 1.7 Photomicrographs of cross-sections of a ring-porous hardwood (ash). (a) slowly grown, narrow growth rings. (b) fast grown, wide growth rings.

the overall density is therefore greater than that of faster grown wood. In ring-porous hardwoods, on the other hand, the position is reversed because in the narrow growth rings associated with slow growth the bands of pores lie closer together than they do in material with wider growth rings and the pores thus occupy a larger proportion of the volume of the wood, resulting in lower density.

The density of wood is of practical importance because it is the best single criterion of strength. Broadly speaking, the mechanical properties of wood are greater the higher its density, as may be expected. Nevertheless other factors, in particular structural features, have an important influence on individual strength properties so this generalization must be treated with some reservation. Strength properties of wood will be considered in more detail in section 1.1.6.

Since wood is a porous material it is possible to impregnate it with resins or plastics and thus to increase its density and modify some of its other properties. In this way low density and inexpensive woods may be up-graded to improve their performance, or materials with new properties may be produced. The production and properties of these densified or 'improved' woods is described in section 1.5.9.

b. Thermal properties. The thermal conductivity of wood is dependent upon three factors: (1) the direction of heat flow with respect to the grain direction, (2) density of the wood, and (3) its moisture content. The thermal conductivity is nearly the same in the radial and tangential directions, but is approximately $2\frac{1}{2}$ times greater in the longitudinal direction. It increases linearly with density, and is greater at higher than at lower moisture contents.

The thermal insulating value of wood is the reciprocal of the thermal conductivity and is important in those applications of wood, mainly in building, where heat insulation is required. It is clear that low density woods are more effective heat insulators than those of higher density. Balsa is used, for example, for insulation of refrigerated stores and ships. Because it is a good insulator wood feels warm to the touch and this is a desirable property in some of its uses – for example, in furniture and flooring.

Table 1.1 gives the thermal conductivity of timber and certain other materials used in building or for insulating purposes, and shows that woods of low density have very good heat insulating properties.

Table 1.1 Thermal conductivity of building materials (approximate values)

Material	*Density* kg/m^3	*Thermal conductivity* $cal. cm/sec/cm^2/°C$ ($\times 10^{-3}$)
Mild steel	7800	120
Aluminium	2700	570
Copper	8900	920
Brick	1600–2000	1.4
Concrete	2400	2.4
Glass	2500–2700	2.0–2.5
Rubber	950	0.32
Cork	240	0.10
Timber, average	400–600	0.3–0.4
balsa	160	0.13
Insulation board	250–350	0.13
Hardboard	900	0.30

The thermal expansion of wood is small in comparison with that of most materials and is of little practical importance. The coefficient of thermal expansion in the longitudinal direction is approximately 3.4×10^{-6} per degree Celsius regardless of the kind of wood, while in the radial and tangential directions values of 25.7×10^{-6} and 34.8×10^{-6} per degree Celsius respectively have been obtained for wood of specific gravity 0.46. However, in practice, changes in temperature of wood are generally accompanied by changes in its moisture content and the resulting swelling or shrinkage in the radial and tangential directions are much larger than the changes due to temperature alone (see section 1.1.5).

c. Electrical properties. Dry wood possesses high electrical resistance, the direct current resistivity of oven-dry wood ranging from about 3×10^{17} to 3×10^{18} ohm-cm. Its resistance decreases rapidly as the moisture content increases up to the fibre saturation point (about 28 to 30% moisture content), when it approaches that of water alone, i.e. 10^5 to 10^6 ohm/cm. Wood is thus a satisfactory electrical insulator in dry situations.

The electrical resistance of wood decreases with increasing temperature and depends to some extent upon the species of wood and the direction in which it is measured. It may also be affected by the presence of extractives or minerals in the wood.

Electrical resistance moisture meters are widely used for determining the moisture content of wood and depend upon the change in resistance (or conductivity) with changing moisture content (see section 1.3.5). They must be calibrated for each species of wood that is being tested, as well as for changes in temperature. Wood which has been treated with preservatives or fireproofing salts, or which contains soluble salts, picked up, for example, from floating logs in sea water or in other ways, may possess abnormally high conductivity in relation to its moisture content, and in such cases electrical moisture meters may give false readings.

The dielectric constant of wood also increases with its moisture content and varies with specific gravity, direction in which it is measured and the frequency of the alternating current. Moisture meters based on measurement of electrical capacity are also in use.

d. Acoustic properties. In the use of wood in buildings its acoustic properties are important in two ways. First, it is generally desirable to restrict the passage of sound from one room to another through walls, partitions and floors. It is not possible to discuss the science of acoustics here, but briefly it may be said that the ability of a material to absorb sound depends on its mass, on the way in which it is fixed, and on the acoustic properties of its surface, i.e. the extent to which the surface absorbs or reflects sound. When timber is fixed so that it cannot easily vibrate the surface has a deadening effect on sound waves. For this reason wood performs satisfactorily as a sound barrier, but the actual level of sound transmission through a piece of wood depends on a number of factors, including its density, thickness, surface condition and the wavelength of the sound waves.

Secondly, the acoustics within a room depend on the design of the room and the degree of sound absorption or reflection at the interior surfaces, and the acoustic properties of these surfaces will clearly affect the conditions in the room. The acoustic properties are important, not only in concert halls and lecture rooms, but also in offices, restaurants, etc., where it is desirable to keep the noise level down. Wood-based materials, especially insulation board, having improved sound absorption qualities have been developed for use in such situations.

Apart from the use of wood in buildings, certain woods are employed for specialized purposes in musical instruments. The qualities required here are quite different from those mentioned above and depend upon the elastic properties of wood which determine its vibrational characteristics. The special quality imparted to notes emitted by wood is very pleasing and is difficult to reproduce with other materials, but careful selection of the wood is necessary to obtain the best results. It has been found from long experience that certain species have the particular qualities that render them suitable for specific parts of musical instruments. For example, the preferred timber for sounding boards of pianos and violins is slow-grown European spruce (generally known as 'Roumanian pine') from certain areas, having a uniform texture, regularity of growth and freedom from defects.

1.1.5 Wood-moisture relationships

Wood, in common with other cellulosic materials, has an affinity for water and, when exposed to a humid atmosphere, takes up within its structure an amount of water which is dependent on the relative humidity of the surrounding air. The interaction between wood and moisture is of practical importance because many of its properties and its behaviour in service are influenced by its moisture content and, particularly, by changes in moisture content due to changes in the external conditions.

Wood takes up water in two principal ways. The so-called 'bound' water penetrates into the cell walls where it is adsorbed on the cell wall components and may be considered to be attached to the polysaccharides by a form of chemical linkage (hydrogen bonding). In so doing it causes swelling of the cell walls. The 'free' water, on the other hand, enters only the pores and cell cavities where it is held by physical forces. Although there is probably no sharp dividing line between these two modes of attachment of water, the distinction between them is important because, whereas the free water has no significant effect on wood properties, the amount of bound water determines the degree of swelling of the wood and affects its strength, electrical and other properties.

a. Fibre saturation point. The fibre saturation point of wood is the moisture content at which the fibres are completely saturated with water, but no 'free' water exists in the coarse, microscopically visible, capillary structure. As indicated above it is impossible in practice to remove all the water in the cell cavities without removing some from the cell walls, but the fibre saturation point is nevertheless a convenient concept. It is approximately equal to the moisture content of wood in equilibrium with an atmosphere fully saturated with water vapour at a given temperature. For most timbers the fibre saturation point lies at a moisture content of about 28 to 30% at normal atmospheric temperatures. It may be mentioned here that the moisture content of wood is usually expressed as the weight of moisture in 100 parts of oven-dry wood. It is thus possible, in very wet wood, to have moisture contents greater than 100%.

b. Equilibrium moisture content. The moisture content of a piece of wood in equilibrium with the surrounding atmosphere is directly related to the relative humidity of the air and increases steadily up to the fibre saturation point as the relative humidity is raised to 100%. The equilibrium moisture content at any given relative humidity of the air varies somewhat from species to species, as also does the change in moisture content corresponding to any given range of atmospheric conditions. The equilibrium moisture contents of some common timbers at 60% and at 90% relative humidity of the air are given in table 1.2.

c. Swelling and shrinkage. It is common knowledge that wood swells when it absorbs moisture and shrinks when its moisture content is reduced. Over most of the range from the fibre saturation point to the oven-dry condition the shrinkage/moisture-content relationship is practically linear; that is, the volumetric shrinkage from the fully saturated condition is proportional to the reduction in moisture content, but at the upper end of the range shrinkage is slightly greater than that given by the linear relationship.

Most woods shrink and swell only about 1/100 to 1/50 as much in the fibre direction as across the fibres and for most purposes longitudinal movement can be disregarded. Swelling and shrinkage take place in the lateral rather than the longitudinal direction because uptake of water causes lateral swelling (increase in thickness) of the fibre walls

Table 1.2 Equilibrium moisture contents and movement of some common timbers

	Equilibrium moisture content in 90% humidity	*Equilibrium moisture content in 60% humidity*	*Corresponding tangential movement %*	*Corresponding radial movement %*
Softwoods				
Scots pine (redwood)	20	12	2.2	1.0
European spruce (whitewood)	18	12.5	1.5	0.7
Douglas fir	19	12.5	1.5	1.2
Western red cedar	14	9.5	0.9	0.45
Western hemlock	21	13	1.9	0.9
Hardwoods				
Beech, European	20	12	3.2	1.7
Oak, European	20	12	2.5	1.5
Ash, European	20	12.5	1.8	1.3
Sycamore	23	13.5	2.8	1.4
Rock maple	21	12.5	1.3	1.0
Teak	15	10	1.2	0.7
Iroko	15	11	1.0	0.5
Afrormosia	15	11	1.3	0.7
Mahogany, African	20	13.5	1.5	0.9
Muninga	13	10	0.6	0.5
Greenheart	16	11	2.0	1.6
Obeche	19	12	1.2	0.8
Ramin	20	12	3.1	1.5
Utile	20.5	13.5	1.6	1.4

Source: Forest Products Research Laboratory Technical Note No. 38, H.M.S.O.
Note: Figures for movement in columns 4 and 5 are given as a percentage of the dimension of the wood in the condition of column 3 (60% humidity).

with little change in their length. Swelling and shrinkage are about 1.5 to 2.5 times as great in the tangential direction as in the radial direction. The various forms of distortion which tend to develop in a piece of timber as it dries or as its moisture content changes in use are due to the fact that the shrinkage of wood, unlike that of homogeneous synthetic materials, differs considerably in the three fundamental directions. The extent to which twist, cupping, or bow develop in a piece of wood during drying depends on the position of the piece in the tree and on any special growth characteristics (such as spiral grain) of the tree.

Table 1.2 gives the dimensional changes (movement) which occur in the tangential and radial directions for a number of commonly used timbers when they are conditioned, first in air at 90% relative humidity and then in air at 60% relative humidity. It will be seen that the movement is considerably smaller in some timbers than in others and the relatively high dimensional stability of such timbers as teak, iroko, afrormosia, muninga and western red cedar is a valuable property of these timbers.

The dimensional changes which occur with changes in the external humidity conditions form one of the major drawbacks to the use of wood in many of its applications. For this reason many attempts have been made to find ways of modifying wood in order to improve its dimensional stability. Chemical methods of reducing the hygroscopicity of wood have been investigated, but generally the treatments are accompanied by undesirable side effects such as increased brittleness of the wood or loss of abrasion resistance. Methods based on impregnation with resins, waxes or polymers show rather more promise and are referred to in section 1.5.9.

Good stability can also be obtained by restraining the movement by mechanical methods. In plywood, in which alternate layers of wood are laid at right angles to each other, the movement of the individual layers is restricted by the firm bonding to the adjacent layers, which tend to swell and shrink in a different direction. Chipboard, which has no pronounced grain direction, similarly possesses a high degree of stability. The movement of both these materials is much smaller than the tangential movement of solid wood, but they swell more in thickness. The dimensional change can be further reduced by bonding resin-impregnated paper, plastic or thin metal sheet to the faces.

d. Moisture content and strength. The mechanical properties of wood will be discussed in more detail in section 1.1.6, but it should be noted here that the strength of wood is influenced to a considerable extent by its moisture content. When wood is dried from the green condition, practically no change in mechanical properties occurs until the fibre saturation point is reached. Below this point most of the strength factors increase almost linearly with a further decrease in moisture content, but the rate of increase is different for the different properties, and in the case of toughness there is generally a small decrease as the moisture content decreases.

For wood at 12% moisture content the average increase in value of some strength properties resulting from a decrease in moisture content of 1% is approximately as follows:

Modulus of rupture	4%
Modulus of elasticity (stiffness)	2%
Maximum compression strength	5%
Side hardness (i.e., hardness measured on the side grain)	2%
Shearing strength parallel to grain	4%

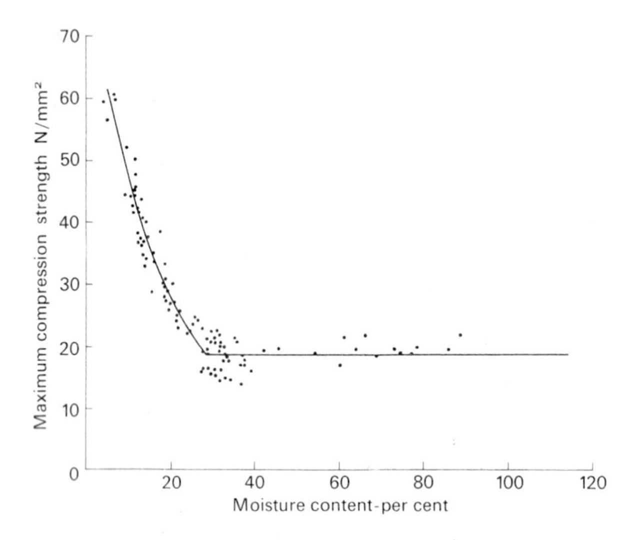

Fig. 1.8 Relationship between compression strength and moisture content (Scots pine).

Curves showing the relationship between compression strength and moisture content indicate clearly the significance of the fibre saturation point, above which an increase in moisture content produces no further decrease in strength.

e. Moisture content and electrical properties. It has already been mentioned in section 1.1.4 that dry wood, like other fibrous materials, is a good electrical insulator, but its conductivity increases rapidly with increasing moisture content up to the fibre saturation point. At this point there is a sharp break in the conductivity/moisture-content relationship (as in the case of strength properties), and at higher moisture contents the conductivity increases much more slowly. Over most of the moisture content range up to the fibre saturation point the relationship between the moisture content and the logarithm of the electrical conductivity is approximately linear. Use is made of this relationship in electrical moisture meters, in which the moisture content of wood is determined from measurement of its conductivity. In using these moisture meters a probe having two sharp prongs is inserted into the wood and the resistance (or conductance) between them is measured. This reading is converted into moisture content by reference to tables which make allowance for the species of wood and temperature. The method gives an instantaneous result and is therefore more suitable for routine control than is the oven-drying method, which necessitates removal of a specimen of wood and determining the loss in weight when all the moisture is removed by drying in an oven. However, the depth at which moisture content can be measured by means of moisture meters is limited by the depth to which the prongs can be forced into the wood, and it is not possible to determine moisture contents at the centre of thick pieces of wood in this way.

1.1.6 Mechanical properties

In the selection of a timber for a particular purpose many factors have to be considered, but in a great many cases the choice depends to some degree upon one or more of its mechanical properties. A basic knowledge of these properties is therefore essential if timber is to be used efficiently.

The fact that wood is a natural product has two important consequences so far as its mechanical properties are concerned. First, there is a natural variation in strength between different specimens of the same timber. Within a single species the range of values found for each strength property is generally rather wide. For example, one piece of Douglas fir may be twice as strong in compression as another. The factors responsible for these variations in strength are discussed in section 1.1.7. To determine the strength properties of a timber it is therefore necessary to test a considerable number of specimens under controlled conditions and to express the results in terms of the arithmetic mean and the standard deviation, which is a measure of the variation in strength.

Secondly, because of its anatomical structure and method of formation, wood is an anisotropic material and its strength depends on the direction of the stress in relation to the direction of the grain. This is self evident, and timber is almost invariably employed in such a way as to utilize its strength properties to the best advantage.

Timbers vary very widely in their mechanical properties, and the most important factor influencing strength is density or specific gravity. In general terms, species with a high density have correspondingly high strength values and vice versa. However, there are irregularities in the relationships due to variations in structure of the woods and it is not practical to select a species for a particular purpose on the basis of density alone, because there may be marked differences in certain specific properties between timbers of the same density. Thus European ash, having the same density and bending strength as European beech, has an energy absorbing capacity about 50% higher than that of beech and consequently has a higher 'toughness'. For this reason ash is particularly well suited to the manufacture of sports goods (e.g. hockey sticks) and tool handles.

The dependence of strength on moisture content of wood has been discussed in section 1.1.5. Because of this relationship it is essential, when comparing the strength of different species of wood, to carry out tests on specimens which have been conditioned to a standard moisture content. Most strength properties are also affected by temperature changes, though the effects of temperature are less marked than are those of moisture content. In general, a reduction in strength occurs with a rise in temperature.

a. Strength testing. The strength properties may be determined either by measuring the strength of timber in structural sizes, in which case realistic values are obtained, or by testing small clear specimens, free from knots and other defects. In view of the need to test a considerable number of specimens because of the variability of the material, the latter procedure is more generally adopted because it is much more economical in material. Procedures are then available for deriving working stresses from the values so obtained by making appropriate allowances for the various strength-reducing factors which may be present.

The more important tests are described briefly below; full details of the test methods are given in B.S 373: 1957, *Methods of Testing Small Clear Specimens of Timber*, and Forest Products Research Laboratory Bulletin No. 50, *The Strength Properties of Timbers*. Standard test pieces are 20 × 20 mm (0.79 × 0.79 in) in cross section, cut parallel to the grain, and are free from knots, splits, etc. Specimens for testing are conditioned at 25°C and 60% relative humidity, and testing conditions (size of specimen, rate of loading) are precisely defined.

Static bending. The test piece, 300 mm (11.81 in) long, is supported over a span of 280 mm (11.02 in) and is centrally loaded. Readings of load and deflection are continued past the maximum load, until either the load has fallen by 90% or a deflection of 6% has been reached. The properties determined from this test are modulus of rupture, modulus of elasticity and work to maximum load. The modulus of rupture is the equivalent stress in the extreme fibres at the point of failure calculated on the assumption that the simple theory of bending applies. It is a measure of the ultimate bending strength of timber and applies only to the size of specimen and loading conditions employed in the test. The *modulus of elasticity* is a measure of stiffness and is of importance in determining the deflection of a beam under load. It is generally considered in conjunction with bending strength, as for many uses stiffness is the controlling factor in design. The *work to maximum load* is determined from the area under the load-deflection curve up to the point of maximum load, and is a measure of toughness. It is important where timber is subjected to shock loads.

Impact bending. A weight of 1.5 kg (3.306 lb) is dropped centrally from increasing heights on to a 300 mm (11.81 in) long specimen supported over a span of 240 mm (9.45 in) until fracture occurs. The resistance of the timber to a suddenly applied load is indicated by the height of the drop causing fracture of the beam.

Compression parallel to grain. The specimen is 60 mm (2.36 in) long and is crushed axially between parallel plates. The maximum load is recorded and measures the ability of the timber to withstand loads when applied on the end grain. This property is important where timber is used as short columns or props. When considering the strength of a long column or strut the critical property is the stiffness of the material.

Hardness. The load required to indent the timber on the side grain by means of a hemispherically headed plunger to a depth equal to the radius of the plunger is determined under defined conditions. This measures the resistance to indentation and is important in rollers, bearing blocks, flooring, etc.

Other strength properties that are determined as required include shear strength parallel to the grain, resistance to cleavage, and abrasion resistance.

b. Strength values. As already mentioned, there are large variations in strength between different species. A heavy, strong timber, such as greenheart, is about eight times as strong in compression parallel to the grain as the very light timber, balsa. Table 1.3 gives average values for the modulus of rupture (bending strength), modulus of elasticity (stiffness) and compression strength parallel to the grain for a number of common timbers to illustrate the range of values commonly met with. More detailed information for most of the timbers in general use will be found in Forest Products Research Laboratory Bulletin No. 50, mentioned above.

It would be interesting to compare the strength of wood with that of other materials, but this cannot always be done in a straightforward manner because different materials are used in different ways. Thus, bricks and concrete are always used under compressive stresses, fibres are used in tension, metals may be used in compression or tension, and for wood the stiffness is often of major importance.

The tensile strength of wood is lower than that of the common metals, but when the lower density of wood is taken into consideration it is found that wood has a remarkably high strength/weight ratio, superior to that of most metals and plastics and comparable to that of high tensile engineering steel, as the figures in table 1.4 show. It should be added, however, that metals are generally free from some of the less desirable proper-

Table 1.3 Average strength values of some common timbers at 12% moisture content (determined on small, clear specimens)

Timber	Density lb/ft³	Density kg/m³	Modulus of rupture p.s.i.	Modulus of rupture N/mm²	Modulus of elasticity 1·000 lbf/in²	Modulus of elasticity N/mm²	Compression parallel to grain lbf/in²	Compression parallel to grain N/mm²
Softwoods								
Scots pine	32	513	12900	89	1450	10000	6870	47.4
European spruce	26	417	10400	72	1480	10200	5300	36.5
Douglas fir	34	545	13500	93	1840	12700	7560	52.1
Western red cedar	23	368	9400	65	1010		5080	35.0
Hardwoods								
European beech	43	689	17100	118	1830	12600	8170	56.3
European oak	43	689	14000	97	1460	10100	7490	51.6
Black poplar	27	433	10400	72	1250	8600	5430	37.4
African mahogany	31	497	11300	78	1300	9000	6730	46.4
Teak	40	641	15400	106	1450	10000	8760	60.4
Greenheart	61	977	26300	181	3050	21000	13040	89.9
Obeche	23	368	7900	54	800	5500	4090	28.2
Ramin	41	657	19400	134	2030	14000	10500	72.4
Utile	40	641	15000	103	1560	10800	8760	60.4
Afrormosia	46	737	19400	134	1810	12500	10350	71.4

Source: Forest Products Research Laboratory Bulletin No. 50, H.M.S.O.
Note: 1 N/mm² ≈ 0.1 kg/mm².

ties of wood, namely low strength across the grain direction, and swelling and shrinkage. Some of the modern fibre-reinforced materials possess very high strength properties in relation to their weight, but they are also very costly.

In designing wooden structures, the stiffness of the material is often of greater consequence than its ultimate strength. It is an interesting fact that the ratio of the modulus of elasticity (stiffness) to the density is very nearly constant for a number of common materials (metals, wood, glass, etc.). This means that beams of any of these materials of equal weight will have very nearly the same stiffness.

1.1.7 Quality, defects, grading

a. Quality. Quality in wood can only be considered in relation to the use to which it is to be put. The various end uses require emphasis to be placed on different properties or features of wood, such as its appearance, strength, durability, impermeability, etc., but in practice the term quality is generally used in connection with appearance or strength or a combination of these. It is closely associated with the presence or absence of defects in the wood.

For many purposes, for example in furniture, panelling, flooring, and joinery (unpainted), the attractive appearance of wood is one of its valuable assets. In this context

Table 1.4 Tensile strength of some woods, metals and plastics (approximate values)

Material	*Density* *kg/m³*	*Tensile strength* *p.s.i*	*N/mm²*	*Strength/Density*
Woods				
Sitka spruce	430	15000	100	0.24
Scots pine	510	17500	120	0.235
Douglas fir	530	19000	130	0.25
Ash	690	21000	145	0.21
Beech	680	25000	170	0.25
Hickory	820	36000	250	0.30
Metals				
Cast iron	7100	20000–40000	140–280	0.02–0.04
Mild steel	7700	60000	410	0.055
High tensile steel	7700	225000	1550	0.20
Copper	8900	30000	210	0.025
Brasses	ca.8400	45000–90000	310–620	0.037–0.074
Aluminium	2700	18000	125	0.046
Aluminium alloys	ca.2700	22000–70000	150–480	0.055–0.18
Magnesium alloys	1800	33000–45000	230–310	0.13–0.17
Plastics				
Low density polythene	910	1000–2300	7–16	0.008–0.018
Polypropylene	900	5000	35	0.039
Rigid PVC	1390	8000	55	0.04
Plasticized PVC	1280	1500–3500	10–24	0.008–0.019
GRP	1600	20000	140	0.087

quality is a subjective term and cannot be expressed quantitatively. The principal factors affecting the appearance of wood are its colour, texture and grain pattern (figure).

From a practical viewpoint colour is important because it may enhance or detract from the decorative value of timber. Pure wood substance possesses no colour of its own, and the wide range of colours that is found, from almost white in some timbers, through various shades of yellow, brown and red, to black in ebony, is due to the presence of natural colouring matters which are present in association with main chemical components of wood. The sapwood of many timbers is paler in colour than the heartwood because during the growth of the tree the colouring matters are deposited mainly in the heartwood. The colour of many timbers changes on prolonged exposure to light–either darkening or fading–and this can be troublesome, especially in the furniture trade.

The terms *grain* and *texture* should be used to refer to two quite distinct characteristics of wood, but they are often confused in everday use. Grain refers to the direction of the fibres in relation to the axis of the tree or the longitudinal direction of a piece of wood, while texture refers to the relative size and variation in size of the different types of cells making up the structure of the wood. Timbers are described as having coarse or fine, and even or uneven texture, and these differences affect not only its appearance, but also its wearing qualities. Timbers having a fine and even texture (e.g. maple) possess greater resistance to wear than those of coarser and less uniform texture. This factor is of particular importance in timbers used for flooring.

The term grain is often used to describe features which are not directly connected with its true meaning, namely the direction of the fibres relative to the axis of the tree, but are associated with texture ('coarse grain', 'fine grain', etc.), width or uniformity of growth rings, or direction in which the wood has been cut.

Many types of grain pattern occur in different timbers and they affect both the appearance of the wood and its strength and other properties. In straight-grained timber the fibres lie more or less parallel to the axis of the tree. This leads to good strength properties, ease of machining and low wastage, but does not give rise to ornamental figure. This is produced when the grain direction is irregular. For example, many tropical hardwoods have interlocked grain. This means that in alternate growth layers the fibres are inclined in opposite directions, producing a ribbon or stripe figure on quarter-sawn (radial) surfaces which is of decorative value. However, timbers with heavily interlocked grain may present considerable difficulty in sawing and planing because the inclination of the grain to the direction of cutting leads to production of a rough surface. Wavy grain is a feature of some timbers. This gives rise to a series of more or less horizontal darker and lighter stripes on longitudinal surfaces, because of variations in the reflection of light from the surface of the fibres. This is referred to as 'fiddle-back' figure. A broken stripe or 'roe' figure is produced when wavy and interlocked grain occur together in the same piece of timber.

Many other types of figure are recognised in trade terminology and logs that produce timber having an attractive figure command a high price in the trade. They are valued especially when they are suitable for cutting decorative veneers. Knots in timber are generally considered to be an undesirable feature, but even they can be of decorative value, as in the knotty pine used for panelling.

b. Defects. Wood, being a natural product, is liable to contain a number of abnormalities which reduce its usefulness for specific purposes and lower its economic value. These are commonly called defects, but it must be appreciated that the extent to which a structural feature, such as grain deviation, is to be regarded as a defect depends upon the intended use of the wood. We have already seen that wood possessing certain grain patterns is valued for its decorative qualities, even though the irregularity of grain direction may reduce its strength.

A *defect* may be defined as any feature which lowers the technical quality or commercial value of a piece of timber, while an *allowable defect* is one which does not exclude a piece from a defined grade.

The more important features which occur naturally in wood and adversely affect its technical properties are knots, inclined grain, reaction wood, and a group of features (compression failures, shakes, resin pockets) which cause discontinuity of the wood tissue.

Knots. These are perhaps the most common type of defect in timber. As the tree grows branches emerge from the stem and as the latter increases in diameter it gradually encloses the bases of the branches. The enclosed portions form knots. The degree of knottiness depends on the species of tree and conditions of growth but, especially in softwoods, knots are formed throughout the length of the stem. If, however, the branches in the lower part of the stem fall off naturally or are removed by pruning, the wood produced in later years grows over the knots and the outer part of the stem consists of clear wood, free from knots.

Knots are of two kinds, depending on whether a branch was alive or dead at the time of its inclusion in the stem. If the branches are alive their tissues are continuous with those of the main stem and the knots so formed are 'live' or 'tight' knots. When a branch dies a stump remains which is gradually surrounded by the tissues of the stem, but these grow round the stump without being connected with it and a 'loose' or 'dead' knot results. Such knots may fall out when the timber is converted or dried.

The shape of knots as they appear on sawn surfaces depends primarily on the plane and angle at which they are cut through during conversion of the log. Various terms are in use to describe knots in sawn timber, and they are also classified according to size, which varies from very small ('pin knots') to several inches in diameter.

Knots adversely affect the appearance and technical properties of wood, and in many species they are the primary cause of lowering of quality. The strength of wood may be greatly reduced by knots, depending on their kind, size and location, partly because of their abnormal structure and also because of the deviation in grain direction in the wood surrounding the knots. They also give rise to difficulties in drying, machining, and finishing (painting and varnishing).

Inclined grain. In many trees the direction of the grain is not exactly parallel to the axis of the stem of the tree; during the growth of the tree the fibres are laid down in a spiral arrangement about its axis. When the log is converted into boards in the usual way the grain of the wood is therefore inclined at an angle to the faces or edges of the boards. Inclined grain reduces the strength of timber and is also liable to cause distortion during drying and tearing of the surface in machining.

Spiral grain is of common occurrence in both softwoods and hardwoods, but the extent of the deviation from straightness (the angle between the fibre direction and the stem axis) is very variable both between trees and within a tree. Interlocked grain, which is a common feature of tropical hardwoods, is a special case of spiral grain in which the direction of the spiral alternates at intervals. For a number of years the fibres spiral in one direction, then the direction is reversed, and so on.

The significance of inclined grain lies mainly in its effect on the strength properties of timber. When the slope of the grain is 1 in 25 there is an average reduction in bending strength of about 4%. With a slope of 1 in 20 the reduction is 7%, with 1 in 15 it is 11%, with 1 in 10 it is 19%, and with 1 in 5, 45%. The stiffness of a beam is also reduced by sloping grain, but to a smaller degree, while the ability of wood to absorb shock is much impaired. It is thus particularly important to use straight-grained timber for such purposes as ladder rungs and stiles, scaffold boards, tool handles and sports goods.

Reaction wood. When trees grow under conditions of mechanical stress, for example the stresses present in leaning stems or the lower sides of branches, or those due to wind pressure, abnormal forms of wood tissue are produced. This is termed reaction wood because it is formed as a reaction to the imposed stress. Reaction wood is of two kinds, namely *compression wood* which is formed in softwoods on the compression side of the stem (the lower or leeward side), and *tension wood* which is found on the upper or windward side of hardwoods where the wood is under tension. Compression wood and tension wood are both associated with eccentric growth and occur on the side of the stem where the growth rings are wider.

Compression wood and tension wood differ from normal wood both in their microscopic features and their chemical composition. Compression wood in softwoods is darker than normal wood, having a reddish-brown colour, and when pronounced its growth rings appear to be composed almost entirely of late wood. It is higher in density than normal wood, and is notable for its abnormally high longitudinal shrinkage, which can lead to serious distortion or splitting during drying. Its strength properties are lower

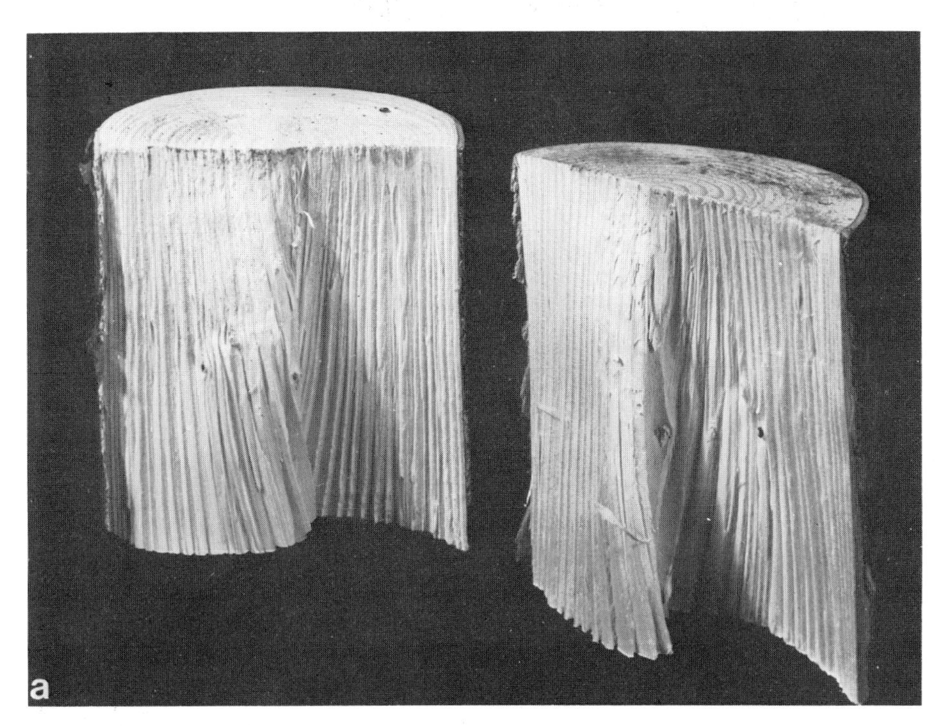

Fig. 1.9 Examples of pronounced spiral grain (a) European spruce, (b) plum.

than those of normal wood, and because of its higher density it is more difficult to nail and work with tools.

Tension wood is liable to occur in a wide range of commercial hardwoods, but is more common in some species than others. It is less easily recognised than compression

Fig. 1.10 Cross-section of stem of Sitka spruce of eccentric growth showing dark bands of compression wood.

wood, but is often paler in colour than normal wood and tends to have a lustrous or shiny appearance. Like compression wood in softwoods it also has a characteristic appearance under the microscope and differs from normal wood in its chemical composition. It also has abnormally high longitudinal shrinkage in drying, causing distortion especially in relatively thin boards. Sawn timber containing tension wood has a fibrous or woolly surface and is generally more difficult to machine than normal wood.

Both compression wood and tension wood vary greatly in the intensity of their development in the wood and they can only be regarded as serious defects when they are strongly developed and affect a considerable proportion of the material.

Other natural defects. There are a number of other defects which originate in the growing tree and which affect the strength of the wood because they introduce discontinuities in the wood structure.

Compression failures are ruptures of wood stressed in compression parallel to the grain, generally through pressures caused by wind or snow, or resulting from careless felling.

Shakes are also ruptures of the wood of living trees, and occur longitudinally, generally parallel to the growth rings. Their cause is uncertain, but they may perhaps be due to internal growth stresses.

Resin pockets or pitch pockets occur in softwoods and a few hardwoods. They are lens-shaped cavities lying along the growth rings and filled with resin, which may flow out when they are sawn through.

Brittleheart is a feature which occurs near the centre of large trees of some tropical hardwood species. As its name implies, the wood of brittleheart is low in strength properties and density, and particularly in resistance to impact. It should be avoided in timber used for structural purposes.

In addition to defects arising during the growth of the tree wood may also be affected by external agencies, of which the most important are the wood-rotting fungi and wood-boring insects. The effects of these organisms will be considered in sections 1.6.3 and 1.6.4. Wood may also be subject to degradation in a variety of ways if drying is incorrectly carried out.

c. Grading. The strength properties of a timber are determined by testing to failure a large number of small, clear specimens under controlled conditions in the laboratory. The strength values so obtained cannot, however, be used directly for design purposes because there are a number of strength-reducing factors which have to be taken into account.

First, the strength, even of small, clear specimens, is subject to natural variation, so that some specimens are appreciably lower in strength and others appreciably higher than the average. To allow for this, the stress below which only 1 in 100 results will fall is determined statistically. Secondly, the test is of very short duration, and since timber can resist higher stresses for short periods than for longer periods, it is necessary to reduce the stress obtained from the test to make it applicable to normal loading conditions. The *basic stress*, applicable to material free from defects, is thus arrived at by making appropriate adjustments to the stress values actually determined to allow for the effects of variability of the material and duration of loading.

In addition to these strength-reducing factors, most timbers also possess defects, as a result of which individual pieces will have widely different strength values. If wood is to be used efficiently and economically as a structural material it is, therefore, desirable that it should be divided into groups or grades so that the stronger material may be assigned to and used at higher stress levels. This type of selection is known as stress grading and it enables material to be placed in grades having stress values which are definite fractions of those that would apply to clear timber.

The strength-reducing effects of such features as knots, inclined grain, fissures and wane (the irregular edge of a board where it meets the surface of the log from which it is cut) vary with their size, location and extent relative to the dimensions of the timber. Stress grading rules have been drawn up for softwoods, specifying the maximum size of each of these features that can be allowed in a particular grade. The application of the rules consists simply in making a visual examination of the surfaces of a piece of wood to locate the feature which would cause the greatest reduction in strength. An assessment of the relative size of the feature enables the grade of the piece to be determined. In Britain, four basic stress grades are specified for sawn softwood, and three for laminated timber. The strength ratios for the sawn timber grades are 75, 65, 50 and 40% respectively. This means that a piece of timber which would satisfy the requirements of 65 grade would have stress values given to it which would be 65% of the basic stress of timber of the same species which was free from all defects. The working stress, or grade stress, of a piece of timber can thus be easily calculated from a knowledge of its basic stress and the grade to which it has been assigned. Visual stress grading is not a difficult operation, but requires considerable experience if it is to be carried out rapidly and accurately. Each of the major softwood producing countries has developed grading rules which, although differing slightly from country to country, are generally comparable in their effects.

Softwood species provide most of the structural timber used in the United Kingdom and consequently much more attention has been given to softwood than to hardwood grading. In addition to grading rules based on defects, which have been developed in some hardwood producing countries, a method of grading known as the *cutting system* is also in use for hardwoods. In this system the grade of each piece is based on the percentage of timber, either free from all defects, or with only acceptable defects, which can be cut from it. Each grade within a set of rules specifies a minimum area of acceptable cuttings in reasonable sizes which can be obtained from a plank, and expresses this as a fraction of the total plank area. Thus, the existence of a serious defect which could easily be cut out, leaving the rest of the piece faultless, would not lower the grade as in softwoods.

d. Machine stress grading. Visual stress grading results in more efficient use of wood, but is neither a very accurate nor a very rapid method of assessing the strength of a piece of wood. Some factors which influence strength, e.g. density, are disregarded. The possibility of some form of mechanical stress grading arose from the discovery that bending strength (modulus of rupture) is closely related to stiffness (modulus of elasticity). The basis exists, therefore, for a method of estimating the strength of a piece of timber non-destructively and continuously throughout its length by applying a relatively light load and measuring the deflection. Machines operating on this principle have been constructed in the United States and Australia and, after a considerable amount of development work, the Australian 'Computermatic' machine has been brought into commercial operation in the United Kingdom. This machine is continuously fed with timber and assesses the basic shape (curvature) of each piece. It imposes a predetermined load at intervals of 6 in (152 mm) along its length and measures the resulting deflection, the load being applied continuously by means of rollers as the timber passes through the machine. This information is sent to the machine's computer which compares it with the basic shape information. The computer grades each section according to its strength and instructs the machine to apply colour coding by means of sprays of different dyes at the appropriate section. As the piece of timber leaves the machine its end is also colour-coded according to the lowest grade recorded over the length of the piece.

The chief advantage to be derived from mechanical stress grading is a greater confidence in the material because the method is so certain and reliable. Lower factors of safety can therefore be used and the yields of usable timber and, particularly, of high grade timber, are much higher than those that can be achieved by visual grading.

1.2 SPECIES OF WOOD

1.2.1 Classification

Forest trees are divided into two classes – gymnosperms and angiosperms. The distinction between these two classes is a botanical one and is based mainly on the way in which the trees carry their seeds. The gymnosperms have exposed seeds (from the Greek *gymnos* = naked), while the angiosperms produce covered seeds (from the Greek *angeion* = container). Gymnosperms are separated into seven orders, of which one, coniferales (conifers), is the source of all the commercially important softwoods. Angiosperms are classified into two sub-classes – monocotyledons and dicotyledons. Monocotyledons are largely annual grass-like plants, but they include some of the palms. All the forest trees that produce commercial hardwoods belong to the dicotyledons.

Thus, the division of commercial timbers into softwoods and hardwoods, which is based on a long-standing custom of the trade, corresponds to the botanical distinction between the conifers belonging to the gymnosperms and the broad-leaved trees (dicotyledons) belonging to the angiosperms.

Both these groups are further divided into a number of *genera*, each of which contains one or more *species*. To characterize a plant or tree accurately it is necessary to quote the name of its genus and species. Latin names are used for this purpose in order to avoid the confusion that might arise from the use of different languages in different countries. Examples of botanical names are *Pinus sylvestris* (Scots pine), *Fagus sylvatica* (European beech), *Tectona grandis* (teak), and so on.*

Botanical names are not normally used in the timber trade, and timber is referred to by common or standard names such as Swedish redwood, African mahogany, etc. In the UK, to avoid ambiguity it is important to adhere to the standard names laid down in British Standard 881 & 589 : 1955. In many instances other names are used in other countries. It has been common practice to apply the names of well-known woods such as oak, walnut, mahogany or teak, with a geographical or other qualification, to timbers that have no botanical relationship with the true oaks, walnuts, etc., but which have a superficial resemblance to them. When names of this kind are used they should be distinguished by quotation marks to indicate that they are not true oaks, walnuts, etc. Examples are 'Rhodesian teak', 'Tasmanian oak', 'Parana pine', 'African walnut'.

In the following paragraphs short descriptions are given of twenty-one commonly used timbers (six softwoods and fifteen hardwoods) and data on a further forty-one timbers are summarized in table 1.6. More detailed information on these and other timbers in commercial use may be found in the publications *A Handbook of Softwoods* and *A Handbook of Hardwoods*, published by H.M. Stationery Office.

The following notes should be read in conjunction with the descriptions of the timbers.

For convenience, numerical data are given in both Imperial and SI units.

Densities of timbers are average values and refer to a moisture content of 12%.

With regard to strength, the hardwoods are compared with one or more of three reference timbers – obeche, European beech, and greenheart – which can be regarded as having low, medium and high mechanical properties, respectively. The more important of these properties at 12% moisture content of these three reference timbers are given in table 1.5.

The term 'movement' refers to the dimensional changes that take place when timber which has been dried is subjected to changes in atmospheric conditions. It is described as small, medium or large, and gives an indication of the dimensional stability of the timber in use.

* Classical scholars may be puzzled by the apparent disagreement in gender between the noun and the adjective in the Latin names of many trees, e.g. *Pinus nigra*, *Quercus alba*, *Fagus sylvatica*, *Populus tremula*. The explanation is that the names of trees, although they often end in *-us*, are not masculine (like *dominus*) but feminine and therefore take a feminine adjective.

Table 1.5 Strength properties of reference timbers (hardwoods)

Timber	Bending strength lbf/in²	Bending strength N/mm²	Modulus of elasticity (stiffness) 1 000 lbf/in²	Modulus of elasticity (stiffness) N/mm²	Compressive strength parallel to grain lbf/in²	Compressive strength parallel to grain N/mm²
Obeche	7900	54	800	5500	4090	28.2
European beech	17100	118	1830	12600	8170	56.3
Greenheart	26200	181	3040	21000	13040	89.9

The term 'natural durability' refers to the resistance of a timber to fungal decay and is important when timber is used in situations where it is liable to become damp. Timbers have been classified into five broad grades based on the performance of their heartwood in contact with the ground under conditions prevailing in Britain. The five grades are as follows:

Grade of durability	*Approximate life in contact with the ground (years)*
Perishable	Less than 5
Non-durable	5–10
Moderately durable	10–15
Durable	15–25
Very durable	More than 25

When timber is to be used in conditions favourable to decay or to attack by insects or marine borers the ease with which it can be impregnated with preservatives is important. Timbers differ greatly in their resistance to impregnation, and the terms used to describe the extent to which a timber can be impregnated under pressure with preservatives have been arbitrarily defined from the results of standardized tests. Timbers are described as permeable, moderately resistant, resistant, or extremely resistant to impregnation.

In the paragraphs below, the chief softwoods are described in sections 1.2.2 to 1.2.7, the hardwoods in sections 1.2.8 to 1.2.22 and some additional woods more briefly in section 1.2.23.

1.2.2 Scots pine or redwood–Pinus sylvestris

In accordance with established custom, timber of this species imported from the European continent is commonly called redwood or red deal (especially in the north) or yellow deal (especially in the south), while timber grown in the British Isles is generally known as Scots pine. Names indicating the origin of imported timber, e.g. Finnish redwood, Swedish redwood, Archangel redwood, etc. may also be used.

a. The tree. The tree has a very wide distribution in Europe and Siberia, ranging from Spain into the Arctic Circle. Timber is imported into the United Kingdom mainly from the Scandinavian countries, USSR, Poland and Czechoslovakia. The tree commonly grows to a height of 100 ft (30 m) or more, with a diameter of 2 to 3 ft (0.6 to 0.9 m).

b. The timber

Properties. The extensive geographical range of the tree covers a wide variation in growth conditions, and the character of the timber varies accordingly, particularly in regard to texture, density and the number and size of knots. The pale reddish-brown heartwood is usually distinct from the lighter coloured sapwood. Average density about 32 lb/ft^3 (510 kg/m^3). Compared with other softwoods, Scots pine is intermediate in strength, being somewhat stronger than the spruces and less strong than Douglas fir and larch. It has medium movement.

In the descriptions of other softwoods their strength properties are generally compared with those of Scots pine. Average values for the principal strength properties of Scots pine at 12% moisture content are as follows:

Properties	*Strength* *lbf/in²*	*N/mm²*
Bending strength	12900	89
Modulus of elasticity (stiffness)	1450000	10000
Compression parallel to grain	6870	47.4

The heartwood is non-durable, and moderately resistant to preservative treatment, but the sapwood is permeable. The sapwood is liable to attack by the common furniture beetle and the house longhorn beetle.

Processing. The timber dries rapidly and well. Owing to its tendency to blue stain it should be dried as soon as possible after conversion. It generally works easily and cleanly in most hand and machine operations, but knots may be troublesome as they are liable to come loose and fall out in planing and sawing. It takes nails well and, with the exception of very resinuous material, can be stained effectively and gives good results with paint, varnish and polish.

Uses. The timber is used for all kinds of constructional work, particularly in house building, while the better grades are employed for joinery, furniture, etc. It has for long been the standard general utility timber of northern Europe and combines adequate strength with light weight and good machining and nailing properties. The sapwood can be readily treated with preservatives and large quantities of the timber are used for railway sleepers, telegraph and transmission poles and pit props.

Home-grown Scots pine is essentially similar to imported redwood, but is generally faster grown and consequently has wider rings, coarser texture and larger knots, and yields a smaller proportion of the better grades than typical redwood from northern Sweden and Finland. Properly dried and graded, however, it is quite suitable for building work.

Scots pine is extensively used for pulping and papermaking and in the manufacture of fibreboard and chipboard.

1.2.3 European spruce or whitewood – Picea abies

Timber imported from the European continent is commonly called whitewood or white deal, while timber grown in the British Isles is generally known as European spruce, common spruce or Norway spruce.

a. The tree. A native of Europe and western Russia, and grows well in Great Britain. Reaches an average height of about 120 ft (37 m), with a diameter of 2½ to 4 ft (0.8 to 1.2 m), but grows to larger sizes in Central Europe. Much of the timber used in the United Kingdom is imported from the Baltic countries.

b. The timber

Properties. Varies in colour from almost white to pale yellowish-brown. The annual rings are less prominent than in Scots pine and there is no clear distinction between sapwood and heartwood. Timber from the Baltic has an average density of about 29 lb/ft^3 (460 kg/m^3). Home-grown spruce is of faster growth and is slightly lower in density. The timber is slightly lower in strength than Scots pine.

The timber is non-durable and is resistant to preservative treatment. The sapwood is susceptible to attack by the common furniture beetle and damage by house longhorn beetles is sometimes present.

Processing. The timber dries very rapidly and well, with little tendency to split, check or distort. It works easily with hand and machine tools and can be nailed and glued without difficulty.

Uses. Used largely for general joinery and carpentry, interior finishing of houses, and packaging. Owing to its lack of durability and the difficulty of impregnation with preservatives it is not generally suitable for exterior joinery or for use in contact with the ground. Smaller trees are used in the round for scaffold and flag poles, masts and pit props and similar purposes. Spruce is extensively used for the manufacture of pulp and paper, including groundwood pulp for which it is particularly suitable.

European spruce grown in the British Isles tends to be faster grown and somewhat lighter in weight than imported Baltic whitewood, but is suitable for many of the same purposes. Much of the spruce planted in the United Kingdom is Sitka spruce (*Picea sitchensis*), introduced from western Canada. The timber is generally similar to that of home-grown European spruce, but may be slightly lower in density and rather coarser in texture, and is inclined to be knotty.

1.2.4 Douglas fir–Pseudotsuga menziesii (also called British Columbian pine, Oregon pine)

a. The tree. A large tree, sometimes reaching a height of 300 ft (90 m) or more, but more commonly 150 to 200 ft (45 to 60 m), with a diameter of 3 to 6 ft (0.9 to 1.8 m). It is a native of western North America, being most abundant in British Columbia, Washington and Oregon. It has been planted extensively in Great Britain and other parts of Europe.

b. The timber

Properties. Douglas fir has a pinkish-brown heartwood distinct from the pale sapwood, with prominent growth rings. The wood is generally straight-grained, but sometimes has a tendency to wavy or spiral grain. Average density about 33 lb/ft^3 (540 kg/m^3). The timber is appreciably stronger than Scots pine, especially in bending, though similar to it in density. Its movement is small.

The sapwood may be attacked by the common furniture beetle, but appears to be attacked only rarely by the house longhorn beetle. The heartwood is moderately durable and is resistant to preservative treatment.

Processing. The timber dries rapidly and well with little checking or distortion, but knots tend to split and loosen. It works readily with hand and machine tools, though a little less easily than Scots pine of average quality. Hard and loose knots may be troublesome in sawing and planing. It is somewhat harder to nail than other softwoods. It is used on a very large scale in North America for the manufacture of plywood.

Uses. Owing to the large size of trees in North America, Douglas fir is obtainable in ample widths and long lengths. This, coupled with its good strength properties, renders it very suitable for heavy construction work in buildings, bridges, etc., including laminated arches and roof trusses. It is also widely used for vat making.

Home-grown Douglas fir is generally similar in properties to the American timber, though not available in as large sizes or free from knots. It is suitable for carpentry, structural work and general building purposes and for packing cases, or for use in the round for transmission poles and pit props. Selected material is suitable for joinery. Douglas fir plywood is very widely employed in building and packaging, a major use being for concrete shuttering.

1.2.5 European larch – Larix decidua

a. The tree. Larch grows in many parts of Europe, including Great Britain, and commonly reaches a height of 100 ft (30 m) or sometimes more. For silvicultural reasons, much of the larch that has been planted in Britain is Japanese larch (*Larix leptolepis*) or a hybrid between this species and *Larix decidua*, but the timber produced by the three species is generally similar. Larch is not imported in any great quantity.

b. The timber

Properties. Larch is one of the heaviest and strongest softwoods, having an average density of 37 lb/ft^3 (590 kg/m^3). The heartwood is resinous and is pale reddish-brown to brick red in colour, sharply differentiated from the narrow, light coloured sapwood. The annual rings are clearly marked. The timber is harder and slightly stronger than Scots pine. Movement is small.

Larch is more durable than most softwoods, but the durability of the heartwood varies greatly and it is classed as moderately durable. Damage by the common furniture beetle and the house longhorn beetle is usually unimportant. The heartwood is resistant and sapwood moderately resistant to preservative treatment.

Processing. The timber dries fairly rapidly with some tendency to distort. It saws and machines fairly readily and finishes cleanly in most operations, but the knots, which are often hard and loose, tend to damage the cutting edges of tools. It takes stains, paints and varnishes satisfactorily, but care is needed in nailing to avoid splitting.

Uses. Larch is stronger and more durable than most other softwoods and is employed primarily for outdoor structures e.g. farm buildings, gates, fences, poles, footbridges, etc., and in boat building. Small trees are used for pit props, stakes and transmission poles. Now that preservative treatment is becoming more general, natural durability is less important than it used to be. For very long life, preservation even of moderately durable timbers such as larch is recommended.

1.2.6 Western hemlock – Tsuga heterophylla

a. The tree. Western hemlock is a large tree growing in western Canada and United States. It reaches a height of 200 ft (60 m) with a diameter of 6 to 8 ft (1.8 to 2.5 m). Considerable quantities are imported into the United Kingdom from British Columbia. It has been planted to some extent in Great Britain.

b. The timber.

Properties. A straight-grained, non-resinous softwood of medium density, averaging about 30 lb/ft^3 (480 kg/m^3). It is pale brown in colour, with growth rings that are less prominent than those of Douglas fir. It is lighter in weight and correspondingly lower in strength than Douglas fir, and is comparable in properties to Scots pine. Medium movement.

The sapwood is liable to attack by the common furniture beetle. The heartwood is non-durable and is resistant to preservative treatment.

Processing. When green the timber often has a very high moisture content and does not dry as rapidly or as easily as Douglas fir, but with care it can be kiln dried very satisfactorily. It works readily in hand and machine operations, though somewhat less easily than imported redwood, and it has little dulling effect on cutting edges. It has fairly good nailing and screwing properties but it is advisable to bore for nails near the ends of boards. It gives good results with stains, varnish, paint and polish, and can be glued satisfactorily. Used in North America for manufacture of plywood.

Uses. The timber can be obtained in large sizes and is used for constructional purposes, joinery, sleepers, vehicle building, etc. Timber from trees grown in Britain is generally similar in properties to the imported product, but the trees are faster grown and comparatively young, and the material has considerably more knots.

1.2.7 Western red cedar – Thuja plicata

a. The tree. Western red cedar is also a tree of western Canada and United States, where it grows to a height of 150 to 250 ft (45 to 75 m), with a diameter of 3 to 8 ft (0.9 to 2.5 m). It has been planted to some extent in Great Britain, where it grows well.

b. The timber

Properties. The timber is reddish brown, non-resinous, straight-grained, somewhat coarse in texture and with fairly prominent growth rings. It is the lightest in weight of the common softwoods, having a density of about 23 lb/ft^3 (370 kg/m^3). The timber is soft and comparatively low in strength, in accordance with its low density. Small movement.

The heartwood is durable, and resistant to preservative treatment. Attack by wood-boring insects is not generally serious.

Processing. Timber in the thinner sizes dries readily with very little degrade, but some thicker planks may prove very difficult to dry. Works easily in all hand and machine operations and has little dulling effect on tools, but sharp cutters are needed to produce a good finish on account of the softness of the wood. Takes nails and screws well, and stains and paints satisfactorily.

Uses. The outstanding properties of western red cedar are its light weight, good durability and good dimensional stability. It is used in the construction of greenhouses, outhouses and bee-hives and for exterior work such as weather boarding and cladding. It is also employed in the production of roofing shingles. The timber has somewhat acidic properties and tends to accelerate the corrosion of metals in contact with it. Precautions should be taken, either by protecting metals from direct contact with the wood or by use of corrosion-resistant metals.

1.2.8 Beech, European – Fagus sylvatica

a. The tree. Beech is of widespread occurrence in Europe approximately between latitudes 40° and 60°N, and in western Asia. It grows well in the British Isles. The tree reaches a height of 100 ft (30 m) or occasionally 150 ft (45 m), with a diameter of about 4 ft (1.2 m), and when well grown has a long clear bole.

b. The timber

Properties. The timber is whitish or very pale brown, darkening on exposure to slightly reddish brown. The practice of steaming beech, common in south-east Europe, changes the colour to pink or light red. The sapwood is not normally distinguishable from the heartwood. The grain is usually straight and the texture fine and even. The density is variable according to growth conditions and climate, averaging 42 lb/ft^3 (670 kg/m^3) for central European beech to 45 lb/ft^3 (720 kg/m^3) for home-grown and northern European beech. It is one of the strongest of home-grown timbers (see table 1.5). Its moisture movement is large.

The timber is classed as perishable, and is permeable to preservatives. It is immune from attack by the powder-post beetle but liable to attack by furniture beetles, and timber in old buildings may be attacked by the death watch beetle.

Processing. The timber dries fairly well and fairly rapidly, but with some tendency to split, check and distort. Shrinkage during drying is high. Working properties vary according to the conditions of growth, but are generally satisfactory. There is some tendency for the saw to bind when green timber is converted. It has very good turning properties. Pre-boring is necessary in nailing. Its steam bending properties are exceptionally good. It takes stains and polishes well. Beech is employed for manufacture of plywood.

Uses. In the United Kingdom beech is used in larger quantities than any other hardwood. Both home-grown and imported timber are used. The timber is heavy and strong, usually straight grained, of plain appearance, and has good turning and bending properties. The largest consumer is the furniture industry and it is also used for joinery, general turnery, tool handles, brush backs, bobbins, domestic woodware, toys, and many other purposes. It is unsuitable for exterior use owing to its lack of durability.

1.2.9 Oak, European – Quercus robur and Quercus petraea

a. The tree. Both species grow throughout Europe south of about 63°N, and in Asia Minor and North Africa. In Great Britain oak is the commonest forest tree, being most abundant in the South and Midlands. The tree reaches a height of 60 to 100 ft (18 to 30 m) with a diameter of 4 to 6 ft (1.2 to 1.8 m), and forms a straight, clear bole when grown under forest conditions, but branches low down when grown in the open.

b. The timber

Properties. The appearance of oak is well known. It is yellowish brown in colour, with a distinct growth ring figure, while quarter-sawn material exhibits a characteristic silver-grain figure. The sapwood is light in colour and distinct from the heartwood. Density, structure and quality are affected by growth conditions. Home-grown oak averages about 45 lb/ft^3 (720 kg/m^3) in density, while timber from central Europe is milder in character and averages about 42 lb/ft^3 (670 kg/m^3). Oak is slightly lower in strength than European beech, and has medium movement. Dark stains are liable to appear on the wood if it comes into contact with iron or iron compounds under damp conditions.

Oak is well-known for its durability and is classed as durable. The sapwood is permeable but the heartwood is extremely resistant to preservative treatment. The sapwood is susceptible to attack by powder-post beetles and the common furniture beetle, and sapwood and heartwood of timber in old buildings are liable to be attacked by the death-watch beetle, especially when decay is present.

Processing. Oak is not an easy timber to dry. It dries very slowly with a marked tendency to split and check, and considerable care is needed to avoid excessive degrade. In kiln drying, mild drying conditions are essential. In general it can be sawn and finished readily, but its working properties vary with density. Nailing is difficult and pre-boring is advisable. It is a very good wood for steam bending. It glues satisfactorily and stains and polishes very well. It is widely used as decorative veneer, which is cut by slicing in order to obtain the most attractive grain pattern.

Uses. Oak is a strong, durable timber with an attractive appearance, which finds many uses both for structural purposes out-of-doors and for decorative purposes indoors. Good quality oak is used in furniture production, panelling, joinery, shop-fitting, boat building, and flooring, while the lower grades are widely used for fencing, gates, wagon construction and other purposes. Its impermeability combined with good bending properties make it very suitable for cooperage.

1.2.10 Ash, European – Fraxinus excelsior

a. The tree. Ash commonly grows to a height of 80 to 100 ft (25 to 30 m), with a clear bole of 30 to 50 ft (10 to 15 m) according to growth conditions. It occurs widely in Europe, except the extreme north, and Asia Minor.

b. The timber

Properties. The timber is white to pale brown, with no clear distinction between heartwood and sapwood. The grain is straight and the texture coarse, with clearly marked annual rings. The density varies according to growth conditions and is generally in the range 32 to 52 lb/ft^3 (510 to 830 kg/m^3), averaging 43 lb/ft^3 (690 kg/m^3), that is, about the same as beech and oak. Fast-grown timber is higher in density and strength than slower-grown timber and is preferred for the more exacting purposes. In strength, ash is comparable to European beech, but is outstandingly high in toughness, being superior in this property to any other home-grown hardwood. It has medium movement.

The sapwood is liable to attack by powder-post beetles and by the common furniture beetle. The heartwood is perishable and is moderately resistant to preservative treatment.

Processing. The timber dries fairly rapidly with little splitting or checking but with some tendency to distort, and end-splitting is sometimes severe. Its working properties are satisfactory. In nailing, pre-boring is advisable except with the less dense material. It stains and polishes well. It has excellent steam-bending properties except when irregular grain or knots are present.

Uses. Ash is highly variable in quality. Good selected material is outstanding for its toughness and its good bending properties, and is used widely for sports goods and for handles of tools such as picks, shovels, axes and hammers. It is also suitable for furniture parts. Ash is employed extensively in road vehicles and agricultural implements for parts where toughness and weight are important, but it should be used with care on account of its lack of decay resistance. It is also used for a range of items in boat building.

1.2.11 Birch, European – Betula pubescens and Betula verrucosa

a. The tree. Birch is a common tree in northern and eastern Europe and grows widely in the British Isles where it is often known as silver birch on account of its silvery bark. It is not a large tree, generally reaching a height of 60 to 70 ft (18 to 21 m), but is occasionally larger. In Scandinavia boles are often straight and clean for 30 ft (9 m), but in Britain it is usually less well formed.

b. The timber

Properties. The wood is white to light brown in colour with no conspicuous structural features, and no clear distinction between sapwood and heartwood. It is usually straight-grained with a fine texture. Its density is, on average, 41 lb/ft^3 (660 kg/m^3) and it is comparable in most strength properties to beech, but superior to it in toughness.

Birch is not attacked by powder-post beetles, but is liable to damage by the common furniture beetle. The timber is classed as perishable. It is permeable to preservatives.

Processing. The timber dries fairly rapidly and well, but with some tendency to distort. As it is very susceptible to fungal attack it should be dried as soon as possible after felling and conversion. Its working properties are satisfactory, though care is required in machining material with irregular grain around knots. It glues well, but when nailing near the edges of material with irregular grain, pre-boring may be advisable. It has good bending properties if free from knots and irregular grain, but these features are often present. It takes stains and polishes well. It is very widely used for plywood manufacture, especially in Finland.

Uses. Birch, mainly from the Continent, is used in furniture and for brushes and brooms, and turnery. When treated with preservatives it is suitable for posts. It is principally used in the form of plywood, which is manufactured in Finland and USSR and imported in large quantities into the United Kingdom.

1.2.12 Maple: Rock Maple – Acer saccharum; Soft Maple – Acer saccharinum

a. The tree. The maples are trees of eastern Canada and USA. Rock maple is also known as hard maple or sugar maple. Both species grow to a height of about 70 to 90 ft (21 to 27 m) with a diameter of 2 to 3 ft (0.6 to 0.9 m).

b. The timber

Properties. Both timbers are creamy white in colour and the sapwood is not easily distinguishable from the heartwood. The grain is straight and the texture fine and even. Growth rings are rather more pronounced in rock maple than in soft maple. Rock maple (about 45 lb/ft^3, 720 kg/m^3) is considerably higher in density and stronger than soft maple (average density about 34 lb/ft^3, 610 kg/m^3) and is comparable in strength to European beech. Both timbers have medium movement.

The maples are non-durable and are liable to attack by the common furniture beetle. Rock maple is classed as resistant, and soft maple as moderately resistant to preservative treatment.

Processing. The timbers dry slowly, but without undue difficulty. Rock maple is harder and more difficult to work than soft maple, but saws and machines well if proper care is taken. Rock maple appears to be a good bending wood, but is not much used for this purpose. Both timbers give good results with stains and polishes.

Uses. Rock maple has good strength properties, and finishes and turns well. Its outstanding features are its fine, smooth texture and its high resistance to wear, and its principal use in the United Kingdom is as a flooring timber. It is suitable for industrial flooring taking heavy traffic and is also used for roller skating rinks, dance halls, squash courts, bowling alleys, etc. It is also suitable for furniture and panelling, and is used for rollers in textile machinery, shoe lasts, sports goods, etc.

Soft maple is softer, lighter and lower in strength than rock maple. It is suitable for furniture and joinery and for turnery, but less suitable than rock maple for flooring.

1.2.13 Mahogany, African – Khaya spp.

The standard name African mahogany covers all species of *Khaya*, but most of the African mahogany used in the United Kingdom consists of *Khaya ivorensis* and *Khaya anthotheca* shipped from West Africa. Trade names indicating the locality or port from which the timber is derived are also in use.

American mahogany is a somewhat similar timber obtained from several species of *Swietenia* growing in Central and South America. The name mahogany, with a geographical or other qualification, is sometimes applied to other woods that are not related to *Khaya* or *Swietenia*, e.g. cherry mahogany, Gaboon mahogany, Philippine mahogany, etc., but the use of such names may lead to confusion, and is not recommended.

a. The tree. Both species of African mahogany grow to a height of about 180 ft (55 m). They have a long clear bole up to 4 to 6 ft (1.2 to 1.8 m) in diameter, generally with buttresses at the base.

b. The timber

Properties. The timber is pink when freshly sawn, darkening on exposure to reddish brown. The creamy white or yellowish sapwood is not always clearly demarcated from the heartwood. The grain is sometimes straight, but more usually interlocked, producing a stripe figure on quarter-sawn material, and the texture is moderately coarse. Brittleheart is present at the centre of some logs. The average density of the wood is about 33 lb/ft^3 (530 kg/m^3) and in strength it lies about half way between obeche and beech. It has small movement.

The sapwood is liable to attack by powder-post beetles and by the common furniture beetle. The heartwood is moderately durable, and extremely resistant to preservative treatment, the sapwood being moderately resistant.

Processing. The timber dries fairly rapidly with little degrade. Its machining properties may be affected by interlocked grain and by woolliness in some material, but generally a good finish can be obtained with proper care. It takes nails and glues satisfactorily, and stains and polishes well. It is extensively used for manufacture of plywood.

Uses. African mahogany is a wood of medium density and pleasing appearance, having good working properties and small movement. It is used extensively in the furniture industry for exterior and interior parts, and in the joinery trade for panelling, general interior joinery, doors, bank fittings, etc. In boat building it is suitable for most parts of boats and is widely used for planking and, in veneer form, in the cold moulding process; it is very suitable for racing craft where low weight is important. It is used also for many other purposes for which a good quality, medium weight hardwood is required.

American mahogany is a high class timber, generally regarded as superior to other mahoganies. Its uses are mainly confined to special purpose high quality work on account of its relatively high cost and limited availability.

1.2.14 Teak – Tectona grandis

a. The tree. Teak is indigenous to India, Burma and South-east Asia, but has been planted in Africa, the West Indies and elsewhere. The tree is very variable in size and form according to locality and conditions of growth. In favourable localities it may reach a height of 130 to 150 ft (40 to 45 m) with a diameter up to 6 to 8 ft (1.8 to 2.5 m), but is generally somewhat smaller.

b. The timber

Properties. The timber is golden brown in colour, darkening on exposure, and is sometimes figured with dark markings. The grain is often straight, but sometimes wavy, and the texture is coarse and uneven. Its density averages about 40 lb/ft^3 (640 kg/m^3), and its strength is comparable to, or slightly lower than, that of beech. It has small movement.

The heartwood is classed as very durable and is reported to be highly resistant to termites. It is extremely resistant to preservative treatment. The sapwood is liable to attack by powder-post beetles.

Processing. Teak dries well, but rather slowly. In sawing and machining it has a variable, and sometimes severe, blunting effect on cutting edges, and tungsten carbide cutters are advantageous. Care is needed to obtain a good finish. Gluing of freshly machined or newly sanded surfaces is satisfactory, but long exposed surfaces may present difficulty. It takes polishes satisfactorily.

Uses. Teak is an outstanding timber on account of its many valuable properties, including durability, strength, moderate weight, relative ease of working, dimensional stability and pleasing appearance, and has a wide range of uses. It is extensively used in shipbuilding, for boat planking (particularly for use in tropical waters), and for interior fittings of boats. It is a good timber for exterior joinery and is also extensively used in furniture manufacture and for garden furniture. It has good resistance to chemicals and has been used in industrial chemical plant.

1.2.15 Afrormosia – Pericopsis elata (formerly Afrormosia elata)

a. The tree. A tall tree, reaching a height of 150 ft (45 m), with a clear bole up to 100 ft (30 m) long above buttresses. Its diameter is about 3 ft (1.0 m) or sometimes more. It grows in West Africa, mainly Ghana and the Ivory Coast.

b. The timber

Properties. The timber is yellowish brown, somewhat resembling teak, but progressively darkening to dark brown on exposure to light. The sapwood is narrow, and is light coloured and clearly distinguishable from the heartwood. The grain varies from straight to interlocked, with a moderately fine texture. It is slightly heavier than teak, with an average density of about 43 lb/ft^3 (690 kg/m^3), and it is slightly higher in strength than European beech. Its movement is small. Dark stains are liable to appear on the wood if it comes into contact with iron or iron compounds under damp conditions.

The heartwood is classed as very durable, and is extremely resistant to preservative treatment. It is generally resistant to insect attack.

Processing. Afrormosia dries rather slowly but very well, with little degrade. Its working properties are generally satisfactory provided that cutting edges are kept sharp, but interlocked grain affects many machining operations. It stains and polishes well. It is employed for plywood manufacture and is usually available in the United Kingdom.

Uses. Afrormosia is a good alternative to teak for many purposes. It is outstanding for its small dimensional movement, high durability and good appearance, combined with good strength properties and satisfactory working and drying properties. It is suitable for a wide range of exterior and interior work such as joinery, furniture, flooring and boat building. It is widely used as decorative veneer as well as in solid form.

1.2.16 Iroko – Chlorophora excelsa

a. The tree. Iroko is widely distributed in Central Africa where it grows to a height of about 160 ft (50 m) and diameter up to 8 to 9 ft (about 2.5 m). The bole may be unbranched for about 70 ft (21 m) or more.

b. The timber

Properties. Iroko is a timber possessing many of the valuable features of teak. It is yellowish brown in colour, deepening to dark brown, with lighter markings. The sapwood, which is 2 to 3 in (50 to 75 mm) wide, is pale in colour. The grain is commonly interlocked and texture rather coarse. When dry it lacks the characteristic greasy feel of teak. In density the wood averages about 40 lb/ft^3 (640 kg/m^3) and it is slightly lower in strength than beech. Hard deposits known as 'stone' occasionally occur within the wood and may damage saws. Its movement is small.

The heartwood is very durable and has been reported highly resistant to termites, but the sapwood is liable to attack by powder-post beetles. The heartwood is extremely resistant to preservative treatment, while the sapwood is permeable.

Processing. The timber dries well and fairly rapidly without much degrade. It has a moderate blunting effect on saws, but this may be severe when stony deposits are present. Machinability is generally good, but may be affected by interlocked grain. It can be nailed satisfactorily and glues well. It takes stains and polishes well when a filler is used.

Uses. Iroko is a strong, very durable timber of attractive appearance and small movement, suitable for many of the purposes for which teak is used, including exterior and interior joinery, boat and vehicle building, bench tops and draining boards. It is a valuable structural timber, suitable also for piling and marine work.

1.2.17 Keruing, Gurjun – Dipterocarpus spp.

Timber of the keruing type is produced by a large number of species of the genus *Dipterocarpus*. It is known by distinctive names according to its country of origin. The most important commercially are:

Keruing from Malaya, Sarawak, Sabah and Indonesia
Gurjun from India and Burma
Yang from Thailand
Apitong from the Philippines

The general character of the timber from the different countries is similar, but gurjun and yang tend to be more uniform in properties than keruing.

a. The tree. Trees vary in size according to species and locality, but are commonly 100 to 200 ft (30 to 60 m) in height and 3 to 6 ft (1.0 to 1.8 m) in diameter, with a clear straight bole up to 70 ft (21 m) long.

b. The timber

Properties. The timber varies in colour from pinkish brown to dark brown, sometimes with a purple tint, and is rather plain in appearance. The sapwood is grey and usually well defined. In comparison with many tropical timbers these timbers are characterized by straightness of general grain direction, which may nevertheless be interlocked. The texture is moderately coarse, but even. Some of the timber exudes resin, which mars the surface, especially when exposed to heat or sunlight. The density of the timbers is mostly in the range 45 to 50 lb/ft^3 (720 to 800 kg/m^3). On average they are somewhat higher in strength than beech. They have medium to large movement.

The timber is fairly durable and it varies from moderately resistant to resistant to preservative treatment.

Processing. The timber generally dries slowly, even at high temperatures, and distortion is often considerable. Its working properties are very variable according to species, and it may have a moderate to severe blunting effect on tools because some of the timber contains silica. Apart from this it generally saws and machines satisfactorily, though resin adhering to tools may sometimes be troublesome. It has adequate nailing properties. Care is required in staining and polishing when resin is present.

Uses. These are useful general purpose timbers of relatively low cost. They are suitable for general construction work and are used for many of the purposes for which oak was formerly used, e.g. for frames, flooring and sides of road vehicles, and in boat building. Some timber is liable to exude resin if used for external joinery exposed to the sun. Resin exudation appears to be more common in keruing than in gurjun or yang.

1.2.18 Ramin – Gonystylus spp., principally G. bancanus

a. The tree. A tall tree with a straight, cylindrical bole, unbuttressed but sometimes slightly fluted at the base, and having an average diameter of 2 ft (0.6 m) or sometimes more. Ramin grows in south-east Asia, especially Sarawak. Timber from Malaya is known as melawis.

b. The timber

Properties. Ramin is a white to pale straw-coloured wood without any outstanding features. The sapwood is not easily distinguished from the heartwood. The grain is straight or shallowly interlocked and the texture moderately fine and even. The average density is about 41 lb/ft^3 (660 kg/m^3), and it is generally slightly higher in strength than beech, but slightly less tough and hard, and weaker in shear and in resistance to splitting. It has large movement.

The timber is perishable and the sapwood is liable to attack by powder-post beetles. It is permeable to preservatives.

Processing. Ramin dries readily with little distortion, but with a tendency to end-splitting. A strong unpleasant odour may be evolved during kiln drying, but does not generally persist after the wood is dry. Working properties are generally satisfactory and the timber glues well, but has a marked tendency to split when nailed. Staining and polishing are satisfactory with a small amount of filler.

Uses. Ramin is a plain, pale coloured wood, slightly lighter in weight than beech and employed for similar purposes. It is a good utility timber for interior uses, and is used in furniture, interior joinery, mouldings, small handles, wooden toys and other domestic articles.

1.2.19 Utile – Entandrophragma utile

a. The tree. Utile, known in France and the Ivory Coast as sipo, is a large tree of west and central Africa. It grows to a height of 150 to 200 ft (45 to 60 m) with narrow buttresses, above which the bole is straight and cylindrical, clear of branches for 70 to 80 ft (21 to 24 m), and up to 8 ft (2.5 m) in diameter.

The related species, sapele (*Entandrophragma cylindricum*) is similar to utile in many of its properties.

b. The timber

Properties. Utile is of a fairly uniform reddish or purplish brown colour, with a light brown sapwood, distinct from the heartwood. The grain is interlocked and rather irregular and the texture is more open than that of sapele. Quarter-sawn material exhibits a stripe figure, but this is less marked than it is in sapele. It has an average density of about 41 lb/ft^3 (660 kg/m^3) and it is comparable in strength to European beech. Movement is medium.

The heartwood is classed as durable, and is extremely resistant to preservative treatment. The sapwood is liable to attack by powder-post beetles.

Processing. The timber dries at a moderate rate with little difficulty. Its working properties are generally satisfactory, but care is required to avoid tearing of interlocked grain. Nailing is satisfactory and it takes glues and finishes well. It is employed in plywood manufacture and is usually available in the United Kingdom.

Uses. Utile is a reddish brown timber, similar in general appearance and properties to sapele, but less liable to distort during drying and in subsequent use. It is suitable for furniture and cabinet making, interior and exterior joinery, and construction work.

1.2.20 Meranti, seraya, lauan – principally Shorea spp.

A large number of species of the genus *Shorea* growing in south-east Asia produce timber known according to its country of origin as meranti, seraya or lauan. Generally, the name meranti is applied to timbers from Malaya, Sarawak and Indonesia, seraya to timbers from Sabah, and lauan to timbers from the Philippines. The timbers vary in colour and density and are conveniently grouped as follows:

1. Light red meranti, light red seraya, white lauan (white lauan also includes other species).
2. Dark red meranti, dark red seraya, red lauan.
3. Yellow meranti, yellow seraya.
4. White meranti, melapi.
5. White seraya, white lauan (this is not the same type of timber as white meranti).

The commercial timber in each of these groups consists of a number of species and therefore exhibits some variability, but there are fairly clear distinctions between the groups. The most important of these timbers are the light red meranti group and the dark red meranti group.

The trees are large, often up to 200 ft (60 m) in height, with well-shaped boles.

a. Light red meranti. This timber, and the corresponding timbers from other countries, varies in colour from very pale pink to mid-red. The grain is usually shallowly interlocked, producing a broad stripe figure on quartered surfaces, and the texture is coarse but even. Brittleheart is commonly present at the centre of logs of some species. The timber is of medium weight, with an average density of about 32 lb/ft^3 (510 kg/m^3) and its strength properties lie about half-way between those of obeche and beech. It has small movement. The heartwood ranges from non-durable to moderately durable and is resistant, or extremely resistant, to preservative treatment.

The timber generally dries fairly rapidly and its working properties are satisfactory. It takes nails and glues well, and stains and polishes satisfactorily when filled. It is extensively used for plywood manufacture.

Light red meranti is a useful medium-weight timber, and is used for a wide range of light structural work and interior joinery, and for general purposes. It has been employed as a mahogany-like wood for interior parts of furniture.

b. Dark red meranti. This timber, and corresponding grades of seraya and lauan, is darker in colour, heavier, stronger and more durable than light red meranti. In density it averages about 42 lb/ft^3 (670 kg/m^3), and in strength it approaches European beech. It dries more slowly than light red meranti, and saws and machines satisfactorily. It is also used for plywood manufacture.

Dark red meranti is attractive in appearance and is somewhat less variable in character than light red meranti. Owing to its superior strength and durability it is suitable for more exacting purposes in construction, exterior and interior joinery, shopfitting and boat building.

1.2.21 Obeche or wawa – Triplochiton scleroxylon

a. The tree. A large West African tree, reaching a height of 150 to 180 ft (45 to 55 m), with narrow buttresses extending up to 20 ft (6 m) up the bole. The bole is 3 to 5 ft (0.9 to 1.5 m) in diameter, and is straight and cylindrical and free from branches up to 80 ft (25 m).

b. The timber

Properties. The timber is nearly white or pale straw coloured, with a wide sapwood which is not clearly distinguishable from the heartwood. The grain is generally interlocked, giving a characteristic striped appearance to quartersawn material. The texture is moderately coarse, but even. Brittleheart is sometimes present at the centre of large logs. The timber is light in weight, having an average density of about 24 lb/ft^3 (380 kg/m^3), and correspondingly low in strength properties (see table 1.5).

The timber is non-durable and the sapwood is liable to attack by powder-post beetles. The heartwood is resistant to preservative treatment, but the wide sapwood is permeable.

Processing. Obeche dries very rapidly and very well with little tendency to split or distort. It is easy to work with hand or machine tools, but sharp cutters are required owing to the softness of the timber. It can be easily nailed but its holding qualities are poor. It glues well, and takes stains and polishes satisfactorily if carefully filled. It is used for manufacture of plywood, which is usually available in the United Kingdom. It is frequently used as the core of plywood which is faced with veneers of other species.

Table 1.6 Properties of commercial woods

Name	Sources of supply	Density	Mechanical properties: Bending strength	Mechanical properties: Stiffness	Mechanical properties: Crushing strength	Natural Durability	Resistance to impregnation	Drying properties: Drying rate	Drying properties: Movement in service	Woodworking properties: Resistance in cutting	Woodworking properties: Blunting effect
Softwoods											
Fir, Silver	Europe	Medium	Low	Low	Medium	Non-durable	Moderate	Rapid	–	Low	Slight
'Parana pine'	S. America	Medium	Medium	Medium	Medium	Non-durable	Moderate	Slow	–	Low	Slight
'African pencil cedar'	E. Africa	Medium	–	–	–	Durable	Extreme	Slow	–	Low	Slight
Pitch pine	Southern USA and Caribbean region	Heavy	Medium	High	High	Moderate	Resistant	Rather slow	–	Medium	Moderate
Pine, Corsican	S. Europe and UK	Medium	Low	Low	Medium	Non-durable	Moderate	Rapid	Medium	Low to medium	–
Pine, Maritime	S. Europe	Medium	Low	Low	Medium	Moderate	Resistant	–	–	Low	Slight to moderate
Pine, Radiata	Australia, New Zealand	Medium	Medium	Low	Medium	Non-durable	–	–	–	Low	Slight
Pine, Yellow	E. Canada and USA	Light	Very low	Low	Medium	Non-durable	Moderate	Rapid	Small	Low	Slight
Yew	UK	Heavy	–	–	–	Durable	Resistant	Fairly rapid	–	Medium	–
Hardwoods											
Abura	W. Africa	Medium	Low	Low	Medium	Perishable	Moderate*	Rapid	Small	Medium	Moderate
Afara or limba	W. Africa	Medium	Low	Medium	Medium	Non-durable	Moderate	Rapid	Small	Medium	Slight
Afzelia	W. Africa	Very heavy	High	High	High	Very durable	Extreme	Very slow	Small	High	Moderate
Agba	W. Africa	Medium	Low	Low	Medium	Durable	Resistant	Fairly rapid	Small	Medium	Slight
Balsa	Central and S. America	Except. light	Very low	Very low	Very low	Perishable	Resistant*	Rapid	Small	Very low	Slight
Boxwoods	Europe, W. Asia, S. Africa, W. Indies, etc.	Very heavy	–	–	–	Durable	–	Very slow	–	High	Moderate
Camphorwood, E. African	E. Africa	Medium	Medium	Medium	Medium	Very durable	Extreme	Slow	Small	Medium	Slight

Chestnut, Sweet	Europe	Medium	Low	Low	Medium	Durable	Extreme	Slow	Small	Medium	Slight
Dahoma	W. and E. Africa	Heavy	Medium	Medium	High	Moderate	Resistant	Slow	Medium	Medium	Moderate (variable)
Danta	W. Africa	Heavy	High	Medium	High	Durable	Resistant	Rather slow	Medium	Medium	Moderate
Ebony	Ceylon, E. Indies, Tropical Africa	Very heavy	Very high	Very high	Very high	Very durable	Extreme	Fairly rapid	–	Very high	Severe
Elm, English and Dutch	UK	Medium	Low	Low	Low	Non-durable	Moderate	Fairly rapid	Medium	Medium	Moderate
Gaboon	W. Africa	Light	Low	Low	Medium	Non-durable	Resistant	Rapid	–	–	Moderate to severe
Guarea	W. Africa	Medium	Medium	Low to medium	Medium to high	Very durable	Extreme	Fairly rapid	Small	Medium	Slight to moderate
Hickory	Eastern N. America	Very heavy	High	Very high	High	Non-durable	Moderate	–	–	High (variable)	Moderate to severe
Hornbeam	Europe	Heavy	Medium	High	High	Perishable	Permeable	Fairly rapid	Large	High	Moderate
Idigbo	W. Africa	Medium	Low	Low	Medium	Durable	Extreme	Rapid	Small	Medium (variable)	Slight
Jarrah	W. Australia	Very heavy	Medium	High	High	Very durable	Extreme	–	Medium	High	Moderate
Jelutong	Malaya, Indonesia	Medium	Low	Low	Low	Non-durable	Permeable	Fairly rapid	Small	Low	Slight
Kapur	S.E. Asia	Heavy	High	High	High	Very durable	Extreme	Rather slow	Medium	Medium (variable)	Moderate
Lignum vitae	W. Indies and Central America	Except. heavy	–	–	Very high	Very durable	Extreme	Very slow	Medium	Very high	Moderate
Makore	W. Africa	Medium	Medium	Medium	Medium	Very durable	Extreme	Fairly rapid	Small	Medium	Severe
Mansonia	W. Africa	Medium	High	Medium	High	Very durable	Extreme	Fairly rapid	Medium	Medium	Moderate
Muninga	Southern Africa	Medium	Medium	Low	High	Very durable	Resistant	Rather slow	Small	Medium	Moderate
Niangon	W. Africa	Medium	Medium	Low	Medium	Durable	Extreme	Fairly rapid	Medium	Medium	Moderate
Opepe	W. Africa	Heavy	Medium	High	High	Very durable	Moderate	Rather slow	Small	Medium	Moderate
Poplar, Black	Europe	Light	Low	Low	Medium	Perishable	Moderate*	Fairly rapid	Medium	Medium	Slight
'Rhodesian teak'	Rhodesia	Very heavy	–	–	–	Very durable	Extreme	Slow	Small	High	Severe

Table 1.6 (continued)

			Mechanical properties					*Drying properties*		*Woodworking properties*	
Name	*Sources of supply*	*Density*	*Bending strength*	*Stiffness*	*Crushing strength*	*Natural Durability*	*Resistance to impregnation*	*Drying rate*	*Movement in service*	*Resistance in cutting*	*Blunting effect*
Sycamore	Europe	Medium	Medium	Low	Medium	Perishable	Permeable	Fairly rapid	Medium	Medium	Moderate
'Tasmanian oak'	Australia	Medium to heavy	Medium	High	High	Moderate	Resistant	Fairly rapid	Medium	Medium (variable)	Moderate
Walnut	Europe, S.W. Asia	Medium	–	–	–	Moderate	Resistant	Rather slow	Medium	Medium	Moderate
Willow	UK	Light	Low	Low	Low	Perishable	Resistant*	Fairly rapid	Small	Low	Slight

* These timbers often contain large amounts of permeable sapwood.

Source: *A Handbook of Softwoods* and *A Handbook of Hardwoods*, H.M.S.O.

Uses. Obeche is a light, soft, and easily worked hardwood, which is widely used for purposes where its low strength and lack of durability are unimportant. Its principal uses are in furniture manufacture, where it is employed for drawer sides, interior rails and other parts, and for interior joinery and model making.

1.2.22 Greenheart – Ocotea rodiaei

a. The tree. Greenheart is a South American tree, growing chiefly in Guyana and part of Surinam. It reaches a height of 70 to 130 ft (21 to 40 m) or occasionally more, and has a long, straight, cylindrical bole, 50 to 80 ft (15 to 25 m) long and about 3 ft (1.0 m) in diameter.

b. The timber

Properties. The heartwood is light to dark olive green in colour, sometimes marked with brown or black streaks. The sapwood, which is 1 to 2 in (25 to 50 mm) wide but may be more in small trees, is pale yellow or green, shading gradually into the heartwood. The grain is straight or interlocked and the texture fine and even. Greenheart is a very heavy timber, having an average density of about 64 lb/ft^3 (1 030 kg/m^3), and has exceptionally high strength properties even when its weight is taken into account (see table 1.5). It has medium movement. The heartwood is very durable.

Processing. Dries very slowly and with considerable degrade, particularly in the thicker sizes. Timber over 1 inch (25 mm) in thickness should be partly air dried before kiln drying. The timber is not unduly difficult to work when its hardness and high density are taken into consideration, but interlocked grain when present affects many machining operations. Its gluing properties are variable, and it polishes satisfactorily.

Uses. Greenheart is a very heavy and hard timber, outstanding in most of its strength properties, and possessing very high durability and resistance to marine borers when all sapwood is excluded. It is available in very large sizes and is therefore suitable for piling, piers, lock gates and dock and harbour work. As sawn timber it has been used for pier decking and hand rails, factory flooring, and in chemical plant, and for other specialized purposes.

1.2.23 Other timbers

Table 1.6 summarizes in general terms the more important properties of some additional softwood and hardwood timbers that are in common use in the United Kingdom. The classification adopted for density and mechanical properties is as follows:

Density

Rating	*Density at 12% moisture content* *lb/ft*^{3}	*kg/m*^{3}
Exceptionally light	Under 19	Under 300
Light	19–28	300–450
Medium	28–40	450–650
Heavy	40–50	650–800
Very heavy	50–62	800–1 000
Exceptionally heavy	Over 62	Over 1 000

Mechanical Properties

Rating	Bending strength		Modulus of elasticity (stiffness)		Compression parallel to grain	
	lbf/in²	N/mm²	1000 lbf/in²	N/mm²	lbf/in²	N/mm²
Very low	Under 7200	Under 50	Under 1200	Under 8000	Under 2900	Under 20
Low	7200–12000	50–85	1200–1500	8000–10000	2900–5000	20–35
Medium	12000–17000	85–120	1500–1800	10000–12500	5000–8000	35–55
High	17000–25000	120–175	1800–2500	12500–17500	8000–12000	55–85
Very high	Over 25000	Over 175	Over 2500	Over 17500	Over 12000	Over 85

Note: 1 N/mm² ≈ 0.1 kg/mm².

1.3 PRODUCTION OF WOOD

1.3.1 Timber resources

In the earliest historic times forests occupied a much larger proportion of the earth's surface than they do today. As a result of land clearing and of devastation of natural woodlands for commercial gain or to supply timber needs in times of war, much of the original forest has disappeared, particularly in the more densely populated areas. However, realization of the importance of forests both for production of timber and for protection of the land has led to extensive replanting in many areas, and even the more populous regions are now quite heavily forested. Europe, including the European region of the USSR, the United States and Japan as a whole all have at least a quarter of their land area under forest.

Broadly speaking, forests of one type or another cover nearly a third of the world's land surface. Forests are classified by foresters under a number of different types (cool coniferous, temperate mixed, warm temperate moist, equatorial rain, tropical moist deciduous, and dry forests) and information is available on the geographical distribution of these main types, but for the present purpose it will be more convenient to record the total forest areas in each of the main continental areas. These are given in table 1.7. Not all of this forest area is productive of timber because of inaccessibility, state of

Table 1.7 World distribution of forest areas

Region	Total forest area (million ha)	Forests as percentage of total land	Productive forest land (million ha)	Composition of forests %		
				Coniferous	Broad-leaved	Mixed
Europe except USSR	168.7	30.4	129.1	53	37	10
USSR	910.0	40.6	710.8	76	24	–
North America	750.2	38.8	425.9	55	32	13
Latin America	871.1	42.6	324.5	1	94	5
Africa	762.7	25.4	377.2	1	98	1
Near East	8.8	1.6	1.2	10	86	4
Far East except China	443.1	44.7	181.0	7	72	21
China	96.4	9.9	–	–	–	–
Pacific including Australia	218.2	27.2	41.1	36	14	50
World total	4229.2	32.2	2190.8	35	60	5

Note: 1 ha ≈ 2.5 acres. (Data from *FAO World Forest Inventory, 1963*).

development of the country, or other reasons, and the table also gives the areas of productive forests in each of the regions. The potential output of timber does not only depend on the productive forest area, but is also greatly influenced by the climatic conditions within the region, which determine the rate of growth of trees. For example, the rate of production of wood is much greater in the tropical rain forests than it is in the cool coniferous forests of northern Europe and North America.

Referring to table 1.7, the following figures relating to forest areas in the British Isles may be compared with the corresponding figures for the larger regions given in the table:

Total forest area	1.9 m. ha (4.7 m. acres)
Forests as percent of total land area	6.1%
Productive forest land	1.4 m. ha (3.5 m. acres)
Composition of forests	52% coniferous
	47% broad-leaved

The world distribution of coniferous and broad-leaved trees, producing softwoods and hardwoods respectively, follows a fairly well-defined pattern.

a. Softwoods. Softwoods are found mainly in the temperate zone of the northern hemisphere, that is in northern Europe, USSR, North America and Japan. These areas are accessible to the heavily populated and economically advanced regions immediately to the south of them, and they supply the major part of the large volume of pines, spruces, firs and other important softwood species that are so widely used in construction work and in industry as well as in the manufacture of wood pulp. Pockets of commercial softwood also occur in the tropical zone, for example Parana pine in Brazil, and large softwood plantations have been established in tropical East Africa, and also in the southern hemisphere in South Africa, Australia and New Zealand.

Supplies of softwood for the United Kingdom market are obtained mainly from northern Europe (Scandinavian countries, northern USSR, Poland, Czechoslovakia) and Canada, together with some maritime pine from Portugal and Parana pine from Brazil, and a relatively small quantity of home-produced timber.

b. Hardwoods. Hardwoods occur in great variety in both the temperate and the tropical zones. The most important hardwoods found in the temperate zone are various species of beech, oak, ash, and birch, and the main producing areas are Europe, except the extreme north, the eastern half of North America, and Japan.

Tropical hardwoods comprise a very wide range of timbers, and tropical forests occur in all countries where rainfall is sufficient for their growth, notably South and Central America, Central Africa south of the Sahara, South-east Asia and the islands of the south-west Pacific, and parts of India. Considerable quantities of tropical hardwoods are imported into the United Kingdom, mainly from West Africa and South-east Asia.

c. Home-grown timber. In 1970, about 7% of the sawn timber (excluding pulpwood, pitwood, and wood-based materials such as plywood and chipboard) used in the United Kingdom was home produced. Of this quantity, about 55% was hardwood and 45% softwood but, whereas the production of hardwood has been falling slowly in recent years, the output of sawn softwood has increased by about 50% in the period 1960–70, and is rising steadily as the extensive softwood plantations approach maturity. Thus, while the United Kingdom market for softwoods will always be based predominantly on imported timber, the home-produced material will play an increasingly important part in it, and is expected to provide an appreciable proportion of the total requirements within the next 20 to 30 years.

Further details of sources of supply of softwoods and hardwoods and of other wood-based materials will be found in section 1.8, Statistics. It may be noted here that in the United Kingdom the quantity of softwood used for manufacture of wood pulp, fibreboard and chipboard is greater than the quantity used as sawn timber, but is generally derived from trees that are too small for production of timber.

1.3.2 Forest management

Forestry is the science and art of forming and cultivating forests, usually with the primary object of producing timber either for use as such or for manufacture of wood pulp or other wood-based materials. It is a highly developed subject and its successful practice requires a high level of knowledge and experience of the planting and care of trees, the suitability of different species in various soils and climates, their resistance to pests and diseases, and many other factors. Since a tree may grow anything from 20 to 100 years or more before it is ready for felling, long-term planning, assisted where possible by forecasts of future requirements for wood and wood products, is essential. Forestry practices vary widely from country to country according to the nature of the forest and the state of development of the country, and it will only be possible here to describe in outline the main operations involved in growing trees and harvesting the timber.

Forests are broadly of two types – natural forests and man-made forests (plantations). Although in many parts of the world the natural forest which covered much of the land area has been cleared, large areas of natural forest still remain. In some parts this still exists as untouched virgin forest, while other areas have been brought under some form of management for production of timber and protection of the forest.

There is an essential difference between the extraction of timber from a forest and the extraction of minerals from the ground. Mineral resources are a wasting asset, and when they have once been exploited they cannot be replaced. A forest, on the other hand, is a living entity and is constantly manufacturing wood. With good management there is no reason why it should not continue to do so for an indefinitely long period, but if it is over-exploited it will sooner or later become exhausted and cease to be productive.

The primary object of forest management is to produce the maximum sustained yield of timber from the forest, while preserving the capital asset. This means that the forest should furnish regular annual out-turns of timber which are the maximum which the soil and climate are capable of producing for an indefinite period of time.

Where natural forests are brought under regular management, a great deal of adjustment and improvement is usually necessary before they are in an organized and fully productive condition. For instance, there may be a great excess of old trees and a corresponding deficiency of young ones, or vice versa, or as in many of the mixed forests of the tropics, there may be only a few marketable species among a large number of inferior species having no market value. Measures have to be taken to remedy such deficiencies, to improve the growing stock, and to bring the forest into an organized condition if regular annual yields and maximum productivity are to be obtained.

The process according to which fellings are arranged and the felled areas are regenerated is called a silvicultural system. As a result of long experience, particularly on the continent of Europe, a number of such silvicultural systems have been evolved and have been applied, with suitable modifications, to forests in other parts of the world.

It is clear that a forest which produces regular sustained yields of timber is superior economically to one which produces intermittent or spasmodic yields. The object to be aimed at, therefore, is a forest that contains trees having a regular and complete succession of age-classes from the youngest to the oldest, so that as each becomes ready for felling it may be cut out in its proper turn and regenerated. This type of forest rarely exists in reality, but it should be regarded as the ideal to be approached as nearly as possible in any scheme of forest management aiming at sustained yields. The arrangement of the forest to meet this objective may take various forms. The forest may be laid out in a series of areas of uniform age, which are clear felled in succession as they reach maturity, the land then being replanted to produce the next crop. Alternatively a uniform forest containing trees of all ages can be developed, individual trees or small groups of trees being selected for felling each year. In this case natural regeneration from seed from the existing trees generally takes place, and the remaining trees also provide shelter and shade for the young trees. A number of intermediate systems are in use, in which blocks, strips, wedges, etc. of forest are cleared at one time and are regenerated naturally or by planting seedlings, and all of these systems possess advantages and disadvantages.

a. Coppice. The coppice system involves reproduction by stool shoots or suckers. When felled near ground-level most broad-leaved species, up to a certain age, reproduce from shoots sent up from the stump. As a rule several shoots arise from each stump, with the result that coppice has a characteristically clumped appearance. In Britain the trees most commonly grown as coppice are hazel and sweet chestnut, but other hardwood trees, for example, oak, ash, sycamore, alder and lime, may also be grown in this way. Coppice is essentially a system for production of small or medium sized material up to pole size, but not for the production of large timber.

b. Forest plantations. Where land has been cleared or forests are to be established on new land, this is done by setting up plantations. This provides an opportunity for the forest to be planned from the outset and decisions have to be made concerning the lay-out of the forest, age distribution, spacing of trees, choice of species and whether the forest is to consist of a single species or mixed species. In Britain most of the plantations consist of softwoods because these are faster growing than the hardwoods and they provide the timber for which there is the greatest demand.

For full information on the establishment and care of plantations, books on forestry practice should be consulted. The operations involved are very briefly as follows.

Collection of seed from well-grown trees: it is very important to sow only good seed.

Sowing in seed beds in a forest nursery, which should be conveniently sited in relation to the forest. The soil requirements and operations in the nursery are essentially similar to those for a garden.

Planting. After clearing and preparing the land the young trees (one to two years old) are planted out in their final site. It is generally necessary to cut the weed growth in young plantations for the first two or three years after planting so that the growth of the young trees is not suppressed. Further attention is needed during the next ten years or

Fig. 1.11 Planting seedlings.

so, chiefly to cut down inferior tree species that often spring up naturally and threaten to injure the more valuable species. In some areas protection of the trees from attack by animals (rabbits, deer, etc.) is necessary. The spacing of the trees is generally closer than that required in the final crop, the surplus trees being removed at a later stage by thinning. Close planting, at least in the case of conifers, results in growth of smaller branches which tend to die off, so that clear wood is produced in the outer part of the stem.

Thinning is necessary to reduce the number of trees growing on a site so that those remaining have more growing space and less root competition, thus assisting in their subsequent development. It also reduces the risk of disease and pests by removing unhealthy and dead trees. Thinning is carried out in several stages, and considerable judgment is required in deciding which trees are to be removed. Thinnings are generally too small for use as sawn timber but nevertheless have some value as poles or for pulping, thus providing some financial return between planting and maturity.

Pruning may be necessary, especially if trees are not closely grown, to improve the quality of the timber by preventing the formation of knots. Broad-leaved trees, if properly grown, do not require much pruning but fast-growing conifers which have persistent branches often benefit from pruning. In the case of Douglas fir, for example, the branches do not fall off when they are dead and it is worthwhile to prune them off at the time when the first thinning is made.

1.3.3 Felling and extraction

Trees may be felled by hand or by means of power tools, the method chosen being governed largely by the size of the tree. When felling by hand, either an axe or a saw or a combination of these may be used. In Britain, felling with an axe alone is confined to small trees, usually of pole size, but in any case not greater than 18 in (450 mm) in diameter. Where there is insufficient space to swing an axe, trees may be felled with a saw alone. Sawing is started on the side on which the tree is to fall and completed on the opposite side at a slightly higher level. Wedges are driven in behind the saw to prevent it from jamming. For felling larger trees it is more usual to cut a 'gullet' with an axe at the base of the tree on the side on which it is intended that the tree should fall. Sawing is then commenced on the opposite side, using wedges behind the saw, and continued until the tree falls.

Power felling, using chain saws, has now largely replaced other methods of felling trees. Chain saws consist of an endless chain running round a guide and having saw teeth inserted in it. They are driven by a petrol engine or electric motor and are usually operated by one man. They may also conveniently be used for cross-cutting and trimming logs in the forest.

In other countries different techniques may be used, especially for felling the much

Fig. 1.12 Tree felling.

larger trees growing, for example, in western North America and in tropical forests. Many large trees in tropical forests grow with buttresses at the base, extending up to 20 ft (6.5 m) or so up the stem. In such cases it is necessary to build a platform above the buttresses in order to fell the tree at that height above the ground.

Timber is a heavy and bulky material and its removal from the forest and transport to the place where it is to be further processed (sawmill, pulpmill, etc.) may present considerable engineering problems.

The removal of logs from the growing site in the forest to a point from which they can be transported by other means, such as road, railway or river, is termed extraction. Tractors and winches are commonly used to move logs in the woods, and special equipment is used for gripping logs and to facilitate sliding along the ground. There are various patterns of timber arches, originating in the USA, which consist of a frame in the shape of an inverted 'U' carried on two wheels or sets of caterpillar tracks. Provision is made for lifting the ends of logs between the forks of the frame by means of a winch so that the small ends trail on the ground.

Overhead cableways are used in some countries, such as Switzerland, where the land is steep and uneven.

Further transport of logs is effected by road or forest railway, or by water. Road haulage is carried out by timber carriages or various types of lorries, and mechanical devices are necessary for loading timber on to lorries. The construction and maintenance of forest roads is a costly but necessary part of forestry operations. Forest railways have been used less in Britain than in other countries, but possess advantages under certain conditions.

Transport by floating in water, either as rafts or as single logs, has also not been common practice in Great Britain owing to the lack of suitable rivers in forest areas, but is widely used in some countries, e.g. Scandinavia, Canada, and some tropical countries. Where it is possible it is generally the cheapest method. Floating of single logs is only possible with timbers that are buoyant, and this rules out many tropical timbers, but heavier timbers can often be transported in mixed rafts containing a sufficient proportion of lighter logs.

1.3.4 Sawmilling

The function of the sawmill is to convert logs into timber in a form ready for direct use or for further processing. The types of machine that are used and the methods of handling vary according to the size and nature of the logs. The principal operations carried out at the sawmill are debarking, breakdown of the log into baulks, battens or boards, resawing these lengthwise in a ripping or edging operation, and cutting them across the grain in a cross-cutting operation. Subsequent operations (sawing, planing, drilling, etc.) are dealt with in section 1.4, Woodworking.

a. Log storage. Logs arrive at the sawmill by land or by water and must be stored until they are required for conversion. It is common practice, particularly at larger mills, to store logs by floating them in water in a log pond. This is particularly convenient in those countries where the logs are transported from the forest by floating them down rivers. The method also has certain advantages over storage on land because excessive drying leading to end-splitting of the logs is avoided, and they are protected from attack by insects and fungi. Where there is a quick turnover of logs, or water storage is not available, logs are stacked in piles on land.

Fig. 1.13 Transporting logs by water.

In tropical countries some hardwood species are very rapidly attacked after felling by insects (ambrosia beetles, or 'pinhole borers') which penetrate the wood causing small holes, often associated with a black stain. To avoid this form of deterioration logs of susceptible species are sprayed with an insecticide immediately after felling. Some softwoods, especially the pines, are similarly liable to be affected by staining fungi (see section 1.6.2) unless suitable precautions are taken.

b. Debarking. In most parts of the world where large-scale sawmilling is undertaken, the logs are debarked before conversion. There are two principal reasons for doing so. First the bark usually cracks in growth and picks up grit and small stones during transport to the mill, and hence causes blunting of saws, and secondly, the offcuts can be sold for pulping if free from bark, but are not acceptable if they have bark adhering to them.

There are several systems of debarking, suitable for different sizes of log. In Scandinavia mechanical rotary debarkers using knives are employed, while in North America jets of high-pressure water and air are used for debarking the large trees. In tropical countries many logs are debarked by hand adzes.

c. Conversion. Conversion is defined as the operation which turns a log into timber. Saws used for this operation are mainly of three types, namely circular saws, band saws and gang or frame saws.

Fig. 1.14 Debarking drum, using high-pressure water and air method.

Circular saws are used in the large majority of small and medium sized American sawmills and in some British mills. The first patent for a circular saw was granted in England in 1777. The early saws were usually turned slowly by water power and were roughly made and inefficient, but modern saws are carefully designed and constructed and are thinner so as to remove less material as kerf, and they run at high speeds. An important development in the design of circular saws was the invention of the inserted-tooth saw, which made possible a much longer life of the saw plate. Various types of teeth can be inserted in sockets on the circumference of the saw plate to suit the species of log that is being converted. The inserted-tooth saw has replaced the solid saw in many circular saw mills.

Saw teeth may be either swage-set or spring-set. Swaged teeth are flared at the cutting edge so that they are wider than the thickness of the plate, thus permitting the plate to clear the sides of the cut when in use. In the spring-set saw each tooth is bent away from the plane of the saw plate, alternate teeth being bent or set in opposite directions. This provides the necessary plate clearance. The teeth of inserted-tooth saws are manufactured to produce a swage type of set, while those of solid-tooth (plate) saws may be either swage-set or spring-set. A wide variety of tooth types and profiles are also available and the type selected depends upon the kind of wood to be cut, and to some extent on the rim speed of the saw and the feed rate.

Table 1.8 gives the specifications for teeth of circular saws for rip-sawing various classes of timbers at a rim speed of 10 000–12 000 ft/min (3 000–3 700 m/min).

For a saw to cut straight and accurately it must run true and its toothed rim must be

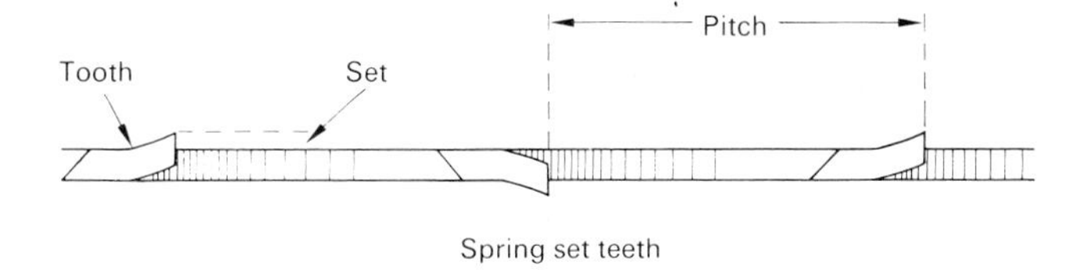

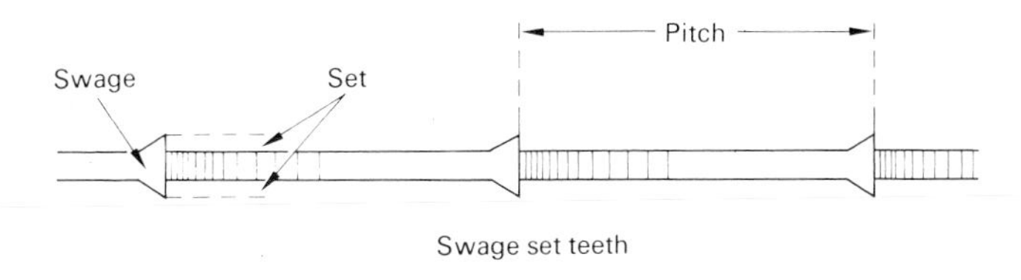

Fig. 1.15 Spring-set and swage-set saw teeth.

tight. Under normal cutting conditions the rim has a tendency to expand more than the centre, due to the effects of centrifugal force and of the heat generated by friction. To compensate for this it is necessary to tension the saw blades. This entails hammering the centre part of the blade to enlarge it – a highly skilled operation – so that the toothed rim is put into a state of tension, even when the saw is in operation.

The wide *band saw* is the most widely used machine for breakdown of logs in large sawmills and is particularly well suited to cutting logs of large diameter, but its use is by no means restricted to large scale operations. Band saws have the advantage over circular saws that they produce a narrower kerf, thus reducing the loss of wood as sawdust.

A band saw consists of a continuous band of steel with teeth on one edge, mounted on two large wheels which may be up to 6 ft(1.82 m) in diameter, one of which is above and one below the cutting area. Power is applied to the lower and heavier wheel, which acts as a flywheel, driving the saw downwards through the log as it is fed on the carriage. Band saws came into use about 1850, but in the early machines difficulties

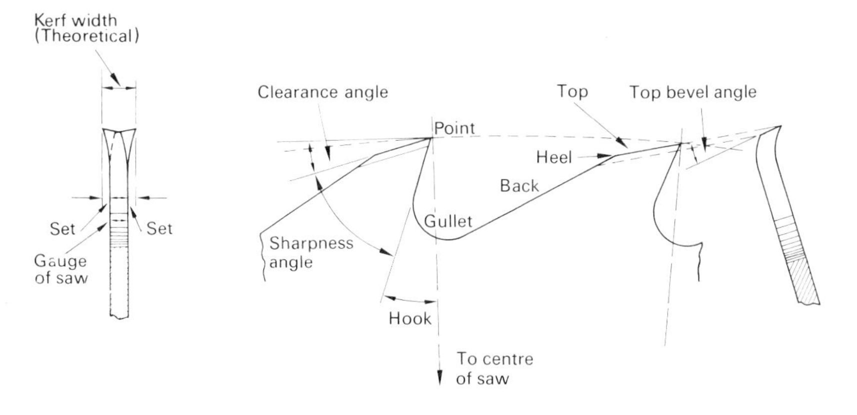

Fig. 1.16 Diagram of rip-saw tooth showing terms used in describing parts and angles.

Table 1.8 Specification for teeth of circular saws for ripping

Tooth shape number	*Timber Density range (lb/ft³, seasoned)*	*Number of teeth*	*Clearance angle*	*Rake (hook) angle*	*Sharpness angle*	*Top bevel angle*
SR48	Less than 35 (560 kg/m³)	48	15°	30°	45°	15°
HR54	35–50 (560–800 kg/m³)	54	15°	20°	55°	15°
HR60	50–65 (800–1 040 kg/m³)	60	15°	15°	60°	10°
HR80	Over 65 (1 040 kg/m³)	80	15°	10°	65°	5°
TC*	Abrasive hardwoods	54	15°	15°	65°	0°

* Tungsten carbide tipped
Source: *Forest Products Research Laboratory Leaflet No. 23.*
Note: The gullet depth to pitch ratio is 0.4 to 0.5, and the set per side is 0.015 in (0.4 mm).

arose from the inability to produce good joints required to form an endless band. The saw is subjected to severe bending and tension stresses as it passes round the two wheels, and highest quality steel and joints are required to give satisfactory service. Band saws usually run at speeds from 6000 to 12000 linear ft. per min. (1 800 to 3 700 m. per min.), the highest speeds being used for softwoods and the lower speeds for heavy hardwoods.

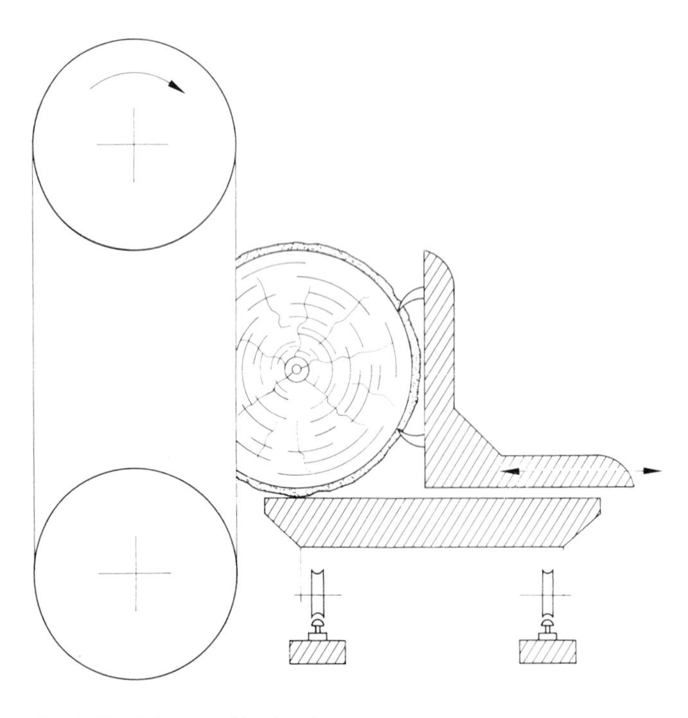

Fig. 1.17 Diagram of log bandsaw.

Like circular saws, band saws must be tensioned to operate satisfactorily, in order to allow for the expansion which occurs in the toothed edge as a result of frictional heat generated in the cutting action.

The *gang saw* or *frame saw* is the machine frequently used in Europe for the conversion of softwood logs of relatively small diameter. It consists essentially of a heavy reciprocating frame in which are mounted a number of parallel saws. As the log passes through the machine it is cut up completely into boards. Gang saws produce accurately cut timber and labour costs are low in comparison with those in other types of sawmill. The principal disadvantages are that the gang saw generally has a lower feed speed, and in order to obtain optimum yields logs require sorting to quite close size limits.

In all types of sawmill much attention is given to mill layout, log carriages and methods of feeding logs to the saws, equipment for handling and turning logs, etc. Information on these subjects, and on maintenance of saws (sharpening, levelling and tensioning) may be found in books dealing specifically with sawmilling.

d. Sawing patterns. The method of conversion of a log depends on the purpose for which the timber is required and there is clearly a variety of ways in which a log can be cut. The objective is to obtain the optimum yield and quality of timber from the log.

Flat-sawn timber is timber that is cut tangentially, i.e. more or less parallel to the growth rings, while timber that is cut radially, i.e. at right angles to the growth rings, is termed quarter-sawn. Quarter-sawn timber is less liable to distort, check and split during drying than flat-sawn timber, and its subsequent swelling and shrinkage are smaller (see section 1.1.5). It generally has greater resistance to wear and is therefore preferred for some uses, for example for flooring. There are also differences in machining properties and in grain pattern, which may be important when the timber is used for decorative purposes.

The optimum cutting pattern for a large log may be quite complex, depending upon the distribution of knots, the quality of the timber in different parts of the log, the taper of the log, etc. Small hardwood logs are generally cut through and through; that is, they are converted into a succession of parallel boards, as this gives the maximum yield of timber at minimum conversion cost, but for large logs and the majority of softwood logs more complex cutting patterns, requiring turning of the log, are often advantageous.

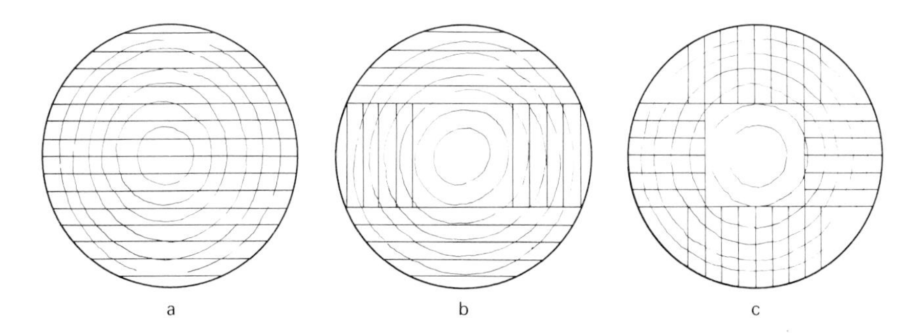

Fig. 1.18 Methods of conversion of logs:
(a) Log sawn through and through
(b) One method of cutting flat-sawn timber
(c) One method of cutting quarter-sawn timber

e. Sawdust. The production of sawdust in all sawing operations represents a considerable waste of material, and efforts are continually being made to reduce the amount of sawdust formed, for example by the use of thinner saws, or to find useful outlets for it. There are a number of uses for clean sawdust but most of them are of low value in comparison with that of the sawn timber. In countries with a large wood pulping industry there have been important developments in the use of sawdust in pulp mills, either alone or in admixture with chips, and sawdust now forms a substantial proportion of the raw material in some pulp mills. In Great Britain a saw has been designed which, instead of producing sawdust, forms chips which are suitable for use in manufacture of chipboard. Developments of this kind may provide a partial solution to the problem of utilization of sawdust.

A number of unconventional methods of cutting wood have been examined, largely with the object of reducing waste. These include the use of the plasma arc, which produces very high temperatures with a nozzle of very small diameter, high speed water jets having velocities up to 800 m/s., and gas lasers which give an intense concentration of energy. None of these methods, however, has so far proved technically and economically successful.

f. Re-sawing. This is the operation whereby the dimensions of unedged or square-edged timber are further reduced after conversion by cutting parallel to the grain. The softwood re-sawing mill is the most important type of sawmill in Britain because a very large proportion of our timber supplies consists of imported softwood in the form of deals, battens and boards, which often require to be reduced in dimensions to the size required by the user.

The machines used for re-sawing are essentially of the same type as those used for conversion of logs, namely circular saws, bandsaws and, more rarely, reciprocating frame saws. They differ from the breakdown saws in size and in details of construction in order to perform their different function. For conditions in Britain the band re-saw is generally the most suitable type. It will cut rather faster (up to 200 ft/min, 60 m/min) and removes less kerf than a circular saw, and is a simpler machine to operate, but preparation and maintenance of the saw is a highly skilled operation.

g. Cross-cutting. Cross-cutting of converted timber is generally carried out with circular saws, a number of types of machine being available for this purpose. The pendulum cross-cut saw consists of a circular saw fixed to the lower end of a long arm hung from an overhead shaft. The saw is driven by a motor at the top end of the arm, and swings like a pendulum towards a bench which is set out in front of it. The bench is provided with a number of stops and cross-cutting is done simply by pushing the wood up against the stop and swinging the saw across. The machine can be dangerous unless correctly used.

The pendulum cross-cut saw has been largely replaced by the pull-over saw which is mounted on an arm which moves in slides across the bench. Pull-over saws are made with automatic feed which drives the saw across the bench at a speed which is controllable. This reduces the labour of pulling a heavy motorized saw across the timber, and speeds up production. Mechanized systems for feeding and removing the wood by means of powered rollers or belts are also in use, but in general cross-cutting is an operation in which insufficient attention has been given to productivity.

1.3.5 Drying

Timber in the log, or when freshly sawn, contains a large amount of moisture, though there is much variation in this respect between species and from one tree to another of the same species. In hardwoods the initial moisture content, conventionally expressed as the percentage of water by weight of the dry wood substance, is commonly around 80%, with little difference between heartwood and sapwood. In softwoods the heartwood generally has a moisture content in the range 30 to 50%, and is much drier than the sapwood, which may have a moisture content as high as 200%.

For most uses it is necessary to remove the greater part of the initial moisture in green timber by a drying or seasoning process to enable the wood to give satisfactory service. In comparison with undried wood, dry wood is lighter in weight, stronger and harder, less liable to discoloration and decay, less likely to split or distort in use, and a better material for painting, polishing or treatment with preservatives. The drying process must be carefully controlled in order to avoid the deterioration ('degrade') which is liable to occur if timber is dried too rapidly or unevenly. Little drying takes place in the log, and timber is generally dried after conversion into boards or planks, and before further processing.

Reference has been made in section 1.1.5 to the ways in which water is held in wood. Up to the fibre saturation point (about 28 to 30% moisture content for most species) moisture is present in the so-called 'bound' condition; that is, it is closely associated with the wood cell walls, while at higher moisture contents 'free' moisture is present in addition, and occupies the cell spaces and pores. As drying proceeds the free moisture first moves to the surface and then evaporates, usually without much trouble, and it is only when the remaining bound water begins to leave the wood that shrinkage occurs.

a. Distortion. The various forms of distortion which tend to develop in a piece of wood as it dries are due to the fact that the shrinkage of wood, unlike that of homogeneous synthetic materials, differs considerably in the three dimensional directions. It is greatest across the grain in the direction of the growth rings (tangential shrinkage), less at right angles to the rings (radial shrinkage), and very small along the grain. The relative shrinkage which occurs in the tangential and radial directions in drying from the freshly sawn state to a moisture content of 12% is indicated for certain types of timber by the following average values:

	Tangential (%)	*Radial* (%)
Most softwoods	4.5	2.7
Hardwoods such as oak and beech	9.0	4.2
Hardwoods such as teak and afrormosia	2.5	1.5
Light-weight hardwoods, e.g. mahogany	Similar to softwoods	

As a result of the difference between tangential and radial shrinkage, flat-sawn boards tend to cup during drying while quarter-sawn (radial) boards retain their shape. If the grain is irregular, spiral, or interlocked, then other forms of distortion (spring, bow, twist) tend to develop along the length of a board.

Longitudinal shrinkage is generally unimportant, except when reaction wood is present (see section 1.1.7). Material containing compression wood or tension wood has abnormally high longitudinal shrinkage and may be troublesome in drying.

In addition to distortion resulting from differential shrinkage in the tangential and

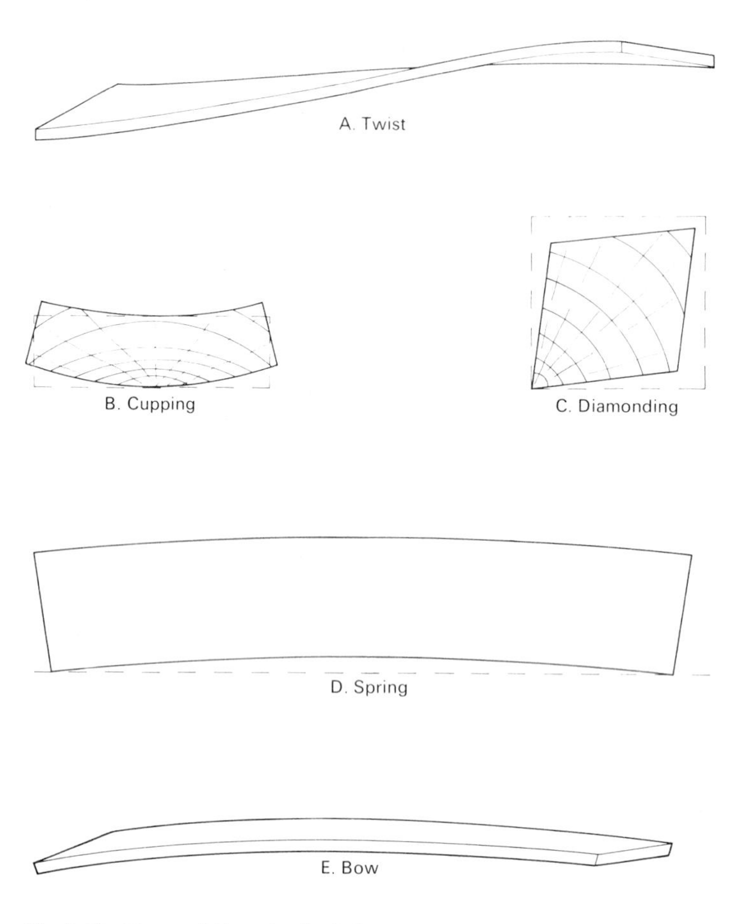

Fig. 1.19 Forms of distortion in timber.

radial directions and from irregular grain direction, stresses may be set up during drying as a result of moisture gradients within the wood. When a piece of wood is dried, the surfaces dry in advance of the interior and therefore tend to shrink first. The surface layers are thus brought into a state of tension, with the result that splitting and checking may develop. This can generally be avoided by careful control of the drying conditions so that drying does not take place too rapidly. The drying rates which different timbers can safely withstand vary widely, depending upon their shrinkage values and the resistance to movement of moisture within the wood. Consequently, different drying schedules are required for different timbers.

b. Moisture content of timber in use. Wood is a hygroscopic material and, when kept in any particular environment, its moisture content will eventually reach a value which is determined by the relative humidity of the surrounding air. Relative humidity is a function of absolute humidity and temperature. At a given absolute humidity the relative humidity is lower, the higher the temperature.

The aim in drying wood must, in general, be to bring it to a moisture content at which it is in equilibrium with the air conditions to which it will be exposed in use. There will be slight fluctuations as the air temperature and humidity vary, but these changes will be relatively small, and so long as the moisture content of the wood is adjusted to the average conditions attained in service, the timber may be expected to give little trouble in use. Table 1.9 gives some guidance as to the average moisture content of timber in a number of environments, together with a few other useful reference points on the moisture content scale. Small variations occur between different timbers but for practical purposes these may generally be neglected.

Some difficulty may be experienced when joinery must be installed in a new building before the drying-out process is complete. In such cases the air in the building will temporarily have a high relative humidity, but will ultimately reach a value corresponding to a moisture content of about 10 to 12% in the wood (in a residential or other normally heated building). In these circumstances the wood should be dried to a moisture content not much above 12% before installation, and allowance for swelling made, where possible, by expansion joints or other means.

Table 1.9 Moisture contents of timber in various environments

Moisture content of timber %	*Environment*
27	Appreciable shrinkage commences at about this point
26	
25	Suitable moisture content for pressure treatment with preservatives, fire-resisting solutions
24	
23	
22	
21	
20	Decay safety line
19	
18	Exterior joinery
17	
16	Garden furniture
15	Aircraft, motor vehicles, ships decking
14	Woodwork for use in situations only slightly or occasionally heated
13	
12	Woodwork in buildings with regular intermittent heating: block floors, furniture, brushes, musical instruments, sports goods, etc.
11	Woodwork in continuously heated buildings
10	Woodwork in situations with high degree of central heating: offices, hospitals, department stores
9	Woodwork in close proximity to sources of heat: radiator casings, mantelpieces, wood flooring laid over heating elements
8	
7	
6	
5	

Source: *Forest Products Research Laboratory Leaflet No. 9.*

Attention is drawn to the 'safety line' shown in the table at 20% moisture content. Timber which is at a moisture content higher than 20% for prolonged periods is liable to fungal decay unless suitably protected, but if it is dried to a moisture content below this level and kept in this condition, decay will not take place (see section 1.6.3).

Timber may be dried either by stacking it in the open air and allowing it to dry under the prevailing air conditions, or by placing it in some form of drying kiln where the atmospheric conditions, and consequently the rate of drying, can be controlled.

c. Air drying. In air drying, timber is stacked in such a way as to allow free circulation of air round and through the stack. It should be stacked on a well-drained site from which vegetation has been cleared and which has been covered with ashes or otherwise treated to prevent renewed growth. As a general rule stacks should not be more than 6 ft(1.8 m) wide because in very wide stacks the timber inside dries slowly. In order to permit ample ventilation the bottom layers of timber should be raised well above the ground. Tall stacks are generally preferred because the additional weight of timber tends to minimize distortion which may occur during drying. A stack should always be provided with some form of roof to protect the timber from heavy rain and from the full heat of the sun.

In building a stack the boards are separated from each other by piling sticks lying crosswise, so that both surfaces of each board are exposed to the air to facilitate drying. It is very bad practice to leave green sawn timber close-stacked for any length of time, because in this condition it is very susceptible to fungal decay or staining. It should be exposed to the air so that drying may commence as soon as possible after it has been converted. Piling sticks should be of clean, dry timber, generally 1×1 in (25×25 mm) or $1 \times \frac{1}{2}$ in (25×12.7 mm) in cross section, and should be arranged in vertical rows in the stack. Some control over the drying rate can be achieved by varying the thickness of the sticks and hence the degree of air circulation through the stack.

A common form of degrade which occurs during air drying is checking or splitting at the ends of boards, especially when certain hardwoods are put out to dry in hot weather. This can be largely avoided by protecting the ends from the sun and air so that they do not dry too rapidly. The best method of doing so is by application of a good moisture-proof coating, e.g. a bituminous paint, to the ends of the boards.

Timber may be stacked for drying at any time of the year, but the rate of drying will clearly be faster in warm, dry weather than in cool damp winter weather. Many hardwoods are liable to split if dried too rapidly during the early stages, whereas softwoods are more tolerant of severe conditions. Consequently, where a choice of season is possible, it is preferable to stack hardwoods in the winter months when the drying conditions are mild, and softwoods in the spring when comparatively rapid drying may be expected.

It should be realized that in Britain the moisture content of timber cannot be reduced by air drying to a value below about 17%. While this is adequate for some purposes (mainly outdoor uses), for many of its uses timber requires to be dried to a lower moisture content, and this can only be done by kiln drying.

d. Kiln drying. The principal advantages of kiln drying over air drying are, first, that kiln drying is much more rapid than air drying, secondly, that it enables lower moisture contents to be reached, and thirdly, that the drying conditions are under control and can be adjusted so that they are appropriate to the timber being dried.

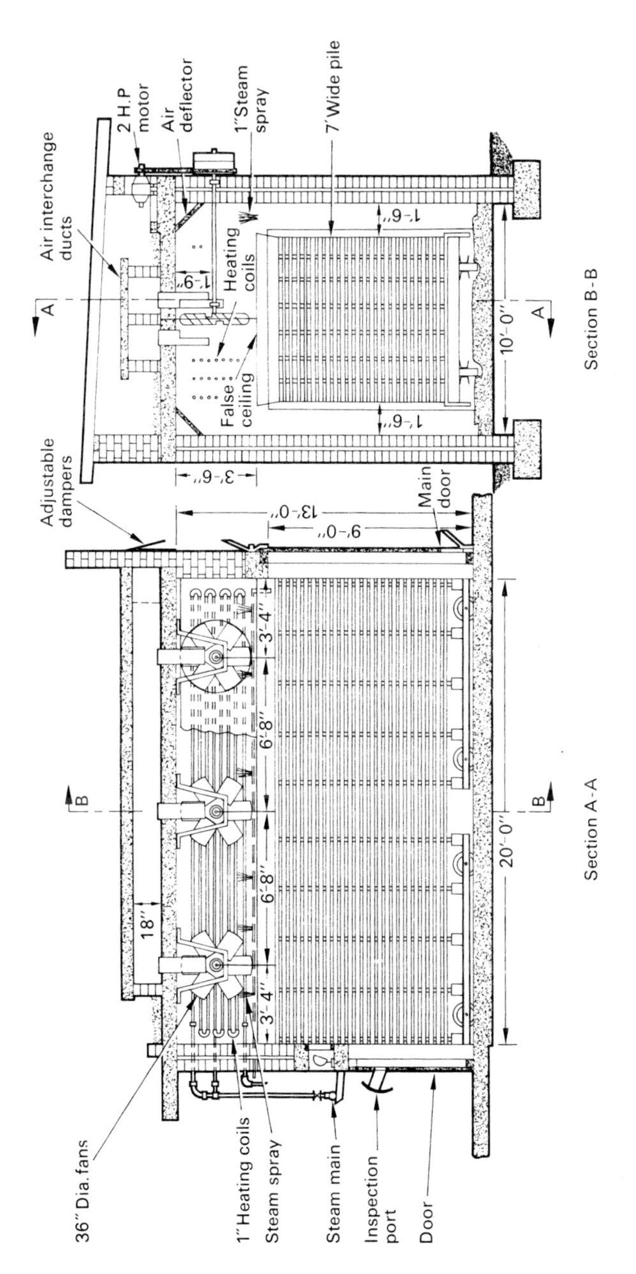

Fig. 1.20 Cross-shaft overhead fan timber drying kiln.

A drying kiln consists of an enclosed chamber into which a load of timber, stacked with sticks separating the boards as for air drying, can be moved on a truck. Provision is made for heating the kiln, generally by means of steam pipes, for controlling the humidity of the air in the kiln by introducing live steam and by allowing moist air to escape through vents, and for moving the air through the stack of timber by means of large fans. A number of different designs of kiln have been developed, varying in the arrangement of the fans and of the heating and humidifying equipment. Some have the fans inside the kiln, situated either below, above, or to one side of the timber pile, and others have the fans placed outside the kiln proper. Each type possesses advantages and disadvantages, but kilns fitted with fans inside the drying chamber and situated above or to one side of the pile of timber have certain advantages over most other types. They are simple in construction and provide a strong, uniform air circulation throughout the whole length of the kiln. This is a factor of considerable importance because, in whatever way the general condition of the surrounding air is controlled, heat must be conveyed to, and moisture removed from, the surface of every board if drying is to continue. Efficient air movement is therefore essential.

Provided that an adequate air flow is maintained through the stack of timber, the rate of drying is determined by the temperature and humidity of the air. These may be controlled by adjusting the heating coils and operating the steam spray and air vents as required. Instruments are available for recording the conditions in the kiln and for controlling them automatically. Alternatively the simple wet and dry bulb hygrometer may be used to determine both temperature and relative humidity. The difference between the dry and wet bulb readings gives a measure of the latter.

During drying the temperature and humidity are adjusted continually so as to maintain a drying rate which is the highest that the timber can safely withstand. Timbers vary widely in the rate at which they can be dried without suffering degrade or discoloration, and a series of drying schedules, suitable for various classes of timber, has been evolved as a result of experience. These lay down the temperature and humidity conditions which are to be maintained in the kiln throughout the drying process. They commence with relatively cool and moist conditions so that initial drying is not too rapid, and change by stages to warmer and drier conditions, the points at which a change is to be made being defined by reference to the moisture content of the wood. Mild schedules are necessary for timbers that are liable to check or warp, but some timbers, e.g. softwoods, are able to withstand quite severe drying conditions. Details of kiln drying schedules suitable for a large number of timbers are given in *Forest Products Research Laboratory Technical Note No. 37* 'Kiln-Drying Schedules', obtainable from H.M. Stationery Office.

In order to follow the course of drying and to control the drying according to a prescribed schedule, it is necessary to have a means of ascertaining the moisture content of the timber at any time. This is done by selecting sample boards at the start of the run, which are tested for moisture and inserted into various parts of the pile. They are removed from time to time and their moisture content determined, either by drying a small piece in an oven and noting the loss in weight or, more rapidly but somewhat less accurately, by means of electric moisture meters.

Fully automatic kilns have been developed, in which the conditions in the kiln are controlled by instruments which measure the condition of the wood continuously, but the instrumentation is quite complex.

1.4 WOODWORKING

There is not always a clear line of demarcation between sawmilling and woodworking operations, but generally speaking woodworking processes may be considered to be those further cutting and machining processes that are carried out on wood after it has left the sawmill, and generally after it has been air dried or kiln dried.

For some purposes, for example for structural work in building, rough sawn timber straight from the mill can be used, but for many applications further cutting to size or shaping is necessary, or a smooth surface which cannot be obtained by sawing is required. This is essential where wood is to be painted or where a good surface finish is required, as in joinery and furniture.

Safety considerations. By its nature woodworking machinery does not lend itself to very efficient safety arrangements and the number of accidents that occur is regrettably high. Much has been done to devise efficient guards and protective devices but there is room for improvement in the standard of safety precautions both by the designers and manufacturers of machinery and by the management and workers in the factory.

In Great Britain anyone setting up a sawmill or factory using woodworking machinery is required by the Factories Acts to post a copy of the Woodworking Machinery Regulations in the factory and to keep an accident book in which all accidents are registered at the time of their occurrence. It is to be noted that the regulations include requirements as to guards and certain other safety devices for woodworking machines, as well as to working space, condition of floors, and temperature. Similar regulations apply in many other countries.

Another provision of the Factories Acts is that no person shall be permitted to work a woodworking machine unless he has been adequately trained or is under the supervision of a properly trained person. It is essential that this provision be strictly observed because a high proportion of the accidents occur to people who are relatively new to the work.

The Factories Acts have been extended to cover the use of grinding wheels for sharpening saws and cutters (Abrasive Wheels Regulations 1970), which can present a serious hazard if not correctly mounted and used.

1.4.1 Sawing

Sawing operations are generally performed either on a circular saw bench or on a bandsaw machine. Some of the features of these machines as applied to breakdown of logs have been described in section 1.3.4 (Sawmilling). The machines used in woodworking factories are similar in principle but of less massive construction. Many different models, with either mechanized or hand feed, are available to suit specific requirements.

The performance of a saw depends on its suitability for the particular class of timber for which it is used and on such factors as the true running and condition of the blade and the shape, spacing, setting and sharpening of the teeth. For efficient sawing the gullet must be large enough to hold all the sawdust produced by the following tooth, and the tooth itself should have the correct angles and dimensions. Each tooth should have a top clearance angle to prevent the top rubbing on the wood, and the correct rake (hook) angle for the species which is being cut. Generally, for dry timber the hook

Fig. 1.21 Circular saw bench.

should be 20° for medium hardwoods and 25° for softwoods. The bite of each tooth in the wood should be between 0.8 and 1.3 mm (0.03 to 0.05 in); the bite can be increased by reducing the number of teeth on the saw (increasing the tooth pitch) or by increasing the feed speed, but care must be taken not to overload the gullet with sawdust.

a. Circular saws. The term circular saw includes saws mounted in a bench and used for ripping, deep cutting or cross-cutting, and also swing saws and other types of saw used for cross-cutting in which the saw is moved towards the wood.

The most common type of circular saw is the general purpose or hand push bench in which the saw blade is mounted on a spindle below the table of the machine and the wood is fed to it by hand and held in position by a fence. Machines are made in various sizes and there are many variations on this basic design, including machines with mechanical feed.

The blade of a circular saw working in a bench is required to be guarded in three ways:

1. Guards are necessary for the part of the saw below the bench. These may take the form either of a completely enclosed frame below the table or of a guard fitted round the portion of the blade below the bench. Serious injuries have been sustained while clearing sawdust or retrieving objects which have fallen inside the frame of the machine. Guards which do not restrict the escape of sawdust from the teeth can usually be provided without difficulty.

2. A guard must be provided for the crown and the front of the saw. This should cover the top of the saw and have a flange on each side, the flanges being deep enough to extend below the roots of the teeth at the top of the saw. The guard must be capable of easy adjustment, to a point as low as practicable at the cutting edge.
3. A riving knife must be fitted at the back of the saw above the bench. This is a blade of high-grade steel set behind and in line with the saw and firmly held in position. Its main purpose is to prevent the two sides of an incomplete cut from closing in on the up-running part of the saw and thus to reduce the risk of the workpiece being thrown back at the sawyer. There are requirements concerning the shape, size and position of the riving knife.

In addition, with saws that are fed by hand a push stick should always be used if the hand that exerts the feeding pressure is liable to approach dangerously close to the saw.

b. Bandsawing machines. Bandsawing machines vary from large band mills with pulleys 8 ft (2.4 m) in diameter carrying blades 14 in (350 mm) wide down to small bench machines with pulleys 12 in (300 mm) in diameter and blades $\frac{1}{4}$ in (6 mm) wide. Cutting is done by the down-run of the blade between the top pulley and the machine table. Bandsaws can cut more quickly and with less waste than circular saws and may

Fig. 1.22 Bandsaw.

be used for practically all types of sawing. The smaller machines can also be used to make curved cuts, and their blades do not require tensioning.

To prevent accidental contact with any moving part of the machine, the pulleys and the whole of the saw blade except the down-running part between the top pulley and the table should be completely enclosed, and the part of the blade between the top pulley and the top, adjustable guide should be protected by a plate in front of the cutting edge of the blade, with a flange on one or both sides. The guards, which are usually of a high standard on modern machines, must be hinged to give access for maintenance.

1.4.2 Planing and moulding

In the surface planing machine, which is a very widely used machine in the woodworking industry, the wood is fed either by hand or by powered rollers along the front table, up to and over a cutter block revolving on a horizontal axis below the table, and is removed from the back table. Each table is adjustable for height so that the required depth of cut may be obtained. The cylindrical cutter block usually carries two, or occasionally three, four or more knives. A cutting angle of 30° to 35° is suitable for planing softwoods and hardwoods of low to medium density, but when cutting against the grain tearing is liable to occur unless a reduced cutting angle is used. Thus, to obtain a good finish when planing timbers with pronounced interlocked or wavy grain (many tropical hardwoods are of this type) it may be necessary to reduce the cutting angle to 20° or 15°. The reduction in cutting angle is accompanied by increased feed resistance and power consumption and should only be made when necessary.

The action of the planer is to produce a series of curved (approximately circular) cuts in the surface of the wood by the succession of knives impinging on it as it passes through the machine. This results in a series of small shallow corrugations, known as knife marks or cutter marks, across the face of the wood, and a planed surface can be described in terms of the number of knife marks per inch. The pitch and depth of knife marks are

Fig. 1.23 Surface planer.

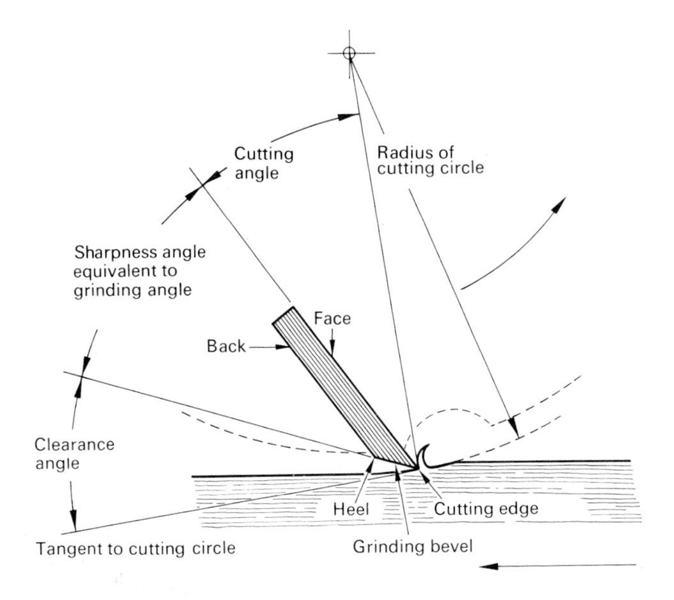

Fig. 1.24 Diagram showing parts and angles of planing cutters.

affected by the speed of roation of the knife block, the diameter of the cutting circle described by the knives, the number of knives in the knife block and the feed speed. Unless the pitch of the knife marks is very small they must be removed by subsequent sanding in order to produce a good surface finish. It is desirable that surfaces that are to be painted or polished should have not less than 16 to 20 knife marks per linear inch (1 knife mark per 1 to 1.5 mm).

Other operations such as rebating, moulding, and tonguing and grooving can be carried out on the planing machine by using cutters of appropriate shapes and suitable set-up of the machine.

To protect the operator's hands, a surface planing machine should be provided with a 'bridge' guard set centrally over the axis of the cutter block and constructed so that it is easily adjustable both in a vertical and horizontal direction. The guard should be of the same length as the cutter block and should be wide enough to overlap the edges of the feed and delivery tables. A sectionalized guard which is rigid when extended is the best type for a wide machine because, when drawn back from the fence, it will not project unduly beyond the side of the machine. As with other woodworking machines, accidents on planing machines can most effectively be prevented by the use of an automatic feeding device which eliminates the need for the hands to approach the cutters.

1.4.3 Other operations

Other operations carried out in woodworking factories include moulding, routing, mortising, drilling and turning. The special machines that have been developed to perform these various operations cannot be described in detail here, and for fuller information reference should be made to books specializing in the machining of wood.

The vertical-spindle moulding machine is one of the most widely used of woodworking machines. It consists essentially of a spindle which projects through a hole in the centre of the machine table and on which a variety of cutters may be mounted, the type, size and shape of which depend on the work to be done. A wide variety of shaping processes can be carried out on this machine using different loose top spindles and a range of cutter blocks and cutters.

Vertical-spindle moulding machines have a long history of serious accidents, largely because of the exposed cutters. Irrespective of the type of work being done, the top of the spindle and the inactive part of the cutter circle should be effectively guarded, the guard being as strong as possible in order to minimize the risk of the escape of flying cutters. No single type of guard or other safety device can deal adequately with the variety of work which can be done on a spindle moulder. Each job must be considered individually, and the most effective protection provided for the particular circumstances.

The high-speed router is essentially an inverted vertical-spindle moulder, and consists of a cutting tool fixed to the lower end of a vertical spindle which is raised and lowered by means of a pedal. The variety of tools which can be mounted and the high speeds at which they run (up to 24000 r.p.m.) permit a very wide range of operations to be carried out quickly and accurately. These include drilling, plain grooving, slot mortising, edge rebating, profiling, and many others.

Mortising machines are of several types, generally using either a hollow chisel or a chain, the links of which have cutting edges. To cause the tool to enter to the required depth into the wood it is moved vertically downwards by means of a long handle, the movement being controlled by means of slides. A hollow chisel consists of an auger revolving in a steel tubular member, with a square outer section, the bottom edges being sharpened to form chisels. When the tool is lowered into the workpiece the auger removes the bulk of the wood and the chisel cuts the corners square. A mortising chain comprises a continuous chain which is pulled at high speed round a guide bar, each link of the chain being equipped with a cutting edge.

Other operations, such as drilling and turning, are carried out on machines similar in principle to those used with other materials.

1.4.4 Joints in timber

Timber is normally available in fairly small cross sections and limited lengths and it is often necessary to build up components of larger dimensions or having complex shapes using these relatively small pieces. Joints are commonly, though not always, required to transmit loads. The choice of a suitable method for making a joint depends on the geometrical arrangement of the joint and on whether or not it has a load-bearing function. Methods of producing larger cross-sections by lamination will be considered in section 1.5.7, while end-jointing, to produce longer or continuous lengths, is described below.

a. Nails and screws. The nail is one of the oldest fixing devices and is known to have been used at least as early as 2000 BC. Later, the Romans built up a nail-making industry on a large scale, possibly with factories in Britain. A large store of Roman nails has been found in Scotland. The modern industry dates from about 1565.

Nails are made of various types of steel, of which plain mild steel, used in wire nails, is by far the most common. They are also produced in copper and aluminium and these metals may be advantageous when nailing some woods which are liable to discolour in contact with iron.

Round wire nails are the most widely used type of nail, but many other designs are available and are suitable for specific purposes. In recent years, new and improved nails

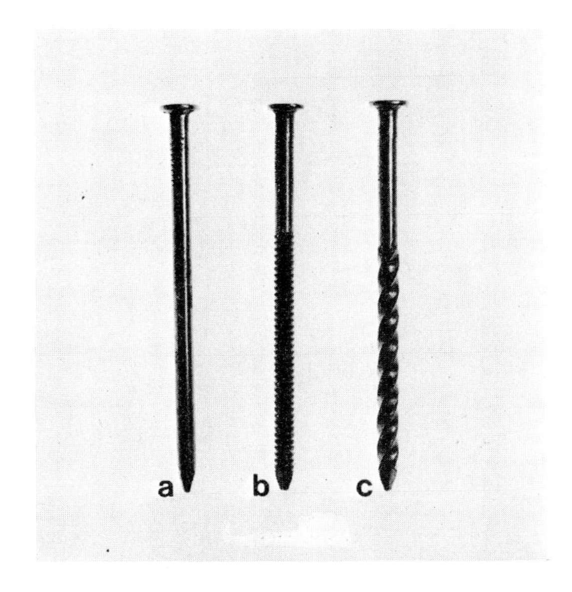

Fig. 1.25 Types of nail: (a) round wire nail, (b) ringed-shank nail, (c) helically threaded nail.

of the ringed-shank (annularly grooved) and helically-threaded types have been developed. They have increased resistance to withdrawal under axial loading, and the helically-threaded nails are particularly effective in the end grain of timber, where only low withdrawal resistance can be achieved with round wire nails.

In the softwoods commonly used in building work nails can be driven without any preparation of the members and, with the use of nailing machines, nailed joints are quick and cheap to make. With many of the denser hardwoods, however, pre-boring is necessary.

Screws perform a similar function to nails. They are not very widely used for structural purposes because of the increased costs involved, but find application in joinery and cabinet work.

b. Bolts and connectors. Bolts have long been used for joining timber members, either by overlapping the timbers or by using timber or steel gusset plates. Bolts require holes to be drilled in the adjoining members, but the joints are easily assembled. The strength of bolted joints falls markedly when the load is at an angle to the grain of the timber and, because the strength of timber is much lower in shear than in compression or tension, the distance of the bolts from the ends of the members is critical for bolted tension members.

Except in heavy timber constructions, bolts are now usually supplemented by timber connectors. These are positioned at the interfaces of the members in a joint and are used in conjunction with bolts. They give a much greater bearing area and, consequently, stronger joints which are less affected by the angle of load to the grain than are joints made with bolts alone. Toothed plate connectors, either single-sided or double-sided, are available in diameters from $1\frac{1}{2}$ to 3 in (38 to 75 mm), and are only suitable for softwoods or light hardwoods since their teeth have to be embedded in the wood. Split-ring connectors and shear plates, which are available in $2\frac{1}{2}$ to 4 in (63 to 100 mm) diameters,

Fig. 1.26 Types of connector joints.
(a) Toothed-plate connector joint
(b) Split-ring connector joint
(c) Shear-plate connector joint

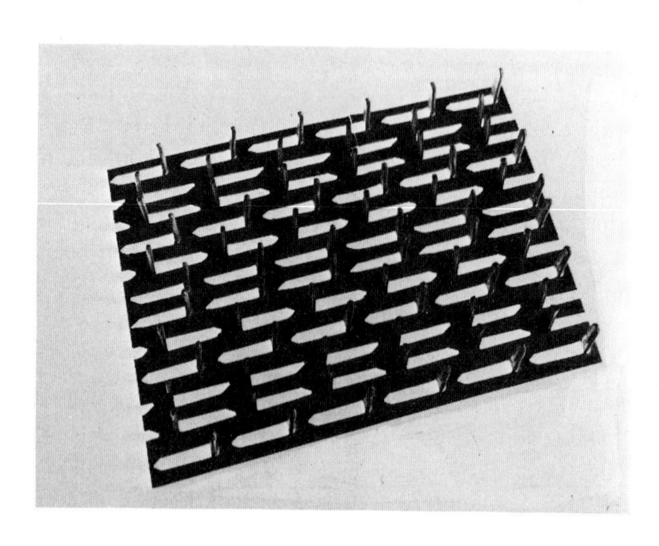

Fig. 1.27 Toothed-plate metal connector.

require grooving of the timber and are suitable for both softwoods and hardwoods. Shear plates and single-sided toothed plate connectors may be used back-to-back in pairs in demountable timber joints, or singly for joining wood to steel or concrete.

c. Metal plate fasteners. One of the most significant recent developments in the assembly of timber structures of softwood has been the introduction of punched metal plate fasteners. These fasteners are manufactured from galvanized mild steel strip, which is cut and punched to form a large number of integral nails or teeth which project at right angles from one surface of the plate. The plate overlaps two or more members at a joint, the nails or teeth being embedded under pressure. Plates are normally fixed on both faces of a joint, all the timber members being in one plane. They are particularly well adapted to the shop assembly of pre-fabricated building components, especially trussed rafters used in house building. Perforated steel jointing plates using separate nails are available too and are suitable for similar purposes.

Plywood can also be used for gusset plates and may be fixed to the timber with either nails or glue. Such joints may be fairly highly stressed, but it is essential that plywood of controlled and known quality be used.

d. End joints. The end-jointing of timber serves three main purposes. It removes the limitations otherwise imposed by the available lengths of sawn timber; this is of particular importance in production of laminated timber. It enables short lengths which are normally discarded to be converted into useful material. Thirdly, it makes it possible to improve the quality of low-grade material by cutting out defects such as large knots and rejoining the pieces.

In this section only end joints formed by gluing the two pieces together will be considered. Other methods of end-jointing are possible, but they are unsuitable for many applications because they involve overlapping pieces or projecting bolts or jointing plates. Before considering the types of end-joint, a factor which affects the strength of all types of joint between wooden members should be appreciated. This is that glued joints between end-grain surfaces of wood are much weaker than those formed between side-grain surfaces. For this reason, and because the area of the surfaces in contact is small, the simple butt joint, in which the ends of the two pieces to be joined are squared and glued together, is the least efficient of all joints. The advantage of the butt joint is that the amount of timber lost in making the joint is negligible, but its practical application is limited to situations where strength is not important. The strength of a butt joint under tension is normally less than one-tenth of that of an unjointed piece of wood.

Two types of end-joint having much greater strength than the butt joint are the scarf joint and the finger joint.

A *scarf joint* is formed by cutting a sloping plane on the ends of the two pieces to be joined and gluing together the sloping surfaces. The strength of such a joint depends mainly on the slope of the joint surface, the joint being stronger the flatter the slope and hence the nearer to the side-grain condition. It has been shown that joints with a slope of 1 in 12 or flatter can have as much as 95% of the strength of the unjointed timber. The corresponding figure for butt joints, noted above, is 10% or less. However, the scarf joint suffers from two important disadvantages. First, it involves wasting a piece of timber equal in length to the joint. It is therefore unsuitable for recovering short lengths of wood, and even when joining longer pieces the wastage may amount to 10 to 15% of the material. Secondly, a simple scarf joint is difficult to locate and because of this the

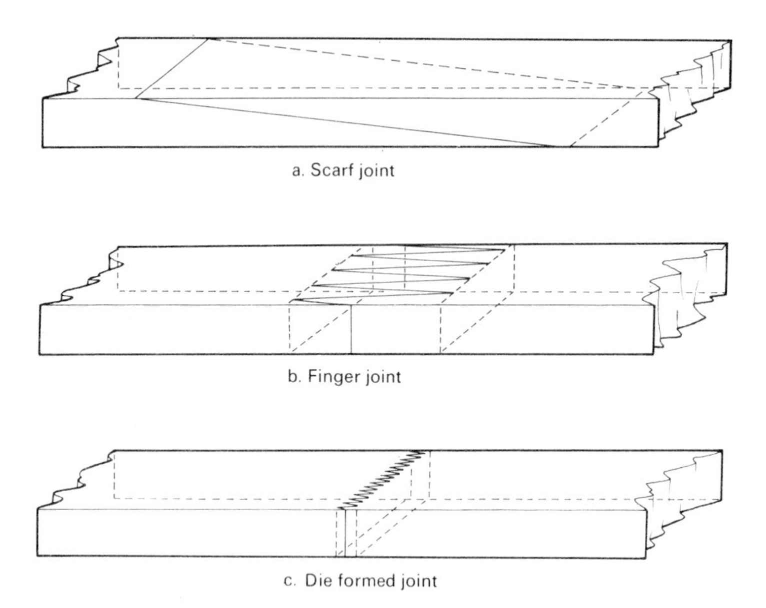

Fig. 1.28 Types of end-joint.

jointing process may be slow. Various modifications have been made to the plain scarf joint by introducing hooks, steps and dowels to aid location. Mechanical scarf joints may also be made, but drilling the members for bolts or screws reduces the joint strength.

These disadvantages are largely overcome in the *finger joint*, which is produced by

Fig. 1.29 Die for making die-formed end-joints.

cutting tapered projections ('fingers') on the ends of the pieces to be joined, and interlocking and gluing the two pieces together. The strength of finger joints depends to a large extent on the geometry of their profile, and the design of the joint can be modified according to the strength and other requirements. The amount of timber wasted in making a finger joint is considerably less than it is in the scarf joint, and the self-locating property of the finger joint makes it ideal for production line work and for the manufacture of laminated timber. Special machines are available for accurately cutting the fingers.

Recently a new type of finger joint, called the die-formed end joint, has been developed. In this case, instead of removing the timber by cutters, a die is forced into the ends of the two pieces to mould a finger joint pattern on them. The fingers are very short–about 0.1 to 0.5 in (2.5 to 13 mm) long–and the strength of a die-formed joint is consequently lower than that of a conventional finger joint, but the joint is nevertheless suitable for many joinery purposes. Development work has been carried out on the production of corner joints by a similar process. The wastage of timber in making die-formed joints is very small and this is particularly important if offcuts and short lengths are to be reclaimed. Die-formed joints can only be made in timbers which are soft enough to be indented by the serrated teeth of the die, i.e. softwoods and low-density hardwoods.

e. Joints used in joinery and furniture. Joints falling under this heading are predominantly those by which two pieces of wood are joined at right angles (or less com-

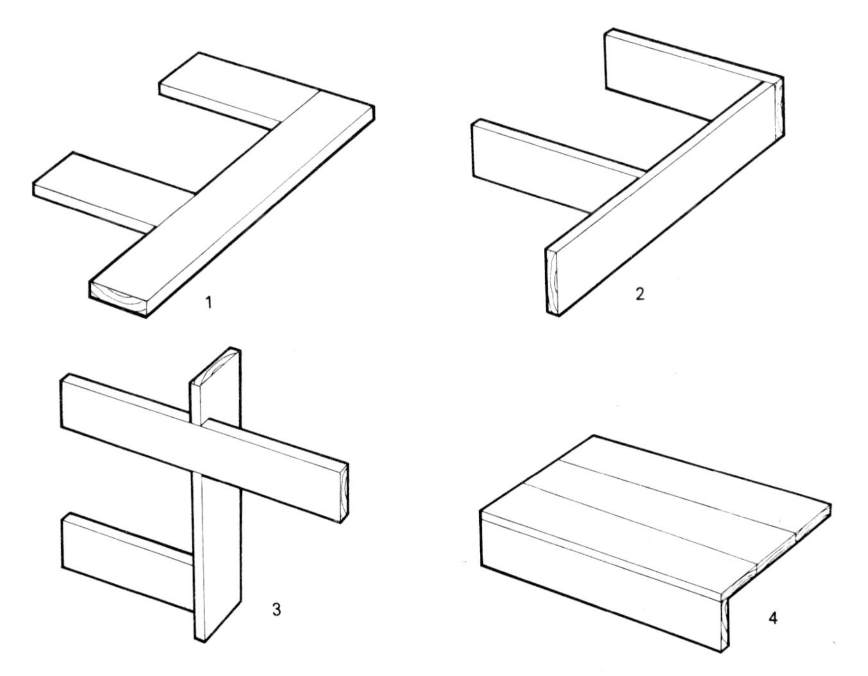

Fig. 1.30 Classes of joint used in joinery.
(1) Square or flat sections joined at right angles as in the head and stile of a door. (2) Flat sections joined at right angles as in a bookcase. (3) Flat sections joined at right angles as in the framing of fitments. (4) Flat sections joined longitudinally as in wide shelving or counter tops, or at right angles.

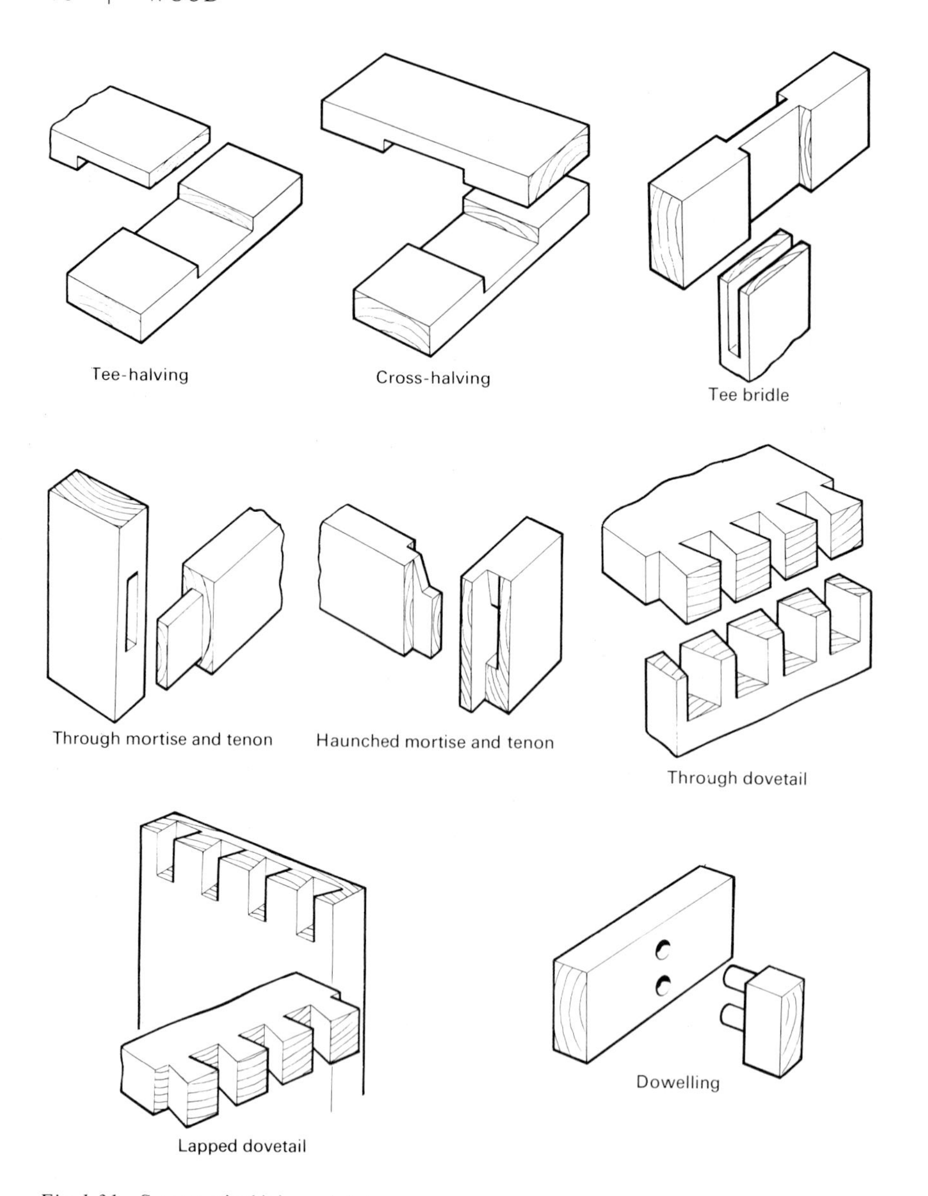

Fig. 1.31 Some typical joinery joints.

monly at some other angle) to each other by machining them to an appropriate profile so that they fit accurately together or interlock, and gluing the surfaces in contact. The proper design of the joint imparts the necessary strength and rigidity to it.

In an industry which has become largely mechanized it might be expected that the use of machinery would have an influence on the design of joints. Although there are a few joints that are favoured because of their ease of manufacture (e.g. the comb joint) it is nevertheless generally true that the nature of the material has more influence on the design and shape of joints than the tools that are used to manufacture the components. In a survey of joints that have been used in joinery from time to time it has been found that there are about 100 which are in quite common use. It might have been thought that with the introduction of machine working and new, synthetic resin glues only a small proportion of them would still be used, but in fact very few of them are considered to be obsolete.

It is not possible here to describe or list all the kinds of joint that are used in joinery or furniture, but the great majority of them fall into one or other of the following four groups: (1) square or flat sections joined at right angles, as in the head and stile of a door, (2) flat sections joined at right angles as in the construction of a bookcase, (3) flat sections joined at right angles, as in the framing of fitments, and (4) flat sections joined longitudinally, as in wide shelving and counter tops, or at right angles. In making joints of any of these types the edges or surfaces to be joined are machined to form projections, grooves, slots, etc., which fit accurately together. Some principles that should be followed in designing a joint are that as much as possible of the strength of the individual pieces is maintained, the actual joint takes place within the section of the wood (in carpentry many joints are formed by lapping the two pieces of wood over each other), the joint must be easy to make and to locate and economical of machining operations and material, the gluing area must be as large as possible, and the joint must withstand dampness if it is to be exposed to the weather.

f. Glues. All joints have to be glued, except in those instances where provision is made for movement, as in fixing wide boards to a base. For practical purposes three main types of glue are used in the United Kingdom.

Animal glues, made from bones and hides (known as Scotch glue) or from fish, are the traditional glues which have long been used in joinery and cabinet making. They are still favoured by some, though they have been largely superseded by the synthetic resin glues. Although they normally have to be applied hot, they are convenient and economical to use and do not stain the wood. They form strong and durable joints in dry conditions, but are not resistant to moisture and should not be used in joints that are to be exposed to the weather or to damp conditions.

Casein glues, which are made from milk products, are prepared simply by mixing the powdered glue with water and are applied cold. They may stain some woods, and are irritating to the skin. They are suitable for interior use, but the joints lose their strength if they are wetted although they regain it on re-drying.

Synthetic resin glues are now very widely used in assembly work as well as for veneering and in manufacture of plywood and chipboard. A number of types are available, based in the main on urea formaldehyde (UF), phenol formaldehyde (PF), melamine formaldehyde (MF) and resorcinol formaldehyde (RF) resins; mixtures of these are also supplied. They are classified according to their resistance to water (cold and hot). PF and RF resins fall in the highest durability class and make joints which have been proved by systematic tests and their records in service over many years to be highly resistant to weather, cold and boiling water, and dry heat. UF resins, which are the cheapest of the synthetic resins, are classed as 'moisture resistant and moderately weather resistant' and joints made with these adhesives will survive full exposure to weather for only a few years, but they are very suitable for use in interior conditions. Their water resistance is sometimes improved by mixing them with MF resins, which themselves fall in a higher durability classification. Unlike animal glues, all of the synthetic resin glues are resistant to attack by micro-organisms (fungi and bacteria).

Each type of synthetic resin glue can be obtained in several varieties, some in powder form and some in syrup form, suitable for different conditions of use. Some glues contain 'extenders' which improve their gap-filling properties. Usually a hardener is supplied separately, though in some cases resin and hardener can be supplied ready-mixed in powder form, needing only the addition of water.

Synthetic resin glues require greater care in the gluing operation than the animal and casein glues, their 'pot' life is relatively short, and they also have a dulling effect on tools. However, they are well adapted to factory use under controlled conditions and they are widely used on account of the variety of types available and their markedly superior weather resistance.

When a joint is assembled the two surfaces should be held together under pressure until the glue sets. This is normally done by means of cramps or by screwing the two components together. For large areas screw or hydraulic presses are usually necessary.

1.5 WOOD-BASED MATERIALS

The sizes in which timber can be produced are limited by the dimensions of the logs from which it is cut. To overcome these limitations, and to improve the properties of wood in specific directions, a number of wood-based materials have been developed. These will be described in the following paragraphs.

1.5.1 Commercial forms of solid wood

A number of terms are in use in the timber trade to describe pieces of timber of different sizes. The following are the more important of these:

Baulks – Pieces of sawn or hewn timber of equal or approximately equal cross dimensions of size greater than 4 × 4½ in (100 × 115 mm).

Planks (softwood) – Pieces of square-sawn timber 2 to 4 in (50 to 100 mm) in thickness by 11 in (280 mm) and over in width.

Planks (hardwood) – Not precisely defined, but generally pieces of square-sawn timber 1½ to 2 in (38 to 55 mm) and over in thickness and 6 to 9 in (150 to 230 mm) and over in width, commonly 8 ft (2.4 m) and over in length.

Deals – Pieces of square-sawn softwood timber 2 to 4 in (50 to 100 mm) in thickness by 9 in (150 mm) to under 11 in (280 mm) in width. Thicknesses less than this are termed boards.

Battens – Pieces of square-sawn softwood timber 2 to 4 in (50 to 100 mm) in thickness by 5 to 8 in (120 to 200 mm) in width. The term batten is also used to describe timber of smaller section used for fixing roofing tiles or slates.

Boards (softwood) – Pieces of square-sawn timber under 2 in (50 mm) in thickness by 4 in (100 mm) and over in width.

Boards (hardwood) – Less precisely defined, but generally pieces of square-sawn timber under 1 to 2 in (25 to 50 mm) in thickness of any length.

Scantlings – Pieces of square-sawn softwood timber 2 to 4 in (50 to 100 mm) in thickness by 2 to 4½ in (50 to 115 mm) in width. Also applied to hardwood timber converted to an agreed specification, such as wagon scantlings.

Squares – Pieces of strictly equal-sided sawn timber of any stated square dimension.

Round timber is referred to as logs, poles (if too small in size for conversion into sawn timber), and bolts or billets (short pieces of roundwood, as cut for pulping, firewood, etc.).

For shipping and sale of sawn softwood timber the unit of volume which has been in use in the past is the *standard* (Leningrad standard). This is equivalent to 165 ft^3 (4.67 m^3). The unit in board measure is the *board foot*, which is used mainly in North America and is the amount of timber in a piece 1 ft square and 1 in thick ($\frac{1}{12}$ ft^3). In practice, the working unit is 1 000 board feet.

In the hardwood trade the standard unit of measurement has been the cubic foot. However, in the UK both the softwood and the hardwood trade have now changed to the metric system in which the unit of volume is the cubic metre. The following are the relevant conversion factors.

$1 \text{ ft}^3 = 0.0283 \text{ m}^3$ $\quad$ $1 \text{ m}^3 = 35.3 \text{ ft}^3$

$1 \text{ standard} = 4.67 \text{ m}^3$ $\quad$ $1 \text{ m}^3 = 0.214 \text{ standard}$

The use of common units of volume for both softwoods and hardwoods enables quantities and prices of timbers to be more easily compared than they were under the earlier system.

1.5.2 Veneer

Wood veneers are produced basically for two purposes – for application as a surface veneer to a solid base material in order to present a pleasing appearance, and for use in the manufacture of plywood.

Veneered panels which display to the best advantage the attractive appearance of wood have been made by furniture craftsmen for a very long time. Indeed, examples of veneered woodwork dating from about 2000 BC have been found in objects extracted from the tombs of the Pharaohs in Egypt, and in Britain craftsmen from the eighteenth century onwards have employed wood veneers with much skill in furniture and cabinet making.

Decorative veneers are produced from such well known timbers as oak, teak, walnut, mahogany and sapele, and also from a large number of rarer woods, mainly of tropical origin, which have an attractive figure when correctly cut. It is partly because of the limited availability and high cost of these species that they are used in the form of veneer rather than as solid wood. It should be emphasized that a well-made veneered panel is in no way inferior to a panel of solid wood, and may in fact be superior to it in its technical properties. By laying the veneer on a suitable base material, for example plywood or chipboard, a panel can be made which is indistinguishable in appearance from solid wood of the same species, and is less liable to distort or shrink in changing humidity conditions than is solid wood of the same dimensions.

For cutting veneers, special types of circular saw were formerly used, and it was possible in this way to cut veneers of thickness down to $\frac{1}{16}$ in (1.6 mm) or even less. However, production of thin veneers by sawing is a very wasteful process because, for every veneer produced, an approximately equal amount of wood is converted into sawdust. This method of cutting has therefore been almost entirely superseded by methods employing a knife, namely slicing and peeling (rotary cutting). For production of decorative veneers slicing is generally preferred because by this method the plane of the cut can be chosen to expose the most attractive grain pattern from the log.

A log is prepared for slicing by first cutting it length-wise into thick pieces (flitches) on the band saw so as to expose the grain pattern. The precise method of opening up the log requires considerable judgment on the part of operator, so as to produce the maximum yield of high quality veneer from the log. With many species the best grain pattern is obtained from flitches that have been cut on the quarter (radial surfaces), but other methods of cutting are often preferred.

Before cutting veneers, the wood of most species must be softened by steaming the flitches or by steeping them in hot water. This operation is carried out in pits situated outside the veneer mill, in which the flitches are heated by steam or by steam pipes immersed in water for a sufficient length of time for the heat and moisture to reach the centre.

Slicing machines are of two types – vertical and horizontal – but both operate on the same principle. The flitch is fixed to the bed of the machine by means of gripping claws (dogs), and the powerful knife, carried on a rigid frame, is driven diagonally across the face of the flitch, cutting a thin slice of wood from it. In some machines the knife is fixed and the bed is movable. As the knife is forced across the flitch the veneer is produced by a shear cut and emerges between the cutting edge of the knife and a pressure bar which runs immediately ahead of it. At the end of each forward stroke the knife returns to its original position and operates a ratchet which advances either the flitch or the knife a fraction of an inch, corresponding to the thickness of the veneer to be cut. In this way a succession of veneers is produced having similar figure or grain pattern and, when applied to panels or furniture, these can be arranged or 'matched' to produce attractive effects.

The loss of wood in cutting a sound flitch on the slicer is small and is restricted to that portion which is gripped by the dogs.

The second method of veneer cutting (peeling) is used primarily for production of veneers for plywood manufacture and is described below (1.5.3, Plywood). However, decorative veneers are also sometimes produced by peeling.

After slicing, veneers are piled in the order of cutting so that the stock can afterwards be properly matched. They are then dried, reassembled, and generally sold in complete flitches.

1.5.3 Plywood

Although wood possesses many valuable properties by virtue of which it is very widely used in construction work, furniture, joinery, boat building and for many other purposes, it is a non-homogeneous material and therefore suffers from certain disadvantages when compared with other structural materials. In particular, its strength properties in the transverse direction are low, it is liable to swell and shrink when exposed to different humidity conditions, and it is apt to split longitudinally. The sizes in which it is available are also limited by the diameter and length of the stem of the tree.

These shortcomings of solid timber are largely overcome when wood is converted into plywood. Plywood is made up of at least three layers of veneer assembled by gluing, its chief characteristic being the crossing of alternate plies at right angles, thus producing a material having much greater homogeneity than the original wood. The high longitudinal strength of wood is distributed in the two major directions, and in certain plywood constructions used for highly stressed components, greater homogeneity and a still more uniform distribution of strength is achieved by laying alternate layers at 45° to one another, but this is more costly and for most purposes unnecessary. Plywood is also very resistant to splitting and can, for example, be nailed close to the edges.

Since the adjacent veneers in plywood are firmly bonded together the tendency of the wood in one layer to swell laterally is restricted by the adjoining layer, whose tendency to swell is in a different direction. Consequently the material possesses a high degree of stability. The average movement of plywood of 3-ply construction of a large number of species on changing from 7% to 19% moisture content is about 0.19% in the direction of the grain of the face veneers and about 0.27% in the transverse direction, and is much smaller than the movement of tangentially cut boards.

a. Manufacture of plywood. The following is a brief general account of the manufacture of plywood. The actual process may vary in different mills and countries, but with few exceptions the process is substantially the same.

Plywood is made from round logs and never from sawn timber. Not all species of wood are suitable for manufacture of plywood, preference being given to those of even texture and medium density (generally between 25 and 50 lb/ft^3 (400 and 800 kg/m^3)). Although light-coloured woods are generally favoured, colour is not of major importance provided it is uniform. European species which are used for plywood include birch, beech, alder, ash and various pines. In North America the most important species is Douglas fir, but other softwoods and some hardwoods are also widely used. Large quantities of plywood are also manufactured from certain tropical hardwoods, either in the country of origin or in mills in Europe (including Great Britain) and Japan. It is interesting to note that some species which contain in their structure small grains of silica, and consequently have a severe blunting effect on saws, can nevertheless be converted into veneer and used for plywood without difficulty. White seraya (*Shorea* sp.) from Sabah is an example of such a species.

Formerly, only trees of large diameter and top quality were acceptable for plywood manufacture, but methods have been evolved for producing high class plywood from relatively small and irregularly shaped logs. For example, birch, which does not grow to large sizes, forms the raw material of the important plywood industry in Finland.

As in the production of sliced veneers, logs for plywood manufacture are generally first subjected to a softening process by steaming or boiling them in large vats for a number of hours or days, depending upon the size and species of log. This process ensures that the log has a high and consistent moisture content throughout, and it softens the wood tissue so that the cutting of veneer is facilitated. Certain species, e.g. birch, Douglas fir, and European pines, can be peeled successfully without pre-treatment, and are taken directly from the log pond to the mill.

Peeling. Veneer for plywood manufacture is almost invariably produced by rotary cutting or peeling. After cutting to a suitable length, the log is mounted on a large lathe and rotated against a knife which runs the full length of the log. A continuous ribbon of veneer is thus produced. As cutting proceeds, a gear from the main drive slowly feeds the knife towards the centre of the log, so that a constant thickness of veneer is maintained. The veneer emerges between the cutting edge of the knife and an adjustable pressure or nose bar. The function of the pressure bar is to prevent the formation of fine splits, known as lathe checks, by exerting compression on the wood as it is fed towards the knife.

Generally, the best quality veneer is obtained from the outer portion of the log. As the knife approaches the centre, knots become more frequent, and other defects may be present at the centre of large logs. The quality of veneer is also influenced by the setting of the lathe and the sharpness of the knife. The best quality veneers are used as face veneers in plywood, those of lower quality being used in the core.

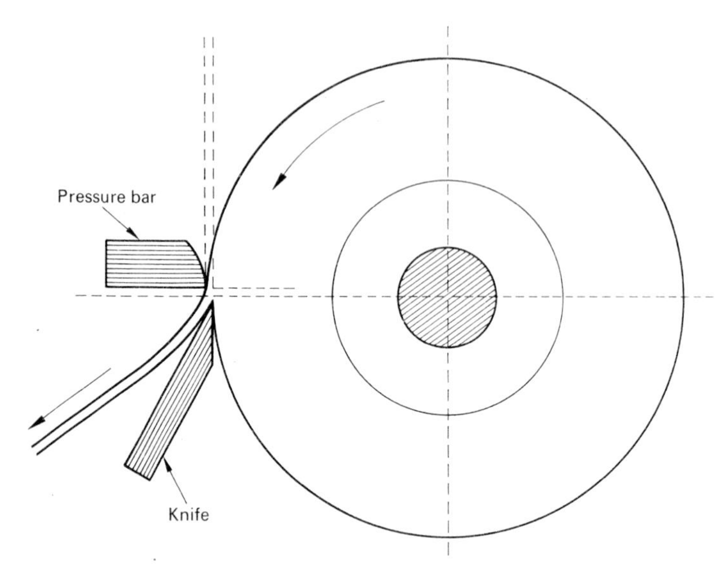

Fig. 1.32 Diagram of rotary peeler.

Fig. 1.33 Rotary peeler.

The thickness of veneers for commercial plywood manufacture is generally between 1 mm and 4 mm, though decorative veneers are often cut to smaller thicknesses. As the veneer leaves the log it is turned through a fairly sharp angle. This leads to the development, especially with thicker veneers, of slight surface checking (cutting checks or lathe checks) on the face next to the log. Careful setting of the nose bar (pressure bar) can minimize these. The veneers thus have a 'tight' and 'loose' side, and should always be laid with the tight side of the face veneer outwards.

In every log a core remains after the rest of the log has been converted into veneer. The diameter of the core varies with the original diameter and soundness of the log. In some factories the cores from large lathes are reduced in length and re-cut on smaller lathes, producing veneers which are used for plywood core stock or other purposes.

The handling of veneers as they are produced by the rotary cutters varies from factory to factory depending on the thickness and properties of the veneer and on the speed of production, which may be very high (up to 600 ft per min. (180m per min.)). The veneer is either wound on reels or run on to a conveyor table where it is cut into widths by a guillotine. The veneer may either be cut into standard widths to suit the press size or, if major defects are present, these may be cut out by clipping to shorter widths which are subsequently joined by edge gluing to produce pieces of the required size.

Veneers are normally dried before gluing. A number of types of continuous drier are in use, and reduce the moisture content to a suitable value generally within the range 5–14%. Excessive moisture in the veneers can cause faulty adhesion, and variable moisture content may give rise to distortion of the board.

Gluing. After drying the veneers, the narrower widths are joined where necessary to form the standard sizes required for pressing, using special machines which hold the glued edges in close contact and at the same time apply heat to set the glue. At this stage also defects such as large knots may be cut out by means of a punch and the resulting spaces filled with plugs of sound veneer of similar colour and texture to that of the surrounding wood.

The prepared veneers are conveyed to the gluing section where the glue is applied and the individual veneers assembled into the pack for pressing. Liquid glues are generally applied by means of mechanical glue spreaders consisting of corrugated rollers which revolve in troughs containing the glue, which must be maintained at the correct consistency and temperature. Accurate control of the glue spread is important because a thin glue line produces the best bond, and the objective is therefore to obtain the thinnest practicable glue line of even spread. In making 3-ply boards the core veneer is generally glued on both faces and then laid with the grain at right angles to the face and back veneers. Similarly, for multiply boards of 5, 7, 9 or more veneers, alternate veneers are glued on both faces and laid at right angles to their neighbours until the requisite number of veneers for the final board has been assembled.

The types of glue used for making joints in wood have been described in section 1.4.4. Glues used in plywood manufacture are of the same general types, but are specially formulated to render them suitable for the particular conditions of use.

Plywood glues are classified according to their resistance to moisture and it is important to use a glue having a suitable durability classification for the particular end use to which the plywood is to be put. For example, plywood used for interior applications such as furniture needs to be well bonded but does not necessarily require to have resistance to moisture, while plywood for exterior use requires an adhesive possessing good resistance to the weather. The following are the four main durability classes for glues:

Type WBP – Weather and boil proof. Glues which make joints that are highly resistant weather, micro-organisms, cold and boiling water, and dry heat. Examples: phenol-formaldehyde (PF) and resorcinol-formaldehyde (RF) resins (seldom used in plywood).

Type BR – Boil resistant. Glues which have good resistance to the weather and to boiling water, but fail under very prolonged exposure to the weather. Joints will withstand cold water for many years and are highly resistant to attack by micro-organisms. Examples: melamine-formaldehyde (MF) and melamine-fortified urea-formaldehyde (MF: UF) resins.

Type MR – Moisture resistant. Joints made with these glues will survive full exposure to the weather for only a few years. They will withstand cold water for a long time and hot water for a limited period. They are resistant to attack by micro-organisms. Examples: urea-formaldehyde (UF) resins.

Type Int – Interior. Joints made with these glues are moderately resistant to cold water but need not withstand attack by micro-organisms. Examples: animal glues, blood albumen, casein, vegetable proteins.

Much imported and all British-made plywood is manufactured with synthetic resin adhesives of one type or other. It should be noted that the synthetic resin adhesives mentioned above (PF, RF, MF, UF) are often blended together in proprietary brands. Some manufacturing countries use protein glues such as those derived from soya and casein products.

Pressing. The packs of glued veneers are pressed in large hydraulic presses having multiple heated platens so that a number of boards can be pressed simultaneously. The press must be of heavy, rigid construction so that pressure is uniformly applied over the whole area of all the boards without deflection of the platens. Sometimes the packs of veneer are pre-pressed cold before the final hot pressing. Loading of the press is carried out either by hand or, more usually, by means of automatic loaders. The pressing temperature depends upon the type of glue that is being used, and rigid control of platen temperature, moisture content of veneer, pressure and time of pressing is essential for production of boards of good quality. The platens are generally heated by steam under pressure, though other methods, e.g. high pressure hot water and heat transfer fluids, have also been used.

After removal from the press the boards are reconditioned by stacking them until their moisture content is uniform throughout. They are then trimmed to size and sanded to a predetermined thickness with drum sanders.

b. Types of plywood. Plywood is made from three or more veneers, and usually has a 'balanced' construction; that is, it is constructed from an odd number of veneers arranged symmetrically relative to the central plane of the sandwich. The tendency of the finished board to distort is reduced by adopting a balanced construction. With very few exceptions the grain of each veneer runs at right angles to that of the veneers on each side of it.

The veneers in a sheet of plywood may either all be of the same thickness or they may differ in thickness, the face veneers being either thinner or thicker than the core veneers. Plywood may also be made with veneers of a single species of wood or it may include more than one species. The strength properties of the plywood will clearly be dependent on the type of construction. In any case the desirability of retaining a

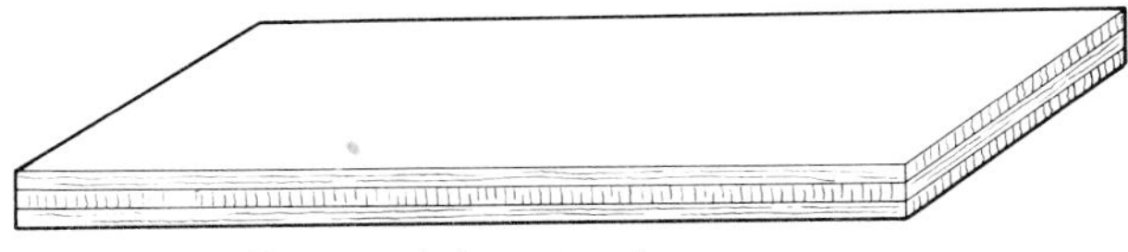

Three equal ply construction

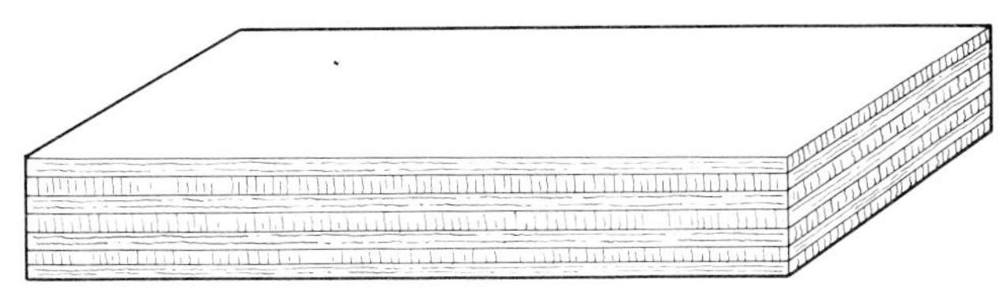

Seven ply construction

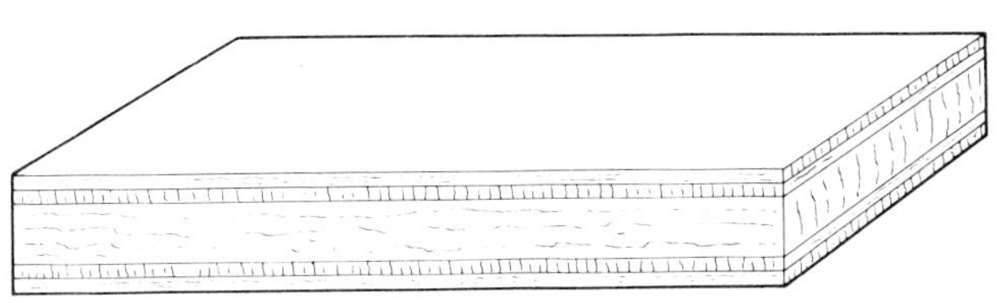

Five ply construction with thicker core

Fig. 1.34 Some examples of plywood constructions.

symmetrical or balanced construction should be borne in mind.

Commercial plywood is normally obtainable in thicknesses ranging from $\frac{1}{8}$ to 1 in (3 to 25 mm), and in many sizes, 8 × 4 ft (2.4 × 1.2 m) being the most common.

About 97% of the plywood used in Britain is imported. Plywood manufactured in Britain is made entirely from imported timber, mainly hardwoods from West Africa and Asia. The principal species used are afara (limba), agba, gaboon, African mahogany (*Khaya* spp.), makoré, obeche, sapele, utile and the serayas.

c. Uses of plywood. Apart from solid timber, plywood is manufactured on a larger scale and has a wider range of uses than any other wood-based material. It is not possible to list all its uses, but the following are the industries in which the greater part of it finds application:

Building work, including doors, panelling, external cladding, partitions, prefabricated units.
Concrete formwork.
Packaging containers – crates, boxes, barrels.
Furniture, including radio and television cabinets.
Shop fitting.
Marine uses – hulls and planking of boats, interior fittings, etc.

Caravans.
Agricultural work.
Transport vehicles.

Plywood is made with a range of surface coatings or finishes. Decorative wood veneers, plastic laminates, or metal facings may be incorporated during manufacture or applied afterwards in a separate operation. If an additional veneer is applied to plywood a compensating veneer should be placed on the opposite face to maintain a balanced construction and prevent twisting.

Plywood is better adapted to factory methods of pre-finishing than solid wood since it can be made in large panels. Application of synthetic lacquers by spraying and accelerated drying can greatly reduce the time taken in finishing. Special surface treatments are available to improve resistance to weather or to wear. These include phenolic resin films and resins incorporating granite particles. To protect plywood from fungal decay or attack by wood-boring insects the veneers may be treated with preservatives before assembly.

1.5.4 Blockboard and laminboard

Blockboard and laminboard consist of a core made from narrow strips of wood placed together with or without glue between the strips, and having outer veneers on each face, the grain direction of which runs at right angles to the grain of the core. In some boards (Finnish blockboard) there are two outer plies on each side of the core, but the grain of all the veneers runs at right angles to that of the core.

In blockboard the blocks forming the core are up to 1 in (25 mm) wide, while in laminboard they are narrower (3 to 7 mm in width) and generally consist of strips of veneer laid on edge. A further variation on blockboard construction, widely used in North America and Japan but seldom seen in the United Kingdom, is battenboard, also known as lumber-core. In this type of construction the core is formed of battens up to 3 in (75 mm) wide in place of the 1-inch blocks.

a. Manufacture. Cores for blockboard are made from boards of well dried softwoods, dressed on both sides to a finished thickness of $\frac{7}{8}$ to 1 in (22 to 25 mm). These

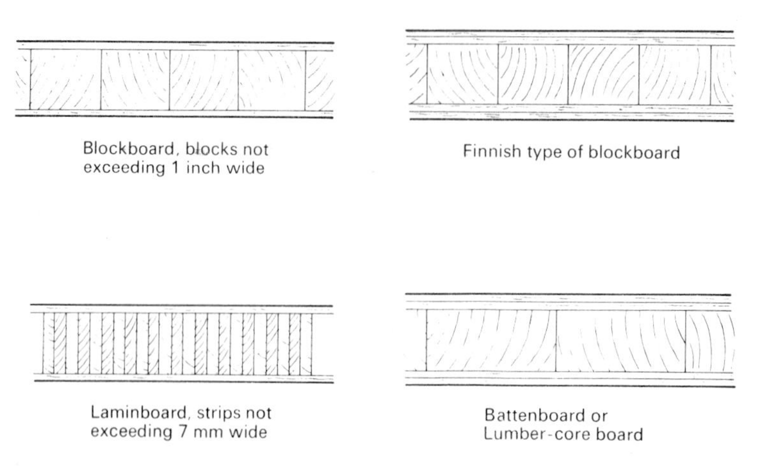

Fig. 1.35 Blockboard, laminboard and battenboard.

are passed through high-speed multiple rip saws to yield strips, the width of which will be the the thickness of the block core when assembled. Provided the saws are kept sharp and correctly set the cut edges will be sufficiently smooth for subsequent operations without planing, but serious defects or knots must be cut out. The strips are fed into a composing machine with a sawn edge uppermost and are arranged in pairs, heart side to heart side, as this minimizes any tendency for distortion. To make up the required length several pieces may be fed in, end to end, care being taken to ensure that the joints are staggered. The strips are glued along their entire length and placed on the flat bed of the machine whilst lateral pressure and heat are applied to set the glue. As quite short strips can be used and the width is cut as required, wastage is very small.

A built-up core is passed through the glue spreader and the outer plies, built up from smaller pieces of veneer by edge gluing if necessary, are placed above and below it, and the whole assembly is pressed in a hot press to set the glue. It is common practice to assemble the core and apply the outer veneers in a single operation.

Laminboard is somewhat similar to blockboard in its construction and properties but differs in the method of manufacture because the core is made from wood veneers rather than solid blocks of wood. The veneers may be sliced or rotary cut and may be produced from softwoods or low density hardwoods, e.g. gaboon. Two methods are in use for preparing the core material from the veneers:

1. The veneers are accurately dried to a specified moisture content in the range 7 to 10%, trimmed to standard widths, and assembled in a framework, the grain of all the veneers lying parallel. The block is built up to a height of about 36 in (0.9 m), then each alternate veneer is passed through the glue spreader and the block reassembled, taking care that the ends of veneers are cut so as to form a close joint without gaps. The block is then pressed in a cold hydraulic press, the pressure being taken up by retaining clamps until the glue has set. The block is fixed to the bed of the saw bench on its side and sawn into slabs of appropriate thickness on a frame saw or band saw. The width of the core stock is made up by gluing two or more slabs edge to edge. Cores are then built up into boards by gluing, assembling between outer plies, and pressing.
2. In Finland, rotary-cut redwood veneers, 2.6 mm thick, are laid with the grain running in the same direction in assemblies which will yield boards about 50 × 50 in by 25 mm thick (1300 × 1300 × 25 mm). After glueing and pressing, the boards are cut in the direction of the grain into strips 11–13 mm wide. The strips are then turned through 90° and assembled into cores, lengths being staggered as in production of blockboards. Double veneers of birch are used for face and back, the inner one passing through the glue spreader. In this method of production veneers are produced on standard plywood-making equipment and little further expensive machinery is required.

b. Properties and uses. Blockboard and laminboard are materials having good dimensional stability and freedom from distortion. They can be made lighter in weight than plywood of the same thickness by using low density species such as pine or obeche for the cores. They can be supplied in large sizes such as 6 ft (1.8 m) by up to 17 ft (5.2 m) and are used for large flat surfaces, such as desk tops and cabinet doors in the furniture industry, where rigid panels are required. Laminboard is used almost entirely in furniture manufacture, but blockboard is also used to some extent in building, shipbuilding and joinery, mainly in interior situations. It should be noted that for use in furniture blockboard is rapidly being replaced by chipboard, which is comparable in many of its properties, but lower in cost.

A large proportion of the blockboard and laminboard used in the United Kingdom is imported from Finland.

1.5.5 Chipboard

Wood chipboard, also known as particle board, is made from solid particles of wood, which are produced mechanically by a cutting operation and are bonded together into a rigid material by means of a synthetic resin binder. Particle boards are also manufactured from materials other than wood, notably flax shives.

The chipboard industry is a comparatively recent development and has grown up almost entirely since the Second World War. It originated in a time of wood shortage, as a means of making a useful product from waste wood, and the early plants utilized largely waste material such as shavings and offcuts from other wood-using industries. Manufacturing methods and equipment were somewhat crude and experimental and many of the products were of relatively low quality.

A great deal of study and development work has been put into the manufacture and properties of wood chipboard, with the result that it has now become a major product of high quality and utility which can be manufactured economically to closely controlled standards. It must be emphasized that it is no longer a cheap substitute material, but a high quality product in its own right which has valuable properties, is easy to use, and has a wide range of applications.

a. Manufacture of chipboard. The aim in manufacture of chipboard is to produce a board with a high strength/weight ratio, with good dimensional stability and a surface suitable for the application of thin veneers. Certain additional properties, e.g. good screw holding, low water absorption, are required for specific applications.

Initially the manufacture of chipboard was little more than an attempt to utilize waste shavings, which were hammer-milled into appropriately sized splinters and particles. In some mills material of this kind is still used for part or all of the raw material. However, it was soon realized that less resin was needed, and the properties of the board were improved, if prepared shavings of controlled size and shape were used.

Originally, chipboard was made as a single layer of essentially uniform density. A consequence of the introduction of specially prepared chips was the manufacture of three-layer boards, in which thinner shavings with a higher resin content are used for the surfaces and thicker chips with a lower resin content for the core. This gives boards with smooth, hard surfaces of higher density than the core, and results in good surface properties and mechanical strength at a lower cost than that of boards consisting of thin chips throughout. Other types of board contain a continuous gradation in chip size from fine on the surfaces to coarse in the core. Such boards have a fine, smooth surface suitable for various surface treatments such as painting or printing.

The principal operations in chipboard manufacture are chip preparation, drying, mixing with resin, forming into a mat and hot pressing.

Chip preparation. As already mentioned, the raw material for chipboard may consist either of factory waste, e.g. planer shavings, or solid wood. In the latter case solid waste (offcuts, etc.) or roundwood from forest trees (thinnings) may be used. The properties of the wood, in particular its ability to be cut into chips possessing a smooth and even surface, are of some importance. The resin in chipboard is a major item in its cost and it is therefore essential to utilize it with maximum efficiency. The smoother the surface and the smaller the proportion of end-grain in the chips, the smaller is the amount of resin lost through absorption. Timbers of low to medium density are preferred and other properties to be considered include compressibility and permeability. In most respects softwoods have some advantage over hardwoods, though some low to medium density hardwoods are quite suitable.

Shavings are reduced in size by means of hammer mills or similar types of equipment. Solid wood is converted into chips by a variety of chipping machines which aim to produce chips of predetermined size and uniform thickness. Chips are hammer-milled to reduce particle width. The shape of the chips is of great importance for the quality of the final board. Long, smooth and slender chips are necessary for production of strong boards, while relatively large, bulky chips on or near the surface of the board impair the dimensional stability and the appearance after veneering.

Chipping machines fall into two groups: disc type chippers in which the cutting plane lies parallel to the plane of rotation of the disc, and cutter head chippers in which the cutting plane is transverse to the plane of rotation. The object is always to cut chips parallel to the grain of the wood, and the main differences between the various types of machine within each group lie in the methods used for determination of chip length and in equipment for feeding and clamping the wood.

Drying. Roundwood is chipped in the undried condition since the best cutting results are obtained by using green timber. It is then necessary to dry the chips so that they have the correct moisture content for the pressing operation. A very wide range of drying equipment has been used, but essentially the wet chips are dried in a current of heated air arranged to ensure that they are fully exposed to the air stream. The moisture content of the chips leaving the drier is usually in the range 5 to 12%, and it is desirable that it should be as uniform as possible.

After drying, chips are screened on vibrating screens to separate them into suitable sizes for face chips and cores.

Resin blending. Most chipboards are bonded with urea-formaldehyde resins, which possess the advantages of low cost and a light colour, and are satisfactory for products intended for indoor use where high resistance to moisture and heat is not required. Where greater durability is sought, for example in boards for exterior use, phenol-formaldehyde or other binders are used. Although these are not yet very widely employed, developments in this direction may be expected, and experimental work has also been carried out on the use of binders based on waste pulping liquors containing lignosulphonates.

A controlled amount of liquid resin mixed with hardener is sprayed on the chips in a chip/resin mixer, which consists of a longitudinal trough with a shaft carrying mixing arms running down the centre. The aim is not to provide a complete coating of resin over the whole surface of the chips, but rather to apply the resin uniformly as small droplets which bond the chips together. Additional strength is imparted to the board by the interlocking of relatively long and slender particles.

Mat formation. The glued chips are formed into a mat by spreading them either on a continuously moving steel belt or on a series of adjacent flat metal plates (caul plates) with or without separating walls. If the mat is laid on a belt or if no separating walls are employed, segments of the mat are cut out with saws before the glue has set to provide separate chip cakes for pressing into board. The importance of uniform distribution of the chips is obvious; any variations in density or thickness of the mat will lead to lack of uniformity in properties of the final board. In manufacture of three-layer boards a layer of face chips is first laid on the belt, followed by a layer of core particles and finally a second layer of face chips. Accurate proportioning of the two types of chips is essential.

Pressing. This is carried out by two basic methods, namely flat pressing and extrusion, and there are important differences in physical properties of boards produced by these two methods.

Flat pressing is the more widely used method and is generally carried out in multi-platen presses, similar in principle to plywood presses. Before entering the hot press the mat is usually subjected to a cold pre-pressing operation. This consolidates the mat and reduces its thickness so that it can be more easily handled and loaded into the hot press. In the hot press heat and pressure are applied to set the resin and bring the mat down to the 'thickness stops'. The platen temperature and pressing time are adjusted to suit the requirements of the synthetic resin adhesive used.

In the extrusion method the particles, coated with resin, are forced through a long heated die by means of a reciprocating ram, the equipment usually being arranged so that the chipboard is extruded vertically. The pressure is applied in the direction of the length of the board, and the resin sets as the board passes through the heated die.

b. Properties of chipboard. Chipboard is essentially a panel product which can be made in large sheets with closely controlled properties and quality. Most of the production is in the medium density range (density between 28 and 45 lb/ft^3 (450 and 720 kg/m^3)). Panels are commonly 8 × 4 ft (2.4 × 1.2 m) in size, but larger sizes are available. Thicknesses generally range from $\frac{3}{8}$ to $1\frac{1}{2}$ in (9 to 40 mm), with $\frac{3}{4}$ in (19 mm) as a common gauge. For special applications thicknesses as great as 3 in (75 mm) are available; such boards often have holes running lengthwise to reduce their weight and are made by the extrusion process mentioned above.

Some differences in properties exist between the more common flat-pressed boards and extruded boards because of the different arrangement of the particles in the two types. In flat-pressed boards the particles tend to lie with their long dimensions in the plane of the board, whereas in extruded boards many of the particles are orientated in a plane perpendicular to the surface of the board. This difference in structure between the two types of board is reflected in their properties, which differ in two principal respects. First, the extruded boards are relatively low in strength, especially bending strength, and for this reason they are usually veneered to provide additional strength to the surfaces. Secondly, when the boards take up moisture flat-pressed boards swell mainly in thickness while extruded boards, owing to the different orientation of the particles, swell in their longitudinal and cross directions. Flat-pressed boards swell and shrink in the plane of the board to about the same extent as plywood, i.e. about one twenty-fifth as much as solid wood, while the change in thickness is slightly greater than that of solid wood.

More detailed information on the properties of chipboard may be found in British Standard 2604 : 1970, Specification for Resin-Bonded Wood Chipboard.

c. Uses of chipboard. Most chipboard is bonded with urea resin and is thus suited only to interior applications. The greater part of it is employed in the manufacture of furniture and kitchen cabinets, doors, partitions and floors, and as a base for veneer and plastic overlays in panelling. Very little board is used in its natural state. Normally it is finished with wood veneer, paper, plastic or even metal laminates, or by painting or lacquering. It has excellent properties as a base material to which veneer can be applied for furniture or other uses, being a homogeneous material with no directional properties, low swelling and shrinkage, and little tendency to distort.

Wood chipboard is still a relatively new material and further developments may be expected in the production of boards having improved properties to render them suitable for particular applications. In particular, production of an exterior grade chipboard having good resistance to the weather is a development that is already taking place and should open up many new uses in building work, concrete shuttering, packaging, etc.

1.5.6 Fibreboard

Fibreboard is distinguished from chipboard by its method of manufacture. It is made from wood fibres, prepared by a pulping process, and not from solid particles produced mechanically, such as are used in chipboard manufacture. The wood fibres are then reconstituted into a rigid board by the action of heat, with or without pressure, with little or no added bonding material. Bonding between the fibres takes place as a result of their felting together and their inherent adhesive properties.

Fibreboards are subdivided into two types – insulation board and hardboard – which differ principally in density. *Insulation board* is defined as fibreboard having a density not greater than 400 kg/m^3 (25 lb/ft^3) (generally less than 300 kg/m^3), and is used for heat and sound insulation. *Hardboard* is a denser material, used for a wide variety of purposes including panelling, partitioning, furniture, vehicle construction, etc., and generally has a density in the range 800 to 1 200 kg/m^3 (50 to 75 lb/ft^3). A superior grade of hardboard having improved strength properties and resistance to water is known as super-hardboard or tempered hardboard.

a. Manufacture of fibreboard. In Britain, fibreboard is made from forest thinnings and other small-sized roundwood, mostly softwood, but some hardwood may be included. In other countries forest material together with solid sawmill waste is used. Bagasse (sugar cane fibre), groundwood screenings and repulped waste paper are also utilized, particularly in insulation board manufacture.

The wood is converted by a variety of methods into a coarse pulp. This differs in two important respects from paper pulp. First, it is less highly refined and generally coarser in nature, consisting in the main not of individual fibres but of fibre bundles, and secondly, little delignification takes place in the process of pulp preparation and most of the wood components are retained in the pulp.

Insulation board. Pulp for production of insulation board is prepared either by grinding of logs in the presence of water by means of stone grinders, or by defibration or grinding of chips, generally after some form of treatment (steaming or water soaking) to reduce the power requirements and improve the quality of the pulp. The pulp is refined as necessary, by passing it between a pair of rotating fluted discs, sizing (water-proofing) agents are added, and a suspension of the pulp is allowed to flow on to the wire of a machine resembling a paper-making machine, but running more slowly. In this way a thick, continuous mat of fibre is formed on the wire. Water is withdrawn from the mat by drainage through the wire, assisted by suction boxes, and the mat is lightly pressed between rollers to consolidate it. A cross-cut saw cuts the mat into lengths, which are fed into a long tunnel drier which removes the water by means of a fast-moving air stream at a high temperature, without further compression. In this way a rigid board of open structure and low density is produced.

Hardboard. Pulp for manufacture of hardboard is prepared almost entirely by defibration of wood chips, and a number of methods are available for performing this operation. The most important of these are the *Masonite* process, the *Defibrator* process, and the use of an attrition mill after steaming. The Masonite process is one of the earliest methods used for defibration of chips, and differs in principle from other methods. It is an 'explosion' process, in which chips are fed into a special type of high pressure digester, called a gun. After closing the gun, steam is admitted at high pressure for a short time. A special type of instantaneous relief valve is then opened, with the result that the charge is blown out of the vessel with explosive violence. The very rapid expansion of steam within the chips disintegrates them into coarse fibre, which is collected in a cyclone, mixed with water, refined and washed.

The Defibrator process, which is a Swedish invention, is based on the fact that lignin, which is the major bonding agent between the fibres in wood, is a thermoplastic material; that is, it softens at high temperatures. Consequently the separation of fibres can be carried out with a lower expenditure of energy and with less damage to the fibres at temperatures above the softening point of lignin than at lower temperatures. The Defibrator is a continuous machine in which chips are first steamed under pressure in a pre-heater, then pass immediately to the defibration zone while still under steam pressure. Here they are separated into fibre by passing between a pair of fluted discs, one of which rotates while the other is fixed. Defibration is thus achieved under optimum conditions, while the lignin is in a plastic condition. The Defibrator process is widely used in Scandinavian countries, where there is a large fibreboard industry, and in other countries in Europe and elsewhere.

A number of mills utilize processes in which softening of the chips and defibration are carried out as separate operations. The chips are first softened by steaming or other treatment, then reduced to fibre in disc refiners or attrition mills. There is considerable latitude for production of pulps possessing a wide range of properties by suitable control of the processing variables.

After refining and screening, the pulp is mixed with a small amount of a waterproofing agent and passes to the sheet-forming machine. This is similar to that used in manufacture of insulation board and produces a thick mat (about 1 in (25 mm) thick) of wet pulp, which is cut into lengths and pressed in a multiplaten hot press at a temperature in the range 180 to 210°C. Under these conditions much of the water in the fibre mat is squeezed out and the remainder evaporates, while at the same time bonding between the wood fibres takes place under the influence of heat and moisture. This results in formation of a dry, rigid board of high density and good strength.

Until recently most manufacturers of hardboard have utilized the wet-forming process described briefly above, in which the pulp fibre is suspended in a large volume of water. However, in order to minimize the problems of supply of large quantities of clean water and of disposal of effluent, so-called dry and semi-dry processes have been introduced. In these methods of manufacture the moist fibre produced by defibration of chips is not dispersed in water but, after mixing with a small proportion of waterproofing agent and phenolic resin, is carried in an air stream to the felting machine where it settles from an air suspension on to a moving belt. After passing through a precompression unit to reduce its thickness it is conveyed to the hot press. The dry and semi-dry processes are claimed to have certain advantages in economy of operation over the more usual wet-forming process, but the quality of bonding achieved is lower in the absence of excess of water, and in order to produce boards of adequate strength it is necessary to supplement the natural fibre bonding by addition of a phenol-formaldehyde resin.

b. Properties and uses. Insulation board has a density in the range 15 to 25 lb/ft^3 (240 to 400 kg/m^3) and is commonly made in thicknesses from $\frac{1}{2}$ to 1 in (12 to 25 mm). Sheets are generally up to 4×12 ft (1.2×3.6 m) in size. Being a low density material, its strength properties are relatively low, but it is not used in situations where high strength is required. It is used primarily for heat and sound insulation, for example for roof and wall linings in buildings, shipbuilding, refrigerated stores, etc. Acoustic tiles made of insulation board are perforated or slotted to improve their sound absorption qualities and are used on ceilings.

Standard hardboard is normally $\frac{1}{8}$ to $\frac{1}{4}$ in (3 to 6 mm) thick and is made in sheets 4 or 5 ft (1.2 or 1.5 m) wide and up to 18 ft (5.5 m) long. It has a density of about 60 lb/ft^3 (960 kg/m^3), but other grades of lower density (medium hardboard, produced in greater thickness), and higher density (tempered or super-hardboard) are also made. The latter are oil-treated during manufacture to give improved properties.

The properties that were initially prescribed in specifications for hardboard were a minimum bending strength and maximum water absorption under defined conditions, and limiting values of these properties were laid down for the various grades. Much attention has been given recently to other strength properties that are important in practice, such as tensile and compressive strength in the plane of the board and perpendicular to it, shear strength, and rigidity.

Most hardboard is subjected during manufacture to a heat treatment after pressing. The boards are suspended for a period of a few hours in chambers through which hot air is circulated. This hardens the surface of the board, probably by chemical cross-linking reactions, thus improving its strength and water resistance. Further modification of properties can be achieved by the use of additives such as bitumen emulsion to increase water resistance, or of flame retardant chemicals, or by the use of drying oils such as linseed oil and tung oil to produce tempered hardboard. An additional technique, now widely used, is to introduce a surface overlay of modified pulp before pressing. This provides a surface of controlled absorption with improved paintability.

1.5.7 Laminated wood

In plywood, the grain directions of alternate layers of wood lie at right angles to each other. This results in a material having approximately equal strength properties in the two principal directions. Laminated members are made by gluing together smaller pieces of timber with their grain directions parallel to each other.

Lamination is essentially a method of building up members of greater length and cross section than those that can be obtained economically from available logs. However, laminated timber possesses certain advantages over solid beams of similar dimensions. It allows strength-reducing defects such as knots to be dispersed throughout the length and cross-section, thus in effect upgrading poor quality material. A beam having a number of widely distributed small knots will have better strength properties than one having a large single knot. Secondly, the thin laminae can be easily and rapidly dried before gluing, whereas drying of timber in large dimensions can only be carried out very slowly and with extreme care. Thirdly, laminating permits the manufacture of curved members such as roof arches, by bending the thin laminae before gluing and assembling.

Laminating was probably first used in Germany in the early years of this century and was gradually adopted on a limited scale in most European countries. At first only casein glue was available to bond the laminations together and the members were therefore only suitable for interior use, but with the introduction of the synthetic resin adhesives this limitation was removed and the field of use extended to include even the most severe environments.

a. Manufacture of laminated timber. The early manufacturing techniques were simple and although individual stages in the operation have been improved, the basic procedure has undergone little change. Sawn boards are first kiln dried to a moisture content suitable for gluing, i.e. below about 16%. If the boards are shorter than the member being produced it is necessary to build up the required length by end-jointing. For this purpose finger joints (see section 1.4.4), which are easy to locate and give adequate strength for the grades of timber normally employed in laminating, are generally used. Special machines have been developed for cutting the fingers and for assembling the joint.

After end-jointing the laminations are planed to a uniform thickness. This is usually between $1\frac{1}{4}$ and $1\frac{3}{4}$ in (32 to 44 mm) for straight members, but less ($\frac{1}{2}$ to 1 in (13 to 25 mm)) for curved members so that they can be bent to the required profile without fracturing. The laminations are glued by means of roller glue spreaders, then assembled in a jig and a bonding pressure of at least 100 p.s.i (0.7 N/mm^2) for softwoods and 150 p.s.i (1.0 N/mm^2) for the denser hardwoods is applied, and maintained until the glue is set. Laminating jigs vary considerably in design and complexity. After removal from the jig the member is finished by scraping off excess glue and planing if necessary.

Most manufacturers of laminated timber use a batch system of production as outlined above, but this is slow, has a high labour content, and tends to be wasteful of timber. A more efficient continuous laminating machine has been developed in the United

Fig. 1.36 Laminated timber construction.

Kingdom to assist in meeting the increased demand that is expected as architects and structural designers appreciate more fully the benefits to be gained from the use of laminated construction. Random-length planed boards with fingers cut in the ends are coated with glue and inserted in the continuous laminating machine, which automatically joins the boards end-to-end, then glues and assembles them into a continuous laminated member, which emerges from the machine and can be cross-cut to the required lengths.

b. Uses of laminated timber. Laminated timber is used primarily for structural purposes in building, either as straight beams or as curved and shaped members in structures such as roof arches and bridges. It also finds application in boat building for stems, framing, knees, and other parts.

Laminated members are produced in large cross-sectional sizes which would be virtually impossible to obtain in solid timber. Members having depths exceeding 80 in (2 000 mm) have been used. In addition to their excellent strength properties, curved laminated members have an attractive appearance and pleasing architectural effects can be obtained by their use. This is an important consideration in buildings such as churches, assembly halls, etc.

Small curved articles such as hoops, coffee table rims, seat rings, tennis racket frames and steering wheels are also made using laminating techniques. These are referred to below.

1.5.8 Bent wood products

Curved wooden members are required for many purposes, for example in the manufacture of furniture, boats, sports equipment and other articles. At one time these were hewn or sawn from limbs of trees having the necessary curvature, but this method of production has obvious limitations and is rarely practised. Curved parts can also be cut from straight boards by band-sawing, but when producing a curved piece of comparatively small radius of curvature it is necessary to saw some part of it across the grain of the wood, so that the strength of the piece as a whole is much reduced.

There are two main methods by which straight pieces of timber can be bent to shape while retaining their strength: one by softening and bending the complete piece in the solid, and the other by bending and gluing together a number of thin laminations or plywood strips to produce a built-up piece of the required dimensions and curvature.

a. Solid bending. In the process of bending a solid piece of timber the fibres on the concave face are compressed while those on the convex face are stretched by an amount depending on the thickness of the piece and the radius of curvature of the bend. With most timbers the strains that can be applied without failure either in compression or in tension are not very great. However, by subjecting certain timbers to heat treatments, such as steaming or boiling, their compressibility becomes very considerably increased and in consequence bends of smaller radius of curvature can be produced without fracture. The degree of curvature that can be achieved is then limited by the maximum permissible extension of the stretched fibres on the convex face, which is appreciably less than the permissible compression. If it is desired to decrease the radius of curvature further the stretched fibres must be supported by attaching a steel strap to the convex face during bending and tightening this by means of end stops or other devices.

The process of bending therefore consists of three stages: steaming the wood to soften it, bending the piece round a form by hand or with the aid of machinery, using a steel supporting strap if necessary, and setting the bend by holding it to shape while the wood cools and loses some of its moisture. It is not necessary to describe in detail the various kinds of machines and equipment that are used to assist in making bends in wood. Füller information on these may be found in the literature references (section 1.9).

The precise mechanism by which wood becomes amenable to plastic deformation when it is subjected to heat and moisture is not fully understood, nor is it known why some timbers are much more responsive to this treatment than others. The structure of the wood is undoubtedly an important factor in determining whether or not a species is suitable for bending, but other factors also play a part. In general, most of the temperate zone hardwoods are good bending timbers, but many of the tropical hardwoods and most softwoods are not well suited to bending in the solid form. The poor bending properties of many tropical hardwoods may well be associated with the presence of interlocked grain, which is a common feature in these timbers.

Information has been gained as a result of laboratory tests on the suitability for bending of a large number of timbers. This is expressed as the minimum radius of curvature at which a reasonable percentage of faultless bends (breakages not exceeding 5%) can be obtained, using clear material of a given thickness (25 mm), both when the timber is supported by a metal strap and when it is unsupported. Some typical examples of the data thus obtained are given in table 1.10.

It should be noted that, even though a species of timber may possess good bending properties, it may prove unsatisfactory in practice for other reasons. For example, distortion may occur during the steaming treatment, the timber may be inclined to check or distort during setting, it may stain in contact with iron or steel, or steaming may be accompanied by exudation of resin or dyes.

Although softening by steaming is the usual method of making wood sufficiently plastic for bending, there are other ways of bringing it to this condition. A number of chemicals are known to have a plasticizing action on wood. One of the most effective of these is ammonia, and a process has been developed in the United States for making wood pliable by immersing it for a short time in anhydrous liquid ammonia at a low temperature. The wood can then be bent or twisted to any shape, and hardens again as the ammonia evaporates. It is stated to have proved satisfactory for bending thin pieces of ash, birch and elm.

If wood of a suitable species is steamed and compressed longitudinally and immediately released, it does not fully regain its original length. The initial compression may be up to about 20% of the original length of the piece, while the residual strain may be about 3 to 4%. The pre-compressed wood is then capable of being bent in the cold state to a fairly small radius of curvature without fracture. Still greater flexibility is obtained if the heavily compressed material is set while still under load. These methods of production of pre-compressed wood, and a device for supporting the wood while it undergoes compression, have been patented in the United Kingdom.

b. Laminated bending. Thin strips of most timbers can be bent to a small radius of curvature without fracture, but owing to their elastic properties will tend to return to their original shape when the bending force is removed. If a number of concentrically bent pieces are securely fixed to one another by gluing, relative movements are impos-

Table 1.10 Limiting radii of curvature for material 25 mm thick of various species

Species	Approximate radius mm	
	Supported by a strap	*Unsupported*
Temperate hardwoods		
Ash, European	64	300
Beech, European	38	330
Cherry, European	51	430
Chestnut, Sweet	150	380
Elm, Canadian Rock	38	360
Lime, European	360	410
Oak, European	51	330
Sycamore	38	370
Walnut, European	25	280
Tropical hardwoods		
Afrormosia	360	740
Keruing	410	940
Mahogany, African	970	890
Mansonia	250	390
Obeche	460	430
Teak	460	890
Utile	910	1020
Softwoods		
Douglas fir	460	840
Larch, European	420	740
Pine, Corsican	860	740
Spruce, European	940	740
Yew	220	420

Source: *Forest Products Research Laboratory Leaflet No. 33.*

sible, and the composite piece retains its bent shape. This method of producing bent objects is known as laminated bending, and possesses a number of advantages over bending of solid timber. Bends of small radius can be built up from thin laminations of any species of wood, and material of relatively poor quality which would be unsuitable for solid bending can be incorporated. Long lengths may be obtained by end-jointing of shorter pieces, and normally no softening treatment is required. On the other hand, in comparison with bending of solid wood rather more technical skill and better equipment are required, and the larger number of operations leads to higher cost of the product. The presence of visible glue lines may in some instances be undesirable.

In laminated bending, relatively thin wooden strips or laminations are assembled adjacent to each other with the grain approximately parallel, and all are bent simultaneously over a single bending form. The surfaces are covered with glue and the laminations are pressed between male and female forms, or fluid pressure is applied uniformly over the whole area of a bend by means of a flexible hose or tube which is inflated with air or water. The method used for setting the glue depends upon the type of glue, but low-voltage heating strips are often used for rapid setting of bends up to about one inch in thickness.

A wide variety of curved articles are made by the laminating method. It is particularly well adapted to the production of continuous bends for articles such as cylinders or drums, hoops, tennis racket frames and steering wheel rims.

Large curved members for use in building are also made by laminating techniques and have been referred to in the preceding paragraph (section 1.5.7, Laminated wood). In this case the laminations are thicker (up to a maximum of 2 in (50 mm)) and it is generally necessary to use timber of better quality than that which is acceptable for the thinner laminations employed in smaller bent articles. The techniques of bending the laminae, clamping, and heating the glue lines to set the glue are also somewhat different when producing large, curved structural members.

It is possible also to make bends in plywood, but the operation resembles the bending of solid wood more closely than the production of laminated bends because the laminations in the plywood are unable to slide over one another. Plywood is bent after softening it by soaking or steaming, and the process is affected by the orientation of the grain of each ply and by the species and thickness of the plywood. Curved shapes made from plywood have many uses in the production of chair parts, travel goods, radio and television cabinets and other items of furniture, containers, etc.

1.5.9 Improved wood

Wood possesses many valuable technical and aesthetic properties, including a high strength/weight ratio, ease of fabrication and fixing, attractive appearance, etc. A great variety of timbers is available covering a wide range of density and other properties, so that it is generally possible by proper selection to find a timber that is suitable for any particular purpose. Nevertheless, all timbers suffer to a greater or lesser degree from certain drawbacks, one of the most important of which is their lack of dimensional stability – that is, their tendency to swell and shrink with changing humidity conditions. This is the reason for the sticking of doors and drawers, twisted window frames, and gaps between floorboards that are familiar to everyone.

Much effort has been devoted to attempts to improve the dimensional stability of wood by a variety of methods. Many of these are of little practical interest either because they adversely affect the mechanical properties or because of their high cost. Four methods of possible potential value will be described very briefly.

a. Acetylation of wood. The high affinity of wood for water is a consequence of the existence of large numbers of hydroxyl groups in the wood structure. If these can be converted into less hydrophilic groups it may be expected that the water absorption, and therefore the swelling of the wood, will be reduced. In the acetylation process the hydroxyl groups are converted by treatment with acetic anhydride and pyridine into acetyl groups which have a lower affinity for water. The pyridine acts as a swelling agent to open up the structure and a catalyst to promote the reaction. Acetyl contents of 20–25% can be obtained without difficulty and a reduction in swelling of about 70% can be achieved, though the acetylation of thicker pieces of wood is very slow. A process for acetylating wood was operated commercially in the United States for a time, but has been discontinued. Apart from its water absorption, the properties of acetylated wood differ very little from those of the untreated wood, and the main obstacles to its use appear to be the cost of treatment and the difficulty of treating wood in large sizes.

b. Polyethylene glycol treatment. Polyethylene glycol is a wax-like solid which is soluble in water in all proportions. In spite of its large molecular size it diffuses readily into the cell walls of wood that is swollen in water, and under controlled conditions it is possible to replace all the water in wood by polyethylene glycol by a diffusion process. In this way the wood is maintained in its swollen condition even after all the water has been removed, and subsequently exhibits very little swelling or shrinkage in changing humidity conditions. Polyethylene glycol treatment has found limited application in the stabilization of wooden carvings and in the preservation of waterlogged objects of archaeological value.

c. Resin-impregnated wood. Wood is a porous material, and if the voids in it can be filled by impregnation with a resin or plastic it is to be expected that its dimensional stability will be improved because the presence of the resin restrains the movement of the wood. The effect is still greater if the resin is able to penetrate into the cell walls, in which the swelling mainly takes place, as well as filling the void spaces.

Phenol formaldehyde resins possess the necessary properties to stabilize wood in this way. They have a strong affinity for wood and are able to diffuse into the cell walls, thereby swelling the wood beyond its water-swollen dimensions. It is not easy to impregnate satisfactorily solid pieces of even the more permeable species of wood with the viscous resin syrup, and the treatment is usually carried out on veneers. The veneers are impregnated by immersing them in the resin solution in an open tank or in a cylinder to which pressure can be applied. They are then pressed together in a hot press for sufficient time to cure the resin and bond the veneers together. The resin plasticizes the wood so that it can be compressed to a greater extent than untreated wood and the product, containing at least 30% by weight of phenol formaldehyde resin, therefore has a high density (1 350 to 1 400 kg/m^3).

This material has very low water absorption and a high degree of dimensional stability. It is also very hard and has good strength properties as a result of its high density, though its toughness is somewhat reduced by comparison with untreated wood. It also has good resistance to chemicals and to decay, and very high electrical resistance, but its cost is relatively high so that it can only be employed for purposes where its special properties are necessary. Much of the material that is made is used for electrical insulation purposes, but it also finds other applications which take advantage of its mechanical properties.

A similar product of lower density in which the wood is uncompressed is made by bonding the resin-impregnated veneers by heat and light pressure, insufficient to compress the wood. This material has properties intermediate between those of untreated wood and the fully compressed impregnated wood, and has a useful degree of dimensional stability.

d. Polymer-impregnated wood. In recent years attention has been given to the possibility of improving the dimensional stability and certain other properties of wood by impregnating it with liquid monomers, which can then be polymerized within the wood to convert them into solid polymers. The polymerization reaction can be brought about by the action of heat or catalysts, or by means of gamma radiation. The latter method possesses the advantages that, owing to the high penetrating power of the radiation, it induces uniform reaction through pieces of wood of moderate thickness and enables polymerization to be carried out at low temperatures. However, special equipment and precautions are needed and the radiation dose must be carefully controlled, because an excess of radiation has a degrading effect on wood.

This method of improving the properties of wood is still in the development stage and is open to many variations, producing materials with a range of properties. The monomer used may be varied; most of the work has been done with methyl methacrylate, styrene and acrylonitrile or mixtures of these. The possibilities also exist of inducing the monomer to penetrate the cell walls, and not merely to fill the void spaces, and of grafting the polymer on to the wood components. Modifications of this nature may be expected to affect the water absorption and swelling properties of the products.

Polymer-impregnated wood, also referred to as wood-plastic composite, has many useful properties, combined with an attractive appearance which needs no polish or other surface finish. It has been used mainly for small articles such as knife handles and brush handles. It would clearly be a suitable material for flooring and other uses in building, but is unlikely to find large scale applications unless a substantial reduction in production cost can be achieved.

1.5.10 Wood-cement products

Building materials consisting of some form of wood mixed with cement have been produced for many years. Sawdust-cement products were quite widely used at one time as a light-weight concrete, but more recently wood-wool-cement compositions have largely replaced them because they have superior strength properties.

Wood wool, which is also used as a packing material, is made by cutting roundwood into thin strands on special cutting machines. Good quality wood is necessary to produce wood wool having satisfactory strength properties. The wood wool is mixed with cement slurry and formed into slabs which are allowed to set in the usual way. The products are useful light-weight concretes with good strength properties owing to the reinforcing action of the wood strands, and are widely used in building work. The alkalinity of the cement confers a considerable degree of decay resistance on the wood in wood-cement products.

It is known that certain species of wood contain substances which delay or even inhibit the setting of Portland cement. Water-soluble or alkali-soluble wood components (sugars, hemicelluloses and phenolic substances) are mainly responsible for this effect. Hardwoods in general are more troublesome than softwoods in this respect and are therefore not much used for wood-cement products. Among the softwoods, larch and the heartwood of Douglas fir are known to cause difficulty. The species that are generally used are the pines, spruces and true firs (*Abies* species). Decay in timber also causes severe inhibition of the setting of cement, in addition to reducing the strength of the wood. It is important therefore to use sound wood that is free from rot.

1.5.11 Sawdust and shavings

When a tree is felled and the timber in it is converted into useful products, a large amount of wood is wasted in the various processes of conversion. This wastage is so great that in many cases not more than half the wood in the standing tree eventually finds its way into the final product. The waste takes many forms: tops and branches, slabs and edgings, sawdust, shavings, offcuts, etc. Part of this material is left in the forest, but uses can often be found for the solid waste from manufacturing processes. Sawdust and shavings are more difficult to dispose of profitably, but the growth of the chipboard industry has provided a useful outlet for clean shavings of suitable species. The most useful way of disposing of waste wood, particularly when it is dry, is often to use it as fuel under boilers, though this is a relatively low grade use, and special burners may be required.

Sawdust is produced in very large quantities at large sawmills and its utilization presents considerable difficulty. Damp sawdust is a difficult material to handle, and any method of utilization must be sufficiently profitable to cover the costs of collection, transport and storage. Sawdust as normally produced is not generally suitable for chipboard manufacture because the particles are not of the desired shape and size and they require a greater proportion of glue for bonding than do specially prepared particles. Developments have taken place in recent years in the use of sawdust for production of wood pulp. Chips are normally used for pulping because the fibres in them are longer and have been subjected to less damage than they have in sawdust, but it has been found possible to use sawdust as part of the raw material for some types of pulp without causing too great a reduction in pulp quality. Substantial quantities of sawdust are now used in pulp mills in North America. A number of relatively small scale uses also exist for sawdust, but in many of these dry sawdust is required. A great deal of effort has been devoted to attempts to convert the organic matter in sawdust into useful products by chemical processing. These will be considered briefly in section 1.7.

The amount of sawdust that is produced can be reduced by the use of thinner saws, but the extent to which this is practicable is clearly limited. Two other developments which aim at producing the waste in a usable form are worth mentioning. The first is a method of chipping away the wood from the outer part of the round log until a square cant is formed. The chips that are produced can be used for pulping or chipboard manufacture. The second is the development of a special saw which produces chips suitable for chipboard manufacture rather than sawdust (see section 1.3.4.).

1.6 WOOD DECAY AND PROTECTION OF WOOD

Strictly speaking, the term decay as applied to wood refers only to fungal decay, that is, attack by the various wood-rotting fungi. However, wood may be subject to other forms of deterioration and these will also be considered here.

Under favourable conditions wood is an extremely durable material. It is well known that woodwork in buildings and furniture will last practically unchanged, apart from purely superficial effects, for many hundreds of years. However, wood is an organic material and in some circumstances is subject to attack by certain biological agencies, of which the most important are the wood-destroying fungi, wood-boring insects and marine borers. Wood also suffers some deterioration if exposed to the weather for long periods or to the action of acidic, alkaline or oxidizing chemicals.

1.6.1 Weathering of wood

As already stated, wood is very stable under dry indoor conditions at ordinary temperatures. Any changes that take place are confined to the surface of the wood, and its physical and mechanical properties remain unchanged for very long periods. The change in colour of wood which takes place when wood is exposed to light is of some consequence where the appearance of the wood is an important factor, as in furniture, panelling and flooring.

All woods change in colour when exposed to light and air. Generally some yellowing occurs and many woods also darken in colour, while a few become paler. Both the main

Fig. 1.37 Darkening of wood caused by exposure to light. Note that the area of the floor ('Rhodesian teak') covered by the mat has retained its original colour.

structural components of the wood and the extraneous components, or extractives, are involved in these changes. The part played by the extractives has been studied in a few cases only. In some of these there is little doubt that the colour changes are due to oxidation of extractives promoted by ultra-violet light. Attempts have been made by furniture manufacturers to minimize these changes by incorporating an ultra-violet absorber in the lacquer, thus preventing the ultra-violet light from reaching the wood. This retards the colour change in some instances, but does not entirely prevent it. A method for stabilizing the colour of wood has also been developed in which the components in the surface layer are modified by a chemical oxidation process.

When wood is exposed to the weather, the effects of light are supplemented by the leaching action of rainwater and the swelling and shrinkage of the surface layers that result from alternate wetting and drying. The combined action of air and light causes some oxidative decomposition of lignin, the products of which are leached out by rainwater. Since the lignin functions primarily as the bonding material between the fibres, its removal from the surface layer of the wood results in the formation of a layer of loose fibrous tissue consisting largely of cellulose fibres. Thus, wood that has been exposed to the weather for long periods acquires a silvery fibrous appearance. In practice, the surface often turns to a silver-grey colour owing to accumulation of dirt and of traces of iron which interact with phenolic compounds in the wood, forming dark iron complexes. The effect may also sometimes be obscured by the growth of moulds and lichens on the surface of the wood.

At high altitudes, where the intensity of ultra-violet radiation is greater, the effects of weathering are rather different. The end result is often a dark brown colour of the wood, which has not been entirely satisfactorily explained.

The effects of weathering are normally confined to the surface layer of the wood, but in large cities they may be accelerated by polluted atmospheres. The presence of sulphur dioxide in the air causes partial hydrolysis of cellulose, so that the fibres are embrittled and are readily flaked or rubbed off. Thus, progressive breakdown of timber can take place in these circumstances.

Wood is subject to a special form of decomposition when it is in contact with actively rusting iron. It is often found that the wood around unprotected iron fittings or fastenings on gates and fences has seriously deteriorated after a period of exposure to the weather. It has been suggested that in changing from the ferrous to the ferric state iron acts as a catalyst for the oxidation of cellulose.

1.6.2 Resistance of wood to chemicals

Although wood does not withstand the action of strong acids, alkalis, or oxidizing agents for long periods without deterioration it possesses very good resistance to most neutral, or weakly acidic or alkaline solutions at moderate concentrations and temperatures. For this reason it has been used extensively for construction of certain items of equipment, such as vats and tanks, filter presses, rollers, paddles and fume ducts, in those industries where relatively mild conditions are met with. These include some branches of the chemical industry, and the manufacture of textiles, leather, foodstuffs and beverages. In particular, as a material for use in contact with weak acids, which are of very common occurrence in these industries, wood possesses advantages in cost and ease of fabrication over most other available materials. It is somewhat less resistant to alkalis than to acids of equivalent concentration.

Because of its good resistance under mildly acidic conditions, wood finds some application as a structural material in chemical works, where it is exposed to corrosive atmospheres which cause relatively rapid deterioration of metals. It is suitable not only for roof structures, but also for supporting structures for chemical plant, flooring, stairways, railings, etc. Its performance in corrosive atmospheres can be improved by means of protective coatings, and in some circumstances timber may prove more economical in practice than alternative materials, in terms of initial cost, length of useful life and maintenance costs.

For manufacture of items of equipment such as vats and tanks there is a tendency for wood to be replaced by other materials, particularly stainless steel and some plastics, largely on account of the greater ease of cleaning the latter.

Individual woods differ considerably in their resistance to chemicals, particularly acids and alkalis, and these differences can be correlated with the chemical composition of the woods, though physical and structural features also play an important part. The softwoods are intrinsically more resistant to attack than the hardwoods because of their higher alpha-cellulose and lignin contents, but some hardwoods of high density and low permeability to liquids also possess very good resistance. In addition to possessing good resistance to chemicals, timber for the construction of watertight vessels (vats and tanks) requires to be straight-grained and free from defects, and to have low permeability. For this purpose softwoods, e.g. Douglas fir and pitch pine, are generally preferred, and the best timber comes from the outer heartwood of large trees. Some dense hardwoods are also used, though they require greater care and accuracy in machining.

For the manufacture of barrels it is necessary to choose a timber that can be bent and for this reason barrels are commonly made of hardwoods, such as oak.

1.6.3 Decay by fungi

All organic materials are liable to decay. Some materials decompose rapidly, while others, including wood, do so more slowly. In every case the decomposition is brought about by the action of bacteria or fungi. In the case of wood, decay is caused by certain kinds of fungi, and bacteria play little part in its decomposition. Nevertheless, the properties of wood may be affected by bacterial action. For example, wood that has been stored in water may have its permeability greatly increased owing to the action of bacteria. Such wood absorbs abnormally large amounts of preservatives or water repellents, and this may lead to difficulties in subsequent painting. In temperate countries fungal decay is by far the most important cause of deterioration in timber, and much study has therefore been given to the wood-decaying fungi and to methods of preventing decay.

Fungi represent a low form of plant life and are distinguished from all other groups of plants in that they do not possess the green colouring matter, chlorophyll, which enables the higher plants to build up their food materials by the action of sunlight on carbon dioxide and water. For this reason they must derive their nourishment from the organic matter present in the dead or living parts of plants or animals. We are only concerned here with decay of wood after felling, but it must be remembered that decay can also take place in living trees, though the fungi responsible are usually different from those that attack timber. An outstanding exception is the standing oak fungus, which is the main cause of decay in the timbers of H.M.S. *Victory*.

The wood-destroying fungi are filamentous in nature and spread over the surface or through the wood by means of extremely fine hyphae (fungal threads). Fungi are reproduced by means of *spores*, which are microscopically small and are produced in very large numbers by fruit bodies, which are usually formed only after the fungus has made considerable growth. The result of this immense production of spores is that sooner or later any piece of timber exposed in the open becomes infected. In order to prevent the growth of the fungus and decay of the wood it is therefore necessary to maintain the wood in a condition in which the spores do not germinate.

In addition to a suitable food material, a fungus requires for its growth oxygen (from the air) and moisture. The important and active wood-destroying fungi are unable to make any growth in the total absence of air, and wood that is waterlogged or kept permanently submerged therefore remains virtually unaffected by decay, though very slow surface decay can be caused by microfungi. It is partly for this reason that logs are often stored by floating them in water in log-ponds.

Fungi also require moisture for their growth. The spores do not germinate except under damp conditions and fungi cannot grow in timber containing less than 20% by weight of moisture. Timber is liable to attack only while it is in an undried, green condition or when it has become damp owing to re-wetting. This fact is of fundamental importance in considering the decay of timber, and explains why woodwork in the interior of buildings, as well as in other situations where access of moisture is prevented, is normally free from decay. Decay of timber in buildings generally occurs as a result either of faulty design which permits penetration of water into timber by direct action of rainwater, condensation, or capillary action, or of inadequate maintenance, allowing deterioration of paintwork. Where it is not possible to ensure that timber will be kept permanently dry, other methods of preventing decay must be employed. These include the use of durable timbers or treatment with preservatives.

a. Recognition of decay. The different wood-destroying fungi differ greatly in the way in which they attack timber, but it is easy to recognise an advanced stage of decay in wood. Decayed wood is often lighter in weight than sound wood, and it generally shows some form of discoloration, being either lighter or darker than normal. This discoloration usually takes the form of blotches or streaks running along the grain. Decayed wood is also softened and can often be detected by probing with a sharp pointed instrument.

The early stage of attack, which is often referred to in the timber trade as 'dote', may be difficult to detect. With experience, it is often possible to establish whether incipient decay is present by raising the wood fibres with the tip of a knife blade and observing whether they break off short, indicating brashness, or whether they separate in a splinter. For a more positive determination of the presence of decay, microscopic examination is necessary.

b. Effects of decay. The principal effects of fungal decay on the properties of wood may be listed as follows:

Discoloration of the wood.
Loss in strength, which may be quite serious even in the early stages of decay. The first strength property to be affected is the toughness.
Loss in weight. Wood in an advanced stage of decay may have lost up to 70 to 80% of its air-dry weight.
Change in smell. Decayed wood loses its fresh smell, particularly when damp.
Increased water absorption.
Loss of calorific value.

c. Types of decay. Three main types of decay in wood may be distinguished:

1. *Brown rots.* Wood which has been attacked by a brown rot fungus eventually becomes a dark brown colour and in the final stages tends to break up into cubical or brick-shaped pieces as a result of cracking along and across the grain. Wood severely attacked by a brown rot can be readily crumbled under the fingers. In chemical terms, the brown rots are distinguished by the fact that primarily the cellulose and associated polysaccharides in the wood are attacked by a hydrolysis process, whereas the lignin is left practically intact. The dark colour of wood that has been attacked by brown rot fungi is due to enrichment of lignin or modified lignin in the decayed wood, resulting from the decomposition of cellulose and hemicelluloses.
2. *White rots.* Although darkening of the wood is sometimes the first indication of a white rot, it eventually becomes very much lighter in colour than normal. This lightening in colour may cover a considerable area, which is often surrounded by narrow dark lines, or it may occur in small patches, causing a white 'pocket' rot. Even in an advanced stage of decay, wood attacked by a white rot does not crumble into powder under pressure, although it may disintegrate into a lint-like material. Most of the decay in hardwoods is caused by various types of white rot. The chemical action of a white rot differs from that of a brown rot in that all the constituents of the wood are attacked, the lignin as well as the cellulose being broken down.

Fig. 1.38 Wood decayed by a brown rot fungus (dry rot fungus, *Merulius lacrymans*).

Fig. 1.39 Wood decayed by a white rot fungus.

The terms 'dry rot' and 'wet rot' have also been used to describe different types of decay, but they are often misleading or ambiguous. Dry rot is a term originally employed to describe any form of decay in buildings, but is now used specifically for the true dry rot fungus *Merulius lacrymans* (see below). This type of rot eventually leaves the wood in a dry, friable condition, but it cannot attack really dry wood. The term 'wet rot' has no precise scientific meaning. It is generally applied to the decay of timber caused by fungi that tend to require a fairly high moisture content for their growth. Wet rot occurs in timber out-of-doors in contact with the soil, and in parts of buildings where flooding with water from defective plumbing or roofs has occurred.

3. *Soft rot.* This is a form of slow decay proceeding from the surface of timber, which has been recognized more recently than the other, more vigorous types of rot. It is caused by microfungi which were first known only as destroyers of cellulosic fabrics. These fungi can develop when only very small amounts of oxygen are present, and decay can take place under water. This type of decomposition is most severe in the surface layers of wood, and can often be found as a softened zone up to $\frac{1}{8}$ in (3.2 mm) thick on the surface of wood that has been immersed in water for a long period. It has been observed, for example, in the superficial layers of wooden slats in water-cooling towers. In comparison with other wood-rotting fungi, decay by soft rot is very slow and is therefore of relatively little economic importance. Another feature of the soft rot fungi is their relatively high resistance to the action of the fungicides used in wood preservatives.

The following are the more important wood-rotting fungi causing decay of timber in buildings in the United Kingdom:

Merulius lacrymans (brown rot), the true dry rot fungus, is the most destructive fungus in buildings in Europe, and is responsible for a very large amount of damage to house property in the United Kingdom and elsewhere. It grows most actively in damp and poorly ventilated conditions where a high atmospheric humidity exists, and does not grow on dry wood or in very wet situations. The fungal strands are able to penetrate brickwork and to pass over inert substances such as stone or metal, and so spread to other timber in the vicinity. It is essentially a fungus of buildings and is not found in the open.

Coniophora cerebella (brown rot), the cellar fungus, normally only attacks timber that is definitely wet and is the principal cause of 'wet rot'. It is the commonest cause of decay in woodwork that has become soaked by leakage of water, and is of frequent occurrence in damp mines. The decayed wood often presents a similar appearance to wood attacked by dry rot, but *Coniophora* more commonly produces dark strands running over the wood, associated with pronounced cracking along the grain. It is less difficult to eradicate than the dry rot fungus.

Polystictus versicolor (white rot) is a common cause of decay in hardwood joinery, in hardwood timber in contact with the soil such as fence posts, in hardwood props in damp mines, and in ash and beech planks stored under insufficiently ventilated conditions.

Poria contigua is a white rot which attacks softwood joinery.

Lenzites sepiaria and *Lenzites trabea* (brown rots) occur on softwood window joinery, but are more common on the Continent of Europe than in Great Britain. Infections can often be traced to imported timber, which had been infected before its arrival in this country.

Poria vaillantii (white rot) occurs in very wet conditions, for example where leakage of water has occurred in buildings, and is probably the most important cause of decay in damp mines, where it rots pit-props very rapidly.

d. Sap stain or blue stain. So far we have considered only those fungal infections of timber which lead to its decay by breaking down the wood substance. Many other fungi can grow in wood. Some of these have no effect on its properties or appearance, while others cause staining owing to their hyphae or spores having a dark colour. In some cases the fungi cause actual staining of the wood cell walls.

Sap stain, or blue stain, is caused by fungi which have a negligible effect on the strength of wood, but reduce its commercial value by disfiguring it so that it cannot be used where a clean, bright appearance is required. It should be emphasized that fungi that cause blue stain belong to an entirely different class from those that cause decay, and blue stain is not the first stage in decay. However, the conditions that favour the development of blue stain also encourage decay.

Blue stain fungi develop in the sapwood of logs or sawn timber of susceptible species when the wood is partially dried. Spread of stain ceases when the wood dries to a moisture content below 25%, but may resume if the wood is re-wetted. The stain varies in colour from bluish-grey to almost black, and is caused by the presence of numerous fine, dark-coloured hyphae in the wood cells. It is of greatest importance in softwoods, particularly the pines, but also occurs in some light coloured hardwoods, e.g. obeche.

Staining is best prevented by rapid extraction and conversion of logs and rapid drying of the surface of the timber. Storage of logs in water is effective in preventing staining and is widely practised in Scandinavia. Anti-stain chemicals may be used where other methods are not applicable.

1.6.4 Attack by insects

In the United Kingdom damage to timber owing to attack by wood-boring insects is less extensive than that caused by fungal decay, but is nevertheless very substantial. Fungal decay and insect attack are often confused and are sometimes both present in the same piece of wood, but it is important to distinguish between them so that the correct measures for prevention of infestation or eradication may be taken.

Insect attack is generally characterized by small holes on the surface of the wood and by tunnels (often filled with bore-dust) within the wood. This dust may be ejected through the holes, forming small mounds beneath or on the surface of infested wood, indicating that adult insects have recently emerged or that live larvae are present inside the wood. Sometimes, if the attack is severe, the wood may be reduced to a powdery condition. It is not uncommon to find that a piece of wood has a very thin skin of sound wood, while the interior is largely reduced to powder. In such cases the external appearance of the wood may not give a true guide to its condition.

a. Life-cycle of wood-boring insects. The life-cycle of insects such as beetles, moths and wasps is made up of four stages: egg, larva (grub), pupa, and adult.

Eggs are laid by the adult insects on the surface of the wood or in pores or cracks. The eggs hatch into larvae.

The larval stage is usually the longest stage in the life of the insect, and it is the larvae which bore into the wood and cause the damage. When the eggs hatch the larvae begin to bore, deriving their nutriment from the wood which passes through their bodies. Their tunnels are often filled with excrement, called bore-dust or frass, which is sometimes characteristic of the particular insect and aids in its identification.

a

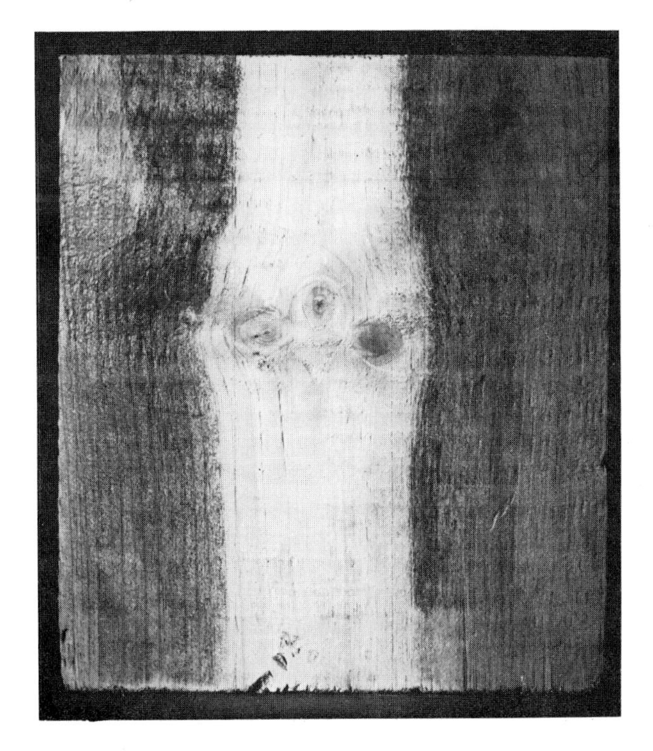

b

Fig. 1.40 Blue stain in pine, confined to the sapwood: (a) cross-section of log, (b) stain in a board.

When fully grown, the larva changes into a pupa (chrysalis), which is a non-feeding, resting stage, during which the intricate anatomical changes from a grub to an adult take place.

The adult emerges from the wood through a flight hole or exit hole. Adults are often short-lived. Many of them do not feed and some fly strongly. They spread infestation by mating and laying eggs. The complete life-cycle in temperate climates may take from one to ten years or more, depending on the species of insect, suitability of the wood, and the environmental conditions.

Some timbers are more attractive to insects than others, and it is apparent that insects are often selective. Little is known about the factors which render wood suitable for attack by insects, but it is known that the protein content of the wood, usually measured as its nitrogen content, is a limiting factor in determining whether it can support the growth of insects. Attack is generally, though not always, confined to the sapwood.

There is a considerable number of species of insect which can attack wood, but in practice damage in the United Kingdom is mainly caused by five types of insect, which are described briefly below.

b. Powder-post beetles. Powder-post beetles are so called because their larvae attack timber and reduce it to a fine powder. In the United States, several families of beetles are included under this name, but in Britain it is restricted to the families

Fig. 1.41 Bore dust from *Lyctus* powder-post larvae on surface of infested oak.

Lyctidae and *Bostrychidae* (usually imported), and by far the commonest species is *Lyctus brunneus*.

The beetles are small (about 5 mm long) and are reddish brown or black. The curved white larvae reach a length of 6 to 7 mm when fully grown. Damage by the larvae is limited to the sapwood of certain hardwoods. Softwoods are normally immune from attack. Examples of susceptible timbers are oak, ash, elm, walnut and hickory from temperate regions, and obeche, agba, afara, mahogany, ramin and seraya from tropical regions, and also many Australian hardwoods. Beech and birch are not attacked.

Lyctus is mainly a pest of sawmills, timber yards and manufacturers' premises where converted hardwoods are stored. It also occurs in veneers and plywood of susceptible species. Trouble may arise when infested wood is inadvertently used in manufactured woodwork, and attack only becomes apparent later when beetles emerge from the wood.

It has been shown that starch, which is commonly present in sapwood, is essential for the growth of *Lyctus* larvae, and methods of producing starch-free wood have been investigated as a means of preventing attack, but have not found widespread use. Much can be done to reduce losses by regular inspection of sawn hardwood stocks and removal of infested material. Control measures include sterilization by heat in drying kilns, fumigation with methyl bromide, and the use of aqueous emulsions of contact insecticides to protect sawn stock.

c. The common furniture beetle. This beetle (*Anobium punctatum*) is widely distributed throughout the British Isles and is the most frequent cause of 'woodworm' damage. Despite its popular name, this insect occurs more frequently and causes more damage in structural timber and joinery than in furniture.

Fig. 1.42 Damage by common furniture beetles in sapwood of pine.

The larvae are curved and whitish, up to 6 mm long, and the beetles, which emerge chiefly in June and July, are 2½ to 5 mm long and are reddish to blackish brown.

Anobium will attack the sapwood of a wide range of timbers, both softwoods and hardwoods, but most tropical hardwoods appear to be immune. Wickerwork and older types of plywood made up with animal glues are particularly susceptible. Attack is favoured by high moisture content, maximum temperatures not exceeding 22°C, wide sapwood, low resin contents and the presence of fungal decay, though the last is by no means essential. High temperatures and dry conditions are unfavourable to the growth of this insect, and the extent of attack is therefore likely to be less in centrally heated buildings.

The length of the life cycle of *Anobium* in buildings is at least three years, and often longer, and infestations are often not detected until several life cycles have elapsed. Consequently, infestations normally increase slowly over a period of years, and a house or article of furniture may be twenty or more years old before the damage becomes obvious. This gives the impression that new wood is not attacked, but this is not the case.

d. The death-watch beetle. The death-watch beetle, *Xestobium rufovillosum*, is also a member of the family *Anobiidae* (furniture beetles), and is well known on account of the damage it causes to the woodwork of historic buildings. It is widely distributed in England and Wales, mainly in the southern half of the country, and there is no record of its occurrence in Scotland. The name 'death-watch' is derived from the intermittent tapping sounds made by the adults during the mating period and believed by the superstitious to be a herald of death.

The beetle is 6 to 9 mm long and chocolate brown in colour. The fully grown larva is slightly larger than the beetle and is creamy white and curved.

The death-watch beetle occurs in old buildings where it causes severe damage to hardwoods, chiefly oak because of its wide use for structural purposes, but damage has also been reported in beech, chestnut, elm and walnut. Attack is almost entirely confined to timber which is in a decayed condition, the decay resulting either from installation of the timber in an unseasoned condition or from its re-wetting in use. Sound timber is very rarely attacked. The insect can spread into softwoods from adjacent infested hardwood. The combined effect of death-watch beetle and fungal decay often produces severe structural damage. If decay is slight or no longer active, the death-watch beetle is the primary pest.

In buildings where death-watch beetle damage is suspected a thorough survey by a specialist should be undertaken before embarking on remedial measures. These may include replacement of affected timber, elimination of conditions favouring attack (chiefly by preventing access of moisture), and insecticidal treatment, all of which may be costly operations, for example in a church roof.

e. The house longhorn beetle. The family of longhorn beetles (*Cerambycidae*) comprises a large number of species, most of which are essentially forest insects. The house longhorn beetle, *Hylotrupes bajulus*, is an exception in that it can breed only in converted softwoods. Damage is confined to the sapwood.

The importance of this insect has increased in recent years. It is of widespread occurrence in Europe where it causes serious damage to structural timber in buildings. In Britain it has become established in an area to the south-west of London, probably because the summer temperatures, which are higher in this area than elsewhere, favour the development of the insect.

The beetle (10 to 20 mm long) and its larva (about 30 mm) are larger than those of the other insects considered here, and its tunnels and exit holes are correspondingly larger.

In buildings, all the softwood timbers such as doors, windows, floorboards and other joinery work can be attacked, but in Britain serious damage is virtually confined to the

Fig. 1.43 Roofing timbers attacked by house longhorn beetles. (a) before and (b) after removal of the surface skin.

structural timbers of roof spaces. Infestation is not easy to detect, especially in the early stages before the oval exit holes appear on the surface of the timber. In order to prevent the spread of this destructive insect, regulations are in force in districts where the risk of infestation is high requiring that all softwood timber used in the roofs of new houses must be protected with a suitable preservative.

f. Ambrosia (pinhole borer) beetles. These beetles, of which a large number of species exist, attack only green timber. They are most common in tropical timbers in which the attack may occur in the living tree, but frequently starts in logs soon after felling and may also take place during extraction and storage of the logs. On conversion and drying, the insects eventually die, and there is no risk of recurrence or spread of attack in seasoned timber. The tunnels remain as an indication of past infestation, but the structural strength of the wood is seldom seriously affected.

The tunnels often have darkly stained walls and the stain may extend into the wood as patches or streaks. These insects cause considerable economic loss on account of the unsightly appearance of wood in which bore holes or tunnels with their associated stain are present. Other properties of the wood are little affected and there is no danger of further damage when the wood is dry. The effects are most serious in some light coloured tropical hardwoods, particularly those that are cut for veneer. Damage is best avoided by conversion and rapid drying of the timber. The application of insecticides to freshly felled logs is also widely practised.

It is important to distinguish between damage caused by ambrosia bettles, which has always ceased in the seasoned timber, and that caused by the insects previously described, which may continue to spread unless active control measures are taken.

1.6.5 Attack by marine borers

Timber exposed to sea water, such as harbour pilings, boats, and timber floated in salt water, is subject to rapid and extensive damage by certain wood-boring animals known as marine borers. They are of two distinct types, belonging to the molluscs and crustaceans respectively. These organisms differ widely in appearance and habits and in the nature of the damage that they cause, and also in the relative effectiveness of preventive treatments. Although a number of species are active in attacking wood it will be sufficient for the present purpose to describe briefly two of the more common types of marine borer.

Shipworm (*Teredo* species). Species of the mollusc *Teredo* have a world-wide distribution, but in Britain attack is sporadic and is mainly important in southern waters. They enter wood as minute larvae which then increase rapidly in size and develop the characteristic worm-like body, at the same time enlarging the size of the burrow. The entrance hole is only slightly enlarged during the growth of the larvae, so that considerable damage may be caused by these animals without much external evidence. The shipworm when fully grown may reach a length of as much as four feet (1.2 m) and its diameter varies from about $\frac{1}{8}$–$\frac{1}{4}$ inch (3.2 to 6.4 mm) to 1 inch (25 mm). The intensity of infestation is related to temperature and is much more severe in tropical waters than in those of the temperate zone.

Gribble (*Limnoria* species). Species of *Limnoria* are widespread in the British Isles and do not show the localized distribution of *Teredo*. They are crustaceans having the general appearance of a wood louse, and bore into wood forming tunnels which are much shorter and smaller than those caused by *Teredo*, seldom penetrating more than

Fig. 1.44 Typical damage to piling by *Limnoria*.

½ in (12.7 mm) below the surface of the wood. However, when they are very numerous the entire surface may become infested, and the weakened surface is eroded by the action of the sea, exposing fresh timber to attack. The damage is thus progressive and can become quite severe.

No timbers are completely immune from attack by marine borers, but some are highly resistant and are commonly used for marine structures. These include greenheart (*Ocotea rodiaei*), basralocus (*Dicorynia guianensis*), pyinkado (*Xylia dolabriformis*), jarrah (*Eucalyptus marginata*), manbarklak (*Eschweilera subglandulosa*), turpentine (*Syncarpia laurifolia*) and a few others. Alternatively, timber that has been impregnated under pressure with preservatives may be used for marine piling and wharf construction. Suitable preservatives are coal tar creosote, creosote-coal tar solutions, and copper/chrome/arsenic water-borne mixtures. Long experience has shown that when timber is thoroughly impregnated with a high loading of one of these preservatives it will last for a very long time in marine conditions. It is necessary to choose a timber which is sufficiently permeable for easy treatment. Another method of protection is to encase the timber in an inert material which prevents the borers from gaining access to it. The material must be resistant to corrosion by sea water. Copper sheathing and concrete have been used in this way.

1.6.6 Natural durability

In the United Kingdom the term natural durability generally refers to the resistance of a timber to fungal decay, because this is the most important cause of deterioration in service. The natural durability of a timber is of importance in situations where the timber is liable to become damp (moisture content greater than 20%), and there is consequently a risk of decay. These include outdoor uses and certain indoor situations where there is a risk of wetting by moisture penetration or condensation. Timber in contact with the ground is particularly vulnerable to decay organisms. On the other hand, when timber is used in situations where it will always be kept dry, natural durability is unimportant for there is then no risk of decay.

Two points concerning the natural durability of timbers should be noted. First, the sapwood of all species is non-durable; that is, it possesses little natural resistance to attack by decay organisms. Secondly, the natural durability of the heartwood varies greatly with the species. Some species are rapidly destroyed by fungal decay, while others are highly durable and give long service under adverse conditions. The difference in life between durable and perishable timbers in conditions where there is a high decay risk is of the order of at least 20 or 30 to 1. Moreover, different samples of the heartwood of the same species, even when taken from the same tree, may show a marked variation in decay resistance. For this reason durability can only be expressed in approximate or average terms.

Standardized methods of measuring the resistance of different timbers to attack by wood-destroying fungi have been developed and are used to classify timbers according to their natural durability. Small specimens of the heartwood of the timber are exposed to attack by certain fungi growing under controlled conditions in the laboratory for a standard period of time, and their loss in dry weight is determined. This loss in weight, expressed as a percentage of the original dry weight, gives a useful indication of the amount of decay that has occurred and provides a basis on which to compare the resistance of different timbers to decay. The results of such tests are correlated with those of long-term field tests and in this way a durability classification has been arrived at, based primarily on the results of field tests. Timbers are classified into five broad grades, based on the performance of their heartwood when in contact with the ground. The five grades are shown in table 1.11.

The average life quoted for each grade refers to conditions in the United Kingdom and relates to material of cross-section 2 × 2 inch (50 × 50 mm). Larger sizes will, of course, last longer and in general the increase will be in direct proportion to the thickness (smallest dimension), and not to the cross-sectional area.

An outline of the chemical composition of wood has been given in section 1.1.3.

Table 1.11 Natural durability classification

Grade of durability	*Approximate life in contact with the ground (years)*
Very durable	More than 25
Durable	15 to 25
Moderately durable	10 to 15
Non-durable	5 to 10
Perishable	Less than 5

Wood consists essentially of the three main cell wall constituents, cellulose, hemicelluloses and lignin, with which are associated varying amounts of a wide range of minor components or extractives. The wood substance itself possesses little resistance to decay organisms, and the high durability of some species is due to the presence among the extractives of components that are toxic to the wood-destroying fungi. Even in the highly durable timbers the sapwood is non-durable, and this is in accordance with the fact that the sapwood contains only relatively small amounts of extractives. A large number of investigations have been carried out on the relationship between the durability of timbers and the extractives present in them, and it has been shown in many instances that substances can be extracted from the wood which have fungicidal properties. These substances are often phenolic in nature, and include the tannins which are of widespread occurrence in hardwoods.

In the descriptions of timbers given in section 1.2, their natural durability classification, as defined in table 1.11, is quoted.

1.6.7 Wood preservatives

Wood may be protected from decay in three ways: (1) by ensuring that the wood is dried initially to a moisture content below 20% and is always maintained in that condition, (2) by the use of wood that possesses sufficiently high natural durability, taking into consideration the conditions of use and the length of service that is required, and (3) by treatment of the wood with preservatives.

In most outdoor uses of timber, e.g. exterior woodwork in buildings, fencing, telegraph poles, railway sleepers, boats, etc., and in some interior woodwork in buildings, it is not possible to ensure that the wood will always be kept dry. While freedom from decay can generally be achieved in such situations by the use of durable timbers it must be borne in mind first, that natural durability is a variable property and some pieces, even of a durable species, may possess insufficient decay resistance for the particular situation and, secondly, that the sapwood even of durable timbers is non-durable. Moreover, the very durable tropical hardwoods such as teak, afrormosia, iroko and jarrah, are too costly for many uses.

It is often possible to obtain a high degree of decay resistance with greater certainty and at lower cost by treating timber with a suitable preservative. By choosing a suitable species of timber and treating it correctly, wood can be made to last almost indefinitely even under severe conditions. To ensure that the treatment is effective it is necessary to employ a timber that is sufficiently permeable to allow the preservative to penetrate to an appreciable depth into it. Timbers vary greatly in their resistance to penetration by liquids, and it is because of the importance of this property in relation to treatment with preservatives that information on resistance to impregnation is given in the descriptions of timbers in section 1.2. Scots pine, beech, birch and ramin are examples of timbers that can be easily treated, while spruce and many tropical hardwoods are very difficult to treat.

Types of preservative. There are three main types of preservative in general use today: tar-oils, water-borne compounds, and organic solvent type preservatives.

The most important of the tar-oil preservatives is coal tar creosote, which is a product of coal tar distillation. It is suitable for use on much exterior woodwork and provides very good protection against wood-destroying organisms if correctly applied. It is also effective against marine borers and is used on piling, dock and wharf timbers, etc. In the UK its quality is covered by British Standards 144:1954 and 3051:1958.

The principal drawbacks to the use of creosote are that its odour is objectionable in enclosed spaces, it is dark in colour and is liable to stain adjacent materials, and creosoted wood cannot be painted. Creosote is cheap and is extensively used for preservation of telegraph and transmission poles, sleepers, fencing and gates, and timber in farm buildings. It is not generally suitable for buildings of other types.

Decay of timber can be effectively prevented by treatment of the wood with aqueous solutions of inorganic salts which are toxic to the decay fungi. Examples of salts which have been used in the past are copper sulphate, mercuric chloride, sodium fluoride and zinc chloride. However, these salts may be leached out of the wood if it is exposed to wet conditions, and some of them have a corrosive action on metal fittings attached to the wood. For these reasons, multi-salt mixtures have been developed, the constituents of which interact with the wood components and with each other to form insoluble products, which are very highly fixed in the wood while retaining their preservative effect. The most effective of these multi-salt formulations is a mixture of copper sulphate, sodium dichromate and sodium arsenate, commonly referred to as the copper/chrome/arsenic type of preservative. Such preservatives are widely used under various proprietary names in the United Kingdom and elsewhere, and some other compositions are in use in other countries. When correctly applied, water-borne preservatives are extremely effective and ensure that timber will remain free from decay for a very long time under any conditions without maintenance. They are free from odour and leave the timber clean so that, when it has been adequately dried, it can be painted satisfactorily. They are used for treatment of timber in buildings, as well as for other purposes, e.g. in packaging, agricultural appliances, motorway fencing, and water cooling towers, where the presence of creosote would be undesirable.

The boron diffusion process is a special method of treatment of wood with a water-borne preservative. A concentrated solution of disodium octaborate is applied, preferably by dipping, to the surfaces of freshly sawn undried timber, which is then close-piled in the wet condition. Diffusion of the boron preservative into the wood then takes place, a process that may require several weeks, after which the timber is dried in the normal way. This method of treatment is inexpensive and possesses the advantage that, provided the initial moisture content of the timber is suitable, complete penetration can be obtained even with timbers such as spruce and hemlock which are difficult to impregnate by other methods. The preservative is odourless, clean and safe, and only one drying period is necessary. However, the process can only be used with green timber, and the preservative is not fixed in the wood, so timber treated in this way should not be used in situations where leaching may occur. Nevertheless it is suitable for most purposes in building construction work where leaching is not a normal hazard.

The organic solvent type preservatives are in general much more expensive than the other types and their cost limits the method of application to brushing, dipping, or the double vacuum process (see below). Their main advantages over the water-borne preservatives are that they do not swell or distort the timber and that the carrier solvent evaporates from the treated wood relatively rapidly. The chemicals principally used are pentachlorophenol and its derivatives, copper and zinc naphthenates, and tributyltin oxide. In addition, some preservatives of this type also contain specifically insecticidal substances such as dieldrin and gamma benzene hexachloride, and others contain water repellent constituents such as synthetic resins and waxes, which restrict the penetration of water into the wood.

1.6.8 Preservation methods

The choice of the preservation method to be used depends upon the decay risk to which the timber is to be exposed and the length of life required. Where the conditions are favourable to decay, for example in timber in contact with the ground, or a very long life is required, impregnation under pressure, which ensures deeper penetration and a higher loading of preservative in the wood than other methods, is the most effective treatment.

Simpler methods of treatment (non-pressure methods) are also available which give a sufficiently high standard of preservation for many purposes.

a. Pressure impregnation. The treatment is carried out by placing the timber, which must be previously dried, in a pressure cylinder which is then evacuated to remove air from the pores of the wood as far as possible. The cylinder is then filled with the preservative and hydraulic pressure is applied for a period long enough to force the requisite quantity of preservative into the wood. Alternatively, where a lower loading of preservative is considered sufficient, the vacuum stage may be omitted and pressure applied directly after introducing the preservative into the cylinder. Provided the timber is sufficiently permeable, a good depth of penetration of preservative into the wood is achieved by these methods of treatment and the treated wood is immune from decay. As has already been mentioned, however, timbers vary very widely in their permeability, and those which are classed as resistant to impregnation cannot be treated satisfactorily, even by using high pressures. When high resistance to decay is required it is therefore important to choose a timber which is sufficiently permeable to permit treatment to be

Fig. 1.45 Plant for vacuum-pressure impregnation of timber with preservatives.

carried out effectively. It may be noted that, among the softwoods widely used in building, spruce, hemlock and sometimes Douglas fir are resistant to impregnation. They can, however, be treated satisfactorily by the boron diffusion process (see above). The preservatives normally applied by pressure impregnation are creosote and the waterborne preservatives.

b. Double-vacuum treatment. This method is particularly suitable for application of low-viscosity organic solvent preservatives. The timber is placed in a cylinder which is evacuated, filled with the preservative solution, and the vacuum then released. After draining the solution from the cylinder a final vacuum is applied to remove excess preservative. The degree of treatment obtained can be controlled by varying the treating cycle. The process is often used for preservative treatment of exterior joinery timber.

c. Hot-and-cold open-tank treatment. Pressure or vacuum treatment requires fairly expensive equipment and is only economic when a large volume of timber is to be treated. The open tank process can be carried out in very simple plant, and consists in immersing the timber in a preservative, which is then heated to 85–95°C and maintained at this temperature for an hour or more. It is then allowed to cool, thus drawing the preservative into the wood, before the timber is removed. The process is a very effective method of treating timbers that are permeable or only moderately resistant to impregnation, and is a convenient and efficient process for preservation of much of the timber used on farms and estates.

Fig. 1.46 Plant for treatment by the double vacuum process.

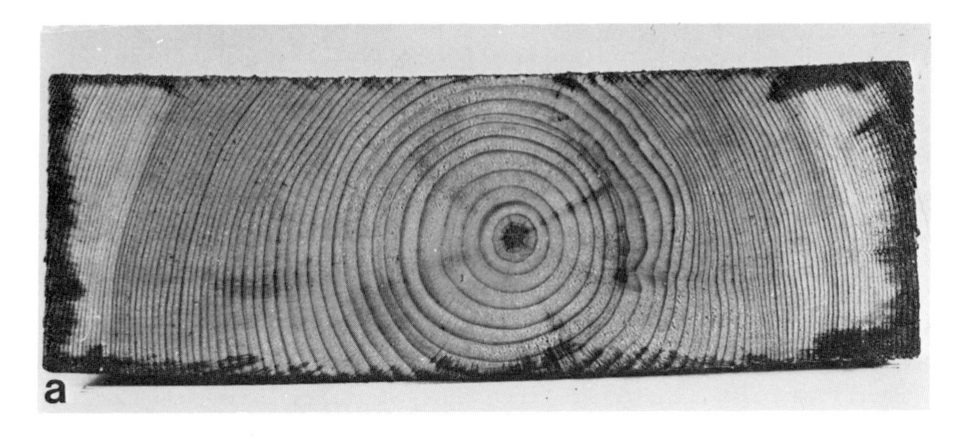

Fig. 1.47 Scots pine joist treated with coal-tar creosote, (a) by 15-second immersion, (b) by vacuum-pressure impregnation. Note increased depth of penetration by pressure treatment.

d. Brushing, spraying and immersion. Where long life is required of timber exposed to adverse conditions, it should be deeply impregnated with preservative, but where conditions are not so favourable to decay, e.g. in timber exposed to the weather but not in contact with the ground or in some interior woodwork, treatment by brushing, spraying or immersion can provide a useful degree of protection. The preservatives that are normally applied by these methods are creosote (for outdoor use) and the solvent type preservatives. With most timbers brush application results in little more than skin-deep penetration and re-treatment may be necessary every few years. Spraying results in a more effective covering of the timber, and there is a greater chance of the liquid penetrating into holes and cracks. Immersion, which may range from a few seconds dipping to days or weeks, provides more complete contact with the preservative and ensures a thorough coating of all surfaces. Treatment of exterior window joinery by dipping is now standard practice. Deluging is a form of immersion treatment in which pieces of timber are propelled mechanically through a tunnel in which an organic solvent preservative is sprayed on to the piece, or through a running stream of the preservative.

1.6.9 Surface treatments

Surface coatings are applied to wood and wood-based materials to improve their appearance or to protect them from deterioration, or for both of these purposes. They are, in the main, organic materials and may be classed as (1) paints, (2) varnishes, and (3) stains and polishes.

One of the major factors affecting the performance of paints and varnishes on wood is the tendency for wood to swell and shrink with changes in humidity conditions. This imposes a stress on any film-forming surface coating and partly explains why the durability of a coating on timber is often lower than that of the same coating on metal. The stress is naturally lower in coatings on timbers having low movement values than in those on timbers with high movement, and failure of coatings can be minimized by devising coating materials that are sufficiently elastic to withstand the imposed stress without rupture. Surface texture and grain pattern also have an important influence on the performance of surface coatings. In general, coatings will have a longer life on timbers with a smooth, even texture than on those with a coarse texture, prominent growth rings or large pores.

a. Paints. Paints consist of a polymer, together with a pigment to impart opacity or colour to the product, generally dispersed in a solvent but occasionally applied without one. The presence of a solvent facilitates the application of the paint but, because the solvent evaporates after application, the film thickness obtained is less than it is when the solvent is absent. Drying oils, e.g. linseed oil, which harden by oxidation are widely employed in paints. It is not possible here to discuss the wide range of polymers used, but many of them have characteristic features which render them suitable for particular purposes. Thus, alkyd resins are satisfactory for non-yellowing finishes, epoxy-based paints have good chemical resistance and toughness, and polyurethanes have high abrasion resistance and resistance to water and chemicals. (See also vol. 5, ch 11.)

A smooth surface is essential for good paint performance on wood, and planing is generally necessary, while sanding in addition is usually beneficial. Before being painted timber should be dried to a moisture content close to that which it will reach in use. Very wet timber may, during drying, cause a paint film to blister.

Provided that they remain intact, paint films are effective in preventing liquid water from entering wood, but they are not generally impervious to water vapour. Where timber is to be used in moist conditions it is therefore advisable to treat it with a preservative, even though it is to be painted, in order to avoid the risk of decay. In exterior joinery, which is a major field of application of paint on wood, preservative treatment of the wood is highly desirable, not only because of the risk of penetration of moisture through the paint, but also in order to protect the timber at the joints, where rupture of the paint film frequently occurs, allowing ingress of moisture.

With some timbers difficulty is experienced due to exudation of resin through the paint. In the case of softwoods, such as the pines and Douglas fir, this can generally be prevented by sealing the knots or resin streaks with shellac 'knotting' or a leafing aluminum primer, but exudation from a few hardwoods, notably keruing, is very persistent and cannot easily be controlled in this way. Here it is advisable to eliminate highly resinous pieces of wood.

b. Varnishes. Clear varnishes are used both on interior and exterior woodwork. For interior purposes they present few problems and a range of varnishes or lacquers is available for use on furniture, table tops, etc., some of which possess special properties such as hardness and resistance to scratching, heat resistance and resistance to solvents. Finishes that are suited to mass production methods of application are used in furniture manufacture.

Varnishes are used on exterior woodwork when it is desired to preserve the natural appearance of wood, but they have considerably lower durability than exterior paints. Varnishes fail by cracking, peeling and delamination, exposing the timber to the effects of water and sunlight and to attack by unsightly mould growth. Furthermore, unlike paints, they do not protect the wood from the effects of ultra-violet light, which causes deterioration of the surface and loss of adhesion to the varnish. Clear varnishes at present only last about 1½ to 2 years. Many attempts have been made to improve the performance of clear finishes on exterior woodwork. One such method which shows some promise is to incorporate in the varnish small particles of inorganic solids which are not themselves degraded by the action of ultra-violet light but which, because of their suitable refractive index, become invisible in the varnish film. Like the pigments in paint films, they protect the wood from the effects of ultra-violet light. Trials with modified varnishes of this type have shown considerable improvement in service life over the usual varnishes, with no appreciable change in appearance.

Other methods of protecting exterior timber are available. The use of boiled linseed oil for this purpose has long been practised. It is fairly satisfactory, subject to two reservations. First, it requires frequent renewal and, secondly, a black mould is liable to grow in the oil, spoiling the appearance of the wood. This can be avoided by incorporating a fungicide (pentachlorophenol) in the oil.

Preservative stains, which can be brushed on to wood and are absorbed by it without forming a film on the surface, provide satisfactory protection, but they modify the colour of the wood. They can be easily renewed when necessary, without scraping the wood. A modified stain which gives good protection, with durability approaching that obtained with a good paint coat, has been developed by the US Forest Products Laboratory. It is based mainly on boiled linseed oil, together with colouring matters, paraffin wax and pentachlorophenol, dissolved in turpentine.

c. Stains and polishes. Both water stains and spirit stains are used for staining light coloured woods used in furniture, interior joinery and flooring. Darker coloured woods or those with an attractive appearance are left in their natural colour. Many woods require careful filling to close the pores before staining and polishing.

1.7 WOOD AS A CHEMICAL RAW MATERIAL

The preceding sections have dealt with wood which is either used as a natural material or is reconstituted by various methods into other wood-based materials which have improved properties or can be made in more useful sizes. However, wood is an organic material and very large quantities of it are also used as a raw material for chemical processing. Brief surveys of the principal chemical processes by which wood is utilized will be given in the present section. Some of them are described in more detail elsewhere in this volume.

Wood waste. Some of the processes for chemical utilization of wood have been devised specifically with the object of finding a useful outlet for part of the very large quantity of waste wood that is produced in the course of timber production. When a tree is felled and the timber in it is converted into useful products, a large amount of wood is wasted in the various processes of conversion. This wastage is so great that in many cases less than half of the wood in the standing tree finds its way into the final product. The waste takes many forms: tops and branches, slabs and edgings, sawdust, shavings, offcuts, etc. In addition to the wastage in conversion, there are other important sources of unused wood. In forest plantations, thinnings are removed at intervals according to a definite plan, and these are generally too small in size for utilization as sawn timber. In natural forests (especially in tropical forests), inferior species commonly grow along with the more valuable ones, and these, too, provide a very large volume of unused wood. Chemists have devoted a great deal of effort to the study of methods by which the organic matter of which this vast quantity of waste wood consists may be converted into useful products. However, the main chemical industries based on wood (pulping processes) operate on a very large scale and require for their efficient working a large, continuous and uniform supply of good quality raw material. They do not, therefore, provide a solution to the problem of the utilization of waste and inferior wood, though they may in some cases be able to absorb a proportion of material of this kind. Other processes which have been proposed more specifically for utilization of waste wood are referred to in section 1.7.5.

1.7.1 Destructive distillation of wood

The destructive distillation of wood to produce charcoal is by far the oldest chemical industry based on wood, and is known to have been practised by the ancient Chinese. The Egyptians, and the Greeks and Romans, made charcoal by carbonization of wood and they also collected the volatile products of distillation and used them for embalming purposes, and for filling the joints in wooden ships. At a later date, with the development of the iron industry, very large quantities of charcoal were produced from wood for smelting iron ores, and this led to serious devastation of forests in some industrialized countries.

When wood is heated out of contact with air, decomposition takes place slowly at temperatures up to about 250°C, but an exothermic reaction commences at about 270°C and extensive decomposition then takes place without further addition of heat, with formation of volatile distillation products and a residue of charcoal. The charcoal retains considerable amounts of the higher-boiling volatile products, which can only be removed by heating to higher temperatures.

The destructive distillation of wood may be carried out either for the production of charcoal alone or for the additional recovery of the volatile by-products, chiefly methanol, acetic acid, acetone and wood tar. In the latter case a much more elaborate plant is necessary for the collection, separation and refining of the volatile products.

In early times, charcoal was produced by slow carbonization of wood in pits or mounds with controlled access of air, and such methods are still in use in various parts of the world. However, simple earth kilns have now been replaced in many countries by brick kilns, or by portable metal kilns which can be set up in the forest and moved about as sources of supply of wood become exhausted. These allow closer control of the carbonization process and give better yields of charcoal, as well as a cleaner and more uniform product.

A modern wood distillation plant with full recovery of by-products is a much more complex installation than the simple charcoal kilns described above. Carbonization is carried out in an iron retort which may operate batchwise or continuously, and the distillation products are condensed and separated. The products consist of (1) charcoal, which is the solid residue left in the retort; (2) pyroligneous acid, a solution of the water-soluble products of decomposition of wood, which is separated by further processing into methanol, acetic acid and other products; (3) wood tar, containing the water-insoluble products, which settles to the bottom of the pyroligneous acid; and (4) wood gas, the non-condensable portion of the decomposition products, which is scrubbed to remove spirit and acid vapours before being burnt as fuel.

For distillation purposes, an important distinction is made between hardwoods and softwoods. Hardwoods produce a higher yield of a dense charcoal and higher yields of by-products, particularly acetic acid, than softwoods. Since charcoal is the major product of all distillation plants, it is the type and yield of charcoal that is the determining factor in the choice of wood, and for this reason hardwoods are generally preferred for wood distillation. The approximate yields of products from hardwood distillation are shown in table 1.12.

Softwoods produce a lighter, softer charcoal and lower yields of by-products, but they also yield turpentine and various oils derived from the resinous components of the woods. The commercial distillation of softwoods is carried out at a number of plants in the United States, which use mainly the resinous pines of the southern and south-eastern states.

In most industrialized countries charcoal from wood distillation has lost much of its former importance since the introduction of coke as a smelting agent, but it finds wide use as fuel in less developed countries. Nevertheless, charcoal still plays an important part in a number of industries, for example in the manufacture of case-hardening compounds for metals, for absorption of impurities from gases and solutions, and in manufacture of carbon disulphide used in production of viscose rayon, though here, too, it is being replaced by coke. Although charcoal production is not likely again to be a major use for wood, the industry continues at a steady pace. The current tendency is to operate plants which produce only charcoal without recovery of by-products, because the economic marketing of the other products of wood distillation is becoming increasingly difficult as they have to compete with the same chemicals available at low cost from the chemical and petroleum industries.

1.7.2 Pulp and paper

The manufacture of wood pulp and the production from it of paper and of cellulose for textiles and cellulose derivatives are dealt with in chapters 2 and 3 of the present volume, and will only be mentioned very briefly here.

Table 1.12 Products from hardwood distillation

Product	*Yield, % of dry wood*
Charcoal (17.5% volatiles)	36
Acetic acid (including formic acid, propionic acids)	6
Methanol + acetone	2
Tar and oils	12
Non-condensable gases	20
Water of pyrolysis and loss	24

The manufacture of wood pulp is by far the largest and most important chemical industry based on wood. In the production of wood pulp, coniferous woods (softwoods) and broad-leaved woods (hardwoods) are both used. These two types of wood differ in their structure, their chemical composition, and in the dimensions of the fibres of which they are largely composed. These differences are reflected both in the processes used for production of wood pulp from them and in the properties of the pulps obtained which render them suitable for different end uses. In the past the softwoods were generally considered more suitable for pulping than the hardwoods, and it is partly for this reason that the pulping industry was at first concentrated mainly in North America and northern Europe where plentiful supplies of softwoods were available. With the development of improved techniques for pulping hardwoods the industry has now spread to many other parts of the world.

The purpose of all pulping processes is to separate the fibres in wood, while retaining as far as possible their form and strength. This may be achieved either by mechanical or by chemical methods, and the choice of process depends upon the type of wood and the purpose for which the pulp is to be used. The principal pulping processes at present in use belong to the broad types described below. Further details are given in chapter 2.

a. Mechanical pulping. In the mechanical processes defibration is accomplished entirely by mechanical methods, without the aid of chemicals. The most important of them is the groundwood process in which billets of wood are ground to fibre by pressing them, in the presence of water, against the surface of a revolving grindstone having a specially prepared face. Pulp may also be prepared mechanically from chips by the use of disc mills. Since in these methods there is very little loss of material, the yield of pulp is high (about 95%). Softwoods, especially spruces, are generally favoured for mechanical pulping, but some hardwoods are also successfully utilized. Groundwood pulp is produced in very large quantities for manufacture of newsprint.

b. Chemical pulping. A number of chemical processes are in use, all of which achieve defibration by dissolving the lignin which cements the fibres together. The wood, in the form of chips, is digested with chemicals, generally at high temperature and under pressure. The principal processes used are the sulphite process, utilizing the sulphites or bisulphites of calcium, sodium, magnesium or ammonium, and the sulphate process, which is an alkaline process in which the active chemicals are sodium hydroxide and sodium sulphide. The sulphate process is particularly suitable for pulping hardwoods and some softwoods (e.g. pines) which are not readily treated by the sulphite process.

c. Semichemical pulping. In semichemical processes, which are of more recent development, the wood is treated with delignifying chemicals under milder conditions than those employed in full chemical pulping. In this way part of the lignin is removed, and the wood is considerably softened so that it can subsequently be defibred by suitable mechanical treatment without excessive damage to the fibres. A variety of processes of this type are in use, producing pulp suitable for certain types of paper and paperboard, in yields generally in the range from about 55 to 80 to 85%.

By far the greater part of pulp production is used for the manufacture of paper and paperboard, the remainder (dissolving pulp) being used for production of cellulosic fibres (viscose rayon) and cellulose derivatives. The latter include cellulose acetate, nitrocellulose, hydroxymethyl cellulose, and other derivatives used in the manufacture of films, plastics, explosives, varnishes, solvents and other chemical products.

1.7.3 Board materials

The principal board products utilizing wood as raw material are chipboard (or particle board), fibreboard, and wood-cement boards, all of which have been described in section 1.5. Of these, chipboard and wood-cement boards are made by bonding particles or strands of solid wood, prepared mechanically, with synthetic resin adhesives and cement respectively, and no chemical processing of the wood is involved.

The manufacture of pulp for production of fibreboard (hardboard and insulation board) resembles the manufacture of pulp for production of paper. However, pulp for fibreboard differs in two important respects from paper pulp: firstly, it is less highly refined and generally coarser in nature and consists, in the main, of bundles of fibres and not of individual fibres, and secondly, the pulps are generally made by defibration of wood chips that have been steamed, without chemical treatment. Little delignification takes place in this process and most of the wood components are retained in the pulp, which is produced in high yield.

Initially, insulation board and hardboard were made almost exclusively from coniferous timbers and the industry developed primarily in those countries where these timbers were plentiful and where experience in pulp and paper manufacture was available. Later it was found that hardwood species, particularly those of lower density, could be incorporated without loss of quality, especially in hardboard, in spite of the shorter fibre of hardwoods. Some mills now operate entirely on hardwoods, for example on species of *Eucalyptus* in Australia. Roundwood (forest thinnings) and solid sawmill waste (slabs, edgings, offcuts) are commonly used as raw material for fibreboard manufacture.

1.7.4 Extraneous components of wood

It has been indicated in section 1.1.3 that, in addition to the main structural components of which the wood cell walls are built up, all woods contain smaller amounts of minor or extraneous components which can generally be extracted from the wood with organic solvents or sometimes with water. In most woods the amounts of these components are so small that their extraction and utilization would not be economic, but there are a few species which contain larger quantities of potentially valuable extractives, and these form the basic of quite important industries.

a. Turpentine and rosin. From an industrial standpoint, turpentine and rosin (known in the United States as naval stores) are among the most important of wood extractives. They are obtained from certain species of pine growing principally in the southern states of USA and in southern Europe. Two methods are used for recovery of these products. First, they may be obtained by tapping the living trees. An incision is made in the bark and outer part of the stem causing the viscous liquid to flow from the tree. It is collected in cups fixed to the stem below the incision, and is subsequently separated by distillation with steam into turpentine and rosin. Secondly, turpentine and rosin may be extracted from the wood, broken down into chips, by means of suitable solvents (petroleum naphtha or benzene). The solvent is first recovered from the extract by distillation and the turpentine is then separated from the rosin by treatment with steam. The process utilizes largely stumpwood from trees that have previously been felled, which is particularly rich in resins and oils. In the United States the products obtained by tapping the growing trees are termed gum turpentine and gum rosin, while those obtained by solvent extraction of the wood are termed wood turpentine and wood rosin. There are some differences in composition between gum and wood turpentine.

Resin can also be obtained from Scots pine by tapping, but experience in Germany has shown that the yield is not large enough for the industry to compete with the imported product from southern France and USA. It is unlikely that Scots pine in Britain would yield more resin per tree than the same species grown in Germany, and it may be concluded that climatic conditions are unfavourable for production of resinous products from home-grown pines.

b. Tannins. Tannin is present in the wood and bark of many species of tree, mainly hardwoods, and a few are used commercially for tannin production. Extraction of tannin from oak wood for use in leather manufacture has been carried out in the past, but oak has a relatively low tannin content (average about 10%) in comparison with other sources. Quebracho wood (*Schinopsis lorentzi*) from South America and wattle bark (*Acacia mollissima*) from South Africa, which may contain 20 to 30% of tannin, are among the important sources of vegetable tannin at the present time. The bark of some softwoods also contains substantial quantities of tannin and has been utilized for tannin production.

c. Essential oils. A number of oils such as cedarwood oil, sandalwood oil, sassafras oil, etc., which are used in perfumery or in medicine, are obtained by steam distillation of the wood or other parts of certain trees.

1.7.5 Wood hydrolysis products

In principle, it is possible to convert wood into a number of useful products by chemical processing, and this might appear to be a profitable way of utilizing the very large amounts of wood waste that accumulate at sawmills and processing plants. The most promising process consists in hydrolysis of the cellulose and hemicelluloses in wood by means of acids to yield a mixture of sugars. Several processes have been devised, and have been operated in different countries, using cold concentrated hydrochloric or sulphuric acid or hot dilute sulphuric acid to bring about hydrolysis of the polysaccharides. The sugars produced may be used directly as a constituent of animal foodstuffs, or they may be fermented to produce alcohol or used as a substrate for growth of a fodder yeast, or pure, crystalline glucose may be prepared from them. However, a hydrolysis plant requires to be large, relatively complicated, and costly, and the value of the end-products is low because they can be produced cheaply from other sources. For these reasons the wood hydrolysis plants that have been set up have not generally been successful, for economic rather than technical reasons. For example, a plant has been operated in the past in Switzerland under heavy government subsidy, but even this plant has been closed. The prospects for successful operation of a hydrolysis plant in Britain at the present time are not considered to be good. So far as is known, the only country in which there is a substantial wood hydrolysis industry is the USSR.

It is clear that no truly simple and economically attractive hydrolysis process is yet available or can be anticipated in the near future. No single product is sufficiently valuable to pay the costs of raw material collection, handling and processing, and if hydrolysis processes are to play an important part in the utilization of waste wood, a fuller utilization of the raw material must be realized.

Other processes for producing useful chemicals from waste wood (e.g. phenols, furfural) have been investigated, but all of them meet the same obstacles, and no economically successful process is in sight. Indeed, it is generally much cheaper to produce these chemicals from other sources, such as wheat straw, oat hulls, etc., or from the petroleum or coal tar industries.

1.8 STATISTICS

1.8.1 World trade in wood and wood-based materials

Table 1.13 Value of World Production (1 000 m. $ US)

	1950	1960	1965	1970
Processed wood (sawn wood, boxboards, sleepers)	10.3	12.9	15.5	18.6
Panel products (veneers, plywood, particle board, fibreboard)	1.0	3.0	4.7	7.2
Pulp products (paper and paperboard)	8.7	13.2	16.9	23.7
All other wood products	3.9	5.8	7.4	8.8
Total	23.9	34.9	44.5	58.3

Source: *Yearbook of Forest Products, 1970.*
Values estimated on the basis of average unit prices prevailing in the respective year.

In Tables 1.14, 1.15 and 1.16 the principal timber exporting and importing countries are listed, with the quantities of the various classes of wood and wood-based materials exported from, or imported by, each country in 1970.

Table 1.14 Exports and imports of softwood logs, pulpwood, and sawn softwood timber, 1970

	Coniferous logs (1 000 m³)	*Pulpwood (1 000 m³)*	*Coniferous sawn timber (1 000 m³)*
Exports from:			
Canada	1 211	2 626	17 337
USA	12 180	201	2 720
France	180	1 238	191
Austria	126	–	3 342
Finland	–	735	4 644
Sweden	617	3 462	6 877
Rumania	–	500*	1 238*
German Federal Republic	156	260	183
Czechoslovakia	188	932	661
Poland	–	412	807
USSR	7 351	5 982	7 980
New Zealand	1 809	–	257
Imports by:			
Canada	1 304	304	399
USA	482	1 065	13 460
France	–	1 244	1 688
German Federal Republic	472	1 773	3 954

* Figures for 1969

Table 1.14 (continued)

	Coniferous logs (1 000 m³)	*Pulpwood (1 000 m³)*	*Coniferous sawn timber (1 000 m³)*
Italy	727	1 677	3 195
Finland	572	1 670	–
Bulgaria	390	–	–
Hungary	637	292	1 027
Austria	–	1 172	–
German Democratic Republic	–	889	1 464
Netherlands	–	354	2 692
Belgium	–	1 048	872
Norway	307	3 499	306
United Kingdom	–	284	8 107
Japan	18 395	1 060	2 667
Australia	–	–	631

Source: *Yearbook of Forest Products, 1970–71.*

Table 1.15 Exports and imports of hardwood logs and sawn hardwood timber, 1970

	Non-coniferous logs (1 000 m³)	*Non-coniferous sawn wood (1 000 m³)*
Exports from:		
Canada	–	395
USA	312	278
France	759	387
Yugoslavia	–	579
Rumania	–	745*
Ghana	765	241
Ivory Coast	2 511	183
Gaboon	1 633	–
Congo (Brazzaville)	431	–
Indonesia	7 834	–
Phillippines	8 622	140
Sarawak	3 182	315
Malaysia	2 075	1 030
Sabah	6 150	–
Singapore	–	721
Japan	–	100
Imports by:		
France	1 532	256
German Federal Republic	1 704	395
Italy	1 910	752
United Kingdom	279	802
Netherlands	400	311
Japan	19 851	306
Korea	2 865	–
Singapore	1 844	300
Canada	305	216
USA	172	791
Australia	204	252
South Africa	–	296

Source: *Yearbook of Forest Products, 1970–71*

* Figure for 1969

Table 1.16 Exports and imports of plywood, particle board and fibreboard, 1970

	Plywood (1 000 m³)	Particle board* (1 000 m³)	Fibreboard (1 000 metric tons)
Exports from:			
Canada	369	–	57
USA	115	–	41
Italy	97	68	–
German Federal Republic	60	229	29
Finland	602	168	152
Sweden	–	92	388
France	100	98	79
Belgium	42	698	39
USSR	281	145	132
Japan	322	–	–
Korea	822	–	–
Phillippines	256	–	–
Imports by:			
Canada	131	40	22
USA	1771	–	174
France	96	163	24
German Federal Republic	213	318	160
United Kingdom	1101	414	289
Netherlands	150	349	157

Source: *Yearbook of Forest Products 1970–71*
* Including flaxboard

1.8.2 Production of wood and wood-based materials

The quantities of wood and wood-based materials produced in the United Kingdom in 1960–71 from home-grown timber or, in the case of plywood, manufactured in the UK from imported logs, are shown in table 1.17.

Table 1.17 United Kingdom production of wood and wood-based materials, 1960–71

Year	Sawn softwood (1 000 m³)	Sawn hardwood (1 000 m³)	Sleepers and crossings (1 000 m³)	Plywood (1 000 m³)	Wood chipboard (1 000 tons)	Fibre-board (1 000 tons)	Pitwood (1 000 piled m³)
1960	185	439	8.2	52.5	34.7	73.9	1 572
1961	183	422	8.8	40.8	48.8	69.1	1 480
1962	177	419	8.5	36.6	72.8	60.3	1 517
1963	201	403	4.8	34.7	90.9	53.5	1 511
1964	253	403	6.8	39.2	123.6	72.3	1 566
1965	245	396	6.8	44.1	147.7	75.8	1 480
1966	234	373	6.5	37.6	162.3	57.8	1 382
1967	240	365	5.1	32.5	147.6	35.8	1 346
1968	265	366	4.4	31.3	179.5	36.8	1 227
1969	282	349	4.4	28.9	190.1	38.0	1 216
1970	297	341	4.8	27.6	178.9	37.4	1 167
1971	310	335	4.5	28.0	159.8	31.3	1 075

Source: *Year Book of Timber Statistics*

Table 1.18 gives the quantities of plywood, particle board and fibreboard manufactured in the major producing countries in 1970.

Table 1.18 World production of panel products, 1970

Country	Plywood (1 000 m³)	Particle Board (1 000 m³)	Fibreboard Hardboard (1 000 metric tons)	Fibreboard Insulation Board (1 000 metric tons)
Canada	1 850	283	193	183
USA	14 119	3 123	1 316	1 083
France	643	1 131*	230	31
German Federal Republic	569	3 778	231	38
Italy	420	920	–	–
Finland	706	380	199	43
Sweden	80	389	650	70
Belgium	–	530*	60	–
Austria	–	479	83	–
United Kingdom	28	281	20	18
Rumania	285†	314	212	–
Spain	228	444	64	–
Poland	205	217*	206	65
German Democratic Republic	–	451	94	–
USSR	2 158	1 981	542	197
Japan	7 008	444	391	92
Korea	847	–	–	–
Phillippines	341	–	50	–
Brazil	169†	270	50	35
Australia	106	264	151	–
World total	32 607	18 127	5 748	2 030

Source: *Yearbook of Forest Products*
* Excluding flaxboard
† Figures for 1969

1.9 LITERATURE

1.9.1 Wood in general

A. J. PANSHIN and C. DE ZEEUW. *Textbook of Wood Technology*. 3rd Edn. New York, McGraw-Hill, 1970.

F. F. P. KOLLMANN and W. A. COTÉ JR., *Principles of Wood Science and Technology. I. Solid Wood*. Berlin, Springer Verlag, 1968.

F. W. JANE. *The Structure of Wood*. 2nd Edn. Revised by K. WILSON and D. J. B. WHITE. London, Adam and Charles Black, 1970.

GEORGE TSOUMIS. *Wood as raw material*. Oxford, Pergamon, 1968.

H. E. DESCH. *Timber: Its structure and properties*. 4th Edn. London, Macmillan, 1968.

Yearbook of Forest Products, 1969–70. Food and Agriculture Organization of the United Nations, Rome, 1971.

Yearbook of Timber Statistics 1970, Timber Trade Federation, London, 1971.

1.9.2 Species of wood

A Handbook of Softwoods. Forest Products Research Laboratory, London, H.M.S.O., 1960.

A Handbook of Hardwoods. Forest Products Research Laboratory. London. H.M.S.O., 1972.
British Standard 881 & 589: 1955. *Nomenclature of Commercial Timbers*.
G. M. LAVERS. *The Strength Properties of Timbers*, Forest Products Research Laboratory, Bulletin No. 50, 2nd Edn. Metric units. London, H.M.S.O., 1969.
B. J. RENDLE. *World Timbers*, vol. 1, *Europe and Africa* (1969). vol. 2. *North and South America* (1969). vol. 3, *Asia and Australia and New Zealand* (1970). London, Ernest Benn.

1.9.3 Production of wood

R. S. TROUP. *Silvicultural Systems*, Oxford University Press, 1928.
R. S. TROUP. *Colonial Forest Administration*, Oxford University Press, 1940.
N. D. G. JAMES. *The Forester's Companion*, Oxford. Blackwell, 1955.
NELSON C. BROWN and JAMES S. BETHEL. *Lumber. The Stages of Manufacture from Sawmill to Consumer*. 2nd Edn. New York. John Wiley, 1958.
V. SERRY. *British Sawmilling Practice*, London, Ernest Benn, 1963.
Kiln Operator's Handbook. A Guide to the Kiln Drying of Timber. London, H.M.S.O., 1961.

1.9.4 Wood working

P. HARRIS. *A Handbook of Woodcutting*, Forest Products Research Laboratory, London, H.M.S.O., 1946.
Y. S. RAO and S. D. RICHARDSON. *Timber Trades Journal*, London, Ernest Benn. 29.1 and 4.3. 1972.
J. RAYMOND FOYSTER. *Modern Woodworking Machine Practice*, London, Business Publications, 1963.
G. RIDDINGTON, M. A. TYRREL and S. D. RICHARDSON. *Timber Trades Journal*, London, Ernest Benn. 25.12. 1971.
S. A. SHEA. *Timber Trades Journal*, London, Ernest Benn. 18.3. 1972.
Safety in the Use of Woodworking Machines. Health and Safety at Work Series, No. 41, Dept. of Employment and Productivity. London, H.M.S.O., 1970.
The Machine Planing of Hardwoods. Forest Products Research Laboratory, Bulletin No. 51. London, H.M.S.O., 1967.
J. EASTWICK-FIELD and J. STILLMAN. *The Design and Practice of Joinery*, 3rd Edn, revised, London, Architectural Press, 1966.

1.9.5 Wood-based materials

A. D. WOOD. *Plywoods of the World*, Edinburgh and London, W. and A. K. Johnston and G. W. Bacon, 1963.
Plywood. Timber Research and Development Association, 1961.
British Standard 1203 : 1954 *Synthetic Resin Adhesives for Plywood*.
British Standard 1203 : 1954 *Synthetic Resin Adhesives for Plywood*.
British Standard 1455 : 1956 *British-made Plywood for General Purposes*.
British Standard 4071 : 1966 *Emulsion Adhesives for Plywood*.
R. A. G. KNIGHT. *Efficiency of Adhesives for Wood*, Forest Products Research Laboratory Bulletin No. 38, 4th Edn. London, H.M.S.O., 1968.
R. A. G. KNIGHT. *Requirements and Properties of Adhesives for Wood*, Forest Products Research Laboratory Bulletin No. 20, 4th Edn. London, H.M.S.O., 1964.

L. MITLIN, Ed. *Particle Board Manufacture and Application*, Sevenoaks, Pressmedia, 1968.

L. AKERS. *Particle Board and Hardboard*, Oxford, Pergamon, 1966.

W. C. STEVENS and N. TURNER. *Wood Bending Handbook*, Forest Products Research Laboratory, London, H.M.S.O., 1970.

1.9.6 Wood decay and protection of wood

K. ST. G. CARTWRIGHT and W. P. K. FINDLAY. *Decay of Timber and its Prevention*, London, H.M.S.O., 1958.

W. P. K. FINDLAY. *Timber Pests and Diseases*, Oxford, Pergamon, 1967.

J. D. BLETCHLY. *Insect and Marine Borer Damage to Timber and Woodwork*, London, H.M.S.O., 1967.

G. M. HUNT and G. A. GARRATT. *Wood Preservation*, 3rd Edn, New York, McGraw-Hill, 1967.

W. P. K. FINDLAY. *The Preservation of Timber*, London, Adam and Charles Black, 1962.

Timber Pests and their Control. Timber Research and Development Association, 1964.

Dry Rot in Wood. Forest Products Research Laboratory, Bulletin No. 1, 6th Edn, London, H.M.S.O., 1960.

1.9.7 Wood as a chemical raw material

B. L. BROWNING, Ed. *The Chemistry of Wood*, New York, Wiley (Interscience), 1963.

W. E. HILLIS, Ed. *Wood Extractives*, New York, Academic Press, 1962.

R. H. FARMER. *Chemistry in the Utilization of Wood*, Oxford, Pergamon, 1967.

L. E. WISE and E. C. JAHN, Ed. *Wood Chemistry*, New York, Reinhold, 1952.

H. F. J. WENZL. *The Chemical Technology of Wood*, New York, Academic Press, 1970.

CHAPTER 2

Paper

2.1 INTRODUCTION

A sheet of paper is made up of millions of microscopic vegetable fibres together with non-fibrous additions. It is made by depositing a mixture of fibres and mineral matter in water on to a continuous moving wire mesh. The removal of most of the water leaves a mat of intertwined fibres and mineral particles, forming paper. By varying the mode of manufacture products can be obtained ranging from fine tissue to thick laminated board, and from writing parchment to heavy blotting paper. This versatility is made possible by varying the source and type of the fibrous raw materials used, by combining them in different proportions and by treating them differently during the chemical and mechanical processes of manufacture.

Because it is made from natural fibres and these are responsive to changing external conditions, paper has a 'nature' of its own. This must be recognized by the user and due allowance made. It is hoped that the reasons for this will become clear from the following pages.

2.2 HISTORY

As the material which enabled ideas to be conveyed more simply and cheaply than before, paper contributed greatly to the advance of culture, science and the arts. A study of the history of paper is of interest because it reveals much of the basic nature of paper and its manufacture, some aspects of which have remained virtually unchanged in principle since paper was invented.

There were earlier 'writing materials' of course. Man must always have had a great need to illustrate what he could not describe in words and it is to be assumed that the very earliest attempts of this kind were scratched in the sand or soft earth with the point of a stick. Later came smooth stones or pieces of bone on which scratches could be made and these were probably the first portable means of communication. The marking of linen or other fabrics followed, which must have improved both the range of ideas which could be expressed and the ease with which they could be taken from one place to another. There are obvious limitations to such methods, however, and a need undoubtedly arose for a material which could be used specifically for writing.

The first of such materials was papyrus. The earliest papyri are judged to date from about 3 500 B.C. although there is no certainty about this. Papyrus is a reed which still grows on the banks of the Nile and it is perhaps interesting that, although it gives its name to a product which is in use even today, papyrus itself is not a paper. It was prepared by peeling the outer layers of the plant and laying them flat on a hard surface. Other layers were then placed on the first at right angles. The whole was then moistened, releasing the natural glue in the plant. The whole 'sheet' was then pressed and allowed to dry in the sun. A final rubbing with a smooth stone or piece of bone resulted in a surface which would accept ink, was reasonably even and which could be written upon.

Papyrus existed as the main, indeed only, 'civilized' writing material for something like 4000 years, and its use persisted long after the practice of writing on animal skins or 'parchments' was introduced about 170 B.C. It is possible that the introduction of parchments came about because of a need for something more permanent than papyrus, since they are first heard of in Pergamos where at that time a very famous library existed. The method of preparing parchment was to wash the skin, stretch it on a frame and treat it with a mixture of lime and chalk to remove the hair, which was scraped off after drying. The skin was then rubbed to provide a smooth and supple surface for writing upon.

True paper, virtually as it is known today, is believed to have first been made in China in about A.D. 105, by T'sai L'un, whose name was for long afterwards revered as the God of papermakers. The actual fibrous raw materials and manufacturing processes of that time did not differ in essentials from those in use today. The vegetable fibres used were obtained from plants and old clothing. The main manufacturing procedures – reducing the raw material to its component fibres, suspending the fibres in water on a mesh, drying and then polishing the mat of paper so formed – were all carried out in a manner primitive but not differing in principle from those in use at the present day.

At this time although trading caravans frequently travelled between the East and the middle eastern countries, China was still largely isolated and it was probably six or seven hundred years before knowledge of paper began to slowly move westward. Papermaking was in fact a 'secret art', guarded with the greatest care. Eventually the capture of some Chinese papermakers at Samarkand led to the establishment of the first paper mill outside China, in Bagdad. This was sometime in the eighth century A.D. but it was another 100 years before manufacture commenced in Egypt and paper began to displace papyrus.

The use and manufacture of paper spread along the southern shores of the Mediterranean to Morocco whence the Moors took it across to Spain; there is the record of a paper mill in operation at Xativa about A.D. 1150. Although a French mill was established at Herault some 40 years later it still seems to have taken some 200 years for papermaking to spread further across Europe through Fabriano in Italy in 1260 to Nuremberg in Germany in 1389. Papermaking entered England via Switzerland and the Netherlands, the first mill being established in Hertfordshire in 1490. In 1532 a mill is believed to have been functioning in Sweden but it took well over 100 years for the craft to be introduced into Norway; the earliest record of a Norwegan mill is in 1698. Yet papermaking was introduced to the American continent slightly earlier than this, in 1690. The length of time between the establishment of papermaking in Sweden and its appearance in Norway, together with the interval of only eight years between this event and the first American mill may well be accounted for by the political situation in France. Many French papermakers were members of Huguenot families and when the revocation of the Edict of Nantes, which had protected the Protestants, took place in 1685, they were forced to leave the country. After a period of many years during which the spread of papermaking throughout Europe had been slow, there was suddenly an influx into countries neighbouring France of both wealthy owners of paper mills and their work people. The result was a more widespread stimulation of the paper trade than had ever been experienced before.

So far, all paper was made by hand. The breaking down of the rags or plants into constituent fibres was still being carried out by a pestle and mortar in the fifteenth century, and it was not until towards the end of this period that these operations were replaced by a rather crude stamping mill in which a number of pestles were raised and dropped by a cam carried on a revolving shaft. A hundred and fifty years later the windmill-powered 'Hollander' beater was invented in Holland in about 1750. All modern machinery to separate fibres and produce their particular character is derived from this.

Like nearly all fundamental inventions, that of the papermaking machine was contributed to by a number of people. The basic machine invented by Louis Robert, a Frenchman, in 1799, established the mechanical principles upon which subsequent, more efficient, machines were built. They were all simply arrangements for producing a continuous wet web of paper which was subsequently dried by pressing out the moisture and finally air-drying in sheets. It was not until methods of improved drying were invented early in the nineteenth century that the foundations of the modern papermaking industry became truly and firmly laid. From that time onward both the manufacture and usage of paper rapidly increased. The number of ways in which paper came to be used also multiplied so that it was no longer a material solely for writing or printing but was becoming of increasing significance industrially and commercially.

2.3 OUTLINE OF MANUFACTURE

To understand why paper behaves as it does in use, why it is especially suitable for some purposes and unsuitable for others, it is necessary to know how it is made and from what raw materials.

Papermaking is a continuous process, conveniently divided into five main parts. Each of these will be dealt with in some detail in subsequent sections but are set out here in sequence so that the relationship of one part with another can be appreciated.

The first operation is the selection and preparation of the raw material. This comprises fibrous vegetable matter with chemical and mineral additions. In theory almost any cellulosic vegetable fibre may be used for papermaking but in practice some have proved more suitable than others, chiefly on account of availability and the ease with which the vegetable fibre itself can be isolated from accompanying extraneous matter. Wood fibre, either from coniferous or deciduous trees, has proved especially acceptable. A wide range of trees growing in the temperate regions of the world has been found to provide suitable fibres and the result has been the growth of the pulp and paper industries which are characteristic of countries such as Sweden, Norway, Finland, Russia, Canada and the United States. Although wood pulps provide by far the greater proportion of fibrous raw material for papermaking, many other fibres are used. They include cotton and linen, hemps and jutes and a fairly wide range of grasses such as the esparto of North Africa and alfa grass of southern Spain. Preparation of fibres at this stage consists of removal of deleterious material and possibly cleansing and bleaching.

'Beating' is the second stage of manufacture and is the basis of the character of the final sheet of paper, because it is at this stage that the individual fibres from which paper is made are separated and then mechanically treated to produce a paper having the required properties. This is done by passing the fibres, suspended in water, between fixed and moving knives. The blades may be varied in sharpness and the pressure between them increased or decreased so that the resultant fibre is of a length and physical type to ensure a paper of appropriate strength, smoothness and absorbancy. The simple 'beating engine' has developed into quite sophisticated machinery, often forming part of a continuous processing line, as distinct from batch beating. Such developments have not altered the principles involved but only increased productivity and the amount of control. Beating is a critical stage of manufacture and one which has well justified the many separate studies which have been devoted to it by the industry. Pulp preparation and beating are becoming more closely associated as one continuous integrated operation.

The third operation takes place at the papermaking machine itself. There are two parts to the machine, the 'wet end' and the 'dry end'. At the wet (ingoing) end an aqueous suspension of fibrous and mineral raw materials in proportions of 100 parts water to one part fibre is delivered to a continuous moving wire mesh. As this travels forward it is shaken slightly from side to side so that the fibres left behind after the water passes through the mesh are intertwined and are not lying in the direction of the machine's flow. Towards the end of the travel the mixture of fibre and water has become a wet web, which though still containing a high proportion of moisture is strong enough to pass from the wet end of the machine to the next stage. This is the dry end of the machine and consists of large revolving heated cylinders over which the web of paper passes. The heat input to the cylinders increases from the wet to the dry end to eliminate further moisture from the web and produce a paper with some 7% moisture at the extreme end of the machine. Depending on the paper to be produced, this may form the final operation in the entire process or it may simply provide a base which is then the subject of finishing, coating or other operations.

The fifth stage of manufacture comprises finishing processes. As paper has become used more widely as a raw material, finishing operations have become more diverse. Finishing may mean the application of a simple coating of mineral matter and adhesive so as to provide a surface capable of accepting the most demanding colour printing; or it can mean the application of an animal 'size' in order to confer a high degree of permanance and the ability to withstand handling over a long period. Impregnation, lamination, and other final operations are included in finishing, so that paper as it is reeled up at the end of the papermaking machine can be considered in two lights, either as a finished product ready for use – as would be the case for instance with a newsprint or book paper – or as a base product which will then be treated by coating, laminating or impregnation. Some of the more frequently used finishing processes, especially coating and sizing, are not longer always separate operations; they are carried out on the papermaking machine itself, by dividing the dry end of the papermaking machine and placing the machinery for coating or sizing about the middle of the drying process. This, while simplifying the papermaking operation yet not altering the principles employed is an example of how, with modern continuous processing lines the demarcation between the five stages is becoming less clear.

2.4 RAW MATERIALS

2.4.1 Fibrous raw material: introduction

The character of the paper being made largely depends on the type of fibre and the method of processing. A vegetable fibre consists of a minute cylinder or hollow cell of vegetable matter which may occur in the tree or plant as single cells or in groups known as fibro-vascular bundles. The fibre walls enclose a canal or lumen and the character of the fibres is determined by the thickness of the walls, the width of the lumen, and the flexibility and length of the fibre itself. The source from which a fibre is obtained is identifiable by the appearance of the fibre and cells when examined under the microscope. If microscopic examination is undertaken on fibres from a pulp, however, allowance must be made for the state of the pulp (i.e. whether beaten or unbeaten). The appearance of any fibre under the microscope is changed by beating and it cannot be expected that fibre taken from a finished sheet of paper will show the same undamaged clearly outlined fibres and cells as fibre taken from untreated raw material.

2.4.2 Fibrous raw material: sources

The manufacture of paper demands a constant and continuing supply of fibrous raw material delivered to the paper mill at an economic cost. For this reason while some fibres such as the mechanical and chemical wood pulps are almost universally used, there is also a continuing search for suitable raw material which may be more readily to hand. The type of fibre present greatly affects the properties of the paper. Thus a paper containing predominantly short fibres with wide lumens will be absorbent and of only moderate strength. Long fibres with narrow lumens and therefore thick walls will be found in paper made with strength and durability in mind. In practical terms the manufacture of paper is not so simple that only one sort of fibre is likely to be used in any paper and the careful mixing of different types of fibres is a large part of the papermakers' craft.

While most analyses of the materials from which paper is made will reveal certain commonly used raw materials (e.g. wood fibres), unusual fibres may sometimes be present, reflecting local conditions of supply, cost and technical suitability. Thus fibres used in papers emanating from the countries with highly developed papermaking industries such as North America, Great Britain, West Germany, Scandinavia and Japan are likely to be of much the same type unless the paper has been made for a speciality purpose. Production outside the chief papermaking regions may very well display a far greater degree of utilization of local resources.

At the present time, with the exception of newsprint, papers are imported and exported within fairly small areas, considered globally; paper is not a sufficiently expensive product to justify being transported great distances, though it may be moved from, say, one European country to another or between Latin American countries. However, with demand for paper growing at about 4% per annum and with alternative calls being made on wood pulp supplies new sources of raw material may have to be exploited.

At present the three main categories of fibre for papermaking are wood pulps, grasses and rags. Of these, wood pulps of various kinds are the most widely used. Not only has a whole industry devoted to the production of wood fibre for papermaking been developed during the last fifty years, but cellulose chemistry has been successfully directed towards finding alternative materials and methods of treatment which will provide any fibre or combination of fibres necessary for a very wide range of products.

Many of the raw materials now being offered are acceptable alternatives to the grasses or rags which have been historically the staple of at least a certain section of papermaking, and which are still used.

Theoretically any vegetable matter is capable of producing a fibre suitable for papermaking and it is true that in time of war or other emergency usable papers have been produced from the most unlikely material. However, certain economic and technical conditions must be met. The supply must be ample for the purpose intended and the growth rate of the vegetable source must allow replacement of the material used for a long period ahead. The tree, plant or grass must yield without expensive chemical or mechanical treatment sufficient fibre; and the fibre itself should be of a satisfactory length, strength and structure.

The vegetable growth abounding on either side of the equator produces fibrous material which neither works easily on the papermaking machine nor provides a strong stable product. On the other hand the spruce and pine of Scandinavia and North America, alfa or esparto grass from Spain and North Africa, sugar cane, bamboo, eucalyptus, all are now being used to produce satisfactory papers. Thus although in theory a wide variety of fibres could be used, in practice the choice is limited, and the following fibres are those which are likely to be encountered most frequently.

a. Pine. Many pines are available for papermaking, the species being distributed throughout the northern temperature zones. The most important sources of supply are

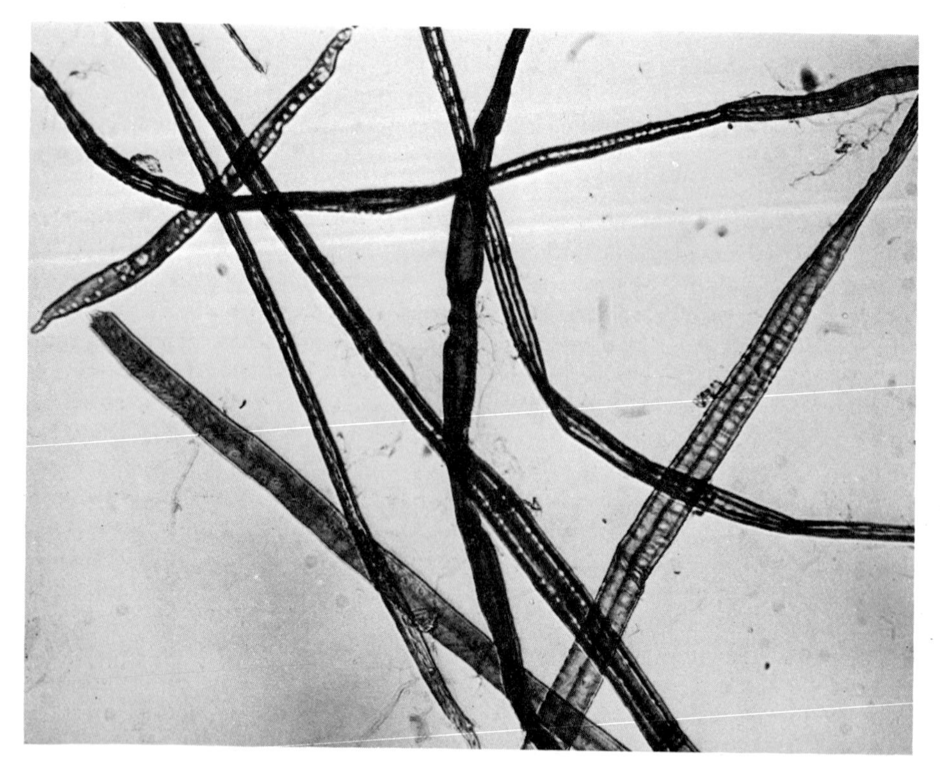

Fig. 2.1 Semi-bleached pine, before beating. Fibres, as seen under a microscope, showing well-defined lumen and characteristic pit marks. Magnification 200.

Fig. 2.2 Semi-bleached pine, after beating. Note how the fibres have become cut and fibrillated during the beating process. Magnification 200.

the North American Continent, Europe and Scandinavia. The growth rate of the tree is from 70 to 120 years to maturity, depending on latitude, with growth naturally becoming slower the farther north one goes.

Two types of pine fibre are distinguishable under the microscope; a broad ribbon-like fibre and one of rather similar nature but thinner and shorter. The broader fibres measure from 2.5 to 3.75 mm in length, and from 0.045 to 0.060 mm in width. The smaller fibres are from 1.5 to 2.5 mm long and 0.020 to 0.025 mm wide. The ratio of length to width is 60 : 1 in the case of the broad fibres and 90 : 1 for the thinner fibres. Both fibres have a wide, well-defined lumen and are distinguished by pit marks. The sides of the fibres are frequently notched. When seen under a microscope the fibres, although naturally twisted, are rather less so than many others, spruce especially, and this is often a quick, though not entirely reliable means of identification.

b. Spruce. Like pine, spruce is well distributed throughout the northern hemisphere and these two timbers probably provide the major part of the wood pulp used throughout the world. Spruce is a less resinous timber than pine and more easily prepared for papermaking. It yields a longer fibre than pine and therefore conduces to a stronger paper. Like pine, there are usually two types of fibre to be found under microscopic examination: broad and thin. The length of the former may range from 2.60 to 4.20 mm and the width from 0.045 to 0.065 mm. The thinner fibres vary from 1.60 to

4.30 mm in length and 0.015 to 0.025 mm in width. Length/width ratio is 70:1 for the broad fibres and 160:1 for the thinner ones. Both fibres have a well-defined lumen, although the walls of the thinner fibres are often thicker and better defined than those of the broad. Spruce may be distinguished from pine by the fibre ends, which are rounded rather than sharply tapered. The pit marks which are a characteristic of both fibres are also less clearly defined in spruce than in pine.

c. Poplar. This is one of the hardwoods which has been increasingly used as demands on timber resources become greater. Better methods of separating the fibres have improved the usefulness of this species. Nevertheless, as the fibres are shorter than those from spruce or pine, poplar is not usually employed by itself except for papers not requiring great strength. Originally it was employed for papers needing maximum absorbency such as blotting paper, but the advances in processing have widened its scope, and especially when combined with other fibres it is likely to be encountered frequently. The fibre length ranges from 0.55 to 2.70 mm and width 0.01 to 0.040 mm. There are also characteristic pith cells which allow them to be fairly easily identified under the microscope. The cells have a length of from 0.40 to 1.30 mm and width of 0.050 to 0.20 mm. Many of the cells display perforations in the walls which are a useful identification feature; the perforations may form a regular pattern or may be irregular, and are small in relation to the size of the cell. The fibres themselves, while coming within the dimensions given, are usually found in three main types. There

Fig. 2.3 Bleached spruce, before beating. Under the microscope the thick, well-defined fibre walls show clearly. Magnification 200.

Fig. 2.4 Bleached spruce, after beating. The fibrillation of the fibres can be clearly seen. Magnification 200.

is a fairly broad fibre, regular in width and with clearly defined walls. This type has blunt, rather quickly tapering ends. There is also a thinner fibre, often with a more angular configuration and long, more gradually tapering ends. The third type falls between the two others dimensionally and is characterized by more noticeable variations in diameter. Taking these variations into account, the average length/width ratio of the poplar fibre is 60 : 1.

d. Eucalyptus. The development and use of eucalyptus as a papermaking raw material is a classical example of the trend towards the use of indigenous fibres to replace imported ones. Research into its use took place because, although at first it did not seem an ideal fibre for papermaking, it was the only tree available in sufficient quantity in Australia and therefore offered the sole opportunity of utilizing a home raw material. It is interesting that, as its use has grown in Australia, the tree has been successfully introduced in other parts of the world. The fibre obtained from eucalyptus is short, stiff, and rather unyielding. Under the microscope it is characterized by a larger variety of pith cells than is common with other fibres. It follows that although eucalyptus is a valuable source of raw material in Australia and in some other parts of the world the fibre should be mixed with other longer types if a satisfactory paper of reasonable strength is to be produced. The fibres themselves are of two main types: a broad ribbon type, and a narrower thicker-walled type. The length of the ribbon type is 0.40 to 0.95 mm and its width 0.020 to 0.040 mm, the length/width ratio being 29 : 1. The length of the narrower fibre, with its noticeably thicker walls, varies from 0.50 to 1.48 mm and its width from 0.01 to 0.025 mm, the length/width ratio in this case being approximately 56:1. Both fibres display fairly extensive pit marks, a regular pattern being especially characteristic of the ribbon-type fibres. The ribbon fibres have irregularly blunted ends and the thinner fibres ends which are more sharply pointed. The pith cells to be seen under microscopic examination fall into two main types. One is large with distinctively wide ends and is completely perforated. Occasionally, but not invariably, the perforations are in a regular pattern. The length of the cell is likely to be 0.35 to 0.70 mm and width 0.20 to 0.40 mm. The smaller cells are of a regular cushion shape, the length being from 0.40 to 0.20 mm and width 0.015 to 0.040 mm. The cells may be perforated or unperforated. Eucalyptus also displays on occasions a number of even smaller irregularly shaped cells, many with large perforations. These can sometimes be mistaken for fibre fragments.

e. Esparto grass. Esparto is the leaf of *Stipa cenacissima*, a grass indigenous to sterile areas of North Africa and southern Spain, the chief sources of supply being Algeria and Tunisia. It is sometimes called alfa grass. Its use is practically confined to Great Britain although in recent years paper mills have been built at the sources of supply. Political and economic factors encouraged the use of esparto for nearly 100 years. At one time it was a cheap and useful return cargo for Scottish cargo steamers

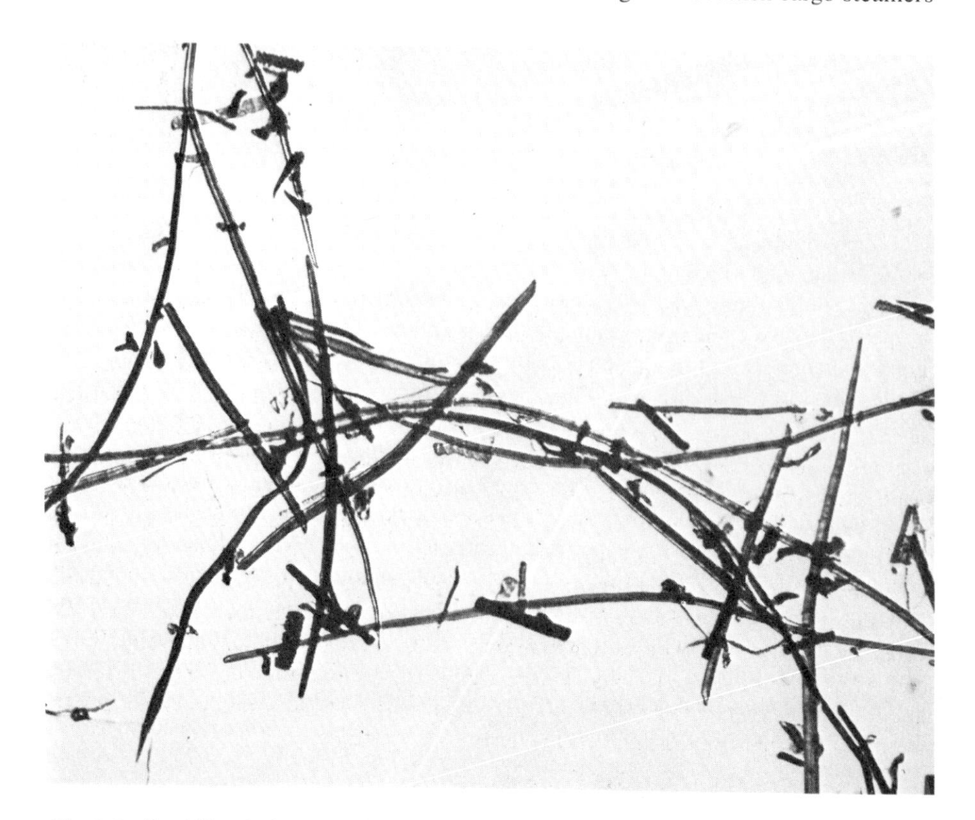

Fig. 2.5 Semi-bleached esparto, before beating. The smooth cylindrical nature of the cells is clearly seen under the microscope, as are the epidermal and 'toothed' cells so characteristic of esparto. Magnification 200.

Fig. 2.6 Semi-bleached esparto after beating. A noticeable number of the cells seem to have been unaffected by the beating process. Magnification 200.

trading to ports in the Mediterranean, but as trade diminished it became profitable to build paper mills where esparto is grown, and wood pulp processing improved so that the grass could be dispensed with. Esparto is no longer cheap or readily available.

Although not now greatly in use in Great Britain esparto grass is included here because it exemplifies many grasses which continue to be utilized in different parts of the world. Fibres are rather short, varying in length from 0.095 to 3.55 mm and in width from 0.004 to 0.016 mm. They are smooth and cylindrical and have thick walls and a narrow lumen. The ends of the fibres usually taper sharply although there are occasional examples of irregular or forked ends. Pit markings are found on many of the fibres, frequently rather sparsely spread out and not as distinct as in the case of wood fibres. Esparto cells are of two types: epidermal cells and the so-called toothed cells. The former are about 0.025 mm long and 0.115 mm wide, and are often pitted and with longitudinal serrations. Toothed cells are rather narrower; length is usually about 0.030 mm and width 0.060 mm. The edges of the cell have the appearance of the teeth of a saw, and are especially characteristic of esparto.

f. Cotton. The use of cotton as a papermaking fibre is deeply rooted in history. It is one of the two fibres (the other being linen) which are most likely to be found in any rag paper. It does not ordinarily come to the papermaker as a prime fibrous raw material such as wood pulp or esparto, but in the form of rag cuttings which have been sorted into various grades. As in the case of esparto, changes in living habits and the development of alternative raw materials have affected the rag market. The displacement of natural by man-made textile fibres has reduced the amount of cuttings available and new methods of processing cotton fabrics have made it difficult and in many cases impossible to isolate the fibre for papermaking purposes.

Cotton fibres have an average length of 28 mm and a width of from 0.01 to 0.03 mm, and are of two main types, a thick-walled fibre and a wider and more ribbon-like type. The width is often not constant and wall thickness increases so that at narrower points a constriction is created. It is also characteristic of the cotton fibre that the lumen may vary in width along its length. Cotton fibres can be distinguished from wood by the fact that they are not pitted, and furthermore have the almost unique characteristic of folding upon themselves and twisting. The natural fibre has a length-to-width ratio of 1 250 : 1.

g. Linen (flax). This other largely-used rag fibre is equally interesting historically. It has been in use both as a textile and in papermaking as long as cotton although in many other ways it differs from it. Sources of supply are geographically fairly widespread throughout both northern and southern temperate and semi-tropical zones. Cultivation is carried on in a number of European countries and in Russia, Canada and the United States, It is extensively grown in India and Pakistan. The plant is an annual, rising to a height of 3 to 4 feet (about 1 metre). Harvesting ordinarily is by means of plucking followed by retting (soaking in water for several weeks) to decompose the tissue which encloses the cellulose. The fibre displays very different characteristics from cotton, however, in that, although long, it is thick-walled and angular. The width may vary along the length and the fibres also show both nodules and swellings; cross-marking and striations are especially characteristic. The lumen is fairly even along the length of the fibre, the ends of which taper gradually to a point. Length is variable but may be from 0.005 to 0.025 mm. The ratio of length to width is 1 200 : 1.

h. Sisal. Sisal is a fibre of the hemp family obtained from the leaves of the plant *Agave sisalona.* Geographically, sisal is indigenous to central Africa, Mexico, East Africa and Java. The plant has a six- to seven-year growth rate, rising to 15 to 20 feet (4 to 5 metres). A feature of the fibre is its variety and the presence of two main types of cells. Sisal fibres are fairly thick walled and ordinarily show a clearly defined lumen under the microscope. They are cylindrical and indicate very little tendency to twist or fold. A saw edge is a particular feature of some of the fibres, whilst small pit or pore marks may also be present. The other main type of fibre is also cylindrical, but is usually broader, with a characteristic lattice pattern in the walls. Two types of cell frequently occur; a cushion type and a rectangular type which often has uneven almost serrated edges. A small spiral filament, in appearance like a minute spring, can often be seen under the microscope. The common fibres have a length of 1.60 to 4.30 mm, and a width of 0.010 to 0.031 mm. The lattice-patterned fibres are usually 1.00 to 1.45 mm long and 0.030 to 0.039 mm wide. The cushion-type cells are about 0.045 to 0.30 mm long and 0.30 to 0.045 mm wide. The rectangular cells vary from 0.060 to 0.15 mm long and 0.15 to 0.024 mm wide. The length-to-width ratio of the commonest fibres is 140 : 1.

Among fibres which are sometimes used for papermaking for special purposes or because they are available locally are the following.

i. Jute. This is the bast fibre of a rapidly growing plant in India. (Bast fibres are found in flax, hemp and jute. They are located in the plant just below the surface of the stem.) Because the jute itself has ready use locally for ropes and sacking manufacture, it becomes available to the papermaking industry either second-hand in the form of used ropes or sacks or as a residue of fibre, after the other uses have been satisfied. It is of medium length, perhaps 2 mm long and 0.02 mm in diameter. The most noticeable features under the microscope are the smoothness of the fibre and the fact that the lumens vary in diameter throughout their length.

j. Bagasse. Bagasse is the residue of sugar cane after the sugar has been extracted, and resembles straw. Having no apparent value, it was burned until it was found to yield an admittedly modest amount of usable cellulose fibre. Its use is confined to sugar-producing areas such as the West Indies or the southern part of the United States.

k. Manila. Manila (or manilla) fibre is obtained from the leaves of a plant of the plantain family which grows in the Philippines. The average fibre length is 6 mm, and the diameter 0.025 mm; and the fibre is even and smooth with a rather large lumen in relation to wall thickness. It is used widely in the manufacture of papers which have to be particularly tough, such as electrical insulating papers, stencils and papers which are intended to be as far as possible untearable, such as security envelopes.

l. Straw. This material is obtained from the stems of cereal plants, such as oats, barley, wheat or rye. It often comes into use in times of emergency, when other fibres are not readily available, although some pulp preparation processes have been designed especially to accommodate it. The yield of cellulose fibre is rather poor, and the fibre tends to knot together, proving unsatisfactory in use. The fibre is smooth, tubular and often displays characteristic pointed ends. It mixes fairly well with other fibres and therefore has a limited usefulness in mills which have a ready supply in addition to their main fibrous raw material sources.

m. Bamboo. Bamboo is increasingly used in countries where the supply is plentiful, such as India. As a raw material it has the advantage of an extremely rapid growth rate and it is reasonably well behaved on the papermaking machine, although it is sometimes difficult to obtain a good yield, owing to the fairly drastic treatment necessary for isolating the cellulose fibre.

2.4.3 Synthetic and experimental fibres

The use of natural fibre in papermaking is partly, even largely, historical. The materials concerned were available when the paper trade was developing from a hand craft into an industry. They were already entrenched as the obvious raw material sources when papermaking was a secret art and there was nothing to disturb their continuing use as the industry developed.

Although natural vegetable fibres are the traditional raw material of the paper industry they suffer from variability, inconsistency of behaviour and the fact that they are sensitive to atmospheric change. If an otherwise suitable raw material could be found which did not have these disadvantages it would obviously be of considerable interest to the papermaking industry, and experiments are constantly being carried out to this end. Experimental work has been done in three main areas: animal fibres, fibrous mineral substances and synthetic fibres.

Papers have been made utilizing leather fibres and also wool fibres. These are protein rather than cellulosic substances. Both have behaved reasonably well experimentally but do not so far appear to offer sufficient advantages over traditional materials to be attractive commercially; when mixed with traditional fibres, however, they may have application for some speciality and security papers.

Mineral substances have offered some promise in pilot and occasionally full-scale production. Both glass and asbestos have been used to form sheet material on a Fourdrinier papermaking machine (see section 2.6.2), the former yielding a heat-resistant 'paper' and the latter one with a certain amount of resistance to fire.

It is in the area of man-made fibres that the greatest hopes lie, however, and a considerable amount of work is currently being done to make a satisfactory paper from synthetic fibres.

The vegetable fibres used until now have certain valuable processing characteristics. The size, shape and density of cellulose fibre is such that it can be suspended in water so that it 'settles down' on the paper machine wire smoothly and in the right length of time. In addition, when wet it has a self-adhesive quality. To compete with natural fibres, a synthetic fibre would have to match these characteristics. Work on papers from synthetic fibres has therefore been directed as much towards developing processing techniques as towards a study of the fibres themselves.

Assuming that any necessary changes in technique can be satisfactorily brought about, the chief benefits to be expected from papers derived from man-made fibre are stability and ease of pulp preparation. Because the fibres are manufactured they can be made to a specification and supplied consistently to that standard. Much of the papermaker's work in adjusting to natural variations will be eliminated, as indeed will most of the present pulp preparation, since the synthetic fibre can be supplied ready to use.

Since synthetic fibres have no central lumen and are largely non-hygroscopic, both the fibre and the paper made from it should be largely impervious to atmospheric change. It should also not deteriorate with age to the extent that paper made from present raw material does.

Although paper made from synthetic fibre is an attractive idea there are certain technical problems yet to be solved; but the development of synthetic materials and new methods of papermaking may eventually allow man-made fibres to compete with the natural product in the paper industry as they already do in textiles.

2.4.4 Broke and waste

In addition to the range of fibrous raw materials already mentioned, there are two others which are important economically. They are 'broke' and de-inked waste.

It has always been necessary for the papermaking industry to make the most economical use of its raw materials because of the high wastage at various stages, and to re-use anything of suitable quality. For instance, fibre which is drawn away with the water coming from the machine wire at the wet end of the machine is not lost but is passed back into the stuff chest (i.e. the chest or tank at the ingoing end of the machine, containing the mixture of fibre (or 'furnish'), mineral loading, colouring matter and water). The same principle applies to all paper waste. The sources for this within the mill include the trimmings from the edges of reels, either during re-reeling or when being sheeted, the fairly large amounts of paper wasted when starting up or stopping cutting machines, calenders etc., and the sheets rejected during the final sorting process, unless they are too dirty. All these waste products are passed through a 'broke' mill which consists of a series of gigantic mincers, and are retained in the mill awaiting a suitable opportunity for use. Broke is not simply fibre; it also contains the loading and colouring matter intended for the original making; broke must therefore be carefully controlled and selected for a specific purpose. Used in this way broke is a perfectly legitimate raw material in its own right, with properties, especially in respect of opacity, which enable it to contribute to a satisfactory making of paper. While paper mills usually try to ensure that broke additions to furnishes are similar to the main fibres being used, there are acceptable reasons for departing from this practice. Hence when making a microscopic examination of paper a small proportion of unexpected fibres may be seen, and such alien fibre may derive from the addition of broke.

Paper waste which is brought into a mill is in a rather different category. This material is generally consumed in the production of coarse papers and boards and its function is more of a 'filler'. It is also subject to much less control and therefore suitable only for non-critical products. Nevertheless, with a great many paper trades so heavily dependent on foreign sources for their raw materials, the availability of a waste product suitable for use in printing or writing paper is obviously attractive. For a great many years such waste had to be roughly sorted and graded; above all it had to be clean and free from any substance which could be carried forward to the final manufacture – most commonly printing ink. In recent years de-inking plants which render the paper free of ink and make it suitable for re-use have been installed. Printing ink is extracted from waste by a fairly simple washing and screening process. The waste is disintegrated in water until, as it breaks down into its constituent fibres, the different densities of the fibres and particles of ink cause them to separate, so that when drained – in some cases centrifugally – a fairly clean fibre is obtained. The fibre is then passed through a number of screens which retain any remaining ink particles and fibre lumps which have not fully disintegrated. After treatment, the fibre is either pressed into sheets for later use or, if the mill is fully integrated, is fed back into the main fibre supply for immediate utilization.

Reclaimed fibre, either from broke or de-inked waste, is a valuable addition to the raw material supply. As mentioned, paper waste often has particular opacifying qualities which contribute to the way in which some printing or writing papers can be used. Waste in one form or another is especially useful in the production of boards for cartons and boxes, where thickness and opacity are more valuable than cleanliness. Whenever used for printing or writing paper, however, its contribution to the strength is below that of the equivalent amount of unused fibre, because of the multiple processing.

2.5 PULP PROCESSING

2.5.1 Pulp preparation

The fibres from which paper is made come from a number of sources and arrive at the paper mill in various forms. There are, therefore, several pulp preparation processes, each appropriate to the fibre and product for which it is intended. They have a number of common purposes, however, although these may vary in importance according to the paper which is to be made.

In general, the need for pulp preparation as a stage in papermaking is based on two separate aims. The first is to produce as pure a cellulose as is necessary for a particular type of manufacture and the other is the removal, to the extent needed, of all the dust, grit or other extraneous matter which is contained in the fibre as it arrives at the mill. Paper includes a very wide range of products, which call for different qualities of fibre and degrees of cleanliness. To take two examples, high-grade writing papers and certain types of paper for industrial usage may both contain rag fibres. This may be their only common feature, and while in the first case extreme cleanliness and a pure and bright fibre are the prime needs, in the second instance the purpose of the rag fibre is strength alone, and providing deleterious matter likely to damage the papermaking machinery or to be dangerous in use is excluded, the papermakers' need will be satisfied.

There are few set rules in papermaking in the sense that a particular paper or board is always made in the same way or that particular fibres will only be used for specific paper. There is general practice which dictates the furnishes (or composition) of and the manner in which most papers are made, but also considerable flexibility both in furnish and manufacturing process.

Over 95% of the world production of paper is made from some form of wood fibre and over 80% of the wood is used is coniferous, largely from spruce, pine or fir. Other papers and the fibre from which they are made are equally vital to their particular purposes, however, and the study of pulp preparation must include these as equally important although much smaller usages.

Wood pulps for papermaking are of two main types, each of which has sub-divisions. The two types are chemical and mechanical wood pulp. The first is so-called because the fibre has been chemically treated to isolate the cellulose, to eliminate unnecessary matter and to whiten the pulp. The second, as the name indicates, is the result of purely mechanical processes which, while removing some parts of the wood which are not needed, have done nothing to cleanse the pulp of its impurities or to improve its colour. Only the cheapest papers and those with a very short life-period are made from mechanical pulp or indeed have any mechanical pulp included in the furnish. These include newsprint and any printing or writing paper of very low quality which is likely to be discarded shortly after being read. Chemical wood pulps of one kind or another provide the fibrous furnish for nearly all the printing, writing, packaging or industrial papers which are made today, and a proportion of high quality chemical pulps may also be found in the very best papers whose main furnish may be, for instance, high grade rag.

The preparation of all wood fibre starts with the cutting down of the tree and its conveyance to the pulp mill. At the mill it is common to store the logs either in the open air, or in some cases, under water, often for several months. The first operation in converting the timber is to remove the bark. This may be done by water jets or steam jets or by mechanical means. After this the processing methods diverge according to whether mechanical or chemical pulp is to be produced.

a. Mechanical pulp. This is sometimes called 'ground wood' – a name that exactly indicates the process. The logs, with the bark removed, and cut to standard lengths, are fed in batches between very large grindstones several feet in diameter which revolve while water sprays play on the logs. The grinding action reduces the logs to very small knots of fibre which are then carried away by water through troughs to a series of screens which trap the larger lumps. An alternative to the grindstone is the chain

Fig. 2.7 Feeding logs into the grinders.

grinder in which a stack of logs passes between two continuous chains each of which has abrasive segments. The result is the same in either case, the mechanical reduction of the logs as nearly as possible to their component fibres. An additional mechanical operation which supplements both the grinders and the cleaning screens makes use of a disc refiner consisting of a fixed and revolving disc within a casing. The reduced fibre passes between the two discs so as to separate the individual fibres and eliminate knots of fibre.

The pulp now consists of a mixture of fibre and water, typically in the proportions of one part fibre to one hundred parts of water. If the pulp preparation is part of an 'integrated' mill, where paper is made from conveniently located wood supplies, the pulp will be fed into the paper mill at this stage, after having been slightly thickened by a reduction in the water content. If the pulp is to be stored or sold it is formed into sheets of a board-like thickness on what is, in fact, the simplest and crudest form of paper-making machine, a screen through which water passes leaving the fibre as a damp mat. The sheets are made into bales of ascertained moisture content (because pulp is sold by weight) and are then ready for storage or sale.

b. Chemical pulps. These are produced from timber which has not only been de-barked and cut to standard logs but which has then been reduced to chips of a suitable size for 'cooking' in an appropriate chemical solution, as described below. The purpose of chemical treatment is to reduce the chips to component fibres, to cleanse the wood of impurities and to produce pulp of a colour good enough for papermaking, all while retaining the natural characteristics of the cellulose. There are a number of chemical processes for wood pulp preparation certain of which are associated with particular types of fibre.

Sulphite process. This entails the treatment of wood chips in a solution of calcium bisulphite and free sulphur dioxide in large digesters at a pressure of up to 100 p.s.i. (7 bar) at a temperature of up to 340°F (170°C). The 'cooking' period may be from 6 to 24 hours according to the type of pulp required, the strength of the liquor used and the operating pressure. The liquor is formed by spraying water on limestone at the top of the digester and at the same time introducing sulphur dioxide gas at the bottom. The lime dissolves, forming calcium bisulphite. Since the liquor contains up to 7% sulphur dioxide, some of which is in a free state, it is acidic and indeed, the process is sometimes called the acid or acid liquor process. When digestion is completed unused sulphur dioxide is drawn off to be re-used and the spent liquor drained away. The process reduces the wood available for manufacture by up to 55% because cellulose matter is dissolved to some extent at the same time as the non-cellulosic. The spent liquor is very largely calcium lignosulphite, an extremely difficult effluent to dispose of. Anti-pollution laws and economic expediency have stimulated the development of a number of uses for the spent liquor which offset the overall cost of the process. The sulphite process is used extensively for the treatment of spruce fibre, though not entirely restricted to it.

Sulphate process. In this process the reagent used is a solution of sodium sulphide which, during the period of digestion, is converted into sodium hydroxide. The time of digestion is usually from 3 to 5 hours at a pressure of approximately 120 p.s.i. (8 bar). About 15% of the active alkali is present as sulphide. Cleansing is more gradual than with the sulphite process, and while it does not result in pulp of very good colour, the strength of the fibre is well retained. For this reason the sulphate process is used for fibre intended for kraft or strong wrapping papers and often for other tough, durable papers which are not required to be fully bleached. If a combination of strength and whiteness is required, a chlorine-bleached sulphate pulp can be prepared and in fact for some uses this is growing in popularity.

c. Mechanical/chemical process. This combines the mechanical or ground wood process and a form of chemical treatment in an attempt, frequently quite successful, to provide an intermediate pulp which will avoid the objectionable characteristics of mechanical grades and at the same time offer a cheaper alternative to the simple chemical pulps. In this process, often called semi-chemical, the wood chips are first treated with steam plus a little caustic soda or sodium sulphite and are then disintegrated mechanically. Semi-chemical pulp, because it still contains a great deal of impurity and ligneous matter, cannot be brought to a pure white shade. Also, because the process does not lend itself to the careful separation of fibres and the retention of their strength, the pulp is unsuitable for any paper of quality or durability. It can serve in medium-quality papers, together with better fibre, as an opacifying agent and is used to some extent as the main furnish of cheap printing and writing papers.

Although wood is by far the most used cellulose fibre, others are extremely important within their own fields.

Wood pulp differs from other fibrous raw materials. Where it is bought-in by a paper mill the pulp is supplied in a prepared state almost ready for use, whereas most other fibrous raw materials arrive in a 'natural' state and have to be treated from the very beginning. Esparto is typical of these natural fibres, and its preparation for the mill is as follows.

The first stage is the removal of the dust and grit. The usual method is to shake the esparto in a large wire mesh cage which is revolving rapidly and within which are spikes or knobs which constantly catch the grass and throw it about. The cleaned grass is then boiled in a caustic solution, since although esparto and most grasses are soft the fibres have a hard non-cellulosic skin which must be dissolved away to release them. This is carried out in a digester or boiler with an internal perforated base and central or peripheral vomit pipes. As the liquor boils it is forced up the vomit pipes and over the top of the grass to sink through it and re-circulate. The grass is digested for approximately 5 to 6 hours in a solution of caustic soda in the proportion of about 12% to 15% of the weight of grass, at a pressure of 45 p.s.i. (3 bar).

Straw. Used for economic reasons or in emergency conditions, straw is very often treated in the same way as esparto but digestion is longer and more severe. Where straw is used as a staple fibre more sophisticated digestion processes specifically designed for it and retaining as far as possible the natural characteristics of the fibre are employed.

Rag fibres. Although probably the very first used for papermaking, they are declining in importance because of cost, the improvement in the quality of wood pulp fibres, and the difficulty in obtaining continuous and adequate supplies of rag of suitable quality. Nevertheless, cotton and linen continue to be used to impart strength and dimensional stability, and fibres such as sisal and manila for industrial purposes where a strong paper with high resistance to wear is required.

All come to the paper mill in the form of rags which have been graded, at least to some extent, by the supplying company. Nevertheless, they have to be sorted again by hand and all those which have had any deleterious treatment such as rubberizing discarded. At the same time buttons, hooks or any trimmings are cut away while a very careful scrutiny is made for materials which may have a mixture of natural and man-made fibres. It is apparent from this that the amount of labour and time which must be devoted to removing all unwanted matter from rags must be set beside their value as a papermaking material.

After the rags have been sorted they are cut into small squares of about 3 to 4 in (100 mm) in a rag cutting machine which rips in one direction while a revolving knife cuts at right angles. Because the materials are second-hand, like the grasses they are full of dust and grit and this has to be removed by tossing in a large wire-mesh revolving drum, as with esparto. When they have been thoroughly cleaned mechanically the rags are boiled in large revolving spherical or cylindrical boilers. Because of the greater variation in the condition of the rag when compared to wood or esparto, it is not possible to be specific about digestion times or the strength of the boiling liquor. Much depends on the dirtiness or greasiness of the rags being treated. The liquor is always alkaline, usually caustic soda or sodium carbonate, or sodium suphite with some alkali, and its strength may range up to about 15% of the dry weight of the rags being treated. The length of time necessary for satisfactory boiling depends very much on the type and condition of the rags. Very clean and grease-free rags might take only two or three hours at a moderate steam pressure of 20 to 30 p.s.i. (1.5 to 2 bar), whereas thoroughly greasy rags might need up to twelve hours or more at very much higher steam pressures. Because rag fibres do not come to the paper mill in a 'prepared' state, alkaline digestion is only the first process through which they may pass. When the fibre is thoroughly clean, it is rinsed in hot water, usually still in the boiler so that the grease which has been extracted is washed away. After draining in boxes with perforated bottoms, the rags are washed and also reduced to their component fibres in a machine very similar to a beater (see section 2.5.3), in which the rags are passed between stationary and moving knives. The machine also contains a drum washer comprising a revolving perforated screen; this removes dirty water and some extraneous solid matter. Unlike the beater, the knives are blunt and the pressure applied during the process is comparatively light. The operation is not intended to produce fibres of a particular character as in the beating process; its purpose is simply to separate the rags into fibres. While this is being carried out the fibre is cleansed of any residual alkali and dirt.

Fig. 2.8 Rags being removed from a rotary cylindrical boiler after cleaning.

2.5.2 Bleaching

The cleansing operations so far described control the general cleanliness of the paper being made. It is also necessary, however, to bring the colour of the paper up to a standard of 'whiteness' suited to the purpose for which it is intended, by a bleaching operation.

Almost all fibre needs bleaching to some extent. The degree of bleaching to be carried out varies with the original colour of the fibre concerned and the shade of whiteness required in the paper. For instance, a number of the fibres received at the paper mill in 'natural' states – the grasses and some of the rags are examples – may require quite extensive bleaching. On the other hand a great many of the wood pulps used in the mill will be supplied already bleached to an agreed standard.

Methods of bleaching, like many other paper making operations, vary from mill to mill in accordance with the type of product made. It might perhaps be not only adequate but also desirable to employ an old-fashioned 'batch' method of bleaching where a comparatively small quantity of highly specialized paper was being made, whilst a mill with a large output of a narrow range of papers would conduct an integrated and continuous bleaching operation. Whatever the process used, however, the principles are similar and the bleaching agents are likely to be the same.

Chlorine or chlorine compounds form the largest group of bleaching agents and the bleach liquor most frequently used is calcium hypochlorite. This is prepared by extracting bleaching powder with water or by passing gaseous chlorine into a mixture of lime and water. The resultant solution must be allowed a period of some hours to settle so that bleaching powder particles or other inorganic residues will not be carried forward into the solution. In actual use the liquor is diluted so as to contain 2% to 3% available chlorine.

The simplest method of bleaching and the one that might well be utilized in a small-scale batch process is to add the bleaching solution to the pulp in a 'breaking engine' or 'potcher'. These are circulatory vats very similar to the 'beaters' mentioned earlier and described in the next section, but with rolls designed not to impart particular characteristics to the fibres but merely to separate them.

A more complex bleaching process and one which is common where a large and continuous supply of bleached pulp is required makes use of 'bleaching towers'. In this process the pulp is pumped upwards through a series of towers which contain bleaching agent, the continuously moving pulp being washed between each application of bleach as it passes from one tower to another. The bleaching tower process is not only suitable for large-scale production by virtue of its size and continuity; it is also a more flexible and controllable process because the degree of bleaching can be constantly monitored to prevent any tendency to under or over bleach.

Whatever bleaching process is used, the time, temperature and reagent concentration all have to be controlled very carefully. A concentration of up to 10% of available chlorine, a temperature of 50° to 90°C and a time of 30 to 60 minutes is usual, but the process can be hastened or intensified by raising the temperature or by adding acids or acid salts. The bleaching action is beneficial since it dissolves pectin which forms part of the plant tissue, but at the same time attack on the cellulose fibre must be avoided.

Although chlorine is the commonest bleaching agent, other whiteners are also used. Sodium peroxide is becoming increasingly popular; it is especially effective for treating mechanical wood pulps, on account of its action on the lignin present. It would be used for 'soft' fibres, such as grasses, only with the greatest possible care, however, and only for a particular reason, because of the danger of degrading the fibres.

2.5.3 Beating

Beating is the stage of paper manufacture where the fibres, which have already been separated, as in the case of some rag furnishes, or prepared in the form of dry pulp, such as chemical or mechanical wood pulp, are mechanically processed to give the characteristics needed to produce a paper to a certain specification. This is done by subjecting the fibres to cutting and abrasion in a beater or one of the later engines developed from it. When brought to the beating stage the fibres consist of minute cylinders or tubes of vegetable matter very largely in a natural and undamaged state. To convert them into a material suitable for producing paper the fibres themselves must be treated so as to ensure a particular character in the paper to be made. It may be necessary to reduce the natural length of the fibre in some cases, or to retain it in others. Some fibres fibrillate during the beating process so that they form a mass, while others maintain a great deal of their original nature. The paper maker takes account of the fibre's character and relates it to the mechanical processing of the paper to be made.

The beating process consists of passing the fibre, suspended in water, between static and moving knives. These are adjustable both in terms of sharpness and the pressures which may be used, and also in the case of more sophisticated beating machinery the actual 'pattern' of the cutting process. The beating engine, or 'beater' in which the action takes place, is a large vat, with a capacity of 5 tonnes or more, with straight sides and round ends, and lined with tile. A longitudinal partition known as the 'mid-feather' divides the centre of the area into two compartments but leaves the ends open, so that the contents of the vat may be continuously circulated. Between the mid-feather and one straight side wall of the beater vat and at right angles to it is the beater

Fig. 2.9 A beater roll, showing the grouping of the knives.

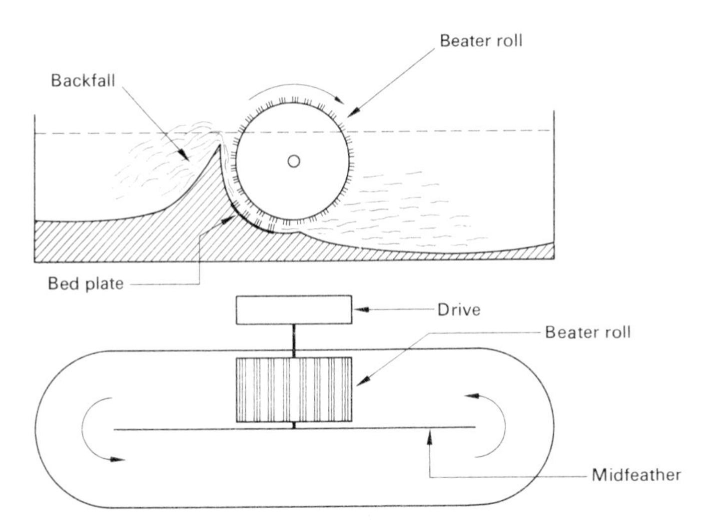

Fig. 2.10 Diagram showing beater action.

roll. This is a large roll fitted with knives, which revolves above a fixed bed plate which also carries knives angularly disposed so that the moving and fixed blades produce a continuous cutting and abrading action. The roll can be raised or lowered and, like the bed plate, it can be fitted with knives with a variety of patterns and keenness of cutting edge. Depending on the sharpness of the knives and the degree of pressure which is exerted by raising or lowering the roll, the fibres can be subject to a very wide range of cutting conditions. The fibres remain separate, however, and it is a principal object of beating to ensure this, and to avoid clumps and knots.

The mixture of fibre and water passes between the bed plate and roll for a period of some hours. The process is progressive, starting with the roll and bed plate separated, so that the initial action on the fibres is simply to separate them from each other. This 'brushing out' period, as it is sometimes called, is also effective in softening the fibre so that it becomes more manageable in the later stages of paper manufacture. When the fibres have been brushed out the beater roll is brought closer to the bed plate, leaving a space between them of only a few millimetres. At this stage the fibres may become cut and abraded and in some cases fibrillation may occur. The character of the fibre, the manner in which it is beaten and the degree of fibrillation are the factors affecting the behaviour of the fibre on the papermaking machine wire and the finished paper. Thus, it might be necessary to produce a paper of high absorbency which does not demand high strength, or one of highest possible strength within the range of the fibre being used, and with a close non-absorbant character.

In the first case the process would be designed to achieve a short fibre, probably not fibrillated to any extent, to give a bulky, absorbent open-textured sheet of paper. The shortness of the fibre would offer little opportunity for the interlocking which promotes strength in paper, but this is not required here. To produce such a short fibre the beating 'tackle' – that is, the sets of knives – would be sharp and the pressure under which the process was carried out would be high so as to result in a clean cutting action resulting in rather short open cylinders of fibre.

At the other extreme is the paper which calls for the greatest possible fibre length combined with a degree of fibrillation which will ensure maximum strength both in terms of tensile stress and folding resistance. Here the purpose of beating would be to draw out each individual fibre without reducing the length, so as to retain its strength and ability to intertwine with other fibres to form a closely knit mat of fibrous matter. The beating tackle would be blunt rather than sharp, and the pressure much less than is required for an absorbent paper.

Generally speaking, the length of fibre that has been found suitable for working on the paper machine wire is from 0.02 to 0.12 in. (0.5 to 5 mm). Reference to the 'natural' fibre lengths (section 2.4) shows that, apart from any characteristics it is intended to introduce during the beating process, the longer fibres will have received a certain amount of cutting in any case, while many of the shorter fibres such as esparto need little or no cutting.

Apart from the physical action of beating in terms of cutting and fibrillating, however, an extremely important feature of the process is the amount of water 'beaten in' to the fibre. The constant passage of the fibres between the bed plate and the roll, especially with blunt knives and rather light pressure, results in a hydrated fibre known to the beater man and papermaker as 'wet' pulp. The opposite effect is produced if the fibre is more sharply cut so that a great deal of fibrillation cannot occur; it is then known as a 'free' pulp. A wet beaten pulp tends to retain moisture on the papermaking machine, so that the fibre remains in suspension for a longer period as it travels forward from the wet to the dry end (see section 2.6). During passage of the pulp, the machine wire on which the paper mat settles moves forward and shakes laterally. A long suspension period allows this action plenty of time and ensures a sheet of paper more even in strength in both the machine (longitudinal) and the cross (transverse) direction and with the fibres so distributed throughout as to result in a close and even sheet. A free beaten pulp, on the other hand, may well be appropriate where these specific characteristics may be safely and correctly sacrificed to speed of manufacture or where an 'open' sheet is sought.

Beaters are becoming less common in paper mills and their use is increasingly restricted to low-tonnage speciality production. They are being replaced by refiners. There are two types of refiner, the conical and the disc. In the case of the former, a cone fitted with longitudinal bars revolves within a shell also fitted internally with similar bars. By adjusting the revolving cone longitudinally, the clearance between the fixed and the revolving bars can be increased or reduced in a similar manner to the raising or lowering of the beater roll. The principle of operation is identical in that, for a pulp which needs only simple preparation, perhaps for a cheap printing paper, the two sets of bars will have a fairly wide clearance; when it is necessary to be more specific about the character introduced into the fibre the bars will be adjusted to carry out more

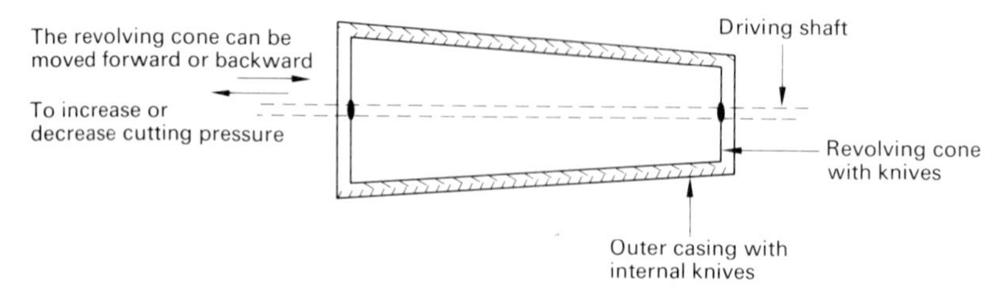

Fig. 2.11 Diagram of a cone refiner.

intensive action. The speed of the revolving cone is very much higher than that of the beater roll and the expanse of cutting edges presented by the fixed and revolving bars so far exceeds that of the beater that pulp preparation can be effected in very much less time. Also, it is a continuous, not a batch, process and can be controlled to a far greater extent than can the beater, so that it is not necessary to rely upon mixing the product from several beaters to equalize the differences in quality between the batches. The disc refiner consists of a central bladed disc revolving at high speed between two outer stationary ones. As with the beater and the conical refiner, clearances are variable to control the character of the fibre being treated. The two types of refiner are sometimes brought together into one, both cone and disc blades revolving on the same axis. A higher degree of flexibility is claimed for the dual refiner, and more finely varied types of treatment are possible. The development of the refiner has been encouraged by the increasing speed of the papermaking process (the refiner is faster than the beater, and operation is continuous) and the improvement in quality control on the part of suppliers of fibrous raw materials to the paper trade. To this may be added the greater use of wood pulp and decline in the proportion of other fibres, except for speciality products.

2.5.4 Loading

Paper made from fibrous raw material alone would be flabby, excessively absorbent and unsuitable for accepting printing or writing. For these reasons loadings–or fillers as they are sometimes known–are added to the furnish. The purposes of loadings are to improve the printing or writing surface of the paper and to modify its absorbency. They contribute greatly to the smoothness of the surface and provide opacifying elements to render the paper suitable for printing or writing. The flatness of the sheet is enhanced, the paper becomes softer and more easy to handle and its dimensional stability is improved.

Loadings are fine white powders prepared from minerals either by simple grinding and screening or by chemical means. Being mineral and therefore non-fibrous they are unaffected by the conditions which cause dimensional changes in fibres, such as variation in temperature or the amount of moisture in the air. On the other hand they add nothing to the strength of the paper. The addition of mineral loading to the furnish is essential to the production of a practical and usable paper, within limits which should not be exceeded if the quality of the product is not to suffer. The loadings in common use are described below.

a. China clay. As well as being a very widely used coating substance china clay (or kaolin, $Al_2O_3 \cdot 2SiO_2 \cdot 2H_2O$) is one of the most extensively used fillers. It is formed originally by the weathering of granite and felspar and is normally obtained by open-cast hydraulic mining. The slurry formed from the mixture of clay particles and water is run over sand traps and screens so that not only are the larger particles of china clay removed but also pieces of grit and other foreign matter. Mica especially is likely to be present and as this can be particularly troublesome in papermaking it must be carefully eliminated. A measure of the quality of china clay is the particle size; the smaller the particle the higher the grade. China clay of large particle size is used for loading and filling the sheet of paper itself and material of finer particle size in coating operations. China clay helps to produce a smooth surface which is receptive to printing ink.

b. Calcium sulphate ($CaSO_4$). Several varieties of calcium sulphate are used in papermaking, most of them natural (e.g. gypsum and anhydrite) but one chemically precipitated. While the usage in different types of paper is fairly wide there is a tendency, especially in the case of the precipitated grade, to use it rather more for writing than for printing papers.

c. Calcium carbonate ($CaCO_3$). This is used fairly extensively in America as a filler in printing papers, but is less used in other parts of the world. There are certain limitations on its use because of a degree of incompatibility with resin size (which is also added to paper) although this is being overcome to some extent with the aid of alternative sizes. A feature of calcium carbonate as a filler is its low acidity, which is an attraction for papers which are required to be highly durable.

d. Magnesium silicate ($MgO \cdot SiO_2$). The particles of this filler are long and narrow, which allows them to lie down more easily with the fibres, and because of this claims are made that magnesium silicate, unlike other fibres, makes a small contribution to the strength of the paper. There appears to be very little evidence to support this. Magnesium silicate is not used a great deal in Britain but is fairly common in European countries and in America. Paper in which it appears as a filler frequently presents a finish so smooth as to appear greasy.

e. Barium sulphate ($BaSO_4$). This compound is used occasionally as a filler, particularly as it is a good opacifying agent. It is more suitable for coating, particularly for photographic papers.

f. Titanium dioxide (TiO_2). This compound is a particularly effective loading where a combination of good opacity and bright white colour is required. Most loadings, while effective for the former, tend to give the paper a dull and leaden appearance unless used with a brightening agent. Titanium dioxide, on the other hand, has a high refractive index and by scattering more light makes the paper opaque without markedly reducing its reflectivity. For economic and other causes printing papers are becoming lighter in weight, and titanium dioxide is proving very useful in making papers which, although thin, can be effectively printed both sides without the print showing through.

g. Zinc sulphide (ZnS). This is a reasonably effective opacifying agent but is somewhat reactive, especially when in contact with copper. It is not used very much in writing or printing paper, therefore, but is fairly extensively employed in the field of industrial papers where it is often mixed with barium sulphite to form lithopone.

Other materials are used for loading, either for speciality papers or because of local availability. Among speciality loadings are those which are added to the furnish for security purposes. One of these is manganese ferrocyanide, often used for security documents such as cheques or bearer bonds. Its tendency to produce a noticeable stain whenever attempts are made to bleach out print or writing on the document renders forgery very difficult.

Although not strictly loadings or fillers in the usual sense it is appropriate at this stage to mention additives – made during the beating stage or as the pulp goes on to the machine wire – which are intended to identify paper to be used for security purposes or to render it resistant to forgery. Obviously such additions are peculiar to the paper concerned and are all closely guarded secrets. In general, however, such protective measures fall into two classes: (a) chemical additions which are known to be reactive to the attempts at forgery to which the paper is likely to be subject, and (b) the use of identifiable additives which establish the authenticity of the paper in the case of dispute or doubt. The latter include specially dyed fibres added to the furnish in a known proportion or the incorporation of metal or mineral particles in a pre-determined pattern, known as 'planchettes'.

Loadings are added to the pulp at the beater stage of manufacture, in the beater itself in the case of mills still using the older methods, or in tanks in which the pulp is mixed prior to going on to the machine wire. Loadings may sometimes be added in the dry state but it is generally considered preferable to add the mineral to water separately and then to mix the liquid and the pulp. It is claimed that by this method a better retention of the loading in the finished paper is obtained; this is not only important from the point of view of economy but also because the process is more easily controlled and so leads to a more uniform quality throughout the making.

2.5.5 Colouring

Even a so-called white paper requires the addition of some colouring matter if it is not to present a dingy appearance, and the demands of fashion create a call for papers with distinctive shades to provide the appropriate background for the printed image or for other purposes. Colouring is therefore not simply a matter of treating papers to provide a range of decorative shades. The colour of a paper, whether it be a effect on its marketability. Papers for printing and display purposes, for instance, could sometimes be produced in vivid colours, but any colour used must not only be attractive but also allow the printed matter itself to stand out clearly. This is why there is a limit to the intensity of colours used in paper to be printed. No such limitations prevail white or a tinted sheet, may frequently be important to its usage and have a considerable where papers are to be used for purely decorative purposes, however, and the shades of papers such as the crepe tissues employed in Christmas decorations or for the manufacture of paper flowers are therefore often very bright indeed.

Shade also becomes important when providing papers for books, particularly text books which may have to be studied for long uninterrupted periods. A bright shade of white might well provide good contrast to the black letter text matter and present a clear image for short periods of reading. It is likely, however, that its brightness would cause a certain amount of eye strain on prolonged exposure, and this must be guarded against. The different types of intensities of artificial light also affect the choice of shade and sometimes the colouring matter itself. The shade of a paper is acquired by reflection of the light to which it is exposed. If it becomes necessary to examine and compare different papers for their colour, therefore, it is essential to compare like with like and in identical conditions. It is difficult to compare different kinds of paper for shade because the light-reflecting properties of the surfaces are not identical. For a similar reason the two sides of the same single-wire paper will be very slightly different, because their surfaces differ slightly. The shade of a paper is a combination of colour, paper type, surface characteristics and lighting and it is with these factors in mind that colouring decisions are made.

The colouring or dyeing of paper is not especially difficult in principle but requires considerable knowledge and experience in practice. It is carried out at the pulp preparation stage, and in its simplest form consists of adding small quantities of blue or red colour to the furnish to correct the rather yellowish shade which is a feature of almost all pulps in their uncoloured state. Papers for critical purposes require careful colour matching and quite extensive laboratory testing is carried out before the type of dye-stuff and the amount to be used is decided upon. This takes into account not only the shade itself but also the paper's usage, length of life required, fastness to light and other features.

Two principal types of colouring matter are available: water-soluble dyes and insoluble pigments. Water-soluble dyes offer the wider range of products and for general use are much preferred to pigments. To be effective as colouring agents, water-soluble dyes which have an affinity for the fibre or for the other constituents of the paper must be used if an economic level of dye retention is to be maintained. Three classes of dyes are in general use: basic, acid and direct. Each class varies in its affinity for particular fibres and in colour and fastness to light. In general, however, basic dyes produce quite strong colourings but have a tendency to fade rather rapidly; the strongest affinity is to unbleached or mechanical wood pulps. Direct dyes, on the other hand, use colours which are less strong but of greater permanence. They are also well retained over a wider range of cellulose fibres than basic dyes. Acid dyes have a fastness to light which is less than that of direct dyes and do not display any affinity for papermaking fibres. They produce bright colours, however, and if used with alum as a mordant can give a satisfactory product where the furnish is entirely or largely of bleached pulps.

Pigments are utilized in special circumstances, sometimes where opacity is important, since apart from their colouring properties they act as a form of loading. The pigments in use include ultra-marine, which is a strong colour with good fastness to light; Prussian blue which also retains its shade well but is affected by alkalis that cause it to fade; and natural earths such as ochres and umbers which may be used in low-shade papers partly for their colouring properties and partly because they are quite good loadings in their own right. One of the problems of using pigments for colouring is the difficulty of maintaining even colour on both sides of the paper. Pigments tend to be drawn from the underside of the paper as it is being formed on the machine wire, so that the shade is paler on the wire side of the sheet.

The demand for bright white papers has given rise to the use of colouring materials which have the property of absorbing the ultra-violet component of light and re-emitting the energy as blue light. These are known as optical bleaching agents. They improve the whiteness of paper where ultra-violet rays are present in some strength in the incident light. At the same time, optical bleaching agents have the drawback of poor fastness to light.

Colouring matter is added to the pulp at the beater stage of manufacture either in the beater itself or in tanks when the loading is being added. The choice of type of dye and amount having been decided prior to the commencement of manufacture, the papermaker's responsibility becomes one of control. This entails two main areas of supervision; eveness of colour throughout the whole papermaking run, and as far as possible equality of shade between the top and wire side of the sheet. Of the two the latter probably requires the greater vigilance because of the constant loss of fibre, coating and colouring matter as the water in which they are suspended is extracted from the wire side of the paper.

2.6 PAPERMAKING

2.6.1 Hand-made paper

The amount of paper made by hand as a proportion of total paper manufacture is insignificant in all industrialized countries, though small hand-papermakers still exist in remote and isolated communities. Nevertheless, the making of paper by hand is of interest because it is the first and oldest method practised, and because it exemplifies, on a small scale, the technique necessary to all papermaking. In industrial communities hand-made papers are sometimes preferred where dimensional stability, strength and durability are required.

The main features of hand-made paper are the use of the highest quality raw materials (rags) and the even distribution of the fibre throughout the sheet. There is therefore much greater dimensional stability and considerably less variability in strength than in machine-made papers. This, together with a generally higher strength and a durability which exceeds that of machine-made papers, means that hand-made paper is peculiarly suitable for purposes such as securities and legal documents or for expensive stationery.

The steps in hand-papermaking are as follows:

1. Preparation of the pulp, as already described.
2. Agitation and warming of diluted pulp.
3. Removal of a layer of pulp on a wire mesh mould combined with agitation of the mould.
4. Drainage of the mat of paper fibre.
5. Removal of the paper.
6. Drying.
7. Finishing.

The preparation of rags for hand-made paper follows the same lines as that already described for machine manufacture except that all operations are on a much smaller scale. A 'beater' in a hand-made mill would probably have a capacity of not more than 100 kg (220 lb.). Also, because of the need to retain the maximum strength of the fibre a refiner would not be used.

After beating, the 'stuff' is emptied into a 'stuff chest', or large tank, fitted with an agitator. More water is added at this stage so that there is an easy flow from the stuff chest through strainers into the vat.

The vat is a lead-lined tank about 2.5 m long by 1.25 m broad and 1.0 m deep. It contains an agitator and steam-heated pipe to keep the contents warm and so improve the drainage of water through the wires of the mould in which the paper is formed. There is a shelf known as the 'bridge' along one long and one short side of the vat, and at one end is a piece of knotched wood known as the 'ass' against which the mould can be rested in a sloping position.

The mould itself consists of a wooden frame covered with a wire mesh. The mesh may be plain or bear a watermark, and may be 'wove'–that is, without lines–or 'laid', which is the term given to papers which appear to have two sets of lines, one set close together and the other at right angles and further apart. Moulds are made in identical pairs and are used with a wooden frame called the deckle which fits over each in turn to form a tray. The sequence of operations is as follows.

The vatman dips the mould consisting of the frame and mesh into the vat vertically and withdraws it horizontally with a mixture of fibre, loading and colouring matter floating in water held within the frame and draining away, partly over the edges of the mould and partly through the wire mesh. As he withdraws the mould the vatman tilts and shakes it sideways with a twisting action. The combined movement carries the surplus mixture of water and solids over the edge of the mould and back into the vat and at the same time interlocks the fibres so that as the water drains away through the mesh they are deposited in an evenly formed mat; the skill with which this is done has a marked effect on the strength and general quality of the paper. The vatman then slips the deckle off the mould, leaving the wire mesh leaning against the ass to drain.

Another worker, the 'coucher', slides a fresh mould along to the vatman and removes the paper from the mould as soon as it is dry enough. The sheet of paper is then covered with a layer of felt, alternate layers of paper and felt being piled in a press until it is filled. Pressure is applied hydraulically to squeeze out most of the remaining moisture.

The next operation, known as 'laying', is to separate the sheets of paper and felt. The paper is then firmly pressed between zinc plates for some hours to remove the marks of the felts.

The paper is then dried by hanging it in a current of warm air, in clips or over bars or ropes; stacking on cloths held in a wooden frame; or, in more up-to-date plants, passing over a heated cylinder.

The paper is finished by applying a thin coating of size – the operation, which is performed in a tub, is called tub sizing. The size is essentially a solution of gelatin, through which the paper is passed. In principle the process is the same as that described in more detail later (section 2.6.6) but because the paper is in sheets and not on a roll, the sheets have to be treated separately. This can be done by passing the sheets through the size tub between continuous felts which are fed between rollers that squeeze out the excess gelatin. The felts are then separated.

Finally the paper is dried, slowly and gently with sheets separated so that the 'skin' of size is preserved, a process which may take up to two days exposure to currents of warm air.

2.6.2 Fourdrinier papermaking

a. Introduction. The first thing that often strikes those who have not seen a papermaking machine, is its size. Even the smallest machine engaged on highly specialized manufacture is unlikely to be less than 50 metres (164 ft) long and to have a width of 2 to 3 metres (6 or 9 ft), whilst the giant machines specializing in newsprint manufacture or perhaps the making of boards for cartons may be as long as 100 metres (328 ft) and have a width of 12 meters (39 ft). Not only is the size of the papermaking machine sometimes a surprise but so also is the speed with which paper is made. The very slowest machines, which may be old or perhaps engaged on some speciality paper, operate at 100 metres (328 ft) per minute and the fastest are now producing paper at up to 1 200 metres (3 947 ft) per minute.

The machine on which most papers are made is the 'Fourdrinier' machine. It takes its name from the inventors, Henry and Sealey Fourdrinier who, together with Brian Donkin, built the first machine early in the nineteenth century. The papermaking machine consists of two sections: the 'wet' end, where the mixture of water and fibre enters the machine, and the 'dry' end where the web of paper has the last of its excess moisture removed and where a number of finishing processes which used to be done elsewhere in the mill are now carried out. The two parts operate as a continuous process.

The variety of purposes for which paper is now used and the wide selection of fibrous raw material available to manufacturers, are such that the actual papermaking machines may differ very greatly. The type of manufacture may dictate the length of machine wire (which corresponds to the mould in hand papermaking) or the angle at which it is

Fig. 2.12 A papermaking machine as seen from the wet end, showing its enormous size.

set between the point where the mixture of fibre and water starts its flow and the point where the web of partly finished paper is taken off. Furthermore, the speed of the machine and the papers it is designed to make determine the extent and disposition of the drying cylinders and their size. Each machine is made for a particular purpose, be it general or highly specialized manufacture; yet the principles involved remain constant.

b. Papermaking machine: Wet end. After beating is completed it is necessary to transfer the contents of the beaters or refiners to the papermaking machine. First, the pulp must pass through a series of revolving screens or mechanical filters to eliminate any particles of matter which may blemish the finished product by causing unsightly spots to appear, or – a more serious flaw – by allowing lumps to get embedded and remain proud of the surface. After the screening, the contents of all the beaters or refiners pass into a stuff chest which contains a head of pulp and water to supply the machine, and agitators to mix the loads from each beater so that any inequalities between them are eliminated.

The mixture of water, fibre and loading material which arrives at the wet end of the papermaking machine contains approximately 1% solid matter. It is in this highly fluid state so that the material can be conveyed from one operation to another prior to the actual papermaking. The purpose of wet end of the papermaking machine is to take this weak mixture and remove the water from it so that the fibre and loading material 'settle down' into the mat of paper. There are three operations: the continual movement forward of the mixture of water and solid matter, the removal of water by gravity and

Fig. 2.13 Papermaking machine wire in operation. The table rollers on which the wire is supported are clearly shown.

suction, and the disorientation of the fibrous matter by sideways shaking of the wire so that the fibres are not all lying in the same direction.

The wet end of the papermaking machine consists of a continuous wire mesh supported on rollers. The length and width of the machine 'wire' (as it is known) and its angle, vary in accordance with the type of paper the machine is designed to make and the economic limits of the intended operation. A newsprint machine, which is obviously highly specialized and designed for producing the maximum quantity of one type of paper, might have a wire width of 12 m (39 ft) and a length of 55 m (150 ft). At the other end of the scale a papermaking machine designed for the production of a wide variety of commercial papers – an operation which might entail a change of quality, furnish, grammage (weight in grammes per square metre) or colour every few tonnes – would need to be smaller if the losses arising from 'stopping time' were to be avoided. Hence papermaking machines are purpose built. The flow of pulp onto the wire may be controlled by partitions or 'slices' which influence the thickness and distribution of the pulp layer, or may – as in the case of machines which have been designed for a single product – pass almost uninterrupted on to the wire.

The slices consist of partitions at the beginning of the wire in line with the flow of pulp. By altering the angle of any one of a series of slices, or of a group, the direction of flow of a part of the pulp may be slightly changed so that an even distribution across the wire is achieved. This is not an operation which has to be carried out at frequent intervals, but is a means of control which is available as necessary. As the speed of paper machines increases and more accurate methods of control become available the use of slices is declining, although they are still found on slow running and speciality machines.

Once the pulp layer reaches the wire, water begins to pass through the meshes, encouraged by the capillary action of the constantly revolving rollers on which the wire rests. To prevent the natural tendency of the fibres to orient themselves in the longitudinal or 'machine' direction the fibres are disoriented by shaking the wire rapidly from side to side as it moves forward, thus simulating to some extent the vatman's art of hand papermaking. Despite this the number of fibres in the finished paper oriented in the direction of flow always exceeds the cross-directed fibres, and it is this that produces the difference in strength and dimensional stability between 'machine' and 'cross' direction of all machine-made papers.

Gravity and capillary action are not sufficient to remove all the water necessary if a paper web of adequate strength is to be formed where it leaves the wet end for the drying cylinders. About two-thirds along the wire a number of suction boxes are provided over which the wire passes. These remove the water more rapidly and it is at this point on the wire that the web of paper really begins to form. Because at this stage the sheet is becoming solid it is here that the water mark – if used – is introduced. The water mark, seen as a clearly outlined design if the sheet is held up to the light, is formed by displacing fibres while the sheet is still slightly soft but is beyond the point of filling in the design so impressed.

The watermark is introduced into paper by means of a revolving wire mesh roll which lies across the machine wire at right angles to the machine direction. This is the 'dandy roll'. The dandy roll has soldered or welded to the mesh the design which is to

Fig. 2.14 A dandy roll. The wheels mounted on the brackets raise or lower the roll as required.

be impressed into the soft surface of the paper. This may be a simple mark which is intended merely to provide a brand name for the paper or, if the paper is to be used for any form of valuable document such as bank notes or share certificates, an intricate combination of lines and shading, difficult to forge. In such cases it is likely that the design, instead of being fixed to the surface of the roll, would be formed by pressing the roll surface between male and female dies. When making the roll the dandy roll manufacturer takes account of a number of factors: the size and speed of the paper-making machine on which it will be used, the type of paper to be produced and above all the amount by which the paper will stretch and shrink between the application of the mark and the final drying process. During manufacture paper tends to become slightly elongated because of the tension under which it travels from wet to dry end. At the same time, as the paper dries the web becomes narrower. If the dandy is to produce a completely circular design, the raised marks on the surface will be slightly oval, with the longer axis at right angles to the flow of paper. The circumference of the dandy and the number of times the design appears on it are calculated to produce a certain number of watermarks in each finished sheet of paper or, as is the case with cheap mass-produced papers, staggered so that no matter how the paper is eventually cut, the water mark will appear several times. If the papermaker has reason to believe that in the case of a particular making it will be necessary to aid the 'closing up' of the sheet for some reason, the paper may be run with a plain dandy with no design on it. It is also claimed by some manufacturers that using a plain dandy in this way is of some assistance in removing water more quickly. A 'laid' paper–that is, one which appears to be lined in the manufacture–is also produced with the aid of a dandy roll. When held to the light the laid paper shows a set of lines close together in one direction and another set at much wider intervals in the cross direction. This is a legacy from the early days of hand-made paper when wires held together in this way formed the papermaker's mould. Today the effect is purely decorative.

The roll itself may be 'free running', revolving by its contact with the wire, or it may be power driven at a carefully controlled speed to harmonize with the speed of the wire. As papermaking machines become larger and faster the use of free running rolls declines because the production of a clear watermark at high speed is very much a matter of care in arranging the dandy roll and wire. If the relationship between them is incorrect there is often a build-up of fibre around the edges of the raised design on the dandy roll and also within the design itself. This can not only lead to blurred watermarking, but if the build-up is excessive a ridge of fibre can form which is too heavy for the sheet to bear. At subsequent processing when passing through couch or press rolls or in the early stages of drying the ridge will fall away or be picked off. Holes may then appear in the paper either at the watermark or in other places where the, by now, hard knot of material has caused damage.

By the time the web of paper has passed the dandy roll and suction boxes, it is reasonably well formed, and its nature, so far as it is controlled by its fibrous furnish, already determined. It is still an over-moist and flabby product and it will not become useful until a great deal more of the water it contains has been removed. Hence at the end of the wire the web must be picked up to go through the drying cylinders. Although now a fully-formed paper, the web is not quite strong enough to be entirely self-supporting. It is therefore carried at this stage by continuous blankets known as 'wet' and 'dry' felts, the wet felt being that in closest proximity to the end of the machine wire and the 'dry' felt that nearest to the drying cylinders. At this point the opportunity may be taken of removing still more moisture by means of suction couches and press rolls. The suction couch is merely a large perforated cylinder to which suction is applied and over which the paper passes, losing more of its moisture on the way. Press rolls may be incorporated at this stage or may be separately installed. They consist of two rolls through which the paper web passes and which nip it to squeeze out the moisture. This has the advantage that by applying pressure to the surface of the paper the sheet is 'closed up' to some extent.

To summarize, the wet end of the papermaking machine accepts a suspension of fibrous and non-fibrous matter in water, removes a great deal of the water by suction, pressure and gravity, disposes the solid raw material as a flat web in as near as possible an evenly distributed state and presents, to the dry end of the machine, a web which although still wet and flabby is recognizably paper.

c. Papermaking machine: Dry end. Much of the efficiency of modern papermaking machines rests on the speed and effectiveness with which paper, having been formed on the machine wire, can be dried. Indeed the evolution of the industry from a craft to a mass production operation could be dated from the addition of drying cylinders to what was then merely a continuous papermaking mould, as described in a patent taken out by T. B. Crompton in 1821. The function of the dry end is to reduce the amount of moisture in the paper from some 17 to 20% to approximately 7%. This is achieved by evaporation. The web of paper is passed over heated cylinders which raise the temperature so that the water in the paper is converted into steam and extracted. Removal of the moisture causes the paper to change its shape so that the drying process is not one of simply removing water.

The size and number of drying cylinders incorporated in any papermaking machine must be directly related to the speed of the machine and the type of end product. Sixteen to twenty-four cylinders with a diameter of 130 to 150 cm are commonly used. The cylinders are normally separately powered in groups, the speed of each group being accurately related to the next. This is necessary because paper contracts to some extent as the moisture is withdrawn. At the same time the web is under tension as it passes over the drying cylinders so that a certain amount of longitudinal stretch occurs, accompanied by contraction at right angles to the machine. The result is a web which is slightly longer and narrower than the paper as it comes off the machine wire. Control of the drying operation is a factor which, with the type of fibrous furnish and method of

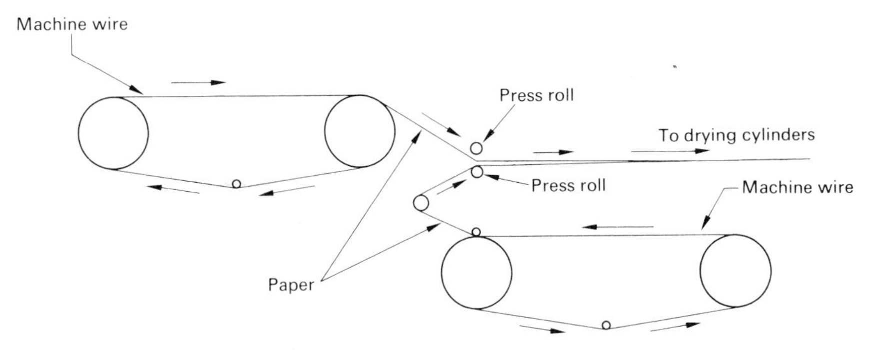

Fig. 2.15 Diagram of a twin-wire machine, a development of the single wire, and itself now further developed to form multi-wire machines.

manufacture, plays a large part in determining whether the paper will be stable towards subsequent changes in temperature and humidity when in use.

Coating and sizing processes can be effectively conducted as part of the drying operation with obvious advantages in terms of continuity and cost. Mechanical treatment to close the surface of the sheet can also be incorporated. It is quite common to find a set of press rolls about two-thirds of the way through the drying process and a great many papermaking machines have calender stacks between the last of the drying cylinders and the final reel-off. (Calender stacks are banks of highly polished rollers, as described in the section on finishing. The paper passes between rapidly revolving rolls under pressure: these impart a smooth finish to the paper.) Considerable latitude is possible in the disposition of drying cylinders, sizing and coating presses, and press rolls; but the principles of operation remain the same.

2.6.3 Machine glazed (M.G.) papermaking machine

The papermaking machines so far described have been Fourdrinier machines or developments of it. The M.G. papermaking machine has a wet end similar to that already described but makes use of a large polished cylinder instead of the series of drying cylinders on the Fourdrinier machine.

The object of the M.G. machine is to produce a paper with a high glaze on one side only, the other side remaining noticeably rough and suitable for accepting adhesives. The sort of papers made on the M.G. machine are poster papers, kraft papers, papers for bag making from sulphites to krafts, and cellulose wadding; in the trade they are described as M.G. kraft, M.G. sulphite, M.G. poster and so on. In place of the drying cylinders of the Fourdrinier equipment, the M.G. machine has a large, highly polished, heated revolving cylinder. The moist web of paper is pressed against the surface of the

Fig. 2.16 An M.G. papermaking machine showing the cylinder and the finished paper being reeled up.

cylinder by rubber covered rolls and dried while it is in contact with it so that within about two-thirds of a revolution it can be drawn away. Pressure between the polished face of the cylinder and the moist paper results in a highly glazed finish on one side of the paper, the side that has not been in contact remaining rough. The M.G. machine is slower than the Fourdrinier and various methods of increasing the speed of operation are used. One way is to introduce small drying cylinders between the machine wire and the M.G. cylinder. By reducing the drying load on the large cylinder it is possible to increase its speed. All modern papermaking machines are 'hooded' to collect the damp air and steam rising from the drying cylinders and convey it away through heated pipes by means of fans. This prevents the air around the drying area becoming saturated and allows the drier to operate more quickly and efficiently – a useful feature in the M.G. process. New methods of paper bag production and the availability of synthetic adhesives has rendered reliance on strong M.G. papers less necessary. Similar changes in respect of poster papers arising from the introduction of sites artificially lit from behind (rear-lit poster sites) and the introduction of plastic sheeting for exposed advertisements are beginning to diminish the use of M.G. poster papers. While papers of this type are still in use it is to be expected that they will decline in popularity in the future.

2.6.4 Board making

A board can be defined as a paper-like product with a grammage of not less than 220 grammes per square metre. This applies to board for printing as well as that used for carton or box making, though to differentiate between a thick 'paper' and a thin printing 'board' is sometimes difficult. In fact, thin board is sometimes made on an orthodox Fourdrinier papermaking machine. Ordinarily, however, boards of any type are made on a cylinder, not a flat wire machine and this gives them characteristics which differ is some respects from those of paper.

Stock preparation is on similar lines to that for the production of paper and the range of fibrous material suitable for board making is roughly the same. As with paper, the bulk of printing board is made from chemical wood pulps and a significant minor proportion of high quality boards is made from esparto. The furnishes of boards for packaging purposes are naturally different and may contain a large proportion of waste fibrous products.

Whatever the furnish the pulp is prepared and screened very much as in papermaking. It is then piped to a mixing box. This has the same function as the stuff chest at the beginning of the papermaking machine; it levels out any variations between consignments of pulp coming from different beaters or refiners. After mixing, the pulp flows through adjustable slits into a vat and it is at this point that paper and board making processes begin to diverge.

In place of the forward-moving shaking wire on which paper is made, there is a cylindrical wire screen which revolves in a vat containing the pulp. As the cylinder rotates it picks up, on its exterior, a layer of pulp from which the water drains through the wire mesh. When sufficient moisture has been removed (i.e. after about two-thirds of a revolution) the by now only moist and sufficiently strong web of paper is transferred from the cylinder onto a continuous moving band of felt. The board machine comprises a number of vats and their associated cylinders, and is known by the number of such units, i.e. a 'six-vat machine' or an 'eight-vat machine'. The layers of pulp from each of the vats are brought together, pressed to form a whole, and passed through the drying process.

There are two types of vat; the uniflow and the contra-flow, the latter being more common. The cylinder of the contra-flow machine revolves against the flow of the pulp and this is said to counteract any tendency on the part of the fibres to orient themselves as a result of a directional movement of the pulp as it deposits on the cylinder. In the uniflow vat the movement of the cylinder and the flow of pulp is in the same direction and its supporters claim a resulting greater regularity of pulp deposition.

Because the layer of pulp is substantial enough to retain appreciable moisture and the fact that it is to be combined with other layers to form the board the removal of water from the web is an important operation. This is done by subjecting the web to pressure or suction at a 'couch' or extraction roll so that it is in a suitable state to receive the next layer. There is, of course, a natural tendency for a thick layer of fibre to retain more moisture than a thinner one, and for the retained water to vary in distribution throughout the thickness of the web.

In the recently developed Inverform process an attempt is made to improve the de-watering process and to produce a board which has similar characteristics on both its surfaces rather than one in which the two sides differ, as is the case with sheets where only one surface has been in contact with the machine wire. In this process the pulp flows between two wires each of which is subject to suction whilst applying a 'nip' to the web. Water is therefore removed from both sides of the web at the same time. There are considerable advantages to be expected from the Inverform process, not only in terms of speed of drying but also in strength, rigidity and general quality.

The principles of water removal after the stages already mentioned are the same as that for paper but because of the weight and bulk of the product it is obviously less manageable and therefore the machinery to handle it is stronger and slower. More important than this is the fact that, because of the composite 'layer' character of board and its bulk, it holds very much more water than paper even after the de-watering processes already described. Whereas paper coming from the Fourdrinier machine is dried very largely by evaporation, with the aid of heat, the drying of board makes much greater use of mechanical extraction by pressure and suction at different stages throughout the whole drying process. The web may pass though a number of press rolls and suction rolls before the residual moisture is removed by heat alone. The principle of graduated treatment is the same as in papermaking, however, and all precautions are taken to avoid too drastic application of pressure or suction early in the water removal

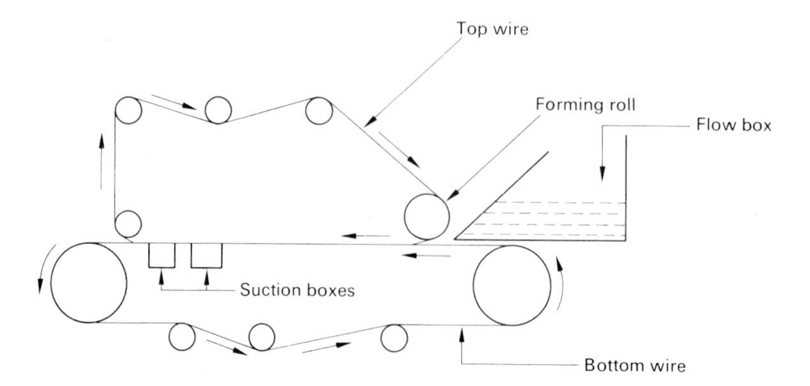

Fig. 2.17 Diagram showing the wire action of the Inverform machine.

process, so that the board will not be distorted, crushed or otherwise damaged while in its early tender state.

After drying, board, like paper, has to be 'finished' by passing through calender rollers as in calendering paper, but it is usually rendered damper by the separate application of water during the operation, partly because board tends to be less resilient than paper and therefore heavy pressure is likely to produce 'plucking', and partly because board has already been subject to more pressure than paper during its drying stage and has usually formed a hard skin on the surface which has to be softened if an even and suitably 'high' surface is to be obtained.

The manufacture of boards differs from paper in one important respect. Paper is usually made for a particular purpose on a machine primarily designed for a special manufacture, i.e. newsprint, kraft, tissues or printing and writing papers. Boards are generally made of a number of layers of thinner stock; and these layers may differ according to the purpose of the end product. The layers from which boards are made may themselves be of different fibre quality, therefore, according to the intended purpose.

The superficial appearance of a board is less reliable as an indication of its quality and purpose than is the case with many papers. Thus a board used for display and with a very short life, would comprise outer layers based on chemical wood and inner layers of mechanical wood fibre, whereas an industrial board might require good barrier layer properties, appearance being less important. In the latter instance the inner layers might well be of different fibrous furnish and made to perform an especial function, while the surface layers are intended to present an attractive appearance only.

2.6.5 Coating

The most widely applied finishing process to which paper is subject is undoubtedly coating. The need for paper to present an even surface suitable for printing has led to a great increase in the amount of coated paper used and also in the variety of processes by which the coating is applied. The development of coated papers is thus directly related to the increasing demands made by the printer, especially the greater use of colour printing not only in high-quality books and magazines but also in newspapers, advertising material etc.

The purpose of coating is to provide a level printing surface by adding to base paper a layer of very fine mineral pigment which may then be smoothed to a higher degree than would be possible with the fibrous, uncoated surface. The mineral pigment used is mixed with an adhesive so that the coating is held fast to the base paper. The chief mineral pigments used are blanc fixe – which is a synthetic barium sulphate – or satin white, a preparation of calcium sulphate and alumina. Because colour illustrations very often gain in appearance by being printed on a paper whose whiteness provides a sharp contrast, a great deal of attention is paid by the papermakers to providing the brightest white shades possible. It is therefore customary to add a whitening agent to the coating mix. One of the most commonly used is titanium dioxide, a brilliant white heavy pigment. Not only does titanium dioxide brighten the shade, but because of its high refractive index it serves as a very effective opacifying agent. High opacity is desirable in coated papers for a number of reasons. A great deal of expensive multi-colour printing is carried out on coated papers of various grades. The lighter the weight of paper on which such printing can be done the more economical the operation. At the same time, the thinner the paper the greater the danger of print on one side of the sheet showing through to the other. The more opaque the paper can be made at any given grammage, therefore, the more effectively can it be used, and while this is obviously important with almost all papers, it becomes especially so whenever the more expensive kinds of printing are being carried out.

The quality of coated paper, in terms not only of cost but also its response to printing (which is a function of the paper, the ink and the printing operation) depends largely on four factors: the furnish of the base paper, that is, the fibre from which it is made; the point at which coating is performed, i.e. on the papermaking machine or as a separate operation; the quality of the coating 'mix'; and the process actually used for coating. As papermaking, despite advances in technical and quality control, is still a craft, these factors may themselves be considered variables. Some years ago, when there was virtually only one form of coated paper, the so-called 'art' paper, the quality of the paper and effectiveness of the printing rested almost entirely on the nature of the raw materials used and the basic skills of the operative. The development of alternative methods of coating paper has blurred what used to be the clearly defined lines separating one quality of paper from another which traditionally were based on the selection of raw materials and coating methods.

The principle of coating is simple enough. A liquid pigment and adhesive, with whitening agent if required, is applied to a base paper. Some of the mixture is absorbed, and the paper is then dried and the surface smoothed by calendering.

Fig. 2.18 Trailing blade coater. The blade is being adjusted as the operator examines the coated surface of the paper.

a. Brush, air blade and trailing blade coating. The original method of application was by brush coating, one side being treated at a time. The web of paper passes over a drum, and the coating is applied by spray or by a revolving brush which is partially immersed in a vat of coating. The coating is then spread evenly on the base paper by a series of oscillating brushes: that nearest the coating vat comprises coarse hair and those successively further away from it progressively finer hair so that the coated surface is smoothed out as the paper moves forward. Surplus coating is sucked from the surface and the paper goes to a drying chamber where graduated increasing heat is applied while the web of paper passes in festoons slowly through the chamber and on to a final reeling machine. The surface of the paper at this stage is still too rough for printing and a separate calendering operation is performed to give the paper its final smoothness.

This method of coating and drying is not fast enough to accommodate the greatly increased throughput of coated papers, neither is it economical in use of time and space. Faster and more economical methods of both coating and drying are now in general use. These comprise both 'on' and 'off' machine installations; in the former the processes are an integral part of the papermaking sequence, in the latter they are separate from it. Either of the techniques to be described may be utilized on or off the papermaking machines and the quality of paper being produced on them may vary according to the intent of the manufacturer. A particular quality or recognizable characteristics are not necessarily associated with a given method of production. So far as coated papers are concerned especially, quality depends far more on the standard of the base paper, the particle size of the coating, the adequacy of the adhesive, and to a very great extent the speed of production and the weight of coating deposited, rather than simply on the mechanical method of production. Two methods of coating other than brush coating are commonly used and one specialized method of manufacture which is nevertheless of increasing importance. The methods are the air blade coater, the trailing blade coater, and cast coating (considered in section 2.6.5*b*).

Body paper to be coated by the air blade process passes through a vat which contains the liquid coating. On emerging, a blast of air removes the surplus liquid and produces a fairly even layer which is further smoothed by passage between calender rolls. The trailing blade coater, on the other hand, passes the web of paper through a vat in such a way as to create a 'lake' of coating in the angle of the web and a blade which is at right angles to the direction of flow. The trailing blade removes the surplus coating,

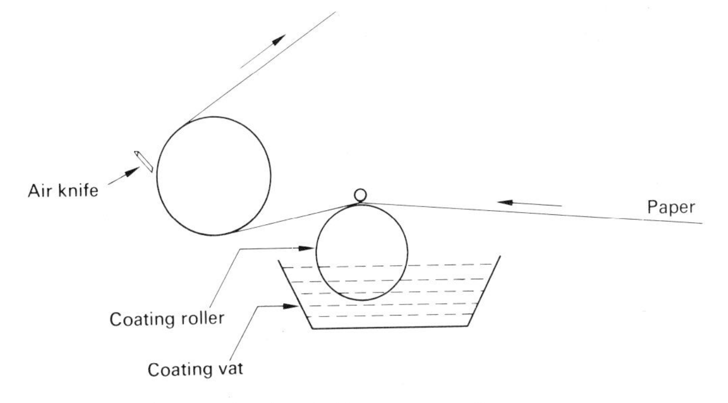

Fig. 2.19 Diagram showing air blade coater action.

leaving only the predetermined thickness necessary, which of course has to be calendered.

Although each of these two methods has its advocates, either will produce an adequate coated paper. The quality is determined by the speed with which the coating machines are operated and the thickness of coating deposited. It is generally reasonable to expect a higher quality of coating from off-machine operation than if the coating is done on the papermaking machine. This is because there is a practical limit to the weight of coating which may be applied satisfactorily during the high speed papermaking operation and close quality control is more difficult, if not impossible, to exercise when coating is only part of the whole papermaking process. When the same or very similar machinery is used as a separate installation, however, coating speeds may be slower, quality control more carefully exercized and a heavier coating applied. This is reflected in the higher price of the paper, which is nevertheless often acceptable for fine quality printing. When coating is done on the papermaking machine, drying is an integral feature of the operation. When off-machine coating is carried out 'festoon' drying is unsuitable. It has been found that there is a limit to the speed with which paper can be dried by this method, partly because of mechanical limitations and partly because of damage to paper which is largely unsupported during its travel yet has surfaces in close proximity to each other. Tunnel driers are used, the paper passing through them at speeds adequate to deal with the production flow. The driers consist of large chambers up to 50 metres in length through which the coated web of paper passes at speeds of up to 1000 metres per minute. The web is largely supported by currents of warm air and the temperature of the tunnel increases towards the end of the process. Apart from the speeds of coating and drying, which are vastly in excess of brush-coating, air knife or trailing blade coating combined with tunnel drying has the advantage of being able to coat two sides of the paper at once instead of one side only.

Paper coating processes have been developed to secure the greatest practical economy and maximum speed. These have to be combined with adequate quality control, however, and although on-machine coating has obvious economic merits a great deal is being done to make off-machine installations, comprising coating and drying machinery as integrated units, so fast and efficient that the difference in cost between the two methods is small. In these circumstances, the choice of method is made for reasons such as the space available, the flexibility required and overall efficiency.

A method of coating which is also coming into use is an adaptation of plastic lamination. The coating is extruded onto the surface of the base paper while it is running at high speed. The method is claimed to be suitable for extremely high speed operation and to be very flexible in its usage of different types of coating materials.

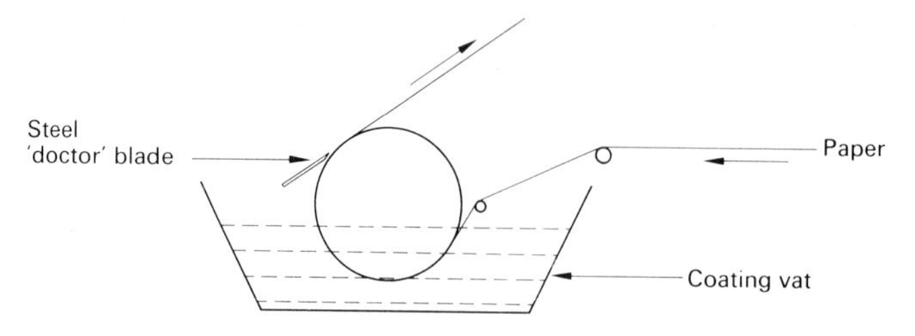

Fig. 2.20 A trailing blade coater.

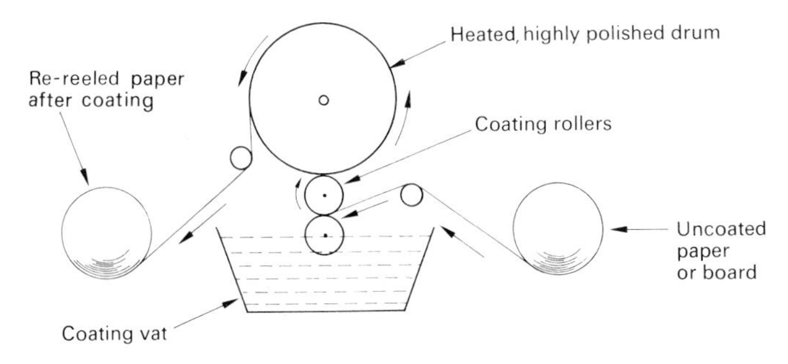

Fig. 2.21 Cast coating machine action.

b. Cast coating. This method differs in principle from the methods just described. Brush, air blade and trailing blade coatings require the application of mineral matter to base paper which then has to be calendered to produce the high finish required. Cast coating is the application of a wet coated paper to a highly polished heated surface and its removal when dry. The paper so coated reproduces the high finish of the surface with which it has been in contact and *so requires no further calendering.*

The web of paper passes between a lower roll and an upper cylinder, both of which rotate. The lower roll revolves in a vat containing the coating mix, a thin film of which is transferred to the base paper. The web then travels around the cylinder where it is pressed into contact with the polished heated metal surface. When dry the paper separates from the cylinder with a highly finished surface, reflecting that with which it has been in contact. Paper produced with a cast coating does not need calendering and has the advantages of retaining its bulk and being receptive to printing ink; calendering reduces the bulk and closes the surface of paper. Cast coating is especially suitable for the multi-colour printing papers necessary in advertising, in producing boards for cartons and boxes, and for luxury and display packaging. The coating can be applied to carton boards which would be too coarse to provide a suitable base for coating by other methods. On the other hand the process is much slower than the others and less suitable for the cheaper qualities of paper.

2.6.6 Tub sizing

As the name implies, tub sizing is the passing of a web of paper through a tub so that its surface is covered with a film of animal size, making it suitable for the reception of writing ink. The meaning of the expression has been broadened and changed to some extent so that it is now a general term for papers which have received a coating of size, on or off the papermaking machine. True tub sizing is a separate operation, carried out on a specific piece of machinery; it is also expensive. The closed surface it produces would make the paper unsuitable for some ordinary methods of printing; it is therefore usually employed only for the best writing papers and perhaps papers such as those produced for scientific instrument charts, which require particularly good surface resistance.

The tub sizer consists of a vat containing a liquid mixture of animal gelatine at 120°F (50°C) with a little alum (which contributes towards the hardness of the sized surface) and formaldehyde. The latter prevents mould growth. The web of paper passing through the vat will either be completely unsized or will have been slightly starch-sized on the papermaking machine. The paper is completely immersed and on passing out of the vat is squeezed between rubber rollers which remove the surplus size. The size-saturated paper then passes over a series of revolving 'spar' driers. These consist of slats through and around which warm air is circulated so that at the end of the machine the paper is dry and ready for sheeting or re-reeling. The size not only provides an excellent writing surface, but also penetrates the base paper, adding to its strength and durability. For economy, gelatine may sometimes be combined with soluble starch, but the strength of the tub-sized paper may then be much reduced.

Starch (derived from rice, maize, tapioca, potatoes etc.) is very often used for sizing papers. It may be added at the beater stage of manufacture or used as a surface size; in the latter case it is applied in a similar manner to surface coatings on the papermaking machine. The use of soluble starches is fairly widespread where a good writing surface is to be achieved at a reasonable cost and no gain in strength and durability is sought. Starch sizing may be carried out on tub sizing machinery, but is equally well and more economically performed on the papermaking machine, as mentioned above.

True tub sized papers ordinarily advertise the fact in the water mark and the expression 'tub-sized, air dried' is the indication that the paper has been fully processed. 'Tub-sized' by itself, however, may mean little more than treatment with one of a number of sizing solutions.

2.6.7 Gummed paper manufacture

The application of gum to paper is allied to coating and is often carried out in a similar way. Early installations utilized brush coating machines and festoon drying chambers but the modern procedure is to transfer the gum by bringing the paper in contact with a roll which is running in a bath of adhesive. The paper may then be dried by passing through a tunnel drier. The adhesive used depends on the intended purpose; it may be gum arabic as in the case of postage stamps or other stickers which adhere on moistening, or one of a number of dextrins chosen to provide a pressure-adhesive or heat-adhesive label. One of the problems in producing gummed paper is 'curling' and the aim of the manufacturer is to produce a paper which is flat and will remain so in all reasonable conditions. Curling is always a problem with papers coated on one side only because of the difference in moisture absorbency between the coated and uncoated side. With gummed paper the problem is particularly marked because, unlike mineral coatings applied for printing purposes, the gummed coating when dry is quite rigid. This means that the gummed side of the paper is not only more resistant to expansion or contraction because of atmospheric changes, but that the more substantial nature of some gum coatings when compared with mineral coatings makes the difference in dimensional change between coated and uncoated sides greater than in the case of coated printing papers. The problem is solved by 'breaking' the coating of gum so that it no longer constitutes a continuous surface. One method of doing this is to pass the gummed paper, after drying, over rollers set at angles of 45° to the direction of movement. By this zig-zag treatment the continuity of the gummed surface is broken, the paper is less intractable and it remains flat in use.

2.6.8 Finishing

'Finishing' embraces a very wide range of operations from simple cutting to impregnation or coating. The operations are conveniently divided into those which may be termed 'routine', i.e. which are necessary to prepare a paper for its eventual usage, and those which add something to the paper, which acts as a base. Among the routine operations carried out between the 'reel up' of the machine and the warehouse, are calendering, cutting, and sorting. These will be described below.

It was noted in the section on the dry end of the papermaking machine (section 2.6.2*c*) that stacks of calenders are often found after the drying cylinders, but for a number of reasons paper mills often install separate calender stacks. Thus, in the case of some older mills, space limitations may prevent installation on the machine itself; or the method of production may make it technically and economically undesirable to include calendering at the machine stage. Also, on some occasions the availability of a separate calendering installation enables a consignment of poorly-finished paper from the papermaking machine to be brought up to standard.

The principles of calendering remain the same, whether the operation is performed on the machine or separately. They are the smoothing and closing up of the surface of the paper. The web of paper is passed between rolls of varying diameter at high speed. The rolls are referred to as 'cylinder bowls'. It is normal for some of the rolls to be of steel and some of highly compressed paper. Because they vary in diameter the paper is burnished during its passage between the rolls. The calender bowls are arranged to subject both sides of the paper to the smoothing operation and paper of high finish can be produced from a base which has come from the papermaking machine with quite a matt surface. If they are part of the papermaking machine the calender stacks must be

Fig. 2.22 A super-calender stack. The speed and pressure is varied from the controls in the foreground.

able to accommodate the whole width of paper being made. If they are a separate installation they must at least be able to process a wide web of paper. It is apparent that the production of a uniformly burnished surface across the whole width of paper is difficult. If a calender stack consists of, perhaps, sixteen bowls, the pressure must be adjusted so as to progressively improve the surface of the paper being treated. Since pressure can only be applied at the ends of each bowl, the bowls themselves must be 'barrelled' so that the pressure on the paper is even across the whole width. The surface of the bowl must be as smooth and uniform as possible, as any blemish will create a mark or variation repeated throughout the whole of the run of a particular web of paper.

Paper varies in compressibility in accordance with its 'furnish', the manner in which it has been made, and the bulk of the paper being treated, and each of these factors has to be taken into account when deciding the calendering pressure. A paper which is 'springy' and 'spongy' will obviously present problems which are absent in the case of a compactly made non-resilient sheet. If too great a pressure is exerted the paper becomes 'blackened'; that is, as a result of pressure combined with the heat frictionally generated, its surface becomes glassy and with a discoloured mottling, not only unattractive in itself but also indicating that it has become so closed as to be unreceptive to printing or writing inks.

The cutting of paper in a mill comes under two headings: the 'slitting' of the machine reel into reels suitable for a reel-fed end-usage or simply to produce reels in the mill which can then be 'sheeted', and the cutting of the paper into commercial sheets. Consider the reels first. Even if a reel of paper of such a size were called for, the reel emerging at the end of the papermaking machine could not be used just as it is wound. When in use, all reels require even tension throughout the reel from the outside through to the core, even tension across the reel, with no 'soft spots' at any point, and accurate width of reel and even and smooth edges. None of these features is present in the machine reel, which must therefore be unwound, slit and re-reeled. The slitting machine ordinarily takes one machine reel and converts it into the appropriate number of smaller reels. Circular revolving knives trim the edges of the reel as it revolves, and additional circular knives divide it into smaller reels. During the slitting process the smaller reels are under constant tension so that the paper is drawn into an even, compact roll which, if necessary, can be used on any reel-fed converting machine, such as a printing press or a machine which produces continuous stationery. The conditions in which many papers on reels are used are highly critical and even small variations between one part of the paper and another can cause tearing, or difficulty in printing, leading to slower production and excessive wastage of paper. It is common in such circumstances to mark the reels to be used according to their position across the machine wire. Because the machine wire is so wide and the papermaking process fast and continuous, there may be small variations in the sheet formation, grammage, dimensional stability, etc., between reels cut from the 'back' of the machine – that is, the side of the drive – and the 'front' of the machine. While these variations may be apparently insignificant they are nevertheless critical when paper is printed at high speed on a reel-fed machine. The marking of reels gives the user the chance of preventing operating losses.

Not all paper is used on the reel, and an additional process is needed to produce the necessary sheets. This is the 'chopping' of the paper by means of a blade at right angles to the movement of the paper from the reel. In this way the sheets are only roughly cut to size. Many printing processes require sheets within limits of about 2 mm, and the swiftly moving cutting machine is unable to produce such close-tolerance sheets. The

Fig. 2.23 Paper being slit from large to smaller reels and rewound. The circular knives in the centre are adjustable for different paper widths.

sheets must therefore be cut by a guillotine, on each edge. Though accurate, guillotine trimming is wasteful of time and labour, and continuous cutting processes, slightly slower but more accurate than the chopping action, are now being developed to increase output while still achieving adequate dimensional accuracy. Cutting is quite a skilled operation and one which is designed to produce minimum waste. For this reason it is common to slit and sheet two or three reels at the same time, the webs being brought together to be slit and chopped simultaneously.

Grain direction – that is, the machine direction of the paper – is important in many printing and converting processes because of the different dimensional changes between machine and cross direction under the influence of atmospheric variations. 'Long grain' paper (paper in which the longer dimension of the sheet is parallel with the machine flow) is often specified. In order to utilize the full width of the papermaking machine, and not to have excessive waste when slitting, it is sometimes necessary to duplex cut or slit reels to different widths which will then be cut into sheets having long grain in different sizes. There is also the cutting of paper for envelope manufacture to be catered for, and as a great many envelope shapes cannot be made economically out of square paper, 'angle cutting' has to be carried out. In angle cutting, the chopping blade operates not at right angles to the movement of the paper but at a predetermined angle to produce sheets of roughly diamond shape so that envelope 'blanks' can be cut with very little further waste of paper.

To understand the sorting of paper and even perhaps the need for it, the nature of paper as a raw material must be appreciated. There are probably few, if any, other raw materials in which a comparatively small, even minute, blemish can have an effect so vastly out of proportion to its initial volume. For example, a lump of hard material, not more than 5 mm square, raised above the surface of the paper, could ruin a printing cylinder. The cylinder would become unusable and the printing machine idle, and the labour costs in effecting repairs and getting the machine going again could add greatly to the total expense. Again, a weakness in the coating of a printing paper no larger than the area already mentioned might cause some coating to come away during printing. Not only would the machine have to be stopped and cleaned but a large number of printed sheets might be blemished. In both cases the defect may be no larger than a thumb nail and a quite insignificant fraction of the whole consignment, yet the quality of the entire amount of paper may be judged by it. When paper is supplied on reels, to be used in the same way, there is little that can be done in the way of intensive examination. Quality control must therefore be of a general nature intended to prevent blemishes rather than to detect them. Sheeted paper can be examined closely, however, and most papers destined for good quality printing or writing purposes are examined sheet by sheet. Sorting of this kind takes place in the mill 'salle', a large well-lit area where specially trained girls scan each sheet with amazing speed, setting aside those that they consider sub-standard. The faults they look for are variations in shade either between sheets or between the two sides of the sheet, 'pinholes' in the surface, specks of any kind, holes in the paper itself, and any hard lumps of material standing proud of the surface. Visual examination is being increasingly replaced by machinery in which electronic scanners traverse the surface of each sheet of paper and reject those which are faulty. The weakness of such machines so far is that although some of them are excellent for detecting sheets which have holes or surface lumps, they have difficulty in assessing quantitatively the effects of specks and 'pin-holes' which would not stop or damage printing machinery, but which lower the quality of the paper being offered.

2.7 THE PAPER MILL AS A TECHNICAL UNIT

The requirements of papermaking are such that paper mills are usually established away from main industrial centres, although some mills which have been established for many years have been overtaken and surrounded by building development. The factors which have influenced the siting of paper mills and continue to do so are chiefly: the necessary size of the mill, access to power, the presence of abundant water, convenient supplies of raw materials, clean air and access to the markets being served. For economic reasons the operation of making paper must be carried out in large units. Not only is papermaking a large-scale operation, it is also a noisy, and in respect of the chemical cleansing and bleaching of fibre which has to be carried out, an odorous one, and papermaking usually takes place in buildings erected for the purpose in an area separated from other industry.

One of the prime needs of papermaking is an ample and continuous supply of clean water. Even a moderately sized mill will use hundreds of thousands of litres of water each day and historically the earliest mills were always established on the banks of rivers or streams. This sort of water supply was adequate for the limited production of 100 years ago or more, but would not serve a modern high-speed paper mill. The water for this is often drawn from a number of sources within quite a large area surrounding the mill, firstly because there are few local supplies adequate in themselves and also because a single source could prove unreliable. It might for instance, drop below the necessary level of flow or even dry up altogether in some seasons. It is not only the volumetric supply of water which is of concern; its cleanliness and quality are also important. Any discolouration of the water and any foreign matter it may be carrying are likely to affect the finished paper, and nearly all modern mills control the quality of their supply with the aid of a filtration plant.

One reason for siting papermills away from industrial complexes is the need for a clean atmosphere. To reduce the possibility of airborne particles getting into the paper an atmosphere as free from pollution as possible is desirable. This can most easily be ensured by locating the mill in an area remote from other industry.

Power, although enormously important, is probably not one of the first requirements to be considered because it can be produced in a number of ways, and coal, oil, and electricity can all be brought without great difficulty to quite remote districts providing the economics of the situation allow. Hydro-electric power generation is often feasible in Scandinavia and North America, and is attractive because the sites are usually rural and the water can be used for the mill processes.

Good access to raw materials is obviously essential, and if these have to be transported from a distance, say overseas, then proximity to a port may be desirable. An indigenous supply of fibrous material (e.g. wood for pulping) is most valuable and for this reason, as well as those mentioned above, well-established papermaking industries are to be found in Canada, Norway and other countries with large timber reserves. Access to markets is, generally speaking, not a great problem since the raw material zones are in the main convenient to the largest areas of industrial development. On a local basis, however, the operators of paper mills seek to have access to adequate transport facilities, nowadays usually provided by arterial road networks.

2.8 VARIETIES OF PAPER

It is not possible here to describe all the many different types of paper and board used in commerce, industry and the home, but some of the more important varieties are considered below. For convenience they can be classified as printing and writing papers; functional papers (other than packaging); and packaging papers. There is a certain amount of overlapping; for example, packaging materials often carry printed information whilst, although bituminized kraft paper (section 2.8.3*d*) is used largely for packaging, it has a functional role when employed as a weatherproof lining for roofs.

2.8.1 Printing and writing paper

Some of the principal varieties of printing and writing paper are described below. A number of these have already been mentioned in connection with paper manufacture, and others are referred to in section 2.8.3.

Antique papers. Bulky printing papers with a matt finish. Furnish may be esparto, chemical wood or, in the case of the cheapest grades, mechanical wood. Mostly wove but a small proportion of laid is also marketed.

Coated printing papers. These may range from expensive brush-coated art paper with esparto furnish through on and off machine-coated papers from chemical wood to cheap coated printings on mechanical base paper. They also include cast-coated papers and boards and highly-finished one-side coated enamel papers.

Bank and Bond papers. These are the 'bread and butter' qualities of writing papers – the distinction between Bank and Bond being their substance only. Bank is the description applied to paper up to 49 g/m^2 and Bond is descriptive of paper above that grammage. They may be of any quality from mechanical wood to a high-grade rag paper – from simple engine sizing to expensive tub sizing and are supplied in white and colours.

Bank note papers. These are always especially made to a very high standard of durability and strength. They would almost certainly contain a high proportion of rag fibre, probably linen with some hemp. To prevent forgery a number of security measures would be incorporated during manufacture. (See 2.5.4.)

Cartridge. This is the term applied to a range of rather strong and rigid printing papers which are used in the advertising field or in book printing where demanding illustrations are used. There has been a growth in the use of the twin-wire papermaking process to make 'cartridges' so that the paper offers similar printing surfaces on both sides of the paper.

Printing made from mechanical wood pulp. The most important of these is of course newsprint which has a high proportion of mechanical pulp and only a small percentage of chemical wood pulp to maintain the minimum strength necessary. Other 'mechanical printings' with varying proportions of mechanical and chemical wood in their finishes are used for magazines and periodicals or any printing where low cost and short life are requisities.

The best *tissue* papers are based on rag and hemp, but cheaper varieties are made from wood pulps. They are required to combine thinness and flexibility with adequate strength, and are used for interleaving and packing delicate goods. *Crepe* papers are similar but thicker materials crimped during manufacture to give greater apparent bulk.

2.8.2 Functional papers

Paper is used for such a wide variety of purposes apart from printing, writing and packaging, that only a few can be mentioned here. Although in some of the applications listed, paper is now meeting competition from other materials, especially plastics, it is likely to retain its popularity for many products which demand, as raw material, a cheap, readily available, easily fabricated flat stock.

A large amount of board and thick paper is used for the *binding* of cloth-covered and paper-backed books. A similar application is to be found in *files* and *document cases*, and in light pack-flat filing cabinets.

Brightly printed board can scarcely be rivalled for pack-flat *toys* and *games*, and for point-of-sale *advertising displays*. These are cut, embossed and creased as required, for final assembly by the user.

Paper *cups*, food display *dishes* and *baking cases* are all made from paper or board. The cups are pressed or crimped, and usually waxed or resin-coated. Baking cases, which must withstand oven heat, are formed by crimping a thin parchment type of paper to the required shape. Fruit baskets and similar pack-flat containers are made from stout board.

Tissue or creped paper is generally used for *toilet* articles such as serviettes, towels, handkerchiefs and toilet paper. These must be adequately absorbent yet strong and soft, and are made from acid-free sulphite pulp, unsized. Paper *underwear*, also a single-use, easily-disposable product, is designed to simulate the conformable nature of a woven fabric. As packed it has the great advantage of occupying very little space.

Plaster board, as the name implies, consists of a core of plaster faced on each side by board. Together, the hard, rigid, but brittle core and the softer, tougher, paper cladding form a laminated product widely used throughout the construction industry for ceilings, walls and partitions. A cellular partition comprising an 'egg-crate' core made of board, faced on each side with insulating or plaster board, has also been used for light walling. A similar idea is the use of resin-bonded and impregnated paper honeycomb as the core of lightweight sandwich structures (see also Volume 3, Chapter 12, section 12.4.4).

Another interesting application is found in the *electrical* industry, in the form of *insulation* for power and telephone cables (where the paper is bitumen- or oil-impregnated), for windings and for interleaving in condensers. The paper used for these purposes must satisfy stringent electrical requirements and be acid-free and mechanically tough. Manila is often used.

In addition to the above, brightly coloured or printed paper is used for such *decorative* and functional purposes as greetings cards, Christmas decorations, artificial flowers (now largely replaced by plastic), calenders etc.

2.8.3 Packaging papers

Packaging papers are required to fulfill a number of functions which are conveniently grouped under the following headings.

Protection of the contents from harmful external influences. These may include dirt and dust, light, moisture and the effects of atmospheric exposure; or protection may be required against shock or impact.

Containment of contents, as in the case of powders or granules, or the provision of a convenient means of keeping a number of objects together.

Separation of objects which could come to harm if in contact with each other. These include items such as glass, or polished objects which could be scratched by contact.

Support or stiffening to objects which should not be allowed to bend.

Provision of a *means of identification* and a way of directing, i.e. addressing.

Advertisement of the product.

Often the packaging paper is required to satisfy a number of these requirements at the same time so that in some cases the paper itself has to be treated or to be used as part of a composite material.

Packaging papers can be classified as follows:

(*a*) Those consisting of paper, pure and simple
(*b*) Papers treated during manufacture to obtain specific protective features
(*c*) Boards
(*d*) Papers which after manufacture are laminated to other materials.

a. Simple (untreated) papers. These include the following:

Krafts. These are of a rich brown shade and are manufactured with strength as the paramount feature. They may be either unglazed or M.G., and are produced from wood pulp by the sulphate process which is particularly effective in maintaining the strength of the fibre.

Bleached krafts. Manufactured by the same method but are bleached to produce a white shade while still retaining the strength required.

Imitation krafts. As the name implies, these are imitations of the true or pure krafts. There is no agreed definition for them. They are usually made from mixtures of sulphate pulp and waste materials and are dyed to simulate the colour of pure kraft. They are distinguished from pure kraft by being noticeably less strong and rigid.

Rope browns or browns. These are generic terms describing dark brown, bulky wrapping papers of rather unattractive appearance. They are made with a large proportion of waste materials in the furnish. Although not very strong their bulkiness is useful for some packaging purposes where it serves as a buffer against shock or impact.

Sulphite. A thin white paper which is almost always machine glazed and is used as a general purpose packaging paper: its name indicates the furnish. It is frequently found in shops, especially the food trade. Tinted M.G. papers used for very light wrapping may be included here. These are described as 'caps' or 'bottle wraps'.

Tissues. These are the other main group, either unglazed or M.G., and of a grammage not exceeding 30 g/m^2. Though light they are relatively strong and are suitable for the most delicate wrapping.

b. Treated papers. There is one property that may be conferred on a fairly wide range of papers when required; 'wet strength'. Wet strength is defined as the ability of paper to retain 30% of its dry bursting strength (see section 2.9) when it has become saturated with moisture. A typical application is to kraft papers, bleached or unbleached, when used in the manufacture of carrier bags. Any paper which is likely to be subject to wet conditions may be treated. The wet strength is obtained by adding melamine or formaldehyde resins to the pulp while it is being prepared for the paper-making machine.

Extensible kraft has a rougher surface then pure kraft and is softer to handle. Manufacture is modified to produce a stretch of about three times the normal for this type of paper so that packaging becomes easier and the paper is capable of providing a moderate 'buffer' against shock. A further development is *creped kraft.* This has a rough appearance and is capable of withstanding considerable twisting and bending without failure.

Vegetable parchment is a paper of similar appearance to greaseproof (see below) but the manufacturing process includes passing the web of paper through a bath of cold 70% sulphuric acid, which partly dissolves the cellulose fibres, forming a gelatinous mass. The acid is then removed by passing the web through vats of water, and neutralizing the residual acid in a dilute soda solution. The final treatment is in a bath of glycerine to prevent the paper being excessively brittle. Vegetable parchment has good resistance to penetration by water or grease and a high degree of chemical purity. It is therefore used extensively for the wrapping of fatty foods such as butter.

Greaseproof paper is manufactured from unbleached sulphite wood pulp. Its characteristic appearance and properties are obtained by especially long and severe beating which results in a pulp which is hydrated; that is, with a high water content which it does not easily part with. The name 'greaseproof' is misleading as the paper is not actually impervious to grease or moisture. It is resistant, however, and suitable for cheap packaging of moist or greasy substances.

The manufacture of *glassine* is similar to that of greaseproof. In addition, it receives super-calendering, making it almost transparent, and because of the closing of the surface it is possibly rather more resistant to grease or moisture.

c. Boards. Among boards used for packaging are box or carton boards and chipboards. The many varieties and combinations of types for particular packaging needs include the undermentioned.

White folding boxboard used for high class cartons such as are required for the perfumery or cigarette trades.

Duplex boxboards which are lined with white or tinted paper.

Chipboard, made largely from waste pulps, is used for cheaper cartons. As with other boards it may be lined with paper.

Strawboard, which is a thick board made almost entirely of straw. It has a characteristically yellow colour with many impurities showing on its surface and is noticeably brittle.

Corrugated boards. These may be either single or double faced; that is, with a flat surface on either one or both sides of the corrugations.

In addition to the foregoing, paper and board may be impregnated or coated with paraffin or microcrystalline wax to increase its stability to resist moisture.

d. Laminated papers. Typical of these is bituminized kraft or 'union kraft'. This consists of a 'sandwich' of two layers of kraft with bitumen in between. The paper is used for the heaviest types of packaging where strength and resistance to moisture is required.

Packaging papers (usually krafts) are often laminated to metal foils or plastic sheeting of various kinds to give protection against moisture, grease or the effects of atmospheric pollution.

2.8.4 Sizes of paper

Newspapers, and to an increasing extent long runs of magazines and books, are printed on rotary presses from continuous reels of paper. Here the size of the reel must be matched to the press, by arrangement with the paper supplier. For writing, and for most other types of printing, cut sheets are required. These are made up in quantities of 500 sheets (1 ream). In the UK the traditional sizes of printing and writing papers, described by names such as 'Double Crown', 'Large Post', 'Double Medium' etc. are giving way to sizes agreed by members of the International Organization for Standardization (ISO) of which Great Britain is one. These sizes, the so called 'ISO sizes', have already been adopted by many other countries.

Information on the sizes, substances and descriptions of papers and boards is contained in BS 4000:1968 *Sizes of paper and board.* This standard also describes the range of internationally agreed trimmed sizes of writing and printing papers promoted by the ISO. The ISO series is based on the metric system of measurement and simply means the rationalization of some thirty different paper dimensions that were previously used. It establishes three standard sheet sizes, being designated 'A', 'B' and 'C' and their subdivisions. Full sheet sizes are coded: A0, B0, C0. Subsidiary sizes obtained from these basic sheets are *suffixed* by continuing numerals, thus A1, A2 etc; B1, B2 etc; C1, C2 etc. Each of these subdivisions is achieved by dividing the immediately previous size into equal parts, the division being parallel to the shorter side, so that the areas of the two successive sizes are in the ratio 2:1. Thus in the 'A' series, A0 (full sheet size) is 841 × 1189 mm; A1 (half size is 594 × 841 mm; A2 (quarter size) is 420 × 594 mm; A3 (eighth size) 297 × 420 mm and so on. For sheet sizes larger than the basic size the letter is *prefixed* by a numeral, thus 2A is twice A0 (i.e. 1189 × 1682 mm). For general purposes A4, A5, A6 and A7 are the most common of the international sizes used, and it is the 'A' range that is probably the best known of the ISO series of paper sizes.

The dimensions given above, and those that have not been specified, are of trimmed sizes. For technical reasons applying to the printing industry – allowance for 'gripping' when machining; provision for trimming or extra trim for 'bleeding' work – untrimmed stock sizes have been introduced from which 'A' sizes can be cut. These stock sizes are designated RA0, RA1 etc. for normal trim, SRA0, SRA1 etc. for extra trim or 'bled' work.

The principal advantage of the ISO system is that every size in the main and subsidiary ranges has the same proportions, facilitating enlargement or reduction of printed matter, diagrams and so on by photographic means. Because it is more rational than the existing multiplicity of sizes current in the UK the ISO system is being widely adopted in the UK paper trade and elsewhere.

2.9 TESTING PAPER

2.9.1 Testing manually

The paper trade together with many users of paper maintain laboratories and trained personnel whose task is to examine the physical characteristics of materials submitted for test. In addition to the instrumental methods (see section 2.9.2) used in the laboratory, it is useful – indeed necessary – to apply quick, simple and easy tests which merely rely on the eye, the nose, the hands and the tongue. Although the maximum information can only be gained by long experience, with a little practice it is possible to become reasonably proficient at judging the general quality of a piece of paper in relation to its price and purpose. A closer look will reveal whether there are obvious surface blemishes, whether specks of dirt are present in the surface and if the finish is satisfactory.

To judge finish by eye alone the paper should be held at eye level horizontally against the light and a sight taken along the surface. This will show not only if the surface is up to standard but also if there are any sub-standard areas, as these will show dull against the light in contrast to the rest. Both sides of the paper should be examined and if a comparison between two sheets is necessary they should, if possible, be held side by side, examining felt side with felt side and wire side with wire side. The two sides of the paper can also be compared for uniformity, texture and colour. Papers to be examined should always be placed side by side and not placed one partly above the other. If this is not done light can be reflected from one surface to the other and a false result obtained.

The nose is a sufficiently sensitive instrument to detect the presence of gelatine size once its rather sharp aroma has been experienced although it is possible the paper will have to be held very close to catch the scent unless a very heavily animal tub-sized paper is being examined. Casein, when used in coated papers, is another aromatic substance which can be recognized if the paper is held to the nose.

The hands and tongue are possibly the best indicators of certain features. The degree of sizing of the paper can be judged by the manner in which it sticks to the tongue. A soft-sized paper when applied to the tip of the tongue sticks noticeably to it and when drawn away displays a mark where the moisture has penetrated the paper. The degree of adhesion and the distinctness of the mark are indications of the amount of sizing the paper has received. The less the paper sticks to the tongue the harder the sizing. Some experts claim to be able to taste the presence of animal size in this way by a distinguishable salty reaction.

By feeling paper between the fingers a number of impressions can be received. Its thickness is the first and most obvious one; finish also can be judged, and with the addition of the visual examination a fairly accurate idea of the quality of the surface. Bending the paper and rattling it when held between both hands gives an indication both of its strength and quality. The harder the sound of the rattle and the more intractable the paper feels, the greater its strength is likely to be and the better the quality. In performing this particular test account must be taken of papers which are soft by nature (blotting or filter papers for instance).

The tearing of paper is possibly the most commonly practised test and the most valuable one. One edge of the sheet is torn across to the extent of about 30 to 50 mm, then the tear is continued for the same distance but at right angles to the original tear. Alternatively, the second tear is made separately on an edge of the paper at right angles to the first; the important feature is to have two tears at approximately 90° to each other. The resistance felt when tearing indicates the strength of the paper. By tearing at right angles a comparison can be made between the machine and cross direction of the paper. The machine direction is indicated by an easily obtained straight tear while the greater resistance to tearing in the cross direction results in a much more jagged and uneven tear. A wide discrepancy in resistance would indicate a unidirectional sheet, with a preponderance of fibres oriented in the machine direction. Such a paper would not be well formed and when being printed or otherwise processed stresses could build up which could lead to distortion and unsatisfactory behaviour. The sound which accompanies the tearing is itself a guide to the furnish of the paper. A soft gentle sound could indicate an esparto furnish or one of the shorter-fibred wood pulps; a harsh sound associated with a strong resistance to tearing would suggest a rag, hemp, manila or specially beaten strong long-fibred wood pulp. The edges of the tear themselves when held to the light show the fibre from which the paper is made and give a fairly clear picture of the furnish, formation and general quality.

2.9.2 Testing by instruments

The tests carried out on paper in laboratories are more complicated, thorough and searching than is possible by the rule of thumb methods just described. In addition, they enable the properties of papers to be measured, compared and recorded so that a body of experience can be built up. Reliance should not be placed on the results of a single test for any given property. Several samples should be tested so that an idea of the variability of the paper may be gained.

a. Weight. Described in the paper trade as grammage or substance, this is the simplest test to be carried out. If a whole sheet of paper is available it may be weighed on a quadrant scale. This will probably be graduated in grammes per square metre although some of the older scales will also show the weight per 500 sheets of paper in a known standard size. If the latter, the weight will probably be expressed as a ream weight; i.e. 20 × 30 in, 45 lb, 500 s. If only a small piece of paper is available recourse will be made to a 'demy' scale which weighs a small square cut from the sheet with the aid of a template. The weight in this case is usually shown in grammes per square metre only.

b. Thickness. To ensure equal pressure in all tests a dead weight micrometer is used for measuring thickness. Large bench models and small pocket instruments are available. Expression of thickness has varied in different countries according to local practice but there is a growing international tendency to state this dimension in micrometres or microns (μm). As the area over which the micrometer is applied is very small more than one measurement should be taken, and preferably six to ten readings on each of a number of different sheets. If a bench micrometer is used it is sometimes possible to measure the thickness of ten sheets at one time and report an average figure for a single sheet. This practice is suitable for smooth papers or boards but is not recommended for soft papers or those with a rough finish, where the compressibility of the paper could affect the result obtained.

c. Bulk. Bulk is not the same as thickness but is a term used in connection with papers such as blotting paper or those made for some types of book printing, as distinct from highly finished, compressed papers. Bulky printing papers are described as 'featherweights' or 'bulky antiques'. Bulk is the inverse of density.

d. Furnish. The surface or cross-section of a sheet of paper may be examined with the aid of a microscope, but to determine the furnish the constituent fibres should be separated. Although for detailed examination high magnification is necessary a great deal of effective and perfectly adequate analysis can be made at a magnification of 100. The first stage is to gently disintegrate a small sample of the paper in a suspension of water, taking care to avoid damage to the individual fibres. There is a marked difference in the microscopic appearance of fibres before and after treatment in the paper mill and it is most important that the general condition of the fibre to be examined should not be altered by lack of care during the preparation of microscope slides. Fibres knot naturally during manufacture, and it is necessary to tease them out for proper identification. As most papers are of mixed furnish it is necessary to carry out as many tests as is practical so that the proportion of each fibre in the furnish can be assessed visually. The fibres are translucent under the microscope and although there are physical features which aid identification much use is made of stains which react differently to particular fibres and so make the examination more reliable.

Hertzberg stain is probably one of the most generally applied and useful. It is made by mixing 25 ml of a saturated solution of zinc chloride with a solution of 0.25 g of iodine and 5.25 g of potassium iodide in 12.5 ml of water at 20°C. The mixture should be allowed to settle for some hours and then turned into a dark stoppered bottle. Hertzberg stain deteriorates and should not be kept for more than eight to ten weeks. When a drop of Hertzberg stain is applied to the fibres being examined the following colour reactions are obtained–cotton fibre, linen, hemp: red; chemical wood, esparto, straw: purplish blue; mechanical wood, jute and some semi-chemical pulps: yellow. The colour reaction depends on the amount of lignin remaining in the fibre, the purer cellulosic fibres giving a purplish blue to red colour whereas the lignous fibres produce yellow.

Phloroglucinol stain is the 'ever ready' means of detecting the presence of mechanical or ground wood pulp. Phloroglucinol stain is made by dissolving 4 g of phloroglucinol in 100 ml of alcohol and 50 ml of hydrochloric acid. The stain when applied to paper produces a red colouration the strength of which can be related to the amount of ground wood present – the deeper the colour the greater the percentage of ground wood. It is a popular way of judging the proportion of mechanical pulp in a paper but is sometimes unreliable because some semi-chemical straw pulps give the same colour reaction and phloroglucinol will also be affected by some dye stuffs. If necessary, therefore, a check can be made by staining with aniline sulphate.

Aniline sulphate, as a solution of 2 g in 100 ml of water containing one drop of sulphuric acid, makes a stain which shows deep yellow when in contact with ground wood, light yellow when in contact with unbleached sulphite or sulphate pulps or pink when applied to a paper with an appreciable esparto content.

In conjunction with the physical appearance (shape, size) stains applied to fibres on a microscope slide or to a piece of the paper itself will allow a reasonable estimate of the furnish to be gained. These tests, together with those for weight and thickness, comprise the basic inspection to which almost all papers are subject either as part of quality control during production or on receipt by the user. Information on many other characteristics is likely to be needed in the case of particular papers or usages; the most important are described below.

e. Strength. Strength is by no means only associated with packaging or industrial papers. Most printing or even writing papers are subject to a certain amount of strain during printing or any other conversion, and apart from this the strength and uniformity are an indication of quality. Four strength properties are associated with paper: *tear strength*, *bursting strength*, *tensile strength* and *folding strength*. Their importance varies with the type of paper.

Tear strength is the resistance to the force necessary to tear through a strip of paper once the tear has been started. A number of instruments have been designed for testing the tear resistance of paper but all operate in a similar manner. A piece of the paper to be tested is held in two sets of jaws which are drawn apart until the paper tears. The resistance to the movement is measured and recorded as the tear strength of the paper. As with other tests, several samples taken from different parts of the consignment should be examined. In this way not only will an idea of the strength of the paper be obtained but also any variations in strength between one part of the sheet and another. The tear test is sometimes called the internal strength test rather than the tear test.

Bursting strength is by far the most commonly assessed strength property, both as a part of production control and also by printers, paper merchants and other users. In fact some users, who do not relate their findings to other characteristics, place too much faith in it. Bursting strength is tested on the 'Mullen' type hydraulic burst tester. This machine holds a piece of paper about 150 mm square in a clamp while pressure is applied by inflating a rubber diaphragm. The pressure applied is recorded on a dial on the instrument, which stops operating the moment the paper bursts. The machine is made in a number of models, some of which are suitable for general use and others for special materials, such as thin paper, strong paper, boards, etc.

Tensile strength is examined by applying a steadily increasing load to a strip of paper, usually about 200 mm long and 15 mm wide, held between two clamps one of which is drawn steadily away from the other until the paper breaks. The force necessary to rupture is recorded. Most modern instruments also record the amount of elongation before rupture. All tensile tests should be done on several samples, in some of which the grain of the paper is parallel to the pull and in others at right angles.

Folding strength is especially important in papers which are likely to get continual handling over a long period, such as bank notes or ledger cards. A strip of the paper to be tested is put in an instrument having two clamps which move towards and away from each other very rapidly so that the paper held between them is continually bent through 180° and then drawn straight, until it breaks. At this point the number of folds is recorded on the instrument.

Strength tests are especially subject to variable results unless the atmospheric conditions are controlled. Both tensile and folding strength increase greatly in conditions of high relative humidity and it is essential that tests which are intended to ascertain quality or to provide comparative information are carried out in similar conditions. Ideally, this entails the use of a humidity-controlled room as is the practice in paper mills. If tests are carried out in uncontrolled conditions a wide discrepancy in results must be expected.

Testing for the degree of *sizing* is fairly frequently carried out. Sizing is a complex characteristic and a very large number of different methods are used to assess it. The Cobb test is the one in commonest use, and indicates fairly accurately the degree to which a paper will absorb water. It is carried out by clamping a short metal cylinder with a cross-section of 100 square centimetres over the paper to be tested. Water at 68°F (20°C) is poured into the cylinder to a depth of 1 cm. After 45 seconds, measured by stop watch, the water is poured off and the paper removed, and the latter allowed to continue absorbing moisture for a further 60 seconds. It is then blotted so that surplus water is removed. By weighing the paper both before and after the operation the amount of water it has taken up can be measured in milligrammes and expressed as a 'Cobb figure'.

For most purposes a hard-sized paper, i.e. one having a low water take-up, is to be preferred, but soft sizing is desirable for blotting paper, filter paper and similar absorbent products. It will also be appreciated that papers are sized to resist different types of penetration. Water-based inks, used in writing, and oil-based printing inks react differently. Hence many tests have been designed which apply to specific needs and are not in common use, but the Cobb test is of general value and is widely understood.

f. Filler content. Fillers or loadings in paper are necessary to achieve printability and ease of handling. They do not contribute to the strength of paper and no more should be present than is necessary to produce the required effect. For instance, a filter paper or a blotting paper would be expected to have a low loading content whereas a paper which was intended for multi-colour printing, and which needed the loading to provide an even printing surface, could properly have a very high proportion of loading. The amount of loading in a paper is determined by burning it under controlled conditions with full access of air and measuring the remaining ash. One gramme of paper is weighed and then ignited in a furnace at approximately 850°C. Only a few minutes ignition is necessary to destroy the fibrous content of the paper. The residue is simply ash. The ash should be completely grey – any traces of blackness indicate incomplete combustion. The test must be performed with great care; for instance, the ash must be cooled in conditions which preclude the absorption of moisture. The ash is weighed and expressed as a percentage of the air-dry weight of the original paper; this gives the quantity of filler or loading, the amount of ash in the fibre itself being negligible.

For great accuracy allowance must be made for the amount of moisture in the filler but this is not necessary for routine quality control.

g. Acidity. Ideally paper and board should be neutral. Ordinarily, however, it is slightly alkaline or acid, usually the latter. Acidity is undesirable not only in a packaging paper, where it may effect the object to be wrapped, but also in printing papers. The drying of certain printing inks is slowed by the presence of acid. This is of especial concern in lithographic printing, where moisture is necessarily present, leaching out some of the acid from the paper. Acidity is also important in papers for industrial purposes – the papers made for electrical cables and windings have to be particularly carefully controlled in this respect. Acidity and alkalinity are measured on the hydrogen ion (pH) scale, the figures of the scale being the negative indices of the hydrogen ion concentration. The neutral point is at pH = 7, whilst a low figure = acid; high = alkaline. Simple and accurate electrical instruments are available for estimating the pH value, or indicators which depend on the pH sensitivity of dyestuffs, such as the BDH universal indicator, may be applied. Any test should be carried out, not on the paper itself but on a water extract, allowance being made for the presence of fibre particles and colouring matter. As is usual with all paper testing, pH determinations should be carried out on several samples.

h. Dimensional stability. The ability of a paper to remain reasonably stable in varying atmospheric conditions depends on a number of factors. The most important of these are its furnish, that is the fibre from which it is made, the manner in which it is made (i.e. whether it is a close or open sheet), the proportion of loading to fibrous matter, since mineral loading is largely unaffected by atmospheric change, and of course, the amount of moisture already in the paper. Dimensional variation differs according to the direction of the paper, the change in the cross direction being five times greater than that which occurs in the machine direction. This effect can be reduced by papermakers but not eliminated entirely.

Tests for dimensional change are simple in essence but require considerable experience in interpretation. Strips of paper are cut both along and across the grain; some of the strips are for control purposes and the others are to indicate changes. They are totally immersed in water for a period of some minutes, sufficient to ensure complete saturation. After immersion the difference in dimension between the immersed strips and the controls is measured and expressed as a percentage. An alternative method is to mark the largest circle possible on the piece of paper to be tested and then to expose the paper to different conditions of relative humidity. If the same circle is then marked again using the same centre, the difference between sizes of the concentric circles will be the measure of dimensional change. Laboratory instruments are also available on which tests of the nature described may be carried out and recorded, such as the Pira expansiometer produced by the Paper and Printing Industries Research Association.

i. Coating. The coating of a paper is tested for three characteristics: smoothness and uniformity of finish; strength; and adhesion to the body paper. Finish is assessed in the manner described for visual examination but the strength of the coated surface has to be assessed mechanically. Such tests are designed to discover whether the coating will break up in use and whether, if it is sufficiently strong in itself, it will break away from the body paper under stress. For these purposes a series of wax sticks, known as

Dennison waxes, can be used. Each wax stick has a different number which indicates a degree of adhesion. The hot wax sticks are applied to the surface of the paper and drawn away sharply at right angles. When cool it will be seen that the coated surface breaks at a certain number and not before. The cavity formed by the stick is then examined to see whether it is the coating only which has come away or if the removal of the coating has also broken into the body paper. In the first case it might well be judged if the 'Dennison' number is low, that the coating is weak, whereas the second type of cavity would suggest that although the coating was adequate the base paper itself was weak.

A more sophisticated test and one which can also be used to judge the way in which the paper is likely to behave when printed is that performed on the Institute Grafic Technologie (IGT) tester. The IGT instrument consists of a narrow drum on which is fixed a strip of the paper being examined. The end of the strip is brought in contact with a small roller which has been inked with a printing ink of known viscosity. The drum is sharply revolved for part of its circumference and a strip of the printing ink is applied to the paper. The composition of the printing ink is controlled, also the pressure between the drum and the inked roller. The action of the drum causes the printing ink to break the coating and possibly break into the body paper. Since the velocity of the drum increases as it revolves, this will occur progressively along the strip of paper. A hard coating or satisfactory adhesion of coating to body paper yield a long strip of printing ink before damage begins, a weaker coating or base paper being indicated by earlier break up. The IGT test is especially useful for coated papers but is also used extensively to measure the surface characteristics of all papers, coated or uncoated, and for the examination of the ink/paper relationship.

j. Finish. Finish may be tested either optically or mechanically. Several tests are in common use and the selection of any particular one depends on the paper and purpose of testing. Optical testing relies on the reflection of light from the surface of paper and the degree to which it becomes diffused. There are a number of instruments which direct a beam of light on to the surface at an angle and measure the amount and nature of the reflection. It is possible to be misled by the results obtained, since the beam is reflected from a flat surface which nevertheless contains small pin holes or pits and these, while small in proportion, may make the paper unsuitable for some purposes. Hence, when examining finish it is often desirable to test by mechanical means also. The instrument ordinarily used, of which there are a number of types, consists of a vertical cylinder mounted on a firm base. A circular piece of the paper being examined is placed on the base within a ring and the cylinder is rested on it. Air is then introduced into the cylinder under controlled pressure, so that it escapes between the ring and the paper–that is, through the channels caused by the natural irregularities in the surface of the paper. It follows that the greater the irregularity of the surface, the quicker the air will be able to escape. The results of such tests are expressed as the number of minutes taken for the escape of a known quantity of air. There are obvious limitations to this method. Once again it expresses the total area represented by the channels without taking account of the manner in which they are distributed over the total surface. The practice in testing for finish is therefore to carry out two or three tests, both optical and mechanical, and to interpret the results in the light of experience.

Microscopic examination of the surface, though valuable, is carried out far more in connection with research and development projects than in routine quality control or comparative examination. It is a most valuable means of compiling information which can be used to improve standards.

k. Colour. Colour should be examined by daylight, preferably a north light in the northern hemisphere and a south light in the southern hemisphere. As daylight is not always immediately available or is variable, colour testing may be done under artificial illumination especially designed for the purpose. A combination of tungsten filament and fluorescent lighting units is used so as not to influence the colours being matched or tested. Without illumination properly designed for colour matching, inaccurate results will be obtained. For instance, papers containing optical dyestuffs to brighten their appearance may in fact appear very leaden and dull under tungsten light whereas a strongly ultra-violet light would cause them to glow. Whether the light source is natural or artificial, papers being compared should be examined with the light coming from above and behind the examiner; in the case of natural light with the back to the window and not with the paper between the examiner and the light.

The samples must be compared wire side to wire side or two felt sides; that is, care must be taken to ensure that the top side of one sample is not compared with the underside of another in the case of single wire papers. The samples being tested or compared must be moved about in relation to each other so as to minimize the effects of small differences in incident light or visual acuity of the observer.

Instruments are available which direct a beam of light on to the surface of the paper, separate the reflected rays into their constituent wavelengths and record them so that a colour curve may be produced. They are effective and can be used as a means of building up information on the relation between mill variables and colour. Such techniques are not claimed to be greatly superior to visual examination for routine commercial inspection.

2.10 THE PHYSICAL CHARACTERISTICS OF PAPER

Because of the nature of the raw materials, the manufacturing method and the requirements of the customer, all paper is a compromise, and the physical characteristics can only be understood with this in mind. For instance, a paper could be manufactured to be outstandingly strong, but the type of fibre used, the treatment it received in the beater and the operation of forming it on the papermaking machine would almost certainly result in a material which would be difficult to handle and which would not run easily on a printing machine. Strength and docility are likely to be opposing features, and a balance must be struck.

A similar situation exists in the case of brightness of white shade, finish and opacity. If shade is not very important, colourings can be used which also act as opacifying agents. Most of the substances which influence shade do not improve opacity, however, and do not help to prevent 'show-through'. Finish is often a part of this problem. If the surface of the paper is low (i.e. rough), its irregularities, although insignificant in writing or printing terms, diffuse the rays of light reflected from the surface in such a way as to present an opaque appearance. If the surface of the paper is high (i.e. smooth) on the other hand (perhaps burnished by the calendering process) the light rays pass right through the sheet. The paper is then translucent and incapable of accepting printing or writing on both sides without marked show-through. Thus, both the brightness of the sheet and its finish oppose opacity. If their influence is countered by the use of loading materials as is ordinarily the case, a balance must then be struck between the need for opacity and the possibility of the paper being too flabby because loading makes virtually no contribution to the strength of the sheet. The 'bulk' of paper, that is its thickness when not compressed, is also a factor directly related to finish. The production of a smooth surface on paper by means of calendering entails considerable pressure and the smooth surface is obtained only at the expense of thickness. Two papers of similar grammage but with different surfaces will have very different thicknesses. It follows that to obtain a paper in a particular thickness but with a high finish it is necessary to specify a heavier weight than for a low finish paper.

Some of the most careful papermaking calculations have to be made when strength, absorbency and a particular finish are all required in the same specification. Strength is associated with length of fibre, absorbency with a short fibre, frequently of a type which is not particularly suitable for producing strong papers, whilst the high finishing of paper closes its surface both by compression of the sheet and by the pressing down of fibres and surface spreading of loading materials to present a 'skin' on the sheet which resists the penetration of its surface by inks or liquids. These requirements are not easily compatible, and the user should be aware of the limitations within which the paper manufacturer must work.

Paper reacts to the atmospheric surroundings, especially in respect of the amount of moisture it takes from or gives up to air around it. This can have an effect on the actual size of the paper as the component fibres swell or contract. A paper which is made with high absorbency for a particular purpose, or one which has an open surface not highly finished, is likely to be particularly sensitive to changes in atmospheric conditions. The causes and effects of this sensitivity merit closer study.

The sensitivity of paper to changes in humidity can be reduced where necessary and economically feasible. The use of fibres which, by their structure, are less liable to pick up or relinquish moisture and their treatment at the beating stage to render them less hygroscopic can result in a paper with the minimum propensity to expand or contract. Complete elimination of dimensional instability is, however, not possible. If a balance exists between the amount of moisture in the paper and that in the atmosphere, so that there is no tendency for the paper to either absorb the moisture from the surrounding atmosphere or for moisture to leave the paper, the paper will, if otherwise well made, remain flat and dimensionally stable. If the paper is now exposed to moist conditions it will become fractionally larger because of the swelling of the constituent fibres. In the case of some papers even a few minutes in an atmosphere containing an appreciable excess of moisture over that in the paper will result in an expansion which can be measured without sensitive instruments. The paper will give up its moisture to a dryer atmosphere equally quickly and will then contract. It follows that exposure to varying conditions over a period of time can result in continual dimensional changes. These are likely to be non-uniform because of small variations in the paper and the interchange of moisture. The stresses applied to the structure of the paper will distort it permanently so that it assumes a wavy condition. As the fibres from which the paper is made take up moisture they expand. Because each fibre is longer than it is broad, it is more greatly altered in circumference than length. There are always more fibres lying in one direction than the other in any sheet of machine-made paper because of the orienting effect of the process. Any sheet of paper will therefore vary more in the cross direction of the sheet than in the grain direction. Although dimensional variation can be reasonably accurately predicted for most known types of paper, it is the constant aim of the papermaker to reduce wherever possible the susceptibility of paper to such changes.

Similar conditions may apply where there are noticeable variations in the temperature to which the paper is exposed. The temperature itself may have little effect but the associated changes in the relative humidity of the atmosphere can quite suddenly adversely effect the stability of the paper. (Relative humidity is the ratio between the amount of moisture in the air and the amount which would be necessary to saturate it it any given temperature.) As the amount of moisture air will hold increases with its temperature, a rise in the latter can create 'tropical' conditions in which the paper may absorb more moisture.

To a certain extent even the tensile and folding strengths of papers can be affected by the amount of moisture in the paper. This is especially so in the case of folding strength, which increases as the paper becomes more flexible because of its moisture content. In the case of a coated paper, however, the conditions which increase the strength of an uncoated sheet may weaken the strength and adhesion of the coating. Although the coating itself provides some insulation between the base paper and the atmosphere, problems can arise in the case of papers which are coated on one side only. Because one side of the sheet is so insulated, any atmospheric change is felt mainly on the uncoated side. If the atmosphere is dryer than the paper, the flight of moisture will cause the paper to contract on the uncoated side, so that curling occurs towards the plain side. On the other hand, if there is an inflow of moisture from the atmosphere to the paper, the curling will take place in the opposite direction because the uncoated side of the sheet will have expanded more than the coated and insulated side. The curl in this case would be with the coated side inwards.

Quite apart from curling and cockling, a difference of 1% in size, which is quite possible, though not visible to the eye, would be sufficient to present problems in printing or converting. Thus in colour printing, if the size varies between one colour working and the next, the 'register' of the print–that is the exact relationship of one colour to another–may be inaccurate. Absorbency and dimensional stability are closely connected and if for a certain purpose an absorbent paper is specified, the likely variation during and after processing must be borne in mind. A paper mill marketing in a highly arid atmosphere, such as exists in certain parts of the United States, would not take as its 'norm' the same conditions as those generally prevailing in the United Kingdom for instance, and mills marketing in humid tropical or sub-tropical atmospheres would make allowance for the high degree of relative humidity to which their paper might be exposed. As a general rule papermakers aim to despatch papers from their mills with a moisture content very slightly above that of the atmosphere in which it will be used. As an example, the moisture content of papers used in the United Kingdom would probably be about 7%. Material destined for other parts of the world with a very different climate would need to be produced with a moisture content appropriate to that climate, and then sealed for transit by being packed in waterproof paper to ensure insulation from atmospheric changes.

2.11 STATISTICS

2.11.1 Production and consumption of paper throughout the world

The per capita consumption of paper in any country is one of the indicators used

Table 2.1 Production of paper and paperboard, 1970

Country	*000 metric tons*
United States	47 599
Japan	12973
Canada	11655
USSR	6704
Federal Republic of Germany	5516
United Kingdom	4979
Sweden	4359
Finland	4258
France	4134
Peoples' Republic of China	3750
Italy	3451
Netherlands	1567
Norway	1421
Democratic German Republic	1158
Spain	1103
Brazil	1085
Austria	1017
Czechoslovakia	819
Switzerland	731
Poland	725

by economists to assess the standard of living. In general the larger the amount of paper used the higher the standard. Printing and writing papers, newsprint and packaging papers, are all included here under one heading to give a total figure of paper consumption for each member of the population.

Table 2.2 Per capita consumption of paper and board, 1970

Country	*kg per annum*
United States	251.6
Sweden	191.0
Canada	180.5
Switzerland	154.0
Denmark	148.0
Netherlands	138.0
United Kingdom	129.0
Federal Republic of Germany	125.0
Japan	121.6
Norway	120.0
Finland	114.4
New Zealand	114.4
Belgium/Luxembourg	112.5
France	94.5
Austria	79.7
Ireland	78.6
Hong Kong	75.0
Democratic German Republic	75.0
Panama	66.0
Costa Rica	65.4

Tables 2.1 and 2.2 show the 20 countries producing the greatest tonnage of papers of all types and the 20 countries with the highest per capita paper consumption.

The disposition of papermills throughout the world and their relationship to the amount of paper produced in each country is shown in table 2.3. Some idea of the state of development reached by papermaking industries in different countries may be inferred by relating the number of mills to the total production. A large number of mills achieving only a modest production suggests an industry which has not yet developed to its fullest potential. A smaller number of mills producing a larger amount of paper at least indicates that a proportion of the countries' papermaking capacity is well established.

Table 2.3 Paper production compared with number of mills, 1970

Country	*Paper production 000 metric tons*	*No. of paper mills*
America		
United States	47 599	809
Canada	11 655	133
Europe		
Albania	8	4
Austria	942	72
Belgium	723	39
Bulgaria	249	36
Czechoslovakia	842	45
Denmark	237	10
Finland	4 058	43
France	4 005	207
German Democratic Republic	1 001	79
German Federal Republic	5 182	229
Greece	140	10
Hungary	251	10
Ireland	112	5
Italy	3 428	600
Malta	—	—
Netherlands	1 592	35
Norway	1 352	44
Poland	889	59
Portugal	193	124
Romania	477	17
Spain	1 044	257
Sweden	4 110	67
Switzerland	688	36
United Kingdom	4 960	198
USSR	6 235	191
Yugoslavia	596	28
Asia		
Cambodia	5	1
Burma	0.5	2
Sri Lanka	8	1
China	3 490	60
Cyprus	—	—
Hong Kong	—	—
India	751	57
Iran	19	9

Table 2.3 (Continued)

Country	*Paper production 000 metric tons*	*No. of paper mills*
Iraq	—	1
Israel	74	5
Japan	11 290	662
Jordan	3	1
North Korea	59	5
South Korea	292	79
Lebanon	26	3
Malaysia	12	2
Mongolia	1	—
Nepal	0.5	200 (small)
Pakistan	135	12
Singapore	1	1
Syria	1	2
Taiwan	361	113
Thailand	52	10
Turkey	131	13
North Vietnam	4	4
South Vietnam	22	5
Latin America		
Argentina	583	100
Bolivia	0.5	1
Brazil	961	156
Chile	254	26
Colombia	191	10
Costa Rica	—	1
Cuba	104	7
Dominican Republic	7	1
Ecuador	7	1
El Salvador	1	1
Guatemala	11	2
Mexico	820	47
Panama	8	3
Peru	110	11
Uruguay	35	12
Venezuela	229	14
Others	—	5
Oceania		
Australia	1013	17
Indonesia	20	6
New Zealand	440	5
Papua–New Guinea	—	—
Philippines	108	18
Fiji	—	—
Africa		
Algeria	38	8
Angola	6.5	3
Cameroun	5	1
Congo (Leo)	1	2
Ethiopia	0.5	2
Ghana	1	1
Kenya	3	1
Libya	4	2

Table 2.3 (Continued)

Country	*Paper production 000 metric tons*	*No. of paper mills*
Liberia	0	—
Malagasa	6	1
Morocco	60	5
Mozambique	1	1
Nigeria	8	1
Republic of South Africa	557	15
Rhodesia	29	2
Swaziland	—	—
Sudan	3	2
Tunisia	4	3
Uganda	—	1
UAR	132	12
Tanzania	—	—
Zambia	—	—

2.12 LITERATURE

CARTER GILMOUN (Ed.). *Paper. Its Making Merchanting and Usage.* London, Longman, 1965. (New edition in preparation: editors. E. Dean and E. Haylock. Longman/ National Association of Paper Merchants).

R. H. CLAPPERTON. *Modern Paper Making.* Oxford, Blackwell, 1952.

Paper Making. Technical Section of the British Paper and Board Makers Association.

JULIUS GRANT. *Laboratory Handbook of Pulp and Paper Manufacture.* London, Arnold, 1961.

DARD HUNTER. *The History and Technique of an Ancient Craft.* [Paper Making] Pleiades Books.

CHAPTER 3

Natural Fibres

3.1 INTRODUCTION

An abundance of fibrous materials occur in nature and, over the centuries, man has examined these materials to ascertain if they could be used as textile fibres from which he could fashion his clothing and furnishings.

The natural fibres vary considerably in their basic properties and the ease with which they can be extracted and brought into a suitable form for textile use. Over a long period of trial and use, the advantages of some fibres have been recognized, while others have been rejected for a variety of reasons. This selection, based on long experience, has resulted in only a few of the many fibres occurring in nature becoming regarded as being really important, with a small number of others finding some limited application.

There are two main types of natural fibre used in textile manufacture, vegetable and animal, classified according to their origin. For the sake of completeness it should be mentioned that reference is sometimes made to a third class, mineral fibres, but only one of these, asbestos fibre, can be said to be used for textile purposes, namely to manufacture a very limited amount of fire-proof fabric. Mineral fibres will not, therefore, be included here.

The vegetable fibres which have been selected for textile use are employed in nature for a variety of purposes. Some are developed by a plant as a means of providing buoyancy for its seeds, to which they are attached, so that as maturity is reached the seeds will be carried in the wind and widely scattered. Others serve to stiffen and brace the stems and leaves of certain other plants. Again, fibres are used to provide strength and toughness in the stems of plants which otherwise would not be able to withstand the storms and high winds to which they are subjected from time to time.

In the case of animal fibres, many of those used by the textile industry form the coverings which provide warmth and protection for certain types of animal, while others are extracted from the coverings or housings which an insect produces as a means of protection during one stage of its existence.

As the fibres occur in nature, they are quite unsuitable for immediate textile use. For example, when cotton fibres are picked from the plant, there are seeds still adhering to the fibres, and these have to be removed. In addition, the cotton plant picks up dirt while it is being cultivated and some of this adheres to the fibres. Further the fibres become entangled and matted during picking and baling, and a considerable number of preparatory operations have to be carried out before cotton is ready to be spun into a yarn. The fibres obtained from the stems of plants, such as flax, are contained in the inner bark of the stem where they are cemented in position by natural adhesives; a whole range of treatments is required before the fibres are freed and in a condition to be used to manufacture yarns. The animal fibres also require a good deal of preparation before they can be used. Wool, for instance, exists in the fleece as a tangled mass of hairs containing substantial quantities of dirt and grease, and the fibres have to be cleaned and arranged in an orderly manner before a yarn can be prepared. Silk must be softened and reeled from the cocoon after a treatment to ensure that the moth inside is destroyed before it can eat its way through the strands which surround it.

All this preparation and treatment to bring the fibres into a form in which they can be used to make textile articles is expensive, and the greater the difficulty of preparation for a particular fibre, the less attractive commercially will that fibre be. It will be appreciated, therefore, that whether or not a particular fibre becomes a major textile material must not only depend upon the usefulness of its basic properties, but also upon the cost of preparation.

The characteristics which are required in a natural fibre to make it attractive commercially for textile use may be listed as follows:

1. To be of any value at all, a fibre must be very long in comparison with its width. Very short fibres are difficult to bind together in the form of a yarn and offer considerable spinning problems.
2. Once the fibres have been twisted together in a yarn they must hold together strongly and resist sliding apart. The fibres must exhibit, therefore, some surface 'frictional' effect, one against another. In cotton, this is provided by a natural twist in a flat, ribbon-like fibre; in flax there are swellings or 'nodes' at fairly regular intervals and a fibre length greater than in cotton; in wool and other animal hairs there are scales on the outside of the fibre which tend to interlock with each other; while in silk, fibre length, slight irregularities in diameter, and a natural gum on the surface of the fibre increase the adhesion within the yarn.
3. Softness to the touch of a piece of fabric is often associated with smallness in diameter of the fibres composing it. For softness of handle and a quality appearance in the fabric, the fibres should be fine and uniform in diameter.
4. The fibres are subjected to severe stresses during the preparative, spinning, and fabric-forming processes, and the fibre must be strong to withstand these, and also provide strength in the final fabric.
5. Flexibility in a fibre is important in order that it will withstand the many bending motions it will have to endure during the preparation of yarns and fabrics, and to provide a fabric with good draping characteristics.
6. A good elasticity is required in a fibre since this provides 'give' in the fabric made from it and ensures long, comfortable wear.
7. The fibre must be free from dyeing difficulties and must absorb and retain dyestuffs uniformly.
8. Reference has already been made to the need for a fibre which is easily extracted and cleaned. If it is difficult to clean and purify, a fibre will not be able to compete in price with others which are offered.
9. An obvious requirement is that the fibre must be available in sufficiently large

quantities to make it a commercial proposition. Each fibre needs its own preparative and manufacturing techniques and these will only be developed if sufficiently large supplies are available, and if any demand which will be created can be satisfied.

In the following sections, detailed information relating to the more important natural fibres will be given. Included in this information will be figures relating to the various properties of an individual fibre and which, to a large extent, determine its usefulness for textile purposes. Information on the tests from which such figures are obtained is given in section 3.9. Figures quoted in relation to the production and consumption of the various fibres are taken from *Industrial Fibres–A Review* which was prepared and issued by the Commonwealth Secretariat in May, 1969.

3.2 COTTON

3.2.1 History

Cotton is still the most important, and most widely used textile fibre in the world. It is also one of the oldest. It is known that the Ancient Egyptians and early civilizations in China used cotton fabrics, and samples of cotton materials have been found in tombs in India dating as far back as the year 3000 B.C.

India is credited with being the earliest region of cultivated cotton. As early as 500 B.C., Alexander the Great transported cotton from India to Egypt and other Mediterranean countries, and there are references to the planting and cultivation of the cotton plant in India in 350 B.C. Over the centuries India was to build up a thriving export trade in cotton and cotton goods to all the important countries of that time.

Cotton growing became established around the shores of the Mediterranean during the time of the Roman Empire, and flourishing cotton manufacturing industries became established in a number of countries in that area.

The importation of cotton fibre, and the establishment of a manufacturing industry for cotton goods, made very slow progress in England. The wool merchants were opposed to the importation of raw cotton into the country and the manufacture of cotton goods, and had succeeded in obtaining legislation which limited them. In 1736, however, the Manchester Act was passed and removed the restrictions, and Lancashire began to make its effort to take the lead in cotton manufacture.

As regards the development of cotton cultivation in the Americas, Columbus had found when he landed in the West Indies in 1492, that cotton was already extensively cultivated there and the preparation of cotton goods well established. Magellan found that the people of Brazil were using cotton for making clothing, and Pizzaro discovered in 1522 that the inhabitants of Peru were wearing cotton clothing. The earliest reference to the cultivation of cotton in North America is to the growing of the plant in the State of Virginia in 1620; references can also be found to cotton-growing in South Carolina in 1664 and in Georgia in 1735.

The systematic, and large-scale growing of cotton in North America was established in the 1700s. At the beginning of that century, the difficulty found in separating the cotton seeds from the fibre on any large scale, and the slowness of the methods available for making cotton goods, tended to discourage the large-scale production of cotton. In the period covered by the years 1730 to 1770, however, a number of remarkable inventions relating to textile machinery were developed in England and offered a means of producing high quality cotton goods more rapidly and cheaply. Further, the 'cotton gin' was invented in 1793, and provided a rapid and reliable method of removing the seeds from cotton. These inventions provided both the means whereby North America could increase her output of cotton and the incentive for her to do so. By the year 1800, North America was exporting 18 m. lb (8 m. kg) of cotton a year to England and by 1811, this had increased to 62 m. lb. (28 m. kg).

3.2.2 Botanical information

The botanical classification of the cotton plant places it in the Mallow family. The particular plants in the cotton sub-division of this family are called *Gossypium*. There are many varieties of cotton plant, but most of them are of no commercial value.

The usually accepted botanical classifications for the important commercial cottons are:

Gossypium barbadense now grown in the West Indies and producing the high quality fibre known as Sea Island Cotton.

Gossypium hirsutum is supposed to have originated in Mexico, and the commercial cottons of the United States classed as the 'Upland' type are a form of *hirsutum*.

Gossypium herbaceum is said to have originated in India, and this type is now grown in India, China, Persia, and Russia.

Gossypium peruvianum is sometimes classed as being *barbadense*, but is more likely to be related to *hirsutum*.

The cotton plants classified as *barbadense* and *peruvianum* were both grown in Egypt during the development of the Egyptian type and, while some people think that the present Egyptian plant was developed by crossing these with native types of cotton, it would appear to be more closely allied to the *barbadense* type.

The fibres grow inside the seed pods of the cotton plant; after the cotton seeds have been sown and begun to germinate, the plant shows above the ground with two leaves of peculiar shape, which later fade and drop off, between which the stalk grows. The first flower buds show in about seven weeks after the seeds have been planted, and these buds take something like two to three weeks to grow and open. In the meantime, the plant has continued to increase in height and width and new buds are formed and bloom. The flowers are creamy-white and turn pink by the end of their first day. During the growing period, the cotton plant likes warm or hot weather, both day and night, and progress is arrested if the nights are cold. A moderate amount of rain is also necessary. On the third day, the flower withers and dies, leaving a small flattened green seed pod which is referred to as the cotton 'boll'.

The boll gradually increases in size and expands, until it has become egg-shaped with dimensions of from 1 to $1\frac{1}{2}$ inches (25–38 mm) in length, and about 1 in diameter. It takes from 45 to 60 days for the boll to expand and mature, after which it splits open into three, four, or five compartments, each filled with a mass of white fibres.

During the period in which the boll is expanding and maturing, hundreds of tiny fibres appear on the surfaces of the seeds. These 'hairs' grow slowly at first, and then begin to increase in length and diameter much more rapidly. After about two weeks of rapid growth, the rate slows up again and finally stops altogether. At this stage the fibres are in the form of thin-walled tubes of cellulose, attached to the seed at one end and closed at the other; the tubes are filled with liquid. When the growth in length stops, the fibre begins to add layers of cellulose from the inside to give strength, and this building-up of layers of cellulose goes on for three to four weeks.

Fig. 3.1 A cotton boll.

As the fibres grow in length and wall-thickness, they become tightly packed inside the seed case until the pressure developed is sufficient to burst open the boll. The fibres stand out when freed and the liquid they contain dries out in the air, causing the cell walls to collapse and leave a ribbon-like structure which is the mature cotton fibre. Each fibre is twisted lengthwise forming convolutions which are characteristic of cotton. In their twisted, flat-tube form, the fibres are in an ideal condition to fulfil their natural function, namely to act as light-weight 'streamers' which will be caught by the wind and carry the seeds attached to them over considerable distances.

There are between seven and ten seeds in each compartment of the boll when it opens, and each seed is covered with a large number of cotton fibres. The entire contents, fibres and seeds, of between 45 and 90 bolls are needed to make up a weight of 1 lb.

All the bolls do not mature at the same time. Bolls may be opening over a period of eight or nine weeks, and the total time between the first flowers appearing and the last bolls opening may be as long as four months.

The ultimate yield of cotton is affected by the conditions of growth of the plant before it begins to flower; a check in its growth at this stage will result in short fibres being formed. The conditions of growth of the plant during the time that the fibres are developing inside the boll influence the quality of the cotton produced; a check on growth then will give thin-walled fibres. There will always be some immature fibres present in the final mass of fibre, and the ratio of immature to mature fibre is an important factor in determining the quality of the cotton produced.

If a mature cotton fibre is examined in cross-section, it will be found to consist of five parts:

1. The outermost layer is the 'cuticle'. This is a thin waxy layer which serves to protect the rest of the fibre.
2. Next comes the 'primary wall'. This is the original thin cell wall of the fibre and consists mainly of cellulose.
3. Inside the primary wall come the 'secondary cellulose layers'. These are the layers which are deposited daily from the liquid inside the fibre while it is thickening up during the period between the fibre ceasing to increase in length and the opening of the boll. These layers form the main portion of the fibre.
4. Going inward still further, the wall of the 'lumen' or central channel of the fibre is reached. This wall is more resistant to chemical attack than the secondary layers.
5. Finally, in the centre of the fibre, is a hollow channel called the 'lumen'. This extends along the whole length of the main part of the fibre.

Examined longitudinally, there is first a base, which is a short cone-shaped portion at the root of the fibre. Next comes the body part of the fibre, which takes up 75% or more of the total length, has thick walls, a central lumen, and is convoluted. There is a final tip to the fibre, this short portion having no lumen, few convolutions, and a considerably smaller diameter than the body part, and terminating in a cylindrical tapered end.

The natural convolutions or twists which it shows distinguish the cotton fibre from other fibres. The frequency of the convolutions varies widely, but an average number is probably 150 half-convolutions to the inch. The direction of twist often reverses. Any agent which swells the cotton fibre will, at the same time, reduce the number of convolutions.

An important dimension, from the textile point of view is the length of the fibre, referred to as its 'staple length'. Cottons are classed according to the staple length. 'Short staple' cotton has a length of between $\frac{1}{2}$ and $\frac{7}{8}$ in (12–20 mm), 'medium staple' between $\frac{3}{4}$ and $1\frac{1}{2}$ in (17–38 mm), and 'long staple' between $1\frac{1}{2}$ and $2\frac{1}{2}$ in (38–63 mm). The shorter staple fibres are inclined to be thicker than the longer stapled, the latter being finer, softer, and more highly convoluted. Fibre fineness is important in determining the final application of a type of cotton. The grading of the various types of cotton produced today in order of increasing thickness of fibre, is roughly Sea Island, the finest, followed by Egyptian, American Egyptian, Delta, Upland, Indian, Chinese and Peruvian.

The present commercial varieties of cotton plant have evolved as a result of careful selection over a period of many years. The repeated plantings, year after year, from the same seed crop can result in some characteristics, present in the original plants, being changed. This is described as the seed 'running out'. These changes in characteristics may be brought about in a number of ways:

1. When different varieties of cotton are growing in the same area, insects may cause cross-pollination. Even if the characteristics of the original type are not worsened by the cross-pollination, mixed pollination and non-uniform cotton fibre from the plantation can result.
2. If cotton growers send their raw cotton to a central establishment for 'ginning' it is possible that the machines are not cleaned properly and freed from seeds from one

grower's cotton before another grower's fibre is loaded into them. Mixing of types of seeds can take place in this way.
3. Poor soil conditions and lack of satisfactory cultivation methods can cause a deterioration in quality of the cotton produced.

For these reasons, the cotton farmers prefer to use new seed each season which is purchased as 'pure strain' seed from growers who produce it under carefully-controlled conditions.

3.2.3 Sources of supply and cultivation

Cotton growing is a major industry in many of the countries of the world. In general, it can be stated that the cotton plant requires plenty of moisture and sunshine during the growing period, followed by a dry period to enable the cotton fibres to mature and be collected. Such conditions exist in a number of countries and in some, where the rainfall is not sufficient, irrigation provides the additional water required. The cotton plant has proved to be able to adapt itself to somewhat varying climatic conditions, but there are certain types of cotton which appear to flourish best in particular surroundings.

The USA provides about one-third of the world's output of raw cotton. Next, in order of amount produced, come the USSR, India and Pakistan, China, Egypt, and Brazil. Lesser quantities are produced in Mexico, Argentina, Peru, and in parts of the South African continent and the Near East.

Most of the US produced cotton is grown in the Southern States. America has a vast growing area; over 2 million farms produce cotton as their main crop and it is estimated that about 500 000 tons (508 025 t) of seed are planted each year.

The climate in Egypt, coupled with good irrigation, enables excellent cotton to be grown, second only in quality and fineness to Sea Island cotton. Great care is taken to preserve the characteristics in the varieties of cotton produced in Egypt.

In India and Pakistan, a wide variety of cottons are produced, differing quite markedly in quality and staple length. The best types are grown in Southern India, and poorer quality cotton is produced in Northern India.

The finest cotton in the world, the so-called Sea Island type, is grown in the West Indian islands. This cotton has a long staple length of from $1\frac{1}{2}$ to $2\frac{1}{4}$ in (38–57 mm), is soft and silky, and is strong. The finest commercial cotton yarns are processed from Sea Island cotton.

In Russia, the Soviet Government is encouraging an extension of cotton production. The aim is to produce, within the country, all the raw cotton needed for home consumption. Both the quality and amounts of cotton grown in the USSR are increasing from year to year.

There are areas in Brazil where the conditions are suitable for growing cotton, and quite substantial quantities are now produced there. Medium and fine qualities of cotton are grown and quite large quantities are supplied to Britain.

In Peru, the major part of the cotton produced is of a type known as 'Tanguis'. The fibres are white, smooth, and of comparatively long staple length.

The cotton produced in China is mostly intended for internal use. Quite substantial quantities of a good average quality cotton, which is white in colour and has a staple length of about $\frac{3}{4}$ in (17 mm), is grown.

In the USA, the cotton seeds are sown in the spring. Mild weather is needed during planting, with a certain amount of rain to germinate the seeds. Cold weather will prevent the seeds from germinating and, if prolonged, can cause the seeds to rot in the ground. Cotton cannot withstand frost. The seeds are usually sown in drills at a depth of about 1 in (25 mm). The rows of plants are usually 3 to 4 ft (0.914 to 1.219 m) apart. When the seedlings have grown to a suitable size they are thinned, or 'chopped' to provide a suitable amount of growing space to the selected plants. How the plant continues to develop until the cotton fibre is ready to be picked, has already been described.

The cotton is picked in the autumn. Ideally, the weather should be warm and dry over the picking season and, for cotton fibre to be at its best, it should be picked before it has been rained on. Rain dulls the fibre and can stain it by transferring colour from the leaves and hulls. Strong winds are also detrimental since they can scatter the cotton from the boll and carry dust on to the fibres. Most of the cotton produced throughout the world is picked by hand and even in the USA, where labour is costly, most of the crop is hand-picked. An experienced adult can pick about 300 lb (136 kg) of cotton plus seeds in a day, and it will be appreciated that, with the enormous output of raw cotton each season in the USA, the provision of an adequate supply of labour during this relatively short period represents a very serious problem. Mechanical picking machines have been devised and used. One such machine consists of a box-like structure on wheels, with an open front, which is drawn along over the row of cotton plants. At the lower part of the front is a series of closely-spaced fingers and, as the structure is drawn along, the bolls, open and closed, are trapped between the fingers and are pulled off the plant and deposited in the box. Unfortunately, the machine also pulls off a consider-

Fig. 3.2 Hand-picking Egyptian-type cotton in the Sudan.

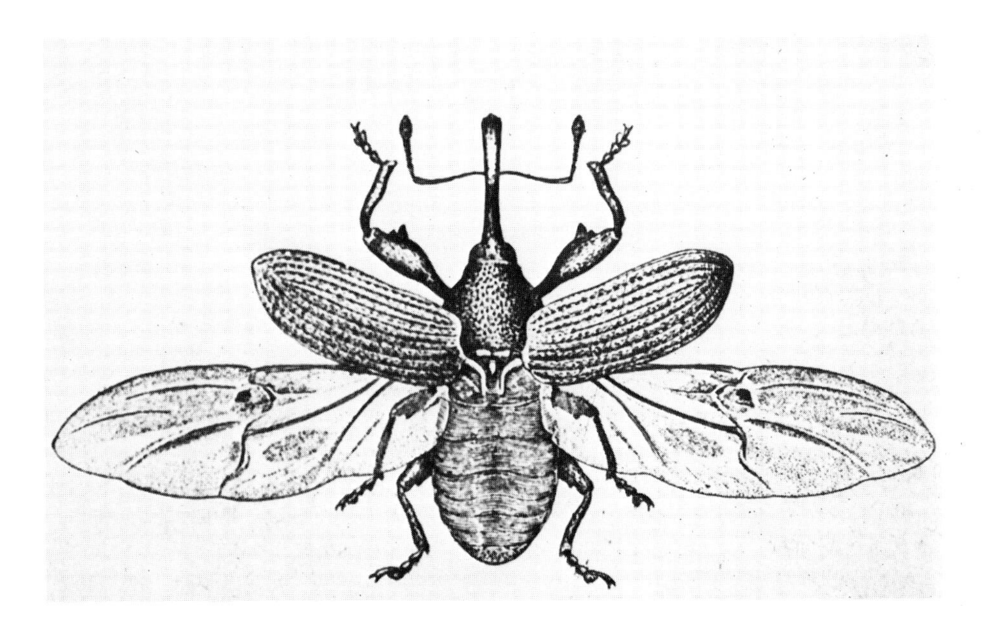

Fig. 3.3 Boll weevil.

able quantity of leaf and branch at the same time, and this becomes entangled with the fibre and is difficult to remove during the processing. Another type of picking machine rides high over the cotton plants and can operate on one or two rows at the same time. Two large shoes are attached to the machine and form a V-shaped opening with each other. As the machine moves along the rows of plants, the gathering shoes guide the branches carrying the bolls into a relatively narrow area and, on either side of this narrow space, are revolving spindles which are sometimes barbed. The fibres projecting from the bolls are caught by the revolving spindles and are torn away from the plant. The fibres are stripped from the spindles by the machine and deposited in a large container. Again, the objection to this machine is the 'trash' it collects at the same time which is difficult to remove from the fibre later. Thus, although hand-picking is costly it is usually employed in order to preserve the quality of the fibre which will be offered for sale.

The cotton plant is very subject to attack by insects and diseases. A small beetle, the boll weevil, is responsible for substantial losses in the cotton crop each year. The weevil feeds on the buds and bolls of the plant and, having penetrated into the boll, it lays its eggs there so that the maggots which hatch out can feed on the boll. The weevil can reproduce and spread at an alarming rate. A good deal of research and investigation has been carried out on the pest, and cultivation methods have been improved and special insecticides used. In India, Egypt and Brazil, considerable damage to the cotton crop is done by another insect, the pink boll worm. The larvae which hatch out from the eggs left by a small brown moth on the cotton plant bore into the bolls and eat the seeds. Cotton attacked by this particular pest is often stained pink. There are other insect pests which feed on the cotton plant in different countries, and adequate precautions must be taken against them.

There are a number of plant diseases which attack cotton. 'Root rot', which develops mostly in alkaline soils, is caused by a fungus in the soil which attacks the roots of the plant and prevents it from receiving an adequate supply of moisture; the leaves turn yellow and wilt, after which the plant dies. Crops resistant to root rot are available, and the soil can also be treated with disinfecting chemicals. Two other diseases arising from the presence of fungi are 'cotton wilt' and 'cotton anthracnose'. Other diseases, caused by the presence of bacteria, also menace the cotton crop, killing any parts of the plant which become infected. Finally a disease called 'rust' is caused by unfavourable soil conditions, such as lack of humus and potash, and failure to provide satisfactory drainage; under such conditions the plants do not develop properly and the leaves curl and wither.

3.2.4 Preparation of raw cotton

When cotton is picked, the seeds are still attached to the fibres and have to be removed. The process by which this is achieved is called 'ginning' and the machine which effects the separation, a 'gin'. The seeds which are broken away from the fibres then go to seed crushing mills where the cotton 'linters', very short fibres still attached to the seed case, are removed before the crushing operation takes place. The linters are much too short in length to be used in the production of cotton yarns, but they do form a valuable source of cellulose used in the manufacture of rayon fibres. The ginning operation is usually carried out by a ginner who has installed the necessary machinery in a central factory serving a growing area.

Difficulties may be experienced if ginning is attempted before the cotton is dry. The mass of fibres can be dried out by being stored in shallow piles in sheds or barns, but many ginners have added drying towers at their factories in which damp cotton can be treated and have its moisture content reduced to the required value. Many cotton producers also consider it worth the extra expense involved in having the seed cotton cleaned and freed from as much foreign matter as possible, since this improves the quality of the cotton and increases its market value. In the cleaning machine, the mass of seed cotton is 'opened' and agitated so that the foreign matter is shaken loose and falls away through a screen in the bottom of the machine. Nowadays, therefore, a ginner usually offers drying and cleaning facilities in addition to the gins which sever the seeds from the fibres.

Two types of gin are in use today, the 'Saw Gin' and the 'Roller Gin'. The saw gin is mostly used in the American cotton-growing districts, while the roller gin is preferred for treatment of the fine, long-staple cottons such as Sea Island and Egyptian, since it is gentler in action and does not produce so much broken fibre.

In the saw gin, a series of circular saws is fastened to a central shaft which rotates below a steel grating in which are narrow slits corresponding to the saw positions. The teeth of the saws show through the slits and catch the seed fibres, which are fed on to the grating, and pull them through the slits. The width of the slits is smaller than the diameter of the seeds and the fibres are torn away from the seeds. The fibres are removed from the saws by means of a revolving brush, and the current of air created by the rotation of the brush carries the fibres down a chute and into a box placed to receive them.

In the roller gin, a wooden roller is covered with leather. Seed cotton, fed on to this leather covering, clings to it and, as the roller revolves, is carried through a gap between the leather surface and a 'doctor knife'. The slit between the two is adjusted to allow the fibres, but not the seeds, to pass through and the seeds are removed, therefore, at the knife.

From the gin, the cotton fibre is usually delivered into a 'press box'. As the fibre falls into this box, it is continually lightly beaten down by means of a mechanical 'tramper'. This compresses the mass of cotton fibre to some extent and enables one box to hold a substantial weight of fibre. When sufficient cotton has been placed in it, the box is transferred to a position under a heavy hydraulic press where the cotton is compressed to form a dense package. Tie bands are then applied and the bale of cotton is ready for shipment.

3.2.5 Properties of cotton

The cotton fibre is a single plant cell, one of the largest known in plant life. All raw cotton has a creamy colour, which is slight in American Upland type, a little deeper in Sea Island, and still deeper in Egyptian cotton. Raw cotton has a natural lustre, probably arising from the structure of its outer casing, and some of this lustre may be destroyed during subsequent chemical treatments.

The lengths of individual cotton fibres can vary considerably, depending upon the type of cotton being examined. Thus a Sea Island fibre may be $2\frac{1}{2}$ in (63 mm) long and a linters fibre can be less than $\frac{1}{4}$ in (6 mm) long. There may also be appreciable fibre length variations within one particular type of cotton. Lengths will be influenced by the conditions under which a particular plant has been grown, and the degree of maturity of the fibre at the time of picking.

Under the microscope, the cotton fibres show as flat, collapsed tubes or ribbons with frequent twists or convolutions along their lengths. In cross-section, most of the fibres have a collapsed tube shape with a central elongated channel or lumen running along the fibre. A few thick-walled, and mercerized cotton fibres have a more nearly round cross-section. The lumen represents quite a considerable portion of empty space in the fibre and it enables the cotton fibre to absorb water by capillary attraction, thus influencing the textile characteristics of the material.

The density of cotton will vary somewhat with the type examined and the number of immature fibres the sample contains. A usually accepted value is 1.54 g/cm^3.

Cotton has a marked affinity for water. Its moisture regain, determined at 65% R.H., is about 8%, and its water retention about 50%.

Cotton is a fibre of moderate strength. Its tenacity is between 3.0 and 5.0 g/denier (27.0 to 45.0 g/tex) and its initial modulus is 30 g/denier (270 g/tex). The absorption of water by cotton produces an increase in tensile strength, and fibres saturated with water can be 20% stronger than dry fibres. Cotton is a relatively inelastic, rigid fibre. After a 2% extension it has an elastic recovery of 74%; after 5% extension, an elastic

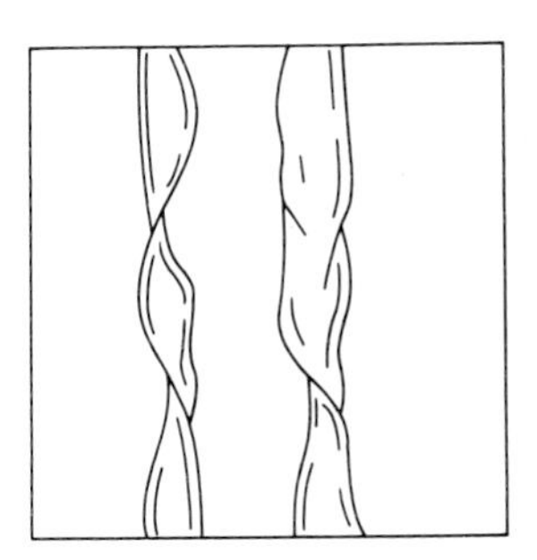

Fig. 3.4 Cotton fibres are flat and ribbon-like and show characteristic twists; cross-sections are collapsed tubes.

recovery of 45%. The extension at break of cotton fibres lies between the limits 5 and 10%. (See section 3.9 for the terms used to describe natural fibres.)

Cotton burns readily, leaving little ash. As the temperature is raised, cotton turns yellow after prolonged heating at 120°C (248°F) and oxidizes and decomposes at 150°C (302°F). Decomposition is rapid round about 200°C (392°F).

Cotton gradually yellows after long exposure to sunlight and loses strength. The deterioration appears to be caused by ultra-violet light and the shorter wave-lengths in visible light. Some protection against deterioration on exposure can be provided by a suitable selection of dyestuffs with which to dye the fibre.

The scoured and bleached cotton fibre used in the preparation of yarns and fabrics consists of about 99% cellulose. Cotton is resistant to most of the chemicals it is likely to encounter in normal use. The fibre is, however, attacked by strong oxidizing agents, including hydrogen peroxide and chlorine bleaching agents, and by acids. Cotton is inert to the common organic solvents. The resistance of the fibre to alkalis is excellent.

Cotton is not attacked by moth grubs or beetles, but it is attacked by fungi and bacteria. Mildews are very troublesome on cotton, and particularly when starch is present on the material, producing rotting and weakening.

Cotton is, by volume of production and use, the world's principal fibre. It is inexpensive, durable in wear, and attractive in handle and appearance. Its strength in both the wet and dry conditions is good, and it will withstand severe wet processing and laundering treatments. The cotton fibre itself is dimensionally stable. A cotton garment may shrink by a certain amount when it receives its first wash, but this is a consequence of the strains introduced during fabric and garment manufacture and not due to any shrinkage of the fibre. Cotton can be dyed easily and the colours are fast to repeated launderings, and particularly when vat dyes are used. Cotton is comfortable to wear, absorbing moisture readily and just as readily giving it up to the air. With such a range of desirable properties, it is not surprising that cotton is the most widely used of all the textile fibres. Cotton is used to make every type of garment and household fabric; it also finds wide application in industrial fabrics, tents, tarpaulins, and tyre cords.

3.2.6 Production and consumption of cotton

Throughout the world, more cotton is produced by far than any other fibre, natural or man-made. The total world production of all the textile fibres, both natural and man-made, has continued to increase over the last 20 years from an estimated 35 144 m. lb (15 941 m. kg) in the season 1951–52, to 52 018 m. lb (6657 m. kg) in 1967–68. For a variety of reasons, cotton production has not kept pace completely with this overall rapid increase, and its share of the world's total fibre production has fallen from about 57% in 1951–52 to the smaller, but yet considerable figure, of 44% in 1967–68.

For many years, the USA dominated the world as the leading producer of cotton, providing something like 40% of the world's output of the fibre. The output was becoming so vast in the USA, and the material was being produced in such ever-increasing quantities in many other countries, that the US government was obliged to introduce legislation in 1965 which had the effect of reducing the acreage being used for the cultivation of cotton. US cotton production fell sharply, therefore, in the 1966–67 and following seasons. There was, however, no corresponding reduction in the output of the fibre in the majority of the other cotton growing countries and, in fact, in some the production was increased. The quantities of cotton produced by the world's main

Table 3.1 Production of cotton in various countries (million lb)

Country	*Average 1951–52 to 1955–56*	*Average 1956–57 to 1960–61*	*1963–64*	*1964–65*	*1965–66*	*1966–67*	*1967–68*
USA	7517	6462	7361	7288	7187	4597	3578
USSR	2870	3322	3871	3968	4206	4445	4445
China	2457	3537	2247	2630	2773	3106	3347
India	1746	1991	2509	2352	2198	2198	2548
Egypt	798	931	974	1111	1146	1003	963
Brazil	796	736	1111	992	1195	978	1314
Mexico	744	997	996	1140	1250	1071	957
Pakistan	655	647	930	840	858	1008	1107
Turkey	324	370	567	717	717	842	873

cotton-growing countries, together with an indication of how their output has changed since the year 1951, are shown in table 3.1. From the table it will be seen that production in the USA fell by 36% between the 1965–66 and 1966–67 seasons, and there was a further reduction of 22% in the following season; the output in the 1967–68 season was only half that of two seasons earlier and it was, in fact, the country's lowest output since 1921. Output in the USSR rose steadily until the 1966–67 season after which it became steady, while the output in China has continued to increase.

Some of the increase in cotton production in certain countries, particularly China, India, Brazil, Pakistan, and Turkey, is a consequence of bigger areas of land being placed under cultivation, but it is also due, in part, to improvements in the yield of fibre per acre in some of these countries. These improvements in yield are the result of better cultural practices and greater mechanization. In the 1962–63 season, the average world yield of cotton per acre was of the order of 285 lb, while in 1966–67 this had increased to 300 lb per acre.

Over recent years, there have also been substantial changes in the pattern of consumption of cotton. Consumption of cotton has increased in Asia and Africa, and has diminished in Western Europe and the USA. In Britain there has been a rapid downward trend, and a level of consumption of cotton has now been reached which is believed to be the lowest for 150 years. This rapid fall can be attributed to a general contraction of the textile industry together with a substantial increase in the use of man-made fibres. Competition from man-made fibres is also given as an explanation for a reduced usage of cotton in the USA and Canada. There has been some greater usage of cotton in one or two of the countries of Western Europe, notably West Germany and Portugal, and quite substantial increases in a number of the developing countries. The USSR and China are also using greater quantities of cotton.

3.3 FLAX

3.3.1 History

Flax may probably have been the first plant fibre to have been used by man to make his textile materials. Specimens of linen fabrics have been found in the tombs of Ancient Egypt, and specimens of mummy-cloth which have been discovered are estimated to be about 5000 years old.

Flax-growing spread over Europe from the Mediterranean area long before the beginning of the Christian era, and many of the linen industries which today exist in a number of countries in Western Europe are of ancient origin. Eventually, linen manufacture spread from Western Europe into England, Scotland and Ireland, much of the skill and know-how being brought in by French and Flemish weavers who were seeking sanctuary from religious persecution.

Before the seventeenth century, only small amounts of flax were grown in England. As in the case of cotton, the wool merchants were too powerful and the development of a liner industry was deliberately retarded in favour of wool. The influx of French and Flemish linen workers was the basis upon which a greater interest in linen did eventually develop, and the increased demand for linen fabrics was met by the importation of flax from other countries, mainly Russia.

Official encouragement by the government in England led to the growth of the linen industry in Ireland, and this began to flourish. However, the invention of new types of machines for processing cotton, and the increase in the use of cotton in England, meant that there was less demand for linen, and the industry in Ireland began to decline. Gradually, the linen industry contracted until only one or two manufacturing areas, such as Northern Ireland, remained.

Today, the flax plant is grown for its fibre content mainly in Europe, this including Russia; it has ceased to be of any real importance in other countries outside this area. By 1939, Russia was growing nearly three-quarters of the total world output, while today she probably produces about the same proportion, with France, Belgium and the Netherlands providing much of the remainder in quantities in that order.

3.3.2 Botanical information

The flax plant is a member of the Linaceae family. It is grown commercially for two purposes, one to produce the fibre from which linen fabrics are made, and the other to produce seed known as linseed from which linseed oil is extracted. The Linaceae family consists of about 150 species which are widely distributed throughout the world in temperate and sub-tropical regions. Of the flax plants, those which are cultivated for seed are bushy in character and have many branches; those grown for fibre are tall and slender and normally branch only towards the top of the main stem. It is not a satisfactory commercial proposition to try and grow plants to yield both seed and fibre.

The only important flax plant grown for fibre is *Linum usitatissimum*. It is an annual with slender, smooth-surfaced, greyish-green stems, growing to a height of about 3 or 4 ft (0.91–1.21 m) and having a diameter between $\frac{1}{16}$ and $\frac{1}{8}$ in (1.5–3.1 mm). The fibres are extracted from the stems of the plant in which they serve to hold the plant erect; they are referred to as 'bast' fibres. The bast fibres are sometimes referred to as 'soft' fibres to distinguish them from the leaf fibres of certain plants, the so-called 'hard' fibres.

The flax strands are constructed of relatively long thick-walled plant cells which overlap one another and are cemented together with natural gums into long strands. It is possible for such a strand to run along the entire length of the stem. The fibres lie along the stem just below an outer surface layer or bark. Beneath them is an inner bark, surrounding the woody core of the stem and pith tube.

The varieties of flax grown for fibre are generally classified into those with white, blue or purple flowers. The white-flowered type is more hardy than the blue, and usually gives a higher yield per acre. The fibre obtained from it, however, is considered

Fig. 3.5 The flax plant.

to be harsher and of lower quality. The purple type is not used extensively for commercial production, but has been used in crossing with the other two in an attempt to obtain improved characteristics.

Flax is usually harvested when the stems are changing from green to yellow, and about one month after the first flowers have appeared. If harvested too early, the yield of satisfactory fibre will be too low; if left too long before gathering, the fibre will lose some of its quality.

3.3.3 Sources of supply and cultivation

The main grower of flax is the USSR, with France, Holland and Belgium also producing substantial quantities. Some flax is grown in Ireland. Outside Europe, China grows a limited amount of flax, and that grown in Japan is all needed for internal use to reduce the amount which otherwise would have to be imported. Egypt produces a certain amount of fibre, and some is exported, and in recent years India has shown interest in flax production. For nearly a century, Australia has taken an interest in the growing of flax and has gained experience in cultivating the plant. When sources of supply were cut off during the Second World War, Australia planted a substantial acreage with flax, but after the war when outside supplies were again available, the acreage was reduced considerably. New Zealand has investigated the cultivation of flax and growing began there in 1936. It was found that better results were obtained in the South Island than in the North Island. Again, due to the special circumstances, there was large-scale production in New Zealand during the 1939–45 period, but after that the acreage under cultivation dropped. Kenya and Uganda attempted to grow flax in 1939 but have since ceased to do so on any substantial scale. The flax plant is grown on quite a wide scale in Canada, USA and Argentina, but mainly as a source of linseed oil, the production of fibre being very small indeed.

The quality of the flax fibre produced is influenced quite considerably by the weather conditions the plant experiences during its growth. In general, it can be said that good climatic conditions are as important as, if not more important than, favourable soils. The plant likes a temperate, equable climate free from heavy rains and frost. A hot, dry summer usually produces a short and harsh fibre, but one having good strength; a moderately moist summer produces plants which yield strong, fine and silky fibres. For this reason, flax grown in the USSR, where the summers are hot and short, gives strong fibres but not ones which can be classed as the highest grade. In Ireland, the climate is, on the whole, suitable for flax, and particularly if a warm, dry spell is experienced during the period towards the end of July. In hot climates, the flax plants tend to branch more and produce stems of only medium length and quality, with increased seed formation.

Flax is not easy to grow successfully. The soil must be right, neither too heavy nor too light. In Ireland it has been found that a good loamy soil, ploughed to a depth of about 6 in, (150 mm) and with a firm clay subsoil, suits the flax plant best. Crop rotations are said to be important. It has been said that no other plant takes more out of the soil than flax, and the flax grower must make sure that his land does not become exhausted. In Belgium, the grower will not sow flax on the same piece of ground more than once in every seven years. In Ireland a 'rest' period between one flax crop and the next is usually about six years.

Before the seed is sown, the soil must be prepared properly. In the Liege area of Belgium, where some of the best flax in the world is grown, the soil is a fairly rich sandy loam, and ploughing to a depth of about 8 in (203 mm) is carried out in the autumn preceding the flax-growing season. In the following April, the ground is broken up by harrowing and rolling, and considerable quantities of artificial fertilizer are applied. The ground is then left to dry for a few days during which whatever weeds there are present begin to germinate. The ground is then given another ploughing to a smaller depth than before and again, after a further wait of a few days, rolling breaks down the soil into a fine tilth and the ground becomes firm.

The flax seed is sown broadcast and subsequently harrowed and rolled in. The sowing rate is about 130 lb (60 kg) per acre. Germination takes place and the shoots begin to appear. The increase in height of the shoots is at first quite slow, but it then speeds up and the height increases at a rate of about 1 in (25 mm) per day for the next 30 to 40 days, after which the flowers are formed. Once the flowers develop the stem ceases to increase in height. Flax can normally be considered to be a three month crop.

When the stems are ready, usually about a month after the first flowers appear, the flax is harvested. Two methods of harvesting are used. In the preferred one, the plants are pulled from the ground; in the other the stems are cut. Pulling gives better results than cutting. With the latter, the fibre deteriorates at the point of cut, and there is also a possibility that weeds will be cut off at the same time and their presence interfere with the further processing of the flax. Hand pulling and machine pulling can be used, but the former is used for the best flax crops.

After pulling the bundles are laid, one crossing over the other, to dry in the sun for 1 or 2 days, and they are then built up into 'stooks' to complete the drying process in from 10 to 14 days. After drying, a good quality flax should have straight, golden-

Fig. 3.6 Flax harvesting.

coloured stems about 3 ft (1 m) in length and of a more-or-less uniform diameter of about $\frac{1}{16}$ in (1.6 mm).

3.3.4 Preparation of the raw fibre

The stems are now ready for the flax fibre to be extracted but, before this is done, the seed cases, or bolls, are removed. This is known as 'rippling', and it used to be carried out by hand, each stem being dragged through a coarse steel comb which broke off the bolls. Nowadays, machines are used, these being similar in action to a thrashing machine but having rotating claws which scrape off the bolls. After seed removal, the stems are sorted by hand according to their length and quality, tied in bundles, and passed on ready for 'retting'.

Retting, or rotting, is the process whereby the fibres are abstracted from the stem. In retting, the stems are submitted to the action of water, fungi and bacteria which decompose the materials surrounding the woody part of the stem but leave the fibres intact. Soaking in water softens the stem, expels air, and extracts the water soluble substances. Micro-organisms present in the flax stems, or in the soil attached to them, then develop rapidly under favourable conditions and break down the stem. Retting is an important process which requires skill and experience to recognize when it is completed. If allowed to go too far, the fibres will be damaged, and yet too short a retting period will not release the fibres.

There are two main methods of retting used, 'dew' retting and 'water' retting. Dew retting is often used for poor quality flax as not being worth the trouble and expense of water retting. In dew retting, the stems are spread out in thin layers on the soil with, as far as possible, the root ends in line and with the top end of one layer overlapping the root ends of the preceding layer. The stems are left for the moisture in the atmosphere to act upon them. The stems are turned approximately once a week, still keeping a similar formation. The retting time can vary between three and seven weeks, depending on the conditions, and the completion is judged by taking a stem from time to time, drying it, and then breaking it at short intervals along its length when, if the retting is complete, the woody material should come away easily. Dew retting is cheap, but, in adverse weather conditions, it can take a long time, and much land space is taken up meanwhile. The main objection to it is lack of control over the process.

Water retting, which is used for the better qualities of flax, is carried out in rivers, ditches, or specially constructed tanks. The bundles of stems should have the stems parallel and their lengths more or less equal. The tie bands which hold the bundles together should not be tight as there is about 10% expansion of the stems during retting. The bundles are usually oval in shape and are immersed under water, usually in a river, and held down by weights, such as stones. As the gas is formed during fermentation, the bundles tend to rise in the water, and more weights must be added. As less gas is evolved, the bundles sink and the weights can be removed, this usually taking place at the end of four or five days. Depending upon the water temperature, retting can take from two to three weeks to complete.

Quicker results, and a better control of the process, are obtained if warm water retting in tanks is used. The concrete tanks usually hold about 6 or 7 tons of stems each and there is a battery of such tanks in a rettery. The proportion of water to stems should never be less than 9:1, otherwise the acids formed during the process will be insufficiently diluted and will destroy the bacteria. The tanks are usually heat-insulated, and the water is kept at around a temperature of 30°C (86°F). During the process, the acidity of the water is checked. At the beginning of the process the water is neutral, but gradually becomes more acid as the process continues. It remains at an increased acid value for something like 10 hours and then begins to climb back towards neutrality. This loss of acidity indicates that retting is completed.

After retting, the stems are rinsed and then spread out to dry in fields, or placed in an artificial dryer. Sometimes excess water is first removed from the stems by passing them between pressure rollers. If they are to be dried outside, the bundles of stems are untied and stood up in cone-shaped formations. After drying, the stems should not contain more than 17% of their weight of moisture, and not less than 12%.

The dried 'straw' now goes forward for 'scutching', the object of which is to separate the fibres from the stems, clean them, and open them up. This is done by beating the stems, a process which was formerly done by hand, using a wooden blade, and is now performed by a machine. In the hand method, the stems were wedged into a notch over a wooden frame and beaten with a flat, broad blade, the straw being turned from time to time. The original scutching machine did exactly the same using a wheel having 12 or 14 pliable wooden beating blades. The more modern machine passes the stems between indented or fluted rollers to 'break' them before passing them forward into a cleaning section. The severe beating action during scutching breaks down some of the fibres into shorter ones referred to as 'scutched tow'.

After scutching, the flax is sold to dealers and some give the material further treatments to improve its quality. The machines used in these treatments give a further cleaning and extract more short fibre, referred to as 'machine tow'. It then goes forward to be spun into yarns.

3.3.5 Properties of flax

The best flax is pale yellow. In scutched flax, the length of the fibre strands can be up to 3 ft (0.91 m), a good strand averaging between 18 and 24 in (457–609 mm). The additional processing in preparation for spinning will break this down further to probably under 12 in (304 mm). Commercial flax is in the form of bundles of individual plant cells held together by cementing materials.

Good flax is soft, lustrous and reasonably flexible. The appearance of the fibres under the microscope is of cylindrical tubes, with swellings or 'nodes' at intervals, and with cross-markings. There is a central lumen which is narrow and can be observed. The cell walls of the fibre are thick and polygonal in cross-section.

Flax is stronger than cotton or wool. Its tenacity can be as high as 6.5 g/denier (58.5 g/tex) and its strength increases when wet, by about 20% compared with its dry value. The fibre is relatively inextensible, with an extension at break of about 2%.

Within its small degree of extension, flax is an elastic fibre, tending to recover to its original length when the stretching force is removed. It has a high degree of rigidity and resists bending.

The regain value for flax is about 12%, determined at 65% R.H.

Flax will burn when ignited. As its temperature is raised, the fibre starts to discolour at about 120°C (248°F) and to decompose at temperatures in excess of that. Flax gradually loses strength on prolonged exposure to sunlight.

Flax will withstand dilute acids in the cold, but is attacked by hot dilute acids and concentrated acids. The fibre has a good resistance to alkalis and the common organic solvents.

Flax resists insect attack and, if the fibre is clean and dry, it has a good resistance to attack by micro-organisms.

Flax is not now used in very large quantities compared with some of the other fibres. The fibre is relatively expensive to produce, but it has a number of attractive characteristics which still make it of value for certain applications. It is available as unusually long fibres and, for this reason, coupled with its strength, it is used for making strong threads and fishing lines. Considerable amounts are used for making sail and tent cloths. It is woven into handkerchiefs, tablecloths and dress fabrics. Flax is a good conductor of heat, and linen sheets made from it feel cool and fresh. Flax absorbs water rapidly and, for this reason, linen towels and glass-cloths are popular. Finally, fabrics made from flax are hard-wearing and durable and, in particular, will withstand repeated launderings.

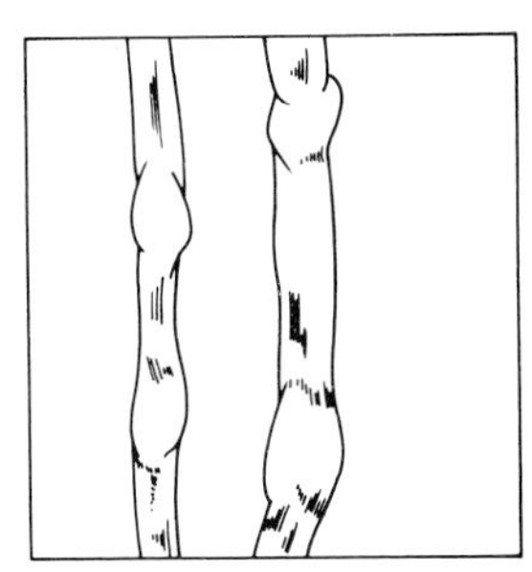

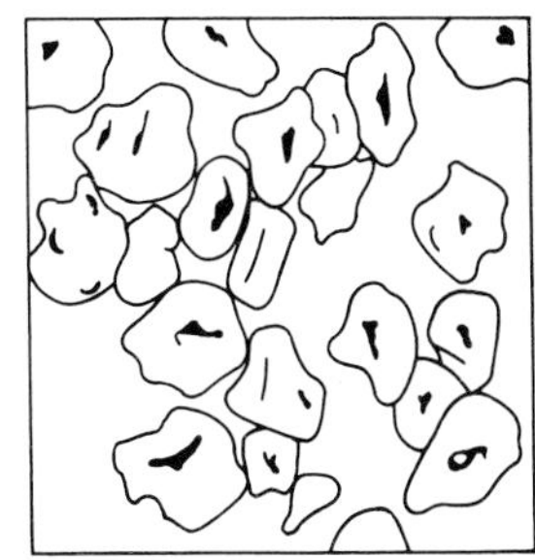

Fig. 3.7 Flax fibres are fairly regular in diameter, with characteristic 'jointed' appearance, gradually tapering to a pointed end. In cross-section they appear as irregular collapsed tubes.

3.3.6 Production and consumption of flax fibre

The world production of flax fibre is about 700000 tons (711235 tonnes) per annum. The production in the Communist countries shows a tendency to increase, but there has been a decrease over the last few years in most of the other countries, and particularly in Western Europe. Table 3.2 shows the output of flax fibre by the main producing countries over the years between 1951 and 1967; the figures in brackets are estimates.

China also produces quite a large quantity of flax fibre, but official figures are not available. It is known, however, that in 1957 production was of the order of 127 m. lb (59 m. kg).

Belgium had always had a reputation for the excellence of the methods used to extract fibre from the flax stems, and considerable quantities of 'flax straw' are exported from France and the Netherlands to Belgium. The amount imported into Belgium in 1965 was as high as 490 m. lb (222 m. kg), but there has been a substantial falling off since that year, mainly in the straw from the Netherlands, and the figure for 1967 is 302 m. lb (136 m. kg).

In the past, Britain has been the biggest importer of flax fibre and still takes in large quantities of the material as shown in Table 3.3, which lists the intake of the main importing countries over the period 1951 to 1967. It will be seen that imports of flax fibre into this country fell sharply in 1966 and 1967, but there was a small recovery in 1968.

Table 3.2 Production of flax in various countries (million lb)

Country	*Average 1951–55*	*Average 1956–60*	*1963*	*1964*	*1965*	*1966*	*1967*
Commonwealth and Ireland	15.7	2.2	–	–	–	–	–
France	82.9	62.7	159.0	190.4	130.0	145.6	138.9
Belgium	76.2	67.2	87.4	103.0	68.4	56.0	38.1
Netherlands	80.6	58.2	73.9	87.4	58.2	44.8	29.1
Egypt	(6.7)	11.2	15.7	18.0	13.4	20.2	13.4
USSR and Eastern Europe	692.0	1140.0	1061.8	970.0	1313.0	1250.0	(1300.0)

Table 3.3 Imports of flax fibre into various countries (million lb)

Country	*Average 1951–55*	*Average 1956–60*	*1963*	*1964*	*1965*	*1966*	*1967*
Britain	91.2	98.3	97.0	95.2	100.0	86.0	66.8
Belgium	29.0	48.0	84.4	83.8	112.0	107.1	114.5
France	39.4	42.8	60.5	42.1	45.7	41.2	37.2
Italy	5.4	32.5	41.2	27.0	34.3	36.5	35.6
W. Germany	36.4	47.7	39.7	32.2	34.5	34.2	29.1
Netherlands	5.0	17.2	11.2	15.7	18.7	20.2	23.3
Japan	6.5	10.0	26.2	21.4	17.7	23.0	29.0
Sweden	5.6	6.7	5.0	8.1	5.6	3.6	4.8
USA	8.1	7.9	3.8	7.9	8.3	7.2	6.7
Finland	2.0	1.3	5.8	5.8	5.8	4.0	2.6
E. Europe	17.5	53.3	49.0	35.2	34.5	47.7	54.0

Britain is the world's largest exporter of linen goods, the USA, being its most important customer. In recent years, however, this trade has tended to fall off, largely because of an increasing use of man-made fibres, and also because of competition from linen made in France, Belgium, Japan and Eastern Europe.

3.4 JUTE

3.4.1 History

The jute plant needs a hot, damp climate, and has always flourished in parts of Asia, mainly India and Pakistan. The fibre obtained from wild plants has been used in those countries for cordage and for making coarse cloth for many centuries. Cultivation of the plant does not seem to have been undertaken until the beginning of the nineteenth century. In the early 1820s attempts were made to spin jute fibre in Scotland at Dundee, and the necessary techniques for doing so were developed and, by 1838, had progressed to the stage that substantial quantities of the fibre were imported from India. By 1850, the jute industry was well established in Dundee, and received encouragement to expand still further until the Crimean War caused supplies of hemp and flax to be reduced in 1853. Dundee has continued to be an important centre of the jute industry.

Since that time, India and Pakistan have increased the numbers of their jute spinning and weaving mills, and both countries now process much of the fibre which comes from the plants they grow. In some Bengal rural areas, jute spinning and weaving are still carried on as a cottage industry.

3.4.2 Botanical information

Jute fibre comes from the inner bark of plants of the genus *Corchorus*, a member of the Lime family. Although there are several species of the plant from which fibre may be obtained, two are mainly used, these being *Corchorus capsularis* L, usually known as 'white jute' and *Corchorus olitorius* L, often referred to as 'Tossa jute'. The two plants are similar in appearance, except in the shapes of the seed pods which are globular and cylindrical respectively, but the most important difference from a cultivation point of view is that *Corchorus capsularis* can withstand waterlogged conditions in its later stages of growth, whereas the other species cannot and is usually grown on higher ground where flood waters cannot reach it. For this reason, *Corchorus olitorius* is sometimes known as 'Upland jute'. The leaves of *Corchorus olitorius* are practically tasteless, whereas the leaves of *Corchorus capsularis* have a bitter taste.

The jute plant requires quite a high temperature, a deep fairly fine soil, and an annual rainfall of more than 40 in (1 m), much of this falling when the crop is maturing. The plant can be grown on all types of soil, but prefers that the soil be not too sandy. If insufficient rain falls during the growing period, the plants become stunted and flower too early. The most suitable growing areas are near rivers where the land becomes inundated with flood water and the soil is renovated annually with silt deposits.

Most jute stems are harvested when about 50% of the plants are in pod, the plants then being at their best as regards the yield and quality of the fibre. While, in some areas, the plants are pulled up by hand and then the root portions are cut away, harvesting is generally done by cutting off the stems close to the ground, using a sickle.

Fig. 3.8 Jute growing in India. (*Corchorus olitorius*).

3.4.3 Sources of supply and cultivation

While very large quantities of jute are grown in India and Pakistan, there are a few other important growers. The production of jute began in Brazil in 1932. At first, the results obtained were not very promising but by using specially selected *Corchorus capsularis* seed, satisfactory production was eventually achieved and good fibres obtained. In Brazil, jute is mainly grown on the banks of the River Amazon in the state of Para. It is usually possible to grow two crops of jute each year, one in the wet season, and the other in the dry season using irrigation. The seeds take about four days to germinate during the wet season and eight days in the dry season; 120 and 130 days covers the development of the plants from sowing to maturity. The cultivation, harvesting and retting are carried out on peasant lines in much the same way as they are in India and Pakistan. The bulk of the jute fibre produced in Brazil is used within the country for the making of sacks, the Government buying all the fibre produced at a guaranteed price.

The production of jute in China has increased since the Second World War. The *Corchorus capsularis* type is grown for the most part. The most important growing area is Chikiang along the Yantse River. Some is grown also in South China in the delta region, and in the Kwangton Province. In China, it is generally considered inadvisable to grow jute for more than two years on the same piece of land; jute is rotated with rice quite often, and sometimes with wheat or sugar-cane. Jute has been grown for many years on Taiwan, having been introduced there from the China mainland.

The cultivation of jute has commenced in Burma in the delta area of the Irrawaddy River. It is hoped to grow jute and padi on the same fields, the padi being planted out when the jute stems are being retted.

There is some production of jute in the USSR, and in Borneo, Malaya, Philippines, Nepal, Iran and Peru.

The preparation of the soil before sowing jute seed is important. The jute roots are more than a foot in length and so very deep ploughing is needed to remove the stubble left from the previous year's crop. The ground is then prepared by ploughing and cross-ploughing five or six times, followed by dragging a heavy log or bamboo 'ladder' across the surface to give a fine tilth. The presence of potash and phosphates appear to check diseases which attack the plant.

The seed is normally sown broadcast, about 10 lb/acre (11.2 kg/ha) for *Corchorus capsularis*, and 6 lb/acre (6.7 kg/ha) for *Corchorus olitorius*. The seeds are very small and are often mixed with dry soil for convenience in handling. After sowing the 'ladder' is again dragged over the soil to cover up the seed, which should be at a depth of about 1 in (25 mm).

When the plants are a few inches above the surface of the ground, they are weeded and thinned to a spacing of about 4 in (101 mm) this relatively close spacing being used to discourage branching of the plants.

As has already been mentioned, harvesting is usually by cutting off the stems just above the ground. The stems are then left to dry for two to three days, then tied into bundles during which the stems are shaken to rid them of as much leaf as possible, and passed forward for retting. In India, an acre yields about 1 240 lb (570 kg) of fibre.

3.4.4 Preparation of the raw fibre

The retting process bears some resemblance to the river retting of flax. The bundles of stems are laid flat in the water, side by side, to form a platform. The platform is then covered with weeds and then heavy logs placed on it to keep the bundles submerged. The retting period varies with the water conditions, the temperature of the water being very important. The retting finally separates the fibre strands from the woody part of the stem. One difficulty associated with the retting of jute is that the thick, lower part of the stem takes longer to ret than the top portion and there is a danger of the top becoming over-retted. This is obviated by standing the bundles upright in about a 2 ft (0.60 m) depth of water for two or three days, before laying them flat in the water to form the retting platform. Often, however, the unretted butt ends are cut off and sold separately as 'roots' or 'cuttings'.

After retting, the fibre strands are removed from the stems by hand, and then cleaned by being spread on the surface of water, when any remaining pieces of stem are removed by hand. The strands are then squeezed to remove excess water, and hung on frames to dry in the sun for two or three days. During the exposure some bleaching by the sun takes place. The strands are then taken from the frame, tied in bundles, and packed ready for shipment.

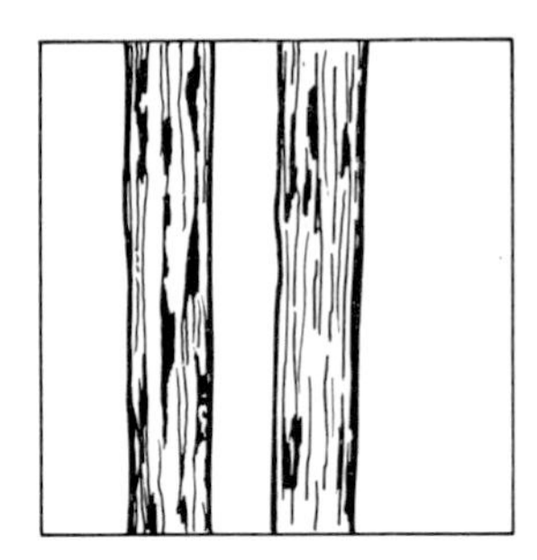

Fig. 3.9 Jute: polygonal cells joined together to form a strand; thick-walled cells with clearly observable lumens.

3.4.5 Properties of jute

Jute varies in colour from yellow to brown. The strands obtained commercially consist of bundles of individual fibres cemented together by natural gums. During retting, some of the fibre ends become loosened and stand out from the strands giving the jute its rough, hairy handle.

The individual plant cells in jute are only about $\frac{1}{10}$ in (2.5 mm) long. The cell sur-surface is smooth and cross-marked here and there.

Under the microscope, a cross-section shows an array of polygonol cells, usually five or six-sided, joined together to form the strand. The cells have thick walls and easily observable lumens.

The tensile property values for jute are less than those for flax, and the individual strands vary quite considerably in strength. Jute is relatively inextensible, having an elongation at break of only 1.7%. It is also relatively inelastic.

The density of jute is about 1.5 g/cm^3. Jute absorbs a considerable amount of water, having a regain value of about 14% determined at an R.H. of 65%.

Jute is very resistant to rotting and this, coupled with its cheapness, good strength and availability, makes it a very suitable material from which to make sacks and packing cloths; probably about 75% of the quantity produced is used for these purposes. It is also used on a large scale for twines, carpet backings, and for webbing. In some of its countries of origin it is used to make lower quality clothing fabrics.

3.4.6 Production and consumption of raw jute

Although jute is being challenged in its traditional use as a packaging fibre by other materials, mainly the man-made fibres, it is still being produced in various parts of the world on a tremendous scale. In the 1967–68 season, the world production of raw jute was 5 898 m. lb (2.675 m. kg). Table 3.4 shows how this production was divided up between the main producing countries. The figures in brackets are estimates.

It will be seen that, in the season 1967–68, India and Pakistan accounted for something like 88% of the world's output of raw jute. The remainder was shared mainly between China, Brazil, Taiwan, Burma and Nepal.

India uses most of the raw jute produced there in her own mills, but in Pakistan, most of what is produced is exported, so that Pakistan is by far the largest exporter of the fibre. The chief importing countries are Britain, Japan, Belgium, France, China and the USA in that order.

Table 3.4 Production of jute fibres in various countries (million lb)

Country	*Average 1951–52 to 1955–56*	*Average 1956–57 to 1960–61*	*1963–64*	*1964–65*	*1965–66*	*1966–67*	*1967–68*
Pakistan	2 162.0	2 271.4	2 350.0	2 130.0	2 544.6	2 560.0	2 688.0
India	1 545.6	1 760.4	2 455.0	2 390.0	1 781.0	2 125.8	2 526.7
China	132.2	230.7	(269.8)	(291.0)	(335.0)	(335.0)	(335.0)
Brazil	50.0	73.9	97.0	112.0	130.0	98.6	82.9
Taiwan	31.4	38.1	22.4	29.1	29.1	31.4	31.4
Thailand	4.5	9.0	15.7	13.4	20.2	24.6	22.4
Burma	–	9.0	26.8	22.4	33.5	29.1	29.1
Nepal	15.7	15.7	45.0	40.4	82.9	179.2	(179.2)
Japan and S. Vietnam	2.2	2.2	4.5	4.5	4.5	2.2	2.2

3.5 OTHER VEGETABLE FIBRES

3.5.1 Ramie

China can be called the home of ramie, where it has been cultivated and used for hundreds of years. Ramie fibre was also used by the early inhabitants of the American continent, mainly to prepare twines and binding materials. The fibre was relatively late in being introduced into Europe and it is only in the eighteenth and nineteenth centuries that mention is made of its use on a commercial scale in England, France and Germany.

Ramie is still produced on a reasonably substantial scale in China, Japan, always an important user of ramie fibre, used to import it from China, but now grows it herself. Smaller quantities are produced in Brazil, USA, Philippines, Indonesia, Ceylon and India.

Ramie fibre is obtained from the stem of a plant *Boehmeria nivea*, which is a member of the Nettle family. This particular plant is the most important of the 100 or so species of *Boehmeria* for the production of fibre; *Boehmeria nivea* is often referred to as 'white ramie'. The plant is a perennial which produces stems which can attain as much as 8 ft (2.4 m) in height and $\frac{3}{4}$ in (19.0 mm) in diameter. The stems have few branches. If the stems are cut during the growing season the root stocks will send up new shoots, and three or even more crops may be harvested in one year. In addition, once the plants have become established, they will go on producing crops for several years without replanting.

The ramie plant is more suited to a warm, temperate climate rather than to the tropics. It flourishes better where a cold season follows the growing period in order to provide a resting time for the plant. It prefers a fairly rich soil able to retain an adequate amount of moisture but it is sensitive to flooding. The plants tend to exhaust the soil and the ground should be fertilized after each harvest, and as much as possible of the waste portions of the crop should be returned to the ground.

As with the other fibre plants, weeding should be carried out in the early stages of growth; later on, the plants will be sufficiently dense to smother the weeds. Male and female flowers form on the same plant. The male flowers mature first and are mostly on the lower parts of the stems. Midway up the stems is a mixture of male and female

Fig. 3.10 Ramie.

flowers, and at the top female flowers only. When the conditions are right, the male flowers scatter pollen, which is blown about, with some alighting on the female flowers.

In China, the stems are harvested when about one-third of the leaves have fallen and when the stems have changed to a yellowish colour. There are then new shoots forming at the base of the plant. The plants are cut with a sickle close to the ground, care being taken not to damage the new shoots.

It will be appreciated that the ramie stems are easily and cheaply grown. This, combined with very useful textile properties would be expected to make ramie a widely applied fibre. Unfortunately the fibres are difficult to extract from the stems satisfactorily by retting, since the gummy material present is not disintegrated by such a treatment, and whatever method of extraction is used, thoroughly clean fibres are not normally obtained.

The method of extraction used is, first to strip long ribbons of bark containing the fibres from the surfaces of the stems. This is usually done by hand. The ribbons are then steeped in cold water for a few hours, and the bark scraped away from the fibres. The bark can sometimes be removed by one scraping, depending upon the skill and conscientiousness of the person scraping, but more often a further short soaking followed by a second scraping is required. The gum content is usually of the order of 20 to 30%; much of this is removed relatively quickly but the difficulty is getting rid of the last 1 or 2%. After the scraping is completed, the fibre, which has been extracted, is hung over poles to dry and bleach in the sun and wind. In some countries, attempts have been made to replace the hand scraping by a machine operation. The machines rely on a beating action to release the fibres.

In Japan, the fibres sometimes come to the spinning mills in gummed ribbon form, and degumming has to be carried out there. This takes the form of treating the strands with caustic soda solution for up to 4 hours, followed by crushing and washing in a machine. The fibres are finally washed by hand and dried, usually in the open air.

Ramie fibre is white in colour and has a lustrous appearance. The fibre is strong and durable in wear, but is not very elastic. Ramie absorbs and gives up water readily; it also dyes without difficulty. Strands of ramie are built up from smooth, cylindrical, thick-walled cells which may reach as much as 18 in (457 mm) in length. The surface of the cell is covered with fine ridges.

The main use of ramie is in twines and threads where the strength and inelastic characteristics make it of particular value. In China and Japan, the fibre is used for making clothing, table-cloths and canvas, and it also proves useful for the construction of fishing nets.

3.5.2 Hemp

Hemp is a bast fibre obtained from the stems of a plant. It is a fibre with a long history, its original home appearing to have been Central Asia, from where its cultivation spread to China. The plant is supposed to have been cultivated in China for more than 4500 years. During the early Christian era, production of the plant spread to the countries of Mediterranean Europe and since that time, has extended to many countries throughout the world. The type grown in the USA originated in China. Russia and the Baltic countries were early producers.

The plant is of the Mulberry family. The plant is grown for a number of purposes. The variety grown mainly for the production of seed from which oil is extracted, is short in height and matures early. The one known as 'Indian Hemp' from which the drug hashish is extracted, is short, profusely branched, and has small dark-green leaves. The one grown for fibre, which is the oldest use of the plant, has slender erect stems which can reach a height of 14 ft (4.2 m), and a diameter of $\frac{5}{16}$ in (7.9 mm). Owing to the presence of the drug, which can be extracted from a resinous juice in the stems and leaves, the growing of hemp is forbidden in a number of tropical countries.

The hemp plant, grown for its fibre, requires a mild, temperate climate with a humid atmosphere and an annual rain-fall of not less than 27 in (685 mm). It likes an abundance of rain over the period in which the seed is germinating and until well established. If the seeds are placed close together, the plants will withstand storms, and will not branch. The plant grows best in a rich, neutral loam soil having a sub-soil which retains moisture well; it will not grow satisfactorily in a sandy, acid soil. The plant can withstand appreciable changes in temperature, although prolonged frost will kill the young plants. The growing of the hemp plant tends to exhaust the ground and it is essential to use fertilizers and put back into the soil any waste material from the plants.

Hemp is an annual, and the seeds are sown in early spring and covered to a depth of about 1 in. Weeding is carried out in the early stages of growth, after which the plant grows rapidly and chokes the weeds. There are male and female plants which look alike until flowering time, after which the male plant turns yellow and dies. The female plant remains dark green for another month or so until the seeds ripen.

The plants are ready for harvesting between the time of blossoming and the formation of the seed. The stems are cut close to the ground using a hemp knife which is like a long handle sickle. The cut stems are spread on the ground to dry. An average yield of dried stems is of the order of 2 to 3 tons/acre (2–3 tonne/ha), and about one-quarter of this is hemp fibre.

The fibre is separated from the stems by retting in a manner similar to that used for flax and then treating as for flax.

In China, the best fibre comes from the Shandun Province of Eastern China. The plant is grown with other crops, such things as wheat and vegetables being sown after the hemp has been harvested. The most important producer in the Western Hemisphere is Chile, where hemp has been cultivated for more than 400 years, for both its fibre and seed. In Chile, the plants are grown on small farms, always using irrigation by reason of the long dry summers. Most of the hemp fibre produced is used locally to make ropes and twines.

The quality of hemp fibre is judged by its colour and lustre. The best qualities are white or pale-grey in colour. The best hemp is grown in Italy and there is one particular variety *bolognese*, which has an excellent colour, a silky lustre and a softness equal to that of flax. Next in quality is that grown in France in the Grenoble area. Russian and American hemps are coarser than those mentioned above but are strong and durable.

Hemp is strong, and is a coarser and stiffer fibre than flax. Strands of hemp may be 6 ft (1.8 m) or more in length. Since the fibre does not bleach well and lacks elasticity and flexibility, it is not used for fine textile fabrics. The individual plant cells are between $\frac{1}{2}$ and 1 in long and are cylindrical in shape with cracks and swellings visible on their surfaces. The cells are thick-walled and are polygonal in cross-section. The central lumen is, however, broader than that in flax cells.

Hemp has been used for a variety of textile applications over its long history, but nowadays it is used mainly for sacking and canvas fabrics, and for ropes and twines.

3.5.3 Kapok

Kapok is a fine vegetable down attached to the inner walls of the seed pods of the Kapok tree, *Ceiba pentandra*, sometimes referred to as the 'silk cotton tree'.

The tree is considered to be indigenous to East Africa, West Africa, India, Ceylon, the Philippines and Indonesia. It can be found in the wild state, giving the best kapok fibre when growing at less than 1 500 ft (460 m) above sea-level.

The kapok tree is large, and can be up to 100 ft (30 m) in height, and has a tall broad trunk. It flourishes best in tropical and sub-tropical countries where there is an abundance of rain during the growing season, and then a dry period for the flowers to appear and the pods to form. The tree flourishes best in a deep, porous, sandy soil; in Indonesia, from which country the finest kapok comes, the trees grow in well-watered volcanic soil.

In Indonesia, the tree starts to flower at the beginning of the dry season in the middle of May. The appearance of the flowers is spread over a period of about a month and the pods mature at different times. The pods are ready to be picked when they are just about to open and when they have changed from green to brown. Men climb the trees to pick the seed pods as they ripen. The kapok floss or hairs, are on the inner walls of the pods and, if the pods are ripe, will fall out in a lump at a touch. The floss is extracted entirely by hand.

Fig. 3.11 The Kapok tree.

The fibres are dried in wire netting containers in the open air. The seeds which are mixed with the fibres, are then removed by beating by hand a mass of the fibres over a screen, the workers usually wearing masks since the kapok fibres are light and float about in the air. In some cases, machine separators are used.

The kapok is finally baled, but not compressed too heavily, as this would tend to destroy the resilience of the fibre.

Kapok is resilient, elastic and light. It remains buoyant after long periods of immersion in water and, when wet it dries quickly; each fibre has a waxy covering which is water-repellent. Individual fibres can be up to 1 in in length, but the majority are much shorter than this. The fibre is creamy in colour. Under the microscope, it will be seen that the cross-section of the fibre varies between circular and oval and is thin-walled with a wide lumen. The fibres are straight with no twist and one end tapers to a point while the other has a characteristic bulbous, root-like end.

The kapok fibres do not make very satisfactory yarns, and their main applications have been as stuffing for such products as life-jackets, clothing for air pilots, linings for rainwear and upholstery.

3.5.4 Sisal

Sisal is a leaf fibre which comes from the plant *Agave sisalana* which is a native of Central America. It derives its name, sisal, from that of the port in Tucaton on the Gulf of Mexico from which it was first exported.

Sisal fibres were used by the ancient Mexicans and Aztecs from which to weave clothing fabrics for themselves. From there, the cultivation of the plant has spread to other countries, including East Africa, Haiti, Brazil, and other parts of South America.

The plant has a short, thick stem from which grows a rosette of leaves. The stem is usually about 6 in (152 mm) in diameter at just above ground level, and its diameter increases to something like 9 in (228 mm) at a height of 1 ft (304 mm) above the ground. The leaves grow from the summit of this stem, and are long, roughly triangular in cross-section, and dark green in colour. The leaves can grow to something like 6 ft (1.8 m) in length and an average width of about 6 in (152 mm). New leaves appear from the centre of the rosette, so that the oldest leaves are on the outside.

The sisal plant can go on producing leaves for many years and, at the end of its life, the stem sends up a long flowering shoot or 'pole' which reaches a height of about 20 ft (6.0 m). This bears white flowers and, when it has flowered the plant produces tiny buds which become small plants and which fall to the ground and take root. After flowering once, the parent plant dies.

Sisal is a water-storing plant and it grows when supplies of water are available and, when dry, practically not at all. The plant requires a tropical climate with a moderate atmospheric humidity. Excessive rain is harmful to the plant and it is damaged by flood water. The sisal plant prefers a dry permeable, sandy-loam soil, containing a certain amount of lime.

The leaves, when they are ready for harvesting, are cut off at the point where they join the main stem using a specially curved sickle knife, care being taken not to damage the younger leaves which are not yet sufficiently developed for harvesting. One leaf, as cut, weighs between 1 and $1\frac{1}{2}$ lb (0.4–0.6 kg). The spike at the top of each leaf is cut away, and the leaves are made up into bundles of about 30 leaves each.

Fig. 3.12 Cutting sisal.

Fig. 3.13 Sisal fibre drying on lines.

After cutting, the bundles of leaves go to a factory for the extraction of the fibre. This must be done as soon as possible after harvesting as the cut leaves deteriorate very quickly. The fibres are extracted by crushing or scraping to remove the green matter surrounding them. This process is carried out using a machine called a Raspador. In principle, this consists of a drum having bars spaced at intervals on its surface, and this drum can be rotated inside a case. One end of a sisal leaf is fed into this machine through a small opening in the case, and the other end is held firmly. As the drum rotates, the leaf is subjected to a combined beating and scraping action by the bars. This breaks away the pulp, which is carried away by a stream of water flooding continuously over the surface of the drum. The extracted fibre is dried on drying lines in the sun, when some shrinkage takes place and the fibre develops a twist.

After drying, the fibre is brushed to straighten it out and remove any traces of waste matter which may remain, together with short, broken fibres of 'tow'. The sisal is then sorted and baled.

Strands of commercial sisal are usually between 2 and 4 ft (0.6–1.2 m) in length. The strands are strong and consist of a number of individual fibres cemented together by natural gums. The strands are rough to handle and not very flexible. If the processing has been carried out satisfactorily, the strands should be nearly white or a pale yellow.

Sisal is a fibre used very successfully in the construction of ropes, cords and twines. It does not flex easily. Some sisal is used to make matting and rugs. Sisal tow is used as a padding material in upholstery.

3.5.5 Miscellaneous vegetable fibres

a. Pineapple fibre. Fibre can be extracted from the leaves of the pineapple plant, *Ananas cosmosus*. The plant is cultivated in a number of countries in the world, chiefly in Hawaii, the Philippines, Indonesia and the West Indies.

The pineapple plant is a biannual or perennial with rather a short life. It produces a short, main stem from which grows a rosette of leaves. The leaves grow to about 3 ft (0.9 m) or more in length and have a fine point. Sometimes they bear spines. The width of the leaf is about 2 or 3 in (50–76 mm).

When the plant is 12 to 18 months old, it produces reddish-purple flowers round its axis; from these the fruit takes about six to seven months to mature. When the plant is grown for the fibre in its leaves, the fruit is removed soon after it is formed to enable the leaves to develop freely. The production of fruit and fibre cannot be combined successfully in the same plant; the fruit requires plenty of sun while the leaves require shade.

A strong, white, fine, silky fibre can be extracted from the pineapple leaves. Its strand length is between 15 and 36 in. There are some difficulties in extracting the fibre. Some leaves tend to die at the tip while still on the plant, and the leaves must be harvested when they are absolutely fresh or the fibres will be difficult to extract. The fibres are extracted either by mechanical methods or by retting the leaves in water. The mechanical, scraping method is carried out by hand by drawing the leaf through a split in a bamboo cane. In the retting method, bundles of leaves are weighted down in slow-running water and left there for five to ten days, depending on the temperature of the water. After retting is completed, the fibres are washed and cleaned, and then hung or spread in a shed to dry; drying in the sun tends to make the fibre brittle.

In the Philippines, pineapple fibre is used for making a fine luxury cloth called 'Pina-cloth'. Some fibre is imported into Spain and used to make choice embroidery. In China, strong, coarse fabrics are made from the fibre, and in Brazil lace and fabrics are made from it. In many countries, pineapple fibre is used for ropes, cords, twines and nets.

b. Kenaf. Kenaf fibre is obtained from the stems of a plant of the Mallow family to which cotton belongs. Its botanical name is *Hibiscus cannabinus*. The plant has been grown in India for thousands of years but little was known of it in Europe until about 200 years ago. The plant flourishes only in tropical or sub-tropical regions, and most of the world's supplies are grown in India, Pakistan and Thailand.

The plant is an annual and is grown from seed and, while weeding is required in the early stages of growth, this can be dispensed with later as the plant grows fast and will smother the weeds. The plant throws up straight, slender, prickly stems which can be 8 to 12 ft (2.4–3.6 m), or even more, in height.

Harvesting of the stems is carried out as soon as they cease to grow in height. The stems are cut by hand as is the case with jute. The fibres are extracted from the stems by water retting, using methods similar to those employed for jute.

Kenaf is a pale-coloured fibre. It has a good breaking strength. It is coarser and less supple than jute but is more resistant to rotting. It has been used, mixed with jute, for the manufacture of hessian and sacks since it is a cheaper fibre than jute. It is, however, not so easy to spin on jute machines as jute fibre and, in spite of its lower price, it is not likely to become very popular for such applications. Kenaf fibre is also used for making ropes, twine, nets and mats.

c. Mauritius hemp. This fibre is obtained from the leaves of a plant, *Furcraea gigantea*, which grows wild in the island of Mauritius, round the Amazon river in Brazil, and in Ceylon.

Fig. 3.14 Mauritius hemp.

The plant is similar in general appearance to the agaves, from one of which sisal fibre is obtained, in that it has a thick stubby stem from which a rosette of leaves grow. The leaves are long and thick, are lighter in colour than those of sisal, and end in a sharp point. They are longer and wider than sisal leaves. The plant appears to have a life of from seven to ten years, and it propagates by scattering 'bulbils' in very much the same way as the sisal plant.

In Mauritius, the first cutting of the leaves is not made until the plants are four years old, and they are then cut every two years. The leaf is easier to cut than the sisal leaf. After cutting, the spiked tip of each leaf is removed and the butts of the leaves are trimmed. The leaves are then made up into 10 lb (450 kg) bundles.

The fibres should be extracted from the leaves as soon as possible after cutting as the juices tend to harden and set. The extraction is by beating, as is the case for sisal, and machines designed to copy the hand beating were built upon the same general principle as those used for extracting sisal fibre. Since all the green extraneous matter is not entirely removed by the beating, the strands of fibre are usually given a short retting treatment for two or three days. The strands are washed and allowed to dry in the open air. The strands are given a brushing after drying.

Mauritius hemp is not so strong as sisal but it is whiter, finer and longer, and is one of the most attractive of the 'hard' fibres. Nearly all the fibre produced in Mauritius goes into the making of sugar bags which are used locally. The fibre is also used for making cloth and twines. It is often used in mixtures with sisal to give ropes and cords a better colour.

d. New Zealand flax. Useful fibres are obtained from the leaf of the plant, *Phormium tenax*, which is better known as New Zealand flax. As the name would suggest, it is a native of New Zealand where it grows wild. Formerly the leaves were collected for their fibre from the wild plant, but some is now cultivated and gives more satisfactory results. The plant has been introduced into the Azores, Argentina and Chile.

The plant has a short, thick rootstock from which the tough sword-shaped leaves grow in a fan arrangement. The leaves can be very long, reaching a length of 14 ft (4.2 m), and widths up to 5 in (127 mm) are obtained.

When ready, the leaves are harvested by being cut near the stem, and are bundled and taken to a stripping mill. A good yield of cultivated leaves is about 30 tons (30.4 t) of green leaf per acre, from which a little over 3 tons (3 040 kg) of fibre are extracted.

The green pulp is stripped from the leaves by means of special machines which combine crushing, beating and scraping actions. The strands are then washed thoroughly, twisted into hanks, and placed on rails for the water to drain away. The strands are then spread out to dry in the open air.

Compared with sisal, the cultivation of New Zealand flax has some advantages. It gives a higher yield of fibre in the leaf, and the fibres are longer. The plant is easily propagated and harvesting is easier. It grows in temperate regions where sisal cannot be cultivated. Its disadvantages compared with sisal are that it requires good land on which to grow and is rather more difficult to clean. The strands are not so strong as those of sisal, but are soft, flexible and lustrous. The fibres are nearly white to pale reddish-brown.

Nearly all the fibre produced in New Zealand is used locally. It is processed into ropes, twines and coarse bagging materials, sometimes being mixed with jute for the latter application.

e. Coconut fibre. The fibre, Coir, is obtained from the husks of the nuts of the coconut palm, *Cocos nucifera*. The nuts are gathered just before they are fully ripe. In Ceylon gathering of the nuts is carried out every alternate month throughout the year. The nuts are picked by hand by men who climb the trees and tap the nuts to determine if they are ready for collecting. The ripe nuts are broken off and dropped to the ground.

The husk of the coconut is made up of a large number of fibres which can be up to 6 in in length. The mills usually buy the husks and extract the fibre from them themselves. This is done by first breaking the husks and then immersing them in water in pits to soften them and enable the fibre to be extracted. The softened husks are then passed between fluted rollers to break down the cementing materials. In Ceylon, the husks are retted in shallow water.

Coir is a coarse fibre. The best fibres are a bright golden colour, while the lower grades are dark in colour. The fibre has a useful strength and elasticity and is used to make ropes, matting and brushes. Ceylon is the only important exporter of coir fibre, but there is production on a fairly substantial scale in India and the Philippines.

3.6 WOOL

3.6.1 History

The name, wool, is reserved for the fibre cut, or plucked, from the coats of sheep and lambs, or pulled from their skins when they have been killed. Wool is one of the oldest textile fibres known and has always been prized as a material from which clothing could be made.

The ancestor of the sheep was probably a rough-haired animal originally breeding on the central plains of Asia. Arabs, moving along the coast of North Africa, probably took sheep with them, and some of these eventually reached Spain. At the same time, it is thought that sheep reached Britain from the flocks owned by certain Asiatic tribes which wandered across Europe.

The early sheep had two coats, an outer coat of long, strong hair which served as a protection for the animal, and an under coat of fine, soft wool which acted as a heat insulator to keep the animal warm. The hairy outer coat was shed each year as the warmer months approached. Over the centuries, selective breeding has changed the character of the two coats. The tough, less-attractive outer coat has been eliminated practically entirely and the soft under coat has been developed. The type of sheep developed in Spain, the so-called 'merino', has a very fine fleece, with practically no sign of outer coat, and requires a warm dry climate in which to flourish. There is no annual shedding of any part of the coat and, in modern merino wool, the presence of 'outer hairs' is regarded as lowering the quality of the material.

In Britain, wool was originally obtained from a small black-faced sheep which roamed wild. In this country, wool has long been valued as a textile material and the domestic rearing of sheep may have begun round about the sixth century B.C. By the time the Romans came to Britain there existed quite a flourishing industry. Under the Romans, this industry flourished even more both as regards the rearing and breeding of sheep, and the weaving of cloth from wool, and continued to expand until the Saxon invasion halted further progress. William the Conqueror put new life into the industry when he arranged for some Continental sheep to be imported into Britain and, from that time, wool production and the manufacture of wool fabrics have become more and more important to the economy of the country. This importance was underlined by Edward III when, in 1350, he decreed that the Lord Chancellor should always sit upon a woolsack, as a perpetual reminder to him of the value of wool to the country, during his deliberations.

Selective breeding to produce the British type of sheep had, of necessity, to be directed towards producing a stronger, sturdier type of animal which could stand up to colder, damper weather conditions than those experienced in the lowlands of Spain. Sheep of the Spanish merino type could not be reared successfully in our climate, and the wisdom of having taken our own independent line in breeding was evident from an attempt made in the reign of George III to rear merino sheep in Britain which failed completely. The sheep which have been bred in this country are stronger and sturdier types and, for many of their characteristics, have become famous throughout the world. Much of the credit for the present-day breeds of sheep in Britain must be given to Robert Bakewell, who was born in 1725 and who evolved specially selective methods of breeding.

It was during the nineteenth century that Yorkshire began to establish itself as the centre of the world's wool trade. The number of mills in operation began to increase rapidly, and the demand for raw wool soon exceeded the home supply. Raw wool was imported, therefore, from the countries of the Dominions which had undertaken the rearing of sheep on a large scale. In those countries, the rearing of sheep constitutes a major part of their economies, and they now supply most of the world's requirements of raw wool.

3.6.2 Origin of breeds and characteristics

The finest, silkiest wool fibres are obtained from the merino type of sheep which, as has already been mentioned, originated in Spain. Originally, the merino flocks were owned by the Spanish crown, nobility and monasteries. There were four main breeds, the 'Escurial', 'Paular', 'Negrette' and 'Infando' types. Of these, the Escurial gave the the finest fibres and belonged to the crown. The Paular sheep were associated with the monastery at Paular, and gave a compact, soft wool. The Negrette and Infando sheep were similar in being somewhat larger animals with folded skins and yielding a shorter fibre. At that time, the exporting of any of these special breeds was forbidden.

Shortly after the middle of the eighteenth century, a number of the Escurial breed of sheep were taken from Spain into Germany. These were located in Saxony and from them was developed the 'Saxony' merino flocks. This breed gives a fine, dense wool and, at the beginning of the nineteenth century, this wool was noted as the best produced in the world. Small numbers of merino sheep were also imported into Austria from Spain and from them were developed the flocks which produce the super fine wool of Hungary.

Merino sheep were also introduced into France towards the end of the eighteenth century. Selective breeding there produced a big-framed type of sheep carrying a heavy fleece of fine fibres; this was known as the 'Rambouillet' breed. This breed has been important in that it has influenced the development of merino breeds in North and South America and in certain parts of Australia.

Merino sheep were introduced into South Africa from Spain round about the year 1790. It was quickly established that the warm conditions of South Africa were suitable for them and the flocks flourished and expanded. More and more land was taken

Fig. 3.15 A merino ram.

over by the sheep farmers to provide feeding pastures and, in a 100 years, the number of sheep in Cape Colony alone had grown to more than 10 million, and has continued to expand since that time. South African sheep produce a wool which is noteworthy for its fineness and softness.

Australia first obtained merino sheep from South Africa in 1795 but, later, other merino breeds were introduced in an attempt to evolve a breed which could withstand the climatic conditions met as the country was opened up. From the breeding experiments carried out have evolved four main types of merino sheep, these being 'Very fine merino' descended from Saxony and Silesian strains, 'Medium merino Peppins' which is a bigger animal evolved by a breeder called Peppin and bred initially from Saxony and Rambouillet types, 'Medium non-Peppins' also a big animal carrying a heavy fleece, and 'Strong merino', which is, again, a large sheep giving a wool of relatively long staple length and having slightly thicker fibres than the others.

Merino sheep were introduced into the USA only in the nineteenth century. The sheep flocks of the State of Vermont were bred from French Rambouillets and this has become an important merino centre in the USA. The 'Vermont' sheep have large bodies with a loose, wrinkled skin carrying a heavy, greasy fleece.

Britain produces no merino wool at all and relies for her supplies on importing from other countries. Over the centuries, however, famous breeds of sheep have been evolved within the country, and these are able to thrive in the climatic conditions which are obtained here. The principal British breeds are 'Lincoln' and 'Leicester'. Both breeds produce wool fibres which are between 10 and 12 in (254–304 mm) in length, but the Leicester wool fibres are the finer of the two. Other breeds producing rather shorter staple length fibres, from 7 to 9 in (117 to 228 mm) are 'Devon Closewool', 'Romney Marsh', 'Cheviot' and 'Border Leicester'. Breeds which produce relatively short wool of excellent quality, being fine, lofty, elastic, and with a good springy handle, are the Down sheep. Chief amongst these is the 'South Down' sheep, but other important breeds are 'Shropshire Down', 'Hampshire Down' and 'Dorset Down'. In Britain, much of the wool produced comes from mountain sheep which roam the mountains and hills of Scotland, Wales and the Pennines. Probably the best known breeds

Fig. 3.16(a) A 'Lincoln' longwool.

Fig. 3.16(b) A 'Leicester'.

of mountain sheep are the 'Scottish Blackface' and the 'Cheviot'. Wool from the former is long and coarse and is made into tweeds and carpets; Harris tweed is made mainly from wool from the Scottish Black face. Cheviot is a medium length wool used mainly in manufacturing suiting fabrics.

In many countries of the world, British breeds of sheep have been used for cross-breeding with merino sheep. British breeds used for this purpose have been Lincoln, Leicester, Romney Marsh, and one or two Down breeds. The object of the cross-breeding has been to develop certain particular characteristics. For instance, sheep bred primarily for their wool are not satisfactory meat producers, and in New Zealand, large cross-bred flocks are providing both wool and meat. In Australia and South Africa, the main producer of wool is the merino type of sheep, but both countries now export quite substantial amounts of wool produced by cross-bred sheep which have been evolved with the ability to withstand particular environmental conditions. In the USA, some cross-breeds of sheep are referred to as 'half-blood', quarter-blood', and so on, the fraction denoting the proportion of merino strain in the animal. A half-blood, for instance, could be a straight cross between a merino and a Lincoln.

3.6.3 Sources of supply of raw wool

The most important wool-producing and exporting country in the world is Australia. It is estimated that about 170 m sheep are reared there, although the climatic conditions experienced in any one year can cause the numbers to fluctuate. Of this enormous number of sheep, something like 75% are merino breeds, and the remainder are cross-breeds.

Australian merino wool is the finest and best quality produced in the world. The wool exported through Port Philip in the State of Victoria is noted particularly for its high quality; it is often referred to as 'P.Ps'. Australian merino wools are usually divided into three main grades, 'fine', 'medium' and 'strong'. The fine grade is white and has a soft, attractive handle; its staple length is of the order of 3 in (76 mm) and the fibres are more crimped, about 25 crimps to the inch, than the other two grades. The medium grade has a slightly longer staple length of about $3\frac{1}{2}$ in (88 mm), and a crimp value of up to 22 crimps to the inch. The fibres of the strong grade are the coarsest of the three but still have a good, soft handle and a satisfactory colour; the staple length is still higher at about 4 in (101 mm), and there are up to 20 crimps to the inch. All these grades, like all merino wools, have good 'felting' properties.

Australian cross-bred wools are also divided into the three grades, fine, medium and strong. The fine grade, of a staple length of the order of 6 in (152 mm), has a good colour, a satisfactory strength, and possesses quite good felting properties. The medium grade is strong and lustrous and has a staple length of up to 10 in (254 mm); it has rather indifferent felting properties. The poorest Australian wool supplied is the strong cross-bred; it has a long staple length of up to 12 in (304 mm), it is lustrous, and it exhibits the poorest felting properties of all the grades.

Sheep are reared in all parts of Australia, but the best wool-producing districts are New South Wales and Victoria, which, between them, are responsible for about half Australia's production of wool.

New Zealand is also a large producer of wool, but mainly as a by-product of the meat industry. Only about 2 or 3% of New Zealand's wool comes from merino sheep, the major portion produced coming from the larger, heavy fleeced, British or cross-bred types. A popular cross-breed, between Lincoln and merino, is known as a 'Corriedale'.

The climate of New Zealand is very suitable for sheep rearing. The North island produces more wool than the South island, but the latter has the reputation of providing the better quality, relatively clean and free from dirt and vegetable matter.

Merino breeds of sheep thrive in South Africa, and particularly in the Cape and Natal districts. South African merino wools are white in colour and are noted for fineness of quality and softness of handle. Some cross-bred flocks are reared in South Africa and the wool is exported.

The main producing countries for wool in South America are Argentina, Uruguay, and in the Tierra del Fuego district at the southern tip of the continent. Probably about a quarter of the wool produced in Argentina is merino obtained from sheep bred from the French Rambouillet type; the remainder of the production is from cross-bred sheep. In Uruguay, good quality wools are obtained from merino and cross-bred flocks. Tierra del Fuego produces wools which are very similar in character to the British Down wools, and most of it is exported to this country.

Large flocks of both merino and cross-bred sheep are reared in the USA.

Wools produced in Britain are classified into four main categories; 'Lustre', 'Demi-lustre', 'Down', and 'Mountain' wools. Lustre wools are obtained principally from the Lincoln and Leicester breeds. Devon, Romney Marsh, Cheviot and Border Leicester breeds provide most of the demi-lustre wools. The best producer of down wool is taken to be the South Down sheep, with Shropshire Down, Hampshire Down and Dorset Down providing wool the fibres of which are probably not quite so fine. The mountain wools are more variable in length, quality and purity than the other types. The sorting of British wool is not at all easy due to the variation in quality which may occur on the same fleece.

Merino sheep have been reared successfully in a number of European countries. It is claimed that wool from the Saxony and Silesian sheep is the finest in the world. Spain rears both merino and mountain types as does France. The Soviet Union produces large qualities of wool, much of it of the coarser type.

In general, India and Pakistan produce coarse, lower grade wools, and carpet quality wools are obtained in Iraq and Iran. Chinese-grown wool is composed of both apparel and carpet qualities.

3.6.4 Preparation of raw wool

Sheep shearing is usually carried out once a year. An expert shearer, using modern power-operated clippers, can shear a sheep quite rapidly and be responsible for producing something like a ton of wool in a working day.

Most of the world's annual supply of new wool is obtained by shearing live sheep, but about 10% is taken from the skins of slaughtered sheep, and is known as 'slipe' wool. The wool is pulled away from the skins, not cut, and before doing so the skins must be treated to loosen the fibres. Two methods of treatment are in common use; these can be referred to as the bacterial and chemical methods respectively. In the bacterial method, the skins are first soaked in water and then hung in a warm, humid atmosphere, under which conditions bacteria attack the basal layer without damaging the actual fibres. After a few days, the appearance of the skins together with a smell of ammonia, indicate that the treatment has proceeded far enough, and the skins are taken down and the wool pulled away by hand. In the chemical, or 'painting' method of treatment, a paste of sodium sulphide and slaked lime is applied to the flesh side of the skin, from which it penetrates through to the wool. Two such painted skins are placed together with the flesh sides in contact and are left in this position for 24 hours for the chemical loosening of the fibres to take place. The wool is then pulled away by hand.

Wool taken from the skins of dead sheep is often regarded as being somewhat inferior to that obtained clipping the live animal. It is mostly used mixed with clipped wool.

The principal wool-producing countries normally sell their wool by auction, this probably underlining the fact that, even with the vast production of today, there is still insufficient to satisfy the world's needs completely. A number of factors will determine for what purposes a batch of wool can be used, and what will be its yield of clean, pure fibre. Such factors will determine the price a prospective purchaser will be willing to pay for the material. Wool is first examined therefore, by experts of long experience who grade it in accordance with its fineness or 'quality', staple length, purity, uniformity and colour.

Fibre quality is very important since, the finer the fibres in a particular batch of wool, the finer will be the yarns which can be spun from it. It was on this latter fact that the 'quality number' system of specifying fineness was based. In this system it is assumed that 1 lb (0.4 kg) of the wool in question is spun into the finest yarn it is possible to make from it; obviously, the finer the yarn which can be spun, the longer, the length of the yarn which can be obtained from the pound of wool. The number of 560-yard (512 m) lengths contained in that limiting fineness yarn, is then taken to be the quality number of that particular batch of wool. Thus, a high quality merino wool might allow a yarn of such fineness to be spun that 1 lb of it would provide an 80×560 yard (73×512 m) length; that particular wool would then be classed as 80s.

Again, 1 lb of a good cross-bred wool may yield 56×560 yards (51×512 m) of the finest yarn which it is possible to spin from it, and this will be graded as a 56s wool. The system has been retained, in spite of modern methods of measuring actual fibre diameter which have been devised, and it is still used in all wool transactions. Merino wools usually cover the range 60s to 80s, cross-breds 44s to 60s, and coarse wools such as are used to make carpets, below 44s. The higher the quality number, the finer the filaments in the wool.

Fibre length is also important in determining the suitability of a particular wool for certain applications. Again fibre length is indicated by terms which are related to the processes through which the wool must pass in being prepared for spinning into yarns. Thus, fibres of a length of $2\frac{1}{2}$ in (63 mm) or more can be processed in a combing machine and this is described as being a 'combing wool'. If the fibre length is less than this, but

Fig. 3.17 Range of wool qualities: (*a*) superfine merino 80s, (*b*) fine merino 70s, (*c*) medium merino 64s, (*d*) strong merino 60s, (*e*) comeback 58s, (*f*) Fine crossbred 56s, (*g*) medium crossbred 50s, (*h*) crossbred 44/46s, (*i*) Lincoln 40s.

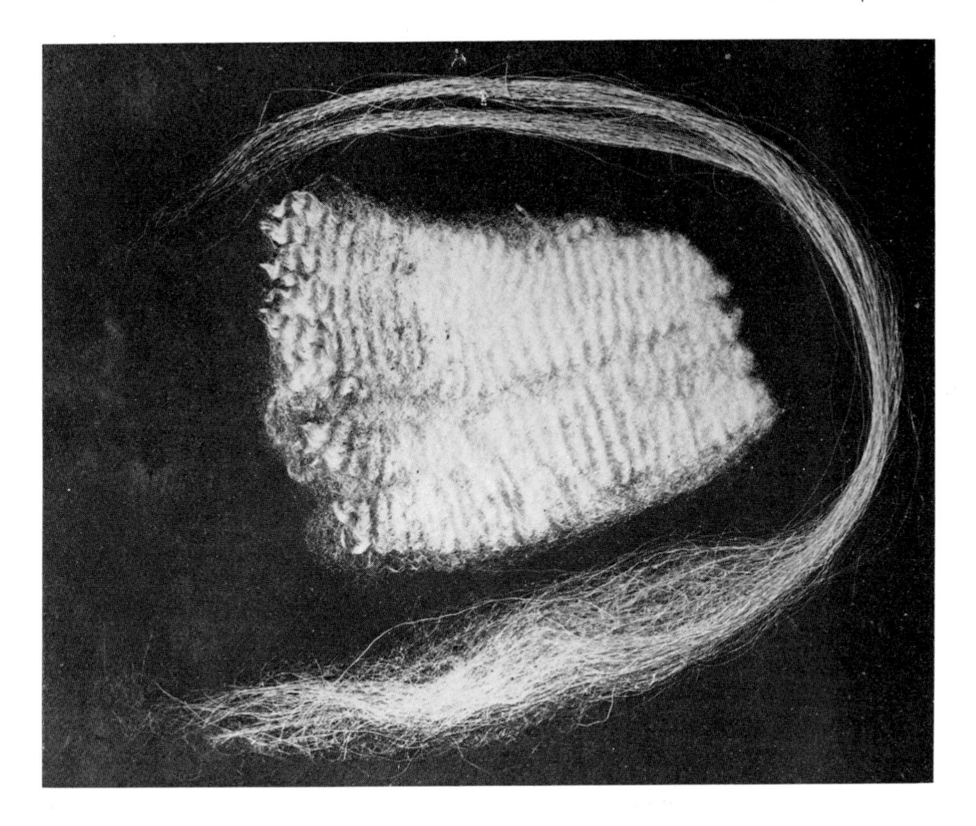

Fig. 3.18 Illustrating the difference between botany merino and coarse carpet wool. The carpet wool is shown surrounding the botany merino sample.

not below $1\frac{1}{2}$ in (38 mm), it can still be processed in a machine known as a French comb, and the material is described as 'French combing wool'. Wools of very short fibre length are not suitable for combing, and in being prepared for spinning, their processing would stop at the carding machine stage. They are referred to as 'carding wools'.

Another factor which has an obvious influence on the price offered is the state of cleanliness of the wool – and therefore how drastic a cleaning treatment will have to be applied – and the yield of good, clean wool. The impurities in raw wool are grease and dried perspiration from the animal, together with vegetable matter, dirt and sand which the fleece will have picked up during the sheep's wanderings. In general, merino wool is dirtier and greasier than, say, a British wool since the fine fibres tend to hold the dirt more tenaciously, and the merino sheep has well-developed grease glands. During scouring, a dirty merino wool can lose up to 50% of its weight. A high percentage of grease in the raw wool will show up as a yellowing or darkening of the fibres, and the grease can be observed at the tips of the fibres. The amounts of sand, dirt and vegetable matter in the wool can be estimated by eye and by squeezing a handful of the material.

The presence of coarse hairs amongst the wool fibres can also reduce the value of a fleece. Some merino sheep which carry a loose, wrinkled skin, may grow coarse hairs in the wrinkles, and these will be in the clipped fleece. Coarse hairs grow on the lower parts of the legs of cross-bred sheep. Some breeds which give a white wool can grow a sprinkling of black hairs in the fleece. Uniformity in the fleece also adds value to it. Fibre length always tends to reduce towards the lower part of the sheep, and the best fleece is one in which the better quality wool extends over the greater part of the fleece.

3.6.5 Re-use of wool

Wool is a relatively expensive fibre and, in addition, there is a need to make as much use as possible of the amount available. Whenever it can be done, therefore, wool fibres from various sources, which are still sound and strong, are re-used in the woollen section of the textile industry. Re-used wool forms quite a substantial proportion of the total world usage of the fibre.

There are three main sources of re-used wool–'soft-waste', 'hard waste' and rags. The first two constitute discarded fibres from the mills during the preparation of yarns. In soft waste the fibres are in a loose state and it includes such things as sweepings from the wool sorting rooms, fibres left in the scouring equipment, waste from the carding and combing machines, and so on. Hard waste is composed of waste from such processes as spinning, twisting, warping, and so on in which the fibres are entangled and bound together. Hard waste has to be pulled loose before the fibres can be used, and this is done in a machine called a 'garnett'.

When using rags as a source of fibre, the rags have to be pulled apart by mechanical action. Soft rags pull apart reasonably easily and yield what is known as 'shoddy' in which the staple length usually varies between ½ and 1½ in. Fiercer mechanical action is needed to disintegrate hard rags and the material obtained from them is 'mungo' containing many short fibres. Rag disintegration is carried out in a machine in which the rags are firmly gripped and fed forward, by means of a pair of powerful fluted rollers, into contact with the teeth of a 'swift' roller rotating at the relatively high speed of around 400 to 700 rev/min.

There can, of course, be difficulties associated with the use of wool re-claimed from rags. The wool fibres in the rags will have already been dyed, and it will be an advantage if the rags can be sorted into matching shades before disintegration. If not, it may be necessary to 'strip off' the dyes using an appropriate chemical treatment after which the fibres can be re-dyed to one shade.

There is also the difficulty that some rags may contain fibres such as cotton and rayon as well as wool, and these other fibres must be removed. The wool may be freed from these other fibres by using a 'carbonizing' process. Two such processes are the so-called 'dry' and 'wet' carbonizing processes. In the former, dried rags are treated with dry hydrochloric acid gas in an enclosed vessel at a temperature of about 90°C (194°F). After a few hours of treatment, usually 3 or 4, the rags are removed, agitated vigorously to break up and shake free the degraded cellulose materials, and then washed. In the wet process, the contaminating fibres are disintegrated in a chemical solution. The rags are steeped for 2 or 3 hours in a weak sulphuric acid solution, after which excess liquid is removed by hydroextraction and the rags dried. The rags are shaken vigorously, washed in a warm soda-ash solution, and again dried and shaken. Although the wet process is more expensive it is more effective when man-made fibres are present in the rags, and it also produces a cleaner shoddy.

3.6.6 Fibre structure

The wool fibre is composed mainly of keratin, a natural protein from which such substances as horn and feathers are also constructed. When the fibre first grows from

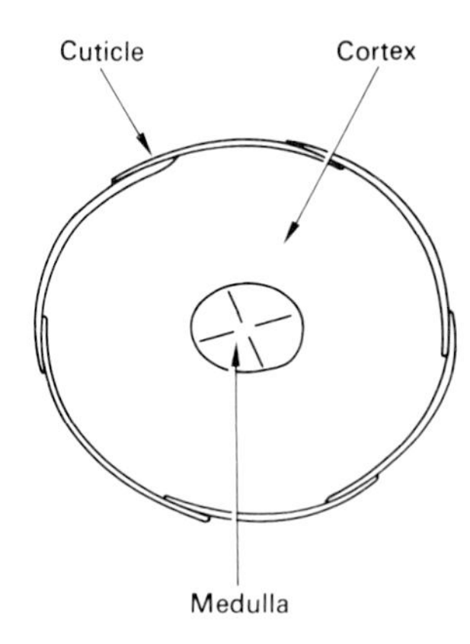

Fig. 3.19 Cross-section of wool fibre.

the skin of the sheep, it has a pointed tip. Once the fibre has appeared from the skin, however, it is a 'dead' material and if the pointed tip is removed, say by shearing, there is no way in which it can grow another point, and so it remains blunt at its end.

A wool fibre is not a solid rod of keratin but is built up in cellular form. It is oval in cross-section, and this cross-section consists of three main parts. On the outside is the 'cuticle', and inside this layer is the main bulk of the fibre, the 'cortex'. Running down the centre of the fibre is usually a hollow core, the 'medulla', which is common in coarse wool fibres and less common in the fine ones.

The cuticle actually consists of two parts, an extremely thin outer skin which transmits water vapour and is known as the 'epicuticle', and under this a layer of overlapping scales which cover the fibre. The free ends of the scales project towards the tip of the fibre. The natural function of the scales is to protect the fibre but, when wool is used for textile purposes, they have a marked and unique effect on the properties of the yarns and fabrics produced from the material. The number of scales is greater in fine wool fibres than in coarse fibres; it can be as high as 2000 scales to an inch in the former, and only 700 to an inch in a coarse fibre.

The cortex is built up from a series of cells which are much longer than they are wide. Their lengths vary between 80 and 100 microns, and their widths at the middle between 3 and 5 microns (1 micron = 0.0001 cm). The cells are pointed at each end and can be described as cigar-shaped. The cells are cemented together in a formation which is parallel to the length of the fibre, and it is they which give the fibre its strength and elasticity. Any stretching force applied to the fibre causes the cells to extend rather than slip relative to each other and, when the stretching force is removed, the cells return to their former shape. Research into the fine structure of the cortical cells has shown that they are built up from a parallel formation of tiny fibrils arranged in much the same way as the cells themselves are in the cortex of the fibre.

The hollow channel running along the centre of the fibre, the medulla, is not always present. It is common in wools obtained from mountain sheep but in fine merino fibres it is often absent or is so small as to be difficult to see. The production of medullated fibres by sheep is probably an inherited disposition; conditions of climate and feeding can also have an effect on the development of medullated fibres. Medullated fibres appear to dye a little lighter in shade than the other fibres. This is probably due to light being reflected back at the walls of the medulla where air and keratin meet, and the fact that the existence of an empty central channel automatically reduces the thickness of the cortex which takes the dye.

Keratin, the protein contained in the wool fibre, is built up from long chain molecules, as are all other fibres (see chapter 4). The chain molecules in wool are folded concertina-fashion and are held attached to each other at intervals by cross-linking bonds. The structure is a stable one and, if it is distorted in any way by the application of external forces, the cross-linking bonds pull it back to the stable form when the distorting force is removed. The combination of folded chain molecules and stabilizing cross-links gives wool its remarkable elasticity. When the fibre is stretched the chain molecules unfold to give a large extension; when the stretching force is removed, the molecules fold up once again and the fibre returns to its original length.

It is possible to break down the cross-links by special treatments. Thus if wool is stretched and, while in this extended condition, it is heated for some time in steam or boiling water, it will, after cooling, be found to have acquired a 'permanent set' in this extended condition. It is believed that moist heat breaks down the strained linkages and then, by continued heating, new links are formed which hold the fibre in this extended state.

In addition to being composed of an elastic material, keratin, the wool fibre has a constructional form which adds to its powers of extension, provides resiliency and warmth in a mass of wool and garments made from it, and enables the fibres to cling together better when they are twisted in the form of a yarn. The fibre has a natural crimp or waviness along its length. This crimp is pronounced in fine wool fibres. A fine merino fibre will have up to 30 crimps to the inch; this number is reduced to five or even less in some of the lower quality wools.

The scaly construction of the cuticle of the wool fibre is both an advantage and a disadvantage from a textile point of view. It does increase the frictional effect between wool fibres and allows strong yarns to be spun from the material, but it is also recognized that this scale formation is the primary factor which enables 'felting' to take place in woollen garments. When a woollen fabric or garment is 'milled' under hot, wet conditions, the fibres are subjected to alternate straining and relaxing movements and under such conditions the fibres migrate in the direction of the root end and become entangled and locked together. In a fabric or garment, this migration can take place between one wool yarn to those adjacent to it and the whole fabric structure will become consolidated. If the milling treatment is prolonged and severe, the identities of the yarns in the fabric will disappear and a thick sheet of entangled fibres will result. In the textile industry, felting can be used to give loose fabric constructions a certain amount of solidity and stability. Fabrics known as 'felts' are made by milling a sheet of loose fibres until they hold together as a fabric without any normal method of fabric construction being employed. The disadvantage of wool's tendency to felt is that 'felting shrinkage' can take place during too vigorous laundering. The motions and conditions generally used during the laundering of garments often simulate those of the milling process, and woollen garments treated in this way can contract in size and lose much of their softness. A mild squeezing, using lukewarm water during laundering, is recommended for soft, open woollen garments and fabrics. Nowadays, much of the wool used in textiles is given an anti-shrink treatment. The object of such treatments is to modify the scale structure on the surface of the wool fibre and reduce the tendency of the material to felt. Anti-shrink treatments are discussed in chapter 7.

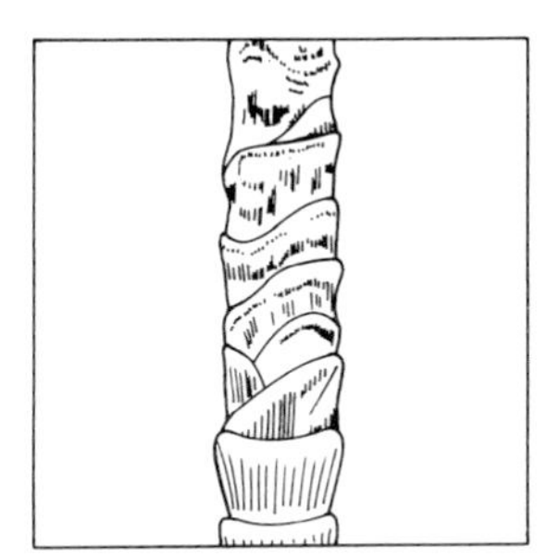
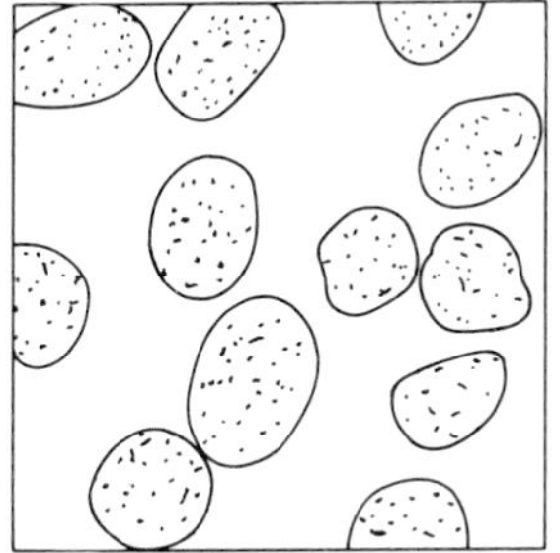

Fig. 3.20 Wool fibres are regular in diameter with a pronounced surface scale structure. Cross-sections are oval in shape.

3.6.7 Properties and uses

The simplest molecular repeating unit in the chains which form keratin can probably best be expressed as:

$$\begin{array}{c}—\text{NH}—\text{CH}—\text{CO}— \\ | \\ \text{R}\end{array}$$

where R is an organic radical.

Under the microscope, the surface of wool fibre is readily seen to be composed of overlapping scales. A good merino fibre shows unbroken scales running the whole of the way across the width of the fibre, but in other breeds and cross-breeds, the scales are more irregular and broken. The fibres are oval in cross-section.

Wool fibres vary quite considerably in both staple length and diameter, depending upon the breed of sheep from which they are taken and, to some extent, upon the conditions under which the sheep are reared. Fine merino wool fibres are usually taken to have a maximum staple length of 5 in (127 mm) with $2\frac{1}{2}$ in (63 mm) being accepted as being a good average length over the whole fleece. Medium wools have fibres up to 6 in in length, and long wools up to 15 in (381 mm). Fibre diameters range from 15 microns for the finest merino to as much as 40 microns for some of the long wools.

Wool fibres have a natural lustre, which varies according to the breeds of sheep from which they are taken. The Lincolns and Leicesters have glossy fibres, and this is referred to as 'lustre wool'. The lustre of fibres taken from some other breeds, including the merinos, is more subdued and these are referred to as 'demi-lustre wools'.

Compared with many of the textile fibres, wool is relatively light in weight with a density of 1.31 g/cm^3. Again, in comparison with some of the other fibres, wool cannot be classed as a strong fibre. It has a tenacity which is no higher than 1.7 g/denier (15.3 g/tex) when the fibre is dry, and when wet this falls to about two-thirds of the dry value. In spite of this, wool is long-wearing, due to its remarkable elasticity. Its extension at break is higher than 40% when dry, and approaching 60% when wet. Wool fibres recover rapidly from stretching and distortion; for small extensions wool has an elastic recovery of nearly 100%. This ability to extend relatively easily and then recover is of considerable textile value. Wool garments can extend when pulled rather than tear, are comfortable to wear, and recover quickly from accidental creasing.

Wool absorbs a considerable amount of moisture. Its moisture regain, determined at 65% R.H. is of the order of 16 to 18%. It can hold as much as 40% of its weight of water. The readiness with which wool can absorb and give up moisture to adjust itself to equilibrium with its surroundings, is of considerable practical value. When a fibre absorbs moisture, it gives out heat, and this effect is marked in the case of wool. A wool garment will thus become warmer as it absorbs moisture from the air.

When wool is held at a temperature of 100°C (212°F) for a long period, it gradually loses softness and strength. At 130°C (266°F) it turns yellow and commences to decompose, while at 300°C (572°F) wool chars and turns black. Wool burns while it is in contact with a flame, but ceases to do so when the flame is removed.

Exposure to bright sunlight produces a discolouration in wool and the material develops a harsh handle. These effects are accompanied by a change in dye affinity and a loss in strength.

In general, wool has a good resistance to acids, and a treatment with dilute acids is often used as a method of removing cotton from wool/cotton mixtures. Wool is decomposed by hot sulphuric acid; it is also sensitive to alkaline solutions. The fibre is dissolved in a few minutes by a boiling 5% solution of caustic soda. The weaker alkalis, such as sodium carbonate, soap and trisodium phosphate are less destructive in their action but, if used at the boil, can cause considerable damage to wool, reduce its tensile strength, and cause it to discolour. Most dyes show some affinity for wool, but those used are often selected for their fastness to acid milling treatments.

Wool is widely used as a material from which to make clothing and domestic textiles. Among the latter can be included such products as blankets and carpets where the elasticity, covering power and crush-resistance of the material are valuable and distinguishing features. Clothing made from wool can be divided into two types of articles manufactured by two distinct sections of the industry, woollen and worsted. In woollen yarns, the fibres are loosely held together with only a light twist and little attempt has been made to obtain an orderly arrangement. Fabrics made from woollen yarns are soft and springy and are warm to the touch. Worsted yarns are spun from well-aligned fibres lying accurately parallel to the axis of the yarn and the fibres are bound together with a substantial amount of twist. Worsted yarns are, therefore, finer, smoother and firmer than woollen yarns, and it follows that the fabrics woven from them are smooth, firm and hard-wearing; such fabrics as fine dress materials and suitings are made from yarns prepared on the worsted system. In the preparation of woollen yarns, only a relatively few processes are involved and one firm will usually undertake all the production stages from raw wool to the final fabric. In worsted manufacture there are too many processes for such centralization to be undertaken. There are probably rather more spindles engaged in the preparation of worsted yarns than in spinning woollen yarns in Britain.

Garments and fabrics made from wool have good wearing properties although, as already mentioned, special treatments have had to be devised to overcome the disadvantages of the material as regards felting, shrinkage and susceptibility to attack by larvae of the clothes moth. Wool is not a strong fibre but its shortcomings in this respect are balanced by the readiness with which it 'gives' and then recovers from sudden snatching. Its abrasion resistance is quite good but not outstanding and nowadays it is a common practice to include nylon fibres in the wool yarn to be used in parts of certain garments which are likely to be subjected to an undue amount of rubbing during wear. Wool fabrics respond readily to steam pressing, but more use is now being made of mixtures of wool with polyester fibres in order to obtain 'permanent creasing' characteristics in the fabrics, and of special treatments which can be given to the fabric, with the same end in view.

3.6.8 Production and consumption of wool

The number of sheep being reared to provide wool throughout the world was estimated to be about 950 m. in 1968. Of this number, there were 170 m. in Australia, 138 m. in the USSR, 60 m. in New Zealand and 59 m. in China.

The output of raw wool on a greasy basis (before being cleaned), by the main producing countries in the years between 1951 and 1968, is shown in table 3.5:

From the table it will be observed that the output in Australia has remained reasonably steady since 1963–64, the drop in the 1965–66 season being explained by a very severe drought. The output of raw wool by New Zealand has shown a continuous increase over recent years, as has that of the USSR. In the USA, fewer and fewer sheep are being reared, and raw wool production is gradually declining. The output of raw wool in Britain has tended to remain steady for a number of years, and this is equally true for Western Europe as a whole.

The main exporting countries for raw wool are Australia, New Zealand, Argentina, South Africa and Uruguay in that order. Of the total of the world exports of 2956 m. lb (1340 m. kg) in 1967, Australia supplied about half. The principal importing countries are Britain, Japan, USA, Italy, France, West Germany and Belgium.

3.7 OTHER HAIR FIBRES

3.7.1 Mohair

Mohair fibres are obtained from the fleece of the Angora goat, an animal which originated in Asia Minor and which takes its name from the district in Turkey with which it was mainly associated. Until early in the nineteenth century, Turkey was practically the only producer of mohair, but Angora goats were then taken to South Africa and the USA, where they have been reared successfully. The USA is now the world's biggest producer and user of mohair fibre, most of the goats being reared in the State of

Table 3.5 World production of raw wool (m. lb – greasy basis)

Country	*Average 1951–52 to 1955–56*	*Average 1956–57 to 1960–61*	*1963–64*	*1964–65*	*1965–66*	*1966–67*	*1967–68*
Australia	1261	1582	1785	1784	1663	1762	1774
USSR	484	690	809	752	787	818	871
New Zealand	434	539	617	623	695	709	728
Argentina	393	414	395	419	430	441	428
S. Africa	268	296	303	296	329	292	307
USA	296	309	281	255	241	236	227
Uruguay	199	181	192	187	183	178	186
Britain	99	116	127	127	129	131	128
Turkey	79	94	95	95	95	97	101
Spain	85	83	86	79	82	82	82
India	65	70	72	72	72	72	72
Brazil	52	62	58	62	64	62	62

Texas. Turkey is the next largest producer of the fibre but is closely followed by South Africa. There is a small production of mohair in Australia.

Mohair fibre grows on the animal in lustrous white locks. The fleeces are graded as 'tight lock' which hangs in ringlets and contains the finest fibres, and 'flat lock' which is wavy and contains fibres of medium quality. Staple lengths in commercial mohair are usually of the order of 5 or 6 in (127–152 mm), although lengths up to 10 in (254 mm) are sometimes obtained. Mohair is smooth and soft to handle, and has about the same degree of fineness as a good cross-bred wool. The fleece grows quite rapidly in suitable surroundings and with the correct diet for the animal. In Turkey, the goats are clipped once annually, but in South Africa and the USA shearing takes place twice a year. The Angora goat shows a tendency to develop a number of short, coarse, brittle hairs in its fleece, and the presence of many of these hairs can downgrade a fleece.

Clean, scoured mohair fibre is usually white and silky. Grading of the fibre is based on lustre, colour, fineness, softness, and the incidence of 'kemp' hairs.

Under the microscope, the surface of a mohair fibre bears some resemblance to that of wool, being covered with scales which are, however, less easily seen than those on wool. The scales are fewer in number, overlap only slightly, and do not stand out so much as those on wool. The fibre is smooth, therefore, to handle. The scales appear to be longer in length than those on a wool fibre of equal fineness, and this offers one means of distinguishing between the two. The cross-section of a mohair fibre is circular with small spots or circles visible. As in wool, the cortical layer in mohair is composed of cigar-shaped cells, and a few of the fibres are medullated.

As regards density and mechanical properties, mohair is very similar to wool. It is usually considered to be the slightly stronger fibre of the two. It is very resilient and fabrics made from mohair have a good resistance to wear. Mohair absorbs moisture readily and can hold nearly as much moisture as wool. Heat, sunlight exposure and chemical properties for mohair are very similar to those for wool, but the fibre is slightly the more sensitive to alkali attack and additional care should be taken during scouring and finishing. Mohair is also probably more easily attacked by bacteria and mildew than wool.

Mohair can be set to shape more easily than wool and is often used in the manufacture of curled pile rugs and imitation astrakhan fabrics. Fabrics made from the fibre are relatively strong and hardwearing and show little tendency to felting shrinkage. Woven dress fabrics are inclined to be a little too stiff to have a satisfactory drape, but open knitted garments, and particularly shawls and scarves, are attractive and durable. Mohair is often combined with wool in lightweight suiting fabrics.

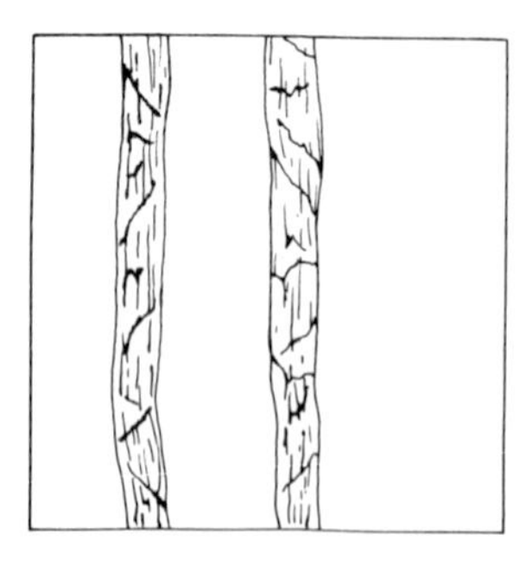

Fig. 3.21 Mohair: reasonably large diameter fibres, striated along their lengths, scales not easy to see.

3.7.2 Cashmere

Cashmere fibre is obtained from the Cashmere goat which is smaller than the Angora goat and is kept as a domestic animal at quite high altitudes in Tibet, Persia, Mongolia, China and Iraq.

The coat of the Cashmere goat consists of two parts, an outer coat of straight, coarse hair, and a shorter fine under coat. Fibres obtained from this undercoat are the ones known as cashmere. The fibre is not clipped but is shed naturally every spring and is gathered by combing from the fleece. Some of the coarser outer hairs are also present in the combings and these are removed by hand as completely as possible, before the material is offered for sale. Each animal yields not more than ¼ lb (0.11 kg) of the fine cashmere fibre each year so that supplies are relatively scarce. China supplies the best cashmere fibre. Britain offers a large market for cashmere from all over Asia.

Good cashmere is white or coloured from fawn to light grey, the coloured fibres being more common than the white. The fibres usually range from 1¼ to 3½ in. True cashmere consists of very fine fibres having a diameter of about 15 microns. With them are mixed some of the coarser hairs, which are up to 60 microns in diameter.

The cashmere fibre is covered with scales which protrude from the body of the fibre more than in the other hair fibres except wool. The cross-section is accurately circular. The cortical layer is composed of cigar-shaped cells, and between the cells are occasional long, narrow spaces which show up under magnification as longitudinal striations in the fibre. The fine cashmere fibres have no easily-visible medullae, but the coarser hairs which are present have well-defined medullae.

Cashmere fibres are chemically similar to those of wool, but the material wets more easily and quickly with water than wool. It is also more sensitive than wool to attack by alkaline solutions, and dissolves readily in a solution of caustic soda.

Cashmere is scarce and expensive, but its softness, warmth and excellent drape, create a demand for its use in the manufacture of luxury articles of wear. It is used in knitted wear, particularly stoles and scarves, and either alone or blended with fine wool in coat fabrics, overcoats and dressing gowns. The mixtures with wool are often lightly milled to produce some slight felting of the wool and so produce a firmer fabric.

3.7.3 Camel

Fibres are obtained from the camel, the best material coming from the Bactrian, two-humped, camels in Central Asia, and particularly North East China. An important Chinese centre for the material is Sinkiang, and a considerable amount is exported from Shanghai and Tientsin.

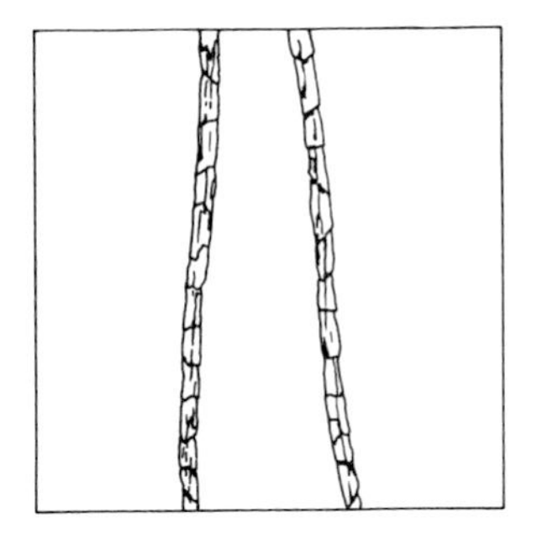

Fig. 3.22 Cashmere is made up of fine fibres of even diameter with clearly visible, unbroken scale structures.

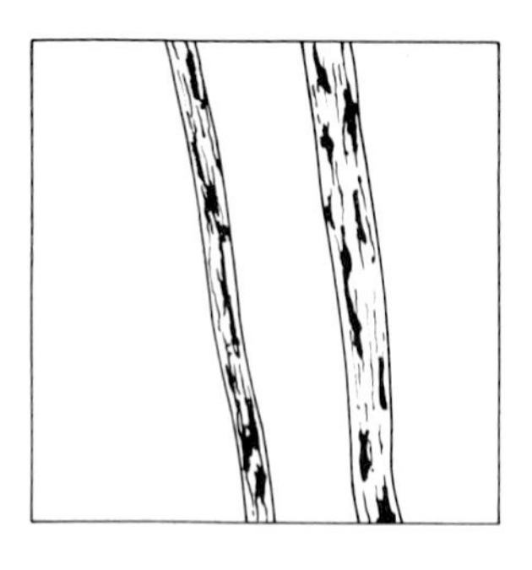

Fig. 3.23 Camel: fibres brownish, with pigment striations and scales having toothed edges which are not easy to see.

The camel has an outside coat of long, coarse hairs, which can be up to 15 in (381 mm) in length, and under it a soft, fine undercoat. These fine fibres vary in length from 1 to 6 in (25–152 mm). The animals moult in June and July, when they are often resting and being well cared for, and the hair is collected and stored. The short fine fibres, and the long coarse hairs are separated as far as possible by hand, and the separation is completed by combing at a later date.

Camel fibres have surfaces which are covered with scales but these are not easy to see. The scales are much shorter than those of cashmere and have toothed edges. The cortical layer is striated due to lines of coloured pigment which give the fibre its characteristic fawn colour. The coarse hairs are markedly medullated. The fine camel fibres have a slightly larger diameter than cashmere fibres but are still extremely fine; they have a round or slightly oval cross-section.

Camel fibres have properties which are similar to those of wool. The material wets out more quickly than wool and, like cashmere, is more sensitive to alkali attack.

Camel fibres are soft and warm, and are used mainly in the manufacture of coats, dressing gowns and knitted goods, usually in the natural colour. The fibres are often blended with a small amount of fine wool to enable a lightly milled finish to be applied. The coarse outer hair is used to make ropes, industrial belts, interlinings and tent fabrics.

3.7.4 Auchenia fibres

Textile fibres are obtained from four species of the Auchenia, or llama-like family of animals. These species are related to the camel. Of these four, two are domesticated, the *llama* being used as a pack animal, and the *alpaca* being bred for its wool. The remaining two are the *vicuna* and the *guanaco* which roam wild and are becoming quite rare. All these animals live at high altitudes in the Andes in South America.

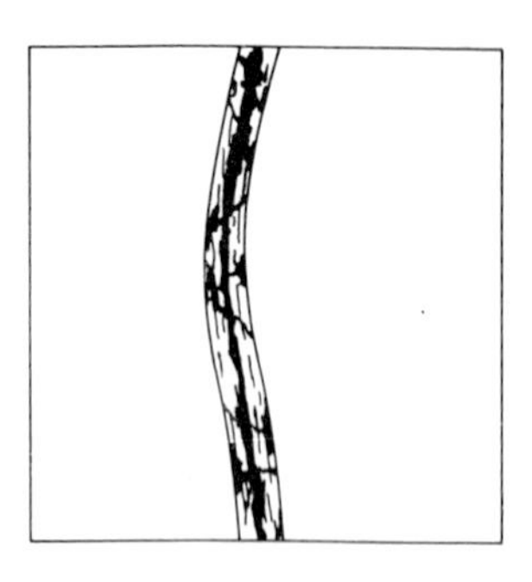

Fig. 3.24 Alpaca: striated fibres with not easily-discernable scales; medulla usually quite clearly marked.

a. Alpaca. From the general textile point of view, the alpaca is the most important of the four animals. The alpaca stands about $3\frac{1}{2}$ ft (1 m) in height and it has a long body. Its hair hangs down on both sides of its body and has a glossy appearance. The alpaca does not shed its hair and, if it is not cut, the hair will continue to grow. It is usual, to shear the animal every second year, when the fibres have reached a length of from 8 to 12 in (203 to 304 mm).

The main centre of the alpaca trade is Arequipa, in Peru, where the fibre is sent after shearing and where it is graded and sorted for colour. Most of the fleece alpaca which is exported, passes through the port of Mollendo.

The alpaca fibres are usually finer than mohair fibres, but are slightly duller. The scales on the fibres are large but not easy to detect, and the material has practically no milling properties. The fibres are striated.

The finer alpaca material is used in its natural colour to make high class knitted-wear. The medium qualities are made into linings, plushes and lightweight suitings. A small amount of fine wool is often added to alpaca to enable a light milling treatment to be carried out and provide a firmer fabric. The coarsest grades of alpaca are used for industrial applications.

b. Llama. The fibres obtained from the llama resemble alpaca but are a little coarser and more likely to contain kemp hairs. The llama fibres are soft and strong, and are often obtained with a staple length of 12 in (304 mm). The colour is mainly brown and the material is not unlike camel hair in appearance. The surface of a llama fibre is covered with scales which are not always easy to detect. The medulla is prominent and is pigmented. Again, llama hairs have practically no milling properties and rely on a mixture with a small amount of wool to provide some firmness in fabrics made from them. Llama fibres are used to make rugs, carpets, and some wearing apparel.

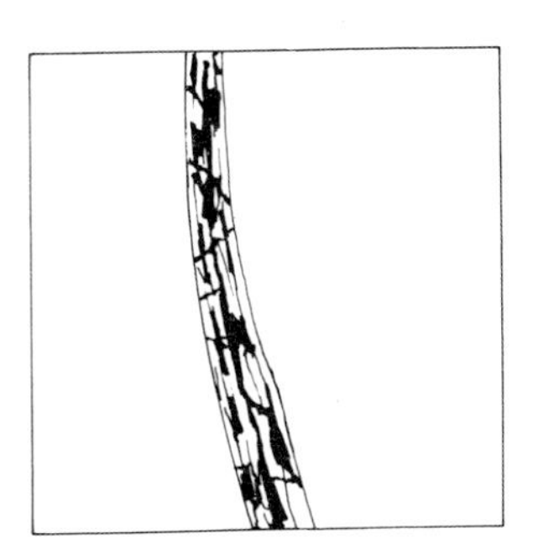

Fig. 3.25 Llama: fibres somewhat similar to those of alpaca, but generally larger in diameter.

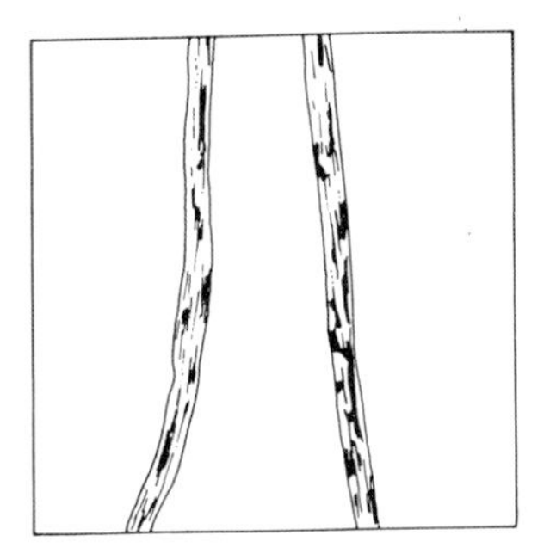

Fig. 3.26 Vicuna: fibres of small and uniform diameter and with evenly distributed colouring; scales not easy to see.

c. Vicuna. The vicuna produces the finest of the wool-like fibres, an average diameter for the fibres being as small as 13 microns. Since it is a wild animal, the vicuna has to be shot to obtain its fleece, which only weighs about 1 lb (0.4 kg). At one period, the vicuna was hunted so much that it came near to extinction, and the Peruvian government was obliged to lay down regulations regarding how many could be killed in any one year. As a result, the amount of vicuna fibre which can be obtained each year now totals about 3 000 lb (1 300 kg) only.

Vicuna fibres are about 2 in long and little more than half the diameter of quite fine wool fibres. The fine fibres which are stripped from the skin of the animal contain a few coarser hairs, and these are removed during processing to preserve the exceptional softness of handle of the material.

The scales on the surface of the vicuna fibre are not easily seen and, as is the case with all the auchenia fibres, the material has little or no felting characteristics. There are striations in the cortical portion of the fibre, and there is rarely any medulla.

Being so rare, and possessing such an exceptional softness, vicuna fabrics are expensive and prized for luxury articles of wear; these include shawls, scarves, dressing gowns and coats.

d. Guanaco. The guanaco also has a fine fleece, but not so fine as that of the vicuna; the diameter of its fibres is about mid-way between those of vicuna and alpaca. The animal is found at lower levels than the vicuna, and chiefly in Argentina. Guanaco is hunted for its skin, from which the fibres are cut.

3.7.5 Rabbit hair

Rabbit hair has long been used in the manufacture of felt hats. The animal has hairs of two kinds, long relatively coarse outer guard hairs, and shorter, finer under hairs. The felt hat trade makes use of the fine hairs, which are heavily milled so that they hold together in the form of a felt. The longer hairs are sometimes used for textiles.

The softest and longest rabbit hairs come from the Angora rabbit. This animal is bred solely for its hair which is used in textiles. It is bred in many European countries and in the USA. The rabbits are shorn two or three times a year.

Commercial angora is graded according to fibre length and the degree of entanglement of the fibres. The best lengths are accepted as being $2\frac{1}{2}$ to 3 in (63 to 76 mm), and an average fibre diameter is about 13 microns.

The angora fibre is covered with scales which have sharp points. Its cross-section is approximately round and its medullary canal is well defined with a characteristic cell formation. The angora fibre is lustrous.

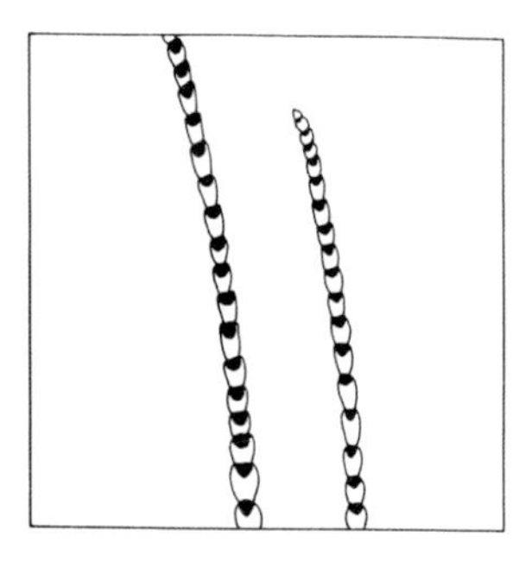

Fig. 3.27 Angora: fibres have a characteristic 'bamboo cane' appearance.

The chief use of angora is in the preparation of 'fluffy' hand-knitting yarns. Here it is used, either alone or in blends with wool or with nylon, to prepare a soft yarn which is subsequently mechanically treated to make the angora fibres stand out from the main body of the yarn.

3.8 SILK

Silk is an animal protein fibre which is generated by an insect as a constructional material from which to make its cocoons. We refer to the insect as the silkworm.

3.8.1 History

The production of silk, or 'sericulture' as it is called, probably originated in China, and there is evidence that raw silk was being produced and converted into fabrics there as long ago as 3000 B.C. China guarded her production secrets very jealously and retained a silk monopoly for something like 3 000 years. Eventually, round about A.D. 300, silkworm eggs, and the seeds of the mulberry tree on the leaves of which the worms feed, were smuggled out of China into Korea, and from there sericulture spread to Japan.

Gradually, the art of producing silk spread westward, first to India, and then to Persia, and the countries round the Mediterranean. It reached Spain and Sicily in the eighth century A.D., and Italy in the twelfth century. It flourished in Italy and the North Italian towns of Milan, Genoa and Venice became famous for the silks they produced. Up to the beginning of the Second World War, Italy was still producing several million lb of good quality raw silk annually. The production of silk was established in France, in the Rhone valley about the year 1500.

The manufacture of fabrics from imported silk was carried out in Britain in the reign of Edward III. Silk fabrics were highly prized and their production began to expand. In 1718, the first silk-throwing mill in Britain was established at Derby, and this was quickly followed by other mills in Congleton, Stockport and Macclesfield. This increasing interest in silk fabrics naturally led to attempts to introduce sericulture into Britain, but none was successful economically. Sericulture can only be successful on a commercial scale if adequate supplies of cheap labour are available as in pre-war Japan, China, India and Italy. For the same reason, sericulture was never established in the USA.

Japan has always made a close study of raw silk production and Japanese methods of sericulture were founded on a scientific basis. Before the Second World War, Japan was the leading silk producer in the world. Now that the material is used in much smaller amounts, Japan's silk production has been reduced considerably, but she still continues to provide the greater proportion of that now required. During the present century, silk production has experienced many set-backs. The difficulty in obtaining raw silk during the First World War, allowed viscose rayon to establish itself and take over a number of applications for which silk had previously been used. Nylon dealt a further blow to silk by ousting the latter material completely in the manufacture of stockings, and in the preparation of many types of fabric. In the period 1939–45, most of the silk-producing countries were at war, and silk production was virtually at a standstill. Since the war ended attempts have been made to revive the silk industry, with some success, but the high cost of production coupled with severe competition from the man-made fibres, will limit the number of applications for which silk will be selected.

3.8.2 Varieties of silk moth

There are two main types of silk moth, the cultivated and wild varieties. The best silk comes from the cultivated type, *Bombyx mori*, and practically the whole of the silk industry is based on this one insect. The moth is bred in silk farms, or 'filatures' as they are called, in Japan, China, Italy and a few other minor producing countries. The silk-work which comes from this moth feeds exclusively on the leaves of the mulberry tree, and consumes enormous quantities of leaves during its lifetime.

However, quite substantial quantities of a strong, coarser silk are obtained in some countries from the cocoons of 'wild' silk moths which live and breed in the open air. The most important class of these is the one which produces 'tussah' or 'tussore' silk. This silk is obtained from several species in the family, Antheraea. The two best-known members of the family are *Antheraea mylitta* which is common in India, and *Antheraea pernyi* which breeds in North China. This family of moths feeds almost exclusively on oak leaves. While tussah silk is produced by a number of different species of insect, the silks obtained are sufficiently alike in character to allow them to be mixed together in use. Silkworms which produce tussah silk are usually appreciably larger than the cultivated variety, and can be up to 6 in (152 mm) in length. They live in the open air, and

Fig. 3.28 Feeding mulberry leaves to silkworms.

produce two lots of cocoons each year, in the spring and autumn. The latter is the time when thorough collection of the cocoons takes place, the spring crop being regarded more as a means of providing silkworms which will contribute to the autumn cocoons. The tussah silkworm leaves one end of its cocoon open except for a thin coating of material which it produces and, when the moth is ready to emerge, it breaks through this coating without damaging the silk filaments from which the body of the cocoon is made.

There are other wild silk moths, the filaments of which are used for textile purposes. In Japan, *Antheraea yama-mai* breeds in the open and feeds on the leaves of oak trees; the moth is often 6 in across, and the cocoon is large. A type of moth, *Attacus ricini*, breeds wild in the American and Asian continents and feeds on the castor oil plant; it produces a good quality white silk. In Africa, a silkworm of the Anaphe family breeds wild and lives mainly on fig leaves. This particular type is interesting in that a colony of silkworms combine to construct a silk nest in which each then produces its cocoon. These are collected in Uganda and parts of Nigeria and used locally to produce silk fabrics.

There are more than 400 different types of moths which produce a kind of silk, but only a few are of commercial value, the rest providing low quality materials which are mostly difficult to collect and process.

3.8.3 Life of the cultivated silk moth, Bombyx mori

The moth lays its eggs in the summer and these are stored in a cool place until the following spring. As soon as a supply of mulberry leaves is available, the eggs are warmed to bring about hatching and, after a few days, tiny silkworms, under $\frac{1}{8}$ in in length, appear. The eggs are usually hatched on trays in heated sheds and, when the silkworms appear, they are provided with a supply of chopped mulberry leaves. The silkworms eat rapidly and continuously and quickly grow in size and weight. They eat nothing but mulberry leaves, which must be supplied in the correct condition of freshness if a satisfactory final result is to be obtained.

The lives of the silkworms are spent in eating, except for four periods, of about one day's duration each, in which they shed their skins, which have become too small for them, and grow new ones. At the end of about 35 days from first being hatched, the silkworm has attained a length of about 3 in (76 mm) and a weight approaching $\frac{1}{4}$ oz (7 g).

When the silkworm is fully grown, it ceases to feed, and looks about for somewhere to construct its cocoon. In the silk farms it is usual to place bundles of straw on the feeding trays, and the worms which are ready to make cocoons climb into the straw and begin to spin. The liquid silk is contained in two glands inside the worm, from which it flows in two distinct streams and emerges through a hole in the head of the insect. As the liquid emerges, it hardens to form two fine filaments which are encased in, and are cemented together side by side by a gummy substance called 'sericin', which is expelled at the same time from two other nearby glands. As the silk filaments are formed, the silkworm moves its head in a figure-eight motion and distributes the material to form the cocoon. Once the silkworm begins to spin, it does not stop doing so until its supply of liquid silk is exhausted. This will take from two to three days, during which time the worm will have extruded up to one mile length of filament, and will itself have shrunk considerably in size.

Once the cocoon construction has been completed, the worm changes into a chrysalis and eventually into a moth. These changes take place over a period of from two to three weeks. If allowed to do so, the moth will then eat its way through the silk filaments to escape from the cocoon. This escape would, of course, break the filaments and prevent them from being reeled off. To prevent this, the cocoons are collected as soon as spinning is completed, and baked or steamed to destroy the insect inside. This has to be done carefully to avoid damaging the silk. The cocoons can then be stored indefinitely until reeling of the silk filaments is to be carried out.

The destruction of the insect inside the cocoon means, of course, that no eggs become available for the next batch of silkworms. In any case, silkworms are susceptible to a number of diseases which either kill them, or inhibit their proper development. Egg producing is, therefore, a separate branch of the industry carried out under carefully controlled conditions. The moth emerges from the cocoon and lives for only a few hours. It eats nothing, and its only objective is to mate and for the female to lay her batch of about 400 eggs and then die. When it has mated, each female is isolated and examined carefully after the eggs have been laid. If disease is present, the moth and its eggs are burned.

3.8.4 Preparation of raw silk

The silk has now to be taken from the cocoons. Each cocoon is covered with about six layers of interlaced fibre which form a protective crust, and this crust must first be removed. If the number of cocoons is small, the crust will probably have been stripped off by hand at the time when the cocoons were being gathered, but if a considerable number is involved, then a 'peeling' machine can be used. In this machine, the cocoons are allowed to come into contact with rotating roughened rods; the projections on the rods catch the outer layers and strip them away. A final hand stripping may be required to complete the operation.

Even when the outer crust has been removed, the cocoon is still hard and solid, since it consists of the layers of silk filament set in hardened sericin gum. In all the operations which constitute the reeling off process, therefore, conditions must be selected which will soften the gum and allow the silk filaments to be drawn away.

The silk fibre as it exists on the cocoon is a double filament, encased in sericin gum, as extruded by the silkworm. We will, however, refer to it as the silk 'filament' in order to avoid confusion at a later stage. Before reeling can begin it is necessary, of course, to find the free end of the filament, an operation which requires skill and experience on the part of the operator. The surface of a tank of hot water is covered with stripped cocoons and a horizontal, rotating, circular soft-bristle brush is lowered to make contact with them; after a period of brushing, the free ends of the filaments become visible. The filament ends are then collected together and a small amount of silk drawn off to ensure that the cocoons will unwind cleanly. The level of the water in the tank is kept constant to ensure that the same brush pressure is maintained. If, during reeling, the silk filament breaks, the cocoon will have to be brushed once again to find the broken end before reeling can be resumed.

Reeling of the silk filament is carried out with cocoons floating on the surface of hot water in a tank, in order to keep the gum softened and allow free rotation of each cocoon. The temperature of the water in the tank is kept constant during the operation. Each filament is so fine that it is usual to reel off filaments together, from say, six cocoons to produce a more substantial thread. The six filaments are taken up from the floating cocoons and then over a rod fixed above the hot water tank. At the same time, filaments

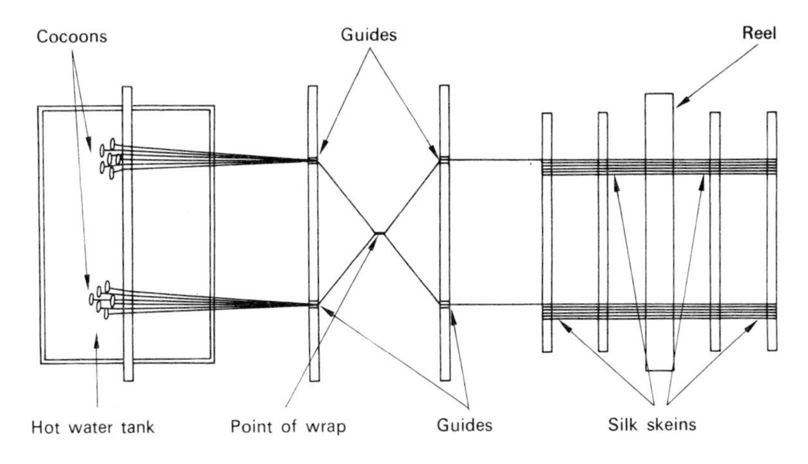

Fig. 3.29 Silk reeling.

are taken up from six other cocoons, floating in a position adjacent to the first six, and again over the rod. By means of guides, these two sets of filaments are led to two individual positions on a reeling machine where they will continue to be drawn off to build up two skeins of silk thread. In their passage from the hot water tank to the reeling machine, the two sets of filaments are allowed to wrap round each other at one point. This wrap is important in that it is a means by which excess water is expelled, and the threads made more compact by the rubbing action.

Reeling is an important process in commencing to prepare a silk yarn. The uniformity and freedom from defects of the yarn which will eventually be made will depend, to a considerable extent, on the skill of the reeling operator. If a filament breaks, then reeling must be stopped immediately and the filament joined up to preserve a uniform thread thickness. When joining takes place, the knots must be secure, but be as neat and inconspicuous as possible. Care must be taken to avoid tangling of the filaments, which will produce thickening, and loops of filaments standing out from the body of the thread.

Tussah silk is more difficult to reel than cultivated silk. The filaments which make up the cocoon are gummed together more firmly in the case of tussah silk and usually have to be given a mild chemical treatment before reeling can be carried out.

The reeling machine winds the silk into skeins. A number of skeins are collected together into lots weighing about 6 lb (2.7 kg) each. These are then packed ready for shipment.

Only a proportion of the silk produced by the silkworm is able to be processed in continuous filament form. Some cocoons may have been broken open by the moth, preparation for reeling produces some silk waste, some cocoons may not allow satisfactory reeling, and others may have been damaged during handling. In addition, a substantial amount of waste silk results from the further processing of the material. All this material is too valuable to throw away, and it is collected and sold for the production of 'spun silk' yarns. The waste silk is composed of fibres of various lengths and it contains dirt and gum. Before it can be used, it is cleaned and degummed, either by boiling in soapy water when the gum is dissolved, or cleaned and partially degummed using a fermentation technique after which about 20% of the gum remains on the silk.

Fig. 3.30 Preparing spun silk prior to it being processed into yarn.

This latter treatment produces what is known as 'schappe' silk which finds specialized uses. After final cleaning and degumming, the silk is passed forward for processing as other staple fibre materials.

3.8.5 Structure and properties of silk

As already mentioned, the silkworm spins two filaments, the material of which is referred to as 'fibroin', and surrounds these by a coating of sericin gum. The presence of this gum gives a rough surface appearance when raw silk is examined under a microscope. In a section through raw silk the two individual filaments can be seen quite clearly, roughly triangular in shape with the angles rounded and one flat face of each

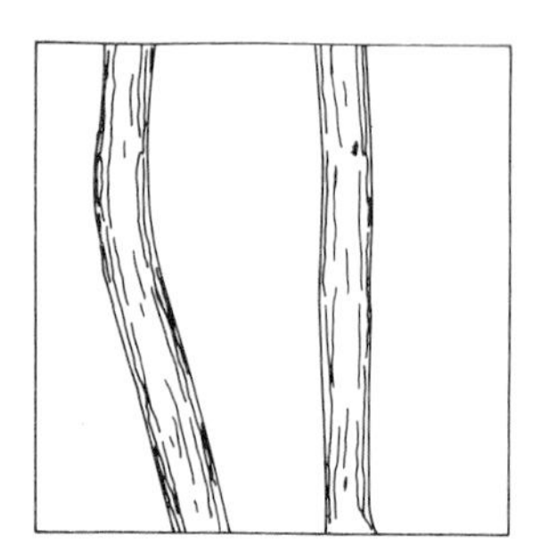

Fig. 3.31 Degummed silk: fibres are rod-like and lustrous with irregular swellings sometimes present.

facing the other. Each filament is coated with gum and the coated filaments are also joined together by gum, giving an overall cross-section which is oval in shape; sometimes the two filaments are spaced a substantial distance from each other. The proportions of filament silk (fibroin) and sericin gum are usually about 75% and 25% respectively. Cultivated raw silk varies in colour between yellow and cream, but much of the colour is removed when the gum is dissolved away.

Wild silk is more variable in filament diameter than cultivated silk. The two filaments are generally wedge-shaped with the short bases facing each other in their gum surround. The colour can be yellow, green or brown.

Under the microscope, degummed cultivated silk shows up generally as fine smooth rods, about 12 microns in width, but there are occasional flattenings. With the removal of the gum, the characteristic silk sheen is developed. The cross-section is a rounded triangle. Degummed wild silk has brownish striations running along the length of the filament and, in the cross-section, these show up as dark-coloured dots in the material.

The density of degummed silk is 1.25 g/cm^3. Silk is a relatively strong fibre, having a tenacity which lies between 3.5 and 4.5 g/denier (31.5 and 40.5 g/tex). The tenacity of the material when wet, is about 75 to 85% of the dry value. Silk shows a good elongation, the value being 16% at break. The elastic recovery of silk when released after stretching is higher than that for cotton and rayon, but not so high as that of wool.

The regain value for silk, determined at 65% R.H., is 11%. The material can, however, hold substantial quantities of water; its water retention value is about 50%.

Silk is able to withstand higher temperatures than wool. It is not affected by temperatures up to 140°C (284°F) but begins to decompose at about 175°C (347°F). Silk does not burn readily and gives a bead of black carbonized matter when contacted by a flame; it gives off an odour similar to that of burnt wool.

Continous exposure to sunlight produces a falling-off in strength of silk.

While the sericin gum is soluble in dilute soap solution the filaments of silk are not soluble. Filament silk withstands the action of boiling water quite well, and certainly much better than wool, but very prolonged boiling tends to weaken it. Its resistance to alkalis is quite good, and better than that of wool, but hot caustic soda solution dissolves it. Silk is resistant to dilute acids and the common organic solvents, but is decomposed by concentrated acids and particularly when they are hot. In moderate concentrations acids cause silk to contract, and this fact has been used to develop crepe effects in the material. The material is attacked by oxidizing agents, and great care must be exercised during bleaching using hydrogen peroxide; hypochlorite bleaches are not used on silk. In general, tussah silk has a rather better chemical resistance than cultivated silk.

Silk has the property of absorbing certain metal salts, such as those of tin, aluminium and iron, from their solutions in water. This property is the basis of the 'weighting' of silk which will be described in chapter 7.

Almost every class of dyestuff which can be used for cotton and wool, can be applied to silk. Neutral-dyeing acid dyes and direct cotton dyes give good fastness characteristics on silk. Dyeing is usually carried out near the boiling temperature.

Silk can be used in a wide variety of fabrics, including knitwear. The fabrics have a characteristic sheen and are soft and warm; they are pleasant to handle and drape very well. The material is relatively expensive and its price can fluctuate, factors which tend to limit its applications. Silk is used in some technical fabrics, mainly in the form of screens.

3.8.6 Production and consumption of silk

In recent years, the world output of raw silk has remained reasonably steady at an average of about 72 m. lb (32 m. kg). Japan is still the largest producer of raw silk. Of the world total in 1967, Japan produced 41.7 m. lb (18 m. kg), China 16.5 m. lb (7 m. kg), and the USSR 6.1 m. lb (2 m. kg). India and South Korea produced relatively small amounts of raw silk, but their output is increasing.

Much of the silk produced in Japan is used internally and Japanese exports of the raw material are falling off. The exports from China and South Korea are increasing. In 1967, the main exporters of raw silk were, China with 9 m. lb (4 m. kg), South Korea 2 m. lb (0.9 m. kg) and Italy 1.2 m. lb (0.5 m. kg).

Prior to 1966, the USA was the largest importer of raw silk but, since that year, the intake has fallen off markedly and is now only a relatively small fraction of the pre-1966 figure. Italy is now the largest purchaser of raw silk, and the main bulk of the imports is from China. In Britain, the silk industry has declined markedly and continues to do so. In 1967, Britain imported only 424000 lb (192323 kg) of raw silk, about 90% of this coming from China.

3.9 TESTING AND IDENTIFICATION OF THE NATURAL FIBRES

There are certain characteristics of fibres which determine how useful the materials are likely to be for specific textile applications. Some of these properties can be examined by means of tests which can be carried out on the fibre, and experience has shown how the results can be related to the behaviour of that particular material in practice. Some of these properties are discussed below. An indication of the method of test used is given, but the reader is referred to the various books which have been written on the subject of textile testing if he desires to know more of the details of such tests.

3.9.1 Dimensions of the fibre

Fibre length, width or fineness and cross-sectional shape, are all of practical importance.

If the staple length of a fibre is too short, then it may not be possible to prepare yarns from such a material at all; it will thus be prevented from being used for normal textile applications. A long staple length means that a long length of fibre is in contact with its fellow fibres in a yarn, and it will be more difficult to drag out of the yarn even when a low binding twist has been applied.

There are always certain inequalities in natural fibres, and a yarn will be more even in appearance, and uniform in properties, the greater the number of fibres in its cross-section. The number of fibres in the cross-section of a yarn of given count depends on the fineness of the fibres composing it. In addition, the finer the fibres, the greater the surface area in contact between fibre and fibre in a yarn and less likely are the fibres to slide apart when the yarn is under tension. Fibre fineness also has an effect on other mechanical properties. For instance, for a given material, the finer the fibre, the more easily will it bend, and the better the drape of a fabric made from it.

The cross-sectional shape of a fibre has an important influence on its behaviour when it forms part of a fabric. Fibres having round cross-sections usually have an attractive

Fig. 3.32 Comb sorter diagram preparation for fibre length.

handle. On the other hand, round cross-sections often give a reduction in covering power when compared with, say, flatter more ribbon-like fibres.

With natural staple fibres, dimensions vary, not only between different types of the same material, but also within the same type. In testing a batch of material for staple length, therefore, it is not only important to determine a length which will characterize that particular batch, but also to gain some idea of the distribution of fibre length. In buying and selling natural fibres, a first assessment of staple length is made by 'hand stapling' methods. A sample tuft of fibres will be taken from a bale and this will be smoothed and drawn out by hand until the fibres are lying parallel to each other in a band about ½ in wide. This band is then laid on a flat, black surface and an estimate made of the average fibre length; in doing this, the thin fringe of fibres at each end is neglected and the measurement made between the two points in the tuft where the ends are reasonably well-defined. A more accurate value for the average staple length and the length distribution can be obtained in the laboratory by taking each individual fibre from the tuft using tweezers, straightening it and recording its length. This is a laborious process, and unsuitable for use in a mill laboratory, and so various instruments, referred to as 'fibre sorters', have been devised. These sorters allow bundles of ditions, the fibres are subjected to alternate straining and relaxing movements and under

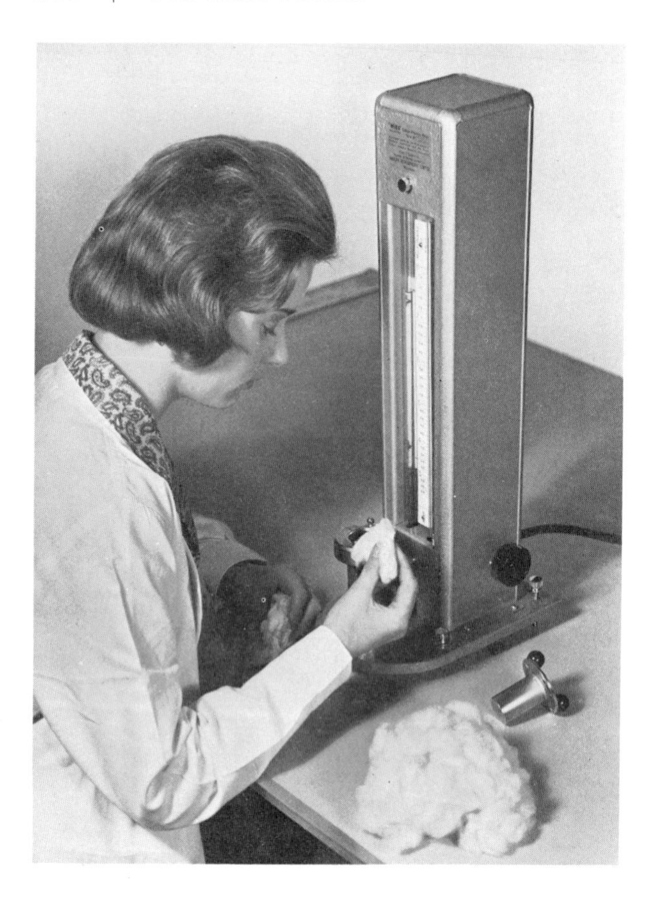

Fig. 3.33 Flowmetre test for fineness and maturity.

of each bundle touching a base line. The equal-length bundles are arranged side-by-side in order of decreasing length to produce an array which is in effect, a diagram illustrating the length distribution in the tuft. From this 'sorter diagram', the average staple length, the 'effective' length (or length of the main bulk of the longer fibres), the percentage of short fibres, and the variability of the material can be calculated.

If all the fibres in a sample had circular cross-sections and these were uniform all along the fibres, then a microscope could be used to measure the diameter of each fibre in a tuft and so determine the fibre fineness. Unfortunately, in natural fibres, these conditions do not exist and optical methods of measurement are not often used. It is more usual to specify fibre fineness in terms of weight per unit length rather than an actual width, since this is independent of cross-sectional shape and variations in cross-section. Weight per unit length is determined by counting out, say, one hundred fibres, weighing them on an accurate balance, and then taking each fibre in turn, straightening it, and measuring its length, when the sum of the lengths divided into the weight will give the information required. Methods of comparing fibre fineness based on the flow of air through a plug of fibres have also been devised. A sample of known weight of fibres is compressed in a small cylinder to a known volume, and an air current under a standard pressure is passed through the plug. The rate of flow is measured by a flow meter which is usually calibrated directly in terms of fibre firmness.

The cross-section of a fibre can be studied by cutting across a fibre and examining this cut face under a microscope. Various instruments for preparing cross-sections are available but a simple method of doing so is to use a thin metal plate in which is drilled a hole of $\frac{1}{32}$ in (0.79 mm) diameter, passing a loop of strong yarn through this hole, and allowing the loop to encircle a small tuft of the fibres to be examined. The loop of yarn is then pulled back through the hole bringing the tuft of fibres with it. When the fibres are firmly wedged in the hole, the ends of the tuft hanging out on either side of the plate are cut off level with the plate surfaces using a sharp razor blade. The cross-sections of the slices of fibres left in the hole of the plate can then be examined by placing the plate under a microscope.

3.9.2 Density

Fibre density is an important property in that it affects the covering power of a given weight of material, influences the way in which a fabric made from the material will drape and, provided it can be measured with sufficient accuracy, offers a means of identifying the material. It is, however, a property which is difficult to measure accurately and in the past, elaborate methods had to be used to obtain precise measurements. A method of measuring fibre density was described, however, in the *Journal of the Textile Institute*, **41**, T.446, 1950, and this has offered a relatively simple means of measuring density and of comparing the densities of a number of fibres.

In this method, a density gradient column is formed in a vertical cylinder using two liquids of different densities. The mixture will have the highest density at the bottom of the cylinder and the lowest at the top, with intermediate values at points in between. Two liquids which are commonly used to determine the densities of fibres by this method are xylene and carbon tetrachloride, and the density gradient is calibrated using small glass floats of 2 or 3 mm diameter whose densities are known. Once the column has been prepared and calibrated, a fibre is dried and then dropped in to the liquid mixture, when it will sink to a depth corresponding to its density, and this can be read off. Since the liquids will expand with increasing temperaure and their densities change in consequence, it is necessary to control the temperature of the column quite accurately. In addition it is necessary to make certain that the immersed fibre is free from air bubbles.

3.9.3 Tensile properties

The load which a fibre will withstand before it breaks is of obvious practical importance. The breaking strength of any material is commonly expressed as the load per unit cross-sectional area required to cause fracture and is usually given in lb/in^2. For a single fibre it is more usual to state 'specific stress' at break, or 'tenacity', which is the breaking load divided by the mass per unit length.

The tenacities of fibre materials included in the literature are often quoted in relation to the 'denier' of the fibres or filaments tested. Denier is the weight in grams of a 9000 m length of the sample being tested and, for a given material, it is a means of specifying the fineness of the fibres of that material. For a staple fibre material, the combined weights and lengths of a number of fibres will provide data from which the denier value can be calculated. For continuous filament yarns, it is simply a question of reeling off a suitable length, and weighing it, when the denier value can be calculated. Thus, if a

100 m length is reeled off and found to weigh, say 0.167 g, the denier of that sample will be:

$$D = \frac{0.167 \times 9000}{100} = 15.03$$

and taking this to the nearest whole number, this would be taken to be a 15 denier fibre. In recent years, a more universal unit, the 'tex', has been adopted and is coming into greater use. The tex is the weight in grams of 1 000 m of the filaments or yarn. It follows that, in order to convert a denier determination to tex, it is necessary to divide the denier by 9. Thus a 120 denier yarn is one of 13.3 tex. When figures are quoted in grams per denier, the value must be multiplied by 9 to convert to grams per tex, i.e. 5 grams per denier equals 45 grams per tex.

The piece of equipment used to study the tensile properties of materials, in this case single fibres, is called an extensometer. Using such an instrument, it is usual to obtain readings which will enable the extension of a fibre to be plotted against the extending force applied; the curve thus obtained is usually referred to as a 'stress–strain diagram'. Such a diagram provides a record of the complete behaviour of the fibre under tension. From it can be determined the tenacity of the material and its elongation at break, the ease with which the material will stretch, and the amount of energy which will be needed to break the fibre.

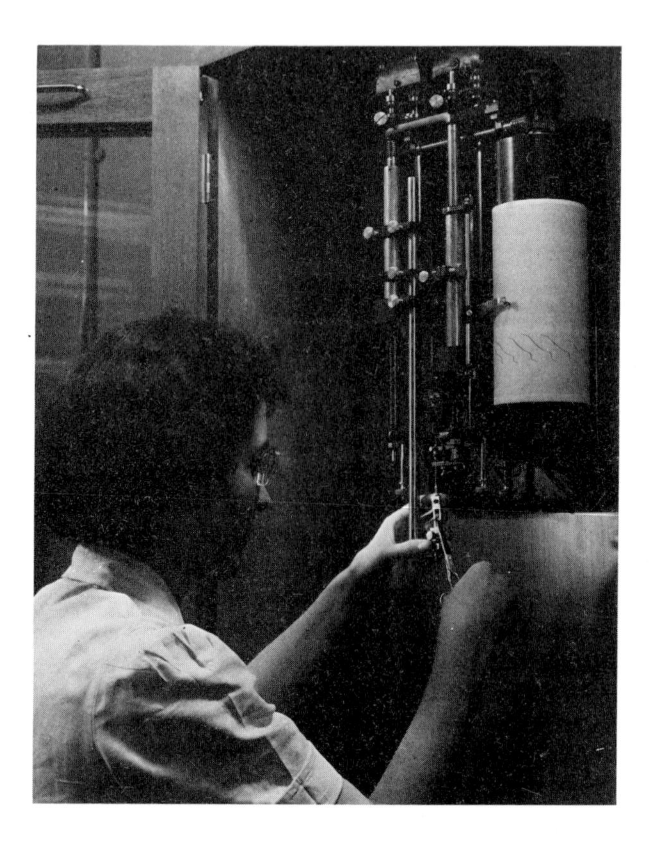

Fig. 3.34 Measuring fibre strength on an extensometer.

Some indication of the elastic properties of a fibre can be obtained from a series of tests carried out using an extensometer. Samples can be loaded by a certain amount and unloaded, reloaded to a still greater amount and unloaded, and so on to provide a range of observations. From these findings, the elastic recovery of the fibre at different values of elongation can be determined. Such findings are of practical value in that they indicate both how the fibre material will behave during the manufacturing processes, and also when it is incorporated in a fabric.

3.9.4 Moisture absorption

The amount of moisture a fibre material will absorb and hold, is of considerable practical importance. A fibre which will readily absorb moisture is often useful in certain types of clothing, underwear for instance. As a general rule, a moisture-absorbing fibre is easy to dye. However, a fibre which absorbs water readily often loses strength when wet but shows an increased elongation at break.

The amount of moisture absorbed by a fibre material is usually indicated by a figure known as its 'moisture regain'. Moisture regain is the weight of moisture taken up from the atmosphere by the fibre, expressed as a percentage of the weight of the fibre after it has been dried to a constant weight in an oven. The amount of moisture a fibre can absorb will be influenced by the amount of water vapour present in the surrounding atmosphere and, in order that the regain figure can have a real meaning, the 'dampness' of the surrounding atmosphere must also be stated. This degree of dampness is indicated by a 'relative humidity' figure (usually shortened to 'R.H.' when test figures are quoted). The relative humidity figure for an atmosphere is its measured vapour pressure expressed as a percentage of the saturated vapour pressure at the same temperature. The vapour pressures are measured using an instrument known as a hygrometer. In this country, moisture regain tests are usually carried out in an atmosphere with a relative humidity of 65% and at a temperature of 20°C (68°F).

3.9.5 Thermal tests

All types of fibre are affected to some extent by heat. In the presence of air, most fibres will burn and it is important, in relation to possible applications, to test how readily they will burn. In addition it is necessary to determine how a particular fibre will react to temperatures it will meet in normal use. Thus, as part of a fabric, it will be subjected to laundering and ironing, and it should be able to withstand these without damage.

3.9.6 Exposure to sunlight

Practically every type of fibre is affected, in some degree, by exposure to sunlight. Some materials deteriorate rapidly and lose strength on exposure while others are more resistant. It is important to carry out exposure tests, therefore, to determine if a material is sufficiently resistance to light, to enable it to be used for curtains and furnishings for instance.

3.9.7 Chemical properties

A knowledge of how a fibre material will react under treatment by various chemicals is important both from the point of view of textile processing, and also the conditions it will be required to meet during use. Modern processing of fibre materials involves the use of bleaching agents, cleaning materials, and special chemicals which may be applied to assist dyeing. In use, fibre materials will be subjected to detergents and alkaline washing agents, and to organic solvents in such treatments as dry cleaning. Tests are usually carried out, therefore, to determine how a fibre will stand up to acids and alkalis in various degrees of concentration and at various temperatures, and to common organic solvents such as trichloroethylene and carbon tetrachloride.

3.9.8 Identification of natural fibres

It is relatively easy to differentiate between the two classes of natural fibres, vegetable and animal. All the vegetable fibres contain cellulose, and all the animal fibres contain protein. If a burning test is carried out, therefore, vegetable fibres burn quite readily with a flame, and leave little ash; the odour they give off while burning is similar to that from burning paper. In general, animal fibres burn more slowly, with little or no flame, and a blackened bead is formed at the end of each fibre; while burning, they give off an odour of burning horn or feathers. The two classes also react differently to treatments with acids and alkalis. Quite low concentrations of caustic soda solutions will attack animal fibres rapidly, but have little or no effect on vegetable materials. In the case of acids, the reverse is the case, the vegetable fibres showing considerably less resistance to acid attack than the animal fibres.

The separation into the two main classes is, thus, relatively simple, but the identification of a particular fibre within a class is much more difficult. Probably the greatest help in identification is obtained from an examination of the fibre under a microscope, although there are certain other characteristics in individual materials which may help to confirm their origin.

Cotton shows under the microscope as a flat ribbon-like fibre, of a fairly uniform width, and with a number of characteristic twists along its length. In cross-section, most of its fibres exhibit a collapsed circular form, rather like a thick-walled deflated bicycle tyre, with a lumen running down the centre. Cotton dissolves in an 80% concentration of sulphuric acid and is decomposed quite rapidly by strong hydrochloric acid and nitric acid. In general, cotton fibres appear to the eye to be rather dull and dead white.

Kapok is softer and silkier than cotton and is creamy in colour. When examined under a microscope, the fibres are flat and smooth and do not show the twists exhibited by cotton. It is possible to observe more or less transparent patches in kapok. A characteristic of kapok is that some fibres have a bulbous, root-like end and, if these can be found, the origin of the fibres is without doubt. Chemically, kapok contains lignocellulose so that, if the fibre is treated with a mixture of phloroglucinol and weak acid, it will turn red while other kinds of vegetable fibre containing normal cellulose will be unaffected.

Flax fibres are stiff, uneven, and often creamy in colour. Under the microscope, the fibres show a fairly regular diameter but with joint-like markings which give them something of the appearance of bamboo canes. The fibres taper gradually to pointed ends. Their cross-sections are irregular and polygonal in shape with small lumens and correspondingly thick fibre walls. Flax is, again, unaffected by boiling in a 5% caustic soda solution. It dissolves in sulphuric and hydrochloric acids but not so readily as cotton.

Ramie fibres are appreciably stronger than those of flax and are something like twice the diameter of flax fibres; ramie fibres are, in fact, the broadest of the bast fibres. The fibres are marked with striations running along their lengths and there are also dark

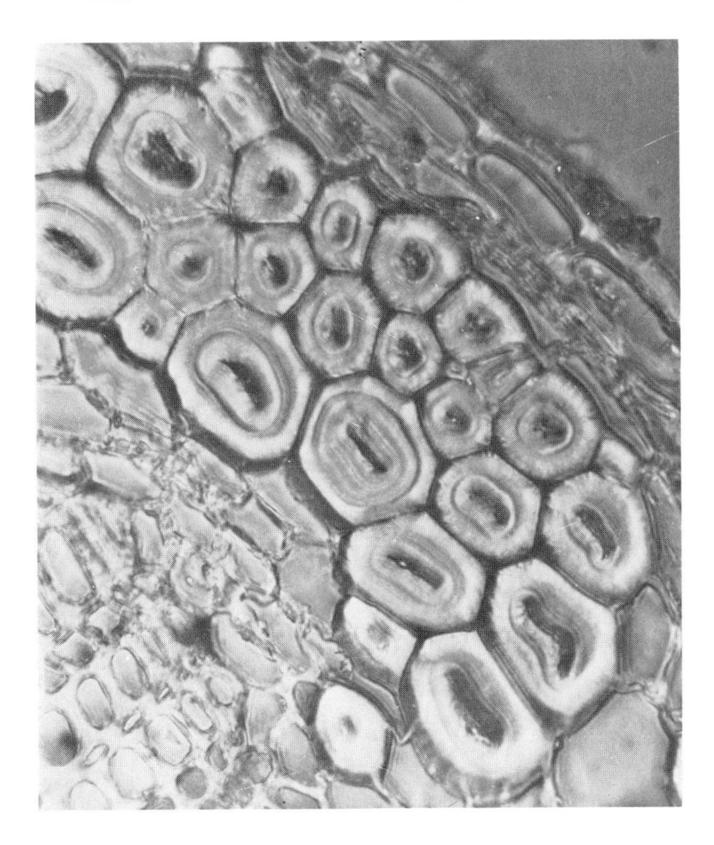

Fig. 3.35 Photomicrograph of cross-section of flax, showing fibre bundles.

lines running transversely across them. Ramie has a somewhat higher resistance to acids than flax.

Under the microscope, jute has a cross-section which shows five- or six-sided cells down which run easily-observable lumens and which have a thick cell wall. Jute also contains ligno-cellulose. Treatment with a weak sulphuric acid solution turns jute a dark brown.

Wool has a springy handle and, in this respect, is completely unlike any of the vegetable fibres. The surface of a wool fibre is covered with scales, and this gives a feeling of roughness if the fibre is pulled between a finger and thumb. Under the microscope, the scales on the surface of a wool fibre are clearly visible and are more prominent than those on any of the other animal hair fibres. Wool fibres also have a wavy or crimped appearance. Individual fibres of sheep's wool are rather more variable in diameter than those taken from goats. Wool will dissolve in a weak solution of caustic soda, and a treatment with hydrochloric acid will cause swelling but will not dissolve it. Wool turns yellow when treated with nitric or sulphuric acids.

Mohair is soft and lustrous and has a characteristic grey-white to grey-blue colour. Under the microscope, the fibres are seen to be fairly large in diameter gradually tapering from the base to the tip. There are striations running along the length of the fibres, but the scale structure is not easy to see.

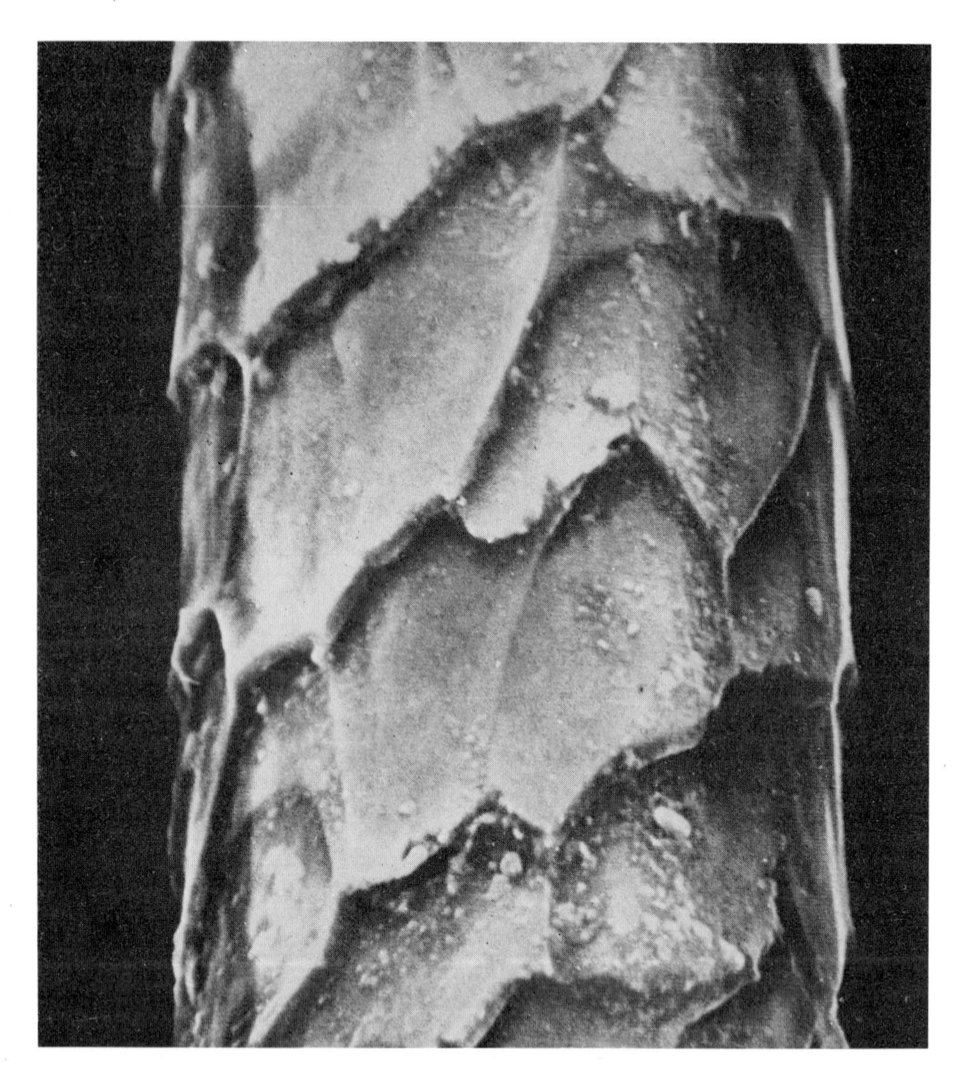

Fig. 3.36 Scanning micrograph of a Romney Marsh wool fibre of 44's quality. (Total magnification × 2 400).

Cashmere fibres are fine and even in diameter. The scale structure is easily seen under the microscope, and the scales are continuous and unbroken. Cashmere fibres are soft and of smaller diameter than most wools.

Camel fibre has a soft handle, there are no crimps, and it is usually light brown in colour. Under the microscope the fibres show a substantial variation in diameter. Patches of pigment can be seen with, here and there, a blood-red spot inside the patch.

The scales on the auchenia fibres are not very prominent. The fibres of alpaca and llama are fairly large in diameter while those of vicuna have a small, even diameter. The diameter of guanaco fibre is intermediate between those of alpaca and vicuna. All the auchenia fibres lack milling properties.

Rabbit fibres are soft and fine and they felt easily under wet conditions. They are identified easily under the microscope by reason of the construction of the medullae which give what seems to be a jointed appearance, again, rather like a bamboo cane.

Silk has a characteristic subdued lustre and a soft handle. Silk burns quite quickly and gives off the characteristic animal fibre odour. There is no mistaking a cross-section through raw silk when examined under the microscope; its dual filament construction is unique. Degummed cultivated silk appears as a solid rod-like fibre of small diameter which is slightly irregular in appearance due to small swellings here and there along its length. The fibres of wild silk are larger in diameter and show occasional twists as well as brownish striations running along their lengths. Cultivated and wild silks can also be distinguished from each other by certain chemical tests. A hot 10% solution of caustic soda will dissolve cultivated silk in a few minutes; wild silk will take something like five or six times as long to dissolve under the same conditions. Concentrated hydrochloric acid dissolves cultivated silk almost immediately but wild silk takes a few hours to dissolve. A cold solution of 10 g of calcium chloride in 100 cm^3 of 90% formic acid will dissolve cultivated silk but not wild silk.

3.10 LITERATURE

P. ALEXANDER and R. F. HUDSON. *Wool: Its Chemistry and Physics*, Chapman & Hall, 1963.

R. R. ATKINSON. *Jute: Fibre to Yarn*, Heywood, 1964.

W. VON BERGEN (Ed.). *Wool Handbook*. Vol. 1, New York, Interscience, 1963.

A. BREARLEY. *Worsted*, Pitman, 1964.

A. BREARLEY. *The Woollen Industry*, Pitman, 1965.

H. B. BROWN and J. O. WARE. *Cotton*, New York, McGraw-Hill, 1958.

P. CARBONI. *Silk: Biology, Chemistry, Technology*, Chapman & Hall, 1952.

ing. Vol 1 of *Manual of Cotton Spinning*. Textile Institute and Butterworth, 1964.

H. HAIGH and B. A. NEWTON. *Wools of Britain*, Pitman, 1952.

J. W. S. HEARLE and R. H. PETERS (Eds.). *Fibre Structure*, Textile Institute and Butterworth, 1963.

P. H. HERMANS. *Physics and Chemistry of Cellulose Fibres*, Cleaver-Hume Press, 1949.

R. H. KIRBY. *Vegetable Fibres; Botany, Cultivation, and Utilization*, Leonard Hill (Books), 1963.

W. E. MORTON and J. W. S. HEARLE. *Physical Properties of Textile Fibres*, Textile Institute and Butterworth, 1962.

W. J. ONIONS. *Wool: An Introduction to its Properties, Varieties, Uses, and Production*, Benn, 1962.

J. G. COOK. *Handbook of Textile Fibres: Natural Fibres*, Merrow, 1968.

A. F. W. COULSON and P. W. HARRISON (Eds.). *Raw Cotton Production and Market-*

H. L. PARSONS. *Jute (Handbook of Textile Technology* No. 4), Textile Institute, 1949

R. ROBSON. *Cotton Industry in Britain*, Macmillan, 1957.

Silk Book, Silk and Rayon Users' Association, 1951.

E. B. SLACK. *Coarse Fibres*, Wheatland Journals, 1957.

A. J. TURNER. *Quality in Flax*, Linen Industry Research Association, 1954.

CHAPTER 4

Man-Made Fibres

4.1 INTRODUCTION

4.1.1 History

A number of the natural fibres, such as wool, cotton and silk, have been used for the making of cloth for thousands of years, but it is only in very recent times that man has learned how to prepare his own fibre materials from substances available to him, and how to process and treat these materials to produce new and valuable textile fibres.

There is evidence that, more than 300 years ago, consideration was being given to possible methods of producing man-made fibres. In the year 1664, in his book *Micrographia*, an English scientist, Robert Hooke, suggested that the ability of the silkworm to spin a fine filament might be imitated by forcing a suitable viscous liquid under pressure through a tiny orifice and allowing it to harden, or set, after it had emerged from the orifice. We have no record of what attempts were made to follow up this suggestion, and nearly two centuries were to elapse before the first recorded practical step was taken towards the development of a technique for producing man-made fibres. In 1842, an English weaver, Schwabe, constructed a machine to produce fine glass filaments by forcing molten glass through tiny holes in a plate, after which the filaments were allowed to cool and harden in the air. The glass filaments which Schwabe produced were too brittle to be of practical value, but he had demonstrated how fine filaments could be extruded through what was the fore-runner of the modern 'spinneret'. His method of forming the actual filaments was now available once a suitable viscous liquid, which could be hardened immediately after being extruded, had been developed.

a. Regenerated cellulose fibres. The investigators of the period were well aware that a very suitable fibre material, cellulose, existed naturally in vast quantities as the main constituent in plant fibres such as cotton and wood. Since, at that time, no solvent for cellulose was known, the problem was how to prepare the viscous solution which could be extruded by Schwabe's method and, having extruded it, how to re-precipitate the cellulose to give a suitable textile fibre.

In 1846, Schonbein was able to form a chemical compound of cellulose which he found could be dissolved in certain organic solvents. He treated cotton with nitric acid to form cellulose nitrate, the explosive material 'gun-cotton' as it became known, and this he found to be soluble in readily-available solvents. Following up this discovery, Audermars in 1855, treated vegetable fibres with nitric acid to produce cellulose nitrate, and dissolved this compound in a mixture of ether and alcohol to produce a highly viscous solution. When a needle was dipped into this solution and then withdrawn it pulled up a fine filament which hardened in the air as the solvents evaporated from it. Sufficient fibre was produced in this way to enable its properties to be examined. It was found to have a reasonably satisfactory strength and flexibility but, being made from such a material, it was highly flammable.

The next important step forward was by a man who was experimenting along the same general lines but with an entirely different objective in mind. In 1883, Swan was attempting to develop an electric incandescent lamp in which he was proposing to use carbon as the material for his incandescent filament. He was looking, therefore, for organic material threads which he could carbonize for this purpose. Taking advantage of the previously reported work, he prepared cellulose nitrate and dissolved it in concentrated acetic acid to produce a viscous solution. This he forced under pressure through fine holes in a plate to produce filaments suitable for carbonizing. In addition to his requirements for carbonizing, he produced sufficient filaments to enable him to make experimental fabrics. He went still further than anyone previously, and treated the material chemically to reduce its flammability.

Not being primarily interested in textile applications, Swan did not carry his work further in this direction, and it was left to Chardonnet to take Swan's findings and devise a practical production process from them. In 1885, Chardonnet nitrated cotton, using a mixture of nitric and sulphuric acids, and dissolved the resulting cellulose nitrate in a mixture of ether and alcohol. The syrupy liquid was extruded through tiny holes in a plate under pressure into water when the filaments hardened to form cellulose nitrate fibres. Later Chardonnet discarded the water bath and allowed the filaments to harden in air by the evaporation of the solvents. The hardened filaments were then given a chemical treatment to de-nitrate them and make them considerably safer to use; this was followed by washing, bleaching and drying. This first commercially-produced man-made fibre had a strong resemblance to natural silk. Chardonnet's initial source of raw cellulose was cotton 'linters', these being the very short fibres left on the cotton seed after the longer textile fibres have been removed, and which are considered unsuitable for the preparation of cotton yarns.

In his search for filaments which he could carbonize, Swan had also made use of a fact established by Schweizer in 1857, namely that cellulose could be dissolved in 'cuprammonium solution', a mixture of ammonia, a copper salt and water, and he found that the cellulose could be re-precipitated with acid after filaments had been extruded, to provide stable threads. He did not pursue this important work to its commercial conclusion, and it was left to Despeissis to work out a production process. His process was the basis of a commercial production unit set up by Bronnert and Company in Manchester. A further discovery was required, however, to make the fibre commercially popular. This was the fact established by Thiele in 1901, that by stretching the newly-extruded filaments, a substantial increase in their strength and durability could be obtained. Employing this modification as part of its process, the Bemberg Company went into the large-scale production of 'cuprammonium rayon'.

In 1892, two chemists, Cross and Bevan, took out a patent covering the preparation of a solution of cellulose which they called 'viscose'. To prepare the solution, they treated natural cellulose with strong caustic soda solution, followed by a treatment with carbon disulphide. The product of these treatments was soluble in dilute caustic soda solution to give their 'viscose' syrup. The development of the fibre-producing process whereby viscose syrup was spun into filaments was largely due to Stearn and Topham.

The extruded filaments were injected into a bath of dilute sulphuric acid to re-precipitate, or 'regenerate', the original cellulose and form textile threads. Topham, in particular, is credited with a number of important practical developments which contributed to the success of the process. These included the ageing, or 'ripening' of the syrup to improve its spinning characteristics, the development of the multiple spinneret, and the centrifugal pot method of collecting the filaments in the form of a 'cake'.

The really successful commercial development of what became known as 'Viscose Rayon' must, however, be credited to Courtaulds Ltd., formerly a silk manufacturing company. They acquired the rights of the Viscose Rayon process in 1904 and, after a period of experimentation carried out at Kew, they opened a factory for large-scale production in Coventry.

A still further method of bringing natural cellulose into solution had been discovered in 1869. In that year, Schutzenberger had prepared cellulose acetate which was soluble in certain solvents. In 1894, Cross and Bevan discovered catalysts which promoted the easy formation of cellulose acetate and opened the way to a satisfactory method of manufacturing the compound. The cellulose triacetate material they produced was soluble in chloroform. It was not until 1903, however, that an economic method of preparing bulk solutions was found possible when it was discovered that, if the triacetate was allowed to break down to some extent to form secondary acetate, this latter material was readily soluble in a much cheaper solvent, acetone.

Up to the beginning of the First World War, many people, among them the Dreyfus brothers, attempted to produce satisfactory fibres from the cellulose acetate-in-acetone solution. It was found that cellulose acetate itself had special characteristics as a fibre material, and it was unnecessary to regenerate the original cellulose to obtain useful textile fibres. The development was interrupted by the First World War in which, however, cellulose acetate was used in another way. It was found that a solution of the material in acetone could be applied to the fabrics of the aircraft of that time to reduce porosity and offer some protection against degradation. Cellulose acetate 'dope' was manufactured for this purpose on a substantial scale by the Dreyfus brothers who, at the end of the war, found themselves left with large stocks of the solution and the specialized equipment to make it. They returned to their original task of developing a satisfactory commercial process for the production of cellulose acetate fibres, and they finally overcame the technical difficulties and commenced commercial production in 1921. The Dreyfus brothers founded British Celanese Ltd., and in 1957 this company was merged with Courtaulds Ltd.

b. Regenerated protein fibres. There are fibre-forming materials other than cellulose existing in nature. Wool and silk are both protein fibres, and it is not surprising that attempts have been made to produce regenerated protein fibres. In 1898 Todtenhaupt produced fibres from casein, the protein in milk. He extruded alkaline solutions of casein with hardening solutions. The fibres he produced had poor strength, particularly when wet, and they lacked flexibility. Only after many years of experimental work was a commercial form of casein fibre evolved.

Other protein fibres have been tried, the vegetable proteins being extracted from maize, soya bean and groundnut respectively. All these reached some type of production stage, but there has never been a large-scale demand for the regenerated vegetable protein fibres and the ones based on the three sources mentioned above have disappeared in the face of competition.

c. Alginate fibres. Another natural material in plentiful supply is seaweed and this was also examined as a possible source of fibre material. In 1883, Stanford succeeded in isolating alginic acid from seaweed, and this compound has been used since as a thickening agent. During the Second World War, Speakman examined the possibility of preparing a fibre-forming material from alginic acid, and he succeeded in extruding fibres from calcium alginate. The fibre has never been used in quantity since it dissolves in dilute alkali solutions. Some production is still carried out, since calcium alginate has found applications where it is used as a support thread, later to be dissolved out of the fabric when no longer needed.

d. Synthetic fibres. So far, this account of the historical development of man-made fibres has been concerned entirely with attempts to take cheap and plentiful natural fibrous materials and refashion them in fibre form. Only one such raw material, cellulose, has really stood the test of time and, even then, the fibres which have been produced from it often leave something to be desired.

In between the two world wars, very considerable progress was made in our knowledge of the nature of fibre-forming materials. This knowledge made possible a rapid rate of development of a whole range of useful man-made fibres. During that period, a number of investigators, among them Staudinger, Mark and Carothers, made fundamental studies of fibre forming materials, and their findings provided a means of progress towards the production of the true synthetic fibres.

In the early 1920s, Staudinger had prepared and studied a material, polyoxymethylene, which he found to have many of the characteristics of natural cellulose. Largely as a result of his work, it became recognised that the chemical process known as 'polymerization', namely the joining up of small molecules to form large ones, could lead to the formation of long chain molecules from which we now know that all fibre materials, natural or man-made, are composed. These chains are held together by ordinary chemical bonds between suitable unit molecules assembled in some kind of regular pattern. During the same period, studies of the X-ray scattering patterns produced by fibre materials established that such materials had a type of crystalline structure, and this 'crystallinity' distinguished them from the resinous substances which were the more usual products of polymerization processes.

In 1928, the Du Pont Company in America decided to undertake an extensive programme of fundamental research into the synthesis and characteristics of large molecules which could be used to build up fibre materials. They selected W. H. Carothers of Harvard University to take charge of this research at their experimental station in Wilmington, Delaware. This work of Carothers and his team has provided us with much of our present-day knowledge of fibre-forming substances. It was this team which provided the first truly synthetic textile fibre, 'nylon'.

The chemical groups selected for study by Carothers were primarily the hydroxyl, —OH, the amine, $—NH_2$, and the carboxyl, —COOH, and a large number of compounds containing such groups were examined. Of those studied, two classes of compound appeared to offer the greatest promise as possible fibre-forming materials, these being the polyesters containing the —CO·O— ester linkage, and the polyamides containing the —NH·CO— amide linkage. In the earlier work, the emphasis was placed on the linear polyesters composed of molecular chains of medium length. These polyesters were prepared from organic acids in the series , $HOOC—(CH_2)_x—COOH$, and glycols in the series $HO—(CH_2)_y—OH$, where x and y are integers of values depending on the positions of acid and glycol in their respective series. All the polyesters examined had crystalline characteristics and all melted at a temperature round about 80°C (176°F). They dissolved quite readily in solvents such as chloroform, but without giving the highly viscous solutions necessary to enable filaments to be extruded from them. It had been shown that the viscosity of a polymer solution was related to the molecular chain length, and polyesters containing longer chains were prepared. These did, in fact, give highly viscous solutions from which filaments could be prepared. After being extended to give additional strength, as in the Thiele modification of the cuprammonium rayon process, these yielded fibres potentially capable of being used in textile applications.

The polyesters prepared by Carothers all incorporated aliphatic radical groups and, from a fibre point of view, they suffered from two defects. They had a low melting temperature of about 80°C (176°F), and a poor hydrolytic stability. No doubt further work would have been done on polyesters of this type in an attempt to make them more acceptable as fibre forming materials, but at about this time, the team prepared another class of compounds, the polyamides, which appeared to be superior as potential fibre materials. Their first polyamide was based on aminocaproic acid and later they prepared one from the interaction of adipic acid and hexamethylene diamine. This latter polymer had a high melting temperature, excellent fibre-forming properties, great strength and abrasion resistance, and a good chemical stability. It was this particular polyamide which the DuPont Company selected for commercial exploitation and which they called 'Nylon'. After a period of about eight years spent in perfecting a manufacturing process, commercial production of the first nylon polymer fibres began in 1939. Nylon was used for war purposes during the Second World War, mainly in replacing silk in parachute fabrics. The first major commercial application for nylon was as a material for ladies' fine gauge hose, for which application its combination of desirable properties placed it in a class by itself, and so much so that the terms 'fine gauge stockings', and 'nylons' became synonymous. This first member of the nylon family, polyhexamethylene adipamide, has become known as 'nylon 6.6'. This nomenclature has been adopted to distinguish between the various members of the family. The figures following the name 'nylon' indicate the chemical groups which are incorporated in the simplest molecular repeating unit of the chain molecule; this is dealt with more fully later on, in the section on fibre structure.

The success of the work of Carothers, and the practical information which had become available as a result of DuPont's exploitation of nylon 6.6, triggered off a burst of investigational activity throughout the world into other possible synthetic fibre materials. Carothers had concentrated his later development work on nylon 6.6, but he had referred to another polyamide, nylon 6, which had similar fibre properties. This material was studied by the German chemical firm, I .G . Farbenindustrie, and successfully developed commercially, when it was marketed as 'Perlon'.

A further important step forward in the development of synthetic fibres was made by the American Viscose Corporation. They had given attention to the possible use of polyvinylchloride, a cable-covering material, as a fibre-forming substance. Its low melting point and poor crystallinity had, however, reduced its possible value for such a purpose, and the American Viscose Corporation investigated possible ways of modifying the material to give it more attractive fibre properties. This they did by preparing a mixed polymer, or a 'copolymer', in which the chain molecules contained molecular units from two substances, one being vinyl chloride, and the other vinyl acetate. This very first copolymer was named 'Vinyon'. Although no very great advantage as regards heat sensitivity was obtained by the copolymerization, the properties of vinyon in relation to those of the two parent materials were of considerable interest, and indicated a further line of investigation which could be followed.

In 1940, the Dow Chemical Company in America succeeded in copolymerizing vinyl chloride and vinylidene chloride to produce a fibre which they named 'Saran'. Although this material had only a moderate heat resistance, its general properties were such as to make it suitable for a number of specialist applications. Still later, in 1956, the Eastman Chemical Company of America succeeded in preparing a copolymer of acrylonitrile and vinylidene chloride, which they marketed under the name of 'Verel'. In addition to a good general chemical resistance, this material was outstandingly non-flammable.

In their work on the polyesters, Carothers and his team had concentrated their attentions almost exclusively on the *aliphatic* polyesters prepared from the aliphatic acids. It occurred to J. R. Whinfield, a chemist employed by the Calico Printers' Association at Accrington in Lancashire, that a material made by reacting a glycol with a symmetrical *aromatic* acid might result in a greater stability, both thermal and chemical. The first investigations along these lines were carried out in 1941 at Accrington by Whinfield in collaboration with J. T. Dickson. They heated together, at a temperature of about 200°C (392°F), terephthalic acid and ethylene glycol, with the latter in excess. When the chemical reaction had proceeded for some time, the whole mass suddenly set solid and, on raising the temperature to 260°C (500°F) remelting of the mass took place. This material was capable of being formed into filaments, had a high melting temperature and, in contrast with the aliphatic polyesters, had a good hydrolytic stability. Once this discovery was made known, and the conditions necessary to prepare the material had been disclosed, other firms, and particularly the DuPont Company, attempted to find alternative reactions of a similar type. While other fibre-forming polyesters were prepared, none proved to be as satisfactory as the polyethylene terephthalate developed by Whinfield, and this material has continued to dominate the commercial fibre-forming polyester field. Manufacture of the material was licensed to Imperial Chemical Industries Ltd., in the United Kingdom, and later to the DuPont Company in America. The fibre manufactured by I.C.I. Ltd. is known as 'Terylene' and by the DuPont Company as 'Dacron'.

During the Second World War, the need to replace supplies of natural rubber lost to this country and America by the occupation of Malaya by the Japanese, led to an intensification of research on synthetic rubbers. Very large quantities indeed of synthetic rubber were eventually produced, and this meant that certain chemicals which had previously only been available in limited quantities were now on offer in bulk. One such chemical, acrylonitrile, was used by the DuPont Company as a material for fibre production, and this was copolymerized, together with small quantities of other polymer-forming compounds, and used to prepare fibres sold commercially first as 'Fibre A', and later as 'Orlon'. By 1945, this fibre was well established for apparel fabrics and becoming increasingly popular. Similar fibres, 'Courtelle' in this country, and 'Acrilan' in America, made their appearance at a later date.

It will be appreciated that, once success had been achieved in the manufacture of the first wholly synthetic fibre, progress in the development of other fibres, some of them with quite remarkable properties, was rapid. This same rapid rate in developing new fibres cannot be expected to continue. The development, and launching on the market, of an entirely new fibre involves the expenditure of vast sums of money, without any guarantee that an adequate profit will eventually be made. In addition, now that a substantial range of the newer fibres is on the market, it is becoming increasingly difficult to provide fibre characteristics which can be regarded as outstanding and hence justifying the financial risk. Nevertheless in recent years two new fibres have made their appearance, these being the so-called 'Spandex' elastomeric fibres, and polypropylene fibres, made from a member of the polyolefin group of polymers.

The elastomers are based on polyurethane, and were developed to replace natural rubber in a number of applications where the natural product is degraded by oxidation, exposure to sunlight, and contact with various substances. Their development was an extension of investigations carried out in Germany during the Second World War and which resulted in the manufacture of a material 'Perlon U'. The modern products represent molecular modifications of that material. The spandex yarns are not cheap to produce, but they appear to have justified their higher cost in certain textile applications in which rubber, or its equivalent, is built into a fabric structure and, in which, failure of this rubber component has meant that a garment, which would otherwise remain serviceable, has had to be discarded.

The availability of polypropylene fibres has been made possible as a result of fundamental research into the mechanics of polymerization carried out by two scientists, Ziegler and Natta. They discovered catalysts for the polyolefin-class of compounds which had an interesting effect on the polymerization reaction. This is discussed later in the chapter (section 4.2) when polypropylene preparation is being described. Using such a catalyst, the unit molecules of propylene were made to link together to form chains in such a way that a powerful binding together of the chains could take place and a strong fibre be produced. Using catalysts of the Ziegler and Natta type, polypropylene is relatively cheap to produce. Imperial Chemical Industries Ltd. obtained a licence to manufacture polypropylene in this country using this new catalytic process, and placed the fibre on the market under the trade name of 'Ulstron'.

4.1.2 Survey of the more important man-made fibres

We have traced the history of the development of the man-made fibre materials which have achieved commercial importance, and a range of these, covered by a wide variety of trade names, is now on the market. In table 4.1 is listed the more important types of man-made fibre in general use, the trade-names under which they are likely to encountered, and their characteristic, and often distinguishing, properties.

Table 4.1 Man-made fibres

Class of fibre	*Trade name*	*Manufacturer*	*Type supplied*	*Characteristic properties*
RAYON *a.* Viscose rayon	Fibro	Courtaulds Ltd.	Staple	Not thermoplastic. Cheap and good processing characteristics
	Danufil	Suddeutsche Chemiefaser A.G.	Staple	
	Fibrenka	Akzo N.V.	Staple	Reasonable strength, but a reduction in strength when wet a disadvantage
	Flisca	Société De La Viscose Suisse	Staple	
	Lenzesa	Chemiefaser Lenzing A.G.	Staple	
	Phrix Rayon	Phrix Werke A.G.	Filament	

Table 4.1 (continued)

Class of fibre	*Trade name*	*Manufacturer*	*Type supplied*	*Characteristic properties*
b. Cuprammonium rayon	Bemberg	J. P. Bemberg A.G.	Filament and staple	Not thermoplastic. Softer drape, less sheen and more silk-like than viscose rayon
	Cupresa	Farbenfabriken Bayer A.G.	Filament	
	Cuprama	Farbenfabriken Bayer A.G.	Staple	
c. High tenacity rayon	Tenasco Super	Courtaulds Ltd.	Filament	Not thermoplastic. Higher strength and abrasion resistance than standard rayon
	Durafil	Courtaulds Ltd.	Staple	
	Fortesco	Dawson Manufacturing Co.	Filament	
	Avron	American Viscose Corp.	Staple	
	Duraflox	Spinnfaser A.G.	Staple	
d. Polynosic rayon (see p. 36)	Vincel	Courtaulds Ltd.	Staple	Not thermoplastic. Higher strength, wet and dry, and lower water absorption than viscose rayon
	Avril	American Viscose Corp.	Staple	
	Zantrel	American Enka Corp.	Staple	
ACETATE	Albene	Société Rhodiaceta	Filament	Thermoplastic. Soft handle, good drape, easy to dye. Comparatively low strength and softening temperature
	Dicel	Courtaulds Ltd.	Filament	
	Celafibre	Courtaulds Ltd.	Staple	
	Lansil	Lansil Ltd.	Filament	
	Rhodia	Deutsche Rhodiaceta A.G.	Filament and staple	
	Silene	Société Novaceta	Filament	
TRIACETATE	Tricel	Courtaulds Ltd.	Filament and staple	Thermoplastic. Soft handle and good drape. Higher softening point and lower moisture absorption than acetate
	Trialbene	Société Rhodiaceta	Filament and staple	
	Trilan	Canadian Celanese Ltd.	Filament and staple	
REGENERATED PROTEIN	Fibrolane	Courtaulds Ltd.	Staple	Not thermoplastic. Low cost and wool like, but very low wet strength
	Miranova	Snia Viscosa	Staple	
ALGINATE	Alginate	Courtaulds Ltd.	Filament	Not thermoplastic. Non-flammable. Low strength. Dissolves in weak alkali solutions

Table 4.1 (continued)

Class of fibre	*Trade name*	*Manufacturer*	*Type supplied*	*Characteristic properties*
POLYAMIDE				
a. Nylon 6.6	Nylon	E. I. DuPont de Nemours & Co.	Filament and staple	Thermoplastic. High melting point. Strong and durable. Good elasticity
	Bri-nylon	Imperial Chemical Industries Ltd.	Filament and staple	
	Blue C nylon	Monsanto Ltd.	Filament and staple	
b. Nylon 6	Enkalon	British Enkalon Ltd.	Filament and staple	Generally as for nylon 6.6 but with slightly lower melting point
	Celon	Courtaulds Ltd.	Filament and staple	
	Perlon	Farbenfabriken Bayer A.G.	Filament and staple	
		Farbwerke Hoechst A.G.	Filament and staple	
		Phrix Werke A.G.	Filament and staple	
		Vereinigte Glanzstoff-Fabriken A.G.	Filament and staple	
c. Nylon 11	Rilsan	Organico S.A.	Filament and staple	Generally as for nylon 6.6 but still lower melting point, and reduced moisture absorption
POLYESTER	Terylene	Imperial Chemical Industries Ltd.	Filament and staple	Thermoplastic. Strong, durable and resilient. High melting point, and heat shapes and permanently pleats well. Good resistance to sunlight
	Dacron	E. I. DuPont De Nemours & Co.	Filament and staple	
	Fortrel	Fiber Industries Inc.	Filament and staple	
	Kodel	Eastman Kodak Co.	Staple	
	Tergal	Société Rhodiaceta	Filament and staple	
	Terlenka	Akzo N.V.	Filament and staple	

Table 4.1 (continued)

Class of fibre	*Trade name*	*Manufacturer*	*Type supplied*	*Characteristic properties*
	Trevira	Farbwerke Hoechst A.G.	Filament and staple	
ACRYLIC	Acrilan	Monsanto Ltd.	Staple	Thermoplastic. Soft, bulky fibre. Good dimensional stability and recovery from deformation
	Courtelle	Courtalds Ltd.	Staple	
	Creslan	American Cyanamide Corp.	Staple	
	Dralon	Farbenfabriken Bayer A.G.	Filament and staple	
	Orlon	E. I. DuPont De Nemours & Co.	Filament and staple	
	Zefran	Dow Chemical Co.	Staple	
POLYOLEFIN *a.* Polyethylene	Courlene	Courtaulds Ltd.	Filament	Thermoplastic. Low density. Low melting point Lack of dye-ability. Inert. Cheap
	Drylene	Plasticisers Ltd.	Filament	
	Nymplex	N. V. Kunstzijde-Spinnerij Nyma	Filament	
b. Polypropylene	Courlene P.Y.	Courtaulds Ltd.	Filament	Thermoplastic. Low density. Strong. Zero moisture regain. Difficult to dye. Medium melting point
	Tritor	Plasticisers Ltd.	Filament	
	Ulstron	Imperial Chemical Industries Ltd.	Filament and staple	
CHLOROFIBRE *a.* Polyvinyl-chloride type	Fibravyl	Société Rhovyl	Staple	Thermoplastic. Non-flammable. Very good chemical resistance. Low softening point
	Rhovyl	Société Rhovyl	Filament	
	Thermovyl	Société Rhovyl	Staple	
	Dynel	Union Carbide Co.	Staple	
	Vinyon	American Viscose Corp.	Staple	
b. Polyvinylidene chloride type	Saran	Dow Chemical Co.	Filament and staple	Thermoplastic. Non-flammable. Good chemical resistance. Moderate heat resistance
	Verel	Eastman Kodak Co.	Staple	
	Kurehalon	Kureha Kasai Co.	Filament	
POLYVINYL ALCOHOL	Cremona	Kurashiki Rayon Co.	Filament and staple	Thermoplastic. Not very heat stable. Higher moisture absorption than other synthetics. Poor resistance to acids
	Vinal	Keowee Mills	Filament and staple	
	Synthofil	Wacker-Chemie G.m.b.H.	Filament and staple	

Table 4.1 (continued)

Class of fibre	*Trade name*	*Manufacturer*	*Type supplied*	*Characteristic properties*
ELASTOMERIC	Lycra	E. I. DuPont De Nemours & Co.	Multi-filament	Similar to natural rubber in extension and recovery, but more resistant to oxidation, exposure to sunlight, and laundering. More heat sensitive than natural rubber. High price
	Glospan	Globe Manufacturing Co.	Multi-filament	
	Spanzelle	Courtaulds Ltd.	Multi-filament	
	Vyrene	Lastex Yarn & Lactron Thread Ltd.	Mono-filament	
MODACRYLIC	Aeress	Union Carbide Co.	Filament	Thermoplastic. Non-flammable. Reasonable tensile properties. Good resistance to sunlight
	Teklan	Courtaulds Ltd.	Filament	

4.2 STRUCTURE OF FIBRES

In order to be able to appreciate what gives a fibre material its characteristic properties, how it may be possible to modify these properties, and why some materials are capable of producing more generally useful fibres than others, it is necessary to know something of the internal structure of fibres.

All substances are built up from units which are called molecules, and the molecules can be further subdivided into atoms. In certain types of organic chemical reaction, conditions have been found whereby unit molecules can be made to link up, one with another, to form long chains. Many hundreds of such units may join up in this way and, quite often, steps have to be taken to limit the number which link together in order to preserve some particular property in the final material. In this branch of chemical investigation, the substances consisting of individual unit molecules which are potential 'links' in the long chains, are referred to as 'monomers'; those produced as a result of the chain formation are called 'polymers'. The chemical reaction by which the change from monomer to polymer takes place is called 'polymerization'.

It has been established that, in the polymer chains, the units are held together by normal primary valence bonds. In order that the unit may be a link in a polymer chain, it must be capable of forming two such bonds, one with a molecule in front, and the other with a molecule behind, and so on along the chain.

There are two main ways by which polymer chains can be built up; one is known as a 'condensation' reaction, the other as an 'addition' reaction. In condensation polymerization, there is more than one chemical entity involved, usually two. Before a reactive unit capable of building up into a chain is formed, an intermediate step must take place. This usually involves a chemical reaction in which a small molecule, often water, is eliminated or 'squeezed out'. Thus, in the preparation of polyhexamethylene adipamide, which we know as nylon 6.6, two compounds, hexamethylene diamine and adipic acid, are reacted together as follows:

$$HOOC(CH_2)_4COOH + NH_2(CH_2)_6NH_2 \rightarrow$$

adipic acid hexamethylene diamine

$$-HOOC(CH_2)_4CONH(CH_2)_6NH_2- + H_2O$$

hexamethylene adipamide water

The hexamethylene adipamide unit thus formed has reactive centres at its ends and is capable of joining up with other similar units to form the polyhexamethylene adipamide chain molecule which we know as that of nylon 6.6. A condensation polymerization is a step-like, relatively slow reaction.

The second type of chain reaction, addition polymerization, consists of the single unit molecules simply linking together without the formation of another substance as an intermediate stage. Thus, the molecule of ethylene gas, $CH_2{=}CH_2$, is potentially reactive and, under special conditions, can be made to link up with other similar units to form a polymer chain, polyethylene:

$$-CH_2-CH_2-CH_2-CH_2-$$

Another example of addition polymerization is the linking up of unit molecules of vinyl chloride, $CH_2{=}CHCl$, to form polyvinyl chloride:

$$-CH_2-CHCl-CH_2-CHCl-CH_2-CHCl-$$

In the case of addition reactions, the chain formation is extremely rapid.

It should not be assumed that all polymer molecules have a simple chain construction. For a number of reasons, some linear chain molecules may also develop side chains or 'branches'. These branches, which are usually short compared with the length of the main chain, may show considerable variation in their individual lengths and in their distribution along the chain. Yet again, there is a possibility that an occasional short link may form between otherwise linear chains; these may occur accidentally during the reaction, or they can sometimes be created deliberately by chemical means when it is desired to give additional stability to the polymer. An example of a material which is slightly 'cross-linked' is wool.

If at least one of the reacting molecules is capable of forming more than two bonds, there is a possibility that a three-dimensional network will be built up. Materials formed in this way are usually infusible, hard and brittle, and of no use for fibre applications.

The illustration shows, in diagrammatic form, the three types of polymer constructions, linear, branched and three-dimensional.

All textile fibres, whether they be natural or man-made, are composed of chain molecules, and the first essential requirement of a material intended for fibre production is that it is built up from chain molecules. Nature has already completed the linking-up, or polymerization, in the fibrous materials she provides. Cotton and wood consist of cellulose which is built up from large numbers of sugar (glucose) units joined together in a chain-like formation. In the case of synthetic fibres, man has learned how to prepare suitable unit molecules, and establish the conditions which will allow them to join up to form the polymer chains thus providing his own fibre-forming material.

It should not be imagined that, because a material contains chain molecules it will be automatically a satisfactory fibre-forming substance. A fibre is of little practical value unless it is strong and will withstand considerable stress. The internal structure

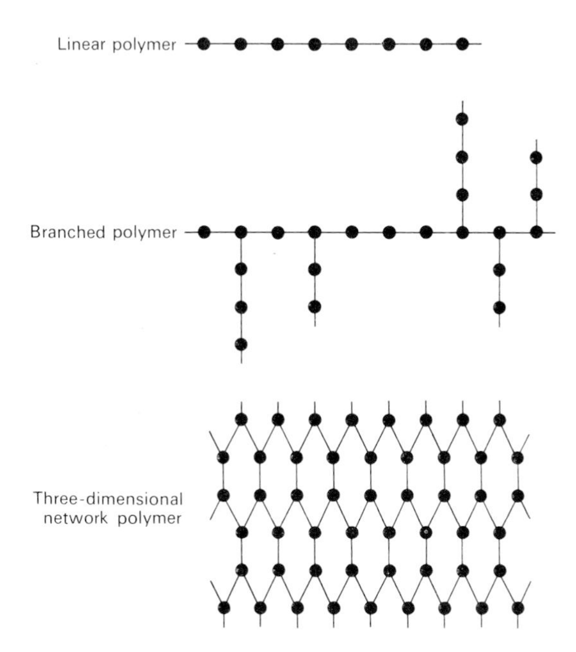

Fig. 4.1 Types of polymer construction.

of a fibre must be such, therefore, that its component parts are bound together strongly and offer considerable resistance to any attempt made to separate them. The more powerful the binding action within a fibre, the stronger and more stable will be that fibre.

As has already been mentioned, the individual units in the molecular chain are held together by valence bonds which are powerful. In carrying out tensile tests on a fibre, therefore, it is unlikely that the chains themselves will be ruptured. There remains the danger that the fibre will give way under stress by the chains sliding apart. The forces which prevent the chains from sliding one over the other are relatively weak and act only over very short distances. The chain molecules can only be attracted to each other with sufficient force, therefore, to give a practically useful fibre if a very large number of these weaker bonds are acting at the same time. This means that a large number of centres of attraction must be in position opposite similar centres in a neighbouring chain molecule, and the two mutually attracting centres must be very close together so that the maximum attraction is exerted. These conditions are met if the linear chains are lying side-by-side in parallel formation, and there are no 'side-chains' standing out from the molecules which would get in the way and prevent the main chains from coming close together. The closer the chains are to being parallel to each other, the more they can pack together, and the stronger the cohesive forces will become. This tight packing produces a high tensile strength in the fibre and also, since the segments of the chains find it more difficult to oscillate under the cramped conditions when an external stimulus is applied, it can influence the melting temperature of the material. This is illustrated by a comparison of the melting temperatures of two polymers, polyacrylonitrile and polymethylacrylonitrile, which are 300°C (572°F) and 115°C (239°F) respectively. In the second of the two polymers, a pendant methyl group has been introduced into the acrylonitrile and has acted as an obstacle to keep the chains from packing tightly together. There has thus been a reduction in the forces which can act between the chains and a greater thermal agitation of the chain segments is possible. This, then, represents a further condition which a good fibre material must satisfy. The chains must be, to some extent, in lines parallel to each other and to the longitudinal axis of the fibre, and they must be able to arrange themselves in close proximity to each other.

When a narrow beam of X-rays is passed through such a closely-packed, orderly array of chain molecules, a sharp diffraction pattern is obtained just as in the case of an inorganic crystal examined in the same way. This oriented state in a fibre material is, therefore, referred to as 'crystalline'. In the natural fibres the orientation of the chains is produced by growth; in man-made fibres the mechanical action of forming the material into a filament introduces some degree of orientation. As will be explained later in the chapter, there are after-treatments which can be given to the newly-formed filaments and which produce a marked increase in their degree of molecular alignments and, it follows, a substantial change in the properties of the resulting fibres.

A fibre does not consist solely of crystalline material. The crystalline state exists in tiny regions referred to as 'crystallites', and these are distributed throughout the fibre and separated by what are known as 'amorphous' regions where the chains are in much greater disorder. From the results of experimental observations it has been possible to estimate the sizes of the individual crystallites. In polymers which crystallize well, such as polyethylene, the range of sizes is estimated to be 100 to 500 angstrom units, where an angstrom unit (Å) is 10^{-8} cm. For linear condensation polymers, such as the polyamides and polyesters, the range is 50 to 100 Å. The crystallites are usually five to ten times as long as they are broad, but the linear dimensions of even the largest of them are still small compared with the length of a fully-extended polymer chain. The chains run, therefore, through more than one crystallite, and the amorphous regions in between them, and 'tie' the structure of the fibre together. There are no abrupt boundaries between the two types of region in the fibre.

Both the crystalline and amorphous regions are important in determining the properties of a fibre. A tight packing of the chain molecules produces a strong cohesive action, and this gives the fibre compactness, high strength and good resistance to abrasion. At the same time it acts as a barrier to the entrance of outside molecules, such as those of water and dyestuffs, and they find it difficult to penetrate into the structure. Thus, in general, a highly crystalline fibre has a low moisture absorption and is difficult to dye. Almost all the reactions of fibres occur through their amorphous regions. It is here that dyes can enter the fibre. Water can penetrate relatively easily into the amorphous regions and, from there, may creep along between the chains to produce some lateral separation of them and a consequent swelling of the fibre. When a substance is heated, its molecules are agitated; in a linear chain polymer, the segments of the chain molecules vibrate more strongly as the temperature is raised. In the crystalline regions, the movements are restricted by the tight packing, but in the amorphous regions, the chain segments are more free to vibrate and, in doing so, can create even greater gaps through which water and dye molecules can enter the structure at these higher temperatures. While too high a crystallinity can result in high strength and stiffness at the expense of difficult dyeing and after-treatments, too great an amorphous content can give a low softening temperature, a high moisture retention which is troublesome after laundering, and a low strength to the fibre, particularly when wet. An abundance of amorphous regions does, however, usually provide a soft and pleasant handle, and an ease in carrying out dyeing and after-treatments. It is obvious that the ratio of crystalline to amorphous contents is an important factor in determining how practically useful a particular fibre will be. This, then represents the third condition which a good fibre material must satisfy. There must be a satisfactory relationship between the amounts of crystalline and amorphous materials present in order to strike a useful compromise between the high strength and toughness of high crystallinity and the pleasant handle and easy processing provided by an abundance of amorphous regions.

Till now, only chain molecules built up from units which are exactly the same have been mentioned. It is possible, however, to build up polymer chains containing two, or even more, different kinds of units, and such units can link up in several different ways. The products resulting from the linking up of different kinds of units are referred to as 'copolymers'. The illustration shows, in diagrammatic form, the main linking-up possibilities for two types of unit molecules. The derivation of the names applied to the three types of copolymer illustrated is obvious from the diagram. One of the advantages which may be gained from copolymerization is that it offers a means whereby the properties of a material built up from one type of unit molecule may be modified in a controlled manner for a particular purpose. Thus, selecting two monomers which, individually, might have different softening temperatures in the respective polymers they provide could, by copolymerization, produce a material having a softening temperature in between the two. Similarly, a graft copolymer may have improved chemical and processing properties when compared with the two materials prepared from chains containing the same chemical units.

Restriction of space has allowed only a brief outline of fibre structure, but it is hoped that sufficient has been included to indicate the importance of the subject in any consideration of the properties of man-made fibres. It should be emphasized that the remarkable progress which has been made since the 1920s in providing the present range of commercial synthetic fibres, has been due practically entirely to the greater understanding of the detailed processes of polymerization which has been accumulated, and of the relationship which exists between the structure of a fibre-forming polymer and its properties.

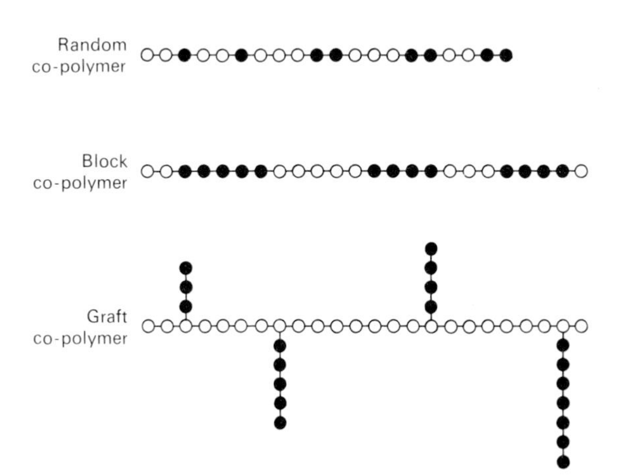

Fig. 4.2 Types of co-polymers.

4.3 FIBRE PRODUCTION

4.3.1 Spinning the fibres

The early investigators decided that the most satisfactory way of preparing man-made fibres was to produce a viscous solution, or syrup, of the fibre-forming material, force this under pressure through fine holes in a plate or 'spinneret', and harden the filaments as they emerged from the spinneret to produce stable fibres which could be taken up on to a suitable 'package'. The same general procedure is still followed, but a combination of a number of individual developments and modifications has resulted in the smooth-flowing, high speed production processes which are in use today.

The viscous liquid, or spinning syrup, to be extruded will be in one of three forms determined by the route which had to be taken to prepare the fibre-forming material and the characteristics of the material itself. It can take the form of a straight solution of the fibre material in a suitable solvent, a solution of an intermediate compound of the material in a solvent when it may be necessary to treat the extruded filaments with a chemical to harden them, or a molten mass of the fibre material. The details of the particular spinning process employed will be determined by whichever type of spinning syrup has to be used.

Remembering the thickness of the average man-made fibre produced, it will be appreciated that the spinning syrup has to be extruded through very tiny holes in the spinneret. It follows that the syrup must be very carefully filtered to remove any particles it may contain and which might block the holes in the spinneret. A type of filter often used for this purpose is the so-called 'candle' filter. It is also possible at the syrup stage, that solid particles of various materials may have been added deliberately. Thus a normally transparent filament can be rendered opaque by adding fine particles of a light-scattering material, such as titanium dioxide, to the syrup. Again, some filaments are coloured at the extrusion stage by adding coloured pigments to the spinning syrup. Any such substances added to the syrup must be in a very finely-divided form and free from agglomerates.

The spinning syrup must also be free from small bubbles of gas or air. If not, there is a possibility that a filament will break when such a bubble reaches a hole in the spinneret. In any case, if actual breakage does not take place, the filament will have been weakened at that point. Allowing the syrup to stand will help towards eliminating the bubbles, and a vacuum is sometimes applied to draw them off.

The spinning syrup has to be forced under pressure through the holes in the spinneret. A metering pump is used for this purpose. The fluid must be pumped without pulsation in a smooth flow, since any pressure fluctuation would give rise to thick and thin places in the filaments, and could possibly generate gas bubbles. For this reason, piston type pumps are not satisfactory and gear-wheel pumps having a large number of gear teeth are normally used. The principle of the gear-wheel pump is quite simple but the few parts which go to make up its construction must be accurately fashioned and close fitting to obtain the necessary uniformity of flow. The pump consists of two gear-wheels meshing with each other and enclosed in a close-fitting housing. An opening drilled in each side of the housing allows access to the space occupied by the gears. As the gears are rotated, each tooth space fills with syrup as it passes the intake opening. When, by continued rotation, the discharge opening is reached, the gears mesh and force the syrup out through this opening. The speed of pumping must be strictly controlled to synchronize with the speed with which the hardened filaments can be taken up, and also in relation to the thickness of the fibres it is wished to make.

The viscosity of the fibre-forming syrup must remain constant; a change in viscosity would result in a different rate of flow through the holes of the spinneret. The high viscosity of the spinning syrup is characteristic of solutions of chain polymers, and arises from the resistance offered by a chain to another sliding over it. In practice, it may be necessary to limit the lengths of the chains in order to obtain a satisfactory viscosity in the syrup, and sometimes even degrade the material to shorten its average molecular chain length.

Some spinning syrups, usually those prepared by dissolving the fibre-forming material in a solvent, are not subject to chemical changes on standing, and these can be stored in airtight containers until required for spinning into fibres. Others must be spun into filaments soon after they have been prepared. In the latter cases, the preparation of the spinning syrups, and the operation of the spinning process, must be carefully synchronized to avoid fluctuations in the properties of the final fibres. In the early days of the production of viscose rayon, for instance, some difficulties were experienced in that different batches of fibre exhibited changes in dye affinity. By tightening up on the details of manufacture and, further, by the careful mixing of several prepared batches of spinning syrup, such difficulties were largely eliminated.

While the spinneret may be regarded simply as a plate in which fine holes are drilled, it does represent a piece of precision-made equipment. The diameters of the holes, through which the spinning syrup is extruded, must be accurately uniform in size to avoid any variations in thickness from one filament to another. The number of holes in a spinneret will determine the number of filaments which will be present in the yarns; this usually varies between 15 and 100. When a 'tow' from which 'staple' fibre is prepared is being made, there will be several thousand holes in the spinneret. These are

Fig. 4.3 Three jets used in the production of viscose rayon. The centre one has more than 10 000 holes in it and is used in the production of viscose rayon staple. The two small jets are for the production of different types of continuous filament viscose rayon yarn.

usually arranged in concentric circles, round the centre point of the spinneret face. The sizes of the extrusion holes will vary with the thickness of fibre it is desired to make, but the most common hole diameters are between 2 and 4×10^{-3}in., (5 and 10 microns) and these are usually drilled to an accuracy of 1×10^{-4}in. (0.25 microns). The diameters of the holes in the spinneret are not necessarily those of the filaments produced; if the take-off rate is somewhat higher than the extrusion rate, then some thinning down of the filaments will take place as they emerge from the holes in the spinneret.

The holes in the spinneret are usually round but, in recent years, other shapes of hole have been used to produce fibres having a variety of cross-sections. Such cross-sections have been selected to produce some special effect when the fibres are incorporated in a fabric. For instance, quite a considerable amount of fibre is now spun with a cross-section which is roughly triangular, or 'trilobal'. When light enters a filament of this shape, some of it is reflected back strongly at various points in much the same way as from the facets in cut glass. Fabrics made from trilobal fibres appear to be covered, therefore, with tiny glittering points, and are selected for their decorative effects. A more practical use of the trilobal cross-section is in the construction of carpets containing man-made fibres. Not only is 'life' added to the carpet, but the strong light reflection also masks the dulling effect which inevitably results from soiling during wear.

The spinneret must be made from a material which will resist wear and deterioration from the spinning conditions used. For instance, the wearing away with time of the material of the spinneret could result in the extrusion holes gradually increasing in diameter with a consequent increase in the thickness of the fibres being produced. Apart from this, the making of a spinneret involves a high degree of skill and the expenditure of very many man-hours, and every care must be taken to preserve its useful life as long as possible. It is for such reasons that expensive, and sometimes not easily worked metals have to be used in the preparation of some spinnerets.

Three main types of spinning process are in use in the preparation of man-made fibres, these being referred to as 'dry' or 'solvent' spinning, 'wet' spinning, and 'melt' spinning, respectively.

a. Dry spinning. This method is used to spin filaments from syrups which can be prepared by dissolving the fibre-forming materials in a suitable solvent. Cellulose acetate, for instance, is soluble in acetone and it is by dry spinning from an acetone solution that its fibres are made. The principle of the method is to extrude the filaments into a stream of warm air, when the solvent will be evaporated and its vapour carried away by the stream of air, leaving the hardened material in the form of a fibre.

The equipment used for dry spinning has a metering pump which forces the spinning syrup through a filter and then through the holes in a spinneret situated at the top of a spinning vessel, or shaft, which usually has a diameter of up to 12 in. and a height of up to 20 ft. An air inlet is provided at the bottom of the shaft and, as the extruded syrup proceeds downward in the form of a number of fine streams, it meets a current of warm air travelling upward and supplied through this air inlet. The current of warm air evaporates the solvent, leaving solid filaments, and air and solvent vapour are taken away through an air outlet situated near the top of the shaft, to a recovery plant where the solvent is extracted ready for re-use. The efficiency of recovery of the solvent is an important factor in the economics of the whole process. After leaving the bottom of the spinning vessel, the hardened filaments, now combined to form a yarn, pass round a guide pulley, as shown in the diagram, before being drawn off by a roller system.

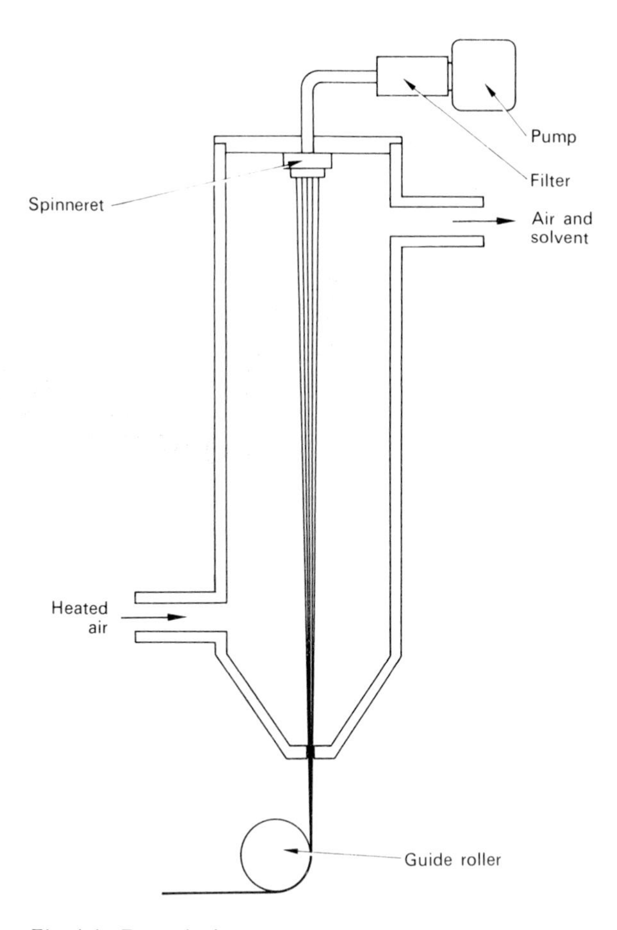

Fig. 4.4 Dry-spinning process.

In the case of dry-spun fibres, the spinneret is usually made of stainless steel and corrosion is not a serious problem. Stainless steel is strong and has an adequate resistance to chemical attack.

In dry spinning, the filaments, now combined to form a yarn, must finally be taken up to form a package. The form of package can be, either a bobbin on which the filaments can be wound free from twist, or with twist on a cop produced by ringspinning equipment. Both these package-forming machines are described in ch. 5.5.6. The yarn can either be passed forward for sale on these packages or, if any further treatments are to be carried out, to the next stage of processing.

b. Wet spinning. Wet spinning is used when the fine streams of syrup emerging from the spinneret have to be hardened into solid filaments by a further chemical treatment to re-precipitate, or regenerate, the fibre materials. Viscose rayon yarns, for instance, are produced by wet spinning. Here, the viscous solution of cellulose xanthate in caustic soda is extruded into a 'coagulating' bath containing a solution of sodium sulphate and sulphuric acid. The filaments emerge from the spinneret under the coagulating liquid, and the illustration shows the general arrangement used.

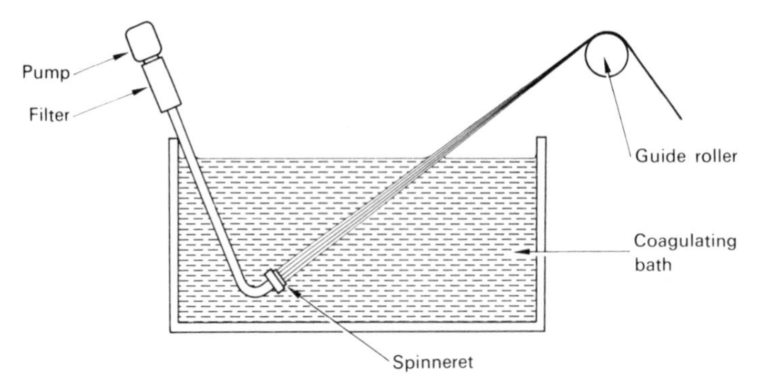

Fig. 4.5 Diagram showing wet-spinning process.

After the wet, hardened filaments, now combined to form a yarn, have left the coagulating bath, they are drawn away by a roller system and fed into some kind of package-forming equipment. A common one, known as a centrifugal pot machine, or Topham box after the name of the inventor, was the first method of taking up viscose rayon yarn employed commercially. A diagram illustrating the principle of the machine is shown. The wet yarn is dropped vertically down an elongated funnel which feeds it, at the lower end, into a rotating pot. This pot is usually about 7 in. in diameter and 6 in. in height, and it revolves at between 8000 and 10000 rev/min. The walls of the pot

Fig. 4.6 Wet-spinning process in action.

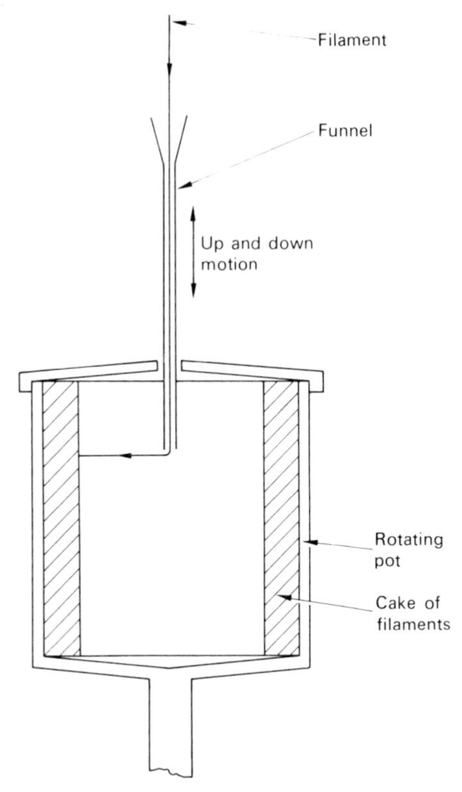

Fig. 4.7 Topham box method of taking-up yarn.

are perforated and, under the centrifugal force produced by the high speed of rotation, excess liquor carried over by the yarn from the coagulating bath is removed, and this can be recovered. While the pot is rotating the funnel feeding the yarn into it is moved up and down, so that the yarn emerging from its lower end is distributed evenly over the inner surface of the vertical wall of the pot. The yarn is thus guided so as to form a thick cylindrical mass referred to as a 'cake'. A cake can weigh between 1 and 2 lb. The rotation of the pot inserts some twist in the yarn, and this serves to bind the filaments together and from a compact yarn. The amount of twist inserted is easily calculated from a knowledge of the peripheral speed of the take-off roller round which the yarn passes after it leaves the coagulating bath (i.e. the linear speed of the yarn), and the rotational speed of the pot. If the linear speed with which the yarn is fed to the pot is l in./min, and the pot rotates at n rev/min, then the twist inserted in the yarn is n/l turns per inch. Thus, with a yarn speed of 250 ft/min (= 3 000 in./min), and a pot rotating at 9 000 rev/min, the twist inserted will be 3 turns per inch.

Another method of taking up the wet spun yarn, bobbin spinning, is an alternative to pot spinning. The yarn is wound on the outside of a perforated metal cylinder without any twist being inserted. The mechanism is situated above the coagulating bath, and the coagulating liquor carried over by the yarn runs back into the bath. The mechanism is so designed that a yarn-filled bobbin can be removed easily and replaced by an empty one without stopping the machine or interrupting the flow of the filaments.

When the cake, or bobbin, has built up to the required size, it is removed and passed forward for the various washing and finishing treatments which have to be carried out on the yarn. This is done while the yarn is in its package form, either on a cake or a bobbin, and there are difficulties associated with obtaining satisfactory penetration of the closely-packed yarn. For this reason, as has already been mentioned, there is now a greater tendency to adopt methods of continuous treatment of the yarn. The yarn is formed by wet spinning, in exactly the same way as described above for the batch process, but from the coagulating bath it is carried forward for its finishing treatments as a single yarn and not as part of a package. One method of doing this is to lead the yarn round a pair of long rotating cylinders the axes of which are inclined at a small angle. As the cylinders rotate, the yarn advances in spiral form from one end of the cylinders to the other. At various positions on the cylinders, the yarn can be sprayed with the appropriate washing or finishing liquid, finally being dried at a heated end-portion of the cylinders before being led away for winding into a package. In another method of finishing a single viscose yarn, the two large cylinders are replaced by a series of small cylinders attached to a frame in descending positions and displaced horizontally in the manner of a series of steps. The diagram shows the arrangement used. The number of wraps of yarn on each of the cylinders is adjustable and, in this way, the time the yarn remains on each cylinder can be set. Each cylinder is under its particular spray, and thorough washing, desulphuring, bleaching and so on, can be carried out on the yarn in single strand form.

Fig. 4.8 'Doffing' out cakes of yarn.

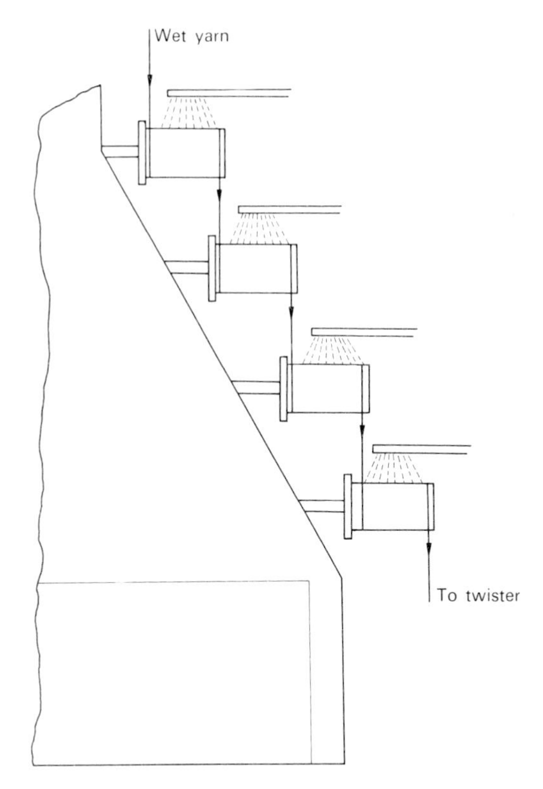

Fig. 4.9 Continuous treatment of viscose yarn.

In wet spinning, the conditions which the precision-made spinneret has to withstand are much more severe than in other spinning methods. In preparing viscose filaments, for instance, a strongly alkaline solution is extruded into a strongly acid coagulating bath, and corrosion of the spinneret is a problem. Originally platinum was used to construct the spinneret, but this gave way to a 70% gold, 30% platinum alloy which gave better service and was easier to fashion. This was eventually replaced by the rhodium-platinum composition which is now used.

In making cuprammonium rayon, wet spinning is again used, but the chemical conditions are less severe than those used in viscose rayon manufacture, and nickel spinnerets can be used.

c. Melt spinning. Melt spinning is used for the majority of thermoplastic man-made fibre-forming materials, although a few are prepared by dry spinning from a solvent.

In melt spinning, the fibre material, in 'chip' form, is fed continuously by gravity from a hopper into a spinning vessel where it falls on to an electrically heated grid and melts. The spinning vessel is usually filled with an inert, oxygen-free gas. As the fibre material melts, it runs through the grid and forms a pool at the bottom of the vessel. Here the temperature is adjusted until the correct viscosity for spinning is obtained. At the bottom of the spinning vessel is a pump which forces the molten polymer through a filter and a spinneret.

The filaments are extruded through the spinneret to meet a stream of cold air, when they solidify. The filaments then pass down through a steam tube to produce stability while they are wound in the form of a package. After passing round a guide pulley, the filaments, now combined to form a yarn, pass to some form of package-forming machine. A diagram illustrating the sequence of melt spinning operations is shown.

In principle, melt spinning is a simple process; the filaments are extruded and, when their temperature has fallen sufficiently, they harden. No solvents are involved and, in consequence, no recovery plant is needed. It was first introduced by the DuPont Company to spin nylon yarns and much effort was put into its development before the smoothly operating spinning process of today was evolved.

4.3.2 After-treatment applied to man-made fibres

If a fibre-forming polymer is allowed to crystallize without any stress being applied, the crystallites it contains are randomly oriented. In such a state, its filaments are not satisfactory for textile purposes, being relatively weak and inelastic. The newly-extruded filaments of all man-made fibres are subjected, therefore, to a stretching or 'drawing' operation such as was first used by Thiele in the cuprammonium rayon process. The flow of material produced by the drawing swings round the crystallites so that their longitudinal axes lie along the direction of flow and, it follows, the length of the filament. It is possible that, in a polymer of low crystallinity, the drawing may allow some additional crystalline material to be formed. The drawing may be introduced as part of the spinning process, as in the modified viscose rayon process to produce polynosic rayon, and in the manufacture of cuprammonium rayon when the dragging effect of the flow of coagulating water stream plus the stretching action of the speeded-up take-off produces a very considerable extension in the filaments. With some polymers, however, the filaments can be taken up more or less as extruded and the drawing carried out at a later stage. The filaments, now gathered together to form a yarn, are built up to

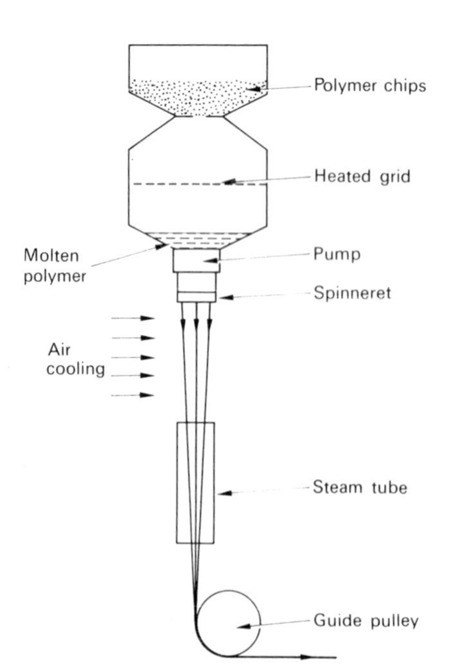

Fig. 4.10 Melt-spinning process.

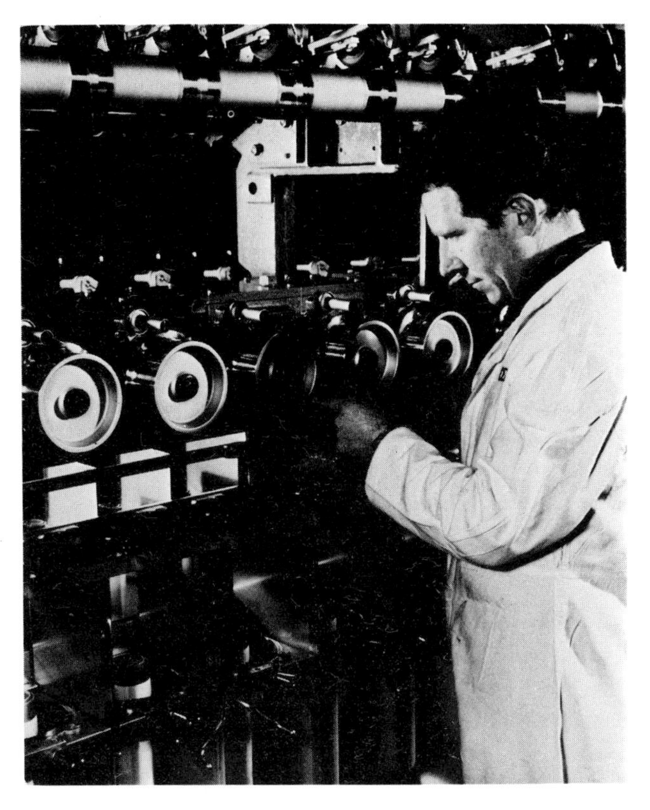

Fig. 4.11 Draw-twisting of nylon yarn. Here the undrawn yarn is being stretched between the feed rollers at the top and the godet rollers in the middle, before being wound on to the cop tube at the bottom.

form a package, and later this package is mounted on a machine known as a 'draw-twister' and the yarn is there subjected to a considerable extension. Some materials, nylon is an example, draw very easily and the operation can be carried out cold; other polymers, the polyesters for instance, draw more satisfactorily when heated.

The draw-twister unwinds the yarn from its spinning package and passes it between two sets of rollers. The second set of rollers has a higher peripheral speed than the first set and the yarn is extended by an amount determined by the ratio of these two speeds. If the nylon yarn is stretched to, say, twice its original length, then it is still inelastic and will not attempt to recover from a small amount of extra extension which may be applied to it. If it is drawn to something like four times its original length, it becomes a strong, compact, elastic yarn which will recover immediately from a further extension which may be applied. The draw-twister is set to produce this stabilizing extension, after which it inserts a regulated amount of twist to bind the filaments together and builds the yarn up to form a package. The draw-twisting of a polyester yarn, such as terylene, is carried out in much the same way as nylon, except that it passes through a heating zone situated between the two sets of stretching rollers. The drawing of a man-made fibre, at whatever stage in the manufacturing process it may be carried out, must always be operated under accurately controlled conditions. Drawing modifies its properties markedly and, in consequence, any laxity in controlling the conditions will result in a variable final product.

The amount of draw given to a man-made fibre intended for general textile use does not necessarily represent the limiting extension which can be applied. Any further extension which may be found possible will increase the tensile strength of the fibre but will, at the same time, reduce its elasticity. A certain amount of 'give' in a fabric is useful for wearing apparel since it can promote comfort in wear. Tough, unyielding characteristics are required, however, in yarns intended to be used in rubber tyres and as part of conveyor belts, and 'hot-drawing' techniques have been developed to provide an even greater extension in certain fibres and yarns. As the name implies, the fibres or yarns are heated and, at the same time, a powerful pull is applied to produce additional extension.

An examination of lists of properties of fibres will indicate that, in general, the synthetic fibres have good electrical insulating properties and, at the same time, a low moisture absorption. They are prone, therefore, to pick up electrostatic charges and, once they have become charged, there is insufficient moisture present to provide a conducting path and allow the charge to leak away. Electrostatic charges are developed when two bodies are brought into intimate contact; it is thought that electrons can leave the surface of one body and attach themselves to the other, depending on the relative 'electro-positiveness' of the two materials. When the two bodies are separated, the one which has picked up electrons has developed a negative electrostatic charge while the other, which has lost electrons, has developed a positive charge. Pressure and friction between the two materials magnify the effect by increasing the area and degree of contact and, in the manufacture of synthetic fibres, there are many opportunities for 'static' charges, as they are called, to develop. Static charges are a nuisance in the manufacture of synthetic fibre yarns, and later in the production of fabrics from them. The charges cause the fibres to adhere to each other and to guides, rollers and machine parts while they are being processed, thus interfering with high speed production and the provision of a uniform product. While the very nature of a synthetic fibre will always mean that static charges are developed, some alleviation of the troubles can be obtained by ensuring that the charges will leak away quickly. The processing machines should be well-earthed and the atmosphere surrounding the machines should contain some water vapour. In addition, the filaments, immediately after extrusion, are coated with a finish which contains an electrically conducting material, and an 'anti-static' finish is retained right up to the final fabric stage.

4.3.3 Staple fibre production

All man-made fibre materials are extruded as continuous filaments but, during the 1920s, the idea was conceived of cutting or breaking the continuous rayon filaments into shorter 'staple' lengths, in which form the natural vegetable and hair fibres are available. The reasons for considering such a step were:

1. A considerable amount of machinery for the preparation of yarns from cotton and wool was available, and there was a natural desire to use this equipment and the specialized skills of the operators to process rayon yarns.
2. It is very difficult to mix fibres of widely differing lengths, and yet it was realized that economic, and in some cases performance, advantages could result from blending the rayons with natural fibres.

3. The handle and appearance of yarns made from continuous filaments and staple fibres respectively, are widely different, the latter being soft and bulky compared with the continuous filament yarns.

Rayon staple fibres were something of a novelty in the 1920s, but began to establish themselves in the early 1930s so that in 1935, both viscose rayon staple and acetate rayon staple were available in quantity.

A wide use of rayon staple fibre was made in Germany and Italy in the years prior to the Second World War. There was an acute shortage in both countries of wool and cotton for clothing for members of the armed forces and rayon staple fibre was used to eke out the supplies of the natural fibres and, in some cases, to replace them. In the first trials, rayon waste was torn up into shorter lengths in shredders, but widely varying staple lengths and tangling which resulted would not allow satisfactory spinning, and an improvement finally came from running the rayon waste through cutters to produce the desired lengths.

The use of staple man-made fibre has increased rapidly year after year, and now accounts for a very large proportion of the annual man-made fibre production. Most of the

Fig. 4.12 Viscose rayon 'tow' passing over a large tension roller, down a water-lubricated funnel to knives which will cut it into the required staple lengths. (The cutter unit cover was removed for the photograph.)

rayon produced is used as staple fibre and a considerable percentage of the synthetic fibre production is supplied in this form.

Basically, the method of making man-made staple fibre is the same as for continuous filament production up to the point where the filaments leave the spinneret. From then on, the two processes differ. The filaments from a large number of spinnerets, probably 100 in the case of viscose rayon, and each having more extrusion holes than the normal continuous filament spinneret, are gathered together to form a thick rope which is called a 'tow'. The tow is then fed into a cutting machine which can be set to cut the rope of filaments into any desired length of staple, in which form it is baled ready for shipment to the various mills.

In the case of viscose rayon the thick tows are sometimes supplied to mills which desire to do their own cutting into staple lengths. In this case, after leaving the coagulating baths, the tows are washed, treated, dried and packed ready for dispatch. By far the greatest quantity of viscose rayon tows are cut immediately they leave the acid bath. It is then necessary to use special materials for the construction of the cutting blades and other parts of the equipment in order to resist the corrosive action of the coagulating bath liquor which is carried over by the tow . The wet tow after cutting must be washed, treated and dried and, from the cutter, it falls on a continuous perforated plate conveyor which carries it through all these treatments. It is then baled ready for shipment.

A simpler version of the above process is used to produce staple fibre from tows formed by dry spinning and melt spinning since no corrosive liquids are involved.

Cotton and wool fibres have the advantage of clinging together naturally and thus

Fig. 4.13 A highly compressed bale of viscose rayon staple fibre being removed from a baling press for wrapping and labelling.

enabling a strong yarn to be spun from them. Cotton is a thin ribbon which is twisted along its length, and these convolutions enable one fibre to entangle with another and cling together by this means, so that only a modest twist needs to be inserted in the yarn during its preparation, to ensure that the fibres will not slide apart when pulled. Wool fibres cling together even more strongly; scales on one fibre interlock with those on another to give a powerful adhesion between the two. The early man-made staple fibre was a straight, smooth cylinder and a substantial amount of twist had to be inserted in the yarn, to prevent the fibres from sliding apart. In recent years, man-made staple fibres have been crimped or waved to improve adhesion, allow more uniform blending with natural fibres, and provide a yarn having a softer handle and greater covering power. Various methods of crimping have been used, but the most usual one is to pass the tow between hot fluted or corrugated rollers, when the filaments are set in a wavy form. The crimped tow is then cut into suitable staple lengths, lubricant and an anti-static agent are added, and the fibres are finally dried and baled.

When the bale of staple man-made fibre is opened ready for spinning, the fibres are subjected to the same operations of loosening, carding and combing as are applied to natural fibres in preparation for spinning into yarn. The object of these particular operations is to arrange all the staple fibres in a parallel formation so that they can be spun into a strong, uniform yarn; in fact, just the formation which existed in the tow before it was cut up and baled. It is not surprising, therefore, that attempts have been made to preserve the state of parallelism which exists in the tow, while dividing the tow into staple lengths. The general method of procedure followed is to pass the tow between two sets of rollers, the second set of rollers rotating at a higher speed than the first set. The filaments in the tow are thus subjected to considerable strain and will break, not all at the same place, but at random positions in the space between the two sets of rollers. By this means a tow of continuous filaments is converted into a uniform 'sliver' of staple fibres and, since the breaking positions are randomly distributed, this sliver holds together sufficiently well to enable crimping to be carried out before the yarn-spinning operations are applied. The more important methods of tow-to-sliver processing which have been operated satisfactorily in practice, are discussed in ch. 5.5.6.

4.3.4 Split-film fibres

In the mid-1930s, another method of preparing fibres from synthetic polymers was devised. In 1936, the German chemical firm, I.G. Farbenindustrie, filed a patent covering the production of fibres from polymers, such as the polyamides and polyvinyl chloride, by first preparing a film of the material, stretching this film in one direction to multiples of its original length, and then cutting up the film along its stretched length into very narrow tapes. This produced strong 'fibres' having a flat cross-section.

Had the position remained that only these flat tapes were available, then the applications of such fibres would have been very limited, but other investigators began to take an interest in the method. It was shown that if the polymer films were stretched in one direction very considerably, the strength of the film along the direction of stretching was increased markedly, and there was a decrease in strength across the film. Further, the film tended to split up into 'fibrils' to produce narrow fibres running along the length of the film and relatively loosely held together across the width of the film. Some mechanical help was required to separate the fibrils; rubbing between two friction surfaces or subjecting the treated film to vibration or violent agitation helped to produce fibrillation.

Various investigators have studied a whole range of polymers which might be used for split-film fibre production, but now practically the whole of the commercial exploitation of such a process is based on polypropylene by reason of its strength, readiness to fibrillate and material cost. The first commercial development of the process based on polypropylene started with the manufacture of twines and ropes. These proved to be very successful and, since that time, there has been a gradual move towards the fine textile fibre fields. It will be appreciated that, provided a sufficently fine fibrillation can be obtained, the method offers a means of producing the equivalent of a spun yarn on a continuous rather than a multi-stage intermittent method.

The process for producing split-film fibres from polypropylene, as it is now operated, consists of the following steps:

1. The polymer is melted and extruded into a film.
2. After extrusion, the film is cooled rapidly.
3. The film passes on to the slitting position where it is divided into narrow tapes.
4. The web of tapes is then subjected to a heating and stretching treatment.
5. There is next an annealing or stabilizing heat treatment.
6. Finally, the tapes are separated for winding on to beams or packages.

A diagram of the practical arrangement used is shown in the illustration. The film is formed by an extruder through a flat film die, after which it is chilled rapidly before much crystallization in the film can take place. Water quenching is the most common method of providing chilling, but some manufacturers prefer to use a series of chilling rollers which provide a closer control over the film. Slitting is usually done using an assembly of razor blades. The life of the blades can be a few days, or only a few hours if heavily pigmented films are being used. The slitter is designed to allow rapid replacement of the blades without having to stop production. The slit film is stretched between a first set of rollers and a more rapidly rotating second set while it is inside a radiant heat, or circulating hot air, oven. For polypropylene, the film is heated to a temperature of about 150°C (302°F) while being stretched 6 : 1 to 12 : 1 depending upon the end use for which the fibres are intended. The setting up of internal stresses in the slit film could lead to subsequent shrinkage, so an annealing section is often included. To anneal the slit film, it is taken through a second oven operating at a slightly higher temperature than the stretching oven, a third set of rollers rotating at the same or slightly lower, speed than the second set of rollers, to maintain control over the split film. The tapes are then taken up on some standard type of winding machine.

A coarse fibrillation for heavy yarns can be obtained by twisting the tapes but, for fine fibrillation such as is shown, a more severe mechanical treatment is required. This can take the form of rubbing between two surfaces, or passing the tapes through turbulent air or liquid streams, or providing some other means of violent agitation. If staple fibre is required, however, the tapes can be cut to the required length and sent forward in this form for processing into yarns. The carding treatment they will receive as part of

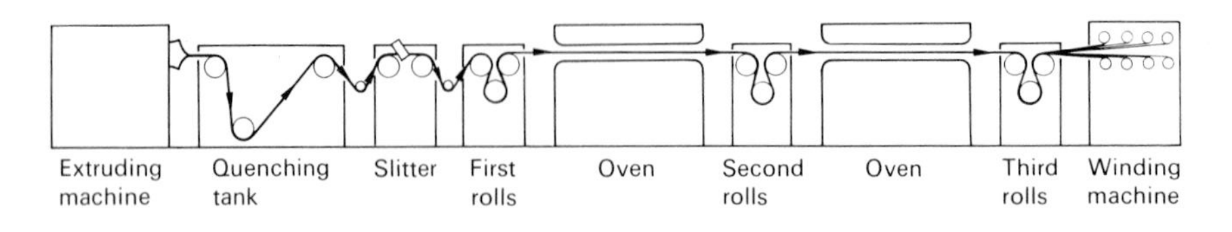

Fig. 4.14 Slit-film production.

Fig. 4.15 Fine fibrillation of a tape.

this additional processing will produce fine fibrillation. It is also possible to use stretch breaking to produce fibrillation.

4.4 THE VARIOUS MAN-MADE FIBRES

The man-made fibres can be divided into three main categories; those based on cellulose obtained from natural sources and usually referred to as the 'rayons', those prepared from naturally-occurring proteins, and the truly synthetic fibres. The cellulose fibres still account for a high percentage of the tonnage of man-made fibres produced throughout the world, while the protein fibres are now produced in only very small quantities. Each of the more important fibres in these categories will be discussed in turn and, in the information provided, figures relating to the properties of the various fibres will be given.

4.4.1 Fibres based on cellulose

The cellulose fibres available commercially are viscose rayon in its three forms, normal, high-tenacity and polynosic, cuprammonium rayon, acetate rayon and triacetate rayon. Viscose and cuprammonium rayons consist of cellulose which has been chemically modified to make it soluble and then regenerated back into its original cellulose form after extrusion, while in acetate and triacetate rayons the cellulose has been left in its chemically modified soluble form.

a. Viscose rayon. Normal viscose rayon fibres are prepared by first treating naturally-occurring cellulose with caustic soda, followed by a reaction with carbon disulphide, to produce a compound of cellulose, cellulose xanthate. This compound is soluble in alkaline solutions, and the viscous liquid thus obtained is extruded in the form of filaments which can be treated with acid to convert the compound back into the original cellulose.

In the preparation of viscose rayon fibre on a commercial scale, the starting material is wood, and the wood selected is usually spruce as it has quite a high cellulose content of a little over 50%. After the bark has been removed, the wood is broken up into chips and then subjected to a purification process by being treated under pressure with a solution of calcium sulphite. This produces what is known as 'sulphite wood' pulp. This material is bleached and washed, and finally formed into thin sheets.

The sheets of sulphite wood are next soaked in an 18% solution of caustic soda for something like two hours, and then placed under pressure to expel the excess liquor, which is collected and stored for further use. The swollen, and still further purified 'soda cellulose' so formed, is broken down into 'crumbs', and these are then transferred to reactor vessels. The crumbs are usually left standing in the vessels to age for two or three days. During ageing, contact with the air produces some oxidation and a consequent shortening of the molecular chains. A quicker ageing can be produced by using a higher temperature but, in that case, precise control is necessary if a satisfactory final result is to be obtained.

At the completion of the ageing process, the crumbs are stirred with carbon disulphide at a temperature of about 25°C (77°F) for a period of from three to four hours, when a compound of cellulose, cellulose xanthate, is formed. This compound is orange in colour. The vessel in which this reaction has been carried out is then evacuated to remove any traces of free carbon disulphide, and the cellulose xanthate is mixed with a dilute solution of caustic soda, water cooling of the vessel being provided. The solution produced contains the equivalent of about 8% of cellulose. If any dulling agents or coloured pigments are desired to be present in the final fibres, they are added to the solution at this stage.

Fig. 4.16 Sheets of wood pulp being loaded into a press ready for steeping in caustic soda solution.

Fig. 4.17 'Crumbs' being emptied into a mercerizing bin from one of the grinders.

The xanthate solution is somewhat unstable and, on standing, a slow hydrolysis begins to take place. This 'ripening' as it is called, results in a solution which will later give easier coagulation, and it is allowed to proceed until a required amount of hydrolysis has taken place. After this the syrupy liquid is conveyed, under pressure, along the pipelines which lead to the wet spinning equipment.

The viscose syrup is extruded through the spinneret into the coagulating bath. The bath vessel is constructed of lead, and it contains a solution of sulphuric acid and sodium sulphate held at a temperature of about 50°C (122°F). The addition of certain salts, in addition to the sodium sulphate, to the bath may sometimes be an advantage, as mentioned below; exact constitution of the coagulating solution is important in that it has an effect on the rate of coagulation, and hence the speed with which the syrup can be extruded, since the two must be kept in balance. Certain properties of the final fibre, dye affinity is one, can also be influenced by the rate of hardening of the filaments in the coagulating bath, and the temperature of the bath must be controlled accurately throughout the hardening process.

The extruded filaments are usually immersed under the coagulating liquid for a distance of between 12 and 24 in., and when they emerge, they are stretched by a specified amount to introduce an additional alignment of the chain molecules composing them. They can then be taken up on to a package by, say, a Topham box, when they will be fed down through the glass funnel of that equipment. During the build-up in the Topham box into a cake form, two or three turns per inch of twist are inserted by the equipment, and this serves to bind the filaments together to form a yarn. After the box is full, the cake is removed. A number of such cakes are placed one on top of another to form a stack having a central bore, and the cakes are held together by clamping plates. Processing fluids can now be pumped up through the central bore of the stack, when they

Fig. 4.18 A 'box-type' machine producing viscose continuous filament textile yarn.

will seep through the mass of yarn to reach the outside. Water is first used to wash away the remaining acid, this is followed by a sodium sulphide solution to remove traces of sulphur, and a final treatment with water is used to complete the purification. Sometimes the cakes may be given a mild bleaching using peroxide or hypochlorite solutions. The cakes are finally hydro-extracted and dried.

Since the yarn is built up quite solidly on the cakes, it is not easy to ensure complete penetration and cleansing during the purification processes, and some manufacturers go to the expense of winding the yarn from the cakes into loose hanks when satisfactory cleansing in that form is assured.

As has already been mentioned, more use is now being made of continuous treatment. After the coagulating state, the filaments are carried forward, as a twistless yarn, to the treatment positions and the cake-forming portion of the process is eliminated.

In manufacturing a material like viscose rayon, it is not easy to produce a yarn which is accurately uniform from day to day. Each stage of the process requires precise control if lustre, physical properties and dye affinity are not to fluctuate. After viscose rayon yarn manufacture began many years of production experience had to be gained before the reliable product we know today could be made available.

Viscose rayon is probably more widely developed than any other man-made fibre, and it is used in practically every branch of textiles. The fibre is composed of cellulose, and its simplest repeating molecular unit may be written as follows:

```
         HC——O
        /     \
   —HC          CH—O—
        \     /
         HC——CH
         |    |
         HO   OH
```

Normal viscose rayon fibres, when viewed under a microscope, appear as shown in the illustration. The fibres are smooth and straight with pronounced striations running along their lengths. Their cross-sections are irregular, having many rounded indentations, and by reason of this indented boundary, each fibre offers a large surface area to any liquid in which it may be immersed.

Normal viscose rayon has a density of 1.51 g/cm^3. Its moisture regain, determined as already indicated, is 12.5% for a R.H. of 65%. This is appreciably higher than the equivalent value for cotton and is a consequence of the looser, more penetrable, molecular structure of normal viscose rayon when compared with that of the natural fibre.

Viscose rayon is not thermoplastic and, when heated, it will not melt. As its temperature is raised, it begins to lose strength at about 150°C (302°F) and finally chars and decomposes as it approaches a temperature of 200°C (392°F). The fibre burns quite readily.

An average value for the tenacity of normal viscose rayon is 2.0 g/denier (18.0 g/tex) (for definitions of 'denier' and 'tex' see chapter 3), which is adequate for the majority of textile applications. What is less satisfactory, however, is the fact that this tenacity falls off quite markedly when the fibre is wet to about 45% only of its dry value. This, again, is explained by the relative looseness of the molecular structure. From a practical point of view, this relatively low wet strength calls for care in laundering garments and fabrics made from normal viscose rayon, and in dyeing to avoid introducing stresses which could lead to irregular colouring.

As regards its resistance to chemical attack, viscose rayon withstands the lower concentrations of cold dilute acids, but it is disintegrated by hot and concentrated

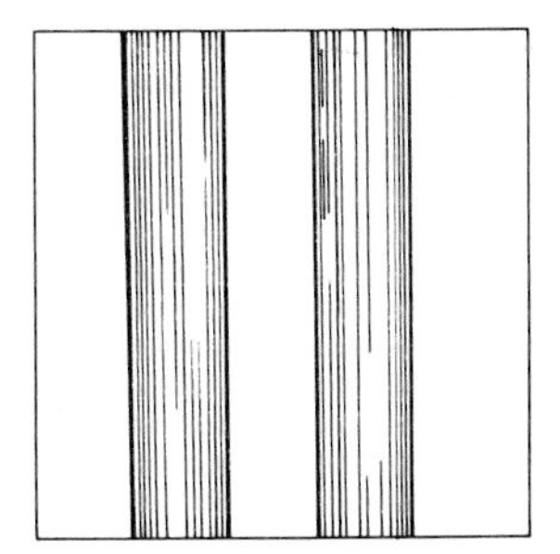

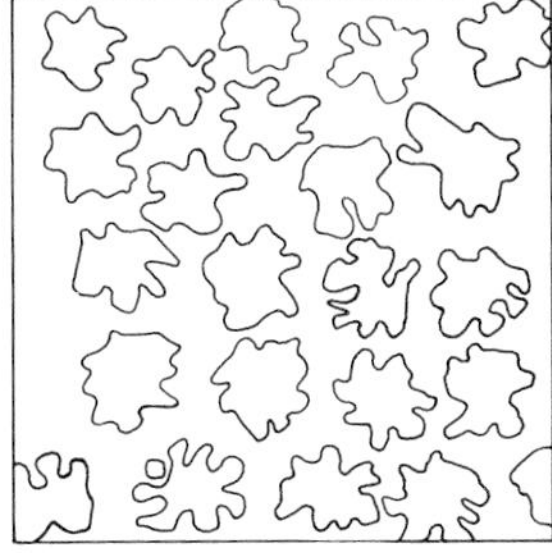

Fig. 4.19 Viscose rayon fibres.

Fig. 4.20 Testing the effects of high temperatures and pressures on viscose yarn.

acids. Cold dilute alkali solutions have no effect on the fibre, and very little when hot. Strongly alkaline solutions first produce swelling of the fibres and ultimate disintegration. In general the fibre is not affected by organic solvents.

Normal viscose rayon dyes with cotton dyestuffs, and more readily than cotton reflecting, once again, its more open molecular structure. The direct, basic, sulphur, vat, azoic and reactive dyes are all suitable for colouring the fibre. Probably the best all-round fastness characteristics are obtained by using the vat dyes and some of the reactive dyes.

Having good processing characteristics, an adequate strength, a pleasant handle, and a good colour, and being relatively cheap to produce, viscose rayon is used in very large quantities indeed, and in most textile applications. There are also substantial outlets for it in blends with natural fibres. It is sometimes used in mixtures with wool to reduce the cost of suiting fabrics; another example of mixing is in combination with cotton for pile fabrics, such as velvet where the backing may be of cotton, and the pile of viscose rayon.

Over the years, attempts have been made to obtain improved tensile properties in viscose rayon, and particularly under wet conditions. Two of these have been successful and two other kinds of viscose rayon besides the normal fibre are available on the market. These are 'high-tenacity' viscose rayon, and 'polynosic' rayon. This latter name was adopted in France to refer to special viscose rayons having improved tensile properties over the normal fibre, and behaving more like natural cotton.

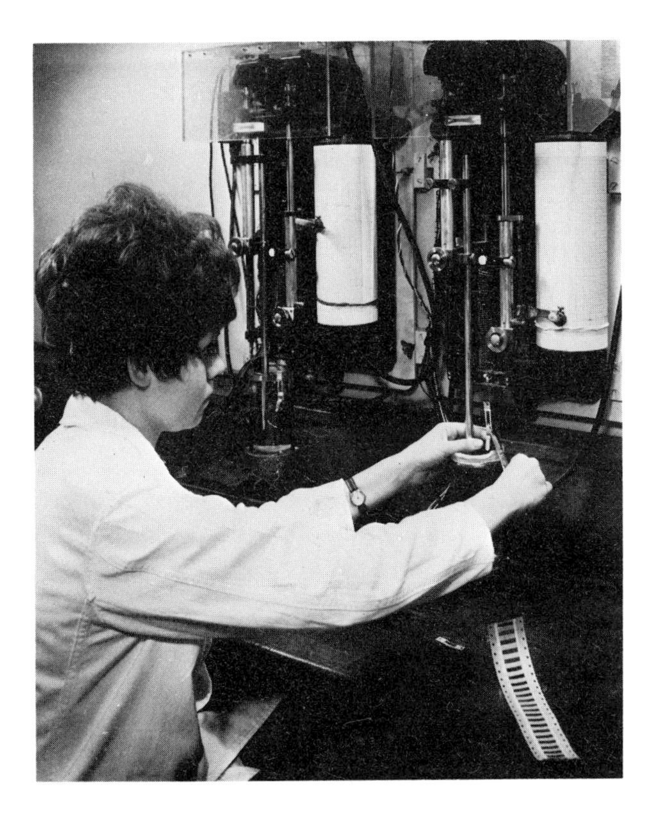

Fig. 4.21 Extensometers used in research laboratories for the tensile testing of filament. As the filament is stretched, its tensile characteristics are drawn as a graph.

High-tenacity viscose rayon fibres are produced by a modified viscose process. The main modifications are:

1. The addition of zinc sulphate to the coagulating bath; this probably applies a skin of zinc compounds to the surface of the fibre.
2. Further modifying the composition of the coagulating bath to slow down the rate of regeneration and produce fibres having longer molecular chains.
3. Stretching the final fibre while it is immersed in hot water to obtain an even better alignment of the chain molecules.

Under the microscope, the fibres of high-tenacity viscose rayon appear as in the illustration. The fibres are straight rods with cross-sections which, while being more regular than those of normal viscose rayon fibres, are imperfectly round, and often bean-shaped.

High-tenacity viscose rayon has a dry tenacity which is more than twice that of the dry normal fibre and, even when wet, its strength is appreciably higher than that of the dry normal fibre. By reason of its strongly oriented condition, high-tenacity rayon has, in addition to extra strength, good abrasion resistance and flex-fatigue characteristics. It has found wide application, therefore, as a tyre-cord yarn and in the construction of conveyor belting. It has found some application in industrial fabrics.

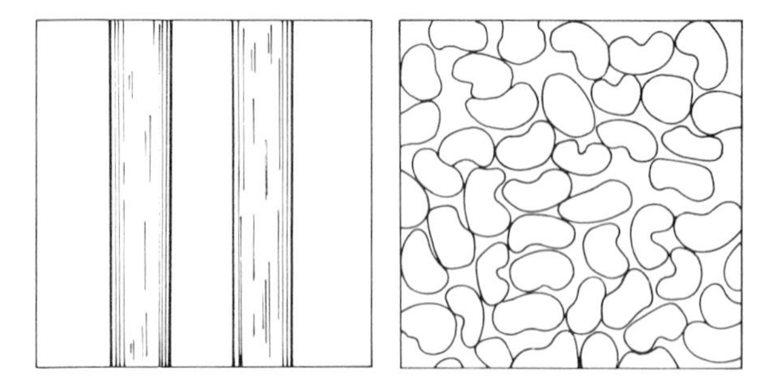

Fig. 4.22 High-tenacity viscose rayon fibres.

Polynosic rayon fibres are again produced as a result of a number of modifications to the normal viscose rayon process some of the more important being:

1. The ageing process at the soda-cellulose stage is omitted.
2. The cellulose xanthate is dissolved in water and not in an alkaline solution.
3. The ripening process is eliminated.
4. The coagulating bath is made up with less acid than in the normal process.
5. While the newly-extruded filament is still immersed under the coagulating solution, it is subjected to a substantial stretching action.

Fig. 4.23 Laying 'Tenasco' on to a drum prior to making test tyre carcases.

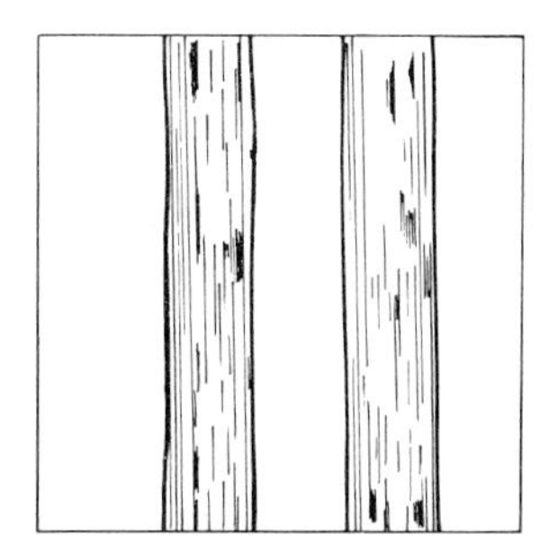
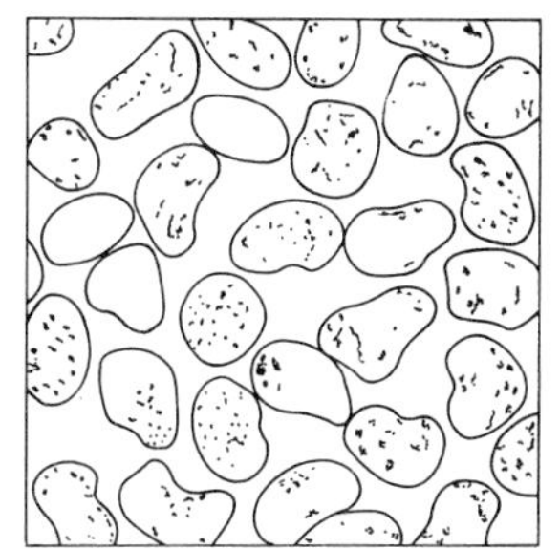

Fig. 4.24 Polynosic rayon fibres.

The result of all these changes is that the fibre is composed of longer, better aligned, chain molecules than is the normal fibre and, in consequence, its properties are closer to those of natural cotton. Its more compact structure gives it a density of 1.53 g/cm^3, a greater stiffness, both dry and wet, than normal viscose, a reduced moisture regain of 11.8% at 65% R.H., and a higher tenacity both dry and wet, than the normal fibre.

Polynosic rayon fibres have the general appearance, under the microscope, as shown in the illustration. The fibres are rod-like, with some pits and streaks, and the cross-sections are of irregular shape but without the pronounced indentations shown by those of the normal fibre. Polynosic rayon dyes with cotton dyestuffs at about the same rate as cotton. It has a crisper handle in fabrics, and a better dimensional stability than normal viscose rayon, and is better able to compete directly with cotton as a dress material.

b. Cuprammonium rayon. The manufacture of cuprammonium rayon is based on the fact that cellulose is soluble in cuprammonium hydroxide, and can later be re-precipitated from solution by treatment with an acid, alkali or other electrolyte.

The starting raw material for the preparation of the fibre can be sulphite wood pulp as in the manufacture of viscose but, preferably, cotton linters are used since their cellulose content is very high. There is nothing incongruous in using cotton linters as a source of cellulose from which to prepare a man-made fibre, since the linters are too short in length for spinning into yarns in the normal way and, by the extrusion treatment, are converted into long fibres which can be spun.

The linters are first kier-boiled with a caustic soda solution to purify them, then bleached using a hot sodium hypochlorite solution, and finally scoured, washed and dried. These treatments are carried out with care in order to avoid breaking down the molecular chains which exist in the original natural cellulose as little as possible. The purified linters are then beaten into a pulp.

Pure copper sulphate is dissolved in water to a high concentration, and to this is added strong caustic soda solution. The two react together to form sodium sulphate, which remains in solution, and copper hydroxide which is precipitated. In practice, the caustic soda solution is added very slowly to the copper sulphate solution, cooling being applied the whole time, until all the copper hydroxide has been thrown out of solution as a light blue precipitate. To this mixture of precipitate and liquid is added the purified linter pulp, and the whole is stirred vigorously for a period of about 30 min, the mixture being kept cool the whole time. The paste thus formed is transferred to a filter press and, under the pressure, the sodium sulphate solution is expelled, and a final washing removes the last traces of the sulphate which would otherwise produce some coagulation of the cellulose solution later.

The moist cakes, usually containing about half their weight of water, are disintegrated and dissolved in concentrated ammonia solution. In order to protect the cellulose against oxidation, protective agents, such as bisulphites and hydrosulphites, are added in small quantities to the ammonia solution. After stirring together for a period of from six to eight hours, the solution is diluted down to an 8 or 9% cellulose content by adding ammonia or caustic soda solutions. If it is intended to have delustrants and pigments present in the final fibre, they are added to the solution at this stage.

The viscous liquid thus formed is filtered several times through fine nickel gauze, and then held for about 90 min under reduced pressure to remove air bubbles. At the same time, some ammonia is also evolved and is collected, and a further thickening of the solution to give the final spinning syrup takes place. Cuprammonium spinning syrup has a reasonable shelf life and can be stored under suitable conditions.

Filaments are prepared from the syrup by wet spinning. The syrup is forced under pressure through the holes in a nickel spinneret which is supported at the top of a funnel down which warm water is flowing. The warm water removes ammonia from the fine streams of syrup and begins the coagulation, which is completed by spraying the emerging filaments with dilute sulphuric acid, or possibly using a bath of the acid. Something like 80% of the copper used is recovered at this stage, and the efficiency with which it is recovered is an important economic factor in the process.

The stream of warm water travelling down the bore of the funnel exerts a downward drag on the newly-extruded streams of syrup, and the flow conditions, and the speed of take-off are adjusted until the correct diameter of filament is obtained. Due to this stretching action, the degree of molecular chain orientation is higher in cuprammonium rayon production than for viscose rayon; the stretching also produces finer filaments. It should be mentioned that ammonia is recovered from the coagulating water stream as a contribution towards keeping processing costs down.

The newly-formed filaments are reeled, washed, neutralized and dried. The only modern innovations which have been made to the process are connected with the continuous washing of the filaments as distinct from that used for filaments built up into a cake form of package.

Cuprammonium rayon is composed of cellulose, and so the simplest molecular repeating unit in its construction is again

```
        HC——O
       /     \
  —HC         CH—O—
       \     /
        HC——CH
        |   |
        OH  OH
```

Under the microscope, the fibres appear as smooth rods, usually of fine denier, with no longitudinal striations. The cross-sections are nearly round with a slight flattening caused by the filaments contacting each other during the slow coagulation process.

Cuprammonium rayon has a density of 1.53 g/cm^3 and a tenacity of 1.8 g/denier (16.2 g/tex), and this is reduced by about 40% by wetting. The orderly alignment of its chain molecules gives a low elasticity. Its moisture regain is 12.5% at 65% R.H.

When heated, cuprammonium rayon does not fuse, but begins to lose strength when a temperature of 150°C (302°F) is reached and to decompose at a temperature a little below 200°C (392°F). The fibre burns quite readily. There is some loss in strength after prolonged exposure to light.

Cuprammonium rayon is disintegrated by hot dilute and cold concentrated acids. It withstands dilute but not concentrated alkali solutions. It is inert to the common organic solvents.

Cuprammonium rayon has a more silk-like appearance, a softer drape, and less metallic sheen than viscose rayon, and its main applications are for lightweight dress goods, underwear, linings and ties.

c. Acetate rayon. So far we have discussed the preparation of rayon fabrics by converting natural cellulose into a soluble chemical compound which can be extruded as a solution and then chemically treated to regenerate, or re-precipitate, the cellulose in a substantially unchanged form. In cellulose acetate rayon, the cellulose is not regenerated, since it has been found that the chemical compound obtained by reacting cellulose with acetic acid has, in itself, fibre characteristics which are of commercial value.

While the preferred starting raw material in the preparation of acetate rayon is cotton linters, high quality wood pulp can be used. The linters are boiled under pressure with strong alkali solution for several hours, after which they are rinsed, bleached with sodium hypochlorite, washed and finally dried as at the start of the cuprammonium process.

In preparation for the treatment with acetic acid, or acetylation, the linters are swelled by steeping in concentrated acetic acid, followed by a treatment with further concentrated acetic acid mixed with acetic anhydride, using equal parts of the acid and anhydride with a quarter part of cellulose. This reaction is carried out in a closed vessel with vigorous stirring and controlled cooling. The reaction is promoted by adding a small quantity of sulphuric acid which acts as a catalyst. The reaction is accompanied by the generation of a considerable amount of heat, and over the five- to eight-hour treatment period, the cooling arrangements must act efficiently to hold the temperature below 30°C (86°F) if a reduction of the molecular chain lengths of the material is to be avoided.

At the end of the reaction period, a viscous, almost jelly-like substance is obtained. The course of the reaction is checked throughout by taking samples at intervals and testing them; at the end of the reaction, the whole of a sample must be soluble in chloroform. The jelly-like substance is cellulose triacetate which is not soluble in a cheap non-toxic solvent like acetone, and it must be partially depolymerized to form secondary cellulose acetate before it can be dissolved in acetone. This is achieved by pouring the cellulose triacetate 'jelly' into cold water and allowing it to stand for about a day; some acetyl groups are lost by this treatment leaving cellulose diacetate, or secondary cellulose acetate. The action is stopped when the conversion has been completed, by adding a large excess of water, when flakes of secondary cellulose acetate are precipitated. These flakes are washed and dried.

The efficiency of recovery of the dilute acetic acid is important to the economics of the whole process, and the dilute acid is extracted and concentrated in a suitable plant ready for re-use.

A number of batches of secondary cellulose acetate are blended together in an attempt to obtain uniform properties in the final product, and a measured amount is then taken from the blender and mixed with acetone, with constant stirring, in a closed

Fig. 4.25 Cellulose acetate flake being passed through squeeze rollers to remove water before entering drying stove.

vessel. After about 24 hours, a clear solution having a viscosity suitable for spinning is obtained. Any delustrants and/or pigments it is desired to include in the final fibre are added to the solution and the thoroughly-mixed syrup is filtered and freed from gas. The syrup is dry spun, the evaporated acetone being collected and recovered as an economic necessity. The final fibre may be given a slight oiling as a protection against mechanical damage during its formation into a package.

The smallest repeating molecular unit in secondary cellulose may be expressed as follows:

```
           CH2OR
           |
          HC——O
         /     \
     —HC        CH—O—
         \     /
          HC——CH
          |    |
          RO   OR
```

where R is either H, present in about 15% of the molecules, or CH_3CO in about 85%.

Under the microscope, secondary cellulose acetate fibres appear as smooth rods with striations along their lengths, with smooth multi-lobed cross-sections.

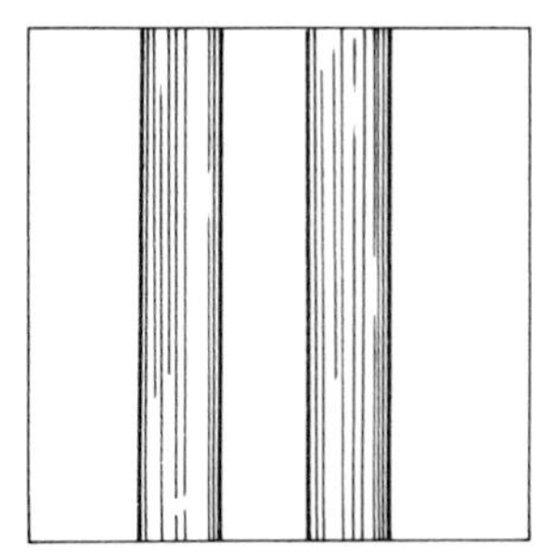
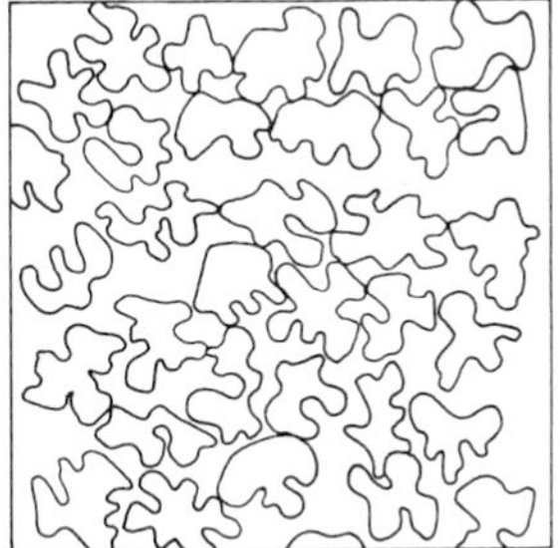

Fig. 4.26 Secondary cellulose acetate fibres.

Secondary cellulose acetate has a density of 1.33 g/cm^3. Its moisture regain is 6.9% determined at a R.H. of 65%. Its dry tenacity is 1.25 g/denier (11.25 g/tex), but this falls to approximately 70% of that dry value when the fibre is wet. The fibre has only moderate elastic properties and will, in fact, flow under a strong sustained pull.

Secondary cellulose acetate is thermoplastic; that is, it can be melted by the application of heat and will re-solidify on cooling. The melting temperature is 240°C (464°F), but softening and deformation start at a temperature of 120°C (238°F). A long exposure to sunlight will produce some loss in strength of the fibre.

The fibre is decomposed by concentrated acids and is soluble in concentrated acetic acid. It is affected by warm dilute alkali solutions, and decomposed by concentrated alkalis. It dissolves in acetone and is swollen or dissolved by a number of other organic solvents.

The best dyeing results on secondary cellulose acetate are obtained using the disperse-acetate dyes, such as those sold under the trade names of 'Duramol', 'Dispersol', 'Cibacet' and 'Serisol'. The dyeing temperature used should not exceed 80°C (176°F) in order to avoid dulling of the fibre.

Secondary cellulose acetate, or 'acetate rayon' as it is usually known, comes second only to viscose rayon, as regards bulk consumption. The material is attractive in appearance, and fabrics made from it have a soft handle, good drape and are quick drying. Their resistance to creasing is superior to that of viscose rayon fabrics. Acetate rayon is used in the production of lingerie, dresses and sportswear, and in blends with other fibres for shirts and suiting fabrics. The material in cigarette filter-tips is usually acetate rayon.

A high-tenacity version of acetate rayon is made and sold. The molecular chains of secondary cellulose acetate have fairly bulky acetyl side groups attached to them and these prevent the chains from binding too close together. When the correct conditions are applied, the chains will therefore slide more easily one over the other. If acetate rayon is stretched in the presence of heat and moisture, an extension of up to ten times can be obtained: this produces a better molecular alignment. The acetyl groups, which aided stretching, can now be removed by treating the stretched material with alkali, and a strong, stable, highly-aligned fibre is obtained. The removal of the acetyl groups results in an appreciable loss in weight, and the price of the high-tenacity fibre is higher than that of the normal acetate. High-tenacity fibre is used in tyre-cord yarns and in tapes.

d. Triacetate rayon. As has already been mentioned, the first stage in the preparation of secondary cellulose acetate is to make cellulose triacetate. Since there was no suitable cheap solvent for the triacetate, the material was once regarded as an unusable product for textile purposes, and the transformation to the acetone-soluble secondary acetate had to be resorted to. A few years ago it was found that a solvent of reasonable cost, in which cellulose triacetate can be dissolved to provide a spinnable solution is methylene chloride, and cellulose triacetate was launched as a fibre material.

The process for preparing triacetate fibres is substantially the same as for secondary acetate up to the stage when cellulose triacetate is formed. The hydrolysis stage, during which the change from triacetate to secondary acetate takes place, is omitted, but it is necessary to wash away the sulphuric acid catalyst which, if allowed to remain, would weaken the fibres. The washed and dried cellulose triacetate is then dissolved in methylene chloride to form a 20% syrup from which filaments can be dry-spun, the solvent being recovered. The filaments are oiled slightly and wound to form a package.

The simplest molecular repeating unit may be expressed as follows:

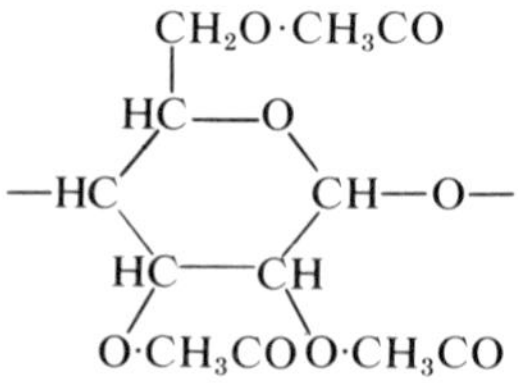

```
          CH2O·CH3CO
          |
         HC——O
        /      \
   —HC          CH—O—
        \      /
         HC——CH
        /      \
      O·CH3CO  O·CH3CO
```

Under the microscope, cellulose triacetate filaments appear as striated rods with smooth-edged multi-lobal cross-sections.

The material has a density of 1.32 g/cm^3. Its tenacity is 1.3 g/denier (11.7 g/tex), and this reduces to about 60% of the dry value when the fibre is wet. Its moisture regain is low, being 4.5%, determined at 65% R.H.

Cellulose triacetate is thermoplastic, and fabrics made from it can be shaped and pleated under the influence of heat. It begins to soften at a temperature of about 200°C (392°F) and melts at 290°C (554°F). There is some loss in strength of the fibre on prolonged exposure to light.

The material is decomposed by moderate concentrations of acids and alkalis; it is inert to weak acids and alkalis and to hydrocarbon solvents.

Fabrics made from cellulose triacetate have a soft, pleasant handle and good draping characteristics. The material is superior to normal acetate as regards softening temperature, which makes the ironing of the fabrics less hazardous, reduced moisture absorption, which enables it to dry more rapidly, and resistance to weak alkaline solutions, which enables it to withstand repeated launderings better. Its main uses are in lingerie, dress and sportswear fabrics. It is also used as a wadding in quilts by reason of its resilience and low moisture absorption. It can be blended with wool, with nylon, and with polyester, for suiting and outerwear fabrics.

4.4.2 Other regenerated fibres.

There are naturally-occurring chain molecules other than cellulose, proteins for instance, and many attempts have been made to manufacture useful fibres from protein materials.

a. Casein fibre. Casein is the protein obtained from skimmed milk. In casein, the chain molecules do not exist in the relatively open straight form as they are in the cellulose materials, but are thought to be curled up into quite tight spirals. In preparing fibres from them, therefore, it is necessary to uncurl the molecular spirals by steaming and stretching and, afterwards, to lock them in this open condition by a certain amount of cross-linking to tie the chains together at a number of points along their lengths. This cross-linking is induced by a chemical treatment using formaldehyde.

Most of our supplies of casein come from the Argentine and, as there is some variation in quality, it is usual to blend together various purchased lots in an attempt to randomize the product. The blend is dissolved in caustic soda solution until a liquid of cream-like consistency has been obtained. This is allowed to stand and 'ripen' until a syrup of viscosity suitable for spinning has been obtained.

The casein filaments are formed by a wet-spinning process in which the syrup is extruded through the holes of a spinneret into a coagulating bath of sulphuric acid to which some formaldehyde has been added. The filaments are then collected in a Topham box, and hardened further by steeping the wound cakes in more formaldehyde or in a solution of an aluminium salt. The cakes are washed and dried. If delustrating and coloured effects are required in the final fibres, these can be obtained by adding the appropriate materials to the syrup before the filtering and spinning operations are carried out.

In a casein fibre, the simplest repeating molecular unit may be expressed as:

$$\text{—NH—}\underset{\substack{|\\ \text{R}}}{\text{CH}}\text{—CO—}$$

The density of casein fibre is approximately 1.30 g/cm^3, and its moisture regain is 14% at 65% R.H. The tenacity of the dry fibre is very low, being 0.75 g/denier (6.8 g/tex). This falls to less than half that value when the fibre is wet.

Casein fibre is not thermoplastic. When held at a temperature of 100°C (212°F) for some time, it turns yellow and becomes brittle; decomposition takes place at 150°C (302°F). The fibres burn when a flame is applied.

Casein fibre resists hot and cold dilute acids very well, but is disintegrated with concentrated acids. It will withstand only a very mild alkaline treatment. It is unaffected by common organic solvents. It will dissolve in a solution of sodium carbonate. The fibre dyes like wool, but takes up dyestuffs rather more readily. It has an affinity for most dyestuffs. The vat dyes are difficult to apply since they require an alkaline bath.

Casein fibre exhibits extreme softness and warmth, and it has a good resistance to light exposure. By blending, it can be used to improve the handle of many materials. Its drawback is a low strength, particularly so when wet, and this makes 100% casein fibre fabrics unsatisfactory. In blends with other fibres, however, and its low price, coupled with excellent aesthetic qualities, may be regarded as offsetting, to some extent, this disadvantage.

b. Other protein fibres. Other protein fibres have been developed and have reached the manufacturing stage. The most notable of these were the fibres prepared from groundnut and soya bean proteins respectively. They are no longer produced and are now only of historical interest. They were prepared in much the same way as casein fibre, namely by dissolving purified protein in an alkaline solution, wet-spinning into an acid bath to regenerate the material, and subsequently hardening the filaments using formaldehyde.

4.4.3 Alginate fibre

Alginate fibre is based on alginic acid, a material which has a structural form closely similar to that of cellulose. Alginic acid is prepared from seaweed.

Seaweed of a suitable type is collected and, after being allowed to air-dry for a period, is taken to the factory where drying is completed. It is then ground up into a powder, which is next treated with a solution of sodium carbonate to extract the alginic acid and form sodium alginate.

The solution is allowed to stand, when the impurities settle out, and a suitable amount of sodium hypochlorite is then added to effect bleaching. The sodium alginate is extracted from the solution and dried to give a pale-coloured powder.

To prepare the spinning syrup, an 8% solution of sodium alginate in water is made. This is pumped through a filter and then extruded through a spinneret into a coagulating bath of slightly acidified calcium chloride solution. The bath converts the sodium alginate of the filaments into calcium alginate. The filaments are washed subsequently, dried and taken up on to a suitable package.

The smallest molecular repeating unit in an alginate fibre material may be represented as follows:

```
            COO·Ca½
             |
            HC——O
           /     \
      —HC         CH—O—
           \     /
            HC——CH
             |    |
            HO   OH
```

Under a microscope, the fibres appear as striated rods with cross-sections having jagged edges.

Alginate fibre is quite heavy, with a density of 1.8 g/cm^3. Its moisture regain, determined at R.H. of 65%, is between 18 and 22%. The tenacity of the dry fibre is 1.2 g/denier (10.8 g/tex) and of the wet fibre only about one-third of that value.

Alginate fibre does not melt and, if heated sufficiently strongly, will decompose. The fibre does not burn.

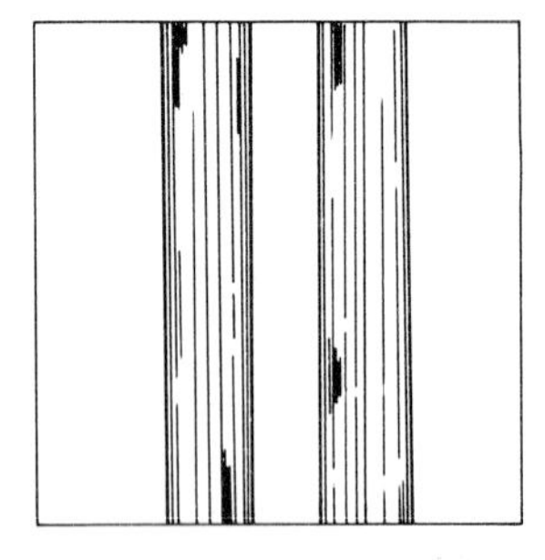

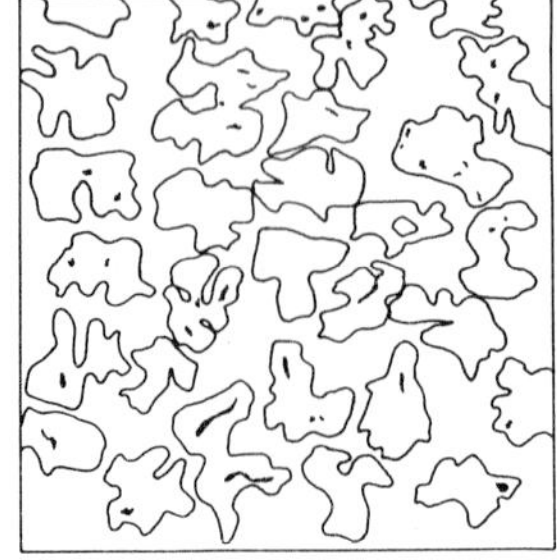

Fig. 4.27 Alginate fibres.

If it is treated with an acid, calcium alginate fibre is converted into the insoluble alginic acid. The fibre dissolves in weak alkaline solutions. Organic solvents have no effect on the fibre.

Alginate fibre is expensive to produce and, by reason of its unusual properties, its uses are limited to speciality outlets. It is soluble in dilute alkaline solutions and so cannot be used for ordinary textile purposes. This property is utilized, however, when the fibre is employed to thread together individual knitted articles allowing continuous operation of the knitting machine. It is also used as scaffolding thread to support very fine wool yarns during the manufacture of certain types of fabric. In both cases, the alginate threads are dissolved away subsequently during the final alkaline scouring of the fabrics. A medical use for the fibre is in soluble dressings.

4.4.4 Synthetic fibres

a. Polyamides. The polyamides were the first truly synthetic fibres to be developed, and the name adopted by the DuPont Co. for the first commercialized member of the family, 'nylon', has now become accepted as the generic term for synthetic linear polyamides.

A number of polyamides have been developed for commercial purposes; some are manufactured in massive quantities, while others find less extensive uses. All are referred to as 'nylons' and a nomenclature had to be devised to distinguish between the various members of the class. The one adopted is based on the total number of carbon atoms in the component, or components, which constitute the repeating unit in the molecular chain of the material. Thus, the first polyamide developed by the DuPont Co., the original 'nylon', has a repeating unit of the following composition:

$$—NH(CH_2)_6NH—CO(CH_2)_4CO—$$

The repeating unit is made up of two components each of which contains six carbon atoms. This particular material is known, therefore, as 'nylon 6.6.'. A polyamide developed later in Germany, and sold under the trade name of 'Perlon', has a one-component repeating unit as follows:

$$—NH(CH_2)_5CO—$$

This unit contains six carbon atom, and the material is referred to as 'nylon 6'. Other polyamides which have been developed commercially, are nylon 6.10, having a composition for its repeating unit of:

$$—NH(CH_2)_6NH—CO(CH_2)_8CO—$$

and nylon 11, made up from a single repeating unit of:

$$—NH(CH_2)_{10}CO—$$

Nylon 6.6. As already mentioned, the first polyamide to be produced commercially was nylon 6.6. This is prepared by reacting together adipic acid, $HOOC{\cdot}(CH_2)_4COOH$, and hexamethylene diamine, $H_2N{\cdot}(CH_2)_6{\cdot}NH_2$.

The starting materials may be prepared in several ways. Most adipic acid is made by the oxidation of cyclohexane to give a mixture of cyclohexanone and cyclohexanol, and this is later oxidized further by catalytic treatment with nitric acid to form adipic acid. The starting compound, cyclohexane, is now usually obtained from petroleum sources. The two-stage oxidation process is preferred since it allows the cyclohexanone and cyclohexanol to be separated from other unwanted hydrocarbon substances before proceeding to the second part of the oxidation process.

Hexamethylene diamine can be prepared from adipic acid by means of catalytic dehydration in the presence of ammonia. This produces a compound, adiponitrile, which can be hydrogenated to form the diamine.

An alternative starting material is butadiene, a compound also of petroleum origin.

The chemical reaction involved in the preparation of nylon 6.6 polymer may be expressed as follows:

$$\underset{\text{(adipic acid)}}{HOOC\cdot(CH_2)_4\cdot COOH} + \underset{\text{(hexamethylene diamine)}}{NH_2(CH_2)_6\cdot NH_2} \longrightarrow$$

$$\underset{\text{(hexamethylene adipamide)}}{—HOOC\cdot(CH_2)_4\cdot COHN\cdot(CH_2)_6\cdot NH_2—} + \underset{\text{(water)}}{H_2O}$$

The terminal groups of hexamethylene adipamide are reactive and hence linking-up to produce the nylon 6.6 chains takes place.

In the manufacture of nylon fibres, two distinct steps are involved, the first being the formation of the polymer, and the second the spinning of the fibres from this polymer. Often, the two steps are carried out in two separate factories.

In preparing the polymer, the adipic acid and the hexamethylene diamine are separately dissolved in methyl alcohol. When these two solutions are mixed together, a mixed compound, usually referred to as 'nylon salt', is precipitated. Nylon salt can be said to have the formula:

$$NH_2\cdot(CH_2)_6\cdot NH_2\cdot COOH\cdot(CH_2)_4COOH$$

Use is made of the low solubility of this compound in methanol to separate and purify it.

The purified salt is dissolved in water, acetic acid being added as a viscosity stabilizer, and the solution heated in an autoclave under a pressure of 250 lb/in^2 (174 500 kg/m^2) and a temperature of 220°C (428°F). An atmosphere of nitrogen is provided to prevent oxidation of the salt. The temperature is raised subsequently to 270° to 280°C (518° to 536°F), the pressure released, and the water removed under a partial vacuum. The total time of treatment is about four hours. During this period, the nylon salt is converted to nylon polymer with the elimination of water. After the treatment has been completed, the molten polymer is extruded, in the form of a ribbon, on to a cooled surface where it solidifies and can be broken up into small pieces or 'chips'.

Careful control is required thoughout the process of preparing the polymer if standard dyeing characteristics are to be obtained, and random mixing of a number of batches of polymer is also carried out in an attempt to achieve uniform dye affinity.

Fibres are prepared from the polymer by melt-spinning, using an atmosphere of nitrogen to prevent oxidation. The step of first preparing nylon chips, which can be remelted to provide the extrusion syrup, is preferred to the direct extrusion of the filaments. In this way, prolonged holding of the polymer in a molten state, when changes in characteristics can take place, is avoided.

Any materials required to be present in the final fibre, such as delustrants and coloured pigments, are added at the melting stage, and the molten polymer is then pumped through the filters and spinneret at high speed to meet a stream of cold air which solidifies the filaments. Subsequently, the solidified filaments are cold drawn to about four times their extruded length to provide the final fibres.

In the nylon 6.6 chains, the simplest molecular repeating unit is:

$$—NH\cdot(CH_2)_6\cdot NH\cdot CO(CH_2)_4\cdot CO—$$

Under the microscope, the fibres appear as smooth rods having circular cross-sections. Reference has already been made to the fact that spinneret hole shapes other than round are sometimes used to produce fibres having a variety of cross-sections; the most common non-round cross-section is the so-called 'trilobal' form.

Nylon 6.6 has a density of 1.17 g/cm^3, and its moisture regain is low at 4.3% determined at R.H. 65%. Its tenacity, both dry and wet, is high at 5.0 g/denier (45.0 g/tex). It has a good elasticity, recovering completely from extensions of up to 8%. Its abrasion resistance is outstanding.

Nylon 6.6 is thermoplastic. As its temperature is raised, nylon 6.6 filaments become tacky at 230°C (446°F) and melt at 250°C (482°F). The material does not burn easily, but hot molten beads can fall from it when it is held in a flame. Long exposure to light produces some falling-off in strength; this can be corrected to some extent by including anti-oxidants in the material. The slight oxidation which comes from long exposure does not cause discoloration of the fibres.

Nylon 6.6 is attacked by concentrated, and hot diluted, acids. It is inert to alkalis, cold dilute acids and the common organic solvents. It is dissolved by some phenolic solutions. The material can be dyed with disperse-acetate, acid, direct, vat and chrome dyes. Its washing fastness is not very good when the disperse-acetate dyes are used.

Nylon 6.6 is a good, all-purpose fibre. It is strong and outstandingly tough. It has a high melting point, which enables domestic ironing to be used without risk of damage. It can be heat-set and shaped under the influence of heat. Garments made from it dry quickly after laundering and are classed as 'drip-dry'. The fibres have a good elasticity and this, together with their strength and resistance to abrasion, make it the preferred material from which to construct ladies' fine gauge stockings. Apart from stockings, the material has a wide range of uses. It is employed in very large quantities for the production of lingerie, ladies' and children's dresses, shirts, socks, knitted outerwear, hand knitting yarns, carpets, overalls and rainwear. It is often used in blends with other fibres, usually to provide additional strength and resistance to wear. Examples of such uses are in mixtures with wool in half-hose and carpets, and with cotton and viscose rayon in body-linen.

Nylon 6.10. An obvious variation on nylon 6.6 was to use another dicarboxylic acid, other than adipic, to react with the diamine. Such changes alter the composition of the

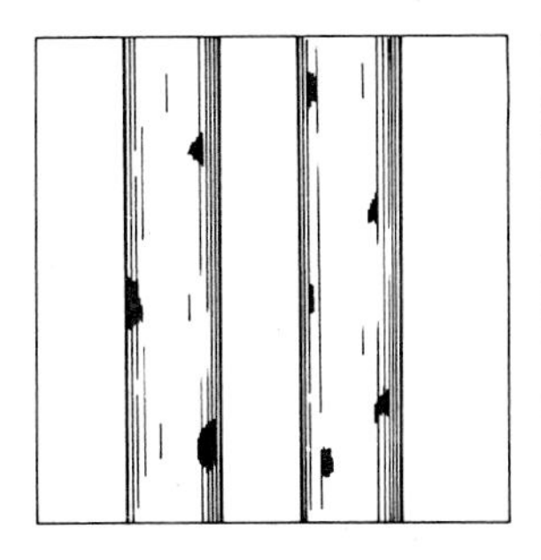
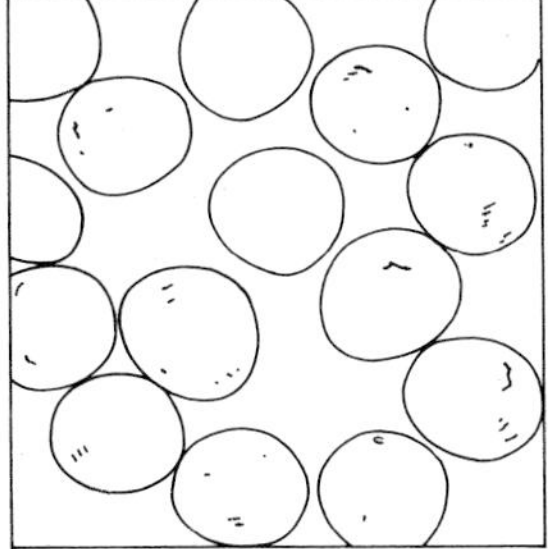

Fig. 4.28 Nylon 66 fibres.

molecular repeating unit in the polymer chain, and vary the number of CH_2 groups, and their distribution relative to the amide linkage. They would be expected, therefore, to make some changes in properties of the final polymer. As an indication of this, the melting points for a range of polyamides made from various dicarboxylic acids and diamines, are of interest. These compounds, with melting points in degrees Celsius, are given in table 4.2.

This table shows only variations in the melting temperature, but other properties are affected in a similar way.

One material produced by a change of this type, which has become of some commercial importance, is nylon 6.10 made by reacting sebacic acid with hexamethylene diamine. This is made, on a production scale, in a manner very similar to that used for nylon 6.6.

Nylon 6.10 has a lower melting temperature than nylon 6.6 and also a lower moisture absorption. It is very suitable for brush bristles and similar applications, since its dry-stiffness and other mechanical properties are retained when the fibres are immersed in water.

The sebacic acid used in the manufacture of nylon 6.10 is usually obtained from vegetable sources, and mainly from castor oil.

Nylon 6.10 is only manufactured on a relatively small scale.

Nylon 6. In the preparation of nylon 6, only one monomeric substance, ε-caprolactam, is used. This can be made to condense with itself to form nylon 6 polymer, poly-ω-caprolactam. The usual commercial method of preparing ε-caprolactam is in three steps. First cyclohexanol is oxidized to produce cyclohexanone, and this is next treated with hydroxylamine to produce the oxime. Finally, the oxime is converted into ε-caprolactam in the presence of 80% sulphuric acid or phosphorus pentachloride in toluene. The route taken to produce the monomer for nylon 6 is somewhat easier than for nylon 6.6 and, in consequence, the former should be a potentially cheaper material to prepare than nylon 6.6.

The preparation of nylon 6 from ε-caprolactam is carried out in a very similar way to that used to produce nylon 6.6. The lactam, with added water, is heated in an auto-

Table 4.2 Melting points (°C) of polyamides made from dicarboxylic acids and diamines*

Dicarboxylic acid		*Diamine*						
Name	*Number of carbon atoms*	*Number of carbon atoms*						
		4	5	6	7	8	9	10
Adipic	6	278	223	250	226	235	205	230
Pimelic	7	233	183	202	196	—	—	—
Suberic	8	250	202	215	—	200	—	—
Azelaic	9	223	178	185	—	—	165	—
Sebacic	10	239	195	209	187	197	174	194

* Source: D. D. Coffman *et al.*, *J. Polymer Sci.*, 1947, **2** 306.

clave to obtain the polymerization reaction, and then the temperature is raised to remove the water. The reaction does not proceed to complete polymerization, and the end product produced is the polymer in equilibrium with about 10% of unreacted lactam. This latter is removed by thorough washing with water. Unless the unreacted monomer is removed completely it can lead to variations in the properties of the final fibre.

Nylon 6 is melt-spun in much the same way as nylon 6.6 and its filaments are stretched subsequently to produce a highly oriented fibre.

The molecular repeating unit in nylon 6 can be expressed as follows:

$$—NH{\cdot}(CH_2)_5{\cdot}CO—$$

When examined under a microscope, nylon 6 fibres appear as straight rods having cross-sections which are accurately round as for nylon 6.6, unless special forms, such as trilobal, have been extruded deliberately.

Nylon 6 has approximately the same values as nylon 6.6 for density, moisture regain, tenacity and elasticity. The two types of nylon have similar chemical properties and dye more or less the same. Nylon 6 has a lower melting temperature – 210°C (410°F). It is said to have a softer handle and be less stiff than nylon 6.6. Nylon 6.6 and nylon 6 are used for similar types of applications.

Nylon 11. This material was primarily developed in France. The starting material from which it is prepared is undecanoic acid, obtained from castor oil. This material is heated to a high temperature, and under reduced pressure, to produce a straight chain acid into which the NH_2 group is introduced with ammonia. Amino-undecanoic acid is polymerized to form nylon 11 under conditions similar to those used for nylon 6, and the molten polymer produced is melt-spun directly without going through the stages of solidifying, breaking into chips and re-melting.

The simplest molecular repeating unit in poly-ω-amino-undecanoic acid, or nylon 11, is

$$—NH(CH_2)_{10}{\cdot}CO—$$

and, under a microscope the fibres appear much the same as the other nylon fibres.

Nylon 11 has a low density, at 1.05 g/cm^3. It is thermoplastic, and has a melting temperature of 185°C (365°F) which is lower than the values for nylons 6.6 and 6. Because of its greater hydrocarbon content, the moisture regain of nylon 11 is very low at 0.9% determined at 65% R.H. and it also has a rather better chemical resistance than nylon 6.6 and nylon 6. It has similar mechanical properties to those of nylon 6.6.

Nylon 11, like all the nylons, is a strong and durable fibre having good elasticity and produces fabrics which can be heat-shaped. It is used for the normal nylon applications and it probably has a slight advantage over the others for certain industrial applications.

Nylon 7. One other nylon, which has been developed in Russia, and which merits some mention, is nylon 7. This material is polyaminoenanthic acid, and the starting materials are carbon tetrachloride and ethylene.

Polymerization is brought about by heating a concentrated solution in water of the aminocarboxylic acid at a temperature of about 250°C (482°F) under nitrogen at a pressure of approximately 225 lb/in^2 (157 000 kg/m^2). The resulting polymer is melt-spun and drawn to produce a high strength fibre having properties comparable with those of the other nylons.

b. Polyester. Since Carothers first studied the class of chemical compounds known as polyesters as possible fibre-forming materials, a whole range of such compounds, based on both aliphatic and aromatic acids, has been examined. Only one of these, polyethylene terephthalate, developed by Whinfield and Dickson, has proved to have a satisfactory combination of desirable textile properties, and has been the subject of major commercial development.

The starting materials for the manufacture of polyethylene terephthalate are ethylene glycol and terephthalic acid, both obtained from petroleum sources. The glycol and the acid can be reacted together directly to form the ester but, in practice, it is difficult to prepare pure terephthalic acid and the dimethyl ester of the acid is used instead. Thus dimethyl terephthalate and diethylene glycol are polycondensed together to produce the polymer.

The process which is used most in practice is in three steps. First, short chain length polyester is produced by reacting the dimethyl terephthalate and ethylene glycol together in an inert atmosphere at a temperature of about 200°C (392°F) and in the presence of a catalyst, with a continuous removal of the methanol which is evolved during the reaction. Next the temperature is raised to 280°C (536°F) and the heating continued at atmospheric pressure for a further 30 min. Finally the temperature is reduced a little to about 260°C (500°F) and the heating continued for a further few hours, a vacuum being applied during this period to draw off any excess glycol, until the polymerization is completed. The molten polymer is then run off in the form of a ribbon, which is cooled and then broken up into chips ready for melt spinning. In this form, the material is in an amorphous state, and its crystallinity is developed only after it has been reheated.

Fig. 4.29 The polymer plant operating floor, where 'Terylene' yarns are produced.

The polyester is melt-spun at a temperature of about 270°C (518°F) after which it is collected in the form of a yarn on a suitable package. To obtain the properties with which we are familiar with in the polyester under the two best-known trade names of 'Terylene and 'Dacron' the yarn has to be extended or drawn up to five times its extruded length to obtain a well-aligned molecular structure; this is done while the yarn is heated.

The simplest molecular repeat unit in polyethylene terephthalate is:

$$—O—CH_2—CH_2—O—CO—C_6H_4—CO—$$

Under a microscope, the fibres appear as smooth rods having round cross-sections.

The density of the polyester is 1.38 g/cm^3 and its moisture regain is very low, being 0.4% at 65% R.H. This low moisture take-up, together with the material's outstanding resistance to creasing, gives to fabrics made from it their 'drip-dry' characteristics.

The polyester is thermoplastic, and can be heat-shaped and 'permanently' pleated. and it has a high melting point at 240°C (464°F). The fibre is strong, with a tenacity of 5.0 g/denier (45.0 g/tex) and this strength is completely unaffected by wetting.

As one would expect of a closely-packed structure, such as is present in this fibre, the material has a good chemical resistance. It is relatively inert to concentrated acids, unless they are strongly oxidizing, to cold dilute alkalis, to hydrocarbons, and to most common organic solvents. There is a very slow attack, which gradually reduces the thickness of the fibre, in hot dilute alkalis. This compact structure has, however, a drawback as regards the dyeing of the fibre, in that penetration by the dyestuff molecules is made more difficult. Disperse-acetate dyes, and a limited number of vat dyes in disperse pigment form, can be used. Dyeing can be eased by the use of 'carriers', or at temperatures higher than 100°C (212°F) in pressurized vessels, both of which 'loosen' the molecular structure of the fibre and permit easier penetration.

This particular polyester gives a strong fibre with good resilience, and it enables durable articles with a good recovery from deformation to be made from it. It is widely used in fabrics for lingerie, dresses, children's wear, ties and socks. To give additional strength, and permanent pleating ability, to a fabric, polyester fibres are often used in mixtures with other fibres; with cotton in shirt fabrics, in suitings mixed with wool, and in sportswear mixed with linen. Other uses are in sewing threads, in industrial fabrics and, by reason of its resistance to sunlight and weathering, in curtains, ropes, nets and sailcloths.

c. Polyacrylonitrile. The monomeric compound from which polyacrylonitrile, best known under the trade names of 'Courtelle', 'Acrilan' and 'Orlon', is prepared, is vinyl cyanide or acrylonitrile. This compound was a raw material for the manufacture of synthetic rubber and, as such, was manufactured in very large quantities and on a scale sufficient to allow it to be offered at a reasonable price for use as the starting material in synthetic fibres.

Although the fibres in this class on the market are often referred to as being composed of polyacrylonitrile, they do, in fact, contain small proportions of other constituents probably totalling up to 10% of the fibre material. These constituents are usually added with the aim of improving the chemical and dyeing properties of the fibre; they are usually vinyl compounds which are closely related to the vinyl cyanide or acrylonitrile to which they are added.

Under suitable conditions, the molecules of acrylonitrile, $CH_2{=}CH{\cdot}CN$, can be persuaded to take part in an addition polymerization chain reaction to produce the polymer:

$$n\mathrm{CH_2{=}CH{\cdot}CN} \longrightarrow -\mathrm{CH_2{\cdot}\underset{\underset{CN}{|}}{CH}-CH_2{\cdot}\underset{\underset{CN}{|}}{CH}-CH_2{\cdot}\underset{\underset{CN}{|}}{CH}-CH_2{\cdot}\underset{\underset{CN}{|}}{CH}}-$$

and in the commercial polymer the number of such repeating units which form a chain is of the order of 2000. In the manufacturing process, the acrylonitrile and the other small quantities of monomeric substances which are added, are emulsified in water, and an oxidizing agent, such as ammonium persulphate or benzoyl peroxide, is added and acts as a catalyst to trigger off the polymerization reaction. The polymer is formed as a powder and, as the polymerization proceeds, this is precipitated and is filtered off and dried.

Unlike the polyamides and polyesters, polyacrylonitrile is solvent spun into filaments. The solvent used is dimethyl formamide and, as the filaments emerge down from the spinneret they meet an upward-flowing stream of heated air or nitrogen which evaporates the solvent and carries it away to the recovery plant. Polyacrylonitrile solutions have been wet-spun, as with viscose, into a coagulating bath containing glycerol but, in that case, recovery of the solvent is much more difficult. The filaments of polyacrylonitrile are stretched while being steamed to give a fibre of good molecular orientation.

In polyacrylonitrile, the simplest molecular repeating unit is $-\mathrm{CH_2}-\underset{\underset{\mathrm{CN}}{|}}{\mathrm{CH}}-$.

The appearance under the microscope varies, to some extent, with the particular method of producing the fibres which has been used. It will be noted that, while the fibres of Courtelle, Acrilan, and Orlon consist of straight rods, the cross-sections of these rods vary. Thus the cross-sections of Courtelle fibres are round but with slightly serrated edges, those of Acrilan are round to bean-shaped, while those of Orlon are peanut-shaped.

The density of polyacrylonitrile fibre is 1.19 g/cm³, and it has a low moisture regain value of 1.5% when tested at 65% R.H. The fibre softens at a temperature of 250°C (482°F) but is tacky at about 200°C (392°F); it will continue to burn when ignited.

The fibre has a tenacity of about 2.8 g/denier (25.2 g/tex) and this is virtually unaffected by wetting. It shows a good resilience and recovery from small extensions. Its resistance to sunlight is good.

Polyacrylonitrile fibre has a good resistance to chemical attack. It withstands acids,

Fig. 4.30 'Courtelle' fibres.

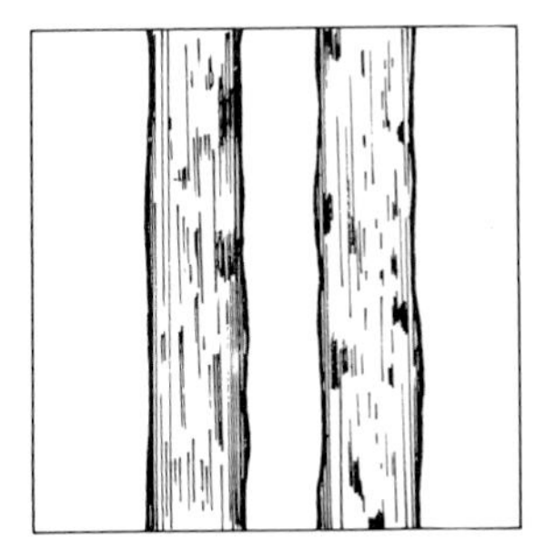
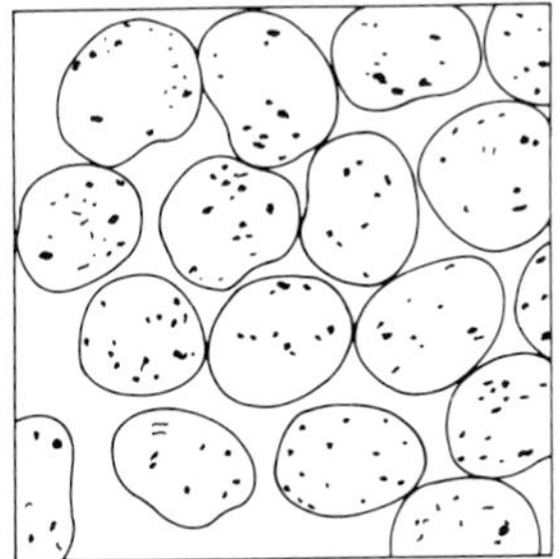

Fig. 4.31 'Acrilan' fibres.

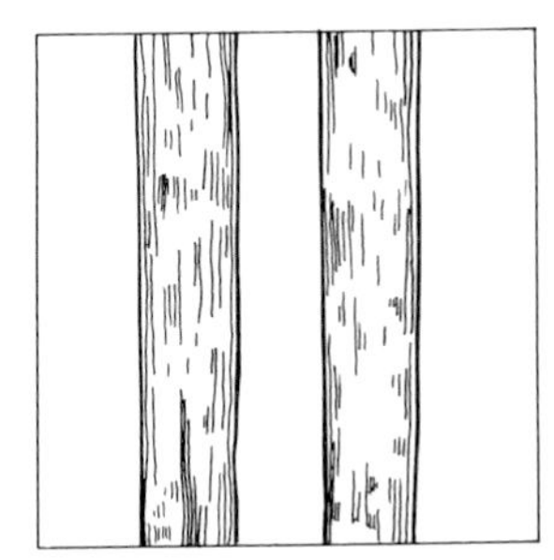

Fig. 4.32 'Orlon' fibres.

even concentrated acids except nitric. It is resistant to weak alkalis but is attacked by hot, strong, alkaline solutions. It is unaffected by the common organic solvents. The same factors which give the fibre a good chemical resistance also make it difficult to dye, and the 100% polymer fibre has been modified, mainly for this reason, in the fibres available on the market. The commercial products can be dyed with basic, acid, and disperse dyestuffs, some fibre materials being specially developed to have a good affinity for a particular class of dyestuff.

Polyacrylonitrile fibres are soft and warm and produce fabrics which are strong, hard-wearing, dimensionally stable, and easily laundered and dried. A feature of this particular type of fibre is the fact that so-called 'high bulk' yarns can be prepared. These yarns are spun from a mixture of two kinds of polyacrylonitrile staple fibre; one has been heat-treated to shrink it so that it will shrink further no more than 1% when immersed in boiling water, while the other is left unshrunk and will contract between 15 and 17% in boiling water. When such a mixed yarn is heated, one of its components shrinks while the other is unaffected; the considerable contraction of the former 'fluffs out' the fibres of the latter to give a bulky yarn.

d. Polyolefins. Polyethylene is the simplest fibre-producing material known. Imperial Chemical Industries Ltd. were the first to polymerize ethylene, $CH_2{=}CH_2$, to form a solid polymeric compound which can be represented as:

$$
\begin{array}{ccccccccc}
 & H & & H & & H & & H & \\
 & | & & | & & | & & | & \\
- & C & - & C & - & C & - & C & - \\
 & | & & | & & | & & | & \\
 & H & & H & & H & & H &
\end{array}
$$

Their method required the use of a high temperature and an extremely high pressure, and they showed considerable ingenuity in developing a commercial process involving the use of such conditions. The ethylene gas was polymerized in a pressure vessel, with a trace of oxygen present as a catalyst, at about 200°C (392°F) and a pressure of 10 tons/in^2 (1.56×10^7 kg/m^2).

To prepare fibres from such a polymer, the molten polymer is extruded and broken up into chips. These can then be remelted and melt-spun into a stream of cold air or into water. The fibrous properties are further developed by hot drawing the material.

The fibre material produced by the high pressure process is built up from branched molecular chains and, in consequence, it has a fairly open structure. This is reflected in its properties.

The discovery of the special catalysts for polyolefins by Ziegler and Natta, discussed later when the manufacture of polypropylene is being described, allowed an alternative method of producing polyethylene to be devised and, moreover, a material having a more compact molecular structure was obtained. Using the new catalysts, ethylene could be polymerized at a pressure of 450 lb/in^2 (314 000 kg/m^2) and a temperature of 150°C (302°F) in a suitable solvent. The polymer formed is removed from the solvent, the solvent being recovered, and broken up into chips for melt spinning and after-treatment in the same way as the high pressure material. The newer process produces fibres which are built up from straight chains and their properties are significantly different from those composed of the branched chains produced by the high pressure process.

The simplest molecular repeating unit in polyethylene is

$$—CH_2—CH_2—$$

Under the microscope the fibres appear as smooth cylinders.

The density of the branched chain polyethylene, i.e. that made by the high pressure process, is 0.92 g/cm^3, while that of the linear chain material is 0.95 g/cm^3, indicating the closer packing of the latter. In both cases, the moisture regain is nil. The tenacity of the denser material is more than 3.0 g/denier (27.0 g/tex) while that of the branched chain polymer fibre is about one-third of this value. Again, the difference in construction of the two fibres is indicated by their melting points; the branched chain fibre softens at about 90°C (194°F) and melts at 115°C (239°F), while the linear chains fibre softens at 120°C (248°F) and melts at 135°C (275°F).

Polyethylene burns with a smoky flame, melting freely while burning. It is very resistant to acids and alkalis, but is soluble in some heated organic solvents – benzene for instance. Both polyethylenes are difficult to dye, although some special dyeing methods have been claimed in patents. Pigment-coloured fibres, in which the colour has been added during manufacture, can be obtained.

High density polyethylene is a strong fibre, but its applications are limited by reason of its low melting point and resistance to dyeing. The applications are those for which its chemical resistance, resistance to weathering, excellent electrical properties, low price, and sometimes its low density are of particular value. Thus, it is made into filter cloths, canvas, awnings, nets, ropes, twines and electrical insulating fabrics. The low density polyethylene is used for similar applications, although its still lower melting point imposes still further restrictions on its use; it has one particular use, however; as a fusible interlining in certain articles of clothing.

Polypropylene. In the normal way, polymer chains built up from propylene possess

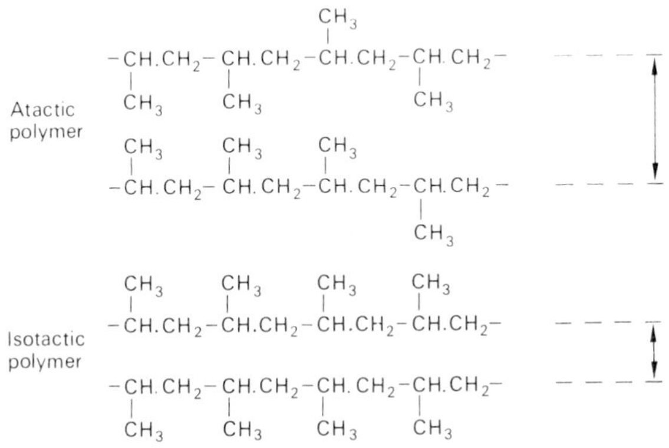

```
                            CH3
                             |
          -CH.CH2- CH.CH2-CH.CH2-CH.CH2-      - - - -
           |        |              |
Atactic   CH3      CH3            CH3
polymer
          CH3      CH3      CH3
           |        |        |
          -CH.CH2- CH.CH2-CH.CH2-CH.CH2-      - - - -
                                   |
                                  CH3

          CH3      CH3      CH3      CH3
           |        |        |        |
          -CH.CH2-CH.CH2-CH.CH2-CH.CH2-       - - - -
Isotactic
polymer
          -CH.CH2-CH.CH2-CH.CH2-CH.CH2-       - - - -
           |        |        |        |
          CH3      CH3      CH3      CH3
```

Fig. 4.33 Chain molecules of atactic and isotactic polymers.

a side chain which prevents one chain from approaching closely to another and results in a lack of cohesion within the material. Ziegler and Natta, to whom reference has already been made, discovered a type of catalytic action which overcame this difficulty. Their catalysts had the important property of moving round the side chains so that they were all extended out on one side of the main molecular chain, leaving a free side which could approach closely to a similar one on another rearranged chain molecule. A polymer in which the side chains extend on both sides of the main chain, is called an 'atactic' polymer, and one in which they extend in one direction only is called an 'isotactic' polymer. From representations of the chain molecules of static and isotactic polymers it is easy to see how the chains of the latter can pack closely together.

To produce the correct action, the catalyst is deposited on the surface of an inert solid which is then immersed in a solvent containing the unit molecules of the monomer. The unit molecules are moving about in all directions and, when one comes into contact with the solid surface, it is held there in such a way that it is pointing in a particular direction. A second unit molecule colliding with the solid surface is held in a similar manner pointing in the same direction, and the catalyst on the solid surface promotes polymerization of the two units while they are held in that way. More collisions take place, and each time the unit is made to face the same way, and thus the chain grows with all the units facing in the same direction. Since polymerization can only take place at the face where the solvent meets the solid, no random formation of polymer is possible throughout the bulk of the liquid, and all the polymer chains are built up in the isotactic form.

The use of a Ziegler/Natta catalyst arrangement allowed an alternative, denser type of polyethylene to be produced; it was essential to prepare any useable kind of polypropylene at all. The process used to polymerize propylene is very similar to that employed for high density polyethylene, namely by using the catalyst on a solid support, immersed in a hot solvent containing the propylene, under pressure. The reaction which takes place produces solid polypropylene. The polymer is purified by dissolving away the short molecular chain material which is present. Filaments can then be produced by melt spinning the polymer, and these are finally cold drawn to give strong fibres.

The simplest molecular repeating unit is:

$$-CH_2-\underset{\displaystyle CH_3}{\underset{|}{CH}}-$$

and, under a microscope, the fibres show as smooth rods with round cross-sections.

The density of polypropylene is very low, being 0.90 g/cm^3, and its moisture regain is nil It has a high tenacity of the order of 7.0 g/denier (63.0 g/tex), and this is the same for the material wet or dry. It is highly resistant to sunlight.

Polypropylene, when heated, softens at a temperature of about 150°C (302°F) and it melts at 165°C (329°F). It burns with a smoky flame, the material melting freely.

Polypropylene has a good resistance to chemical attack, resisting all the acids but concentrated nitric acid. It is very resistant to alkalis. The fibre swells when immersed in ethers and esters and is dissolved by high boiling-point hydrocarbon solvents at the boil, and also hot chlorinated hydrocarbons. By reason of its high chemical resistance, the fibre is difficult to dye, although several special methods for which good results are claimed have been covered by patents. Coloured fibres, in which coloured pigments have been added to the melt just prior to melt spinning, are available.

Polypropylene is a high strength fibre of medium melting point and relatively low cost. It can be heat-set and permanently pleated. The fact that it has such a low density and will float on water is used in some applications. Its uses are restricted somewhat by its relatively low melting point and poor dyeability. It has given excellent service in certain applications, such as ropes, fishing gear and carpets, and attempts are being made to expand its use in fabrics. Since the fibre is resilient and absorbs no moisture, it is being used in blankets and as buoyant stuffing for quilts. It has proved to be very satisfactory in filter cloths, and in fabrics for garden furniture.

e. Polyvinyl chloride. Vinyl chloride, the monomer from which polyvinyl chloride is made, can be prepared in several ways. In one method, the compound is made from the interaction of hydrogen chloride with acetylene in the presence of a catalyst; in another dichlorethane is hydrolysed by reacting it with caustic soda.

Vinyl chloride, under the action of heat, polymerizes readily in an addition reaction to give polyvinyl chloride. A catalyst is used to initiate and help along the reaction. The polymer is dissolved in a mixture of acetone and carbon disulphide and, after filtering and de-aerating, is dry-spun against a stream of warm air from which the solvents are eventually recovered. The filaments are later stretched to produce polyvinyl chloride fibres.

The textile uses of polyvinyl chloride fibre are very limited indeed by reason of its low softening temperature. It begins to shrink at about 76°C (169°F) and to soften at temperatures only slightly in excess of 100°C (212°F). At 180°C (356°F) it begins to decompose. It is completely non-flammable.

Polyvinyl chloride has a molecular repeating unit which can be expressed as:

$$-CH_2{\cdot}\underset{\displaystyle Cl}{\underset{|}{C}H}-$$

Under a microscope, the fibres show as straight rods with nearly-round cross-sections.

The commercial polyvinyl chloride fibres have a density of 1.4 g/cm^3, a moisture regain of nil, and a tenacity of about 3.0 g/denier, (27.0 g/tex) wet or dry.

Polyvinyl chloride fibres have an excellent chemical resistance. They are inert to concentrated acids and alkalis, and to alcohol and petroleum fractions at room temperature. They swell in some hot organic solvents such as chloroform and trichloroethylene, and are dissolved in a mixture of acetone and carbon disulphide.

Polyvinyl chloride fibres have been used in mixtures with less-shrinkable fibres to produce fancy puckered effects in fabrics; when the fabric is heated, the polyvinyl chloride fibres shrink markedly to give the desired effect. The greatest outlets for the fibres are where its resistance to chemicals and weathering can be utilized. It is used, therefore, in some industrial fabrics, filter cloths, awnings, and nets, and as a wadding material.

f. Polyvinyl alcohol. No polyvinyl alcohol fibre is made in the United Kingdom. The main producer is Japan, in which country it is claimed that wide use is made of the fibre.

The preferred commercial method of making the polymer is by the hydrolysis of polyvinyl acetate. The acetate itself is prepared by passing acetylene into concentrated acetic acid in the presence of a catalyst. Vinyl acetate readily polymerizes by heating in the presence of an oxidizing agent such as benzoyl peroxide.

Polyvinyl alcohol is water-soluble, and filaments are made from the solution by wet spinning into a coagulating bath of ammonium sulphate solution. Since a water-soluble fibre is of little commercial value, the spun filaments are given a chemical treatment with formaldehyde to make them insoluble. It is believed that some manufacturers have adopted other methods of producing insolubilization.

Polyvinyl alcohol is built up from simple repeating units of:

$$\begin{array}{c} -CH_2\cdot\underset{\displaystyle OH}{\underset{|}{CH}}- \end{array}$$

Under the microscope recently produced fibres show some striations along their length with cross-sections which are nearly round but with some indentations.

The fibre has a moisture regain of 5% when tested at 65% R.H., and is produced with a range of tenacities, some being as high as 6.0 g/denier (54.0 g/tex). Polyvinyl alcohol has a useful softening temperature of about 220°C (428°F) but care is needed in ironing fabrics made from the fibre since shrinking and hardening can begin to take place at temperatures below this.

Fabrics made from the fibre may be dyed with disperse-acetate, basic, acid, direct, vat and azoic dyes, but none of these is taken up quite so readily as by cotton.

Polyvinyl alcohol fibres are said to be incorporated in all types of clothing, and in many domestic and industrial fabrics. Many of the products are prepared from the fibre in blends with other fibres.

g. Polyurethane elastomers. Earlier in the chapter, when outlining the history of the development of synthetic fibres, reference was made to work carried out in Germany during the Second World War which resulted in a fibre, 'Perlon U'. This was a condensation polymer prepared by reacting together butane-diol and hexamethylene diisocyanate. The modern polyurethane elastomeric fibres, usually referred to as 'spandex' fibres, are composed of materials based on this type of reaction and using a dihydric alcohol, such as ethylene diglycol, and an isocyanate. During the polymerization, some cross-links are formed between the main molecular chains and it is these cross-links, which are always acting to pull the material back into its original shape, which provide the highly elastic properties in the polyurethane fibres.

There are two stages in the preparation of the fibre material. In the first one, a selected glycol is reacted with an isocyanate at a temperature of about 100°C (212°F) to produce what is known as linear polyurethane 'gum'. In the second stage, this gum is heated under pressure to a high temperature, when some cross-linking takes place; this is equivalent to the vulcanization process for natural rubber. The material is then extruded into air as fairly coarse filaments which, while retaining their individual identity, tend to adhere together. At least one company making spandex fibre extrudes the material in the form of one single thick filament.

The illustration shows cross-sections through three well-known brands of spandex fibre, these being 'Lycra' made by the DuPont Co., 'Spanzelle' made by Courtaulds Ltd., and 'Vyrene' made by Lastex Yarn and Lacton Thread Ltd. It will be observed that Lycra has a peanut-shaped cross-section, the filaments adhering together, Spanzelle has a more irregular cross-section, with the filaments again joined together and Vyrene is a solid monofilament.

Spandex materials are appreciably stronger than natural rubber, and have an extensibility and recovery from stretch similar to those of the natural material. They are, however, more heat sensitive and can be discoloured more easily, by the use of chlorine bleaches for instance. The synthetic materials are a good white and can be dyed quite readily. They are superior to natural rubber as regards resistance to sunlight, repeated launderings if the temperature used is not too high, perspiration, mineral oil and solvents. They are appreciably more expensive than natural rubber filaments but, by reason of their greater strength in materials incorporating elastomers, a lower spandex content than with natural rubber can be used.

Spandex yarns are used as uncovered, wrapped or core-spun yarns in a variety of textile applications where high elasticity is important. Such uses include support garments and swimwear, and developments in relation to 'comfort' fabrics are in hand, where the inclusion of spandex fibres will provide additional 'give' in dress and suiting fabrics.

h. Copolymers. Information relating to four copolymers which have become well known as 'Dynel', 'Teklan', 'Vinyon' and 'Saran' fibres respectively, is given below.

Acrylonitrile/vinyl chloride copolymer (an example is 'Dynel'). The proportions of the two monomeric materials in the final polymer used to prepare fibres is of the order of 40% acrylonitrile and 60% vinyl chloride. The fibre material is made in much the same way as is used to prepare polyacylonitrile. The copolymer is soluble in acetone from which it is dry spun into a stream of warm air, the solvent later being recovered, and a final stretching gives a good alignment to the molecular chains of the material.

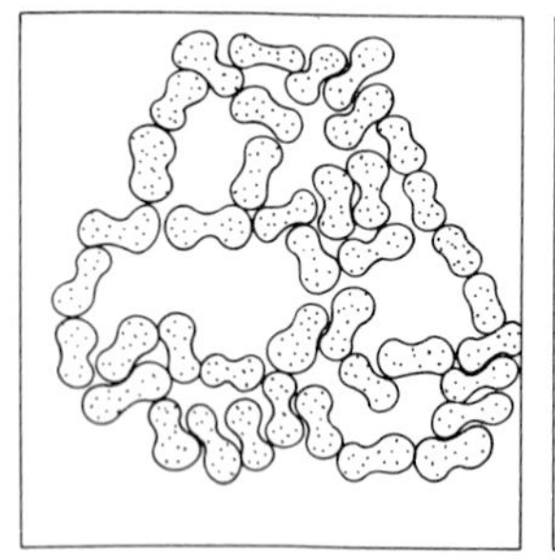
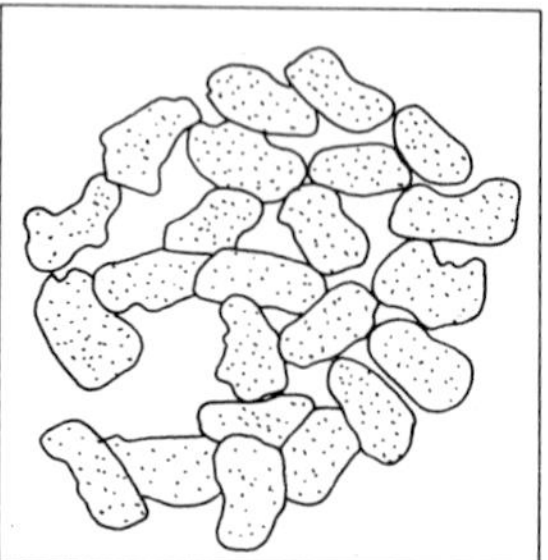
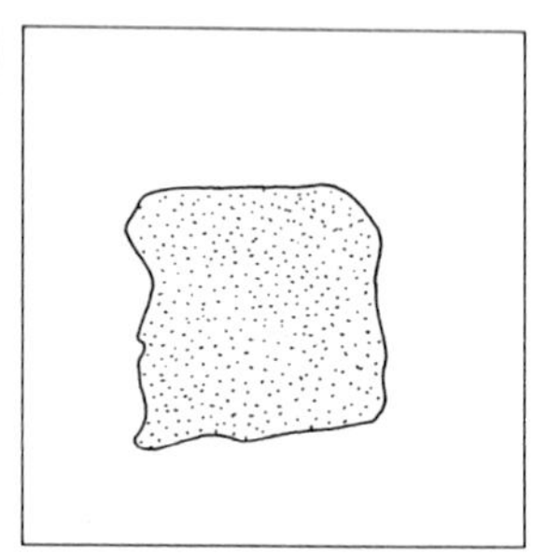

Fig. 4.34 Cross-sections through 'Lycra', 'Spanzelle' and 'Vyrene' spandex fibres.

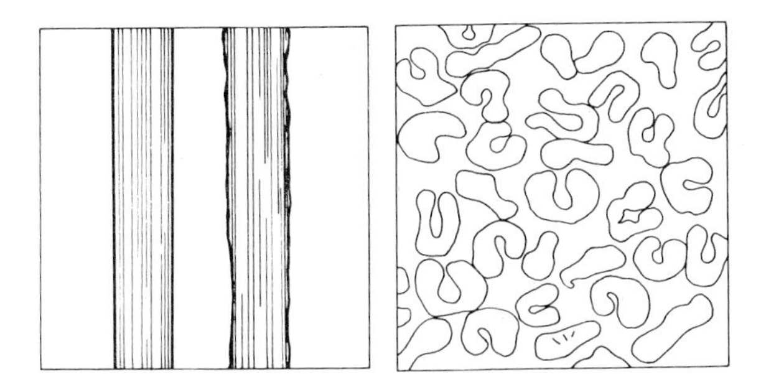

Fig. 4.35 Acrylonitrile/vinyl chloride co-polymer fibres.

Since the chains of the copolymer are built up from two compounds, the simplest molecular repeating unit can probably best be expressed as:

$$-CH_2-\underset{\displaystyle CN}{\underset{|}{CH}}- \cdots\cdots\cdots -CH_2-\underset{\displaystyle Cl}{\underset{|}{CH}}-$$

the first group indicating the acrylonitrile component and the second the vinyl chloride. chloride.

Under the microscope, the fibres show as smooth rods with striations running along their lengths, while the cross-sections are elongated, smooth and often folded.

The density of the fibre is 1.3 g/cm^3. The moisture regain, under a test condition of 65% R.H., is very low at 0.6%, and this is reflected in the fact that the same value of tenacity is obtained for the fibre wet or dry, at 2.3 g/denier (20.7 g/tex).

The fibre is comparatively heat sensitive and care must be taken in ironing fabrics made from it since they begin to shrink at a temperature approaching 140°C (284°F). Fabrics can be thermally-shaped.

This copolymer is soluble in some organic solvents, acetone and warm cyclohexanone for instance, but it is inert to concentrated alkalis and most acids, although decomposed by hot concentrated nitric acid. It can be dyed with disperse-acetate dyes at elevated temperatures, and with some basic, acid and vat dyes.

The fibre is said to be wool-like and to provide warm fabrics which are comfortable to wear. It is used to make apparel, furnishing and industrial fabrics.

Acrylonitrile/vinylidene chloride copolymer (an example is 'Teklan'). This material

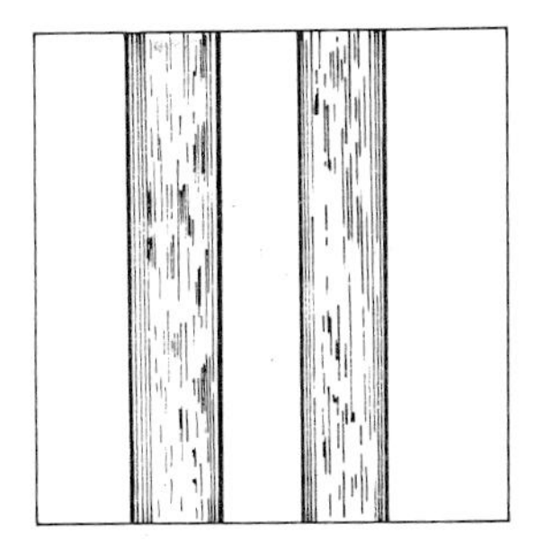
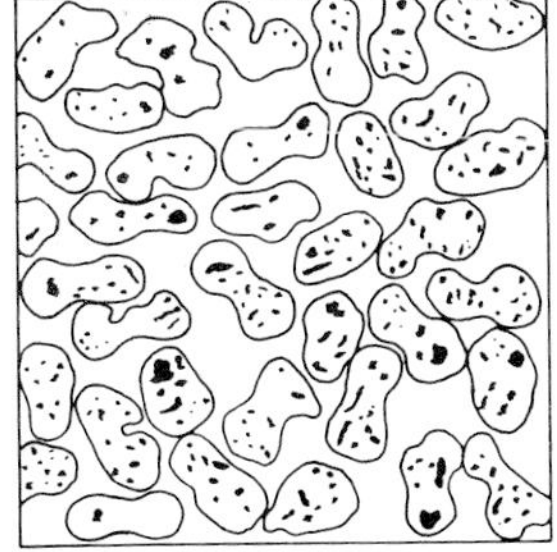

Fig. 4.36 Acrylonitrile/vinylidene-chloride ('Teklan') co-polymer fibres.

is again made in a similar manner to polyacrylonitrile and is dry spun into filaments from an acetone solution.

Under the microscope the filaments are generally rod-like with cross-sections which vary between oval and peanut-shaped.

The density of the fibre is 1.33 g/cm^3, and the moisture regain at 65% R.H. is low at approximately 1.0%. This low value of regain ensures that the wet and dry tenacity values are the same at about 2.6 g/denier (23.4 g/tex).

The fibre is non-flammable and when heated it first shrinks at a temperature of 170°C (338°F) and becomes rubbery and non-fibrous at about 200°C (392°F).

The fibre is dissolved by acetone, and softened by ethyl acetate. It has a good resistance to acids and alkalis in all concentrations. It is strongly resistant to photo-degradation. The material can be dyed at the boil using selected disperse and basic dyes; the dyeing can be speeded up by using a suitable carrier, such as diethyl phthalate.

The fibre is suitable for making a range of fabrics, to which it gives non-flammability, chemical resistance, durability, softness and easy-care characteristics. It has been used in night-wear, dresses, pile fabrics, overalls, filters, nets and curtains.

Vinyl acetate/vinyl chloride copolymer (an example is 'Vinyon'). This copolymer is composed of approximately 11% vinyl acetate and 89% vinyl chloride. The two monomers can be copolymerized together using heat and a suitable catalyst. The resulting material is soluble in acetone with which a suitable syrup for dry spinning can be prepared. The filaments are subsequently wetted and stretched.

The simplest molecular repeating unit can be represented as:

$$-CH_2-\underset{\displaystyle O{\cdot}COCH_3}{\underset{|}{CH}}- \cdots\cdots -CH_2-\underset{\displaystyle Cl}{\underset{|}{CH}}-$$

and the fibres produced are smooth with circular cross-sections.

The material has a density of 1.33 g/cm^3, a moisture regain of nil, a tenacity wet and dry, of about 1.2 g/denier (10.8 g/tex) and a good elasticity. It is non-flammable but heat-sensitive, shrinking at a temperature of 76°C (169°F) and melting at 127°C (260°F). It is virtually unaffected by exposure to light.

Cold concentrated acids have no effect on the fibre, but there is some degradation with hot concentrated acids. It will withstand caustic soda at temperatures of up to 50°C (122°F) and in quite high concentrations. Aromatic hydrocarbon solvents swell the fibre and acetone and chloroform dissolve it.

The fibre is not strong and has limited uses. It can be blended with other fibres for some applications, but its main use is as a mixture with other fibres, which it bonds together when heated, to produce special felts and non-woven fabrics.

Vinylidene chloride/vinyl chloride copolymer (an example is 'Saran'). The two monomers can be copolymerized by heating in the presence of a catalyst. Vinylidene chloride, $CH_2 = C{\cdot}Cl_2$, is unstable and decomposes an exposure to air and light. The copolymer is melt spun into a stream of cold air, after which it is stretched to produce orientation.

The chief constituent in the molecular repeating unit may be expressed as:

$$-CH_2{\cdot}\overset{\displaystyle Cl}{\overset{|}{\underset{\displaystyle Cl}{\underset{|}{C}}}}-$$

The fibre has a density of 1.7 g/cm^3, and its appearance under a microscope is as a smooth accurately round rod. Its moisture regain is nil, and its tenacity, wet and dry, is about 1.7 g/denier (15.3 g/tex). It has a good light resistance.

The material has an excellent chemical resistance. It withstands cold concentrated acids, and is fairly resistant to them when hot. It withstands high concentrations of alkalis. The fibre is swollen and softened by some chlorinated solvents such as chloroform, and some aromatic hydrocarbon solvents have an effect on it. It is unaffected by aliphatic hydrocarbon solvents.

The fibres of the material are usually fairly coarse, and are used for upholstery fabrics, and particularly in public transport. Awnings, screens, filters and webbing are also made. Fabrics made from the copolymer are easy to clean and they do not stain easily.

There are difficulties in dyeing the fibres satisfactorily and colouring is usually incorporated as pigments added at the spinning stage.

4.5 TESTING AND IDENTIFICATION OF THE MAN-MADE FIBRES

In chapter 3, reference was made to a number of tests by which certain important characteristics of the natural fibres could be examined. These same characteristics are also important in the man-made fibres in determining their practical usefulness and can be studied in much the same way.

When examining the effect of heat on man-made fibres, it should be remembered, however, that many of them are thermoplastic and will melt when an appropriate temperature has been reached. Before the actual melting temperature has been attained, a number of the thermoplastic materials will show signs of becoming much more pliable and 'sticky', characteristics which can be troublesome during subsequent processing and use if they occur at too low a temperature. An accurate determination of melting temperature can be of value in identifying a particular fibre. It follows that the temperature tests on man-made fibres should be even more thorough and extensive than on the natural fibres.

In general, the electrical properties of natural fibres are of no very great interest, but some of the man-made materials have outstanding electrical characteristics and could be selected for electrical applications, when properties such as dielectric strength and specific resistance will be of obvious importance. High values for these electrical properties are also invariably accompanied by a low moisture absorption, and this means that such a material can develop troublesome electrostatic charges, which will tend to persist, during processing and in use. A study of the electrical properties of the man-made fibres is thus important.

When considering means of identifying the various man-made fibres, there are a number of tests which can be carried out to determine if they belong to the cellulose class. Cellulose fibres are soluble in cuprammonium hydroxide. They are also soluble in an 80% sulphuric acid solution, but insoluble in xylene and tetrahydrofuran. Where the appropriate equipment is available for studying the infra-red spectrum of a material, the spectrum of cellulose is characteristic. Having decided that a particular fibre contains cellulose, it is necessary to rely on the characteristics of the individual fibre materials to decide which of the rayons it is. One test which can be used is to stain the fibre with a commercial product, Shirlastain A. The various rayons react differently to this material and take on the following shades when stained:

Viscose	– lavender
Cuprammonium	– blue
Secondary acetate	– greenish yellow
Triacetate	– unstained

Normal viscose rayon burns rather like cotton with a bright flame, and leaves little ash. Under the microscope the filaments show up as striated rods and usually with serrated cross-sections. The fibres break quite easily when wet. It is not easy to differentiate between normal viscose rayon and the high-tenacity and polynosic types. There are differences, however. The cross-sections of the fibres of high-tenacity viscose are different in shape from those of normal viscose and they are more highly oriented than the normal material, showing a higher birefringence when examined in polarized light and shorter arcs in the X-ray diagram. Polynosic rayon fibres also exhibit a relatively high birefringence, are more resistant to alkali attack than the other two, and have different tensile properties.

Cuprammonium rayon is softer, more silk-like and, usually, composed of finer filaments than viscose rayon. It can be identified by an examination under a microscope and by its reaction to staining tests.

Cellulose acetate fibres, both in the secondary acetate and triacetate forms, are thermoplastic and can be melted if heated at the necessary temperature. The triacetate has the higher melting temperature. The acetate group can be detected easily by chemical tests in both types of fibre. There are various tests which will allow the two types of acetate fibre to be distinguished. They react differently to staining tests, an 80 : 20 acetone: water mixture will dissolve the secondary acetate but only swell the triacetate, and the triacetate is soluble in methylene chloride and in chloroform while the secondary acetate is insoluble in them and merely shows swelling.

Casein fibre burns rather like wool and gives off an odour similar to that of burnt wool. The fibre usually contains some formaldehyde which is liberated on warming and can be easily detected chemically. In contact with concentrated nitric acid the fibre turns yellow. When treated with concentrated sulphuric acid, regenerated protein fibres swell and develop diamond-shaped flaws after a minute or two of contact. When stained with Shirlastain A, casein fibre shows an orange-brown colour at room temperature and black at the boiling temperature. Protein fibres have a characteristic infra-red spectrum.

Calcium alginate fibres are non-thermoplastic and are non-flammable. They have a high density and their wet strength is very low. Under the microscope, they exhibit characteristic jagged-outline cross-sections. They dissolve readily in dilute alkaline solutions.

Amongst the synthetic fibres, nylon 6.6 and nylon 6 are soluble in cold 90% formic acid solution, but nylon 11 is insoluble. The three types of nylon are all soluble in warm glacial acetic acid and in 80% sulphuric acid solution, but nylon 11 dissolves less readily than nylons 6.6 and 6. Nylon 6.6 is insoluble in 4.25N hydrochloric acid but soluble in 5N strength acid, nylon 6 is soluble in both, and nylon 11 is not soluble in either strengths. When stained with Shirlastain A, nylon 6.6 develops a pale lemon yellow shade, nylon 6 a definite yellow, and nylon 11 is unstained. When heated strongly, nylon 6.6 gives off cyclopentanone, which can be detected chemically. Nylon 11 is distinguished amongst the family by its low melting temperature and its low density.

Polyester fibres are insoluble in a 90% formic acid solution, in acetone and in 80% sulphuric acid solution. They are unstained by Shirlastain A and, in fact, exhibit a rather negative behaviour towards many of the usual straightforward identification tests. The fibres melt at 240°C (464°F) and will burn, but not very readily, with a luminous flame. The material has an unusually high refractive index and the fibres show a very high birefringence.

The acrylic fibres will dissolve in warm dimethylformamide and in concentrated solutions of certain highly-soluble salts such as calcium thiocyanate. The acrylic fibres burn quite readily with a luminous flame and the emission of black smoke which smells rubbery. The fibres are soluble in 80% sulphuric acid solution, but are insoluble in 5N hydrochloric acid, glacial acetic acid and cold 90% phenol solution. Under the microscope, the shapes of the cross-sections will give an indication of which of the acrylic fibres is under examination.

The polyolefin fibres have a waxy feel and, after burning they give off a waxy vapour. They are soluble in boiling xylene and they will float on water. It is not easy to distinguish between polyethylene and polypropylene chemically, and they are best identified through their physical properties. Polypropylene is the lighter fibre of the two and will float if immersed in isophorone, while polyethylene will sink. Polypropylene has an appreciably higher melting temperature than polyethylene. If the facilities are available, polyethylene and polypropylene infra-red spectra can be examined and these will definitely distinguish one from the other.

The polyvinyl chloride fibres are outstandingly non-flammable but are very sensitive to heat. They shrink appreciably when immersed in boiling water. They are soluble in tetrahydrofuran and in mixtures of acetone and carbon bisulphide, but are insoluble in xylene at room temperature, in 80% sulphuric acid solution, and in 90% phenol solution at room temperature.

Polyvinyl alcohol dissolves in warm 20% sulphuric acid solution and, if a 0.1% iodine solution is added to the acid solution of the fibre thus obtained, it will turn a deep blue. Often, formaldehyde can be detected as being present in the fibre. Support for the identification may be obtained from an examination under a microscope of cross-sections of the fibres.

The elastomers are distinguishable from the other synthetic fibres by reason of their rubbery characteristics. When burnt, the synthetic material lacks the characteristic smell of burning natural rubber. The various makes of elastomeric fibre can be distinguished by means of an examination under a microscope of their cross-sections. In addition all the makes, except Lycra, disintegrate when immersed for 24 hours in a 10% caustic soda solution at a temperature of 50°C (122°F); Lycra retains its strength but turns a pale orange colour.

Dynel burns while in contact with a flame but is self-extinguishing when the flame is removed. Dynel is insoluble in xylene, 5N hydrochloric acid solution, 80% sulphuric acid and glacial acetic acid, but it swells and shrinks in a 90% phenol solution; it is soluble in acetone and dimethylformamide. Saran is thermoplastic, is non-flammable and has a high density. Saran turns black on immersion for a few hours in morpholine at room temperature. The modacrylic fibres, such as Teklan are non-flammable. They are insoluble in xylene, tetrahydrofuran and 80% sulphuric acid, but dissolve in acetone. Their chlorine contents, which can be determined chemically, will distinguish the modacrylic fibres from the true acrylic fibres.

4.6 PRODUCTION AND CONSUMPTION OF MAN-MADE FIBRES

Up to 1968 the world's rayon production continued to exceed that of the synthetic man-made fibres. In 1967 rayon production totalled 7536 m. lb, (3149 m. kg) the largest contributors being the USA, 1 388 m. lb, (629 m. kg) and Japan, 1 154 m. lb, (522 m. kg) while Russia contributed 870 m. lb, (374 m. kg); Britain and West Germany were the two largest contributors in Western Europe, having outputs of about 530 m. lb, (240 m. kg) each. In the same year, 1967, the world production of the synthetic fibres was 6 300 m. lb, this being equivalent to 86% of the rayon figure and 84% larger than that for clean wool production.

The output of synthetic man-made fibres has continued to expand year after year in a spectacular manner and now far exceeds the figure for rayon production. Table 4.3 shows how world production of the various synthetic fibres has increased since 1962 up to the present time. In addition, the table includes estimates of what the production figures are likely to be in the years 1975 and 1980, assuming that the present trends will be continued.

Table 4.3 World production of the various synthetic man-made fibres (m. lb)

Material	*1962*	*1965*	*1967*	*1971*	*Estimated* *1975*	*1980*
Polyamide	1 347	2 252	2 897	4 054	5 600	6 787
Polyester	446	1 007	1 659	4 614	8 109	13 574
Polyacrylic	370	891	1 193	2 598	4 300	6 474
Others	218	366	560	1 344	1 994	2 643
Total	2 381	4 516	6 309	12 610	20 003	29 478

The largest percentage increases in output of the synthetic fibres in recent years have been in the USA and Japan. Rapid increases have also been made in West Germany, Britain and Russia.

The table indicates that the three leading synthetic man-made fibres are the polyamides, polyesters, and the polyacrylics. The production of polyamide fibres, the first truly synthetic fibre to be developed, has increased substantially year after year, but the most spectacular increase in production has been that of polyester fibres. By 1971 the production of polyester fibres had exceeded that of the polyamide fibres and this trend is likely to continue in the immediate future. By the year 1980 it is estimated that polyester fibre production will be approximately twice that for the polyamide fibres.

The greater part of production of polyamide fibre has always been in the form of continuous filament rather than as staple fibre. Polyester fibre production has principally always been as staple fibre, but the trend would now appear to be for a greater relative percentage of the production to be in the continuous filament form; by 1980 it is estimated that the total will be divided approximately 50 : 50 between the two methods.

Practically all the production of the polyacrylic fibres has been in the staple fibre form, and little interest has been shown in the material as continuous filaments.

The USA is the largest producer of all three main types of synthetic fibres. The output and usage of the polyacrylic fibres are increasing rapidly in Britain, and the use of the polyacrylics is also growing fast in Japan.

4.7 LITERATURE

F. D. LEWIS. *Chemistry and Technology of Rayon Manufacture*, Reigate, Lewis, 1961.

ROWLAND HILL (Ed.). *Fibres from Synthetic Polymers*, London, Elsevier & Cleaver-Hume Press, 1953.

P. LENNOX-KERR (Ed.). *Index to Man-made Fibres of the World*, Manchester, Harlequin Press, 1964.

R. W. MONCRIEFF. Man-made Fibres, 4th edn. of *Artificial Fibres*, London, Heywood; 1963.

R. ROBSON. *Man-made Fibres Industry*, London, Macmillan, 1958.

C. Z. CARROLL-POREZYNSKI. *Manual of Man-made Fibres*, Guilford, Astex, 1960.

C. Z. CARROLL-PORCZYNSKI. *Natural-polymer Man-made Fibres*, London, National Trade Press, 1961.

H. R. MAUERSBERGER. *American Handbook of Synthetic Textiles*, New York, Textile Book Publishers Inc., 1952.

K. INDERFURTH. *Nylon Technology*, New York, McGraw-Hill, 1953.

Research Dept., American Viscose Corp. *Rayon Technology*, New York, McGraw-Hill, 1953.

S. B. MACFARLANE (Ed.). *Technology of Synthetic Fibres*, New York, Fairchild, 1953.

B. V. PETUKHOV. *Technology of Polyester Fibres*, Oxford, Pergamon Press, 1963.

R. L. WORMALL. *New Fibres from Proteins*, London, Butterworth, 1954.

J. W. S. HEARLE and R. H. PETERS (Eds.). *Fibre Structure*, Manchester and London, Textile Institute and Butterworth, 1963.

CHAPTER 5

Production of Yarns

5.1 INTRODUCTION

It is not possible to weave a rough assembly of staple fibres into a fabric, since such an assembly would lack cohesion and the necessary strength to withstand both the fabric-forming operations, and the treatment the fabric would receive during wear. The insertion of only a few turns of twist into an assembly of fibres has a marked effect on the manner in which they will cling together, and hence on the strength of the yarn thus produced. The strength of a yarn is determined in part by the strengths of the individual fibres composing it, and also by the way in which these fibres cling together. Some fibres, such as wool and cotton, have a natural tendency to cling together and only a moderate amount of twist is needed to produce a strong yarn. Other fibres, which are straight, and smooth, are inclined to slide easily one over the other, and an increased amount of twist is required to bind them together in the yarn.

At the same time, the fibre assembly must be 'drawn out' to provide the correct thickness in the final yarn, and to produce an orderly arrangement of the fibres which will enable them to take advantage of the maximum degree of binding produced by the amount of twist which is inserted. This drawing, or drafting, of the mass of fibres must be done evenly in order to avoid fluctuations in thickness which would introduce weaker spots in the yarn, and also detract from a uniform appearance in the final fabric.

The conversion of a mass of fibres into a yarn in which a more orderly arrangement has been achieved and a binding twist has been inserted is known as 'spinning', and the range of operations which produce yarns suitable for the manufacture of fabrics constitutes the spinning processes.

In order to avoid misunderstanding, it should be mentioned here that there are two types of fabric, which find a certain amount of application, and which are made direct from the loose fibres without first producing a yarn. One of these is wool 'felt' in which a treatment is given to a sheet of loose wool fibres, when the particular characteristics of the wool cause the fibres to become matted and locked together to form a type of fabric. Another, the so-called 'non-woven' fabric, is a fairly recent development. Here, a layer of virtually any type of fibre is bound together to form a fabric using some locking means such as an adhesive, or by punching a barbed needle through the layer at many points over its surface to produce such a state of fibre entanglement that the fabric holds together strongly. Both types of fabric are discussed in greater detail in chapter 6. They constitute special methods of fabric formation of limited application. In the majority of textile uses, the loose fibres are first spun into a yarn from which the fabric can then be fashioned.

5.1.1 History

It is not known whether wool felt or cloth was the first fabric to be produced by man. Samples of cloth dating back to about 6000 B.C. have been found and this presupposes that the preparation of yarns would have been known by then. Coming nearer to more recent times, it is known that the spinning and weaving of a whole range of natural fibres were carried out by the North American Indian tribes long before Columbus discovered the country.

It is thought that the earliest technique of spinning fibres into yarns was by rolling a mass of the fibres between the palms of the hands, and using the feet to anchor one end of the yarn being produced. Another early method was by rolling the fibre mass between the hand and the thigh with the hand moving away from the body.

Once it had been established that the insertion of twist gave such a marked increase in strength to the yarn, one can imagine that thought would be given to more satisfactory means of twisting than the rolling by hand methods which had been used. The use of sticks or spindles was a significant step forward in the spinning of yarns. Probably sticks were first used on which to wind the finished yarn which had been prepared by hand, and then it was realized that a stick was a means whereby actual twist could be imparted to the yarn. Thus, if a yarn has been lightly twisted by hand and is wound on to a stick, and then pulled over-end off the stick, an extra turn of twist will be added each time the yarn slips off the end of the stick.

Another method of imparting twist to a yarn using a stick was to fasten the fibre assembly, or loose 'rope' of fibres, to the top of the stick, allow the stick to hang down on the end of the rope which was held in one hand, and spin the stick with the fingers of the other hand. The twist developed would run back along the rope to the supporting hand. The stick would rotate for only a short time, and it is not surprising that the next development was the attaching of a 'whorl' to the stick or spindle to act as a fly-wheel and keep it spinning for a much longer period, as well as helping to keep the spindle upright on the end of the yarn. Examples of ancient 'drop spindles' fitted with whorls have been found. The spindles were of wood, and usually about 10 in (254 mm) in length, and tapering from the bottom to the top. The whorl was of clay, stone or metal, and the spindle was inserted through a hole at its centre so that the whorl became wedged at a point just above the bottom of the spindle. If the spindle and whorl were heavy, in order to obtain a big rotational momentum, then only heavy yarns which could support that weight could be spun, and a further development was to allow the bottom of the spindle to be supported by a smooth shell or hollowed-out stone and thus reduce the pull on the yarn. To use the simple spindle, a short length of yarn was fashioned by hand and attached to the top of the spindle, either by being tied in position or wedged in a notch cut in the wood. One hand was used to hold the fibre mass at the upper end of the hand-formed yarn, and the other hand to give a twist to the spindle and set it rotating. Twist was developed and ran back up to the supporting hand. While it was doing so, both hands were used to feed out the requisite fibre content from the mass, and some of the twist was allowed to run back between the hands to produce a short additional length of yarn. This newly formed portion of yarn was then fed forward below the hands and the cycle repeated again and again until the spindle stopped rotating. The yarn was then

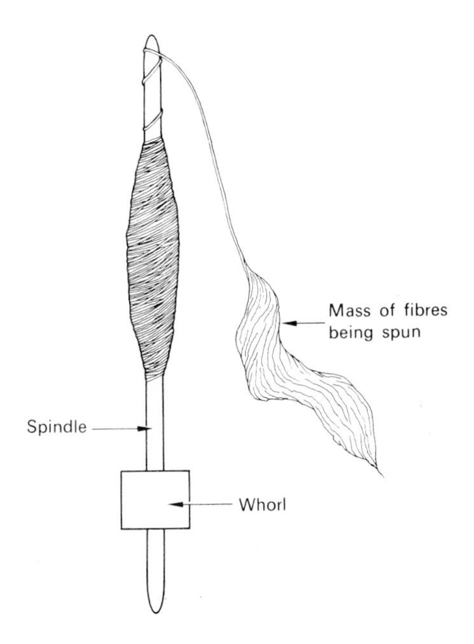

Fig. 5.1 Drop spindle and whorl.

detached from the top of the spindle, wound on to the body of the spindle, and then attached once again at the spindle point, when the complete operation was carried out again, and repeated a number of times until sufficient yarn had been built up on the body of the spindle.

The next important spinning development was the adoption of a wheel to rotate the stick spindle. This was probably first used in India in about the fifth or sixth century A.D. The spindle was supported horizontally in bearings attached to a vertical post. The whorl was replaced by a grooved pulley attached to the spindle, and the spindle was rotated by means of a driving band passing round the pulley and round the circumference of a large wheel supported vertically. A short length of the hand-spun yarn was

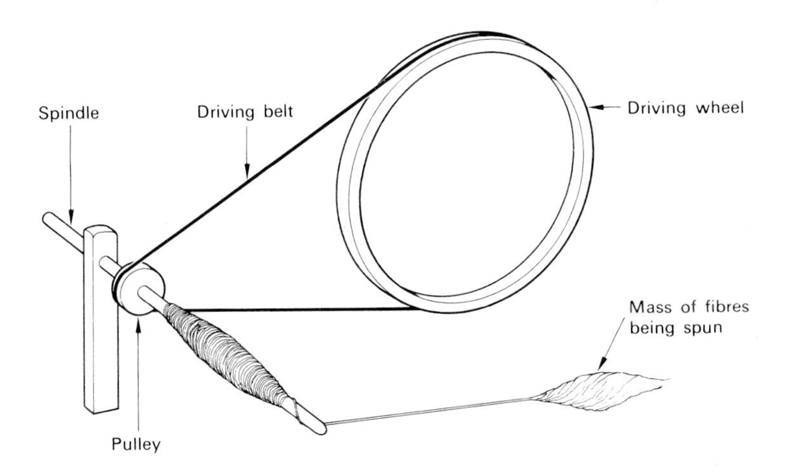

Fig. 5.2 Spinning, using a driving wheel.

attached to and wound on, the spindle, the last few turns spiralling off its pointed end. The mass of fibres was held in the left hand in front of the pointed end of the spindle and slightly to one side, and the large wheel was turned by the right hand. As the spindle rotated, yarn slipped off the end of the point, each revolution inserting one turn of twist. At the same time, the fingers of the supporting hand were used to feed forward a suitable number of fibres from the mass into the yarn being twisted. When a suitable length of yarn had been prepared, the wheel was stopped, reversed for a few revolutions to take off the yarn which was spiralling to the point of the spindle, and the supporting hand moved to the side of the spindle, when the turning of the wheel in the original direction wound the yarn which had been prepared on to the body of the spindle. The operating cycle was then repeated the number of times required to fill the spindle with yarn. By using the wheel method of twisting, spindle speeds of the order of 3000 rev/min could be achieved and the production of yarn speeded up accordingly.

The invention of the spinning wheel with which we are familiar took place in Germany in the sixteenth century. The essential parts of this equipment are shown in the illustration. They consist of a wooden 'flyer' rigidly attached to one end of a metal spindle, a bobbin to take the finished yarn which is a loose fit on the spindle and can rotate on it, and a driving band which passes twice round the circumference of a large driving wheel and once only round each of two pulleys, one on the end of the bobbin and the other on the free end of the spindle. The spindle is supported in bearings in an upright piece of wood. The bobbin pulley is smaller in diameter than the spindle pulley and so the bobbin rotates faster than the spindle as the driving wheel is turned. As the spindle rotates, the flyer is carried round with it to trace out a path round the bobbin. The driving wheel is rotated by means of a foot treadle, leaving both hands of the opera-

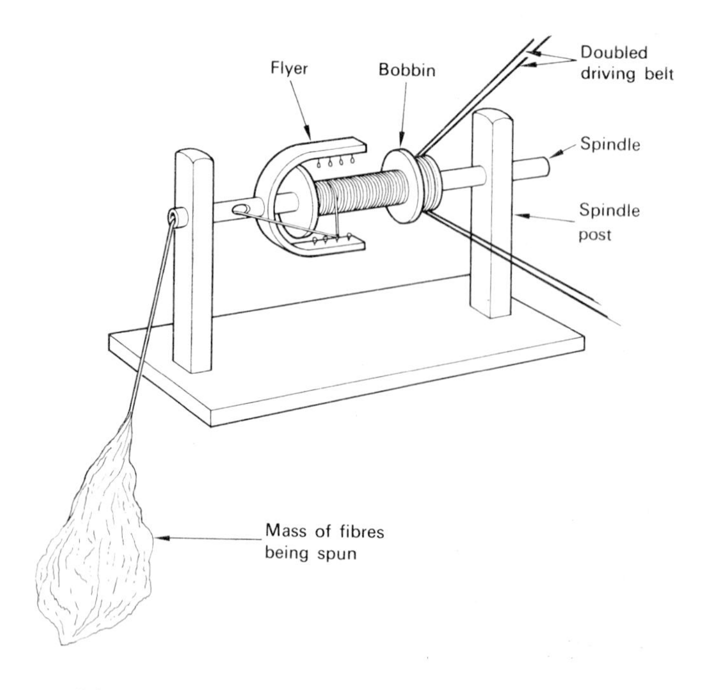

Fig. 5.3 A spinning wheel.

tor free to manipulate the mass of fibres and feed forward the requisite amount of fibre into the yarn being formed. To use the equipment, a short length of yarn is passed through a hole in the flyer end of the spindle, round one of the hooks on the flyer, and is then attached to the bobbin. When the operator starts to treadle, twist is inserted in the fibre assembly being released by his hands and, since the bobbin is rotating faster than the spindle, the yarn thus prepared is wound continuously on to the bobbin, the rate of wind-up being determined by the difference between the two rotational speeds. As yarn build-up on the bobbin continues, the diameter of the yarn mass will increase, and yarn will be taken-up at a faster rate. To compensate for this, an easily-operated adjustment allows the tension in the driving belt to be altered so that some slippage takes place. Much skill on the part of the operator is required if a uniform yarn is to be produced.

The automatic spinning of yarns by machine was developed in the early 1700s. The development was probably started by a man named Paul who, in 1738, first provided a means of drawing out, or attenuating, a mass of fibres by machine using two pairs of nip rollers, the second pair rotating at a higher speed than the first pair. This was followed, in about 1768 by the 'Spinning Jenny' devised by Hargreaves, and for the first time, a number of yarns could be produced at the same time. In the original machine there were eight spindles, with considerably more in later versions of the machine, all driven by cords from a revolving drum which was itself rotated by means of a hand-wheel. The machine consisted of a frame in which the spindles were mounted at one end and, on top of the frame and capable of sliding along its length, was a carriage. This carriage was provided with two clamping bars which could either grip and hold the rovings firmly

Fig. 5.4 Hargreaves' 'spinning jenny' in action.

or be opened to allow the rovings to slide through. The rovings (the mass of fibres until it has been thinned ready for spinning), which had previously been prepared on a hand-wheel, were wound on to packages, one for each spindle, and supported in the lower part of the frame. Each roving was led from its package, through the clamping bars of the carriage, and on to the spindle tip. At the commencement of the spinning operations, the carriage was drawn back a short distance away from the spindles leaving a length of roving between it and each spindle. The clamping bars were now closed to grip the rovings and the carriage slowly drawn still further away from the spindles to draw or attenuate the roving, the drum being rotated meanwhile to turn the spindles and insert sufficient twist to keep the roving intact. The motion of the carriage was then stopped but the rotation of the drum was still maintained to insert the necessary amount of twist to form the yarn. It was now necessary to wind the yarn which had just been spun on to the body of its spindle. The rotation of the spindles was reversed for a moment and the carriage moved still further out to take up the slack thus created. Then a 'faller wire' was lowered on to the yarn, and this guided each yarn on to the side of its spindle so that, as the drum was again rotated and the carriage slowly pushed back to its starting position, the yarn was wound on to the body of the spindle. The last inch or so of each yarn was coiled up the length of its spindle to the tip, the faller wire being now lifted out of the way, the clamping bars on the carriage were opened and the cycle of operations could then be repeated until the supplies of roving were used up. The frame was, therefore, an intermittent means of applying draw, twist and winding-on in that order.

In 1786, Arkwright invented an improved type of spinning frame. In it, he incorporated Paul's roller system of drawing the roving and, after the last pair of rollers, the untwisted roving was fed into the guide-eye in one leg of a 'flyer', which rotated round a bobbin to insert twist, and then wound up on the bobbin. In this machine, therefore, drawing, twisting and winding-on took place at the same time.

A considerable number of people became interested in frame spinning, and the spinning frames were improved in certain respects and continued to be used until well into the nineteenth century. Eventually, the 'Spinning Mule' was designed by Crompton. The design of the mule was based on a combination of ideas from Hargreaves and Arkwright, its hybrid derivation giving rise to its name. It is said that

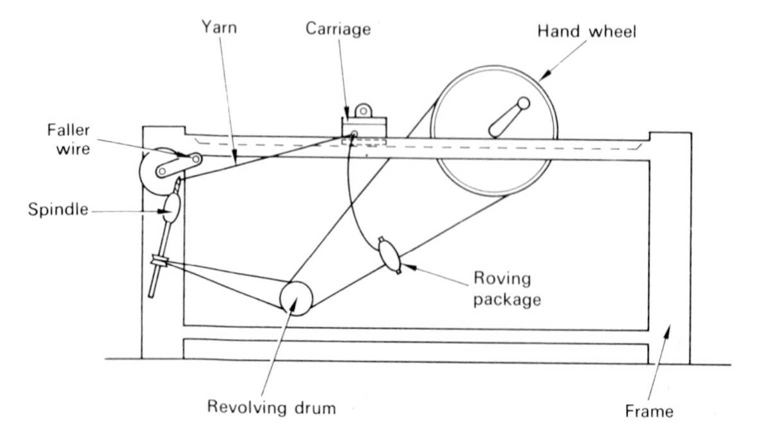

Fig. 5.5 Diagram of the working parts of Hargreaves' 'spinning jenny'.

Crompton embarked on its development out of exasperation with the failure of the earlier machines to produce satisfactory fine warp yarns. The cycle of operations was hand-controlled in Crompton's machine, and it was not until 1830 that Roberts succeeded in making the whole machine automatic.

The mule machine gave a degree of quality and fineness in the yarns it produced which had not been possible to achieve before that time, and it formed the basis on which the technical superiority and prosperity of the textile industry in Britain were founded during the nineteenth century.

5.1.2 General principles of yarn preparation

In general, the natural fibres are supplied for spinning into yarns in the form of a tangled mass. For some of the fibres, cotton for instance, the material is transported in the form of highly compressed bales and, when the bale wrappings are cut, the fibres come from the bale in the form of large hard lumps.

These tangled masses of fibres also contain all the dirt and waste materials which are inevitably associated with their origin. Thus, in addition to the natural fats and waxes which the plant develops, cotton also contains such impurities as dirt, sand and leaf which have been picked up during growth and harvesting. Similar impurities are present in the stem and leaf fibres, in addition to portions of vegetable matter, from their plant of origin, not removed during the extraction of the fibres. In wool, and many of the other hair fibres, there are considerable quantities of grease and dried perspiration, together with dirt and vegetable matter which the animal has picked up during its wanderings. Silk fibres are covered with silk gum which roughens the fibre surfaces and makes the fibres less pliable.

The first stage in preparing the raw fibres for spinning is to pass them through a series of preparative processes which loosen the mass and make it possible to extract, at least those impurities which would interfere with the spinning process. These two objectives are not achieved in one single process, but in a series of processes which are complementary to each other. Experience gained over the long period since spinning was mechanized has shown that the individual fibres must be, as it were, 'coaxed' into their correct positions relative to each other if the best yarns, and it follows the best fabrics, are to be produced. A whole range of processes are involved in preparing a textile yarn and, even today, an individual spinner will have his own ideas of the best combination of processes he should use to produce a yarn which he considers to be satisfactory.

The object of the preparative processes is to deliver the fibres in the form of a clean, fluffy mass ready for the actual spinning treatments. The first of these is the carding process which removes any traces of impurity which may still remain in the mass of fibres and, at the same time, the very short fibres which would detract from the strength and appearance of the final yarn. If the fibres are to be spun into very strong, fine yarns, then they must be arranged accurately parallel to one another, and a combing treatment is given to the fibre arrangement, in which fine needles are passed along it; this treatment also makes sure that the shorter fibres are removed.

The final stages of processing before the actual spinning are referred to as 'drawing'. In this range of treatments, the fibre assembly is attenuated, or drawn out, with the fibres sliding one over the other by a controlled amount to reduce the thickness to that finally required in the yarn and also to arrange the fibres in an even more accurately parallel form.

After passing through all these stages, the material is now clean, uniform in composition, and aligned with its individual fibres parallel to each other, and is ready for twist to be inserted to form the final yarn. The amount of twist inserted will be determined by the intended end use; for a soft, pliable yarn such as is used in hosiery manufacture, only a light twisting will usually be required; for a hardwearing, smooth fabric such as a worsted suiting material, the twist will be greater.

5.2 EARLY PREPARATIVE PROCESSES

5.2.1 Opening machines and processes

In processing cotton, complete cleaning of the material is left to a later manufacturing stage when the fibres will be held in a form which will allow thorough purification to be carried out with less damage to the material itself. This is also true for many of the other fibres. There may be impurities present, however, such as dirt, sand and vegetable waste, which would interfere with the spinning of the fibres to form yarns and these, of course, must be removed. For many of the fibres, therefore, opening processes have been devised to loosen the mass of fibres and thereby enable such impurities to fall away.

A further aim in carrying out the opening processes is to mix and blend the fibres together. With cotton, for instance, each bale will be different from the others as regards the quality of the material it contains and this would be discernable in the yarns made from the various bales. The effect of these differences is minimized during opening and cleaning by feeding small amounts of cotton, taken from each bale in turn, into the machines. Another method sometimes used is to stack layers of cotton taken from the various bales one on top of the other, and then extract the material from a vertical section through the stack as it is fed into the machines. Correct blending is a matter of long experience. Most yarns are made from a blend of several cottons, and the quality of the final yarn is determined, to quite a considerable extent, by the care taken in carrying out this operation.

As has already been mentioned, the highly compressed materials, such as baled-cotton, come away in large, hard lumps when the bale fastenings are removed, and the first machine used in the opening processes is known as a 'bale breaker'. The hard

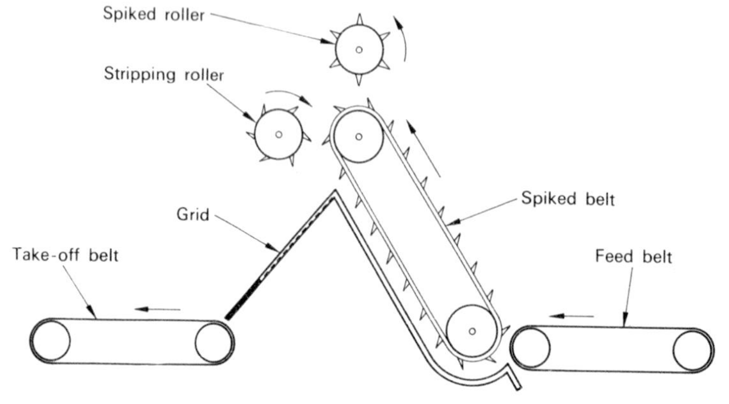

Fig. 5.6 Bale breaker.

lumps of fibre are fed forward to make contact with a continuous belt carrying spikes which is travelling in an upwardly-inclined direction at the point of contact. The spikes rip away small portions of cotton from the mass and carry them upward. At the top of the belt is a spiked roller rotating in a direction opposite to that in which the belt is travelling and separated a small distance away from the belt surface. If too much cotton has been ripped away by the belt spikes, some of it will come into contact with the spiked roller and be knocked back towards the lower part of the belt and be fed forward once again. The cotton which passes under the spiked roller, is removed from the belt spikes by means of a stripping roller rotating in the same direction, but at a higher speed, as the belt, and passes over a grid to the next machine. In passing through the bale breaker, the cotton is agitated violently and loosened, and some of the impurities it contains are freed and fall through the grid below. The cotton is fed from the bale breaker into a 'hopper opener'. This machine works on very much the same principle as the bale breaker but is somewhat less drastic in action. It does produce a further loosening of the fibres and frees a further amount of impurity which falls through the grid at the bottom of the machine.

The use of other types of opening machine, and the number of further opening operations performed, will depend on the preference of the individual spinner concerned. The cotton will often first be taken from the hopper opener to a 'porcupine opener'. This consists of a series of metal discs spaced along a driven shaft, with small plates secured to the edges of the discs. This construction rotates inside a perforated housing which just clears the plates on the edges of the discs. The cotton is fed into this machine by means of a pair of feed rollers and is struck by the rapidly-moving plates on the discs which pull away small tufts of fibres. These are thrown against the outer housing and the impurities which have been loosened by the action pass through the perforations. The further cleaned cotton is carried round inside the housing to a point where it can leave the machine and be carried on to the next stage.

The material can then be passed into a 'vertical opener'. This consists of a series of circular plates spaced along a vertical shaft, each plate having striking blades attached to its edge. The plates increase in diameter up the shaft and the blades are bent so that their ends form an upward spiral of increasing diameter. The whole of this assembly rotates inside a perforated conical housing. An upward air suction is provided to assist the cotton to travel, and as it passes upward between the edges of the plates and the inside of the housing, it is struck by the blades, and loosened further, any impurities present being thrown through the perforations in the housing.

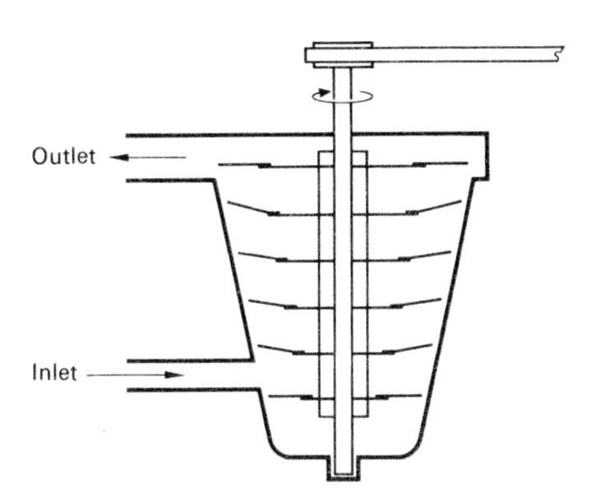

Fig. 5.7 A vertical opener.

The cotton is usually transferred from machine to machine by being drawn by suction through trunking. Along the bottom of the trunking are plates or bars set on edge, and the heavy impurities falling from the loosened cotton lodge between these and can later be removed. At the end of the trunk is a mechanical arrangement for extracting the cotton from the air which is carrying it. The cotton is sucked on to a rotating cylindrical screen where it collects and is carried part way round the inside of a housing until it reaches a pair of rolls which remove it from the surface of the screen in the form of a loose sheet and drop it on to a conveyor.

Other opening machines can now be used until the cotton is in the form of a clean, loose 'lap' or sheet which the spinner considers to be satisfactory. All types of opening machine provide a beating action and means of separating the dirt as it falls away; suction also helps to loosen the fibres and allow the impurities to be extracted.

The loosened fibres are finally delivered in the form of a lap which can conveniently be rolled up into a cylinder for transport to the next spinning stage. During opening, mineral oil will probably have been applied to the material to reduce dust and 'fly' and help in reducing fibre breakage.

5.2.2 Cleaning processes

As has been mentioned already, raw wool and spun silk are scoured before they are passed through any of the spinning processes.

The usually applied method of scouring loose wool is one which involves passing the material continuously through a number of cleaning tanks in a long machine. The machine normally consists of four tanks. The first contains a fairly strong soap solution in which a mild alkali, such as sodium carbonate, has been dissolved, the temperature

Fig. 5.8 An automatic blowing room. Here the bales of cotton are opened and fed by hand on to the lattices which carry the cotton to the blowers. The material is then fully opened and heavy foreign matter removed before forming into laps ready for carding.

being maintained at about 51°C (124°F). The next tank contains a weaker solution of soap and alkali, the third soap solution only, and the fourth and final clean rinsing water. The temperature is progressively lowered between the first and last tanks, the final temperature being about 40°C (104°F). The wool is propelled gently through the tanks by means of swinging forks, and passes between pairs of squeeze rolls in between the individual tanks to ensure that as little as possible of the dirty liquid is carried over into the next tank. Only a gentle motion of the propelling forks can be tolerated during scouring; any more vigorous action under the warm, wet conditions could cause felting and entanglement of the wool and prevent the spinning operations from being carried out successfully. Scouring by this method removes nearly all the dirt and grease and improves the appearance of the wool.

In another method, often referred to as the 'steeping' system, the scouring is carried out in two stages. First, the wool is steeped in cold or lukewarm water for a period to remove the soluble impurities, and then the four-bath system described above is carried out but with milder solutions. It is claimed that this variation gives an improved handle and colour and a material which has a better general appearance.

A relatively small proportion of raw wool is cleaned by solvent extraction of the impurities; this method is claimed to damage the wool less than the others. The wool is first treated with a solvent such as white spirit or carbon tetrachloride to dissolve out the grease and fats which are present, and then this is followed by washing in hot water at about 60°C (140°F), or steeping in lukewarm water for a period, to remove the water-soluble impurities.

The gum is usually removed from spun silk by boiling the material in a strong soap solution. This treatment first softens the gum and then dissolves it.

5.3 CARDING

5.3.1 Carding action

A further process, known as 'carding', contributes to the thorough opening, cleaning and mixing of the mass of fibres. The carding action breaks up any remaining hard tufts of fibre which may have managed to pass through the previous processes, removes the very short fibres, and helps to complete the cleaning of the fibres. It is also an effective means of mixing the fibres thoroughly in order to enable uniform yarns to be spun.

The carding action can be described as the pulling apart, or 'teazing', of the mass of fibres, to give a light, fluffy construction, in between two wire bristle 'brushes' with one set of bristles moving over the other. One of the brushes may be stationary and the other moving, the two may be moving in opposite directions, or both may be moving in the same direction but one much faster than the other. In all three cases, the fibres are dragged apart by the bristles and, as the process is continued, the fibres become distributed in a thin, attenuated condition over the surfaces of the brushes. In this condition impurities are freed to fall away from the fibres, and are also helped to do so by the sliding action of the wire bristles through the mass of fibres. In the days of hand spinning, carding was carried out using two actual wire bristle brushes held in the operator's hands, and the success of the carding operation depended on his skill and judgement.

In modern machine carding, the equivalent of the wire brushes is referred to as 'card-clothing'. This is made with a foundation of leather, or stiff laminated cotton or linen

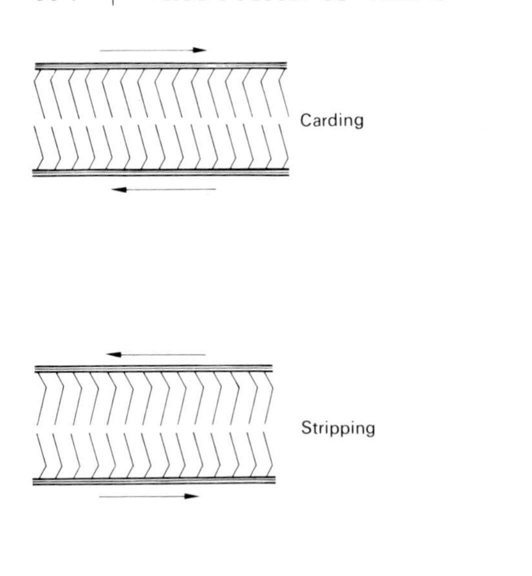

Fig. 5.9 Card clothing as found in modern machine carding.

fabric, covered with pointed wire staples or 'dents' having their points inclined in one direction. The bent shape of the staples not only gives a more effective carding action, but also allows the points to 'give' without jamming against the corresponding points of the clothing travelling in the opposite direction, should a hard piece of foreign material be encountered in the fibre mass. The drawing also shows how the directions of inclination of the two sets of points, and their movements relative to each other, should be arranged to produce the carding action, and then how these conditions should be changed when it is desired for one card to strip all the fibres from the other at the conclusion of the carding process.

For long staple wool fibres to be used for manufacturing worsted fabrics, and for most cotton and flax fibre materials, carding is only one of a series of processes used to prepare for spinning, and the carding process is followed by other treatments before the final yarn twist is inserted. In using the shorter staple wool to make woollen yarns, however, carding is the final operation before the yarn is spun, and must leave the fibre mass in a soft, fluffy condition which is characteristic of a woollen spun yarn. It follows that wool intended for woollen yarns must be very thoroughly treated and mixed during the carding process. For this purpose there are a number of carding positions provided on each carding machine, and the material passes through a number of machines, being removed or 'doffed' from the carding points of one machine before being passed on to the next.

There are two main types of carding machine in common use, the 'roller card' and the 'flat card'. In general, the roller type is used in the woollen, worsted, cotton-waste, rayon staple and flax-tow spinning industries, while the flat type is used in the spinning of cotton, rayon and spun-silk yarns.

5.3.2 Roller cards

In the roller type of carding machine, the card clothing is fastened on all the surfaces of a number of rollers of various sizes used in the machine.

The layer of fibre is carried into the machine on an endless moving belt where it meets an assembly of rollers covered with the clothing, arranged as regards rotational speed and direction of rotation to take the material from the belt and distribute it as a thinner layer over the surface of a card-covered roller referred to as a 'taker-in'.

Operating near to the taker-in roller is another roller, called the 'angle stripper', the function of which is to strip the fibre layer from the taker-in and transfer it to the cylinder, a large clothing-covered roller often referred to as a 'swift'. It will be appreciated that, in order to obtain these transfers of fibre material, the wire points must be facing in the right directions at each point of transfer, and the angle stripper must be rotating with a higher surface speed than that of the taker-in, while the surface speed of the swift must be higher than that of the angle stripper. These preliminary transfers from roller to roller have ensured that the fibre material deposited on the surface of the swift is in the form of a moderately thin layer, and the carding action can now begin.

Carding is effected by a series of pairs of rollers, one of the pair being called a 'worker' and the other a 'stripper' arranged over the upper surface of the swift as shown in the illustration.

The carding action takes place at the worker, which is rotating at a very low speed compared with that of the swift. The separation between the two sets of points is fixed and is quite small, so that only a definite thickness of the fibre layer can get through, the remainder being held and carried by the points of the worker. This fibre is stripped from the points of the worker by a more rapidly-rotating stripper roller and this in turn, is stripped by the even more rapidly-moving swift; the fibre is thus presented once again to the worker for further carding. The fibre layer is thus well carded at this position before it gets through on the surface of the swift to the next pair of workers and stripper

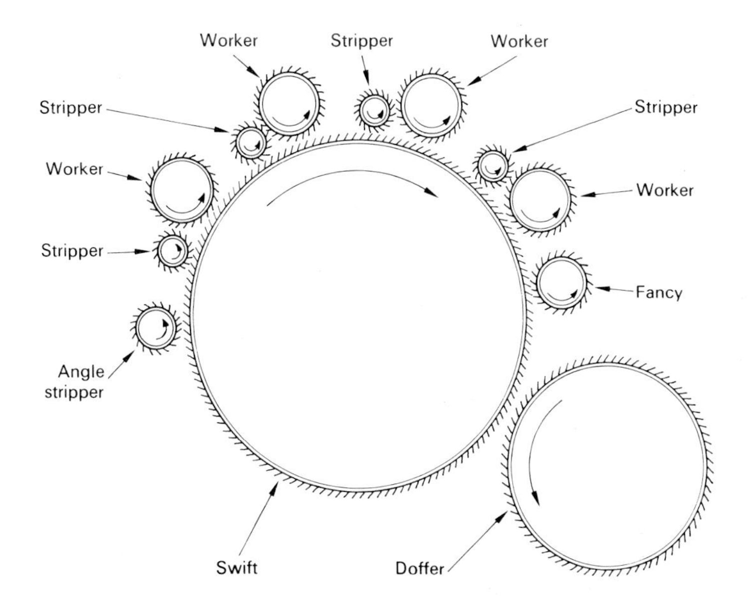

Fig. 5.10 The roller carding machine.

rollers where the process is repeated, and so on through the various pairs over the upper surface of the swift. At each position the fibre layer becomes more and more open and evenly distributed, and thoroughly mixed.

The physical size of the swift limits the number of carding positions which can be arranged round it, and in order to obtain a sufficient carding treatment for a wollen yarn, it is necessary to use a succession of such machines, the fibre material being stripped or doffed from one and passed on to the next. The doffing cannot be done by a simple stripping action, since stripping can only take place if the stripping roller is rotating at a higher speed than the one carrying the fibre material. This would mean that, in a line of machines, each machine would be operating at an appreciably higher speed than the one immediately preceding it. Not only is this impracticable mechanically but it would also attenuate the fibre layer so severely that satisfactory carding could not be carried out. To overcome this difficulty, a high speed 'fancy' roller operates near the surface of the swift, its points penetrating in between those of the swift. The function of the fancy is to brush the fibres up to the tips of the points of the swift, where they are loosely held and are easily taken off by a slowly rotating roller, the 'doffer'. Since the doffer is rotating slowly, it will be appreciated that the layer of fibre which it collects is very much thicker than that on the surface of the swift, and so further carding can take place in the succeeding machines without any risk of too great an attenuation of the fibre layer taking place. The fibre layer is transferred from the doffer to the swift of the next machine by means of an angle stripper as has already been described.

When, at the end of the carding treatment, the fibre layer has to be removed from the doffer of the last machine, this is done by a reciprocating comb which operates with a succession of downward strokes, and peels the fibres in the form of a thin web, or sliver, from the points of the doffer.

It should be mentioned here that, in woollen spinning, no more processes are carried out in between carding and spinning and the sliver which leaves the carding machines is too thick to spin a satisfactory yarn, even though a certain amount of drawing or attenuation takes place during the actual spinning. To obtain a suitable amount of fibre for each spindle of the spinning machine, therefore, the sliver is cut along its length into a series of narrow strips, each of which can be taken up on its own bobbin and used as a roving from which to spin a yarn. The machine which divides up the sliver is called a 'condenser'. The type of condenser in most common use is referred to as a 'tape' condenser. This consists of a pair of nip rollers, the surfaces of which are cut into equally-spaced channels, in which run accurately-fitting tapes, usually made of leather. The trapping, under pressure, of the sliver between the tapes and the rollers gives a scissor-like action which cuts up the sliver into strips. The strips then pass between two moving leather belts, which not only convey them forward to bobbins on which they can be wound, but also by reason of a sideways reciprocating movement of the belts relative to each other subject the strips to a rolling action and thus convert them into rovings.

A similar type of condenser action on the sliver can be produced by covering the final doffer roller in the carding machine with rings of card clothing round the roller and using leather bands to fill in the spaces between the clothing rings. A doffer roller treated in this way will only take the wool from the swift in strip form. This type of arrangement is referred to as a 'ring condenser'.

During the manufacture of fairly coarse yarns from cotton waste of too short a staple length for normal cotton processing the material is treated in much the same way as in woollen processing. Carding is the final process before spinning. The carding is, however, less thoroughly carried out than is the case with woollen preparation. The equipment used normally consists of two parts only, a 'scribbler' card and a 'finisher' card.

As has already been mentioned, in worsted and flax spinning the carding operation is followed by other preparative processes before the actual spinning takes place, and the main purpose of the carding, in those cases, is to open up the material thoroughly and pass on the loosened fibres in the form of a sliver. For worsted processing, the carding machines consist of two or three cylinders, or swifts, together with the attendant rollers as already described. Flax tows in a relatively clean condition are usually passed once through a finisher card, but dirtier tows also receive a preparative treatment using a breaker card.

5.3.3 Revolving flat carding machine

In this machine, the fibre mass is fed forward in the form of a lap until it comes into contact with a toothed small roller, the 'licker-in', which is rotating rapidly. This strikes the material away from the feeding mechanism in tiny tufts on to the card-clothing points of a large cylinder, which has an even higher surface speed. The material is carried round on the wire points of the cylinder, and under the points of a moving, endless 'chain' of flats, each of which is surfaced with card clothing, which runs over the upper part of the cylinder. The carding action takes place here since the flats are moving slowly against the high speed of the cylinder. Any fibre which adheres to the flats is stripped off as each flat reaches the front of the machine. The carded fibres are removed

Fig. 5.11 High production card conversions fed by a continuous lap conveyor.

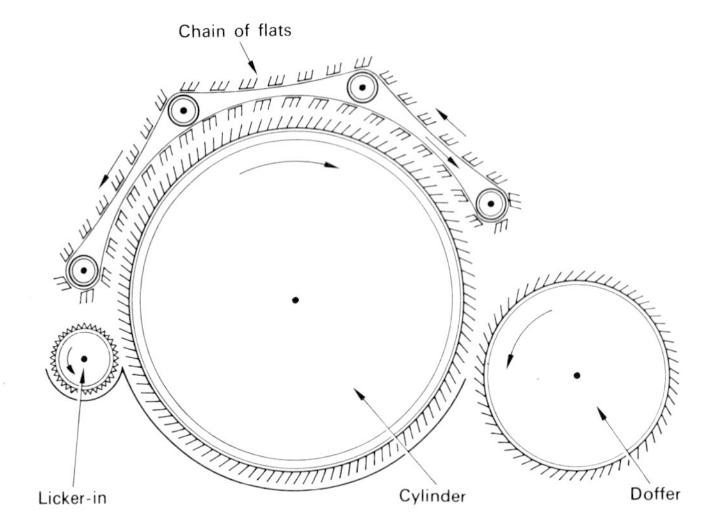

Fig. 5.12 Revolving flat carding machine.

from the cylinder by means of a doffing roller in a similar manner to that used on the roller card, but generally without the use of a fancy roller since there appears to be little difficulty in transferring the bulk of the fibre from the cylinder to the doffer.

From the doffer roller, the fibre material passes, in the form of a lap, through a

Fig. 5.13 Sliver formation and coiling into cans after carding.

vibrating comb, and is then narrowed to form a thicker sliver which is coiled into tall narrow cans ready for the next stage of processing.

5.4 COMBING

5.4.1 Combing action

To obtain fine, smooth, strong yarns, such as are used in the manufacture of worsted suitings for instance, it is necessary to remove the short fibres and arrange the longer ones in an orderly manner parallel to each other so that, when twist is inserted, the maximum binding power will be obtained and no fibre ends will protrude from the body of the yarn. Such conditions are not provided by the carding process, the sliver of fibre material which it delivers being a loosened, but still a jumbled, mass of fibres. A further process is necessary, therefore, to achieve the required objectives, this being known as 'combing'.

As the name implies, combing is the drawing of an arrangement of pins or points through the fibre mass, or alternatively the dragging of the fibre mass over such an arrangement of pins. The moving points straighten out the fibres and pull them into a more or less parallel order, at the same time dragging out the short fibres which are not held firmly in position. Originally the process was carried out by hand using specially-constructed combs which were drawn through the fibre mass.

The first man to attempt machine combing is said to have been Cartwright who, in about 1790, took out patents for a wool-combing machine. This machine was a rather crude affair, but it did demonstrate that machine combing was possible and spurred on others to work on the project. The development work was carried out during the early part of the nineteenth century and it was round about 1850 that combing machines began to take the form which we know today.

5.4.2 Lister comb

This machine was devised by Lister in 1850. The modern machine of this type is used mainly for combing the long animal fibres.

The sliver, suitably oiled, is fed forward by a feeding device in the direction of a swinging frame which, in its first position, is moved close to this feed. Two jaws on the

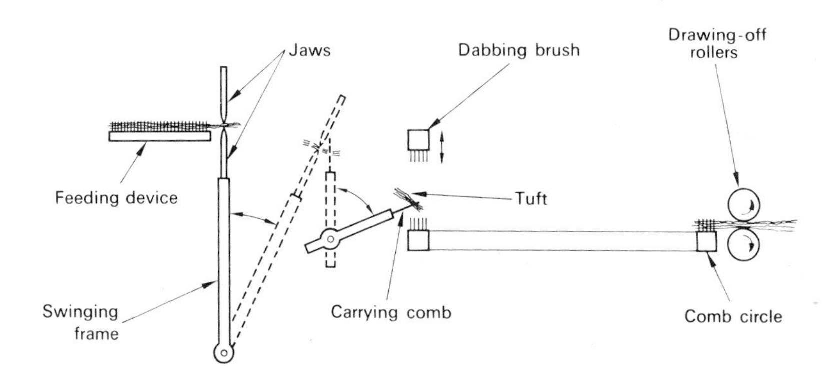

Fig. 5.14 A Lister combing machine.

swinging frame now close on the fibres which are projecting from the feeding device and, when the frame swings away to its second position, it pulls off a tuft of fibres, the trailing ends of which have been combed by being dragged through the pins of the feeding device. When the swinging frame is in its second position, a 'carrying comb' on the end of a lever swings up to it, and the comb embeds itself upward through the tuft of fibres. The nip of the swinging frame now opens, and the swinging frame moves back to its original position adjacent to the feeding device, leaving the tuft of fibres in the carrying comb. While this return journey of the frame is taking place, the carrying frame swings in the opposite direction to the rim of a heated 'comb circle' which carries rows of vertical pins and, at the end of this movement, the fibre tuft is over these rows of pins. A 'dabbing brush' now comes down and presses the fibre tuft well down into the pins of the comb circle and, as the latter revolves, it gets a firm enough grip to pull the tuft out of the carrying comb. It will be appreciated that, at this stage, the half of the tuft which has been combed by being drawn through the pins of the feeding device is projecting outside the comb circle, and the uncombed half is inside the circle.

The tuft impaled on the revolving comb circle now reaches the opposite side of the machine, and its fibre tips are caught in the nip of a pair of 'drawing-off' rollers, when the fibres are drawn through the pins of the comb circle, thus completing the combing of the entire tuft. Fibres which are not sufficiently long to be trapped in the nip of the drawing-off rollers continue to be carried round to another position of the machine where they are removed by a set of 'lifter knives'.

By a repetition of the cycle as described, the comb circle is kept constantly supplied with fibre tufts, and the drawing-off rollers deliver a continuous sliver of long, aligned fibres. In order to keep the tufts under control over their widths, the jaws of the swinging frame, the carrying comb and the dabbing brush, are all curved to conform to the curvature of the comb circle.

The comb circle rotates at about one revolution per minute and an average rate of production is about 65 lb (29.4 kg) of combed fibre per hour. The machine gives excellent results in the combing of long wools and hair fibres.

5.4.3 Noble comb

This machine was invented in 1853 by Noble. In its present form, it is the most popular combing machine used in the worsted industry in Great Britain by reason of its satisfactory rate of production, and also the fact that its essential parts are easily changeable so that it can operate on all staple lengths of wool except the very long and the very short.

The action of the Noble comb is simple in principle but the construction of the modern machine is less simple in order to ensure that the various parts act positively and in a strict sequence. The machine requires a minimum of servicing and gives long, trouble-free, operation.

The main parts of the machine are shown in the illustrations and its method of operation will be understood by referring to them. The fibre slivers to be combed are supported round the cylindrical portion of the machine and rotate with it in a horizontal plane. The main parts of the machine are three heated comb circles, these consisting of one 'large circle' having an inside diameter of about 42 in (1 066 mm) and two 'small circles' of about 16 in (406 mm) diameter mounted within the large circle at either end of a diameter with their outer edges close to the inside edge of the large circle. These three circles rotate in a clockwise direction with their surface speeds approximately the same. Each circle has rows of pins mounted vertically upright on its upper face.

Fig. 5.15 Noble combs in operation. The slivers are clearly visible on the outer circle.

As the large circle rotates, the oiled slivers are carried round and, one after the other, are fed forward until they slightly overlap the rows of pins of both the small and large circles at their point of closest proximity. A reciprocating 'dabbing brush' now comes down and presses the fibres firmly down into the pins of both circles. Further rotation of the two circles of different diameters causes the two sets of pins to separate and drag the fibre mass apart. Those which are held by the small circle have been combed by having passed through the pins of the large circle, and the combed ends project from the outside of the small circle; those held by the large circle have been combed by the pins of the small circle and their combed ends are on the inside of the large circle. It only remains for these to be drawn out of the two sets of pins in the correct directions to complete the combing of the fibres.

The fibres are drawn out of the rings by vertical 'draw-off' rollers, the outside pair taking from the inside of the large circle, and the inside pair from the outside of the small circle. Leather belts working round rollers, as shown in the diagram, first brush the projecting fibres sideways at the appropriate positions so that they are gripped by the roller nips, and then help to convey them to a take-off point where the two streams combine, and also join a stream coming from the second small circle which is operating at the same time and in exactly the same way. The combined stream passes down a rotating funnel to insert just sufficient twist to hold the fibres together, and then through a 'coiler' funnel to be coiled into a can.

The short fibres left in the small circles are lifted out by a suitable device and dropped into a chute which allows them to be taken away. No such device exists on the

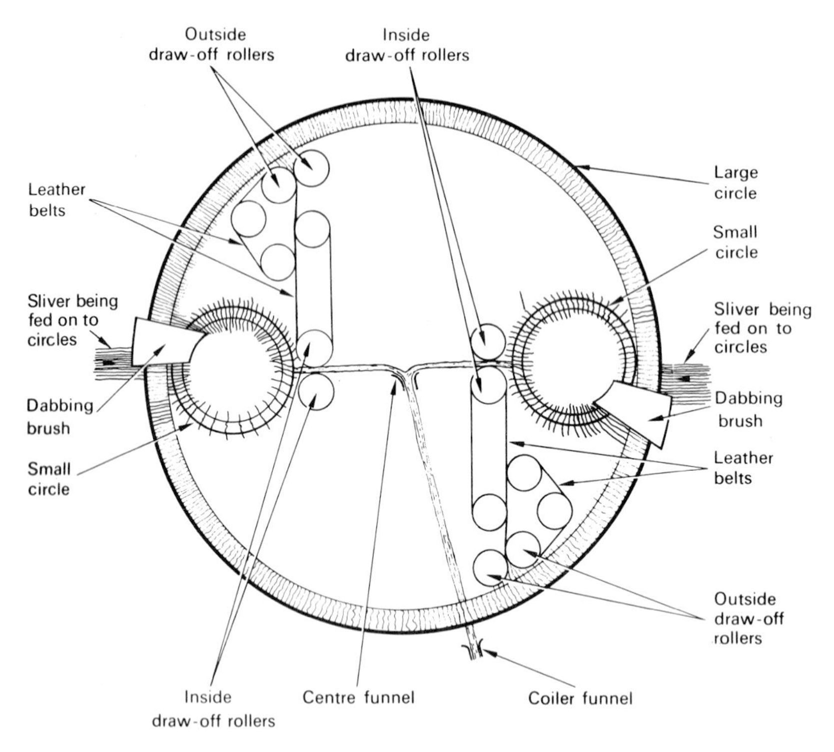

Fig. 5.16 Diagram showing working sections of a Noble comb. View from above the machine.

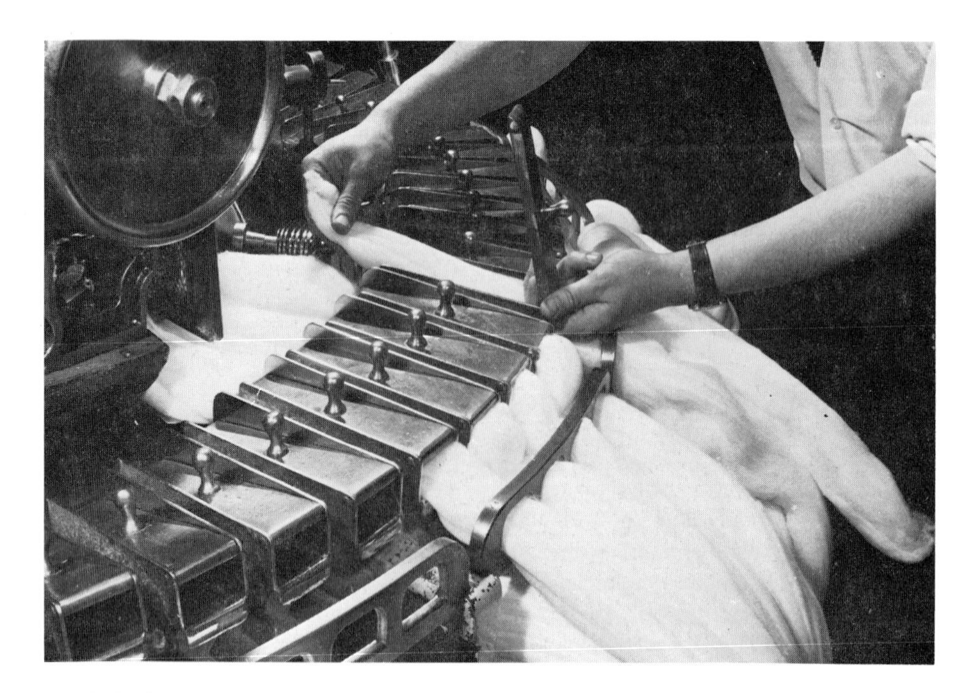

Fig. 5.17 Preparing the creel of a Noble comb with slivers of man-made fibre.

large circle and the short fibres held in the pins remain there until the combing action eventually drags them into the pins of the small circles.

5.4.4 Rectilinear comb

The rectilinear comb is used for combing all kinds of fibres of relatively short staple length, such as cotton, short staple wools and waste silk. The original design was devised by Heilmann in 1846 and the general basic design is still retained in the modern machines of today. The use of the rectilinear comb is often referred to as 'French' combing.

This combing machine is rather more complicated in construction than, say, the Noble comb, since nearly all its parts are subjected to reciprocating movements, which must be accurately synchronized if satisfactory combing is to be achieved. The machine is not heated during operation, and is pleasanter for the operator to handle for this reason. No special feeding arrangements for the sliver are necessary, as in the Noble machine, and the slivers can be taken direct from cans. The rectilinear comb has a lower production rate than the Noble comb, but some increase in speed has been obtained in more recent models.

One modern machine of this type is the Nasmith comber, and the operation of this particular machine, which makes use of the same general principles as all the rest, will be described. The carded sliver of fibres is fed by a feed roller into the nip of two 'nippers', as shown. The 'comb cylinder' then revolves and the needles, mounted in an inclined order on a portion of its circumference, pass through the fringe of fibres and

Fig. 5.18 A rectilinear comb.

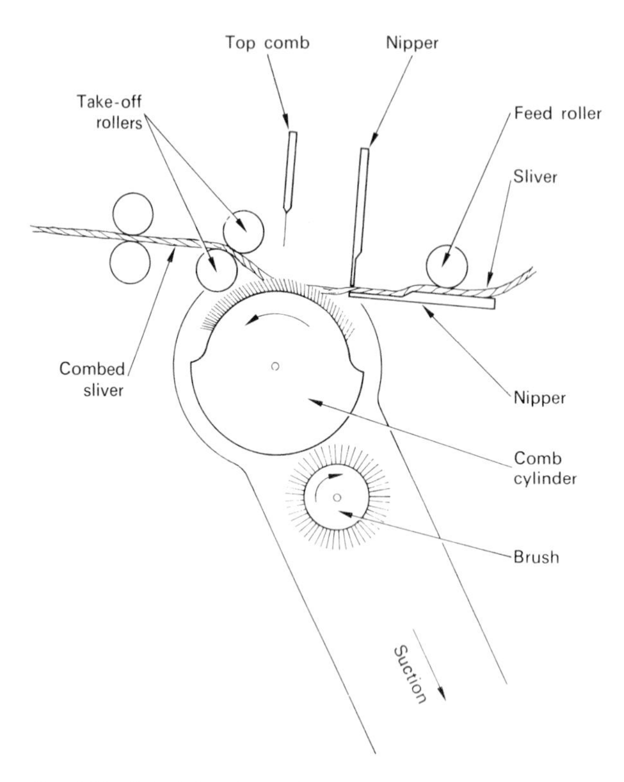

Fig. 5.19 Principles of combing as seen in the Nasmith combing machine.

comb them. The very short fibres which have been combed out adhere to the needles and are carried round by the cylinder until it reaches a position where they are stripped off by means of a revolving brush and carried away by suction to a container at the rear of the machine.

When combing of the fringe is completed, the nippers open to release it, at the same time moving to the left, as depicted in the diagram, close to the take-off rollers which grip the fibre tufts. While in this position, the 'top comb' descends and its pins penetrate the fringe of fibres. The rotation of the take-off rollers pulls off the fibre fringe as a tuft and, in doing so, the fibres are dragged through the pins of the top comb, thus completing the combing of the tuft.

Immediately before gripping the fibre fringe, the take-off rollers are put into reverse and fed towards the oncoming fringe a short length of the combed sliver prepared by the previous cycle. When it grips the new fringe, therefore, the two overlap by a specified amount and, in this way, the machine delivers a continuous combed sliver.

In the modern machines, a number of such units as described are mounted side-by-side and are driven by the same mechanism.

5.4.5 Flax hackling machine

In the flax stem, the bundles of fibres are not independent of each other, but are joined across in various ways to form a type of network surrounding the core of the stem. In spite of the earlier preparative processes, this network remains, to some extent, intact and the task of the combing, or as it is called in the case of flax 'hackling', process is to break this down so that finer yarns may be spun. This requires considerable force, and much more than would be needed than in the case of the combing of wool and cotton, so that the development of the flax hackling machine has been somewhat different from that for the combing machines used for the other fibres.

In the hackling machine, the flax bundles are combed first at the root end by one half of the machine, and then at the tip end in the other half of the machine. The principle upon which the machine is based is shown in the diagram. The flax is clamped in the holder, which is supported in a channel. The channel moves up and down, lifting and lowering the flax bundle as it does so. The flax bundle hangs down, root end first, in between two travelling belts of points made up from single rows of points fixed to strips of wood which are, in turn, attached to endless leather belts.

In operation, the flax is lowered in between the two belts of points. After a dwelling period at the lowest point of the stroke, the channel rises and withdraws the flax. An automatic device now moves the holder one position along the channel, its place being taken up by another holder and new bundle of flax. On the next downstroke of the channel, the original bundle is lowered between the next pair of belts along the machine, and so on until the first half of the machine has been traversed. The first pair of belts used has widely-spaced points so that the bundle is only subjected to a gentle combing and straightening at the first pair, but the density of the points gradually increases in subsequent pairs and the bundle is thoroughly combed.

When the half-way stage has been reached, the flax bundle is taken from its holder and reversed so that the tip end is now hanging down. The second half of the machine is a replica of the first so that, in completing its passage through the machine the remaining half of the bundle is also combed.

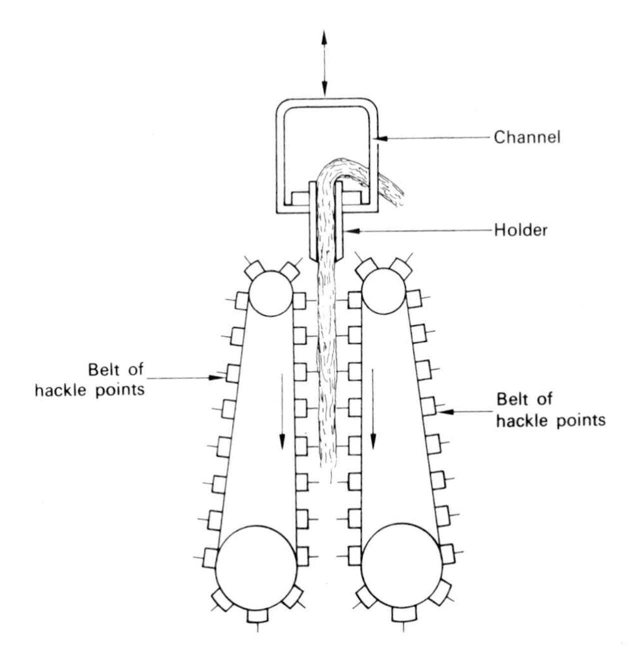

Fig. 5.20 Flax hackling machine.

At the lowest position of the belts of hackle points is a stripping mechanism, which is not shown in the diagram. This removes the short and broken fibres which have been combed out of the bundle and carried round on the points of the belt. These short fibres drop into a bin provided for the purpose. The amount of short fibre removed during a thorough hackling action can be considerable, and the aim should be to obtain an adequate degree of combing consistent with economic and end-use considerations.

5.5 DRAFTING

5.5.1 Drafting action

When describing the hand-spinning of yarns, reference was made to the thinning down of the mass of fibres performed by the fingers of the operator, whereby the mass was drawn out to the correct thickness for spinning, and the fibres distributed as evenly as possible to reduce the tendency for there to be thick and thin places in the final yarn. This thinning down action is called drawing, or 'drafting'. The present-day machine drafting aims to achieve the same objectives as those of the hand-spinner, but in a more controlled and reliable manner and at a much higher speed.

After combing has been completed, the resulting sliver is far too thick and bulky to enable a suitable yarn to be spun from it. In addition, the fibres it contains are by no means lying in an accurate parallel order and it would not give the strength and smoothness required in the final yarn until the degree of orderliness had been improved. Finally, however carefully the combing operation has been carried out, the sliver will have irregularities along its length which, unless removed, will be carried over to appear in the yarn to be spun.

The combing process is followed by a drafting process, therefore, in which the sliver is drawn out to its correct thickness. At the same time, the sliding of the fibres over each other straightens the individual fibres even more and pulls them into a more nearly parallel order. Finally, although it is doubtful if drafting can reduce the inequalities in a single sliver, as we will see later, a number of slivers can be drafted at the same time and, since thick and thin places are likely to be randomly distributed in any one sliver, they are more likely to be cancelled out by such a procedure.

5.5.2 Roller drafting

As has already been mentioned, machine drafting using a system of rollers was first attempted by Paul in 1738, but it would not appear to have been a practical success in the form in which he applied it. It was left to Arkwright to include a workable system of roller drafting in his spinning frame in 1786.

The basic idea of roller drafting is simple. The sliver to be drafted is fed into the nip formed between one pair of revolving pressure rollers, and then through a further pair revolving at a higher speed than the first pair. By this means the sliver is thinned out and is increased in length. Thus, if the second pair of rollers is revolving at, say four times the speed of the first pair, the draft produced will be four and the drafted sliver will be four times as long as it was initially, provided no slippage takes place at the nips. Modern drafting equipment works on this simple principle but, as Paul found out, there are problems to be overcome before good drafting can be carried out.

To a considerable extent, the uniformity resulting from the drafting treatments will depend on how free from irregularities is the starting sliver. There is, however, a tendency for the drafting process to introduce an even greater amount of irregularity. This arises mainly from the difficulty of controlling completely the passage of the fibres between one pair of draft rollers and the next. In the simplest drafting system the separation between the two pairs of rollers cannot be less than the lengths of the longest fibres present, otherwise these would be gripped at both ends and would probably be broken by the differences in rotational speed between the two pairs. If the fibres were all of the same length, then roller drafting would be easy to carry out. The two pairs of drafting rollers would be separated by a distance which was slightly greater than the length of the fibres. For each fibre, therefore, as soon as its trailing end had left the nip of the back pair of rollers, its leading end would enter the nip of the front pair and its travel would be accelerated relative to the fibres still not clear of the back nip. Each fibre would be under complete control the whole time, except for an instant between release from the back pair, and gripping by the front pair. Drafting would, therefore, be perfectly smooth and controlled. Unfortunately, the fibres in the sliver are by no means equal in length and, in fact, short fibres usually out-weigh the long fibres in number. During roller drafting using two pairs of rollers, therefore, once the short fibres have passed through the back nip, they are completely out of control and held suspended only by the long fibres surrounding them. They are not being drafted in that position and tend to move forward in bunches leaving thin places mainly composed of long fibres behind them. It will be appreciated, therefore, that when there is an excess of short fibres, the straightforward roller drafting process can introduce substantial irregularities and, since a higher speed change between the back and front pairs of rollers will draw out the long fibres even more without affecting the short fibres, any attempt to increase the draft at each stage serves to introduce even greater inequalities.

A further consideration in roller drafting is, that in order to retain accurate control and reproducibility of results, it is necessary to ensure that slipping does not take place within the roller nips. Usually the bottom roller in each pair is driven and its rotation is transmitted to the top roller which is resting in contact with it. Various methods of loading the top roller are used to give a good, positive nip on the sliver, and the roller surfaces are treated to increase the grip exerted on the fibres. The bottom rollers are usually of steel and often fluted, and the top rollers mostly have plain surfaces but are constructed of materials selected to give the necessary amount of friction.

Finally, it should be mentioned that drafting roller systems must be well constructed and carefully maintained since any imperfections in the equipment itself can also contribute to the overall inequalities introduced. Thus any variations from a smooth roller drive, and vibrations in the equipment, will affect the amount of draft obtained at any one instant.

Long experience has shown the maximum amount of draft which should be allowed at each drawing stage to provide, ultimately, an acceptable yarn. Many stages are involved, and it is better to draft a little at a time, and use a considerable number of stages, if inequalities are to be kept to a minimum.

5.5.3 Doubling on the drawframe

One method of reducing inequalities is to use the early stages of drawing to mix the fibres of a number of slivers and to allow mixing to take precedence over attenuation, or drawing, before any substantial amount of drafting takes place. A number of slivers, which may be as high as eight in the case of cotton, are fed into the first roller nip which is usually the first of four pairs. If eight slivers are being fed in at once, then

Fig. 5.21 Slivers being fed into a Draft-O-Matic machine where they will be drafted into one sliver of uniform weight per linear length.

it is usual to arrange that the total draft is also eight and the issuing combined sliver has about the same thickness as one of those originally fed in but is, of course, eight times as long. By doing this, no attenuation has been obtained, but the resulting sliver represents a better mixing of the fibres, the fibres composing it are arranged in a more parallel form, and there is less irregularity, than in the eight starting slivers.

It is not, of course, necessary to have the number of 'doublings', as they are called, equal to the draft. In the discussion we have used eight doublings and a draft of eight but, if preferred, either one may exceed the other. If the draft is greater than the number of doublings used, the sliver delivered will be thinner than any of the eight starting slivers and if the draft is the lesser of the two, then the material delivered will be actually thicker than any one of the starting slivers. The combinations of numbers may be used as desired. When the doublings equal or exceed the draft number, then the main objective being pursued is blending since no attenuation takes place; if the draft is greater than the number of doublings, the material is being reduced in thickness while some mixing is taking place. Usually the ratio, draft/doublings, increases as the material progresses through a drawing set by slight increases in draft together with fewer doublings. The number of stages used to produce the total draft necessary will depend on a range of factors, such as variations in fibre length in the starting fibre material, the efficiency of the drafting systems used, and the characteristics required in the final yarn to be spun. It must also be remembered that, the greater the number of doublings and drawings stages used in an attempt to produce a final higher quality yarn, the higher the cost of drafting, and a compromise between the two must be reached.

5.5.4 Short fibre control

In the last two sections it has been mentioned that, in passing the fibre material from one pair of drafting rollers to the next, control is lost over the short fibres. If an improvement is to be made, then some additional means of control must be introduced between the two pairs of rollers which must exercise control over the short fibres while still allowing the longer fibres to be drawn freely. Various such devices are in use and one or two may be mentioned to indicate how the problem has been tackled.

For drafting long wools, flax, and other long-fibred materials, 'gill drawing' is sometimes used. The short fibres are held in their proper place, until they are gripped and drawn forward, by friction against the pins or 'teeth' of vertical combs, known as 'gills'. Between the back and front draw rollers is a series of gills, rows of steel pins mounted in brass rods, extending across the machine, which are lifted to penetrate the sliver soon after it leaves the first pair of rollers, and travel along with it. Just before it would touch the second pair of rollers, each gill drops down to a lower level and is transported back at high speed to its starting position for the cycle to be repeated. It is arranged for the gills to travel at a slightly higher speed than the surface speed of the first pair of rollers so that, at the position in which the points penetrate through the sliver, the latter is under a slight tension and penetration is made more easily and definitely. The gills thus act as a conveyor to carry the sliver over the space between the two sets of rollers and, at the same time, apply sufficient friction to prevent the short fibres being drawn forward as a bunch until they reach the second pair of rollers and are drawn thereby against the friction. Gill drawing is usually confined to the early stages of drafting when attempts are being made to achieve a greater uniformity in the sliver and to arrange the

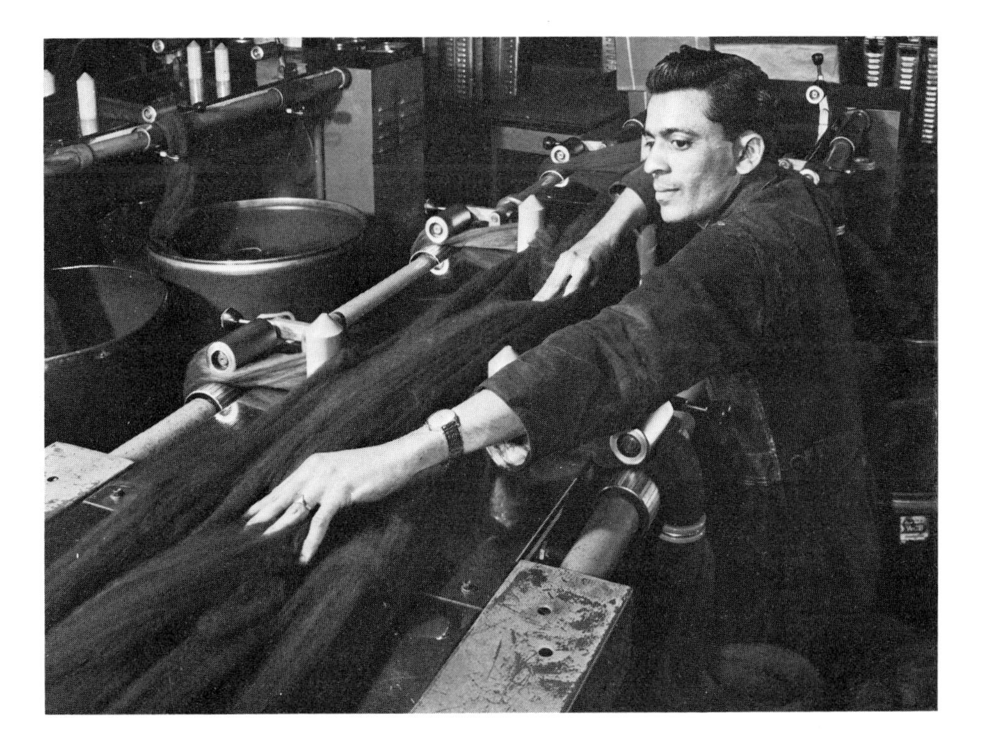

Fig. 5.22 Introducing a new end of sliver into a gill box.

fibres in a more parallel formation. The machines using gill drawing are noisy and there is a limit to the speed at which they can operate. They do, however, fulfil their purpose very well.

For what is known as the 'Continental' system of preparing worsted yarns, a 'porcupine' method of drawing is often used. Here, again, the fibres are controlled by friction against a set of pins, but in this case, they are fixed in inclined rows on a revolving brass cylinder. This is mounted just before the second pair of draw rollers and close to it, so that it is under the sliver with its pins penetrating through the material. The surface speed of the porcupine cylinder is about 5% slower than that of the first roller pair, so that nearly the whole of the draft applied is between the porcupine and second pair of rollers. Carrier rollers are provided between the porcupine and first roller pair to keep the fibre material at the right height for the pins of the porcupine roller. In the Continental system, porcupine drawing is used at all the stages right up to the one at which the material is sufficiently thinned down ready for the spinning operation and is referred to as a 'roving'. This means that no supporting twist can be inserted at any of these stages, as this would prevent the porcupine pins from travelling through the material. It follows that, as the fine roving stages is approached, the material requires very careful handling indeed if separation is to be avoided.

In the 'English' system of worsted processing, the method of exercising short fibre control in the later stages of drawing is by inserting a small amount of binding twist into the fibre strand. The short fibres are then held in position by friction against the fibres surrounding it. Twist is inserted by means of a bobbin and flyer arrangement, operating in much the same way as that used to spin yarns and described in section 5.6. When such a twisted assembly of fibres is being drawn, carrier rollers are located in between the pairs of draw rollers and they support the material in this position and also ensure that untwisting does not take place which would release the short fibre material from constraint. A more recent development, the Ambler Superdraft system, allows control of the short fibres, by means of twist, to be exercised close up to the surfaces of the second pair of draw rollers. With this extra degree of control, very much bigger drafts can be used without sacrificing uniformity in the final yarn, and hence a reduction in processing costs is achieved. In the system, a special condenser guide leads the roving into the nip of two tensioning rollers. These rollers are of small diameter, and the nip is formed between one narrow roller which runs inside a close-fitting lateral groove in the other; these rollers rotate at a little higher speed than that of the first draw-roller pair, thus applying a small tension to the roving. From the tensioning rollers, the roving passes into a 'flume', a small channel of rectangular cross-section which narrows towards its delivery end and which leads right into the nip of the second pair of draw rollers. The twisted roving is moulded into a rectangular cross-sectional form as it passes between the tensioning rollers, and this helps to lock in the twist. The tapered rectangular flume also maintains this regular cross-sectional shape and prevents the roving from rotating and losing twist until the main draft between the second pair of draw rollers and the tensioning rollers has been applied. In this way, the short fibres are held in contact with the surrounding material but the pressure between the tensioning rollers is not high enough to prevent the longer fibres from being drafted in the normal way by being drawn forward out of the twisted roving.

A few years ago, developments were made which automatically adjusted the rate of feed of a sliver or roving through a drawing section in accordance with the thickness of the material entering. This means that such devices increase or decrease the amount of draft applied if a thick or thin place enters the section. This levelling out of the strand of material – the device is referred to as an 'autolevelling' mechanism – has allowed a considerable reduction in the number of doublings used to provide a given degree of regularity in the final yarn. The first, and best-known, of these devices, was the Raper Autoleveller which is made by Prince-Smith and Stells Ltd. In this device, the sliver or roving passes between a pair of measuring rollers just before it enters the nip of the first pair of draw rollers, and the variations in thickness of the material are recorded in an instrument. This instrument is adjusted to the speed of travel of the material, and it allows an interval of time to elapse for the material to travel from the first pair of draw rollers to the position at which the greatest amount of draft occurs, before it acts. The instrument energizes a variable speed mechanism, and the speed of the first draw rollers is adjusted in accordance with the reading which was transmitted by the measuring rollers. By this means a thick place is given more drafting than a thin place. The long-wave variations are thus levelled out to quite a considerable extent, a result which would only be achieved without the device by using a larger number of doublings.

5.6 SPINNING

5.6.1 Principle

The previous processes have cleaned the fibres, thoroughly blended them together to eliminate inequalities and produce uniform properties and, in most cases, arranged them so that the individual fibres are lying parallel to each other. The fine roving thus produced is now ready for the final spinning operation which will convert it into a strong yarn suitable for the making of fabrics.

The roving is still a little thicker than is required for the production of the final yarn, so that the spinning process must include three operations:

1. The roving has to be drawn out, or drafted, still more to reduce it to its final thickness.
2. Twist has to be inserted to bind the fibres together and thus provide strength in the yarn, the amount of twist employed being appropriate to the end-use for which the yarn is intended.
3. The twisted yarn has to be wound on to a bobbin or tube to form a package.

As we have seen, the spinning machines designed to perform these operations have reached their present states of perfection as a result of developments made over a period of more than two centuries. They are of two main types, intermittent and continuous operating machines.

The intermittent type of machine is the mule, which was referred to in section 5.1.1, and which represents the oldest machine-spinning method in use. The machine carries out its complete operations on one length of roving before tackling the next, and all subsequent lengths. Thus, the drafting operation is first carried out on a measured length of the roving, then the necessary amount of twist is inserted, and finally this short length of newly-spun yarn is wound up before proceeding to repeat the cycle of operations on the next length.

The continuously-operating machines are of more recent origin. There are three main types, these being the 'Flyer', 'Ring' and 'Cap' spinning frames. In all three types, the drawing, twisting and winding operations are carried out simultaneously. The roving passes through a system of rollers to effect the drawing, and it then passes down to a guide, or its equivalent, which whirls at high speed round a bobbin or tube to introduce twist, while the bobbin, which is itself rotating at a lower speed, takes up the twisted yarn to form a package.

Cotton yarns are normally made using either mule spinning or ring spinning. The stem and leaf fibres are mostly processed into yarns using flyer spinning, and a modified flyer machine is used in the 'throwing' of silk, when two or more continuous filaments are twisted together to form a stronger, more robust yarn. Spun, or waste, silk is cut up into staple lengths and spun into yarns in much the same way as cotton fibres.

For wool, all four types of spinning method can be used. Woollen yarns contain relatively short fibres and are usually spun on a mule, the use of the draw rollers being omitted. For worsted yarns, ring spinning is the most popular of the methods used, but each method has its own characteristic actions and may provide a slightly different treatment more suitable to the staple length and structure of a particular type of wool. Thus the flyer machine, which operates at a relatively low rate, is often used for long fibre and lustre wools suitable for hosiery and carpets. Ring spinning is suitable for soft, smooth yarns made from fine merino wools. Cap spinning is often used for fine crossbred and merino wools; its rapid action produces a rather more 'hairy' but good type of yarn. Mule spinning is particularly suitable for the more delicate soft and full yarns required for hosiery applications. When using the mule machine for worsted yarns, the drawing of the roving is carried out by means of a roller system; for woollen yarns, the draw and twist are produced on the actual spindle of the machine and the draw rollers are not used.

5.6.2 Mule spinning

As already mentioned, mule spinning is an intermittent process in which the operations are carried out on a short length of yarn before proceeding to process the next length, and so on until a package of suitable size is obtained. The essential parts of the machine are shown in the diagram. This diagram shows only one spinning section of the machine; in practice, there are a number of sections side-by-side on long frames down the length of the mill and the same operations are carried out at the same time on all the various sections.

The machine has a fixed frame which supports the packages of roving and, in front of these, a group of three pairs of rollers rotating at progressively increasing speeds to

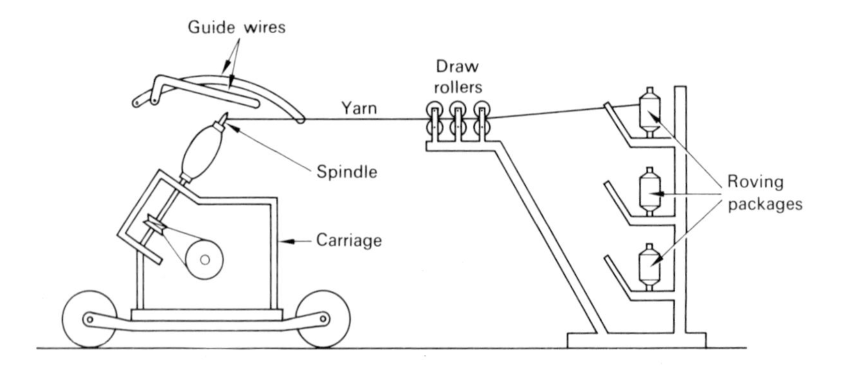

Fig. 5.23 Diagram of a mule spinning machine.

carry out the drawing of the roving fed between them. There is also a moveable carriage which is capable of running to-and-from the fixed frame on a track, the length of travel being about 6 ft (1.8 m). On the carriage is the spindle and means of rotating the spindle; round the spindle is a bobbin to take the final yarn, and the spindle itself is inclined towards the direction of the fixed frame.

The machine deals with about two yards of yarn at a time. The cycle of operations starts with the carriage close up to the draw rollers. The roving is fed from the package, through the draw rollers, and is attached to the spindle. The rollers draw the roving and feed it towards the spindle. At the same time the carriage retreats away from the rollers and the spindle revolves slowly to insert a certain amount of supporting twist in the roving. For fine yarns, the draw rollers cease to rotate and feed yarn when the carriage reaches a distance of a few inches away from the end of its travel; the small extra distance moved by the carriage to complete its travel causes the roving to be extended by the pull from the spindle point.

By the time the carriage reaches the end of its travel, the drawing of the roving is complete and so the spindle speeds up to insert the necessary amount of twist in the two yard length of roving suspended across the machine. By reason of the inclined position of the spindle, each revolution it makes allows one wrap of roving to slip off the spindle point and so inserts one turn of twist into the suspended material without interfering with the yarn already wound on it. The insertion of twist causes the yarn which is thus formed to contract and, in order to compensate for this decrease in length, either the draw rollers are arranged to deliver a small additional amount of roving, or the carriage moves back a short distance.

When the twisting of the yarn is completed, the spindle reverses its direction for a few turns to enable the guide wires to come into operation. One wire lifts the yarn out of its straight path to insert the necessary amount of tension into it, while the other wire presses the yarn down to guide it on to the bobbin round the spindle at the point where it left off at the last run-in.

When the guide wires have moved into their respective positions, the carriage commences its run-in back towards the fixed frame, the spindle rotating at the appropriate speed in the reverse direction, to wind the yarn on the spindle bobbin. When the carriage again reaches the draw rollers, the guide wires disengage and the whole cycle is repeated.

When very short staple fibre yarns are being spun on the mule machine, no draw rollers are used, as already mentioned. In that case, a pair of delivery rollers feed in the roving and what draw is inserted is provided by stopping these delivery rollers when the carriage has made only about two-thirds of its travel. As the carriage continues its journey, the roving receives a certain amount of draw by the spindle point while, at the same time, some draw twist is inserted. After this, the spinning and yarn take-up are as already described.

5.6.3 Ring spinning

Ring spinning is a high speed continuous method which has become popular. In ring spinning, the packages of roving are supported on the upper part of the machine and the roving is led downward towards the spindles through sets of draw rollers which provide the necessary amount of draw. The take-up means is a tube which fits tightly on a belt-driven vertical spindle which rotates at high speed. Round the spindle and the tube it carries is a steel ring supported in a ring holder which stands out from a ring

Fig. 5.24 A ring spinning frame.

lifting rail. A small, light 'traveller' runs in tracks in this steel ring, and can move easily in the tracks by reason of lubrication which is applied to the ring.

The ring rail carries the ring up and down along the length of the tube on the spindle.

The roving from the draw rollers is threaded through a yarn guide above the spindle, through the traveller, and is then attached to the tube on the spindle. In operation, the roving is drawn by the rollers and then carried round the tube on the spindle by the traveller, to be wound up on to the tube. The pull of the yarn as it is fed on to the tube causes the traveller to slide round the ring at high speed to guide the yarn on to the tube and, by reason of the slight drag which it imparts, to provide an even tension in the yarn. The up and down motion of the ring imparted by the ring rail ensures an even distribution of the yarn on the tube.

The amount of twist inserted in the roving to form the yarn in ring spinning is determined by the rotational speed of the spindle and the rate at which the roving is fed through the machine.

5.6.4 Bobbin and flyer spinning

In this machine, the roving packages are again at the top, and the roving is fed downward through a draw roller system. The drawn roving is then threaded through an 'eye' on the end of one arm of a flyer which is mounted above, and rotates round, a vertical spindle which carries a bobbin. The flyer is shaped like an inverted 'U', and it

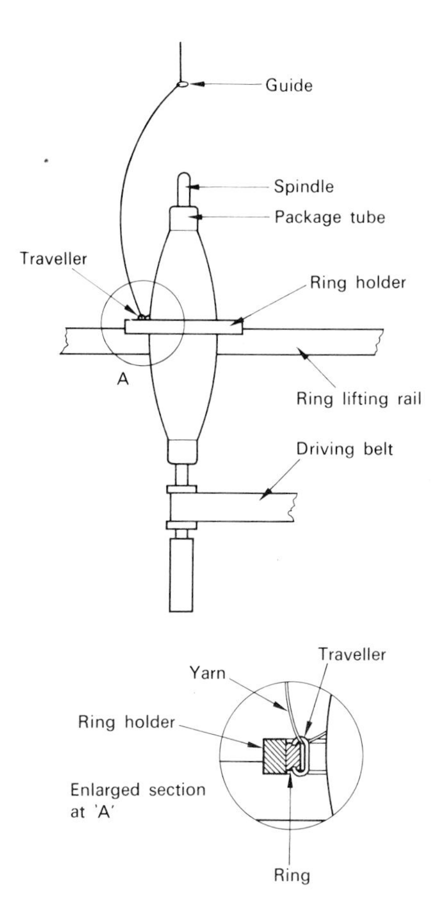

Fig. 5.25 Ring spinning action.

serves to twist the roving to form a yarn and to wind this yarn round the bobbin, a lifter rail carrying the bobbin up and down to allow a uniform distribution of yarn over its length. In this machine the spindle, and the flyer which is attached to it at its upper end, are rotated by a belt drive, while the bobbin is a loose fit on the spindle and is pulled round by the drag of the yarn against the friction imparted by a felt washer between the bottom of the bobbin and the surface of the lifter rail. The speed of rotation of the flyer and the checking action of the felt washer, determine the amount of twist which will be given to the yarn.

Mention has already been made of the fact that bobbin and flyer spinning is used extensively for flax yarns. In spinning fine flax yarns, and yarns to be used for the construction of the warp in weaving linen fabrics, 'wet' spinning is often employed in order to soften the fibres and enable them to slide more readily one over the other during drawing. In such a case, the roving passes through a bath of hot water in its passage from the roving packages to the draw rollers on the spinning machine. In addition to obtaining an improvement in quality in the yarn by wet spinning, it has also been found possible to operate the spindle at a higher speed when compared with dry spinning. There is, however, the disadvantage of having to dry the wet spun yarn, an operation which has to be carried out without much delay if a risk of rotting of the material is to be avoided. A good deal of water is thrown out of the yarn by the high speed rotation of

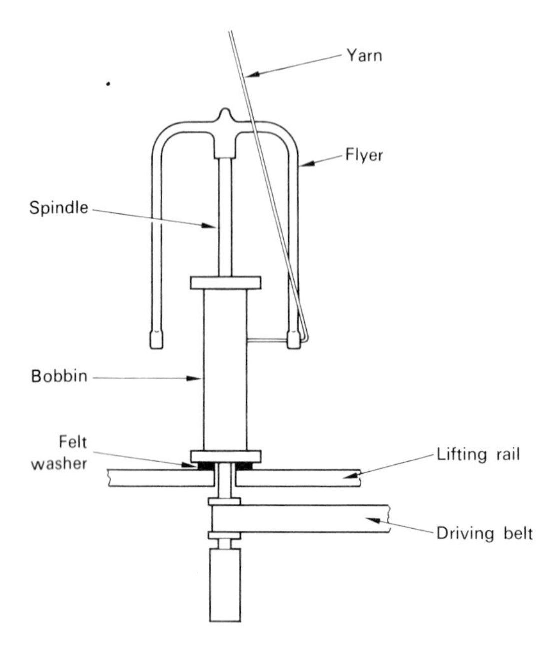

Fig. 5.26 Bobbin and flyer spinning mechanism.

the spindle, but drying has to be completed by winding the yarn into hanks which are then suspended inside a heated room.

5.6.5 Cap spinning

In cap spinning, the yarn is guided on to the bobbin by the edge of an inverted metal cap which is supported round the bobbin. The bobbin is rotated and the yarn licks round the edge of the cap to insert twist, and also by reason of the friction which is developed, to adjust the take-up speed. The action is, therefore, similar in principle to that of ring spinning with the traveller replaced by the edge of the cap.

5.6.6 Tow-to-yarn spinning

Reference was made in ch. 4 to attempts which were started round about 1930 to go direct from viscose rayon tow to spun yarn in an attempt to eliminate the cutting into staple fibres followed by the carding, combing and drawing operations. The principle of the methods which were devised was to extend the tow beyond the elastic limits of the filaments composing it so that breakage at random positions occurred, reduce the thickness of the broken tow by the desired amount by drawing, and then twist the attenuated tow to form a yarn.

After many years of development work, two practical processes were evolved. The first type is known as a direct spinning method, in which the tow is reduced to staple fibres by drawing and then immediately twisted to form a yarn, all in one operation. The principle of the method is shown in the diagram. The tow is first tensioned and then passed over a top back roller and round a middle back roller. It is then carried by a leather belt to a condensing device and after passing through this it goes into the nip of the front rollers, and from there to a yarn twister. The leather belt is carried by a

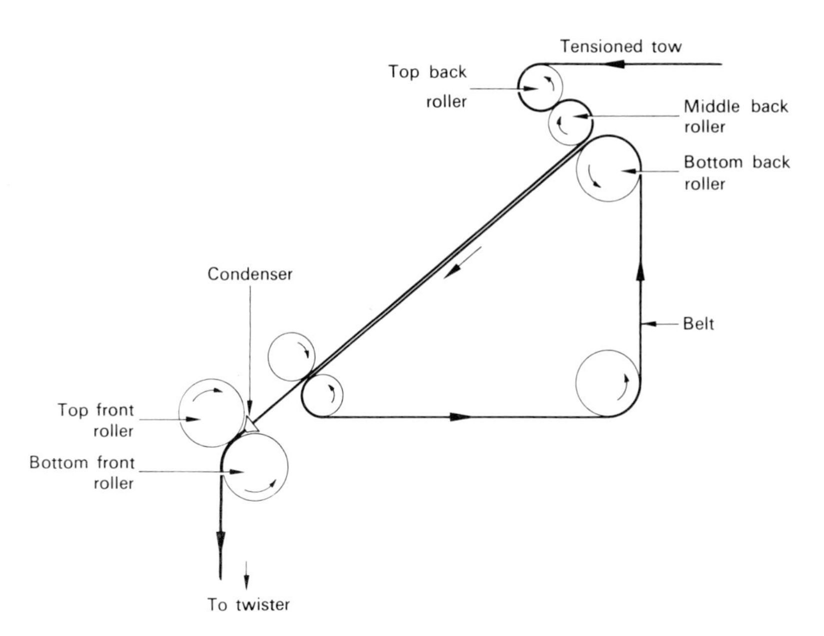

Fig. 5.27 Detail of direct tow to yarn spinning.

roller system as shown and one of these rollers acts as the third member of the back roller system. The back rollers, and the front rollers, are operated under considerable nip pressures to obtain a tight grip on the tow, special surfaces being employed to avoid slippage.

In operation, the front roller system rotates at a higher speed than the back roller system and the tow is thus subjected to considerable stress which causes each filament to break randomly at its weakest point in between the two roller systems and also draws the broken tow. The tow, now weakened by breaking and drawing, is supported in its travel by the leather belt. Once the tow has passed through the nip of the front rollers it is immediately subjected to twisting.

The yarns made by this method are characterized by having a high shrinkage when first wetted, and this is usually of the order of 11%. The reason for the high shrinkage is that the fibres have been stretched to their limit during the preparation of the yarn, and they retain this extension until first relaxed in water. This high shrinkage is used to obtain fancy effects in fabrics by combining such a yarn with a low shrinkage yarn and afterwards wetting out the fabric.

The second method can be used to process heavier tows than the one just described. The tow is again extended between sets of rollers running at different speeds and, in the final stretching section, it passes between two additional rollers which have intermeshing 'breaker bars'. As it passes between the bars of these additional rollers, the tow is sharply kinked, and the extra extension at this point causes random filament breakage. The broken tow is then fed into a stuffer-box where crimp is inserted in the staple fibres into which it has been divided up, after which it is collected ready for spinning.

5.6.7 Up-twisting

This method of inserting twist into a yarn was first used to twist together the filaments forming a silk yarn, and the process was described as 'throwing'. A number of

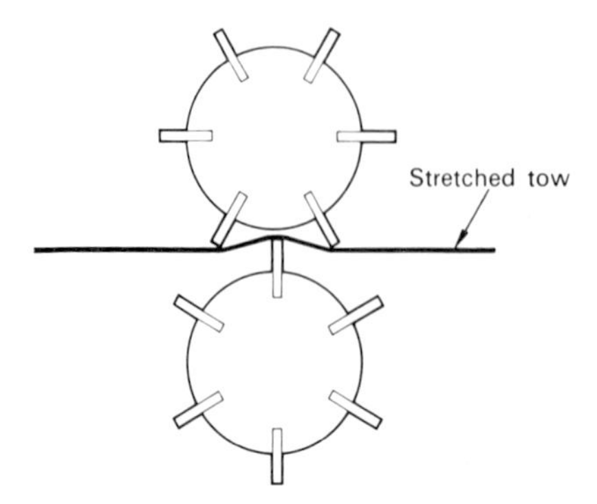

Fig. 5.28 Principle of breaker bars in action.

ends of raw silk were wound on to a bobbin and, in transferring these to a take-up package, the up-twister inserted the requisite number of turns of twist to bind the filaments together. The same type of operation could be used to combine two or more such 'thrown' yarns when a heavier final yarn was required. The up-twister has since been used to process yarns made from man-made filaments and, generally, for the twisting of fine yarns.

The up-twister uses a flyer, in much the same way as does the bobbin and flyer spindle but, in this case, the yarn is drawn away from the bobbin through the flyer instead of on to the bobbin as is the case in bobbin and flyer spinning. The bobbin, on which are wound the untwisted filaments, is fixed on a vertical spindle and the two rotate by reason of the belt drive applied at the bottom of the spindle. At the top of the spindle is a freely-rotating light-weight flyer made from wire with a guide at the end of each arm. Above the spindle and its flyer, is a horizontal spindle carrying a bobbin and with a traversing guide to distribute the on-coming yarn evenly over the surface of this take-up bobbin. The bunch of filaments is led from the lower bobbin, through both eyes of the flyer, and up through the necessary guides to the horizontal take-up bobbin.

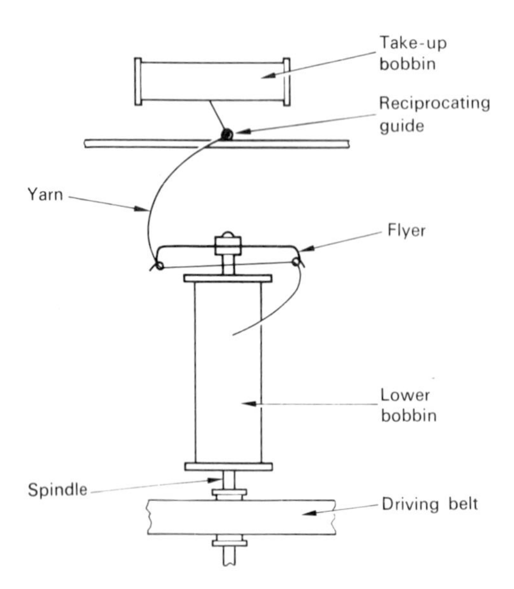

Fig. 5.29 The up-twister spindle action.

In operation, the lower bobbin carrying the untwisted filaments has a higher rotational speed than the take-up bobbin and the filaments are literally 'thrown' off the bobbin surface faster than the upper bobbin can receive them. This would cause a certain amount of slack to be produced, and this slack is taken up by the flyer which is obliged to rotate. The operation inserts twist in the filament assembly before it reaches the take-up bobbin, the amount depending on the difference in speeds between the upper and lower bobbins.

5.6.8 Doubling

The type of yarn provided by the various kinds of spinning systems which have been described, have their component fibres all twisted about a common axis. Such yarns are referred to as 'singles' yarns. For a variety of reasons, it may be desirable to twist together two or more of such singles yarns to produce a combined 'doubled' yarn. The reasons for doubling may be to obtain an extra uniformity, the thick places in one balancing out the thin places in the other, or to obtain a smoother, stronger yarn than a singles one of equivalent count, or to combine different singles and doubling twists and so obtain various changes in appearance and properties of the final fabric.

To double two or more yarns together any of the spinning systems which have been described may be used without, of course, using the drafting facilities provided. It is usual to apply the doubling twist in the opposite direction to that present in the singles yarns since, by that means, the various twists become 'balanced' and by suitably selecting the respective twist values, a stable combination showing no inclination to untwist may be obtained. When the singles and doubling twists are in the same direction, the combination is by no means stable, and such yarns are made deliberately so that, when forming part of a fabric, their 'twist-liveliness' causes them to distort and give special crêpe fabrics.

Doubled yarns are expensive to produce and are only used when an increased cost can be justified. Doubling is an extra process and this puts up the cost of producing the yarn. In addition, two fine singles yarns are required to produce a final yarn of medium thickness, and fine singles yarns are always more expensive to produce than coarser ones.

It may be mentioned here, that there are applications, sewing threads for instance, where two or more doubled yarns are combined by a further twisting. Such yarns are often referred to as 'cabled' yarns. In preparing these yarns, the conditions can be selected to give high strength and uniformity combined with little tendency to extend under pull.

5.6.9 Development work in progress

The yarns produced by existing methods of spinning are of excellent quality and well suited to the final applications for which they are intended. The processes applied in their preparation represent something like two centuries of experience and, over the years, small changes have been incorporated in them to produce the yarns which have proved so satisfactory in practice. The spinning of a yarn is, however, a relatively slow process, involving quite slow preparation of the fibre arrangement and the lengthy operation of inserting the necessary amount of twist.

Development work on spinning is being carried out in the major textile-producing countries, and an examination of the patents relating to the subject which have been filed in recent years shows that it has been concerned with a speeding up of yarn preparation, bearing in mind the standards of quality and performance to which the user has become accustomed.

The lines upon which the various patented methods are based are varied. All appear to acknowledge one important fact, namely that the speed limitation in the present methods of applying twist to a yarn arise from the necessity of having to rotate the yarn package itself in order to insert the twist. It is necessary to rotate the yarn package through a complete revolution to insert a single turn of binding twist, and the rapid rotation of even a small package calls for the expenditure of a considerable amount of driving power. All the methods being developed try to avoid this by continuously feeding the loose fibres into some type of device which will make them twist round each other and thus join up to form a yarn which can be withdrawn from the equipment at a steady rate.

One method being examined is to feed air carrying the fibres down a side tube into a main tube, the two tubes being joined tangentially. The flow of air is provided by suction applied to the main tube and, as the air and fibre enter from the side tube, they are subjected to a whirling action. This action wraps the filaments round each other and the yarn thus formed is withdrawn from the main tube through a small hole in a direction opposite to that of the flow of air. A high speed twisting action can be achieved by this means but there would appear to be a limiting speed beyond which it is difficult to pass. Experimental work done has not yet produced really compact yarns and there has been quite a substantial loss of fibre in the exhaust air.

In a further method being examined, a stream of air carries fibres into a type of 'cage' which can be rotated rapidly. The air and fibre stream is fed into the top of the cage which is constructed from a number of fine needles downwardly-inclined towards the point of a central needle, running along the axis of the cage, and making light contact with it. The motion of the cage twists the fibres together and the yarn thus formed is withdrawn from the point of the central needle. The fibres are straightened and aligned

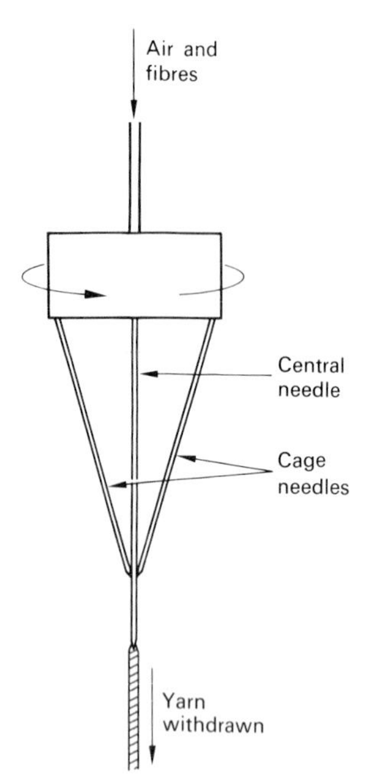

Fig. 5.30 Principle used in the needle cage method of spinning.

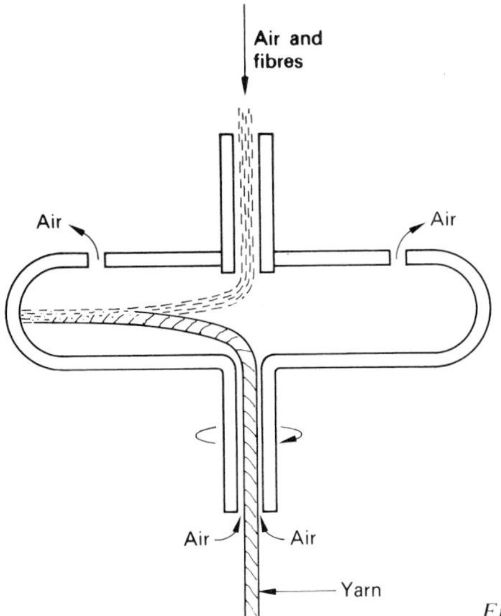

Fig. 5.31 Spinning, using a centrifugal pot system.

as they slide past the needles. The needle cage can be lightly constructed and driven to rotate at very high speeds. Up to the present only 'softer' and hence, weaker yarns than those made by more conventional spinning methods have been produced with equipment of this type.

Substantial progress towards the commercial development of a centrifugal-pot type of high speed spinning method has already been made, and machines have been constructed and are being used. The type of construction which appears to be most favoured at the present time is one in which the fibres are carried by an air-stream and laid on the inner wall surface of a pot which is rotating at high speed. The fibres are held on the inner wall of the pot by centrifugal force created by the spin and, for this reason, the device will not function correctly at low rotational speeds. The fibres are drawn off the inner surface of the pot through a tube lying along the axis of the pot and, in doing so, they are twisted round each other to form a yarn. This method can be used to spin yarns at high speed. The yarns thus produced up to the present have tended to be rather more bulky and a little weaker in strength than conventional ring-spun yarns but the method is very promising and it is likely to be the first of the really high-speed spinning methods to be used on a commercial scale.

5.7 TEXTURED YARNS

In the early days of man-made fibre manufacture, attempts were made to introduce variety into the fabrics, and obtain some of the characteristics of natural fibre fabrics, by cutting up the extruded continuous filaments into short lengths and spinning them into yarns as staple fibres. The development of the synthetic fibres, some of which were capable of being heat-set into a crimped or wavy-configuration, enabled the so-called 'texturizing' processes to be devised and, in some respects, gave to continuous filament yarns the characteristics of staple fibre yarns, as well as properties which cannot be achieved in any other way.

The filaments of untextured synthetic continuous fibre yarns lie straight and parallel in the fabric and it follows that such fabrics will be inelastic, 'lean', lack softness of handle, and only able to retain a little moisture. The various methods of texturizing introduce permanent distortions and waves into the individual filaments so that they now behave more like the fibres in a wool fabric. The appearance and properties of the fabric change considerably. The filaments now refuse to lie flat in contact with each other, so that the fabric appears to become more opaque and bulky, and has developed a much greater softness when handled. The crimps in the filaments add flexibility to the fabric so that it drapes and hangs better, and, by providing innumerable points of contact between filament and filament throughout the fabric, moisture can be held much more strongly by capillary attraction. By making a fabric from crimped yarns, a tremendous increase in its extension and recovery is imparted to it, in marked contrast to the 'deadness' of straight filament fabrics. All these effects can be varied in a controlled manner by selecting the amount and type of crimp in the filaments, and texturizing provides an extra degree of freedom in designing the final fabric.

Textured yarns can be divided roughly into two general types, 'stretch' and 'bulk' yarns. Stretch yarns are produced to impart the maximum degree of elasticity to the garment made from them so that it will cling to the body of the wearer whatever his or her movements may be. Thus ladies' hose and tights, men's half-hose, and swimwear are excellent applications for stretch yarns. The bulk yarns again usually provide elasticity in the garment, but they are made with softness and warmth of handle primarily in mind. They are very suitable for use in the manufacture of all types of knitted garments.

There are various methods of producing textured yarns, all having their own particular characteristics, and these are described in the following sub-sections.

5.7.1 Crinkle yarns

One of the early methods of texturizing produced what are sometimes referred to as 'crinkle' yarns. When a piece of knitted fabric is unroved, the yarn taken from it retains, at least for a time, the crinkles which represented the shapes of the knitted stitches in the fabric. If the fabric has been knitted from a synthetic continuous filament yarn and subsequently heat-set, then the crinkles will be present permanently in the unroved yarn and a type of textured yarn will have been obtained. A texturizing process was based on these operations, special machines being devised to knit long tubes of fabric which could then be heat-set and unroved. More recently a machine has been designed by John Heathcoat & Co. which continuously folds up the yarn in crimps, which correspond to the knitted loop shape, and heat-sets the crimps in the yarn before taking it up as a package suitable for direct knitting. One point must be

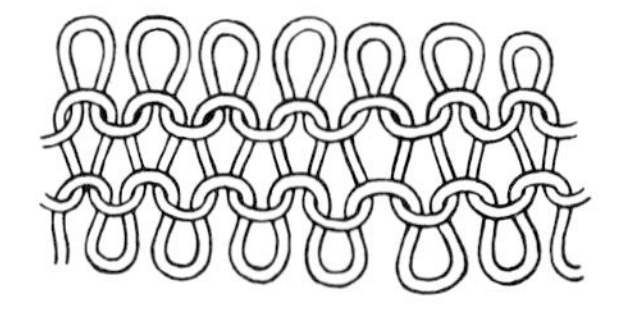

Fig. 5.32 Preparing a crinkle yarn.

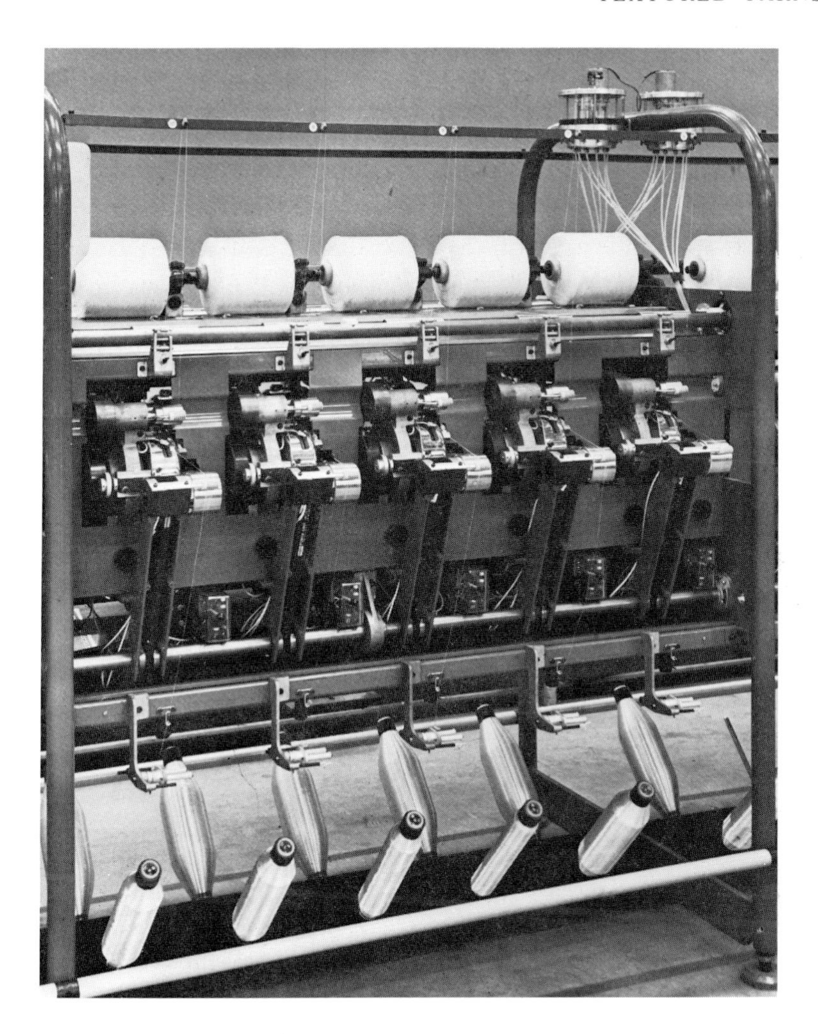

Fig. 5.33 A crinkle machine.

watched; the knitting machine which will eventually be used to produce fabric from such a yarn must construct a stitch size which is quite different from that used to make the crinkle yarn. If the two sizes are more or less the same, then the final fabric will be very uneven and show what appear to be holes or unfilled spaces.

Originally, the crinkle yarns were intended for the construction of the welts of ladies' fine gauge hose to provide an additional extension and recovery in this part of the garment. Interest in the yarn for this application began to fall off but, more recently, crinkle yarn manufacture has been revived to satisfy a demand for its use in knitted outerwear and dress fabrics. Here, the interest is not so much in the stretch properties of the yarn, for which the process was first devised, but in the bouclé type of appearance which it gives to the fabric. It should be noted that, in a crinkle yarn, the yarn as a whole is crimped and the continuous filaments still remain in their original relative formation. There is, therefore, no loss of lustre and, in fact, the tops of the crimps provide light-reflecting spots which give an added sparkle to the fabric. Crinkle yarns

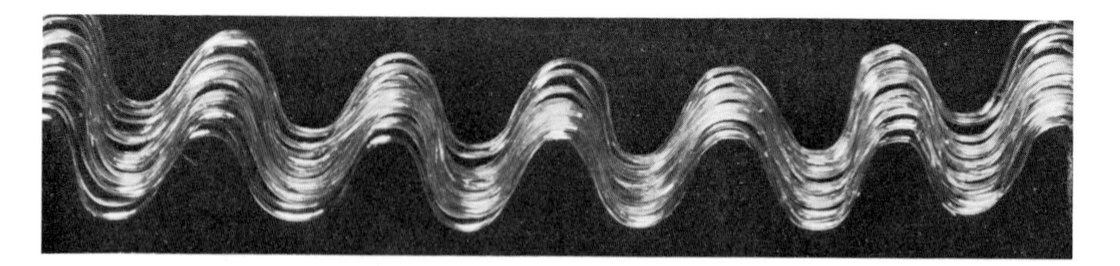

Fig. 5.34 A sample of crinkled yarn.

have been produced from yarns having trilobal cross-sections to enhance the sparkle in the fabrics still further.

5.7.2 Air-crimped yarns

An interesting process, which was originally devised by the DuPont Company, is the use of an air blast to produce a type of bulked yarn known as a 'Taslan' yarn. The filament yarn is fed into a specially-constructed air nozzle where it is subjected to turbulent flow conditions which separate the filaments and whip them about. When the yarn leaves this nozzle, some of the filaments stand out from the main body of the yarn in a 'snarled' or twisted condition and increase the bulkiness of the yarn. The treatment simply increases the bulk and covering power of the yarn without giving it any increase in extensibility. Such a yarn can be woven satisfactorily to produce fabrics having a degree of opacity suitable for a number of applications.

It is not easy to prepare completely uniform yarns using the air-crimping technique and, in addition, the provision of large quantities of compressed air is expensive. Development work has continued with these two points in mind, and new types of jet which are more controllable and, at the same time, are more economic in the use of air under pressure have been devised. Work has also been carried out on the use of the method as a means of intimately combining two or more yarns fed into the jet at the same time to produce novelty effects. Changes in the relative rates of feed of the various yarns, together with controlled variations in the crimping conditions, allow a whole range of fancy effects to be obtained.

5.7.3 Knife-edge crimping

If a heated thermoplastic filament, or assembly of filaments, is pulled with a sharp bend round an edge, the portion of filament in contact with the edge will suffer compression, and the other side of the filament remote from the edge will be extended. When the filament cools down, the stresses introduced into it by this treatment at the edge are 'frozen' in place, and it then requires only a moderate subsequent degree of heating to allow them to act, if the filament is in a free condition. They produce, in effect, a distorted filament with one side being appreciably longer than the opposite side, and the only stable form which the filament can take up is a spiral. If a multi-filament yarn is treated in this way, and is heat-treated subsequently, each filament will develop a helical crimp and provide considerable elasticity and bulkiness in the yarn.

The edge-crimping, or 'Agilon' process, was developed independently by British Nylon Spinners Ltd., in this country, and Deering Milliken Research Corporation in the USA, and both companies reached an agreement regarding the exploitation of the

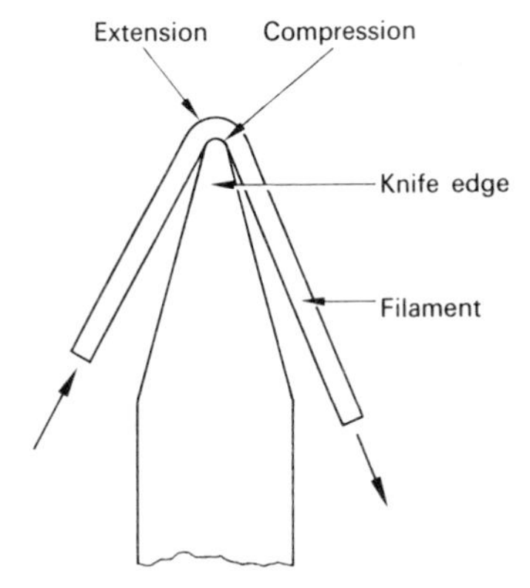

Fig. 5.35 Knife-edge crimping.

method and the use of the name Agilon. Both companies have carried out a considerable amount of development work on the process. The main outlet for Agilon yarn has been in the manufacture of fine gauge stretch stockings and tights. The ability of the process to be applied to fine denier yarns, and the fact that the crimped yarns produced are balanced and torque-free, makes Agilon yarn particularly suitable for such applications.

5.7.4 Stuffer-box crimped yarns

The stuffer-box method of crimping synthetic continuous filament yarns was developed by the Bancroft Corporation in the USA. The thermoplastic yarn is fed by means of a pair of rollers, which grip the yarn, into a heated tube or stuffer box. At the outlet end of the tube is a plug, the weight of which restricts the escape of yarn from the tube. Due to the entering yarn acting against this restricting weight, the yarn is caused to fold up concertina-fashion along the length of the tube and the heat applied to the tube causes the filaments to become set in this saw-tooth form. By reason of the backing pressure provided by the yarn continuing to be supplied from the feed rollers, the plug at the top of the tube lifts slightly to allow some of the crimped yarn to escape from the tube, when it can be wound up on a package. It will be appreciated that the downward pressure of the plug can be selected to ensure a steady flow of yarn along the tube at a rate which will provide the correct form of crimp, satisfactorily heat-set, in the yarn. The yarn produced by this method is bulky, has a good extensibility and recovery, and is balanced and torque-free.

Originally, the stuffer-box process was applied to one denier of nylon yarn only. Further development work has resulted in improvements being made to the process and a wider range of deniers of nylon and other thermoplastic fibre materials can now be treated. The yarns are used to produce a whole range of knitted fabrics and garments.

Two other types of crimping machines, based on the stuffer-box method, have since been devised. One is the 'Spunize' processed developed by the Spunize Company of the USA, which processes a whole sheet of yarns from a warp beam or creel at one time through a type of stuffer box. The other is the 'Pinlon' process developed by the Klinger Manufacturing Company in Britain, in which a single thermoplastic yarn is compressed into a stuffer box of modified design, and which operates at high speed. Both processes were devised with the crimping of heavy denier yarns, such as are suitable for carpet manufacture, in mind.

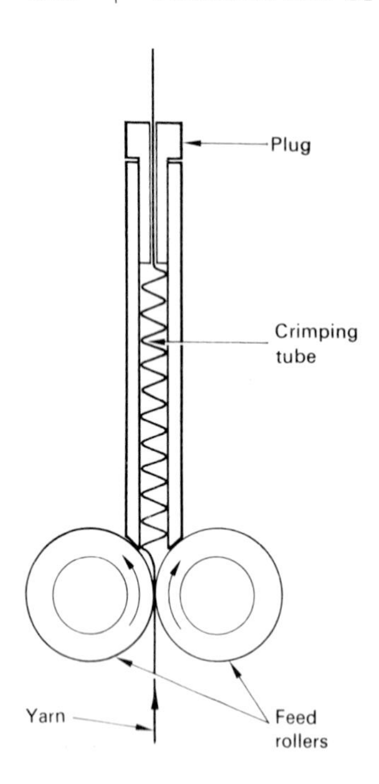

Fig. 5.36 Stuffer-box crimping.

5.7.5 Conventional twist crimped yarns

This was one of the early methods of producing stretch yarns. The continuous filament yarn was first highly twisted and then heat-set in that twisted condition. By this heat-setting, the tendency for the yarn to untwist was 'killed'. The set yarn was now untwisted by an equal amount in the opposite direction, the result being that the filaments were untwisted once again as at the commencement of the processing but with a tendency to get back to their twisted state. In this condition, the individual filaments stood out from the body of the yarn to give additional bulk and, if the yarn was released, it immediately contracted by reason of the filaments trying to take up the twisted form in which they were set. The contracting force for a yarn of even moderate denier was quite strong if the heat-setting had been carried out satisfactorily.

Being so twist-lively, these yarns are difficult to handle during their conversion into fabrics, and the tendency to return to the twisted state is so strong that there is a danger of the fabric made from them becoming distorted. It is necessary, therefore, to balance the yarns by lightly folding together two yarns having twist liveliness in opposite directions, the initial twists inserted during their preparation being in one direction in one of the yarns, and in the opposite direction in the other. Alternatively, single yarns can be knitted but during the knitting of a piece of fabric one row of stitches must be formed from yarn with twist liveliness in one direction and the next row from yarn of opposite twist liveliness, and so on throughout the fabric to effect a balance.

The conventional twist process was a slow method of producing a stretch yarn. The insertion of such a high initial twist into a yarn required a considerable amount of time. The packages of twisted yarn had then to be removed from the twister and subjected to heat-setting. Finally, the set yarn had to be brought back to the twister for an equal amount of twisting in the opposite direction before being wound on to a package ready for dispatch. The yarns made by this method were, however, of excellent quality and imparted powerful stretch properties to the fabrics made from them. The method is still used on occasions to prepare yarns for swimwear and for the manufacture of a type of medical stocking known as a 'support hose'.

5.7.6 False-twist crimped yarns

It was discovered that the sequence of operations necessary to produce a conventional twist stretch yarn could be carried out in a continuous manner by making use of what has become known as the 'false-twist' technique. In this method, the yarn to be treated is run between two sets of rollers, the first set feeding the yarn into the crimping zone, and the second set taking the yarn from the zone and delivering it for winding on to a package. In between the sets of rollers the yarn passes through a twist tube or spindle which is being driven at a high rotational speed. By virtue of this rotation the length of yarn between the feed rollers and the spindle becomes highly twisted in the direction of rotation of the spindle. This twisted length now moves forward and when it reaches the delivery side of the spindle, the spindle rotation takes out the twist which it had inserted on the approach side. Thus, by using one direction of rotation only of the twist spindle, the yarn is first highly twisted in approaching the spindle and then untwisted on leaving the spindle and, if a heater is inserted in the approach section, the twist can be set in the yarn and the sequence of operations used to make a conventional twist crimped yarn can thus be carried out in one single continuous operation.

Since the false-twist technique was first introduced, a considerable amount of development work has been carried out, mainly with the aim of speeding up the process and obtaining processed yarn which is uniform from spindle to spindle. The early twist spindles were straightforward tubes driven by belts and were capable of operating

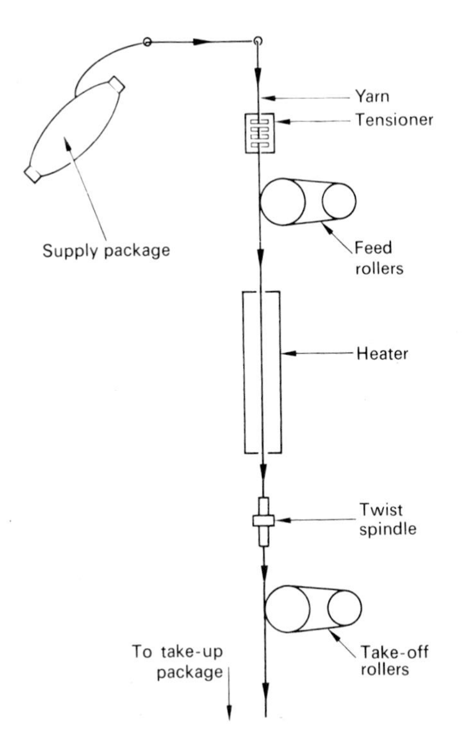

Fig. 5.37 False-twist crimping.

at speeds of up to 20000 rev/min. This enabled yarns to be processed at a linear speed of the order of 20 ft/min. (6.0 m/min.). Improved bearings were designed and enabled speeds of up to three times this value to be obtained. Finally, the idea of rotating the spindle by allowing it to make contact with a rapidly-driven large diameter wheel, or wheels, where it was supported in position by magnetic attraction, was evolved. The modern false-twist spindle machines use this idea, in the main, and speeds in excess of 500000 rev/min. are now possible. It should not be imagined that the achievement of higher and higher spindle speeds has been the only problem in developing the present false-twist spindle machines. High speed twisting is of no value unless the twist so developed can be heat-set in the yarn. Longer and more efficient heaters have had to be developed without increasing the height of the machine beyond that which enables the operator to perform his duties without difficulty. Other problems, such as easy threading-up of the machine, the use of large yarn packages, and providing the maximum number of twist spindles in a machine of a given size, have been met and studied. The modern false-twist crimping machine operates speedily, reliably and economically, and the majority of textured yarns now used are prepared by this method.

The use of false-twist yarns imparts a high degree of extensibility to a fabric knitted from them and there are applications where so great elasticity is important. There are others, however, where softness and bulkiness are more important than high extensibility, and suitable yarns can be prepared by the false-twist technique by applying a second, partial heat-setting to the final yarn to reduce the amount of twist-liveliness present and set the filaments in their open, separated condition. This has been done very successfully with polyester yarns by loosely winding the twist-lively yarn on to a package and then subjecting this package to heat-setting conditions. Most of the makers of false-twist machines are incorporating a second heater stage in their designs so that the twist-lively yarn which is produced can run over a heater as it travels towards the package-forming mechanism. By this means the elasticity of the yarn is reduced and bulkiness developed in one continuous process. Whether such a continuous process can be so controlled as to give results which are the equal of those obtained by using a separate second-setting operation, remains to be seen.

5.7.7 Self-crimping yarns

A type of self-crimping, or rather self-bulking, yarn has long been in use. In it, staple fibres consisting of two types of material, or two differently treated fibres of the same material, are spun together. The two types of fibre in the material differ with regard to the amount they will shrink when the composite yarn is subsequently subjected to some type of heat treatment as, say, during the finishing operations applied to a garment. If one of the components shrinks markedly, then the other, less-shrinkable component will buckle and stand out from the body of the yarn as contraction takes place, thereby providing an increased bulkiness and softness. If one type of material only is used, the differences in shrinkage are provided by drawing one set of fibres more than the other during manufacture, the greater the amount of draw used, the more the shrinkage on subsequent heating.

Self-crimping filaments are now being produced by extruding two different materials through special spinnerets at the same time, the important point being once again that one material will shrink more than the other when the yarn containing such filaments is later subjected to a heat treatment. The two materials can be extruded so that two filaments, joined together side-by-side are formed, or in such a way that one material partially or wholly surrounds the other in the same filament. When heat is applied, the differential shrinkage causes the filament to take up a spiral or folded form, and a number of such filaments in a yarn produce an increase in bulk and softness of handle.

An important point to notice about bi-component filament yarns is that they are prepared by the fibre manufacturer himself as part of his normal production. This could represent a pattern for the future in the man-made fibre industry, namely the supply of potentially-bulkable yarns by the fibre manufacturer without the need to apply any further crimping treatment before the yarns are converted into fabrics.

5.7.8 Development work in progress

Some progress has been made towards the development of textured yarns based on twist but which differ from the twist-crimped yarns already mentioned. The most promising of the machines capable of producing such yarns is the 'Turbo Duotwist' machine developed by the Turbo Machine Company of the USA. In this machine, two single yarns are twisted round each other and are heat-set in that condition by being passed over the surface of a heater. The two yarns are then separated and wound on to two separate packages, each yarn having developed a helical crimp. In operating the machine, the two yarns are twisted round each other over a certain length and, as the two set yarns are pulled apart, the short length of twist runs back into the oncoming yarns, thus enabling a continuous process to be operated simply by separating the two yarns. Initial difficulties were mainly associated with the wandering about of the point of separation, leading to instability in the system, but ingenious methods of overcoming this have been devised and a number of machines based on the principle are in use.

Methods are also being investigated whereby a whole 'sheet' of yarns in a warp formation are given a texturizing treatment at the same time, and others in which a 'microcreping' treatment of crimping and heat-setting is given to a fabric itself without the yarns which compose it having been previously subjected to a crimping process. Both

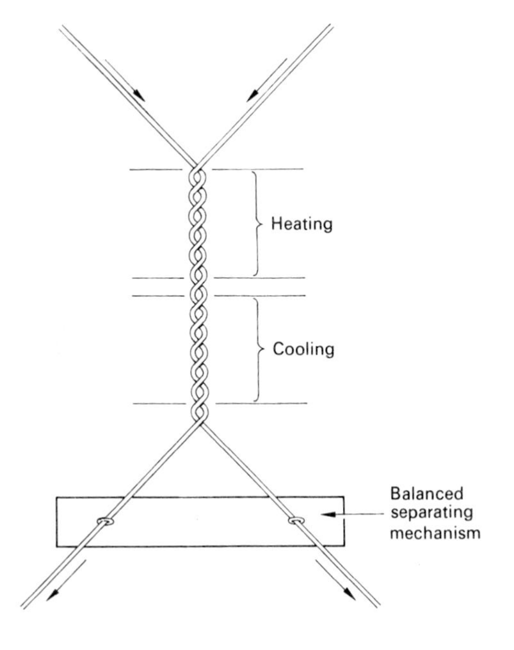

Fig. 5.38 Duo-twist crimping.

these methods are claimed to give new effects in the final fabric and capable of being varied to provide various fabric appearances.

5.8 FIBRE BLENDS

It has long been the custom to blend or mix together, natural fibres. In the preparatory processes carried out before spinning a cotton yarn, for instance, it is usual to take material from a number of bales, a little at a time from each, and mix them together thoroughly during processing. The mixing minimizes the effects of the differences between the materials in the various bales to give final yarns which are more uniform in both properties and appearance. Blending of natural fibres is also carried out for economic reasons. Cheaper types of fibre may be blended with more expensive ones to reduce the overall cost of the yarn and, in some cases, eke out the usage of the better class material. These considerations are important and emphasize the need for blending in the cases of both cotton and wool. In addition, two dissimilar natural materials, such as wool and cotton, have long been used in blends to produce a cheaper yarn and yet one which retains many of the characteristics of the more expensive of the two components.

Natural fibres are often blended together to obtain some particular technical or aesthetic effect. Wool is often mixed with the rarer, and more expensive animal fibres, such as vicuna, to provide the felting properties which the rarer fibre lacks and thus produce a firmer, more stable, final fabric. Wool blends intended for the manufacture of blankets may contain long lustrous fibres together with a shorter wool having good milling properties, the former giving a good pile, and the latter adding 'body', in the final blanket. A carpet yarn may be a blend of strong, springy fibres, combined with finer crimpy fibres to contribute a greater covering power. Fibres dyed to different colours may be spun together to give a mixture effect in the final fabric.

When man-made fibres became available, blending was again used to gain economic and performance advantages. In the early days of man-made fibres, blending was sometimes overdone and unsuitable, and unnecessarily complicated, combinations were devised. These were short-lived and, since that time, a few well-tried mixtures have survived and are still used.

The development of man-made fibres, and particularly the synthetic fibres, offered many fibre characteristics and properties which are not obtainable in any of the natural fibres. At the same time, these particular characteristics of the man-made fibre were not associated with some of the other properties which are present in the natural fibres and which have come to be valued for certain textile applications. Thus, while some of the synthetic fibres have tremendous strength and resistance to abrasion, factors which make for a long wear-life in a garment, they lack the soft handle and comfort in wear provided by the natural fibres such as wool and cotton. An obvious development was the blending of natural and man-made fibres in an attempt to combine the advantages of the two materials and cancel out their defects. Some combinations of natural and man-made fibres have been very successful both from the point of view of cheapening a yarn and thus widening its field of application, and also in providing improved technical effects. The rayons have been used to blend with certain natural fibres to enable cheaper, and yet serviceable, fabrics to be produced. Combinations of wool and nylon have provided improved resistance to abrasion and wear while still retaining the comfort in wear of the natural material. Cotton/polyester and wool/polyester blends have provided comfort in wear combined with 'easy care' characteristics and permanent pleating properties. It would be possible to compile an impressive list of blends of natural and man-made fibres which have been tried and proved successful in practice.

There are three main methods of blending materials together which have been used:

1. The materials are mixed together in loose staple form during the preparative processes for spinning yarns. Satisfactory results are obtained, however, only if certain factors are taken into account. The staple lengths of the two components should be roughly equal if satisfactory rovings and final yarns are to be prepared. In addition, the fibres from the two sources should have nearly the same diameter. As we have seen, staple fibres are subjected to carding, combing and drawing processes before the final twist to form the yarn is inserted. Unless equalities in length and diameter of the fibres are provided, the mixture will not behave as a homogeneous material and migration and separation of the components may take place. Very large quantities indeed of cotton and viscose rayon fibres are blended together, and the rayon manufacturers supply materials of various staple lengths and fibre diameters, which can be selected to be compatible with the various grades of cotton. A third factor is that the two components in the blend must be present in the correct proportions, as has been shown by experience. Uniformity of staple lengths and fibre diameters allow good yarns to be spun, but other factors also contribute to the strength of the final yarn. There have been cases where only a small proportion of a strong fibre has been combined with a weaker one and has had the effect of increasing the ease with which the fibres can slide one over the other, and has, in fact, reduced the overall strength of the yarn.
2. A mixture yarn can be prepared by doubling together two yarns made from different materials. For a successful mixture to be prepared in such a way, the two singles yarns should approximate to each other in character. For instance, a hard, abrasion-resisting nylon yarn doubled with a soft wool yarn can cut into the wool by reason of the movement of the two component yarns relative to each other during wear; in the early use of natural/synthetic fibre combinations, wool yarns doubled with nylon yarns to increase the resistance to the rubbing action of the shoe at the heel of a man's sock, were often unsatisfactory for this reason.
3. Finally, two types of fibre material may be combined by weaving yarns prepared from them into the same piece of fabric. Very often, the warp is formed from a yarn of one material, and the weft from the other. Quite interesting effects can be obtained by this method of combination. Substantial differences in the shrinkage characteristics of the two yarns can be arranged to give a crêpe effect on subsequent heat-treatment, and novelty colour effects arising from the different dye affinities of the two materials can also be obtained.

5.9 LITERATURE

G. R. MERRILL. *Cotton Carding*, 3rd Edn., Ann Arbor, Michigan, Edwards Bros., 1955.
G. R. MERRILL. *Cotton Combing*, 3rd Edn. Ann Arbor, Michigan, Edwards Bros., 1955.
G. R. MERRILL. *Cotton Drawing and Roving*, Ann Arbor, Michigan, Edwards Bros., 1956.
W. E. MORTON and G. R. WRAY. *Introduction to the Study of Spinning*, 3rd Edn., London, Longmans, 1962.

G. R. WRAY. (Ed.). *Modern Yarn Production from Man-Made Fibres*, Manchester, Columbine Press, 1960.

W. A. HUNTER and C. SHRIGLEY. *Opening and Cleaning; Manual of Cotton Spinning*, Vol. 11, part 11. London, Textile Institute and Butterworths, 1963.

J. K. CLEGG. *Practical Cotton Carding*, Manchester, Columbine Press, 1958.

T. F. GRIFFIN. *Practical Worsted Combing*, London, National Trade Press, 1953.

T. F. GRIFFIN. *Practical Worsted Carding*, London, National Trade Press, 1957.

A. E. DEBARR and H. CATLING. *Principles and Theory of Ring Spinning: Manual of Cotton Spinning*, Vol. V, London, Textile Institute and Butterworths, 1956.

G. A. R. FOSTER. *Principles of Roller Drafting: Manual of Cotton Spinning*, Vol. IV, part 1, London, Textile Institute, 1958.

S. A. G. CALDWELL. *Rayon Staple Fibre Spinning*, London, Emmott, 1953.

A. BREARLEY. *Worsted*, London, Pitman, 1964.

J. W. RADCLIFFE. *Woollen and Worsted Yarn Manufacture*, London, Emmott, 1950.

H. WALKER. *Worsted Drawing and Spinning: Part 1, Drawing*, London, Textile Institute, 1954.

CHAPTER 6

Production of Fabrics

6.1 PREPARATORY PROCESSES

Before a yarn, as it is received from the spinner, can be used to make a fabric, it often has to be subjected to one or more preparatory processes. These processes include the arranging of the yarns in a form which will make them easier to use in fabric manufacture, such as the winding on to cops or cones, giving them special treatments designed to produce particular effects, or forming them into 'sheets' or warps which, as will be seen later in the chapter, are essential for carrying out a number of the methods of fabric formation.

6.1.1 Reeling

The reeling process consists in winding the yarns from packages, such as bobbins or cheeses, on to a rotating frame, usually of wire and capable of turning about a central axis, which is known as a 'swift', from which the yarn can then be removed as a hank. Reeling is used for yarns which are to be wet processed, by such treatments as scouring, bleaching or dyeing, as an alternative to carrying out these treatments in a less satisfactory manner while the yarn is in package form. Since the yarns in a hank are free to receive the full treatments, the wet processes can be carried out with uniformity and with less risk of faults developing. Reeling is also carried out to prepare hand knitting yarns in a suitable form for sale. Reeling is an ancient art used for centuries in the preparation of skeins, or hanks, of silk. The machines which have been devised for reeling other textile yarns are based on the methods used in silk reeling.

The winding cages, or swifts, are usually supported near the top of the reeling machine, with the packages from which the yarn is being taken fixed on a creel near the base of the machine. The yarn is attached to the swift which, as it rotates, pulls it up under an accurately controlled tension from the supply package, and through a traversing guide which distributes it evenly across the width of the swift. If no wet processing is to be carried out on the hank of yarn, then a normal traverse is used. Here, the yarn is only traversed intermittently in one direction across during winding so that a series of parallel hank portions are formed. The idea behind this is to avoid any appreciable variation in the length of each wrap which goes to build up the hank.

If the hanks are to be wet-processed, when they may be agitated quite appreciably, there would be a risk of the individual yarns becoming entangled if a normal traverse

Fig. 6.1 Doffing a continuous-type spinning machine producing high tenacity rayon yarn. Here, full bobbins are removed and replaced with empty bobbins.

were used. In such cases, it is usual to employ a rapid traverse backward and forward across the whole width of the hank, it being common to traverse the whole width for each revolution of the swift. By doing this, each succeeding turn will be laid at an angle across the preceding one and the risk of entanglement during wet processing will be reduced. Hanks intended for wet processing are tied to hold the individual windings in position. It is usual to use four tie-bands spaced round the hank, and these bands are normally of coloured yarn which are interlaced in and out across the hank to divide it into a number of sections, say four or five. The two ends of the lacing yarn are tied together in a way which leaves the band slack so as not to interfere with the wet processing at that point. The starting and finishing ends of the yarn in the hank are also tied in such a way that the joint will be found easily.

The swifts usually have a telescopic construction so that their diameters can be adjusted to reel off different sizes of hank; reducing the swift to a smaller diameter after reeling also enables the finished hank to be removed easily. A swift must be light in construction, since it will be whirled round at speed, but it must also be sufficiently rigid not to bend under the combined tensions of a considerable number of wraps of yarn, as this would produce differences in hank size.

The control of yarn tension is important if hanks which are uniform in length and which give level results are to be obtained. Gate or disc tensioners are built into the machine, and a number of the machine makers have developed their own devices for giving more precise tension control. Efficient stop motions, to stop the head when a yarn breaks or tension is reduced beyond an acceptable amount, are also incorporated in the machines, and counters which can be set to stop the head when a pre-determined length of yarn has been wound into the hank are often included.

With such devices forming part of the reeling machine, the operators' time need not be expended on adjusting the reeling conditions, but in tying, doffing and replacing supply packages. With these operations in mind, the number of swift heads for which an operator is responsible should be carefully balanced with the reeling speed being used, to obtain the most efficient production.

A reeling machine does not usually consist of more than 40 reeling heads and, in Great Britain, the heads are generally in multiples of 12. The swifts operate at speeds which are usually between 250 and 500 rev/min. Each swift can be stopped independently of the others, and is ordinarily so arranged that, in the stationary state, it can be swung down to a convenient height at which the tying and doffing can be carried out.

When the hanks have received their wet-processing treatments, they have to be wound back into package form before they can be used for fabric manufacture. The swift carrying the processed hank is suspended in brackets attached to the winding machine so that it can rotate freely. The driven bobbin on which the yarn is to be wound supplies the pull on the yarn, and the latter is transferred from the swift to the bobbin.

6.1.2 Coning

Many fabric manufacturers prefer to receive their yarns wound on to cones. The cones are usually made of compressed paper with their outsides smooth and highly polished except for deliberately roughened portions which are provided as a help to keeping the yarn wound on them in place. The fact that the yarn package surface is conical helps the yarn to draw off the cones easily and smoothly at high speed.

The performance of the additional operation of coning adds to the cost of the yarn but many fabric manufacturers, and particularly knitters, consider it to be justified for the trouble-free subsequent operations it allows. Not only does the yarn come away freely from the cone but other features can be given attention during the coning operation. The yarn can be passed through a slub-catcher set to hold the yarn when a knot, previously tied, or a thick place is encountered. The knot which must be made to join the yarn at such points should be as neat and small as possible and placed on the nose of the cone so that it will not be caught by the yarn being taken from the cone during fabric production. The coning operation also provides an opportunity for oil to be applied to the yarn when this may be required subsequently. Good coning should allow free delivery of the yarn without any plucking, a uniform package density which is firm enough to withstand handling but carries the yarn in a relaxed condition, a uniform application of oil right through the package, all knots on the nose of the cone, and a tail at the base of the cone which can be attached to the start of the next cone to provide a 'magazine' feed.

Coning is a critical operation requiring close attention and control. Coning machines are precision built with devices incorporated to help in exercising close control. The machines are usually built in sections which can be bolted together and operated by a single motor. They are usually supplied with oiling rollers over which the yarn runs. The yarn path is normally as straight as can be arranged.

The packages from which the yarn is being taken for winding on to the cones are usually in a creel near the base of the machine. The oiling attachment consists of a stain-

Fig. 6.2 An automatic cone winder.

less steel roller turning in a trough partly filled with coning oil which is easily scourable. A thin film of oil is picked up by the roller and transferred to the yarn running over its surface. The oiling rollers are driven from the main shaft through a reduction unit and the amount of oil transferred to the yarn is regulated by the speed with which the oiling roller is allowed to revolve.

The yarn tension is usually regulated by means of a gate tensioner, although special tension controllers are sometimes provided. The slub-catcher is usually fixed above the tensioner and its blades are adjustable to suit the various yarn diameters. The running yarn holds up a dropwire which is counter-balanced to avoid adding tension to the yarn. When the yarn breaks, or becomes too slack for any reason, the drop-wire falls and operates a stop motion to close down that particular winding head. It is also possible to have a size-stop device which stops the head when a cone of pre-determined size has been wound.

The yarn is distributed uniformly over the cone surface by means of a traverse motion which gives a reciprocating action to a moveable yarn guide. The motion is provided by a cam revolving inside a housing which is driven through gears from the spindle rotating the cone. The traverse guide rod is supported in a frame which is pivoted so as to move away from the spindle as the yarn package builds up. The pressure of the traverse frame is distributed uniformly along the length of the cone by means of a pressure roller. By carefully controlling the release of pressure and tension separately, a low density, free-delivery cone can be wound. Two types of cone which can be wound are the pineapple cone and the straight cone. The pineapple cone is produced by the use of an attachment which gradually reduces the length of the traverse as the cone builds up.

Tension control is an important and quite difficult problem in coning. Using fine denier and low twist yarns, it is usually more difficult to maintain a uniform tension throughout the package than with coarser yarns. If the yarn is wound too slackly on the cone, there is a danger of it sloughing over the end during handling. If the winding is too tight, the wound mass will be irregular in shape and the yarn will not draw off properly in use. With yarns such as nylon, a very tight winding can contract the cone so much that it grips the supporting spindle of the machine and the package cannot be removed when completed.

6.1.3 Weft winding

The winding of weft yarns or threads on to the bobbin or pirn, which can be inserted in the shuttle of the loom, must be carried out with precision and accuracy if difficulties during weaving are to be avoided.

Two methods of winding are used. One is to wind directly on to the correct type of bobbin on the ring frame, while the other is to produce a large package in the more normal way and later wind the bobbins or pirns from this. In either case, the winding speeds used are high, of the order of 12 000 rev/min.

In winding directly on to the bobbins, a number of precautions not usually considered to be of great importance in normal ring spinning must be taken. The smooth release of the weft thread as the shuttle flies across the loom requires that the yarn is wound in exactly the same way on each bobbin. The exact height at which the winding started was difficult to control when bobbins pressed on to tapered winding spindles to obtain the necessary grip were used, the down-pressure used varying quite widely from spindle to spindle. In addition, wear on the bore of the bobbin was rapid using this method of

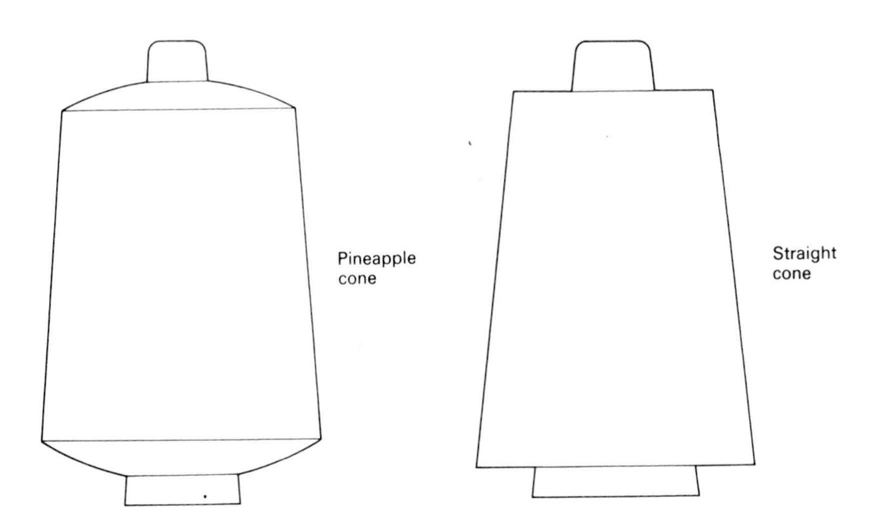

Fig. 6.3 Types of cones.

fixing, and this produced even more height variation and could also lead to the bobbin being supported unsymmetrically. At the high rotational speeds used, perfect balance is essential, and the bobbin must be made, and maintained, as a precision piece of equipment. It could also be mentioned here that the bobbin is designed in such a form as to hold the thread in position and prevent it from slipping while it is being subjected to high acceleration and deceleration forces as the shuttle is propelled across the loom. Further, each bobbin is built so that it can hold a small reserve of yarn when its main length is empty; this is required because, during weaving on modern looms, the fact that a bobbin is empty and has to be replaced is indicated by a 'feeler' coming into contact with the surface of the bobbin itself, and the small reserve of thread prevents the weft from running out completely.

Tapered spindle holders were eventually replaced by other methods of gripping the bobbin, culminating in the modern spring-loaded clutch which is now used. Other problems in direct winding have been overcome by modifications to the bobbins and equipment.

While, at first sight, it would appear that direct winding would offer considerable cost advantages, in that an extra winding process is eliminated, the method has some

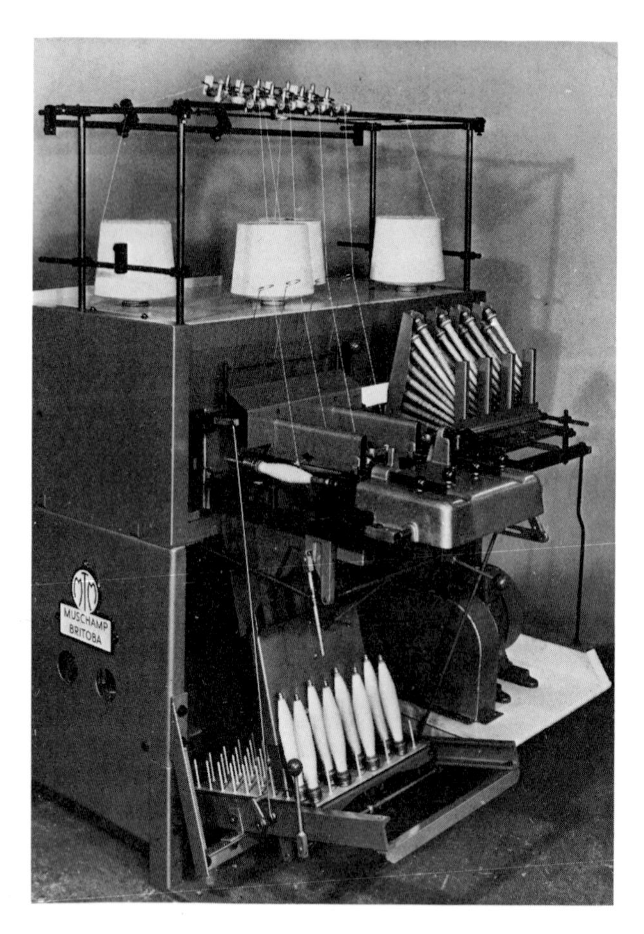

Fig. 6.4 A pirn winder.

disadvantages compared with the formation of a large package and then winding from this, usually referred to as 'pirn winding'. The weaving bobbin is quite a small package, and additional work is involved in dealing with a very large number of small packages; compared with the spinning of large packages, many more operations are required to deal with the extra doffing which is involved, and the spindles are idle for a longer total time. It is also expensive to stock the weft threads on bobbins in the large quantities required by this method, and it takes time for the operator, who fills the 'battery' on the end of the loom with full bobbins, to find the loose end of the yarn as the bobbin is taken from the container, compared with feeding in a wound pirn which has the thread end free. In direct winding, the thread is not laid down so densely on the bobbin as with pirn winding, and not only is there a risk of slipping taking place during weaving, so causing havoc, but also there is less yarn on the package and the shuttle runs out quicker. Probably the most serious objection to direct winding is the fact that it does not allow the thread to be cleared of spinning faults which are not acceptable in the woven fabric; rewinding does allow the opportunity for such faults to be eliminated before the weaving takes place.

Modern pirn winding is an automatic process, and many types of such machines are now available. All the machines perform the same sequence of operations. A continuous supply of pirns is contained in a hopper and is fed into position, and a traverse mechanism controls the correct building of the package so that slipping will not take place during the shuttle action. When the pirn is full, it is automatically ejected, an empty one is put in its place, and winding is restarted. The full pirns are then automatically delivered to a suitable container or on to a pin board. The thread is taken from large packages, cheeses or cones, each of which is capable of filling a large number of shuttle pirns or bobbins, and often magazine creeling is used with the end of the thread from one package being attached to the beginning of the thread from another to give a continuous supply.

Rewound pirns offer more trouble-free weaving and, although the modern pirn winder has reached a high state of perfection, research continues to be carried out with the object of providing still further improvements.

6.1.4 Warping

One of the processes which has to be carried out on the yarns supplied by the spinner before the majority of the different forms of fabric can be manufactured is that known as 'warping'. Processes such as weaving, warp knitting, lace making and carpet making, all make use of a 'warp' of threads. A warp consists of a large number of yarn lengths all arranged regularly side by side to form what may best be described as a long 'sheet' of threads, with each thread being under the same tension so that one is not extended more than its fellow threads in the sheet.

The appearance and characteristics of the fabric which will finally be produced will be influenced, to no small degree, by the excellence with which the warp is made, and the methods of warping used have been designed with this is mind. Once a warp has been arranged with all the threads in position and under a uniform tension, the threads are maintained in this condition by being wound on to a long tube or cylinder which we will refer to as a 'weaver's beam'. This beam with the warp wrapped round it is attached to the fabric-making machine, when it is slowly unrolled at a speed matching the rate at which the fabric is constructed, and so releases the sheet of threads as required. The oldest method of preparing a warp is by hand warping. This is still used occasionally when sample patterns are being tried out and when only a short experimental warp is required. In hand warping, it is usual to employ an upright board carrying two vertical rows of pegs standing out from its surface. The packages of yarn to be used to make the warp are mounted in a convenient frame, called a 'creel', which allows the yarn to be drawn off the packages easily. One 'end' of the yarn is taken from each package to form an untwisted rope of yarns, and this is attached to the top peg in one of the vertical rows. Yarn continues to be drawn off the packages by hand, and the rope is then taken backwards and forwards between the two rows of pegs, descending one pair of pegs at a time, until the necessary length of yarns is in a zig-zag formation on the pegs. The yarns are now cut, and a new rope drawn off and again attached to the same starting peg, and the same procedure carried out using the rows of pegs. The cycle is repeated as needed to provide the number of yarn lengths required in the warp. The large rope of yarns is then removed from the pegs and rolled up to transport it. The warp is finally transferred to a weaver's beam after the individual yarns have been arranged in order by being drawn between the teeth of a comb, or 'reed', to separate them.

One of the earliest mechanical methods of preparing a warp was by the use of the upright 'mill'. By the use of this mill, not only was quicker warping possible, but also longer warps, containing a greater number of threads, could be prepared. The main part of the machine is a large reel, some 15 ft (4.5 m) in diameter, which can be rotated round a vertical spindle. The reel consists of a number of ribs held in position by a frame and spaced out round the circumference of the large cylinder. This reel rotates slowly and, as it does so, it winds round its rib formation the yarns drawn from a number of packages in a creel. In travelling to the reel, the threads are separated by being passed between the teeth of a reed, so that they are arranged side by side on the surface of the reel. At the same time, a driven cam arrangement raises the reed so that the spaced threads are wound spirally on the surface of the reel. The cycle is repeated until the necessary number of yarn ends have been wound on the reel, when the complete warp can be drawn off and transferred to a weaver's beam.

Another method of preparing the warp is by 'section warping'. In this method, the first warp is made up of a number of short sections set side by side. Each section is made separately on a short beam which has a flange at each end. When the various sections have been completed, they are keyed side by side on a long shaft, and the ends of yarn are then drawn off as a complete warp on to a full width weaver's beam. The difficulty in using this method is to make certain that the various sections will deliver threads all under the same tension. A number of devices have been developed which help in providing a more uniform tension across the warp.

A method known as 'beam warping' is similar in principle to section warping but uses, in place of the small sections, a number of 'warper's beams' of about the same width as the final weaver's beam. A very limited number of ends of yarn are wound spaced along each warper's beam, and a number of such wound beams are then used to provide the correct number of threads in the final warp which is wound on to the weaver's beam.

The most widely used warping machine is the 'horizontal warping mill'. Again, sections of warp are prepared, but these are wound side by side on one reel from which they are taken, suitably tensioned, to the weaver's beam. The reel is usually about 5 ft (1.5 m) in diameter and of the order of 5 ft in length. The various sections are arranged on the surface of the reel so that one overlaps the end of the previous one in a carefully

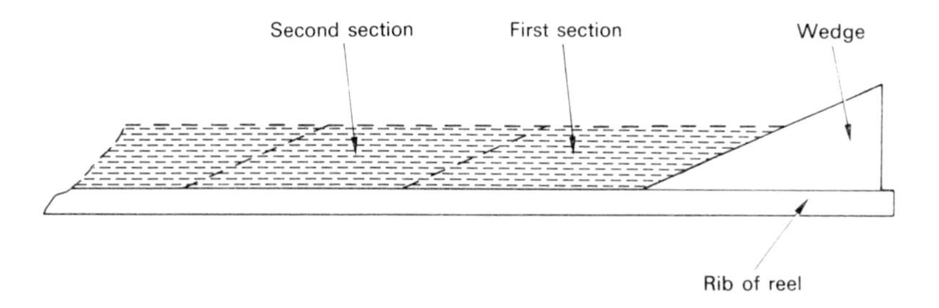

Fig. 6.5 Winding of warp sections on a horizontal warping mill.

arranged manner. The arrangement is obtained by fixing a wedge on the end of each rib of the reel, and the warp section is guided by a moving reed so that its edge is built up on the incline thus provided. When the required length of warp has been wound in that section, the yarns from the supply packages are cut and a fresh section begun at a specified distance along the reel. The same procedure as for the first section is then followed but, instead of the wedge, the incline is now provided by the first section which was wound. Sufficient sections are added in a similar way until the full width of the warp has been obtained. All the threads are then drawn off, under a suitable uniform tension, from the reel and on to the weaver's beam. It has been found advantageous to select the slope of the wedges in accordance with the count of yarn to be wound; it is usual to use a steeper slope for coarser yarns.

6.1.5 Warp sizing

As will be seen from the sections which deal with the various methods of fabric manufacture, the threads composing the warp are subjected to a good deal of stress and abrasive action during fabric preparation. Very often, therefore, the threads in a warp are strengthened and made smoother by coating them with a material referred to as a 'size'. The size normally contains an adhesive, an oil to prevent the size from becoming brittle and cracking when dry, and an antiseptic material to prevent mildew developing. The various firms which carry out sizing usually have their own size recipes. Normally the size adhesives used on wool yarns are derived from protein sources, such as gelatine, albumen and casein, while those for cellulose fibres are starches and gums. The most common antiseptic compound included is zinc chloride. Very often sizes based on compounds such as polyvinyl alcohol are now used, and particularly on yarns containing the synthetic fibres. It should be remembered that, at some later stage, the size will have to be removed, so that no component of the size which is not easily removed by normal cleaning methods should form part of the composition.

Size is normally applied by passing the warp sheet through a trough containing the size solution and then between nip rollers to remove excess solution. The yarns are then heated to dry the coatings and the warp taken up on the weaver's beam.

The amount of size applied to the warp yarns will depend on the type of yarn being used. A small amount will hold surface fibres in position and give some extra resistance to abrasion, whereas a heavy sizing may be required to give additional strength to weaker yarns. The size must penetrate into the yarn and not simply form a coating on the surface, and wetting agents are often added to the size bath to help in getting a more thorough penetration.

In recent years, considerable advances have been made in the design of sizing machines and these now contain all the necessary facilities for preparing and sizing the warp. Thus, modern machines allow winding from a creel to form a suitable warp, and during their passage from creel to beam, the threads are sized and dried. Machines employed in the USA to handle long runs of fabric use a beam warping method of forming the warp on the weaver's beam, the intermediate warper's beams being assembled at the back of the machine. The threads from these back beams then pass to a position where the size solution is applied (excess solution being removed by squeezing), around heated drying cylinders, through a device which separates any threads which may be sticking together, and finally through a reed on to the weaver's beam. The drying conditions must be accurately controlled; sufficient drying must be given to prevent the threads from adhering to each other, but not carried out to an extent which will cause the dried size to crack and break up.

6.2 WEAVING

6.2.1 Woven fabric construction

Weaving is probably the oldest method of combining yarns to form a fabric. The fabric is constructed from two sets of yarns or threads, one set running along the length of the fabric and known as the 'warp', and the other running across the width of the fabric and known as the 'weft'. The weaving process interlaces these two sets of threads with each other, and the frictional forces between the individual threads at the points of cross-over hold them tightly in position so that a stable fabric, capable of being cut without fear of fraying taking place, can be produced. The illustration shows, in diagrammatic form, the construction of a plain woven fabric, the warp threads being filled in black to distinguish them from the weft threads. In practical weaving terminology, the warp threads are usually referred to as 'ends', and the weft threads as 'picks'.

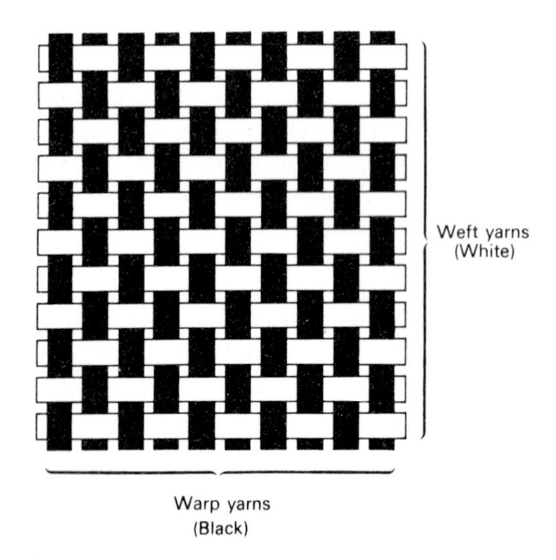

Fig. 6.6 Plain woven fabric.

The warp threads are the most important of the two sets used to construct a woven fabric. When we come to discuss the actual weaving operations, it will be realized that the warp is subjected to very considerable strains. The warp must be strong, therefore, to withstand these strains; and in order to give strength to the final fabric. When a piece of fabric is cut up to make a garment, it is usual to ensure that the cut pieces are so arranged that the warp threads are directed along the direction of most strain during wear.

The weft threads are used to fill in the spaces between the warp threads. If a strong warp is provided, then much weaker yarns can be used in the weft; they must be strong enough, of course, to withstand the action of the shuttle during weaving. In general, the weft threads are more bulky than the warp threads in order to provide a satisfactory 'covering power' in the fabric.

By varying the way in which the two sets of threads are interlaced, different patterns of fabric can be produced by the weaving operation. The fabric illustrated is a plain weave and is the simplest, and probably the oldest, type of woven fabric made. In this fabric, a length of weft thread passes over each odd warp thread and under each even one. With the next weft thread the opposite takes place, so that it now passes over the even warp threads and under the odd ones. In this way, a firmly interlocked and stable fabric is produced with both of its faces identical in appearance.

The plain weave, with weft threads passing alternatively under and over the warp threads, is probably the strongest and most serviceable method of combining the two sets of yarns to form a fabric, but perfectly satisfactory fabrics having different weaving patterns may also be produced by varying the way in which the two sets are interlaced. A simple variation would be to take the weft threads under two, then over one, then under two, then over one, of the warps threads, and so on throughout the piece of fabric, when quite a different fabric pattern would be obtained. Or again, other patterns can be obtained by moving the starting point of the weaving sequence one place to right or left for each successive insertion of a weft thread. On the other hand, it may be desired to have a fabric showing no weaving pattern at all; in such a case a random pattern-less appearance would be obtained by arranging for the weft threads to pass over and under different numbers of warp threads throughout the fabric. There are many ways in which the weaving sequences may be altered. Each will produce a fabric having its own particular appearance and, by reason of the different methods of interlacing, its own characteristic properties. Add to all this the fact that the sets of yarns can be changed as regards relative thickness, smoothness and colour, and it will be appreciated how great are the patterning possibilities available at the loom.

6.2.2 The hand loom

The weaver of ancient times used very simple weaving equipment. The ancient Egyptians attached their warp threads side by side, along a rectangular wooden frame, and the weft threads were threaded in and out across the sheet of warp threads in much the same way as is used for hard darning. At various times, throughout the world, another method has been used in which one end of the warp sheet is attached to some fixed object, such as a tree, with the other end fastened to a leather belt worn by the man carrying out the weaving. By this means he was able to exercise some control over the tension in his warp threads while he interlaced them with the weft threads by passing thin bobbins of thread in and out of the warp threads. These hand methods had one advantage in that the weaver could produce whatever pattern, and combination of coloured yarns, he desired. This method of fabric production was, however, extremely slow and, over the centuries, methods of carrying out the various weaving operations in a quicker manner were sought. These eventually resulted in the invention of the weaving machine or 'loom'.

The basic idea behind the development of the loom was the elimination of the laborious hand weaving in and out of the weft threads across the sheet of warp threads by substituting a method lifting in one operation all the warp threads under which the weft thread must pass and lowering all the threads over which the weft thread must pass. If a length of weft thread is now passed across the space between the two sets of separated warp threads, and the two sets are then returned to their initial undivided positions, the same interlacing as would have been done by the slow hand weaving is obtained by these short rapid movements. The operation of a plain loom will be more easily understood by referring to the diagram. The operator of a hand loom sits in front of the fabric roller upon which he can roll the fabric a few inches at a time, as he produces it. The warp is brought to the machine on a weaver's beam which is then located at the back of the loom. The warp threads are arranged in order on the beam, and this order is maintained as the warp sheet is drawn through the loom by passing the individual threads through the spaces or 'dents' of a reed situated near the front of the loom. The reed is formed by a series of steel pins standing up vertically from a rod running across the loom, with a further rod passing over the upper tips of the pins so that the threads cannot be pulled out of their dents accidentally.

Fig. 6.7 A hand loom.

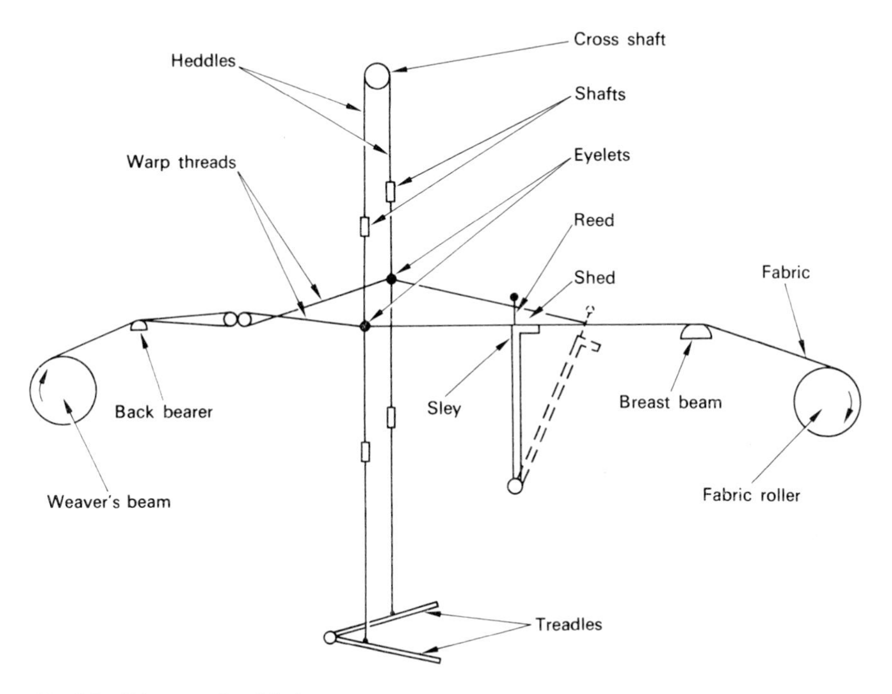

Fig. 6.8 Diagram of a plain loom.

From the weaver's beam, the sheet of warp threads is taken over a cross beam, indicated in the diagram as a 'back bearer', in order to support the sheet at the correct height in the loom.

Passing over the 'cross shaft', situated at the top of the loom, is the 'harness' which controls the movements of the warp threads. Each individual warp thread is taken through its own eyelet forming part of a 'heddle' (also named a 'heald') which hangs down from a 'shaft'. For a plain weave, only two such shafts will be used, one shaft having suspended from it the heddles through which the odd threads are passing, and the other carrying the even thread heddles. It will be obvious from the diagram that the depression by the foot of the operator of one of the treadles situated at the bottom of the loom will cause one shaft and its suspended heddles to lift and pull up, say, all the odd threads of the warp, and a space will be left between the odd and even warp threads which is known in weaving terms as a 'shed'. It is through this shed that the weft yarn is passed. When the foot is taken away and the treadle allowed to come back into its starting position, the odd warp threads will drop to complete the warp sheet once again, thus trapping the length of weft yarn alternately under the odd warp threads and over the even ones. By this simple action, one 'pick' of a plain woven fabric has been obtained rapidly and easily. If the other treadle is now depressed, the even threads will be lifted, when the passing across of the weft yarn will produce another pick of the plain fabric, and so on until the requisite length of fabric has been woven.

In order to pass the weft yarn across the loom inside the shed a small pirn of the yarn is held in a cavity inside a torpedo-shaped piece of hardwood called a 'shuttle'. This shuttle is projected through the shed and, as it flies across the small pirn unwinds under the tension to leave a trail of yarn behind it. Originally the weaver threw the shuttle from side to side of the loom. This limited the width of the piece of fabric to the distance he could throw the heavy shuttle through the relatively small shed, and it also meant that both his hands were full occupied. In 1733, the 'fly-shuttle' was invented by Kay, and this flew faster across the loom and in a straighter path than in the older method and could be operated by only one hand by the weaver. Kay provided two boxes, one at each end of the 'batten' or 'sley' shown in the diagram, and each of these boxes was fitted with a hammer block or 'picker' sliding freely on a metal rod. The two hammers were linked together by a cord held in one hand of the operator. He could jerk this cord in either direction, so suddenly pulling forward one of the hammers to strike the shuttle and propel it across the surface of the sley into the opposite box. Jerking the cord in the opposite direction caused the shuttle to make its return journey.

When the weft thread is laid across the warp and the shed closed, the new pick is a little distance away from the previous weft insertion and has to be 'beaten up' to move it close up to the previous pick and build up a compact fabric. The diagram indicates that the reed, supported on the sley – which is itself held on two uprights which are pivoted to be able to swing forward – is pulled by the free hand of the operator towards him. This movement performs the beating up, the vertical needles of the reed being sufficiently stout to withstand the quite considerable pressure exerted. As soon as the reed is back in its original position, the shuttle is impelled back to lay down another weft thread. When a number of such picks have been performed, the short length of fabric thus produced is wound up on the fabric roller.

6.2.3 The power loom

The principle upon which the modern power loom operates is much the same as in the loom described above, but the weaver's operating movements are replaced by mechanisms.

Patents for the first power-operated loom were granted to Cartwright, a Leicestershire clergyman, over the period 1785 to 1786. Cartwright set out to copy the weaver's movements, arranged to operate in the correct sequence, by mechanisms drawing their motive power from a revolving shaft. By the use of cams, he was able to convert rotary motion into the push-pull action of the weaver. Originally, the driving power was obtained from a bar running across the front of the loom which the weaver pushed up and down, but later a crank was substituted for the weaver's hands so that other means of providing driving power could be used. The original power looms were made entirely of wood, each type of wood for its various parts being selected as regards its suitability for that particular purpose. Thus the shuttles were of box-wood which is capable of taking and retaining a very smooth finish, and the sleys of apple-wood which has good resistance to wear by friction.

In modern power looms, care must be taken to select the designs of mechanisms, and materials, which will continue to give trouble-free service. The beam carrying the warp is attached to the loom in much the same position as shown in the diagram of the plain loom above, and its let-off motion and take-up of the fabric roller are designed to operate at exactly the right rate for the weaving which will take place. The warp can be drawn from a braked beam, or there are types in which a mechanism rotates the warp beam in response to tension applied to the warp by the fabric roller. The latter is arranged to turn a tiny amount each time a pick of weft is inserted in order to keep the same tension in the warp sheet.

The healds are usually made from specially prepared cotton or flax cords, in which

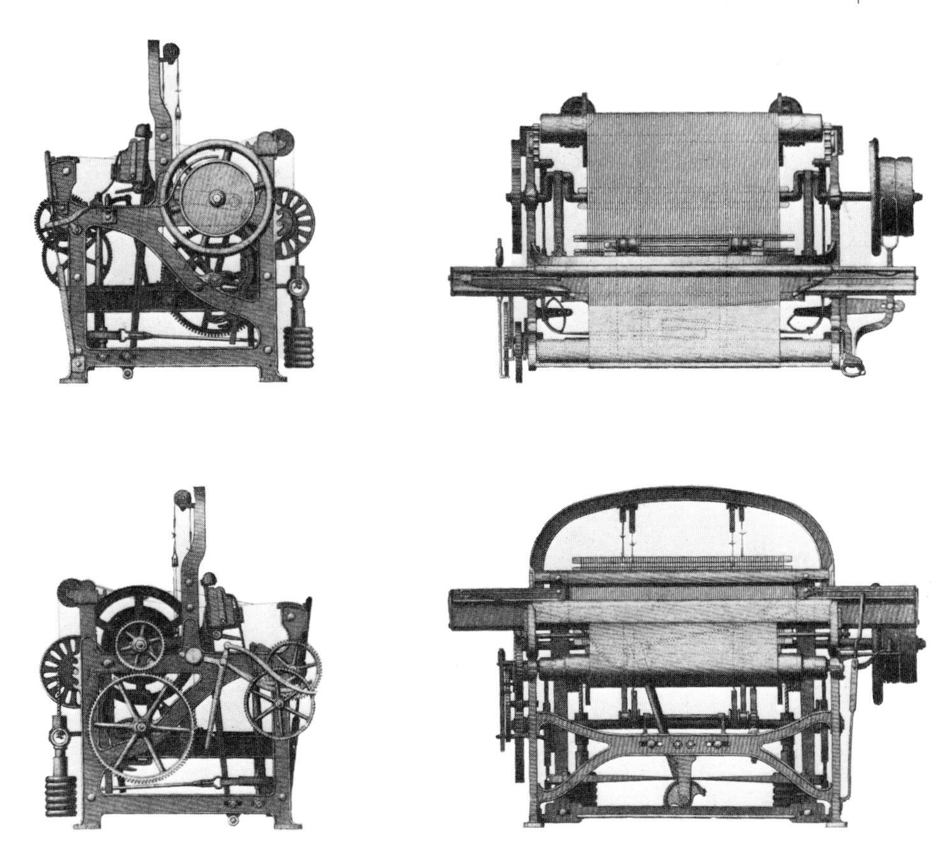

Fig. 6.9 A power loom, dated 1851, showing back, top, side and front views.

metal eyelets are inserted during their preparation, or from metal. Fibre healds are lighter in weight than metal ones, and are preferred when a large number of shafts are in operation on the loom. Metal healds are often made from two wires twisted together to leave an eyelet in the middle and loops at each end to enable them to be attached to metal rods, or comprise flat strips of wire with holes punched in to serve the purpose of loops.

Threading up the individual warp threads through the heald eyelets and reed is a lengthy operation. It is usually carried out more conveniently by transferring the warp beam, healds, and reed to a separate frame and using two persons to carry out the 'drawing in'. One person sits behind the healds and places each weft thread in turn over a hook pushed through the eyelet by the person sitting in front of the healds. On completion of the drawing in, the warp beam, heald shafts and reed are transferred back to the loom. If a warp beam runs out, and it is wished to make no changes in the weaving pattern, then ends from the new warp beam can be tied to the corresponding old ends, either by hand or using equipment developed for the purpose, and then pulled through the healds and reed and on to the fabric roller.

Up to the moment we have only discussed plain weaving and, in fact, the loom depicted previously is only using two shafts and would be producing plain woven fabric. It was indicated earlier that fabrics of various patterns can be produced by selecting different ways in which the warp and weft threads interlace. This calls for more complicated threading up and splitting of the warp sheet during 'shedding', and the whole question of how the various types of warp shed are produced on the power loom is discussed in the next section of this chapter.

The shuttles used on the power loom vary in size with the size of the pirn of yarn contained in them, the nature of the yarns used, and the width of the loom. The shuttles must be able to withstand the severe treatment they receive in use and be made of a wood capable of being given a very smooth finish. The pirn of yarn fits on a spindle inside the shuttle and facilities are provided for its easy replacement. The shuttle is projected across the loom by means of a 'picker'. A rotating shaft has a stud on its surface and when this turns round to a certain position it depresses a lever very suddenly. This causes an upright arm to swing over at high speed and project the shuttle along the sley to the other side of the loom. The upright arm then returns to its former position under the influence of a spring. A similar picking device on the other side of the loom is provided to return the shuttle to its starting side.

On each side of the loom, the shuttle rests in a 'shuttle box' ready for projection and it passes into a similar box when it reaches the other side. Each pirn of yarn inside a shuttle must, of necessity, be very limited in size and its supply of yarn is exhausted in a relatively short time. Formerly, a new pirn was inserted by the operative but now special shuttle-boxes are used whereby this is done automatically. In addition, similar boxes are used to contain a number of shuttles with differently coloured weft yarns when coloured patterns are being woven. In such cases, multiple shuttle boxes are provided on both sides of the loom and mechanisms ensure that the correct section is in line with the sley when that particular shuttle is to be projected through the shed. One type of box used is the circular box usually consisting of six sections; the control mechanisms can rotate the box 1/6th of a revolution forward or backward as required, thus changing the shuttle to be used. Another type is the rising box; this is divided up into sections by shelves on which rest the shuttles and a rod operated by the control mechanism pushes this up until the required shuttle is in line to be projected.

The beating-up of the last pick of weft to the previous pick inserted, and referred to as the 'fell' of the cloth, is by an automatic movement of the sley carrying the reed. The pick must be given a smart blow to slide it into position, and the beating-up must be synchronized accurately with the shuttle cycles. The reed consists of a large number of flat wires, arranged with their edges facing along the loom, held between top and bottom rods or 'baulks'. The ideal would be for a space or dent in the reed to be provided for every warp thread, but this is not practical as an extremely fine reed would be needed and could not be made sufficiently robust. In addition, such a fine reed would not allow knots to pass through. It is usual, therefore, to use a coarser reed and pass two or more ends through each dent.

6.2.4 Methods of forming the shed

There are three types of loom, based on the method employed to form the shed and, it follows, the complexity of the fabric patterns which can be produced on them. The types are the tappet loom, the dobby loom, and the Jacquard loom.

In the tappet loom, cams are fixed to a rotating shaft, and these turn round at every revolution to make contact with a roller or 'bowl' projecting from the treadles. The cam presses against the bowl and depresses it and the treadle so that the shedding action takes place. There may be two treadles only, lifting and lowering two shafts to weave a plain fabric, or there may be a larger number of treadles, each connected to its own shaft, and lifting and lowering selected portions of the warp to produce more complicated patterns. It will be appreciated that the operation of shedding is not simply a question of raising or lowering the treadles. The cams must be located, and of a shape, to obtain shedding at exactly the right moment with respect to the other motions in the loom, and to provide just the right size of shed, not too large to impose undue strain on the warp threads, and not too small to cause end breakages as the shuttle is driven through. There must also be a period of dwell of the shaft movement to allow the shuttle to pass through.

The dobby shedding actions occupy a position midway between the patterning capabilities of the tappet and the Jacquard looms with a control of up to 48 shafts, Dobby mechanisms can be divided into lever and wheel types and these can be further subdivided into positive or negative forms depending on whether or not the mechanism positively depresses the shafts as well as lifting them. For heavy fabrics it is sometimes necessary to use a positive type instead of allowing the shafts to fall under their own weight after lifting.

In the lever type of dobby, a pattern chain runs over a grooved cylinder and, when a boss on the chain comes under a lever it lifts it. Associated with this lever are two 'draw knives', and the lifting action causes one of these to engage with hooks which raise a 'jack' lever and cause the appropriate shaft supporting the healds to lift. When the second draw knife is allowed to engage a hook, in response to the arrangement on the pattern chain, the levers and heald shaft are lowered. The wheel dobby operates on a similar principle but with a system of wheel gears substituted for the lever system.

Complicated woven patterns are produced on the Jacquard loom. In this loom, each individual heald is attached to a lifting wire or cord, the raising of which is controlled by the Jacquard mechanism. This mechanism was devised by a Frenchman, Jacquard, in the mid-1700s and the same principle is still used on various types of machine for making textile fabrics. The essential working parts of the device for controlling four warp threads as a sample of those included in the warp sheet are shown. In this case, four horizontal needles are supported at one end by a needle board, passing through appropriate holes in the board and projecting a short distance on the other side, and at the other end by springs enclosed in a spring box. Passing through eyelets in these needles are four vertical hooks. The lower ends of these hooks are bent back and pass through a grate, the holes in which are shaped to prevent the hooks from turning sideways; at the lower ends of the hooks, the healds are attached. The healds are held vertically by means of small weights hanging below the heald eyelets through which the warp threads are passed. The upper ends of the hooks terminate at a lifting 'griffe' which lifts at every pick and which carries a series of 'knives', one for each hook, and each hook can be lifted by its knife or be pushed on one side to miss its knife and thus remain depressed. The means whereby the hooks are pushed aside to miss lifting by the knives is the essence of the Jacquard device.

The springs on the ends of the horizontal needles keep them pressed forward to project through the needle board and, in this position, the top ends of the hooks will engage with the lifting knives and they, together with the healds and the warp threads passing through, will be raised. If, however, a stiff card is pressed against the back of the needle board, this will push the horizontal needles back and the hooks will miss the lifting knives. By punching holes in the card at appropriate positions, some of the needle ends can pass through while others will be held back; it follows that selected warp threads

Fig. 6.10 Jacquard loom, dated 1810, showing, left, the chain of punched cards.

can be made to lift and others left down. A card must be provided for each pick of the weft thread until the pattern is completed, when the presentation must be repeated to form the next pattern. For this reason the cards, after being prepared, are laced together to form an endless chain and pass over a square drum which makes a quarter turn at every pick and thus brings a new card into the operating position behind the needle board. The square drum is, of course, of honeycomb construction so that the needles can pass through without meeting any obstruction.

The threads are usually in groups of eight with the necessary equipment for a group repeated many times across the machine.

6.2.5 More recent loom developments

There are some drawbacks in using a shuttle for weaving. It must be robust and yet light enough and small enough to be projected through the warp shed. The amount of weft yarn it can carry is, therefore, very limited and the pirn must be changed frequently, often involving the use of complicated equipment for doing so if there is to be no undue loss in weaving time. Many attempts have been made to eliminate the shuttle in recent years, and some devices are being used. The ideal would be a near-weightless method of carrying the weft yarn across while drawing it from a large package which could remain stationary on one side of the loom.

There are three main types of 'shuttleless' loom, employing as a means of conveying weft, rapiers, grippers or jets of air or water. All three have a number of features in

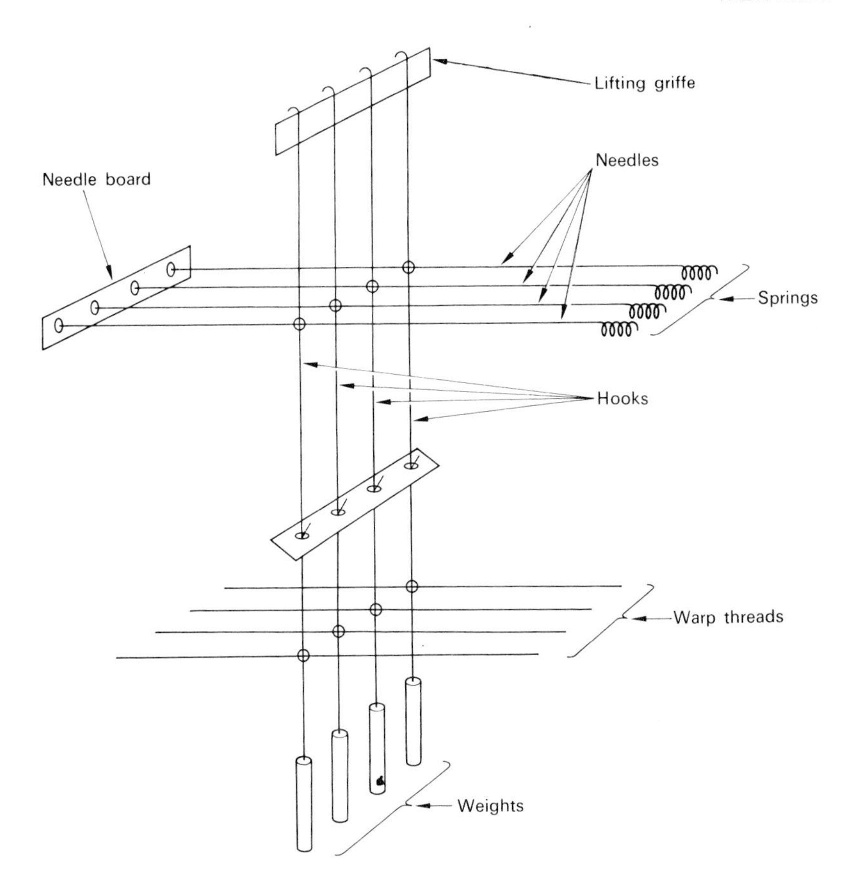

Fig. 6.11 Operation of the Jacquard mechanism.

common. All take yarn from a suitable large package, such as a cone or cheese, all give a high speed of weft conveyance and all need only a small warp shed, thus reducing the strain on the threads. They are not ideal, however, since they produce ragged selvedges, often require complicated and expensive supporting mechanisms, and usually require more floor space than the standard loom.

In one of the rapier types of loom, a long thin needle, operating from one side, has an eye near its tip through which the weft yarn is threaded. This needle carries the weft thread across the shed and a mechanism grips the yarn on the other side and holds it so that, when the needle is withdrawn, a double loop is left across the shed. In another type, two rapiers are used, one from each side of the warp shed and these meet in the middle. One rapier carries the weft thread half-way, at which point it is transferred to the second rapier which completes the pick on its return journey.

In one of the gripper types of loom, a number of very small gripper-shuttles, or 'bullets' loaded with weft yarn are used. These are 'fired' across the shed, paying out yarn as they go. When each gripper-shuttle reaches the other side, the yarn tail is cut, and it falls on to a conveyor and is taken back to the other side to be reloaded ready for use once again.

In the third method, a small high pressure pump directs a jet of air or water from a fine nozzle through the shed. The weft yarn passes through the centre of the nozzle and is carried across by the stream. The yarn is caught and held at the other side and the tail end is cut by cam-operated shears. A water jet gives more reliable results than an air jet but, even with this, as yet only fine and light yarns can be carried over, and these must be such that they do not become water-logged as cotton and wool do.

Very high speeds are possible with shuttleless looms, and they produce far less noise than the shuttle type. The water jet loom, in particular, is practically noiseless, producing only a low humming sound as it operates.

6.3 WEFT KNITTING

6.3.1 Weft knitted fabric construction

Another method of forming a yarn or yarns into a fabric is by weft knitting, usually referred to simply as 'knitting'. A knitted structure can be made using a single yarn. It consists of rows of loops of yarn, called 'courses', running across the width of the fabric and so formed that each course hangs from the course above it and supports the next course which follows. The rows of loops running at right angles to the courses along the length of the fabric are called 'wales'. The knitted stitch is formed when a loop of yarn is drawn through a previously-made loop and, during knitting, the loops are formed across the fabric.

In relation to wearing apparel, knitting and weaving are complementary to each other. Woven fabric is relatively rigid and not liable to be distorted, and is ideal for the construction of certain types of garment. On the other hand if wearing apparel is required to stretch and then recover so as to cling closely to the shape of the body, then knitting will be selected as a means of fabric manufacture. It will be appreciated that, in the plain knitted fabric illustrated, the loops in the wale direction are fairly well extended as they are held in the fabric, and the extension in length of the fabric on pulling will be limited. In the course direction, however, the well-rounded loops can be distorted considerably by pulling and give very large width extension to the fabric. As will be seen later, the art of knitting has advanced so much that various constructions having even greater width extensions, and length extensions too, than plain knitting, can be used.

Knitting is not so old an art as weaving and carpet-making, but hand-knitting was probably carried out in the pre-Christian era in the Middle East countries. The earliest

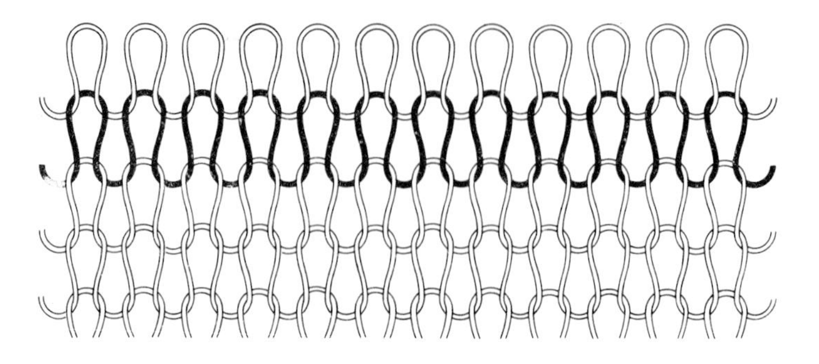

Fig. 6.12 Plain knitted fabric.

Fig. 6.13 A manual stocking frame, dated 1770.

references to hand knitting in Britain were in documents in the fifteenth century A.D. and, by the time of Elizabeth I, hand knitting was a thriving industry.

Machine knitting originated during Queen Elizabeth's reign by the invention of the 'stocking frame' by the Rev. William Lee, and he built his first machine in Calverton in Nottinghamshire. As the Queen would not grant him a charter or patent, it is said on the grounds that the hand knitters would lose their means of livelihood, Lee went to France to develop his ideas. The machines he devised founded the knitting industry in the Rouen district. It is said that Lee's brother subsequently brought back to Britain a number of the machines, and machine knitting eventually became established in London and the Midlands. By the seventeenth century, the machine knitting industry had developed very considerably, with its main areas in Nottinghamshire, Leicestershire, and in certain parts of Scotland. Since that time more and more complicated machines have been devised and these are capable of providing a whole range of garments and knitted structures, each structure having its own characteristic properties and uses.

In the illustration of plain knitted construction the yarn in one course has been blacked in to show the loop formation more clearly. Plain knitting is widely used in the manufacture of various types of fully-fashioned garments and in the making of socks and stockings. In such garments a good elasticity is obtained in the width and a limited amount of extension in the length.

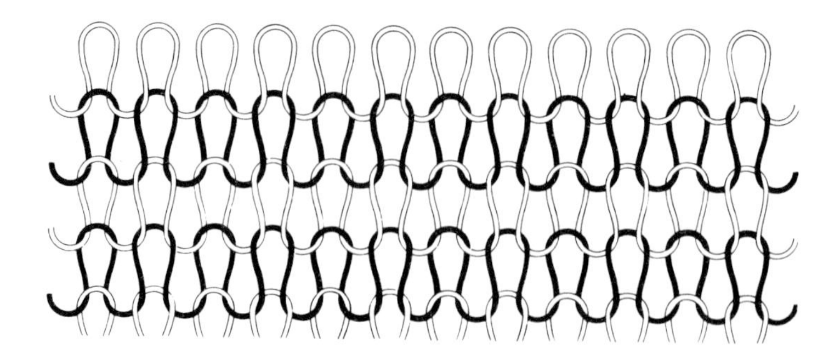

Fig. 6.14 Knitted rib fabric.

Additional contraction in width and an increased extensibility and recovery in that direction can be obtained by using what is known as a 'rib' construction. It will be observed that certain vertical rows of loops are meshed so that the stitches lie in two planes. The diagram shown is of a 1/1 rib fabric in which alternate vertical rows of loops are knitted in the front and back respectively. Again, the blackened-in courses will indicate more clearly how some wales stand out to the front and the others retreat to the back. If such a structure is released from tension, it will close up in width and form a series of ribs running along the length of the fabric. The width contracts to about half the width of a plain fabric and its extensibility is doubled in consequence. Ribbed fabrics are often used at the waist and cuffs of garments to give additional cling. Ribbed fabrics need not necessarily be 1/1; more wales can be meshed to the front than to the back and 3/1, 4/1 and 5/1 are common, the first figure representing the consecutive wales occupying the front position.

'Purl' fabric produces the same general effect as a ribbed fabric but across the fabric. The loops across are knitted to the front or back. The illustration is of a 1/1 purl fabric, and the white yarns are knitted to the front of the fabric, and the blackened yarns to the back. A form of horizontal pleats across the fabric is thus produced and the length stretch of the fabric is thus increased.

A 'float' stitch is made when a needle retains its old loop by not being allowed to rise and take a new loop. This produces a miss-stitch which shows up as a short horizontal 'floating' thread, as shown in the blackened-in course of yarn. This type of stitch is used to produce designs.

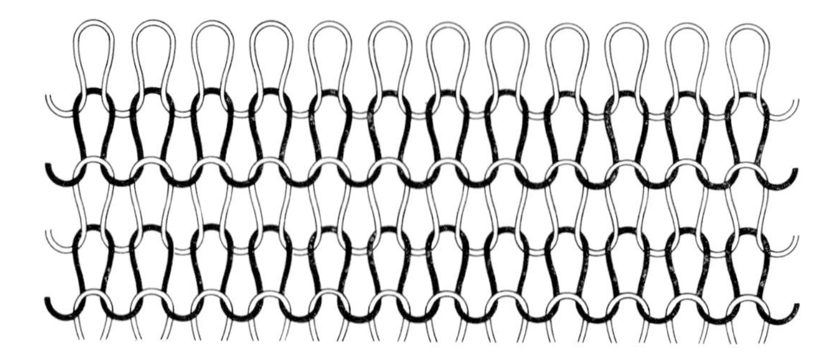

Fig. 6.15 Knitted purl fabric.

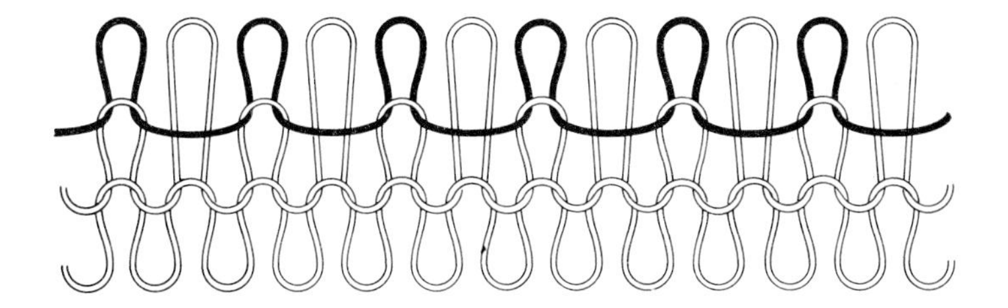

Fig. 6.16 Float stitches in a knitted fabric.

There are many other kinds of design stitches used. One of these is a 'tuck' stitch, and is produced when a needle is lifted to take a new loop while the old loop is prevented from being cast off. Tuck stitches tend to reduce the length of the fabric and to increase its width, and are sometimes used for this purpose. They also reduce the tendency of the fabric to ladder, and are used to make ladder-resistant fine gauge ladies' hose.

Mention should also be made of the 'interlock' types of fabric. This is a double 1/1 rib fabric composed of two ribs interlocked together. Two sets of needles are used, one set knitting in a vertical position and the other in a horizontal position, and these work in conjunction with each other to produce the interlock construction. Such fabrics are used to produce cut underwear garments and, more recently, to make double jersey interlock outerwear garments.

Many other different types of stitches can now be machine-knitted and the reader is referred to the list of books appended if he wishes to follow the matter up further.

6.3.2 Basic weft knitting mechanisms

Two types of knitting needle are used in machine weft knitting, the bearded needle and the latch needle. The bearded needle is the older one of the two and was devised by the Rev. William Lee and used on the first knitting machine. This type of needle is still used on many modern kinds of knitting machines. The top of the needle is bent over to form a hook and this hook terminates in a pointed end, known as the 'beard'. Opposite to the point of the beard is a groove cut into the stem of the neddle. Referring to the illustration, the last loop of yarn knitted into the fabric is still round the stem of the needle

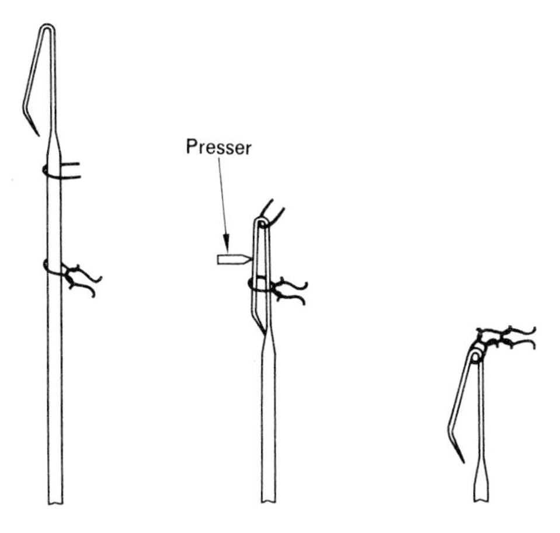

Fig. 6.17 Knitting action of a bearded needle.

with the needle raised. Yarn to form the new loop is folded round the needle stem also, but in a higher position. The needle now descends so that the yarn for the new loop is under the beard in the 'eye' of the needle. The old loop of the fabric also slides up the needle stem but, before it reaches the beard, a presser comes forward and presses the hook so that the point of the beard enters the groove. As the fabric loop travels up the stem, therefore, it passes over the outside of the hook. The presser moves back and the hook opens. Continuing the downward movement further, the fabric loop slips over the top of the hook and the new yarn is drawn through it to form a new fabric loop. The needle now rises to its first position and the cycle is repeated, the new fabric loop now formed taking the place of the old loop.

The latch needle knits by a simple up-and-down motion, no help being required in closing the eye of the needle from a presser. The eye of the needle is closed by a latch which moves freely and swings into position as shown. The old fabric loop holds the latch open, then when the needle rises it slides down on to the stem of the needle. At this upper needle position, new yarn is fed into the hook of the needle. The needle now descends to its lowest position and as the old fabric loop slides up the stem it pushes up the latch, the end of which engages on the tip of the hook and closes the eye of the needle. The old loop continues up the needle but, as the latch is now closed, it passes over the outside of the eye and a new yarn loop is drawn through the old one. The needle then rises, the new loop takes up the original position of the old one, and the cycle is repeated.

In knitting machines using both types of needle, subsidary knitting elements, called 'sinkers', work in conjunction with the needles. The sinkers are specially-shaped strips of metal which are moved into their various positions by the machine to help in the knitting action. In latch needle machines, the sinkers perform the duties of holding down the old loops of the fabric to prevent them from lifting by friction against the rising needles, and also provide a ridge over which the needles draw the loops. In bearded needle machines, the sinkers measure out the yarn and help to form loops of that length, and also again prevent the fabric loop from lifting with the needles.

The machines for producing weft knitted fabrics are of three main types, straight bar machines, circular machines and flat bar machines.

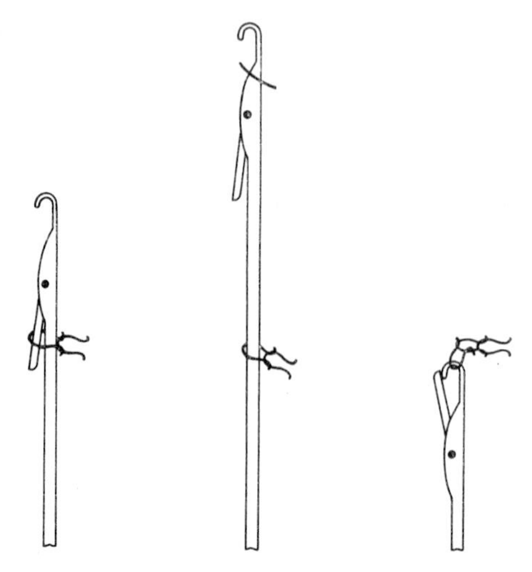

Fig. 6.18 Knitting action of a latch needle.

a. Straight bar machines. The straight bar knitting machine is the direct descendant of the Rev. William Lee's original knitting frame, using the same principle and type of needle – the bearded needle – as he employed. It has been improved, of course, in a number of respects, and the man who was responsible for the improvements and laid down the pattern for the modern machines of this type was William Cotton of Loughborough. He filed a number of patents over a period of about 20 years from 1845 onward and built machines which were, and still are, referred to as Cotton's patent knitting machines.

Straight bar machines produce shaped pieces of fabric which can later be seamed together to form fully-fashioned garments. The machine responds to the settings on a pattern chain and can increase, or decrease, the piece of fabric in width at any course by means of fashioning 'points' which come down on to the needles and lift selected loops for transfer to other needles, either further out, or nearer in, to the centre of the line of needles, whichever is required. The modern straight bar machine is a very long machine divided into sections, each knitting its own piece of fabric, and all performing the same actions at the dictation of the patterning chain.

This type of machine is used for knitting plain or ribbed fashioned underwear and outerwear. It was used extensively for the knitting of ladies' fully-fashioned hose but, since seam-free stockings became fashionable, it is now only used in limited numbers for that purpose.

Fig. 6.19 A tuck and interlock knitting machine.

b. Circular machines. These machines are built in a range of diameters from about $3\frac{1}{2}$ in (87 mm) for the manufacture of stockings and socks, and up to about 30 in (0.76 m) or more for the production of fabrics for outerwear and underwear. Latch needles are used in circular machines, and these are able to slide up or down in grooves cut round the cylinder of the machine and referred to as 'tricks'. A projection standing out from the bottom of each needle engages in a cam track of the required shape to ensure that, as the cylinder revolves, the needles are raised and lowered to perform the motions depicted in the illustration, and thus perform the knitting action. The packages of yarn to be knitted are grouped round the top of the machine, and the machine is usually provided with a number of 'feeding' positions spaced round the circumference.

c. Flat bar machines. In flat bar machines, the needles have individual movement and can slide each in its own groove or trick cut in a flat bar. The machine operates relatively slowly, but its capabilities for producing patterns make it popular. Many textured surface designs can be made and it can also be operated in conjunction with a jacquard mechanism to produce practically any type of coloured pattern. Flat bar machines are either hand operated or power operated. The hand operated machines are used to produce certain types of knitted goods, such as fully-fashioned bulky sweaters. The domestic type of knitting machine is of the flat bar kind.

6.4 WARP KNITTING

6.4.1 Warp knitted fabric constructions

An entirely different method of knitting is called warp knitting. In this method, at least one thread is provided for every needle, and the threads are arranged to run along the length of the fabric as does the warp in a woven fabric. The diagram shows the simplest warp knit construction using a single set of warp threads. One of the threads has been blacked in so that its path is more easily visible. It will be noted that the thread follows a zig-zag path along the fabric, a loop being formed at each change of direction. These loops are intermeshed with other loops in adjacent warp threads which follow a similar path along the length of the fabric. The use of single sets of threads, or single bar

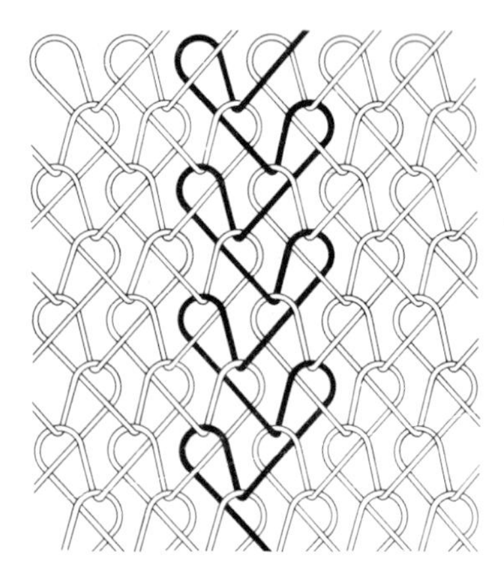

Fig. 6.20 Simple warp knit construction.

structures, is not very common and the majority of warp knit fabrics are produced using at least two sets of warp threads to obtain an improved stability and greater covering power in the fabrics. In recent years, warp knitting production has increased tremendously for a number of reasons. It can produce fabrics of good stability which can compete with woven fabrics in a number of applications – dress and shirt fabrics for instance. An attractive feature of warp knitting is its very high speed of production, made possible by the fact that the threads move very little horizontally during loop formation. Speeds as high as 1 000 courses per minute are claimed for some machines. It,is possible to use a wide range of yarn counts on a given gauge of machine. The warp knitting industry now provides a very big outlet for continuous filament yarns such as nylon, polyester and the rayons, but cotton and staple fibre rayon yarns are also used in substantial quantities.

6.4.2 Warp knitting machines

In warp knitting the threads are wound on to suitable beams which are mounted on bearings usually above and behind the knitting line. The threads are taken through a point bar to keep them in position, round a tension rail, and then through the eyes of the guides which control the motions of the threads relative to the needles. As knitting proceeds, a further supply of warp threads may be obtained by the tension in the sheet rotating the warp beam, or the beams can be positively driven at the correct rate.

The warp knitting machine was developed to make use of the bearded type of needle, and bearded needle machines are still much in use. In this machine, the guides,

Fig. 6.21 A warp knitting machine.

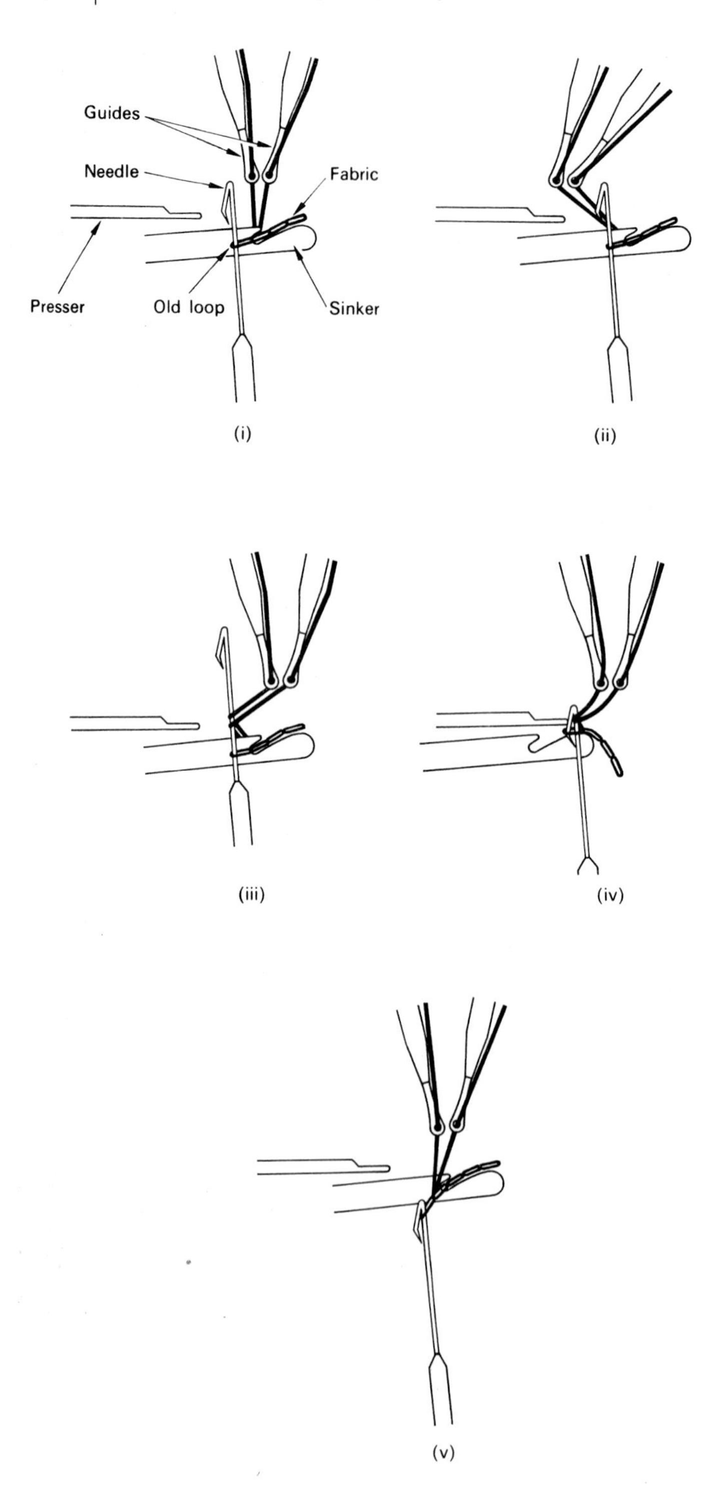

Fig. 6.22 Warp knitting action.

which control the movements of the warp threads are usually mounted in 'leads' and these are fixed on a horizontal bar to give the 'guide-bar assembly'. The knitting action will be understood from reference to the diagrams (i) to (v). The elements shown are the bearded needle, the presser which closes the eye of the needle at the appropriate moment, two guides (since two sets of warp threads are being knitted), and the sinker which serves both to hold down the fabric and provide a means of 'knocking over' when this is required. The sequence shown starts with the guides in front of the needle and the old fabric loop on the lower part of the needle stem. The guides then move back beyond the needle and the needle begins to rise, the fabric being held down in the 'throat' of the sinker. The guides now move to one side a slight amount to wrap the threads round the needle and then swing forward again. By this time the needle has reached its full height and the threads are round its stem. The needle now descends until the newly wrapped threads are in the eye, when a presser moves forward to close the eye. The needle pauses for a fraction of a second to allow the sinker to draw back for the wider part of the sinker to force the fabric up so that the old fabric loop passes round the outside of the closed needle eye. When the needle reaches its lowest position the knock-over has been completed and the sinker has moved forward so that its throat can again hold down the fabric as the needle rises to its starting position. The cycle is then repeated. This takes place simultaneously at every needle across the machine and at very high speed. The bearded needle machine is not restricted to the use of two guide bars, although two bars are employed to make very large quantities of fabrics for such uses as shirtings, sheets, nightwear, outerwear and curtains.

The speed with which warp knitted fabrics can be produced is a very attractive feature of this method of fabric manufacture and it is not surprising that attempts have been made to attain still higher speeds of production. One of the obstacles in the way of achieving this was the type of needle available. Both the bearded and latch needles, such as are used in weft knitting, had disadvantages for very high speed operation. Bearded needles need a presser to close the eyes, and the spring steel from which they are made soon fatigues at such high speeds of opening and closing. When latch needles are used at such high speeds, the inertia of the latch moving up and down becomes a limiting factor and also, in the case of latch needles, the larger eye formed when the latch is closed imposes an increased strain on the loop as it passes over the head of the needle. A special needle was developed, which has proved to operate very satisfactorily at these exceptionally high speeds. Its main part is made from a fine gauge steel tube formed, at its upper end, in the form of a hook. A tongue of smaller diameter tubular steel fits inside the main part and can slide up and down in it to close or open the hook of the needle. This tongue is power-operated independently and the vertical movement of the needle can be reduced considerably with practically no strain imposed on the fabric loops. Very high speed knitting can, therefore, be carried out and yet the high quality of the fabric is maintained.

6.4.3 The Raschel machine

This warp knitting machine makes use of the latch needle. The older type is usually a single-bed machine and it can be used to produce most of the fabrics which are made on the bearded needle machine. It is, however, slower in action than the bearded needle type, and is usually coarser in gauge. It is used, therefore, practically entirely for making special types of fabrics, such as nets, outerwear fabrics, hairnets and veilings.

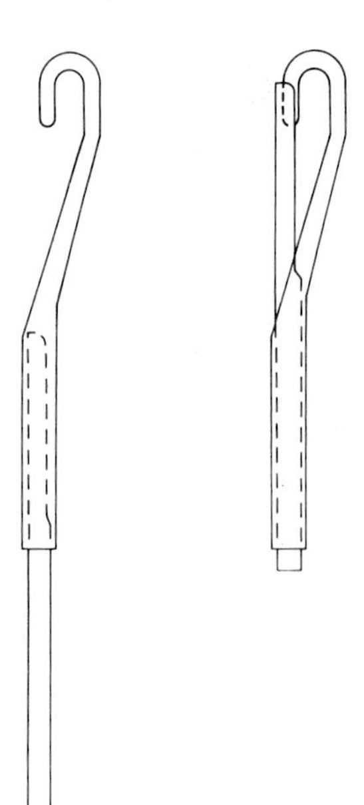

Fig. 6.23 High-speed warp knitting needle action.

Using two needle-beds operating at the same time allows a type of rib fabric to be made, but the fabrics are mainly produced for such applications as underwear, outerwear, scarves and shawls.

Raschel machines, used in conjunction with patterning mechanism, can produce lace-like fabrics, and these are now manufactured in very large quantities. This side of the Raschel activity has been associated very closely with the development of continuous filament synthetic fibres. The high strength of such yarns has enabled fine background nets to be made against which the heavy patterning yarns stand out and this has become acceptable as a form of lace.

6.5 LACE-MAKING

6.5.1 Lace fabric construction

Lace fabrics are made with decoration and adornment mainly in mind, although there are practical applications for such fabrics, as when open-work and net curtains are required to give a partically-obscured view, for instance.

In manufacturing a woven fabric, each pick of weft yarn is beaten-up close to the previous pick to give a solid, covering effect, as well as forcing the two sets of yarns, warp and weft, into sharply-kinked forms at the points of intersection so that substantial frictional forces are brought into play. It is these forces which contribute to the stability of a woven fabric. Considerable difficulties are experienced, therefore, if attempts are made to produce an 'open' woven fabric in which the threads are separated by a substantial amount. If an open structure is to be satisfactory, some other means of holding the threads in place at the points of intersection is required. Lace fabric uses such an alternative method of producing a stable fabric capable of recovering its original shape when released from distorting forces. In lace, the weft threads do not run at right angles to the warp threads, or interlace by passing under and over each other, as in a woven fabric, but are made to twist round the warp threads and provide the necessary gripping action in this way.

By using this simple idea, very attractive lace-patterned fabrics are produced. Such fabrics have been highly prized for their beauty and decorative possibilities for many centuries.

In recent years, the development of the synthetic fibres has provided new opportunities and even greater decorative possibilities for the lace-maker. These new materials allow him to use still finer yarns while retaining an adequate strength in his fabrics, and their freedom from shrinkage in wear and laundering have given the lace fabric new standards of stability of form and resistance to distortion.

6.5.2 Hand-made lace

Lace has been made by hand for many centuries, and two types have been produced, needle-point lace and bobbin lace.

Needle-point lace is made using only a needle and thread, and is very much akin to embroidery. This type of lace is thought to have originated in Venice when a foundation net fabric was first made and a design embroidered on this. Later the use of a foundation fabric was given up and, from about the year 1530 onward, Venetian lace was made with the pattern and background created by needle and thread at the same time. In preparing needle-point laces, the design was first drawn on a piece of parchment and then the out-

Fig. 6.24 Venetian point, dated mid-seventeenth century.

line pricked out using a needle which made a pair of closely spaced holes. Two needles threaded with coarse and fine threads respectively were used alternately to build up chains of stitches of different lengths connecting up the holes. The chains thus formed the outline of the pattern. The spaces within the outlines could then be filled in with a fine network of yarns to give a fine, delicate appearance, or made solid using a heavier yarn in button-hole stitching. Once a number of such design units had been prepared, they were usually joined together using very fine stitches to form the final piece of lace.

Bobbin lace is often referred to as 'pillow' lace, from the fact that it is supported while being made, on a pillow or cushion. The first step in making lace of this type was, again, the drawing of the design on parchment and this was then outlined with close-set pinholes. The parchment was then placed on the pillow and pins inserted through the holes and into the pillow so that they were set upright. The threads being used were wound on small bone pegs or bobbins about 5 in (75 mm) in length and shaped rather like a pencil; each bobbin had a groove at its 'neck' from which the thread was drawn. By manipulating the bobbins relative to each other, the threads could be worked round and between the pins, and intertwined and plaited to produce a stable design.

Hand-made lace was in great demand in this country from early Tudor times. By the middle of the eighteenth century, the demand so far exceeded supply that workers were organized to work in groups in factories. A number of lace-makers would work side by side along a bench; the hours were long and the conditions under which the work was carried out often left much to be desired. Even then, not enough could be produced to satisfy demand.

6.5.3 Machine-made lace

Nottingham has always been famous for its lace, and the first attempts to produce

Fig. 6.25 Bobbin lace, dated late nineteenth century.

lace on a machine were made in that city. These attempts were made by adapting the hosiery knitting frame for the purpose. Some forms of lace were produced on the knitting frame, but it became obvious that nothing which could rival the hand-made fabric could ever be produced in this way.

The first man to recognize that a machine which was entirely different from the knitting frame was required, was Heathcote and he set out to design his 'bobbinet' machine. This name was given to his machine since the threads which crossed and tied together the warp threads were supplied from bobbins. Heathcote solved the problem of how to construct a strong net fabric using a series of twists instead of an easily-distorted 'net' formed from knitted loops. Heathcote's patents were granted over the period 1808–09 and he started his first factory at Loughborough in Leicestershire. His machines were capable of producing 18 in (0.46 m) wide strips of lace net and, by 1810, a substantial work force was employed in making the net and embroidering patterns on it. Heathcote gradually increased the sizes of his machines so that net 30 in (0.76 m) wide, then 36 in (0.91 m), and finally 54 in (1.37 m) was being manufactured. He and his factory prospered, until an association of displaced craftsmen, called the Luddites broke into his factory and destroyed all of his machines. Heathcote was a man of character and perseverance, and he started right from the beginning to construct new machines and set up another factory, but this time at Tiverton in Devon. Tiverton still remains an important centre for making lace net.

In 1823, Heathcote's patents expired and anyone was free to use his method of manufacture. Factories were established in Nottingham where facilities for handling fine yarns and finishing fine fabrics already existed. Soon, Britain was supplying the world with plain lace nets.

Some idea of the basic principles upon which the manufacture of machine lace depends may be gained from the four diagrams (i) and (iv). A and B are two warp threads hanging vertically downward and C is a bobbin thread attached to a bobbin which is able to swing in and out of the plane of the paper. The sequence starts with diagram (i) in which the bobbin thread C is swung back. In diagram (ii), the two warp threads are taken to the opposite sides and cross in the middle, C being still held back. The bobbin thread C now swings forward through the crossed warp threads A and B, as shown in diagram (iii). In diagram (iv), the warp threads A and B have now returned to their original positions, leaving the bobbin thread C wrapped round them. It will be appreciated that, if the bobbin thread C is under strong tension, and the warp threads A and B are not tensioned, then the warp threads will wrap round the bobbin threads. If the tension conditions are reversed, the bobbin threads will twist round the warp threads.

Heathcote's machine was capable of producing the foundation lace net fabric only and it was left to Leavers of Nottingham to devise a machine method of developing the lace pattern as well as the groundwork in one process. His first machine was devised and built round about the year 1813, and it says much for his ingenuity that the modern machine, still referred to as a Leavers machine, uses essentially the same principle of operation as he laid down. Leavers was successful in developing his machine and, from that time, Nottingham became the main centre in the country for machine-made lace.

Almost any kind of pillow type lace can be made on the Leavers machine and today the machine has reached such a stage of development that it is difficult to distinguish between the hand- and machine-made articles. Leavers' first machine was made of wood, was quite small in size, and was hand operated. Sometimes the worker owned his

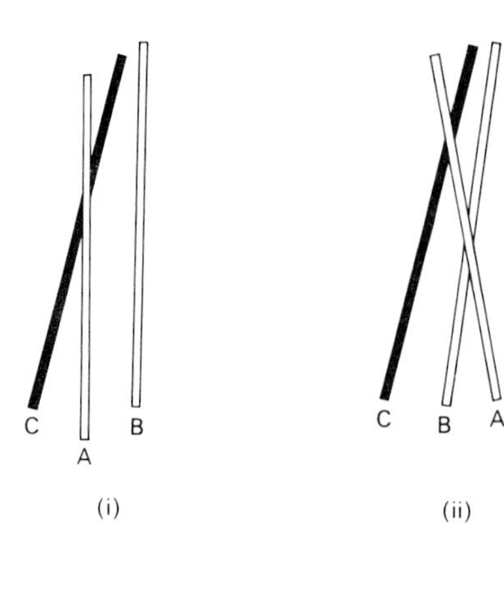

Fig. 6.26 Principle of machine lace making.

machine and operated it in his own home, but more usually a contractor hired it out to him, supplied him with yarn, and collected and marketed the lace he made. As the years went by, the machines became larger and more complicated and finally were power-operated, firstly using water-power and later steam-power, and the lace factory came into existence.

The Leavers machine uses a warp or warps prepared in much the same way as for weaving. The bobbin yarns are wound in special 'bobbins' used only in lace-making. These bobbins consist of two thin brass discs riveted together with a circle of rivets round a central square hole through both discs, and with the discs separated by a short distance. The bobbin yarn is wound in the space between the two discs, the yarn being supported on the circle of rivets. The bobbins are wound, about 120 at a time, by being threaded on to a square rod which is rotated about its axis, thus drawing the yarn from packages on to the bobbins. The yarns are wound under a controlled tension, with each bobbin taking exactly the correct length of yarn. When the bobbins are full, they are taken from the shaft when they will have been found to have expanded in thickness due to the outward pressure of the wound yarn. They are, therefore, stacked one on top of the other in a press, to restore the original thickness. The compressed bobbins are given a treatment with live steam for about one hour and then cooled.

The filled bobbins are finally fixed each in its own thin metal 'carriage'. A circular

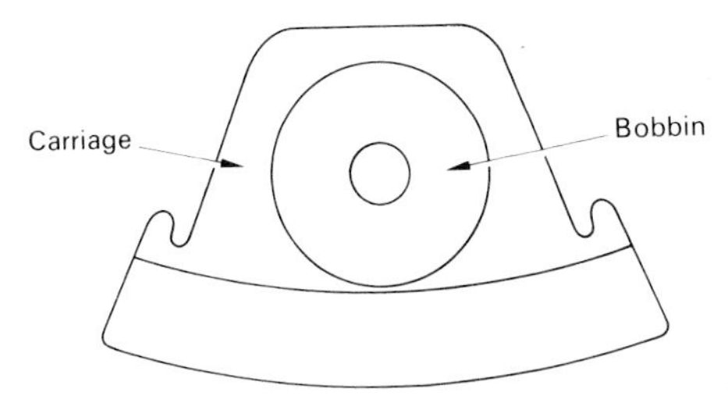

Fig. 6.27 Bobbin and carriage for lace machine.

hole in the centre of the carriage takes the bobbin, and a fine spring locks the bobbin in position. The two shaped depressions, one on either side of the carriage, are designed so that the 'catch bars' of the lace machine can fit in them and control the movements of the carriage. The lower rim of the carriage is thinned down so that it can run between the 'combs' on the machine. The yarn is drawn from the bobbin through a small hole at the top of the carriage.

The diagram is a section taken at right-angles to the longitudinal axis of the Leavers machine. As already mentioned, the machine uses a set of warp yarns, and a bobbin yarn. The bobbins swing between the 'teeth' of two curved combs which are fixed fac-

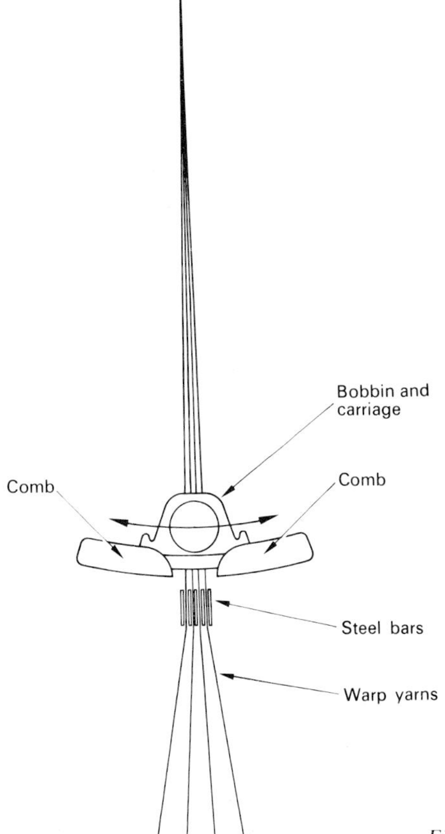

Fig. 6.28 Section through a Leavers' lace machine.

ing each other with a space in between as shown in the diagram; they always pass from one comb to the other like pendulums, being guided by the slots of the combs. The warp beams are supported in slots below and behind the machine and the warp threads are spaced by coming up through a 'sley' after which they pass through a series of holes in steel bars and into the space between the combs. The steel bars are actuated by a Jacquard mechanism and may move either to the right or the left as the patterning mechanism dictates. After each movement of the steel bars, to cause the warp and bobbin yarns to plait together, a 'point' bar moves up and compresses the twist of the previous plait to form another tiny portion of the lace fabric. The take-up mechanism at the top of the machine now takes up this newly-formed portion of fabric, and the cycle of movements is repeated. In the lace-making action, the carriages are supported by the threads wound on the bobbins, and the front and back catch bars of the machine engage and disengage as required to maintain the pendulum-like swinging of the carriages and the bobbins they carry. The carriages are travelling at their highest speed when the bobbin thread is vertical, and there is a slight dwell at each end of the swing to allow the crossed warp threads to be tied firmly by the bobbin threads.

The men who operate the Leavers machines are referred to as 'twist hands' since the method of forming the lace is by twisting yarns together. Many ancillary operations follow the actual lace-making. Each piece of lace is carefully examined and any requiring mending is passed on to the mending room. Clipping, involving the cutting out of motifs, and trimming the scalloped borders of decorative edging, is done by hand using a pair of scissors. Narrow laces are made side by side to form a large width of fabric, and the linking threads are subsequently withdrawn to separate the bands and allow them to be wound on to cards. Crimped yarns are now often used to give a three-dimensional effect provided by the extra bulkiness of the yarns.

6.5.4 Other types of lace

a. Schiffli embroidery. The domestic sewing machine was invented about the year 1824 and this was used at an early date to embroider a design on a foundation net fabric. A single needle was used and, since the machine had to be operated by hand, production was slow and uneconomic. Many people conceived the idea of using a number of needles on one machine all operating together to produce repeat patterns, but the first one to prove successful was the Schiffli machine, developed in Switzerland during the 1850s. This same type of machine is still used for embroidering fabrics. The machine is based on the action of the normal sewing machine, and the bank of needles work simultaneously through the net fabric suspended vertically in front of them. As the needles penetrate the net, a separate shuttle operates for each to catch the loops of thread in the normal way. The design is developed by moving the net as required and not the needles, and the movements of the net are controlled by an operator using a pantograph with which she follows the enlarged design drawn on paper. Very attractive patterns can be produced by this means, and the lace made is often referred to as 'needle-run net'.

b. Guipure lace is made by embroidering using a cotton yarn on a silk fabric and subsequently dissolving away the silk to leave the pattern motifs joined together by strands only and without a background mesh.

c. Barmen lace, often referred to as braid lace, is made on a different type of machine called a Barmen machine. The action of the machine is related to the movements of bobbins guided into interweaving paths by the grooves in a circular plate. It might be described as mechanized pillow lace production. Very fine to very heavy yarns can be used, and this type of machine is used to produce edgings of all descriptions.

6.6 CARPET-MAKING

6.6.1 Carpet construction

A carpet is a textile 'fabric' in which a pile stands up vertically from a backing cloth. Originally the tufts of yarn which form the pile were knotted into the backing fabric and then locked in position by the weaving process. Many carpets are still made in that way, but others rely on the insertion of the tufts into a previously woven backing fabric and the anchoring of the tufts in position by a variety of means.

In the East, carpets were originally designed and constructed for wall decoration, as prayer mats, and as saddle cloths. Nowadays the name carpet is associated solely with floor coverings. Persian carpets have long been prized for their beauty of design and their excellence of construction; the Persian carpets are essentially floral in design and much use is made of curves rather than angular lines. On the other hand, Caucasian and Chinese carpets are essentially geometric in design.

6.6.2 Hand-made carpets

Hand-made carpets have been made for many centuries. Originally the warp was stretched between two heavy pieces of wood, one being supported in an elevated position, such as on a tree, and the other resting on the ground. It is thought that the warp originally consisted of threads of highly-twisted wool; later on these were replaced by linen and cotton threads and, in some early Eastern carpets, the warp is composed of silk threads. Once the warp sheet was in position and under the correct tension, the foundation or backing fabric could be woven by interlacing the threads with a weft yarn as already described in the section on hand weaving. After every pick, or two picks, of weft yarn had been inserted, the carpet-maker tied in a short length of pile yarn to the warp threads in a line across the sheet. The insertion of one or two further picks of weft yarn then locked these tufts in position. This procedure was repeated as weaving progressed until the whole area of the foundation fabric finally produced was covered by tufts of pile yarn standing out from the surface of the foundation fabric.

Eventually, much the same procedure was followed but using a hand-loom. The two rollers, or beams, one carrying the warp sheet and usually referred to as the 'chain warp', and the other to take the carpet as it was made, were supported in the frame of the machine, the warp beam at the top and the carpet beam at the bottom. The warp ends were held at the correct spacing by being passed through a reed. Two rods passed across the warp sheet just below the warp beam, and the warp threads were passed alternately in front of one rod and behind the other, thus dividing the sheet into two layers to form a shed across which the weft threads could be passed. The pile yarn was then tied in by the operator, round a pair of warp threads, one from the back part of the sheet and the other being the adjacent thread in the front part. The knots were pulled tight and the pile yarn cut to the required pile length. Two weft picks were then woven in the foundation fabric and beaten up by the teeth of a heavy comb to lock the knotted pile yarns in position.

Two main types of knot were, and still are, in use; these are the Turkish knot, and

Fig. 6.29 A carpet hand loom.

the Persian knot. The method of tying-in used in each case can readily be seen from the diagrams.

6.6.3 Machine-made carpets

a. Wilton carpets. The forerunner of the present-day Wilton carpet was produced on what was known as the 'Brussels' loom which was introduced into Britain at the beginning of the eighteenth century. This loom was the first on which a pile carpet could be woven mechanically and the first to which a Jacquard patterning mechanism was applied. The loom was first operated by man-power, but later round about 1850, other

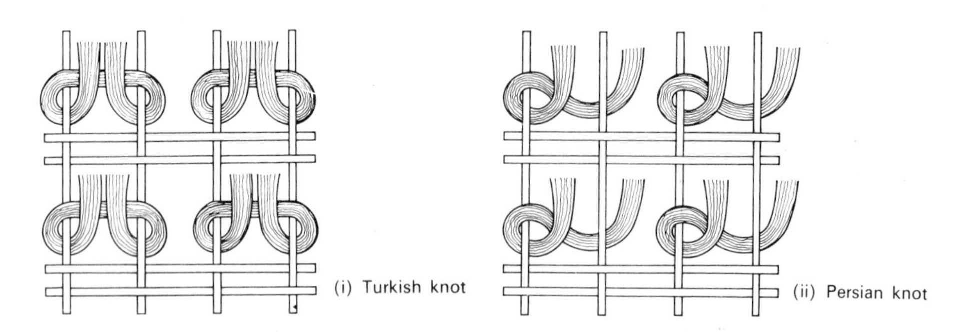

Fig. 6.30 Types of knots for hand-made carpets.

means of driving the mechanism were applied. In the Brussels carpet the tufts were in the form of loops and were packed tightly all over its surface to give a neat, dense appearance and hard-wearing characteristics. Later on, the loops were cut to produce a cut-pile carpet which then became known as the 'Wilton' carpet. Probably the only other main distinguishing feature between the modern Wilton and the old Brussels carpets is the fact that, in the latter, two picks of weft of the foundation fabrics are inserted to lock the tufts into position whereas in the Wilton three such picks are provided.

In manufacturing a Wilton carpet, two separate warps are used, one being composed of strong cotton or jute threads which is used in conjunction with a weft of a similar yarn to form the foundation fabric, and the other, usually of wool threads, to form the pile. The foundation, or 'chain', warp is wound on to a beam and so is the pile warp if a plain Wilton carpet is being made; with patterned carpets the pile threads come from bobbins arranged in creels. The function of the pile warp is simply to form a pile and, when the various coloured threads are not drawn to the surface of the carpet for this purpose, they are lying 'dead' within the body of the carpet and are known as 'stuffers'. The stuffers add to the thickness and luxurious feel of a Wilton carpet; the fact that they are packed away all through the body of the carpet uses up a very considerable quantity of pile yarn and the Wilton construction is a costly one. A good quality Wilton carpet will contain something like 90 to 120 tufts to the square inch (14 to 19 to the square centimetre), and the more intricate a pattern is in terms of the number of coloured pile yarns used, the more stuffers there will be and, it follows, the more expensive the construction.

Operated by the Jacquard mechanism, sheds must be provided for both the foundation warp and the pile warp. When the pile warp threads are lifted, 'wires' are inserted across the loom to hold up the pile warp threads to the length of loop required, and three picks of foundation weft are inserted to hold the loops round the wire. The wire is, in fact, a strip of metal standing on its edge and is inserted by the machine, along a groove in a 'transfer slide'. The construction of the carpet proceeds and, after the three picks of foundation weft have been introduced, another wire is inserted to hold up its row of loops. Twenty or more of such wires are used before any attempt is made to withdraw the first one which had been inserted; this is done to avoid 'robbing' of one loop by another under the tensions applied in the pile yarns. At one end of each wire is a sharp blade and, as the wire is withdrawn at its appropriate position this cuts the loops to give a cut-pile carpet. As the wires are withdrawn, they are moved back to the initial position so that they can be re-inserted once again.

Originally, Wilton carpets were woven on standard looms of only 27 or 30 in (0.68 or 0.76 m) in width. At the beginning of the present century, looms having widths up to 12 ft (3.66 m) were constructed. These wide looms introduced extra difficulties. The long wires were highly flexible and difficult to control and they had to be withdrawn from the carpet construction quite slowly in order to avoid generating an amount of heat which would scorch the loops as they were being cut.

b. Axminster carpets. Machine-made Axminster carpets were so-called, not because they had any association with the town of that name, but because they resembled in appearance the hand-knotted carpets originally made at Axminster.

The machine used to manufacture Axminster carpets involves the use of two distinct mechanisms, one to weave the warp and weft of the foundation fabric, and the other to insert in one operation a full row of tufts of pile yarn across the machine. This latter mechanism inserts just sufficient yarn to form a pile tuft, and practically no pile yarn remains hidden in the body of the carpet as in the Wilton construction. Axminster carpets are, therefore, cheaper to produce than Wiltons.

The more common type of Axminster machine makes use of long spools upon which the various coloured pile yarns are wound in order. A full spool is then taken to a threading machine upon which the free ends of yarn are pulled through a row of tapered tubes, something like hypodermic needles, with one tube for each yarn. A spool, with its tubes and frame, is then taken to the carpet machine and fixed across a pair of endless chains operating at the upper part of the machine. Other spools and attendant tubes are attached to the chains in a similar way, rather like the rungs in a flexible ladder.

The action of the carpet machine is as follows. As the chains bring a spool and tubes into position, the tubes are lowered so that their ends penetrate through the warp sheet of the foundation fabric. The tubes then withdraw a specified amount, leaving behind the second leg of a pile tuft. A knife now cuts through all the pile yarns, leaving a row of tufts across the fabric. Two or three picks of weft thread in the foundation fabric are inserted by means of a long needle entering the warp shed from one side, and these lock the tufts firmly in position. Since no part of the pile yarn contributes to the thickness and weight of the body of the carpet, sometimes a special set of stuffer yarns is included in the structure to produce a more substantial carpet.

A considerable number of pile yarn spools is used in making an Axminster carpet and long feed chains have to be provided and accommodated above the loom. The preparation of the spools is lengthy and attempts have been made to find alternatives to the spool system. One of these is to employ a set of 'grippers' to feed the pile yarn into the carpet structure. The grippers are made of steel, about $\frac{1}{16}$ in (1.6 mm) thick and,

Fig. 6.31 View of spool setting for an Axminster carpet.

Fig. 6.32 An Axminster carpet loom.

viewed from the side, the shape of each resembles the neck, head and beak of a bird. The grippers are threaded in line on a shaft in front of the yarn carriers through the slots of which the individual pile yarns are threaded. The 'beaks' of the grippers can be made to close and open by a special mechanism. In operation the shaft rotates and the grippers swing forward to grip the pile yarns protruding from the carriers. A small amount of pile yarn is now withdrawn, either by the grippers moving back a short distance, or the grippers remaining stationary and the carriers moving back. A comb now descends and prevents any further yarn from being pulled through the carriers and knives cut off the tuft lengths of yarn. The grippers, still holding the ends of the tufts, now swing once again to pass through the upper part of the foundation warp shed and lay the tufts where they can be locked in position by the weft yarns. The gripper's jaws now open to release the ends of the tufts, and the grippers return to their starting position. A 'rake' comes into action and bends the tufts up into a vertical position.

c. Chenille carpets. This method of carpet-making was invented in 1839 but is now only used to a very limited extent.

Chenille carpets are produced using two distinct weaving operations. The first one is carried out to produce the 'fur' and this fur is used in the second weaving operation to form the pile. In producing the fur fabric, the warp consists of groups of a number of cotton warp threads spaced about $\frac{1}{2}$ in. (12 mm) apart. A wool weft is then inserted by a shuttle, and weaving proceeds in the normal way to produce a fabric composed of narrow bands. The fabric is now placed on a cutting machine where it is cut into warp-wise strips in between the bands so that each strip has its group of cotton warp threads with wool weft threads standing out from the sides. The strips are then steamed and passed in the grooves of a heated cylinder to press them into a V-shape, in which shape they become set.

In the second weaving operation, the fur strips are used as weft in weaving the carpet above the foundation structure in which the fur is bound securely.

Since the preparation of a Chenille carpet involves two weaving operations, the method is not cheap to operate, and this has contributed to the decline in the use of carpets of this type.

6.6.4 Tufted carpets

The principle of making carpets by 'tufting' has been in use since about 1930, first using a single tufting needle and later multi-needle machines. The first wide tufted carpet machine was made and used in the USA about 25 years ago, and this was followed by various models capable of being operated at higher and higher speeds. By reason of its speed of production, a tufted carpet is cheaper than a woven one, and its use has extended rapidly in the USA and the various European countries.

In a tufting machine, the pile yarn is taken directly from the cones mounted in a magazine-type creel so that long, uninterrupted runs are possible. The yarn is taken from the cones through guide tubes, which are usually about $\frac{1}{4}$ in (6 mm) in diameter, the threading being carried out with the help of compressed air. The guide tubes lead their individual yarns to the tufting position and keep the yarns separated and in correct order. From the tubes, the sheet of yarns is taken through the nip of feed rollers and then each yarn is threaded through its own tufting needle.

The needles are arranged vertically in a row across the machine. All the needles move down at once to penetrate through a previously-woven foundation fabric, taking the yarns with them, and a looper passes between each yarn and its needle to form the loop. The needles now rise to repeat the action, a presser bar coming into operation to prevent the foundation fabric from rising with the needles, and as fresh loops are produced these accumulate on the neck of the looper, sliding forward with the motion of the foundation fabric. The first loop formed has now travelled along the looper until it has reached a position where it is cut by a blade to form the carpet pile. A loop pile carpet is made in a similar way but without the cutting blade, the loops being allowed to slip off the end of the looper.

When the tufts are formed, they are simply held in the foundation fabric by friction and not locked in position by a weft thread. To hold them in place, therefore, the back of the fabric is coated with an adhesive, such as a latex compound.

Originally only plain tufted carpets were produced, but designs have now been introduced by various means, including colour printing of the pile, using various pile heights and, by giving the needle bar a lateral movement to produce a wavy effect.

6.7 NON-WOVEN FABRICS

Before a fabric can be prepared by interlacing methods, such as weaving or knitting, a yarn must be prepared. The combination of yarn preparation and fabric formation, involving a lengthy series of expensive operations, adds considerably to the cost of the final garment. It is not surprising, therefore, that methods of forming a fabric from a mass of the loose fibres without going through the yarn stage, have been sought, and in recent years much progress has been made in the development of so-called 'non-woven fabrics'. All such fabrics depend on the preparation of a thick layer or web of loose fibres which can then be treated to bind the fibres together and so produce a type of coherent fabric. Non-woven fabrics will always be regarded as being of a lower grade than woven or knitted fabrics, but they are perfectly satisfactory for applications where softness and absorbency, coupled with a moderate strength, can be utilized, and their use is gradually extending.

6.7.1 Preparing the fibre layer or web

Before the loose fibres are bonded together to form a fabric, they must be arranged in an even layer of suitable thickness. The carding machine, mentioned in connection with the processes carried out on the fibres in preparation for the spinning of a yarn, is a suitable method of preparing a fibre web. In another method, a random distribution of the fibres can be obtained by blowing the loose fibres into a large chamber and allowing them to settle on a belt moving across the lower part of the chamber. A non-woven fabric made from a random arrangement of the fibres will have an equal strength in all directions. It is possible, however, to align the loose fibres so that they are all lying in one direction, when the fabric made from such a web will exhibit a maximum strength in the direction of alignment, and a minimum strength at right angles to it. It is usual to superimpose a number of thin webs, one on top of the other, to build up the correct final thickness, and this does provide an opportunity either to have all the fibres facing in one direction to give the directional strength effect, or to cross them in alternate layers and so obtain a final fabric which has equal strengths in the length and width directions.

The web can be formed of one type of fibre selected for its characteristic properties with a particular type of application in mind, or fibres of different materials can be mixed so as to 'tailor' the properties required. Thus, strength and abrasion resistance can be allied to softness and absorbency by using mixtures of natural and synthetic fibres in the webs.

6.7.2 Mechanically-bound fabrics

One of the most successful methods of binding together the web of fibres is by 'needle-punching'. The machine upon which this punching process is carried out has an oscillating beam carrying rows of barbed needles mounted close together. The web of fibres is supported on a slotted table and the needles are pushed up and down into the web until the whole surface has been covered. As the needles descend, the barbs carry down with them some of the fibres, and when the needles lift they leave these fibres embedded in the body of the web. Continuation of the needling leads to an even greater entanglement and the web becomes more dense and takes on the characteristics of a fabric. The needles are oscillated up and down at high speed, rates of up to 900 penetrations per minute being attained.

6.7.3 Adhesive-bound fabrics

Other methods of binding the web of fibres together use adhesives for the purpose. These can be operated in various ways. One of them is to prepare a web of mixed fibres, one of which, at a suitable stage, can be made to act as an adhesive material. The bonding fibres are evenly distributed throughout a fairly thick web, and the layer is then hot pressed, when the special fibres partially melt and become sticky to adhere strongly to the other fibres present. Release of the pressure and cooling down leaves a non-woven fabric firmly held together by the bonding fibres. Often a synthetic fibre having a relatively low melting point is used as a bonding material.

For a variety of reasons, it may be considered advisable to use bonding fibres which do not themselves become tacky on heating but which can be coated with a substance which becomes tacky either on heating or being treated chemically. Whatever method of adhesive bonding is used, there must be no risk of the bonding material re-melting and freeing the web fibres under the conditions the fabric will experience in practice.

6.7.4 Felts made from animal fibres

The oldest type of non-woven fabric is felt, made from wool or other animal hair fibre such as rabbit fur. As with the non-woven fabrics already mentioned, the process starts with a loose web of fibres, built up to a suitable thickness, but the binding action is obtained from the fibres themselves. As already mentioned in ch. 3 when animal fibres were being discussed, wool and certain other hair fibres possess the unique property of being able to 'felt' together, when treated under suitable conditions of dampness and heat, which enables the fibres to interlock with each other. The property arises from the scaly structure of the surfaces of the fibres which gives them a differential frictional effect. Under certain conditions, rubbing of the fibres causes them to creep amongst the mass of fibres in the direction of the root ends, so that finally they become so intermingled and interlocked together that the web has the strength expected of a textile fabric.

The old method of felting the web of fibre, and one still used, is to damp it with soap solution and pound it with wooden mallets. The strokes of the mallets produce compression, followed by relaxation as the mallet heads draw back for the next stroke, and this action, particularly if the soap solution is warmed to about 40°C (104°F), causes the web to felt quite rapidly. Care is taken to see that the whole area of the web is treated so that uniform felting is obtained. The pounding mallets may be operated in two ways; they can be lifted by the machine and then released to fall on to the web under their own weight, or they can be positively driven throughout by being attached to a crank shaft and thus enabled to exert a constant pressure at the instant of impact.

Rough felts are used for quite a wide variety of domestic and industrial purposes. The finest and best felts are used for hat-making, rabbit fur being preferred for this purpose.

6.8 STRETCH AND BULK FABRICS

As has already been indicated in ch. 5, the act of crimping a man-made thermoplastic filament gives to the yarns made from it many of the characteristics of a spun yarn but retains the undoubted advantages of using continuous filaments. Fabrics made from crimped or 'textured' yarns have the high strength, abrasion resistance, toughness and easy-care properties of the fibre material, while the crimp formation gives stretch, bulk, softness of handle, and moisture retention of a fabric made from spun yarns. In addition, by selecting the processing conditions while the crimp is being inserted, changes in the handle and texture of the fabric can be made at will.

Fig. 6.33 Artificially crimped wool emerging from a crimper and being conveyed away by a scotch feed.

The fabrics are prepared either from yarns which have a latent crimp which must later be developed in fabric form, or from fully crimped yarns which are held out in a partially crimped condition by the restraints imposed by the fabric structure. In both cases, the application of heat and moisture is required to obtain the full crimp, when the fabric must be in a free, relaxed state in order to shrink as the crimped yarn contracts. It is probably for this reason that textured yarns have not, up to the present, made any marked impact on weaving and warp knitting and, so far, weft knitting has offered the most important outlet for them. The open, rounded loops of a weft knitted fabric provides conditions where this contraction can easily take place and the full benefit of the texturizing be obtained. Fabrics knitted from crimped yarns are made appreciably larger in size than is finally required in order to allow for the shrinkage which takes place during crimp development.

There are two types of crimped yarn in use. One type can extend very considerably when pulled and will return quickly to the unstretched condition when released from the pull; 'stretch fabrics' are made from such yarns. The other type is one in which the crimp has been partially set out by a heat treatment to give a very limited degree of stretch but an enhanced bulkiness, softness of handle and attractive appearance; fabrics made from these yarns are 'bulk fabrics'. Stretch garments are designed to cling to the body or limbs of the wearer, this being their primary function rather than providing bulk and covering power. Garments of this type are stockings, men's and boys' socks, children's socks and tights, and swimwear. Set or bulk yarns are finding an increasing use in knitted outerwear and underwear. The bulkiness is developed in the garment during after-finishing, but the shrinkage which takes place in bulk fabrics is small, compared with stretch fabrics, and the garments or fabrics can be knitted nearer to the finished size.

The bulk fabric market is dominated by the set polyester and, to a lesser extent, the textured acetate yarns, the latter finding quite a substantial market by reason of cost considerations. Set polyester yarns are used in large quantities for double-knit women's dresses and suits, for women's slacks, shorts and sweaters, and for men's sport shirts and sweaters, and their popularity for other types of garments is still increasing. Their popularity arises from the clearness of stitch, cleaness of appearance, durability and long wearability, resistance to creasing, softness of handle, and the easy-care properties, which they confer on the garments made from them. They are just beginning to be tried in warp knit fabrics and in fabrics for tailored men's wear such as suits, sports jackets and trousers where permanent creasing and a slight amount of 'give' in the direction of strain are important considerations. Applications of this type are considered as offering very big outlets for bulked polyester yarns in the future.

Something like 1 m. lb weight of man-made fibres are processed for stretch or bulk fabrics each week in Britain. The method of crimping in greatest use is false twist and this method probably accounts for about 80% of the textured yarns produced, with the stuffer-box method providing 10% and the other methods making up the remainder.

6.9 ROPES, CORDS AND TWINES

Textile fibres are used in the manufacture of 'cordage' structures, which include the various kinds of ropes, cords and twines. The industry uses substantial quantities of the natural vegetable fibres but only very limited quantities of the natural animal fibres. The usage of the man-made fibres increases year after year. The coarser grades of the natural fibres are, in general, more suitable for cordage manufacture.

Being more concerned with the vegetable fibres than the others, cordage manufacture is divided up in accordance with the types of these fibres which are employed, and each section has developed processing equipment demanded by the characteristics of the fibres which have been selected. The hard fibre section of the industry, which makes use of the leaf fibres, produces coarse yarns, and agricultural and industrial twines, cords and ropes. The soft fibre section uses the bast fibres from which to fashion finer yarns, packaging twines and cords. Both sections of the industry now use some quantities of the man-made fibres, these usually being processed in the continuous filament form.

Cordage structures are built up by means of a series of twisting, folding or plying, and sometimes braiding, operations, such as are employed in the manufacture of the standard textile yarns, and the methods of preparing the fibres in their correctly aligned order for these operations are similar to those already described in ch. 5. The main characteristics required in all cordage products are, high tensile strength, compactness with no looseness in the structure, a good pliability and an ability to avoid distortion of form while in use. The tensile strength developed will be determined by the type of fibre used, the accuracy with which the fibres are aligned along the axis of the cord and the amount of twist which is inserted to bind the fibres together and prevent them from sliding apart. Compactness is obtained by carefully controlling the successive twisting operations; the fibres are first twisted together to form yarns, yarns are combined by twisting together into strands or cords, and cords are twisted together to form ropes. The excellence of the compacting will be determined by the amount of twist inserted, the tension applied in the components of the structure while the twisting is being carried out, and the mechanical methods employed to guide and press the components into their correct positions while they are being combined by twisting. Pliability is influenced by the properties of the fibres used in the construction, and the degree to which one component can slide and adjust its position relative to the others. Resistance to distortion of the cross-section of the cordage structure during use is a function of the compactness which has been achieved during manufacture.

In the cordage industry, yarns are first prepared and the various products are built up from these. A twine is taken to be an assembly of parallel yarns twisted and compacted into a structure of considerable length; it is used mainly for tying and binding. Ordinarily, a twine is an unbalanced structure, and, under certain conditions, can lose some of its twist. A cord is a system of yarns twisted or braided together into a structure of considerable length with effective compacting of its components and of a balanced construction to obviate the tendency to untwist and lose compactness; cords are generally under $\frac{3}{16}$ in (4.8 mm) in diameter. A rope is formed from a series of strands, built up from twisted yarns, and usually with a diameter greater than $\frac{3}{16}$ in (4.8 mm). (Ropes are usually sold on the basis of circumference rather than diameter, so that, for example, a 'two-inch' rope is approximately $\frac{5}{8}$ in in diameter.) A rope is structurally, balanced and is produced using alternate directions of twist at the succeeding yarn, strand and final rope stages. A rope formed from three strands twisted together is referred to as a plain or 'hawser-laid' rope; one formed from four strands is a 'shroud-laid' rope.

6.9.1 Rope production

The process of twisting together the strands to build up a rope is referred to in the cordage industry as 'rope-laying'. As already mentioned, a rope is produced by twisting together three or four previously-prepared strands. The rope twist is opposite in direction to that inserted in the strands, and the effect of this rope twist is to take out some of the twist in the strands. The rope-laying process should, therefore, not only twist the strands into a rope, but also provide some means of restoring the twist which is removed from the individual strands. The term relating to the twist inserted in the rope is 'afterturn', and that for the restoring or retaining twist in the strand is 'foreturn'.

The machines used to lay ropes are based on three different principles of operation. In the first, the strands are individually rotated at the end where they are fed into the rope structure, while the rope twist is applied from the opposite, take-off, end. The second type of machine twists the individual strands and, at the same time, rotates them round each other to produce the rope twist, all from the same feed-in end. The third, and simplest method, is to over-twist the individual strands during their preparation so that, during the formation of the rope, the strands can be simply twisted together without any compensating strand twist being added, the over-twist being lost in the normal rope-laying operation. The overtwisting can cause damage and weakening of the strands, and the method is only employed to make smaller ropes and cords, and ropes which do not have to meet a critical performance specification. The other two methods are used to

Fig. 6.34 A large modern machine for making 8-strand plaited ropes for ships' moorings.

make ropes from hard fibres, the second one having the advantage that the emerging rope can be coiled as it is produced. Originally prepared by hand methods, ropes were first made by a machine, patented by Cartwright, in 1792; this machine design was based on the second of the methods referred to above, namely with foreturn and afterturn inserted from the same end.

The first part of the rope-laying process consists in preparing the strands. As the yarns are twisted together to form the strand, the necessary compactness is obtained in the strand by dragging the component yarns through a compression die or tube, known as the 'strand tube', as they are being bound together by the twist. At the same time this tube serves to guide the individual yarns into their correct relative positions in the strand structure. The yarns for the strands are taken from an assembly of bobbins supported in a creel and, in some creels, the pins supporting the bobbins are sufficiently long to allow two bobbins, one above the other, to be supported on each pin with the yarn from the inside of one bobbin being tied to the outside of the yarn from the other, so that a longer continuous run of the yarns can be obtained. From the creel, each yarn is taken through its appropriate hole drilled in a 'register plate' of curved cross-section, and then into the strand tube. In practice, the strand tube is located as close as possible

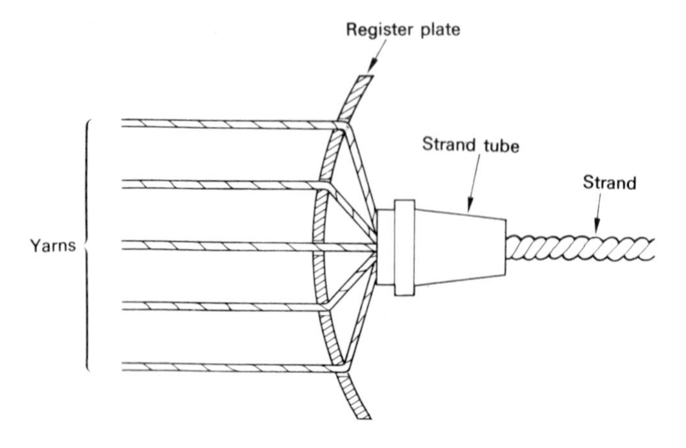

Fig. 6.35 Diagram showing rope laying.

to the register plate, in order that the yarns will have to bend sharply and any protruding fibres standing out from the yarns will be sheared off thereby. Twist is inserted in the strands using a flyer principle, which is referred to in ch. 5, and, since the tensions used are high, hauling capstans are also employed. Various types of strand-forming machines are available.

An older method of forming the strands, and which is still used for certain kinds of heavy strand, is one known as the 'ropewalk'. The equipment consists of a trolley, or 'traveller', which runs on rails along the whole length of the ropewalk, and which carries twisting hooks driven by the motion of the traveller, brackets or 'carriers' which can be made to extend over the traveller rails and support the strand as it is formed, and a rope drive system which causes the traveller to move and, at the same time causes the twisting hooks to rotate. The driving rope is a moving endless rope extending the whole length of the ropewalk, and this is made to rotate a pulley on the traveller to propel the trolley forward, while rotation is also transmitted to the twisting hooks by means of a train of gears operated by the pulley. To produce the strands, the requisite number of yarns are taken from a creel and attached to the twisting hooks. When the rope drive is set in motion, the traveller moves along its rails, and the twisting hooks insert twist to form the strand which then rests, as it is produced, on the projecting carriers. When the necessary length of strand has been produced, the machine is stopped and the strand is cut away and fastened to a post. The other end is removed from the traveller and is also

Fig. 6.36 Doing the 'ropewalk'.

secured to a post, leaving the strand held out under tension between the two posts for a period. Meanwhile, the traveller has been returned to its starting position when the cycle can be repeated. At first sight, the method may appear to be old-fashioned and somewhat clumsy, but it is relatively speedy, simple and flexible in operation.

Having prepared the strands, these are twisted together, in a direction opposite to that used to make the strands, to form the rope. The critical part of the process is the combining of the strands correctly at the point at which they enter the compression tube or die. The die is made of cast iron or steel, and it can either be cut from a single block of the metal, or fashioned in two parts which are then rigidly clamped together. The machining of the die is most important; the correct contour of the bore is sometimes determined by producing a sample piece of rope using a wooden die and noting where the incoming strands wear away this temporary die when a satisfactory rope is being produced.

The machines used for rope-making operate on the same general lines as those used to produce the strands, and consist of four main parts. These are the rotating flyers for the strand bobbins, the compression die, the hauling capstans and the winding mechanism to form the rope into a package. Often, a ropewalk method of preparing the rope is used.

6.9.2 Cord and twine production

Cords and twines are of three general types: laid, twisted, and plaited or braided. The difference between laid and twisted cords and twines is in whether foreturn is provided or not. Twisting is simply a twisting together of the component yarns, while laying is the provision of foreturn during the twisting operation, this producing a better balanced structure. Plaiting or braiding is the interlocking of crossed yarns.

The production machines used to manufacture cords and twines operate in much the same way as those employed in making strands and ropes but, since smaller diameters and weights are involved, much lighter and speedier mechanisms are used. Laid twines and cords usually have a hard, rounded cross-section and high strength, good abrasion resistance and a tendency to resist bending. Twisted twines are softer and less resistant to wear. Braiding machines, used to manufacture plaited cords, are of two main types. One is the older 'maypole' type, and the other a more modern high speed type. The maypole type has two plates which are geared together. The upper plate has cut in it a wavy groove which guides the spindles of the yarn packages in a pre-determined pattern. Each spindle is made to rotate, and travel at the same time along the grooved path, as the machine is in operation, so that the yarns become criss-crossed and overlapped, in true maypole manner, to produce a braiding action. At the same time, the completed braided cord is drawn away by a hauling capstan and wound on to a reel. In all the machines used, the tightness of the braid is determined by the hauling-off speed, the twist in the starting yarns and the amount of interlocking used. Cords are referred to as being hard or soft braided, depending on the rigidity of the cord which has been produced by the braiding action.

6.9.3 Use of man-made fibres

Nylon was probably the first man-made fibre material to be given serious consideration by the rope-making industry. A number of factors stimulated this interest, amongst which were nylon's high strength, its elasticity and 'give' under conditions of shock-loading, and its outstanding abrasion resistance. Later on, other man-made fibre materials were used, amongst them being the polyesters and polyethylene, the former being substituted for nylon when a reduced stretch was required, and the latter when buoyancy and resistance to wetting were important. An important point about all the thermoplastic man-made fibre materials is the fact that ropes and cords made from them can be given a heat-setting treatment to produce a very stable structure. Generally, the stabilizing treatment is given to the finished coils of rope. For applications such as driving ropes, and cords for belting, the ropes and cords are often heat-set while they are under tension to give stability under the conditions they will meet in use.

A relative newcomer to the cordage field, but one which will no doubt play a prominent part in the future, is polypropylene fibre. It is strong, rotproof, floats on water, does not absorb moisture, and is cheaper than many of the man-made fibres which have been tried. Ropes have been prepared from polypropylene in its many available forms – monofilament, multifilament, staple fibre and fibrillated material – and each has shown excellent promise and specific characteristics which may allow it to be selected for particular applications.

More recently, Imperial Chemical Industries Ltd. has been working in collaboration with the rope-making industry, on a new type of rope which they have called 'Parafil'. This construction would appear to use continuous filament yarns lying in parallel array along the axis of the construction and encased in a sheath made from a thermoplastic material. Such ropes are claimed to compete with steel ropes in fixed-rope applications. They could be of particular value in marine applications where the permanent mooring of large and heavy floating structures is required, since the new ropes would be adequately strong and yet free from corrosion difficulties, very little maintenance being required.

6.10 TESTING OF TEXTILE FABRICS

The testing of textile fabrics is carried out for process control purposes, to indicate how processes can be improved and developed, to check the product which is being supplied to the consumer, and to aid research which may be in progress. Tests on fabrics are many and varied and it is not possible to do more here than indicate the types of tests in use. The reader should consult the various books which have been written on the subject for more detailed knowledge.

6.10.1 Checking the fabric construction

A check on fabric 'count' means, in the case of a woven fabric, measuring the number of warp threads to the inch and the number of weft threads to the inch. In the case of knitted fabrics, the corresponding determinations are wales per inch and courses per inch. If a fabric is very densely constructed and there is some felting on the surface, it may be difficult to see the exact structure, and the most reliable method of examination to use is to unravel the fabric and make the count in this way. Usually, however, it is only necessary to obtain a magnified view of the structure to observe the individual yarns, using a magnifying glass, or a microscope, or by projecting an image of the fabric on to a screen by means of a slide projector. Counts can then be made between accurately spaced marks on the fabric or in an area of an opening of known dimensions in a screen placed over the fabric. The construction figures are usually indicated by a combination of numbers such as $a \times b$, where a is the warp threads per inch and b the weft threads per inch.

The weight characteristics of a fabric can be expressed in two ways; the weight per unit area, usually given in ounces per square yard, and the weight per unit length, given as ounces per running yard. The former is easily understood, but the latter requires some explanation since the weight of a unit length of fabric will obviously be affected by its width. In order to appreciate the value quoted it is necessary, therefore, to know the agreed standard width, which is usually 54 in. (1.3 m) in the case of wool fabrics. The weight values are determined by cutting out samples of accurately known dimensions and weighing them. The effects of moisture content can be allowed for either by conditioning the specimen in a standard atmosphere until a steady weight is obtained, or, more usually, drying the specimen in an oven and adding the official regain figure.

The thickness of a piece of fabric is a difficult dimension to measure since all textile materials are easily compressed. Cloth thickness testers consist of a solid 'anvil' on which the sample of fabric is placed, and a flat presser foot which comes down on to the fabric. The separation of the top of the anvil and the bottom of the presser foot under a standard pressure is indicated on a dial. Only relatively rigid fabrics can be tested with any degree of reliability in this way. For soft fabrics and pile fabrics, it is probably better to measure a cross-section cut through the fabric under a microscope.

When the warp and weft threads are interlaced in a woven fabric they follow a wavy path as they pass over and under one another. A measure of this waviness is known as the 'crimp percentage'. The value is of considerable practical value. When a fabric is subjected to abrasion, the tops of the crimps take the brunt of the abrasive action and the yarns with high crimp are the ones mostly abraded. The crimp percentage can have an effect on the shrinkage of the fabric since the fibres may swell and the amount of crimp be increased. During testing for tensile strength, the crimp in the threads in that direction decreases and that in the threads at right angles increases. Finally, a variation in crimp can give rise to faults in a fabric; the cause of crimp variation is lack of control of tensions in yarn preparation and weaving. The tests for crimp percentage are based on an accurate measurement of the length of the fabric compared with that of a thread taken from the fabric in that direction when all the crimps have been straightened out. The differences in the various methods of test are mainly related to the way in which the yarn crimps are straightened out without stretching the yarn itself.

6.10.2 Properties depending on the amount of air held by the fabric

The amount of air held in the spaces in a fabric has a pronounced effect on some properties of that fabric.

Porosity is the amount of air space in the fabric and is determined by measuring the total volume of the fabric and calculating the actual volume occupied by the fibres; the difference between the two expressed as a percentage of the total volume is a measure of the porosity of the fabric. The total volume of the sample of fabric is determined by measuring its area and thickness. The volume occupied by the fibre is the weight of the sample divided by the density of the fibre material taken from the tables of densities of dried fibres. The porosity determination is of value with regard to its effect on the permeability, warmth, and lightness of a fabric.

Two types of permeability of a fabric are of interest. The first is the permeability to air, which is desirable in articles of clothing and allows the body to 'breathe'. The second is permeability to water vapour, of particular interest in underclothing in keeping the body comfortable and fresh. Air permeability can be measured by forcing air under pressure through the fabric at a constant rate and measuring the back pressure which is developed, or it can be determined by driving air at constant pressure through the fabric and measuring the rate of flow.

The permeability to water vapour can be measured by various methods. One is to place the fabric over a vessel containing concentrated sulphuric acid, expose to a standard atmosphere for a specified length of time, and measure the increase in the weight of the acid. Another method is to use the fabric to cover a vessel containing water at body temperature and expose to a standard atmosphere for a definite time, when the decrease in the weight of water will be related to the water vapour permeability of the fabric. A simple method is to expose two similar vessels containing water to the same atmosphere, one of which is covered with the sample of fabric, and after a suitable time to compare the loss in weight through the fabric with that from the open vessel.

The ability of a fabric to resist or allow the passage of heat is also associated with its porosity and permeability since the heat insulating properties of a fabric depend considerably on the amount of air locked up in the construction and not so much on the conductivity of the actual fibre material. One method of measuring the heat insulating value of a fabric is to place the sample between two plates, fitted with thermocouples and at different temperatures, and measure the rate of flow of heat. In another method, a body is held at a constant temperature by an electrical heater and is wrapped in the fabric to be tested. The amount of energy required to maintain the body at a fixed temperature is determined. This method allows a good comparison to be made between samples tested on the same equipment.

6.10.3 Properties in relation to water absorption

The ability of a fabric to take up water is of obvious importance. In this respect, two factors are important, one being the total amount of water absorbed, and the other the rate at which the absorption takes place.

Various tests to study the uptake of water have been devised. One is a 'sinking time' test in which a number of squares of the fabric under test are cut and dropped on to the surface of water contained in a series of vessels. The average time for the squares to sink is observed using a stopwatch. In a 'wicking' method, strips of the fabric are suspended with an inch at their lower ends dipping into water. The heights to which the water, usually coloured, rises after various intervals of time are noted. A further method depends on placing the fabric in contact with a porous plate which is continuously wetted and determining the increase in weight of the sample, and in another a sample of fabric is immersed in water under standard conditions after which it is then taken out, squeezed and weighed. Spray methods have also been used. Water under a given pressure is sprayed down on to the sample for a minute, excess water is removed and the soaked sample is then weighed, after which it is dried in an oven and weighed again. The percentage water absorption is calculated on the dry weight.

For certain types of fabric, resistance to wetting is important. One test to determine this property consists in allowing drops of water to fall on a sample of the fabric which is inclined at an angle, and noting the time for the surface of an insulating plate supporting the sample to become sufficiently wet as to complete an electrical circuit. Hydrostatic pressure tests are also used in which water under a slight pressure is held by the sample fabric and the time noted for a specified number of drops to appear on the free surface.

The strains introduced into a fabric during its manufacture and treatment can cause shrinkage to take place during laundering. A shrinkage test is carried out on squares cut from the fabric which are marked by some indelible means at points in warp and weft directions which are known distances apart. The squares are then laundered under standard conditions, dried and re-measured. The shrinkage is expressed as a percentage of the original measurements.

6.10.4 Handle and draping qualities

The property of 'handle' which is attributed to a fabric ultimately depends on the judgement of an expert. Handle is a very complex quality to which a whole range of properties contribute. Thus weight, surface friction, ease of distortion, compressibility and ease of recovery after distortion all contribute to the quality referred to as handle.

Surface friction is determined in tests by noting the resistance offered by one piece of the fabric sliding over another piece of the fabric under standard conditions of loading and movement.

Flexibility, or ease of distortion, can be measured by folding a strip of the fabric back on itself and measuring the height of the fold when it is subjected to various loads. In another test, a strip of the fabric is fed forward along a plane surface, the end of which slopes down sharply at 45°. As the fabric strip projects beyond the supporting plane it bends, and the distance it projects before its tip touches the 45° incline is a measure of the fabric stiffness. In another test, the so-called 'heart-loop' test, a strip of the fabric of standard width is folded back on itself and clamped so that it hangs in a heart shape. If the fabric were completely limp, it would hang straight down, but the stiffness pulls it up into the shape shown in the diagram. The stiffness is then taken to be inversely proportional to the height of the loop as indicated. Other test methods have been based on the work done in bending the samples into various forms; one end of the sample is attached to the pan of a balance and the force required to bend the fabric is determined in this way.

The Fabric Research Laboratories in the USA have developed an instrument which has been called the 'drapemeter'. This consists of a circular disc of about 5 in (127 mm) diameter on which is placed a circular specimen of about 10 in (254 mm) diameter cut from the fabric under test. If the fabric is perfectly stiff, then there is no draping and the projected area of the draped specimen is the same as the area of the flat specimen.

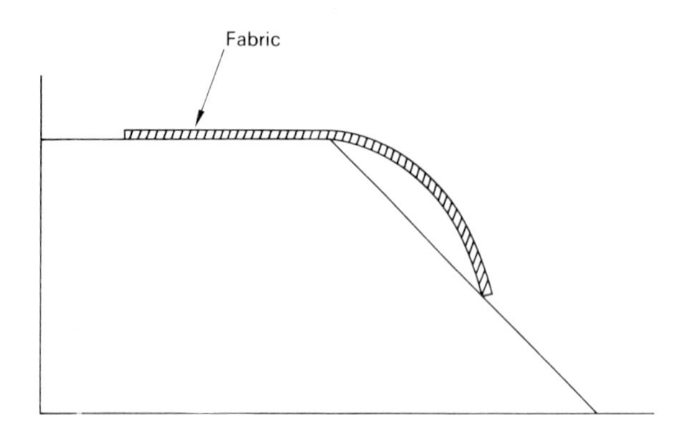

Fig. 6.37 Determining flexibility, using an inclined plane.

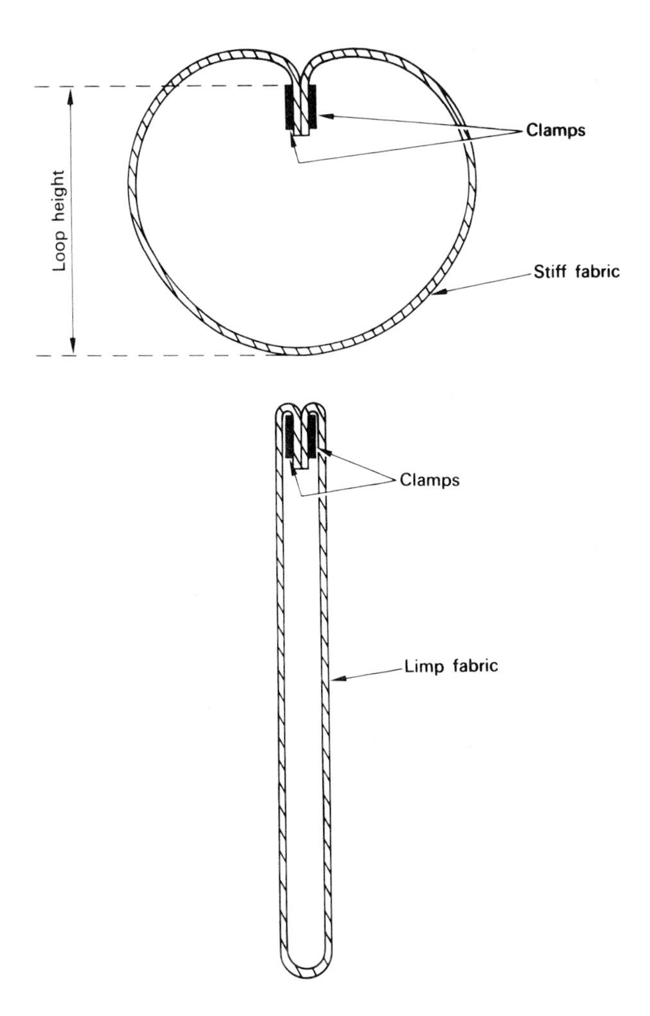

Fig. 6.38 'Heart'loop' test for fabric stiffness.

Fabrics are not perfectly stiff, and so some draping of the specimen takes place to give an irregularly shaped projected ares which is smaller than that of the specimen. Suppose that A_D = area of specimen, A_d = area of the disc, and A_s = projected area of the specimen. Then the 'drape coefficient' (F) is taken to be

$$F = \frac{A_s - A_d}{A_D - A_d}$$

The thickness tester, referred to earlier, can be adapted to measure the thickness of a fabric under various loads. If the thickness is plotted against the load, a small curved portion is obtained on the graph, but this soon straightens out to give a straight line. The hardness of the fabric can then be calculated from the readings for two points on this straight line portion.

The resilience of a fabric may be calculated by first measuring the deformation of the fabric using an increasing load and then the deformation which remains with a de-

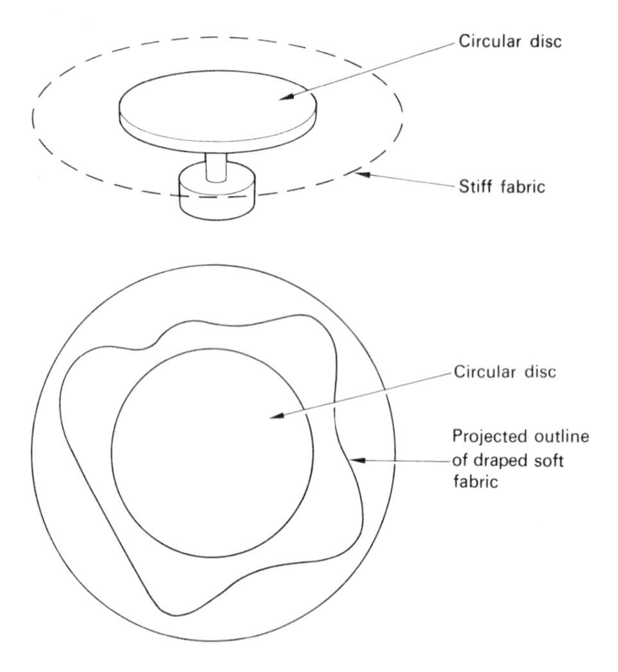

Fig. 6.39 Drapemeter method of measuring fabric stiffness.

creasing load. The two curves are then plotted and the area under each curve determined using a planimeter when:

$$\text{Resilience} = \frac{\text{area under curve with decreasing load}}{\text{area under curve with increasing load}}$$

expressed as a percentage.

The crease resistance of a fabric can be estimated by cutting strips of fabric 4 cm long and 1 cm wide in both directions from the sample, folding them double, and placing a 500 g weight over a strip of spring steel over the fold. After 5 min under this pressure, the strip, which is now 'V' shaped, is suspended over a thin wire. Three minutes are allowed for recovery and then the distance between the ends of the inverted 'V' is measured. This measurement is related to the degree of crease resistance of the fabric. Other, more elaborate pieces of equipment, all more or less operating on the same principle, are available.

6.10.5 Wear and abrasion resistance

Wear is a rather indeterminate property since the conditions under which it takes place are so varied in practice. Probably abrasion is the main factor in wear and attempts have been made to measure this property. For such tests to be of value they must be very much accelerated in comparison with the abrasion which normally takes place in the wearing of a garment and, for this reason, the conditions under which they operate may not entirely approximate to those of normal use. With all abrasion tests, the results which they provide can only be comparative, and the order in which they place a range of fabrics can be no better than a general guide.

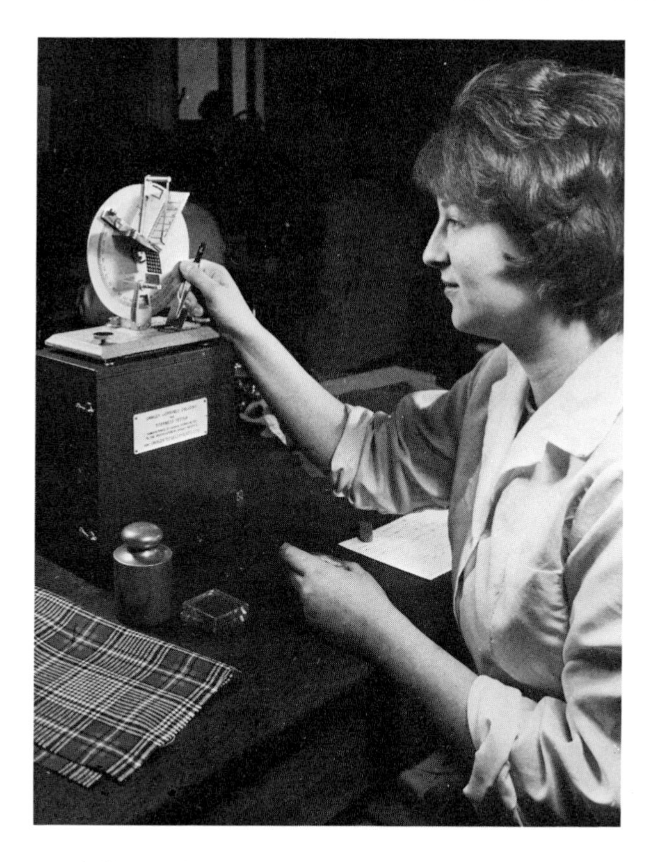

Fig. 6.40 Fabric being tested for crease-resistance by the Shirley Institute method.

All the abrasion tests depend on the movement of the fabric under test either against itself or some standard abrading material. The designers of the tests have their own ideas of the type of movement which will give results as nearly as possible in line with experience in practical wear. It is known that the character of this movement influences the final result. Other important factors are the nature of the abradant, the pressure of the abradant, the tension of the sample, how thoroughly lint is removed and prevented from clogging up the abradant, and what determines the end point of the test. In the case of the latter, some tests determine the number of strokes to produce a hole in the sample, others measure the loss in strength after a standard number of rubs, while others rely on the loss in weight or thickness which a number of rubs produce.

6.10.6 Strength and stretch tests

Strength tests are a useful means of checking the quality of a fabric in comparison with other fabrics. Stretch is desirable for a number of applications. The combination of stretch and strength is an indication of the work necessary to break a sample; in reality a measure of the toughness and durability of a fabric.

The strength of a fabric can be determined using any one of a whole range of machines which are available. In any one of these the specimen is gripped between jaws which are then separated to extend the specimen. This extension can be carried on

Fig. 6.41 Abrasion testing to assess wear resistance of fabrics.

until fracture of the specimen takes place. In carrying out the test, it is usual to plot the load applied to the specimen against the elongation it produces. This curve indicates the behaviour of the specimen from no load and elongation up the the values at which it breaks. From this curve much useful practical information regarding the elasticity, the work of rupture, the yield point and other mechanical properties of the fabric, can be obtained.

The standard tensile strength testing machines are also used for tearing strength tests. A cut is made in the specimen and the force required to complete the tearing apart is measured.

Bursting strength tests are often carried out on fabrics, and particularly on knitted fabrics. In one such test, the specimen is clamped in a ring and then pulled down on a ball until bursting takes place; the machine indicates the pull which caused rupture. In another, the specimen is clamped over a rubber diaphragm and air is pumped in to inflate the rubber, and this is continued until the fabric bursts, when the pressure to cause rupture is recorded.

6.11 LITERATURE

J. B. SMITH (Ed.). *Technology of Warp Sizing*, Manchester, Columbine Press, 1964.

J. READ. *Elementary Textile Design and Fabric Structure*, London, Textile Institute, 1951.

G. A. BENNETT. *Introduction to Automatic Weaving*, Manchester, Harlequin Press, 1948.
J. B. AITKEN. *Automatic Weaving*, Manchester, Columbine Press, 1964.
H. SPILBEY (Ed.). *The British Wool Manual*, Manchester, Columbine Press, 1969.
V. DUXBURY and G. R. WRAY (Eds.). *Modern Developments in Weaving Machinery*, Manchester, Columbine Press, 1962.
A. T. C. ROBINSON. *Rayon Fabric Construction*, Skinner, 1951.
W. WATSON. *Textile Design and Colour* 6th edn., London, Longmans, 1954.
B. L. HATHORNE. *Woven Stretch and Textured Fabric*, New York, Wiley, 1964.
A. CROSLAND. *Modern Carpet Manufacture*, Manchester, Columbine Press, 1958.
G. ROBINSON. *Carpets*, London, Pitman, 1966.
J. CHAMBERLAIN. *Hosiery Yarns and Fabrics*, Vol. II, City of Leicester College of Technology & Commerce, 1949.
H. WIGNALL. *Knitting*, London, Pitman, 1964.
C. ROTENSTEIN. *Lace Manufacture on Raschel Machines*, National Knitted Outerwear Assn., 1954.
G. REICHMAN (Ed.). *Principles of Knitting Outerwear Fabrics and Garments*, National Knitted Outerwear Assn., 1961.
J. CHAMBERLAIN. *Principles of Machine Knitting*, London, Textile Institute, 1951.
D. F. PALING. *Warp Knitting Technology*, Manchester, Columbine Press, 1965.
K. HARDING. *Lace-Furnishing Manufacture*, London, Macmillan, 1952.
D. E. SCHAB. *Story of Lace and Embroidery*, New York, Fairchilds Publications, 1957.
F. M. BURESH. *Non-Woven Fabrics*, New York, Reinhold, 1962.
J. B. GOLDBERG. *Fabric Defects*, New York, McGraw-Hill, 1950.
Identification of Textile Materials, 4th edn., London, Textile Institute and Butterworths, 1958.
R. MEREDITH and J. W. S. HEARLE. *Physical Methods of Investigating Textiles*, New York, Interscience, 1959.
J. E. BOOTH. *Principles of Textile Testing*, London, Heywood, 1964.
J. H. SKINKLE. *Textile Testing*, New York, Chemical Publishing Co., 1949.
J. LOMAX. *Textile Testing*, London, Longmans, 1956.

CHAPTER 7

After-treatment of Yarns and Fabrics

7.1 VEGETABLE FIBRES

7.1.1 Scouring

While the preparative processes carried out on the mass of fibres prior to spinning remove much of the dirt and waste material, some still remains in the yarn and can be carried over into the final fabric. In addition, there are other impurities which cannot be removed by mechanical methods, such as the natural fats, waxes and colouring materials, and to these must be added those substances which are either added to help processing, such as oils and starches, or picked up during processing, such as grease and oil stains and soiling with various materials. All these substances have to be removed to enable uniform dyeing and finishing results to be obtained and give the final textile fabric or garment its clean, attractive appearance. Nearly all these impurities can be removed by scouring. The natural colouring materials in the fibres are not so removed, and have to be destroyed in another process referred to as bleaching.

The principles on which scouring is based are simple, namely to treat the material with suitable chemical solutions which will either dissolve the contaminating substances or convert them into soluble products, or loosen their hold on the fibres so that they can be washed away. This must be done in such a way that removal of the impurities is uniform and thorough, and streakiness in the final fabric is avoided.

There are many chemical substances which can be used to attack and remove the impurities, but they have to be selected not only with their cleaning properties in mind but also taking into account their effect on the fibrous material itself. Thus, alkaline compounds are satisfactory cleaning agents, and particularly in hot or boiling solution form. Such compounds can be used on the cellulose fibres, such as cotton, flax and so on, since these are resistant to alkaline attack, but wool and silk are sensitive to alkalis and such treatments must be applied with caution. Again, acid solutions are useful in removing certain impurities, and here the positions are reversed. Caution is required in treating the cellulose materials with acids, which can degrade and 'tender' them, whereas the animal fibres, such as wool and silk, are reasonably resistant to acids. If the cellulose fibres are treated with acid, the concentration of acid must be kept low and the treatment carried out at a cold, at most warm, and never at the boiling, temperature. After treatment, the acid must be washed out completely, or neutralized using an alkaline solution; if the cellulose fibre material is dried containing even a small amount of acid, its strength and wear properties will be markedly affected. Thus, the cleaning agents used in scouring should be as effective in cleaning as possible, but their selection, and the methods by which they are applied, must be such that they will have the least harmful effect on the fibre material.

To obtain thorough scouring, the material to be cleaned must be moved about in the scouring solution or, if it is more convenient and desirable to keep the textile material stationary, then the solution must be circulated through the material. Only by inducing relative motion between the two can the impurities be loosened or attacked and removed from the fibre material. The different scouring machines which are used, provide this relative motion in various ways.

When vegetable fibre materials are scoured in yarn form, the method of scouring used will be determined, to quite a considerable extent, by the fineness of the yarn. Thus very fine yarns are best processed in package form on cops or cheeses, or even while wound on perforated beams. For more robust yarns, the preferred form of the yarn is in skeins, and the most convenient equipment used is the kier; the same equipment is also used when scouring is carried out on the loose fibre. The material is packed into a vertical steel cylindrical vessel, the kier, and boiling sodium carbonate, or caustic soda, solutions or a solution of both, sometimes containing a little soap or detergent, is pumped up and sprayed over the material. It percolates down through the material and passes through perforations at the bottom of the cylinder into a container portion. It can be taken from here, heated once again to the boiling temperature, and resprayed over the material at the top of the kier. This cycle is repeated a number of times over a period of several hours and, at the completion, the dirty liquid is drawn off to waste and the textile material washed with clean water. Sometimes, the kier scouring is repeated to achieve a complete cleaning, when the textile material in the cylinder is turned upside down so that the lower layers are now at the top. The material is now hydro-extracted to remove excess liquid, and dried.

Scouring in a kier is suitable for treating a large amount of textile material but, for smaller lots, other means of scouring are used. In one, the cleaning liquid is contained in a long, rectangular tank, and the skeins are suspended from rods supported across the sides of the tank and hanging down into the tank so that most of the material is covered by the liquid. At suitable intervals, the rods are slid along the length of the tank, dragging the skeins through the liquid. At the same time, the skeins are adjusted round the rods to make certain that every part of the skein receives equal treatment.

A variation of this treatment in a tank is to hang the skeins from a framework which can be lowered into the tank so that the skeins are totally submerged. A pump, located at one end of the tank, now circulates the scouring liquid continuously through the skeins. At the end of the scouring, the liquid is run off and replaced by clean water. Finally, the skeins are removed, hydro-extracted and dried.

Drying of the skeins after scouring is usually done by conveying them through an enclosed chamber or room in which hot air is circulated. This not only dries the skeins but also gives the yarn a soft, open form.

For scouring yarns in package form, the material can be wound on to perforated tubes and the cleaning fluid pumped through the yarn alternately from inside to outside and then outside to inside. If small packages of yarn are being treated these can be threaded on to long perforated tubes and a similar procedure followed. Yarn wound on to perforated beams can also be treated in the same way. A final washing, removal of excess moisture, and drying follows in each case. In all package scouring, care is required to ensure that the yarn receives uniform treatment throughout. The density of winding on the package and the circulating conditions must be selected to give even processing.

Cotton fabrics are usually scoured in a kier in very much the same manner as has already been described for skeins of yarn. The pieces of fabric are sewn end-to-end to form a long 'rope' and this is then packed into the kier, often with the assistance of a mechanical device which distributes the fabric in such a way that no easily penetrable 'channels' are formed down which the scouring solution could penetrate preferentially and leave much of the mass improperly scoured. The scouring solution which is pumped over the fabric should contain a sufficient concentration of the cleaning chemicals to ensure that there is still some alkali present at the end of the treatment. The kier treatment can, with advantage in the case of cotton, be carried out under pressure so that the alkaline solution is at a slightly higher temperature than the normal boiling point of water. The top of the kier is sealed and the solution is withdrawn continuously from the bottom of the kier, passed through a heater and sprayed over the top of the mass of fabric. The fabric is usually treated twice in the kier, being withdrawn and washed in between the two boils. After the second treatment, the dirty solution is removed, and the fabric washed, usually in a roller washing machine.

The roller washing machine contains two large-diameter horizontal rollers supported over a trough, the rollers being made from wood or hard rubber. The fabric to be washed passes between the nip of the rollers at one end and into running water at the back of the machine. It is then guided along the bottom of the machine to the front and, after going round another roller in this position, is again fed through the nip rollers but slightly to one side of its original entry position. It then follows a similar path through the machine, again entering the nip to one side of the last entry position. Guides keep the fabric strands from becoming entangled. In this way, the fabric gradually works its way across the pair of nip rollers and, by the time it reaches the ends of the rollers and emerges from the machine, it has been washed in running water and mangled a considerable number of times, the exact number being determined by the width of the machine.

While much use is made of the kier method of scouring cotton fabric in Britain, a continuous method of treatment for long runs of the same fabric is more popular in the USA. Very long lengths, 1 000 yd (914 m) or longer, are made up by joining pieces of fabric together end to end, and these are then passed continuously through a series of machines. The fabric is first impregnated with a solution of caustic soda by being passed through a trough of the solution, and then goes through nip rollers to squeeze out excess of the liquid. Its temperature is then raised to boiling point by a heating device after which it then passes into a storage device known as a 'J-box'. This storage device has a large capacity and it takes some considerable time for the hot, impregnated fabric to work its way through. By the time it has done so, the chemical will have completed its cleaning operation, and the fabric can then be led into a roller washing machine where the loosened and dissolved impurities, together with any caustic soda remaining, will be washed away.

7.1.2 Bleaching

To complete the cleansing process and give the yarn or fabric a good, white appearance, it is necessary to follow scouring by a bleaching treatment.

Vegetable fibres are not weakened by a treatment with chemicals which release active chlorine, and can be bleached using an acidified solution of sodium hypochlorite.

Yarns are usually bleached by circulating the bleaching solution through them in some simple form of container or tank.

To bleach fabrics, two methods are usually employed. In the first one, the washed fabric is fed down in rope form into a tank or pit, which is lined with glazed tiles or some other substance which will resist chemical attack, and which has a false perforated bottom over a well containing the hypochlorite bleaching solution. A pump draws up this solution and sprays it over the top of the fabric when it percolates through the fabric and returns to the well. The circulation is continued for several hours.

In the second method, the lengths of fabric are sewn end to end and passed into a trough holding the bleaching solution and through a mangle. The fabric becomes saturated with the solution and, after passing through the nip of the rollers it falls down to form a pile in a tank, lined to resist chemical attack. The tank is then covered to exclude light and the impregnated fabric left for several hours. Under these conditions most of the chlorine available from the solution is used up quite rapidly, and the fabric is not damaged if left for a considerable period in contact with the now-weakened bleaching solution. When the bleaching action is completed, the fabric is passed from the tank through a roller washing machine using running water, then through another washing machine containing a cold, weak solution of either hydrochloric or sulphuric acid of about 0.5% concentration, and finally through another washing machine where all traces of the acid must be washed away. The acid treatment, or 'souring' as it is called, is necessary to give a clear white, and also to remove certain metallic compounds which could otherwise give trouble in later processing.

Just as continuous scouring assemblies have been devised in the USA for long runs on one particular type of cotton fabric, so similar arrangements have been made for continuous bleaching. The two plants are similar in construction and the fabric passes through the scouring section and into the bleaching section in one continuous operation. In the continuous bleaching plant, there is an increasing tendency to use hydrogen peroxide as the bleaching agent since this leaves no harmful residues which cannot easily be rinsed away.

The bleaching action of an alkaline solution of hydrogen peroxide is much more rapid if the treatment is carried out at a temperature in excess of 100°C (212°F). To obtain such a temperature it is, of course, necessary to carry out the treatment under a pressure higher than normal atmospheric pressure. A reaction vessel was designed for this purpose by Imperial Chemical Industries Ltd. and is known as the 'Vaporloc' machine. Using a reaction temperature of the order of 130°C (266°F), the bleaching is completed in under a minute and an economic continuous bleaching process is thus possible. The Vaporloc machine has been developed with a pressure-resisting slot through which the fabric, which has been impregnated with the bleaching liquid, is fed into the reaction vessel in which it is led along a circuitous path while being subjected to the high temperature treatment; some of the rollers over which it passes inside the vessel are driven to reduce tension in the fabric and avoid stretching. After being treated, the fabric emerges from the vessel through the slot.

Another chemical compound which has been used successfully as a bleaching agent for cellulose materials is sodium chlorite. This compound has no weakening effects on the fibre materials and produces a good white. It is, however, liable to attack the materials of the equipment in which the bleaching is carried out and protective coatings must be used. For this reason, sodium hypochlorite is used instead of the chlorite whenever this is possible.

This section on bleaching would be incomplete without some reference to so-called 'fluorescent bleaching'. The chemicals used for fluorescent bleaching are not bleaches at all since they do not modify or destroy the impurities giving rise to colour in the textile material. These added chemicals have the property of fluorescing blue under the influence of the ultra-violet light in daylight and if applied to a yarn or fabric which is slightly yellow, the mixture of yellow reflection and blue fluorescence produces what appears to be an accurate 'white'. The fluorescent compounds were originally used in this way in Germany about 35 years ago. The first examples were by no means wash-fast, but subsequent research has produced improvements and the compounds are now held on to the fabrics rather in the manner of 'colourless dyes'. Fluorescent bleaching agents are now used on quite a substantial scale by the bleachers. They do not allow chemical bleaching to be eliminated but can be used as an adjunct to the normal bleaching processes. Thus, if the textile material is in a state such that severe bleaching is required to produce an acceptable white, weakening of the material can be avoided if a normal chemical bleach, together with the application of a fluorescent bleaching agent, are used.

7.1.3 Mercerizing of cotton

In 1844, Mercer experimented with the treatment of cotton yarns and fabrics with concentrated solutions of caustic soda. He noticed that the treatment resulted in a swelling of the cotton fibres so that they changed from flat ribbons to rounded rods, and the twists or convolutions along the fibre lengths were reduced, or even eliminated. The effect of such a treatment was to shrink the yarn or fabric, increase the dyeing capacity, and sometimes increase the strength. In 1899, Lowe showed that, if tension was applied to the material while it was still in this swollen condition, then a substantial increase in lustre was obtained. The technical process we know as 'mercerization' represents a combination of the two, namely the application of tension either to prevent shrinkage which would otherwise occur, or even to produce a slight additional extension, while the yarn or fabric is being treated with caustic soda solution of sufficient strength to produce maximum swelling of the fibres. After such treatment, the individual fibres reflect light in such a way that is more nearly specular, and the yarns and fabrics assume a very attractive lustre, while the fabrics develop more 'body' due to the closing-up action of the swollen fibres. The mercerization process produces no deterioration of the cotton material.

In practice the cotton material is treated with a strong solution of caustic soda in water for about one minute and the cotton is stretched while saturated with the solution. The material is then washed thoroughly. The material tries to shrink under the swelling influence of the caustic soda and, if allowed to do so, no increase in lustre will be obtained.

Cotton yarns are usually mercerized in skein form. In the type of machine used, the skeins pass round pairs of powerful steel rollers supported parallel to each other at an appropriate distance apart, and under the rollers is a tank. Over the rollers is a pipe which can spray the caustic soda solution down on to the skeins. The rollers can be driven, so moving the skeins, and there is a means whereby the separation between the rollers in a pair can be varied at will.

The rollers are set in motion, and caustic soda solution of 22 to 30% concentration at room temperature, is sprayed on to the moving skeins of yarn. At the same time, the distance between the rollers is reduced slightly to slacken the yarn in each skein and allow it to take up the caustic soda solution. The spray is then stopped and the rollers moved apart gradually to stretch the skeins, now fully impregnated with the solution, to their original, and possibly greater, length. At the same time, washing water is sprayed over the yarn, and washing and maintenance of tension are continued until practically all the caustic soda has been removed. The caustic soda solution and the washing water which fall into the tank are run off to be purified and re-used. The skeins are now removed from the machine, treated with a weak acid solution to neutralize any traces of alkali which may be left, and finally washed and dried.

Cotton fabrics are treated with the caustic soda solution while they are in open width on a powerful 'stentering' machine in which the fabric is anchored by its edges as it travels through the machine. The fabric is first impregnated with caustic soda solution, and any excess is squeezed out by nip rollers before its selvedges are attached to the clips of the stenter chains to hold the fabric out in width. For the first few feet of travel the clips diverge to produce some width extension in the fabric and thereby counteract any contraction in width which may have taken place during the impregnation treatment. From that point, the clips run parallel to each other, and water is sprayed down on to the fabric to wash it free of the caustic solution. By the time the fabric has travelled the 50 or 60 ft (15.25–18.3 m) which represents the length of the stenter, it is practically free from alkali and shows no further inclination to contract. The fabric is then taken from the clips, treated with dilute acid in a washing machine to neutralize any caustic soda which may be left, and then thoroughly scoured and dried. The solutions which have been used are collected for purification and re-use. In order to obtain an effective mercerization treatment, the fabric must be in contact with the caustic soda solution for a period of up to 45 seconds. A long stenter can operate at a fair speed, but a shorter machine must be run at a lower rate or the fabric held back on cylinders before being passed into the shorter stenter. Precautions are necessary when the fabric to be mercerized contains other fibres in addition to cotton, since they may be affected by the caustic soda treatment.

The impregnating rollers and stenter can be replaced by a tank treatment for mercerizing cotton fabric. The fabric is taken up and down, over and under a series of hollow drums which are sufficiently buoyant in a tank of caustic soda solution to press against each other and thus exert some pressure on the fabric passing round them. Two such tanks are used, the first one impregnating the fabric with the mercerizing solution and the second for washing. The fact that the fabric remains pressed between the drums, prevents it from shrinking during the treatment. The fabric is finally neutralized, scoured and dried in suitable equipment.

A recent development based on the mercerizing treatment is to allow the cotton fabric to shrink freely during the treatment. Little improvement in lustre is obtained, but the fabric becomes more elastic and is used to make various types of 'stretch' garments.

7.1.4 Dyeing and printing

The object of the dyeing and printing processes is to obtain a more colourful and attractive effect on yarns and fabrics. Dyeing is used to produce a single shade of colour on the yarn or fabric, whereas printing is concerned with producing colours in pattern form on the fabric, and sometimes on yarns.

Previous to 1856, only natural colouring materials were available for colouring textile materials. These were by no means easy to apply and fluctuations in purity arising from their methods of extraction made it difficult to obtain precise and repeatable results. In 1856, W. H. Perkin discovered synthetic dyestuffs, his first being based on aniline. Later he added more aniline dyes to his range, and other investigators also developed new aniline dyestuffs. At a still later date, further investigation showed that useful dyes could be prepared from coal-tar, and since that time, dyestuffs research has continued to be carried out in this country and on the Continent on a large scale. The dyer now has probably more than 3 000 synthetic dyes, which have been tried and proved, which he can use. Probably a further range of equal size is also known but, as yet, they are not easy to apply and obtain satisfactory fastness results.

a. Dyeing. Cotton is dyed, in yarn form, as hanks, warps, or wound on to a beam. Most cotton yarns are dyed in machines developed for the purpose but, where smaller quantities are involved, some 'hand' dyeing is still carried out. Hank dyeing is one of the oldest forms of cotton dyeing, and large quantities of cotton yarn are still dyed in this form, in spite of the additional cost of the subsequent winding on to packages which must be carried out before the yarn can be used. Large quantities are dyed by suspending the hanks from rods extending across the top of a rectangular tank which contains a solution of the dye. Usually, four men work in pairs from each end of the tank for the first 15 min or so of dyeing time to ensure that the yarn receives a thorough turning. The yarn is turned, a rod at a time, either by each pair of men pulling the yarn over by

Fig. 7.1 Pressure beam dyeing; a beam being removed from the bath.

hand, or by using a stick pushed through the hanks. Three such turns are usually carried out to ensure even penetration. Mechanical methods of performing similar operations are now also used.

Warps of cotton can also be dyed. This is done in a tank fitted with three or four horizontal guide rollers, one of each pair being at the top of the tank and the other near the bottom. At the front of the machine is a pair of nip rollers. Tapes are threaded over and under the guide rollers and through the nip rollers. The tank also has steam coils, which can be of the open or closed type, for heating the dye solution. The tank is partially filled with dye solution, and the end of the warp is attached to the tapes. The tapes convey the warp through the system and, as the threads pass through the solution they take up the dye until nearly all the dyestuff has been absorbed. To prevent uneven dyeing, the warp is often run through the machine twice, the dyeing materials being added, half at a time, at the beginning of each run. A number of such tanks, with squeeze rollers in between, can be used with, say, dye solution in the first two and washing materials in the others.

Another method of dyeing cotton yarn is to keep the cotton stationary and pump the dye solution through it, and this is now the favoured method used. Such a method is clean in use and can give good penetration and levelness. One such method is to use a single open vessel, with the yarn either on perforated spindles or in a perforated cage. The dye solution is then pumped through the yarn. On the Continent closed machines are favoured with the cotton wound in thick layers, pressure pumps being used. Another system uses low pressures and much thinner layers of yarn. In circulation type machines, the yarn packages are threaded on to perforated spindles. Yarns in warp form wound on beams are also dyed by circulating the dye solution. Hanks and warps can also be placed in perforated cages through which the dye solution is made to circulate. Experience in packing the cages correctly is needed if 'channelling' is to be avoided. Since circulation methods give more rapid dyeing, it is not easy to avoid uneven results; very often materials are added to the solution to slow down the action, and dyes which give particularly level colouring are selected, in an attempt to obtain even dyeing.

The dyeing of cotton fabrics, or 'piece' dyeing as it is called, can be carried out by two methods. In the first, little dyeing takes place while the material is actually in the dyebath, and in the second, the complete dyeing takes place within the dyebath. Typical machines using the first method are the 'jigger' and 'padding' machines, and those using the second method are the 'winch' and 'continuous' dyeing machines. All classes of dyestuffs can be applied to cotton fabrics in the jigger machine. The machine consists of a V-shaped trough in which are guide rollers. Above the trough are fixed two rollers on which the fabric can be wound alternately as it passes, in open width, backward and forward through the dye solution in the trough. A steam pipe runs along the bottom of the trough to provide heat. Saturation of the fabric takes place during its short period of immersion and the actual dyeing develops while the fabric is on the rollers. With very heavy fabrics, penetration of the dye solution is helped by using a pair of squeeze rollers. The fabric is kept moving backward and forward. Conveniently, when dyeing is completed, the fabric can be made to run into a second jigger where it will be rinsed. Finally, the moisture is extracted from the fabric either by means of nip rollers or a centrifuge. The final drying is usually by passing the fabric, in open width, over revolving steam-heated cylinders.

In the padding machine, which works on a similar principle as the jigger, a small

Fig. 7.2 Jigger dyeing: taking a cutting from a roll to be matched to a pattern.

trough is used to contain the dye solution, and the solution is replaced from a cistern as it is taken up by the fabric. In one form of the machine, the fabric is guided through the dye solution, while in another a large diameter bottom roller dips into the solution and the fabric is impregnated by running over this roller. In both cases, the impregnation takes place over a short interval of time and no actual dyeing takes place during that interval. Nip rollers remove excess of the dye solution after impregnation. The most popular type of padding machine has two or three rollers giving one or two immersions in the padding solution, but models are available which give more immersions.

In the winch dyeing machine, a square or rectangular vessel, fitted with steam heating pipes, contains the dye solution. The fabric is moved gently in the solution by being passed over a slowly revolving cylindrical frame placed above the vessel. The fabric can be dyed either in open width form or as a rope. The winch machine is generally used for loosely formed fabrics such as lace-curtains and knitted constructions, and for delicate fabrics which could be distorted and damaged in a jigger machine.

In a continuous dyeing machine, a long rectangular tank is fitted at its top and bottom with a series of guide rollers over which the fabric runs in its passage through the dye solution. The tank also has steam heating and is connected to a feed tank which is steam-heated too. The fabric enters through a pair of nip rollers and is dyed in a single passage through the solution, after which it leaves the tank through a second pair of heavier nip rollers. It can then be arranged for the dyed fabric to pass through another similar tank where it will be cleaned. A number of units can be coupled together so that dyeing, washing and drying are carried out in one continuous run.

As already mentioned, the dyer has a whole range of dyestuffs from which he can select the one he desires to use. His choice will be influenced by a number of factors. Cost is of obvious importance and he will not select a dye which is expensive to apply unless the final application warrants it. He must take into account the application for which the fabric will ultimately be used. Thus, yarns must often be dyed to withstand the bleaching or mercerizing processes which will be carried out on the fabrics made from them. The dyer will have to take note of any special finishes such as water-proofing or crease-resistance treatment which may be applied to the fabric and which could cause serious changes in shade. There are applications where the fabric is exposed to weathering and strong sunlight, in awnings, curtains and upholstery, for instance, and the dyes must not be affected seriously by them. Articles such as shirts, dress materials and coloured handkerchiefs will be washed frequently, and the dyes used must be fast to washing. There are applications for which cost and colour are the major considerations and no high degree of fastness to light or washing is required; examples are mattress covers, carpet backing, linings for bags, and many uses where the fabrics are unexposed. The dyer must select the correct dye from the list offered to him in order to satisfy the uses for which the fabric will be employed. The 'direct' cotton dyestuffs are the most widely used for dyeing the material and they are easily applied. A considerable range of direct colours is available, and they are soluble in water and applied simply by immersing the cotton yarn or fabric in a solution and boiling, when the dye gradually passes into the material. Dye take-up may often be assisted by adding common salt to the dye solution. With some of the dyes, the so-called 'basic' dyes which give brilliant colours but lack light fastness, the fibres may have an indifferent affinity for the dye and will not take it up satisfactorily from solution. There are colourless substances, however, referred to as 'mordants', which have an affinity for both such dyes and the fibre material, and the fabric can first be treated with one of these, after which it will take up the dye from a solution satisfactorily. The 'vat' dyes are relatively expensive but give bright colours which are fast to washing. They are not readily soluble in water but, by reacting them with a reducing agent such as sodium hydrosulphite, a so-called 'leuco' compound is formed which is soluble in dilute solutions of an alkali and with which the cotton fabric is readily dyed. The full colour will then be developed by exposing the treated fabric to air, or by chemical means, to affect oxidation which enables the dye to change back to its original form. Since, in its original form, the dye is insoluble, a fabric dyed in this way is particularly fast to washing.

Linen is mainly used in its white state, something like 10% only being dyed for dress goods, brocades, damasks, curtains, and furnishing fabrics. What is dyed is usually processed in fabric form. Similar methods as for cotton are used to dye linen, but it is a little more difficult to dye than cotton since it is a harder fibre and penetration is thus not obtained so readily. All linen fabric is dyed in open width to avoid creasing.

Hemp is not usually dyed; if it is, direct dyestuffs are normally used. To get bright colours, it may be necessary to use basic dyes, having first mordanted the material.

All the direct dyes, as for cotton, may be applied to sisal. Basic dyes may also be used, but they give relatively poor light fastness and penetration is not easy.

Ramie is not usually dyed; if it is, most of the cotton dyes can be used.

All the cotton dyes can be applied to jute, which reacts to dyeing in very much the same way as cotton. By reason of its type of application, jute is usually dyed with the bright, but cheaper, colours and the more expensive ones, such as the vat dyes, are practically never used. Where very fast colours are required, the 'sulphur' dyes are often used. Further information on dyestuffs is given in volume 4.

b. Printing. The art of colour printing dates from ancient times. Block printing, in which the pattern was carved on a block of wood which could be smeared with a coloured paste and pressed on the fabric, is known to have been in use in China more than 2000 years ago. Block printing is still used to a small extent, when exclusive patterned fabrics are required for instance.

Roller printing is the main method used today. The process was devised by Bell in 1783 and, although a number of improvements have been made over the years, the principle he devised is still the basis of modern roller printing. The process is simple and will be easily understood from an examination of the diagram. A large cylinder is covered with a number of thicknesses of fabric to provide a resilient backing for the printing. The fabric to be printed is fed on to this cylinder on top of a further, endless, blanket to increase the resiliency and a 'back grey' cotton fabric to protect this blanket from becoming stained during the printing. The printing paste is contained in a trough into which dips a 'furnishing' roller which picks up paste and transfers it on to an engraved 'printing' roller. The fabric to be printed moves forward with the endless blanket and the back grey cotton fabric as the cylinder revolves until it makes contact with the printing roller and the pattern is impressed on it. A number of rollers can replace the

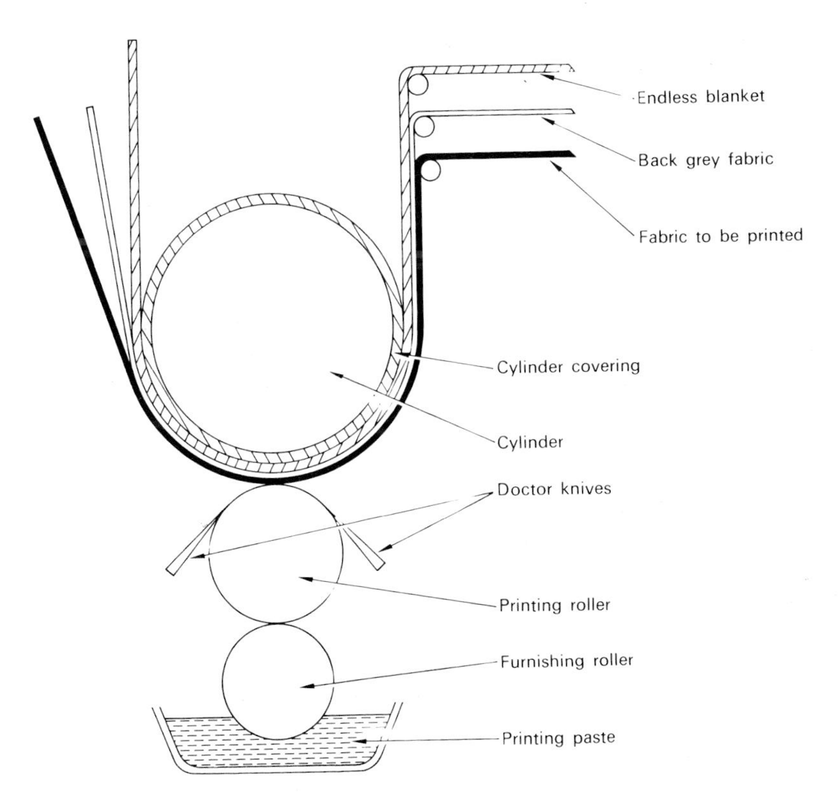

Fig. 7.3 Roller printing action.

single printing roller shown, each contributing its own coloured portions of the pattern. In order to obtain a clear printed pattern, the cotton fabric to be printed is sometimes brushed to remove loose fibres and then sheared by using a machine which operates on a principle very similar to that of a lawn mower. The movements of the fabric and the printing rollers must be such that no slipping takes place, otherwise an indistinct pattern will be printed. When a number of printing rollers is used, each contributing its own colour to the pattern, the 'register' is extremely critical, and various refinements which have been added to the modern machines give help in this respect. When the printed fabric leaves the cylinder, it passes through a drying and steaming chamber to 'fix' the printed pattern, while the back grey cloth is led away for washing ready for further use. The endless blanket continues to be used until it, too, becomes soiled and it is removed for cleaning. Without the steaming, which swells the fibres in the printed fabric and allows the printed dyes to penetrate into them, most of the colours would be only loosely held and would wash off the fabric during the first cleaning. In the steaming chamber, the printed fabric travels up and down, over and under a series of rollers to increase the length of time it is exposed to the steam and so obtain a better fixation. In many cases, it is considered better to dry the printed patterns by passing the fabric over heated plates before it enters the steaming chamber. After steaming, the fabric is thoroughly washed to remove any colour which may be only loosely held, and any other chemical substances which may have been employed.

The above method is suitable for printing on to white or light-coloured fabrics, but not on fabrics of deeper shades since the fabric colour would interfere with the printed colours. A process known as 'discharge printing' is used in such a case. To produce a white pattern on a dark fabric background, the fabric is printed with a paste containing a bleach. On passing through the steam chamber, the bleach attacks the background colour of the fabric to produce a white pattern. There are classes of dyestuffs which are not attacked by the bleach. These may be added to the paste, along with the bleach, so that, after printing and steaming, bleaching and renewed dyeing take place to give a coloured pattern on the background colour.

Another technique which is sometimes employed is referred to as 'resist' printing. Here the white fabric is first printed with a paste containing chemicals which will prevent dyestuffs being fixed in position, and then dye is applied over all the fabric. On drying and steaming, the dye will become fixed except where the pattern has been printed and this will wash out to leave a white pattern on a coloured fabric.

Roller printing is only suitable when the expense of engraving the printing rollers is justified by long runs of fabric. For smaller quantities 'screen' printing is used. Here the pattern is cut from a sheet of suitable material and the colour is brushed through on to the fabric. The cut stencils have now been replaced by silk or wire mesh screens held in a wooden frame. The 'stencil' is then formed by filling in appropriate portions of the screen with an insoluble varnish, leaving the pattern portions open. Separate screens are required for each colour. The screen is laid on the fabric supported on a long table, colour is applied using either a brush or a spray gun, and then the screen is moved to its next position on the fabric where that portion of the pattern is to be reproduced, and so on until the length of fabric has been completed. The fabric is then steamed to fix the pattern.

There is yet another method of printing called 'pigment' printing. Pastes containing dyestuffs can only be used satisfactorily if the fabric to be printed has an affinity for those particular dyestuffs. Pigment printing was developed to remove this limitation.

Fig. 7.4 Screen printing.

The paste contains a water-insoluble pigment, such as is used in paint manufacture, together with a synthetic resin which acts as a binder. Such a paste is applied in the manner already described, the fabric is dried, and it is then heated for a short time at a temperature of the order of 150° C (302° F). This hardens the resin which thus binds the pigment on to the surface of the fabric to give a permanent coloration.

7.1.5 Mechanical modification of handle and appearance

Some fabrics are required to be stiff and crisp to handle, but generally softness is preferred and a good drape. The fabric producer can go some way towards providing both softness and satisfactory drape by using softer, low twist yarns, and allowing a certain amount of looseness in the fabric construction. Even then, the chemical substances used in scouring, bleaching and dyeing can produce a certain amount of stiffness; a fabric dried under tension also develops some stiffness.

Stiffness and unsatisfactory draping characteristics can be eliminated to a considerable extent by mechanical treatments given to a fabric using a machine referred to as a 'calender'. Calendering can produce a remarkable degree of softness in a cotton fabric and also add lustre to its surfaces; cotton treated in this way is thus made more attractive to the user. In a calender, the fabric passes under high pressure between a pair of nip rollers, or 'bowls'. More complicated calenders are used in which there are several rollers situated one above the other to enable the fabric to be nipped under pressure a number of times in one run. Two metal rollers are never run one against the other in calendering as this can cut and damage the fabric passing between. One roller is metal, while the other is covered with compressed paper or cotton to allow a certain amount of 'give'.

The lustre obtained even by multiple calendering may not be sufficient to satisfy the customer, and an additional polishing action is produced by using 'friction calendering'.

Fig. 7.5 A five-bowl hydraulic calender.

Here one of the rollers of the calender is driven at a higher speed than the other to produce sliding over the surface of the fabric.

An even greater lustre can be obtained by using a Schreiner calender. Lustre, in a normal cotton fabric, tends to be destroyed by the broken surface produced by the interlacing of the threads in weaving and the twisting of the fibres in the yarns. Schreiner discovered that these irregularities could be masked by impressing on the fabric surface a series of closely-spaced parallel ridges, something like 300 ridges to the inch, when the fabric acquired a high lustre viewed from certain directions. This produced a silk-like appearance in a plain cotton fabric and the process became popular. The machine in which the treatment is given consists of a paper- or cotton-covered roller and a hollow engraved steel roller which can be heated from inside by electrical or other means to a relatively high temperature. The fabric first runs through a water-spray box to dampen it and is then passed slowly between the two rollers pressing together under high pressure. The treatment results in a big increase in lustre of the fabric surfaces.

In some types of calender, the fabric is not pressed between rollers, but is carried round in contact with a polished steel cylinder and held in position by a heavy blanket. A damp fabric drying in contact with the polished surface develops a reasonable amount of lustre. Two cylinders are normally used so that both surfaces of the fabric can be brought, in turn, to press against a polished surface.

While drying a fabric against a polished surface promotes lustre, it also produces some stiffness and rigidity. An attempt is made to obtain an acceptable compromise between the two by first developing as high a lustre as possible and then subjecting the fabric to a mild mechanical softening treatment. A suitable softening machine is one consisting of two rows of small diameter rollers, one above the other, the surface of each roller, being covered with rounded projections. A fabric drawn between the two rows has its surfaces lightly disturbed and is made softer. In a second type of machine, rollers having spirally grooved surfaces are arranged round a cylindrical framework. Fabric passing over these roller surfaces is softened in the manner obtained with the other machine.

Linen fabrics are valued for their lustrous and smooth, thready appearance and a special type of finishing machine is used to enhance these characteristics. The process used is known as 'beetling'. The beetling machine has a row of heavy wooden 'fallers' arranged side by side above a horizontal roller on which many layers of the linen fabric are wound. The fallers are lifted, one after the other, and dropped under their own weight to pound the fabric. The roller, on which the fabric is wound, rotates slowly during the pounding and moves slightly from side to side to distribute the blows evenly. This treatment may go on for several hours and, after a suitable time, the fabric will be re-wound on to a second roller so that the inside is now on the outside and the pounding by the fallers carried out once again. This treatment develops a good lustre. Instead of fallers dropping under their own weight, some machines have beaters which are controlled by cams.

It will be appreciated that an improvement in lustre produced on the fabrics by mechanical treatment, such as those described, is not maintained after washing. Washing causes the cellulose fibres to swell and, by this means, the surface of the fabric becomes disturbed. In recent years, attempts have been made to produce more permanent finishes by such means as incorporating synthetic resins in the material. Some of these attempts show promise and the investigations are continuing.

7.1.6 Chemical modification of handle and appearance

Chemical treatments can be used to soften or stiffen a cellulose fabric as required. Softening is produced by the application of 'softeners' during finishing, such compounds being based on fats or oils and of a form which allows them to be applied easily. The early softeners used were not fast to washing and only gave a temporary effect. Research has continued, however, and compounds are now available which have fatty or oily characteristics and which bind on to the fibres rather in the manner of a colourless dye. The effect they produce on the fabric, therefore, is much more permanent. Some of these softeners can be added to the dye-bath and applied to the fabric during the dyeing process.

The properties of cellulose materials, including the degree of stiffness, can be changed quite markedly by the application of synthetic resins to them. To obtain any real and lasting effect, it is necessary to deposit the resin inside the fibre itself and this fact dictates the method of treatment which is used. Urea-formaldehyde resins are often used. However, there is no way in which the fully-formed resin can be introduced into the fibres since it is insoluble and, in dispersion or suspension form, the particles are still too large to enter even fully-swollen fibres. The expedient has been adopted, therefore, of conveying the ingredients of the resin, which are water-soluble, into the fibres, together with a suitable catalyst, drying at a relatively low temperature, and then 'curing' the treated fabric at a temperature of about 150°C (302°F) for a short time to form the resin inside the fibres. It is usual to impregnate the fabric with a solution of a urea pre-condensation product, which is water-soluble, formaldehyde, and an acid catalyst, and then dry it, and cure it for a period of about 3 min when a hardened, insoluble resin is formed inside the fibres. Loose resin which may have formed on the surfaces of the fibres is then removed by washing.

A resin treatment, such as has been described, is the basis of crease-resisting treatments given to fabrics made from vegetable fibres. The one developed by the Tootal Broadhurst Lee Co. is based on the use of a urea-formaldehyde resin, but more recently other resins have been used, among them melamine-formaldehyde. Similar treatments can be given to linen to increase its resistance to creasing.

It is not considered advisable to bleach cellulose fabrics which have been treated to obtain crease-resistance, using sodium hypochlorite or any compound yielding active chlorine. The chlorine is absorbed strongly by the resin and is not removed in a normal wash. This retention of the chlorine produces a yellowing of the fabric and also, during storage and particularly hot-ironing of the fabric, hydrochloric acid is formed and this produces deterioration and weakening of the cellulose fibres.

There are parts of cellulose garments, such as collars and cuffs, where a very stiff, permanent finish is required. Such finishes were once provided by the use of starch and gums, but these substances were easily lost during laundering. One improved method which was devised and used was to prepare a solution of cellulose, as is done in the

Fig. 7.6 A section of plant showing where the application of chemical finishes, such as crease-resist, to fabrics is done.

manufacture of viscose rayon, and treat the fabric with this solution. An application of acid regenerated the cellulose inside the fabric to produce a permanent stiff finish.

In modern permanent stiffening, synthetic resins are used in much the same way as has been described for improving the crease-resistance of fabrics, and various resins which are suitable for the purpose are available.

Mention should also be made of the 'Trubenizing' method of producing permanent stiffness, used mainly in the manufacture of collars for men's shirts. A fabric is made up of alternate threads of cotton and cellulose acetate in either the warp or the weft of the fabric, or even in both. If such a fabric is now moistened with a mixture of acetone and alcohol, the cellulose acetate is softened and, when hot pressed, it flows in between the cotton threads so that, on cooling, it sets and cements the cotton threads together. Trubenizing is also used to laminate fabrics together. By hot-pressing two or more of these special fabrics together, the cellulose acetate forms a firm bond between the layers. There are certain other man-made fibre materials which can be used in place of cellulose acetate as bonding substances, and some of these are used, particularly in preparing heavy laminated fabrics.

7.1.7 Improving dimensional stability

During the production of a woven or knitted fabric it is impossible to avoid introducing strains into the yarns which compose it. The yarns become temporarily stretched and the newly-formed fabric is unstable and has an area and shape somewhat different from a fully-relaxed fabric. Wet processing will allow the material to relax and regain its true, stable dimensions but, during certain of the finishing operations, such as smoothing and calendering, the fabric again becomes extended, usually in the length direction. If left in this form, garments made from such a fabric will shrink and distort during laundering, and the problem is to eliminate this tendency to shrink without impairing the finish which has been provided on the fabric. Two types of machines have been developed for the purpose, and the fabrics which have been treated in them are referred to by the trade names of 'Rigmel' and 'Sanforized' shrunk fabrics.

In the 'Rigmel' process, the fabric to be treated is held in contact with the upper surface of a very thick blanket bent over the surface of a roller. Because the blanket is thick, its outer surface, on which the fabric is held, is extended more than the inner surface in contact with the roller, and the difference in length between the two can be made even more pronounced by using a still smaller diameter roller. As the blanket is drawn from the roller to follow a straight path, its outer surface contracts to the same size as its inner one, contracting the fabric which is held in contact with it and, at this instant, a hot iron plate makes contact with the fabric and sets it in this contracted condition.

'Sanforizing' is carried out in much the same way but, at the stage when the fabric contracts, it is trapped in this state between a blanket and the surface of a large steam-heated metal cylinder when the setting takes place.

In both these methods of pre-shrinking to produce a stable fabric, the amount of pre-shrinking required is first determined by thoroughly washing a sample of the fabric. Knowing this figure, the thickness of blanket in relation to the diameter of the roller which will be required to produce that amount of pre-shrinking can be calculated. The methods are not suitable for treating open, knitted fabrics. In that case, the finisher takes steps to adjust the fabric to its stable dimensions, by overfeeding fabric on to the stenter, and adjusting the separation of the stenter clips, for instance.

Investigations have shown that cellulose fibres swell on wetting and the yarn formed from them shrinks correspondingly in length. On drying out, the reverse takes place, but recovery is not quite complete. Here, then, is another cause of dimensional changes in the fabrics, and attempts have been made to reduce the swelling properties of the material. Substances have been used to impregnate the fabric which, when heated to a high temperature, liberate formaldehyde and this combines with the cellulose material in the presence of a small amount of acid catalyst provided, to modify the fibre so that its swelling is reduced appreciably. A similar effect can also be obtained by treating the fabric with a urea-formaldehyde or melamine-formaldehyde resin, as used to improve crease-resistance. A cotton fabric can thus be treated to improve its crease-resistance and shrink-resistance at the same time.

7.1.8 Water-proofing

The water-proofing of very heavy fabrics such as tarpaulins and industrial covers has to be cheap and able to withstand rough usage. It is usual, therefore, to apply a thick coating of a substance containing tar products. For garments, such as raincoats, the coating must not detract from the appearance of the coat, and an acceptable handle must also be retained.

In many types of water-proof garment, it has been usual to coat one side of the fabric with rubber. In some of these, the natural rubber coating is being replaced by synthetic products such as polyvinyl chloride. In the case of all such coated materials there is one defect, namely that the coatings are impervious to air, and the wearer often experiences discomfort in consequence. It should also be mentioned that fabrics of this type suffer some deterioration on making contact with oils and greases.

The ideal raincoat fabric should be permeable to air to give a greater comfort in wear and yet able to withstand heavy showers, and much investigational work has been done with this object in view. Earlier fabrics of this type were made by first treating the material with an aluminium salt and then with a soap solution, when an insoluble, water-repellent aluminium soap was deposited in the material to give the desired effect. Later, a one-bath method of producing the aluminium soap deposit from aqueous emulsions was developed. Treatments such as this have proved to be effective, but they will not withstand dry cleaning, and more expensive methods using complex substances which have fatty water-repellent components can be used to overcome the difficulty. The fabric is impregnated with a solution of one of these substances which can later be broken down by being heated to a relatively high temperature to leave the fatty material firmly fixed in the fabric. The treated material is then rinsed and dried. Not only do such treatments give a water-proof finish which is fast to dry cleaning, but they also make the fabric softer and more pleasant to handle.

Mention should also be made of cotton 'shower-proof' fabrics. The so-called 'Ventile' fabrics are made by closely-weaving special cotton yarns in which the fibres are nearly parallel to each other so that, on wetting, the swelling of the fibres closes up the spaces in between and resists the passage ofwater. Fabrics are also treated with silicone compounds to obtain a degree of shower-proofing. The compounds are applied to the fabric in emulsion or solvent solution form, the fabric is then dried, and a final 'curing' treatment for a few minutes at a high temperature given. The fabrics become shower-proof by this treatment and the finish is fast to dry cleaning and washing.

7.1.9 Flame-proofing

There are cases where a good degree of flame-proofing in a garment is necessary to eliminate the risks of serious accidents taking place. Unfortunately it is not easy or cheap to make a fabric completely flame-poof. Quite a number of flame-proofing treatments do exist but they are not widely used as they are expensive to apply and they also tend to take away some of the softness and warmth of handle from the fabric. The main difficulties are, the large amount of a suitable compound which has to be incorporated in the fabric to ensure a sufficient degree of flame-proofness, and the fact that, when the compound has been introduced, it must withstand the effect of washing. Processes involving the use of ammonium borate, ammonium phosphate, and certain metallic compounds, give a reasonable degree of flame-proofing but are not wash-fast. Research has continued, however, into more permanent treatments, and into the use of combinations of substances one of which will, for instance, reduce the risk of bursting into flame while another will tend to prevent afterglow. Recently progress has been made using complex phosphorus organic compounds which, when applied at the same time as a melamine-formaldehyde resin and heated for a short time at a high temperature, are held firmly in the fabric to give a good resistance to flaming and afterglow, and are fast to washing. The cost of such a treatment is high and, moreover, the handle of the material is affected adversely. Attempts are being made to improve the process in this latter respect.

7.1.10 Improved resistance to bacteriological attack

Many cellulose fabrics are used in the open air, and are exposed to conditions which favour bacteriological attack. When attacked, they lose their strength and will eventually fall to pieces. This is one of the reasons why fabrics made from synthetic fibres are often selected for outdoor use, and why the synthetic materials are selected for twines and ropes where strength must be maintained.

It is possible, however, to increase the resistance to rotting of cellulose materials. A treatment with a solution of a copper salt has long been used to provide additional protection against attack. More recently, copper and zinc soaps have been offered on the market and are claimed to allow an even more effective treatment.

The latest treatments recommended again make use of synthetic resins, to form a barrier round the fibres against micro-organisms. Treatment with melamine-formaldehyde in the manner used to improve crease resistance is effective. If, at the same time, an antibacterial compound, say a mercury compound, is added, then the surface of the material is made much more resistant to mildew, and the life of the fabric is prolonged considerably by a combination of these treatments.

Mention should also be made of developments which are aimed at keeping garments fresh and free from body odours. Bactericidal agents are applied to the fabrics, to destroy the bacteria and organisms which attack perspiration and result in unpleasant odours being given off. The latest ones are fast to washing.

7.2 ANIMAL FIBRES

7.2.1 Scouring

Although a number of the naturally-occurring impurities will have been removed during the preparatory processes carried out on the loose fibres, oils which have been added to aid in carrying out the processes, and dirt which must inevitably be picked up during manufacture, will be present in the yarns and fabrics and must be removed by scouring.

Animal hair fibres are scoured in yarn form when they are intended to be used for hand-knitting or in the manufacture of carpets and certain types of hosiery. Much of the yarn is scoured in hank form. In some small firms, the hanks are sometimes scoured by hand, when they are soaked in the cleaning solutions followed by squeezing and finally rinsing in warm water. Most hanks are scoured, however, by being conveyed by means of conveyors or endless tapes through a series of tanks fitted with squeezing devices at each end of the tank. The conveyor type may have a means of spraying the cleaning solution on to the hanks before they enter the tanks. The cleaning solution usually contains sodium carbonate and a small amount of soap while, in the final rinsing tanks, a little synthetic detergent may be added. Synthetic detergents may be added to the cleaning tanks with advantage if the water used is hard since they help to eliminate uneven deposits of lime and magnesium compounds which may cause trouble when the dyeing is carried out subsequently. The temperature of the solution in the first cleaning tank is of the order of 50°C (122°F), and this is gradually reduced through the series of tanks to about 43°C (110°F) in the final one. The scouring operation is continuous and usually occupies several hours, extra cleaning materials being added as required.

Woven fabrics are usually scoured in rope form in a machine referred to as a 'dolly'. This consists of a large trough, above which is mounted squeeze rollers, the lower one only of which is driven, with the other free-running in slotted bearings. Guide rollers convey the fabric through the cleaning solution in the trough and then between the squeeze rollers. Three or four pieces of fabric are stitched together end-to-end and are scoured together. The dirty solution expressed by the squeeze rollers falls into a small trough immediately under the rollers and may either be run away to waste, or returned to the trough at the discretion of the operator in charge of the machine. The actual compositions of the cleaning solutions and the precise procedure adopted may vary between one works and another. A temperature near to 43°C (110°F) is usually employed, with cleaning solutions containing sodium carbonate and soap, and a small amount of synthetic detergent which is added immediately prior to rinsing. The final rinsing is important and must be thorough to avoid subsequent dyeing and finishing difficulties.

When the fabrics are likely to crease easily, they are not scoured in rope form but cleaned in open width in wider machines. These operate in a similar manner to the dolly machine but the removal of the impurities usually takes longer.

Knitted fabrics are usually scoured in a dolly machine, but using a shallower trough and lighter squeeze rollers. Sometimes a winch machine is used to scour the fabrics just before dyeing.

Knitted garments, including socks, are often scoured in one of two types of machine. In one, the garments in the cleaning solution are beaten by falling wooden 'bumpers'; in the other they are contained in a totally-enclosed cylindrical cage which rotates about an axis inclined to the vertical to agitate the contents, and into which the cleaning solution is introduced. Usually, a single scouring followed by a rinse in warm water is employed. Again, sodium carbonate and soap are used, with synthetic detergent added with advantage at the rinsing stage.

Difficulties can be experienced with some wool yarns and fabrics during scouring by reason of their tendency to 'curl' and 'cockle' when first wetted. With yarns, this can lead to a hank becoming badly entangled, and with fabrics a distortion in shape and surfaces. The defects may be due to high twist in yarns made from certain grades of wool, and in fabrics to unequal shrinkages in the warp and weft threads when they contain different fibres or are made with different degrees of twist. The defects may be obviated by taking advantage of the permanent setting properties of wool (to which reference has been made earlier) before scouring is carried out. When the material is to be scoured in yarns formed into hanks, the hanks are held round spindles in a frame in a stretched condition, and then the frame, with the hanks, is immersed in boiling water for about 30 min, after which the frame is removed and the hanks are allowed to cool down, still in the frame, before the tension is removed. In a similar way, woven fabrics are subjected to a 'crabbing' operation before being scoured; the treatment is again with boiling water while the material is under tension, but followed by a steaming treatment in addition. In the crabbing machine, the fabric is wound tightly on an external beam and then drawn evenly under high tension round guide bars in a trough containing boiling water and then on to a roller above the trough. A similar passage through boiling water in a second trough ensures a uniform treatment since the outer layer in the first passage now becomes the inner layer in the second passage. The fabric is allowed to cool while tightly wrapped on the roller. In the cases of both yarn and fabric treated in this way, the material has received a permanent set and will not be distorted in any subsequent treatments if the setting temperature is not exceeded. The final steaming treatment given to fabrics, and referred to above, is done in order to ensure that a high-temperature setting has been given. The fabric is beamed, under tension, on to a perforated metal cylinder covered with a few layers of cotton cloth to avoid staining, a canvas cover is placed in position, and steam under a pressure of 40 p.s.i. (28 000 kg/m^2) is blown into the cylinder which is rotated meanwhile. After cooling, the fabric is beamed under tension on to a second such cylinder for the treatment to be repeated.

7.2.2 Degumming of silk

The silkworm spins the silk threads surrounded by a protective layer of sericin, or gum, and, in order to develop the characteristic lustre and handle of silk, it is necessary to remove this layer of gum. This operation is referred to as 'degumming', and also as 'boiling-off'. The degumming action can be carried out on the material in the form of hanks of yarn or in fabric form.

In degumming, the silk is held in an olive oil soap solution at a temperature of 90°C (194°F) for something over one hour when the gum begins to swell and leave the filaments. The material is then transferred to a second bath made up with the same materials but with a rather lower soap content. The raw silk loses about one-quarter of its weight in the degumming process. An alternative process is to remove the gum using a hot foam produced by blowing air through a hot soap solution. Since the silk remains almost stationary during such a treatment it is claimed that there is less risk of damaging the silk filaments than in the boiling-off method.

With tussah silk, degumming by boiling-off with soap solution is usually preceded by a treatment at 95°C (203°F) with a solution of sodium carbonate in water.

For some types of yarn, only a portion of the protective gum is removed to give a different type of handle in the final fabric made. In 'souplé' silk, for instance, only about 8% loss in weight is aimed at, the remainder of the gum being 'fixed' so that dyeing may be carried out. The raw silk is treated for about 1½ hours at a temperature of 40°C (104°F) in a soap solution. This softens the gum and, after rinsing, the silk is treated for a further 1½ hours in a weak solution of cream of tartar at 85°C (185°F). This treatment gives a partial degumming of the silk. A higher acidity than this is sometimes preferred and sulphuric acid is used.

'Ecru' silk is one in which a still smaller amount of gum has been removed from the raw silk. The raw silk is treated in a very weak soap solution which does little more than soften the gum. After drying, the filaments have a harsh handle.

7.2.3 Carbonizing of wool

The sheep, from which the wool is obtained, come into contact with various types of bushes and plants, and vegetable matter from these, such as burrs and seed pods, become entangled with the fleece. This vegetable matter is so tightly held by the wool that it is not removed by any normal scouring treatments and a 'carbonizing' treatment is applied to eliminate it. The carbonizing process depends on a chemical treatment which preferentially attacks the vegetable material and makes it so brittle that it can be powdered by crushing and thus enabled to fall out of the wool. The chemicals used are mineral acids applied at a relatively high temperature when the cellulose in the vegetable material is degraded to hydro-cellulose.

Carbonizing is sometimes carried out on the loose wool, but it can be employed more easily and economically on the woven fabric. In that case, it is usually carried out just before dyeing takes place. The carbonizing agent used is usually sulphuric acid, but aluminium and magnesium chlorides are occasionally employed, particularly if carbonizing is carried out on fabrics which have already been dyed.

The impregnation of the fabric is performed in lead-lined tanks containing the sulphuric acid solution. A complete and uniform penetration of the solution into the fabric is necessary and the presence of a wetting agent helps in this respect. Surplus acid solution can be expelled from the fabric by passing the latter in rope form between sets of squeeze rollers or by hydro-extracting. More recently, excess acid has been removed by passing the impregnated fabric over a 'vacuum slot' in open width. The fabric is then dried at a temperature of about 55°C (131°F) and then passed on to a continuous baking machine where it is heated to a temperature of approximately 110°C (230°F) to effect carbonizing. After carbonizing, the degraded cellulose is crushed by passing the treated fabric between rollers and is then shaken out by agitation. Finally, the fabric is rinsed with water, usually in a dolly machine, and thoroughly neutralized using sodium carbonate.

Faults can arise in the subsequent dyeing and finishing operations from unsatisfactory carbonizing. The treatment must be uniform since carbonizing can affect the dye affinity of the wool material. In addition, if the neutralization of the acid present is not complete, this too can give imperfect dyeing results. Streakness arising from uneven distribution of the sulphuric acid, or uneven baking conditions, cannot be rectified by any treatment which can be given to the final fabric. Unevenness due to unsatisfactory neutralization of the acid in the fabric can usually be removed to an appreciable extent by boiling the fabric.

7.2.4 Milling of wool fabrics

As has already been mentioned, wool and certain other of the animal hair fibres have the property of being able to 'felt' under certain treatments, and the preparation of felt fabrics has also been described. A moderate, and controlled degree of felting is often developed during the finishing of wool fabrics by a process known as 'milling'. Milling is applied to most woollen fabrics to provide fullness and what is described as 'body' to the material. The milling action produces a certain amount of shrinkage and thickening of the fabric to give it an appearance and handle which is considered desirable; the appearance changes because the woven structure becomes less distinct, and the handle of the treated fabric becomes firmer. Worsted fabrics are less frequently milled and, even then, the treatment is only a mild one.

Milling is carried out using either soap or acid, the acid usually employed in the hat and felt-making trades is sulphuric acid. In the main, milling machines are of two types, namely stocks and rotary machines. Milling stocks are used mostly in the hosiery trade and in the manufacture of felt. The machine consists of a tank in which the fabric is contained and a series of heavy wooden hammers which pound the fabric. The fabric is first impregnated with the soap or acid solution and then pounded by the hammers, which may be allowed to fall under their own weight and strike the fabric, or may be positively driven by being connected to a crankshaft. Fabrics requiring to be only lightly milled are often treated in the stocks, and this type of machine is often selected for the processing of worsted fabrics.

The rotary type of milling machine is shown in diagrammatic form. It consists of a trough, in the upper part of which are located squeeze rollers, a 'throat', a 'spout' formed of a fixed lower part and a moveable lid connected to the top squeeze roller and able to be weighted at its end, and a guide roller over which the fabric passes. The fabric passes through the machine as shown in the diagram and is compressed in the spout from the end of which it eventually escapes, this action producing the milling. The fabric is first placed in the machine as shown, and the machine put into operation. A 5% soap solution is now sprayed in to secure even distribution, and sufficient is added so that, when the fabric is pressed, a small amount of foam appears. The fabric is then milled until the requisite amount of shrinkage, measured by the distance between two marks sewn into the fabric, has been obtained. If necessary, during the milling time, a small additional amount of soap solution is used. The shrinkage in length of the fabric is controlled mainly by the amount of weight placed on the spout, while that in the width is influenced more by the pressure of the rollers and the dimensions of the throat. After milling the fabric is washed with warm water. usually in a dolly machine. If acid milling is to be carried out, the fabric is first impregnated with the acid solution, passed between squeeze rollers to remove excess solution, and then milled. After milling, the fabric is rinsed, probably using a small amount of sodium carbonate in the water, and finally washed.

In all methods of milling used, the amount of moisture present is important. If there

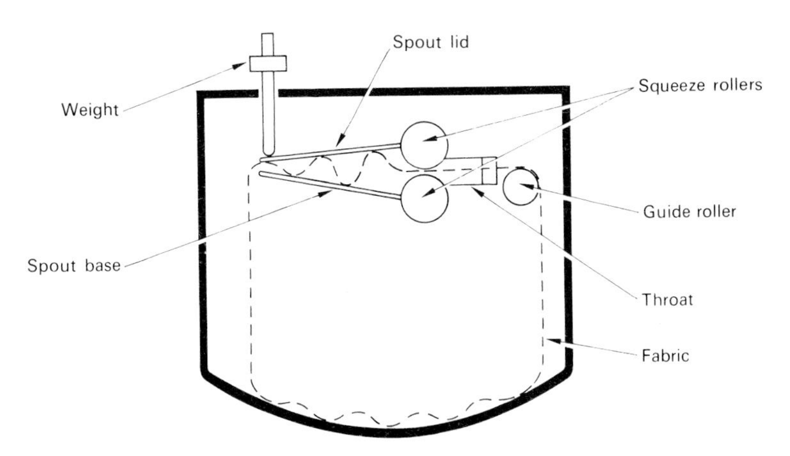

Fig. 7.7 Diagram of a rotary milling machine.

is too much, the fabric is spongy and milling is slow; if too little, then the fabric surface is poor.

Fine wools mill more rapidly than the coarser wools, and the more scaly the surface of the fibre, the better will it mill. The amount of crimp in the fibre and the fibre length, together with the structures of the yarn and fabric, also have an effect on the milling rate. The amount of cotton or man-made fibres present in the blends with the wool also affects the rate of milling; more than 20% of such fibres present usually inhibits milling. An increase in the amount of twist inserted into a yarn reduces the rate of milling by restricting the movements of the fibres, and the tighter a fabric is woven, the longer will it take to mill, for the same reason.

7.2.5 Bleaching

Wool intended for dyeing is seldom submitted to a bleaching process, but bleaching is essential for all-white wool yarns and fabrics. The methods employed involve the use of sulphur compounds which are reducing, such as sulphur dioxide or sodium bisulphite, or of the peroxygen compounds such as hydrogen peroxide or sodium perborate.

The oldest method of bleaching wool, and one still in general use, is by 'stoving', when the damp material is exposed to sulphur dioxide gas in an airtight chamber. The yarn or fabric is draped over poles in the chamber and sulphur is then ignited to produce the gas; sulphur dioxide may be provided directly from cylinders as an alternative method. The treatment can occupy several hours, after which any gas which may be left is blown away and the wool material is rinsed. The method is cheap and simple, but has the disadvantages of giving an objectionable odour to the wool and a bleaching effect which is not permanent.

If a sodium bisulphite method of bleaching is used, the scoured wool yarn or fabric is treated in a cold solution of the compound which has been made acid by the addition of sulphuric or acetic acid. The time of treatment varies, and can be as short as one hour.

There are advantages, in permanence of the bleached whiteness and freedom from odours, in using hydrogen peroxide as a bleaching agent for wool. It can also be used on mixtures of wool with cotton, or with rayon, without producing any harmful effects. Such a treatment is, however, more expensive than one using either sulphur dioxide or sodium bisulphite. Wool yarns are usually bleached with hydrogen peroxide in wood or stainless steel vessels, as hanks. The bath is made up by diluting, up to 100 times, hydrogen peroxide solution commercially available; sodium peroxide may be used instead when supplies of the liquid are not procurable. In the latter case, the alkalinity of the sodium peroxide is neutralized with acid to reduce attack on the wool. The temperature of the bleaching bath is held at about 45°C (113°F) and the period of immersion is up to six hours depending upon the type of wool being treated and the effect desired.

Wool fabrics are bleached with hydrogen peroxide in winch machines or in jiggers. In the latter case, the fabric can be passed through the solution as in jig dyeing or, after being impregnated with the solution, it may be wound up on one of the rollers, covered with a damp cloth to prevent rapid drying, and left for several hours to complete the bleaching.

Peroxide solutions are sensitive to catalytic influences and substances such as copper and iron must be excluded. The catalytic action they produce can give wasteful decomposition of the peroxide, and tendering of the wool can also be experienced. In general, peroxide bleaching is a relatively safe process, but damage to the wool can occur if the bath temperature is too high, too alkaline solutions are used and the time of treatment is unnecessarily prolonged.

If silk is of such a colour that bleaching is considered necessary, then it can either be treated by sulphur stoving or in a hydrogen peroxide bath. In stoving, skeins of damp silk yarn are hung from rods in the stoving chamber, left in contact with the gas for several hours, and then rinsed and dried. If peroxide is used, the silk is usually steeped overnight in the solution.

7.2.6 Dyeing

In the same way as was discussed in the section on vegetable fibres, the dyer concerned with yarns and fabrics made from the animal fibres, and particularly wool, must know for what applications the material is intended, and what treatments it will subsequently receive, in order that he can select the dyes he should use and the methods he should employ. Thus, if wool is to be dyed in the yarn form, worsted yarn must not be allowed to shrink in length to produce an increase in thickness and softness, since the worsted characteristics of the fabric which will be eventually made from it would be marred thereby. Worsted yarns are dyed, therefore, in a stretched condition, and the method of treatment is selected accordingly. On the other hand, wool yarns intended for knitting are required to be soft to handle, and some contraction in length and a certain amount of untwisting are encouraged during dyeing. In the case of most wool yarns, the important factor is the shade on the outer surface of the fibres, but for carpets where the tufts are to be cut, and the cut ends viewed, the interest is mainly in the appearance of the cross-sections of the fibres. Most wool material is, however, dyed in fabric form since a stock of white fabric can be kept and pieces dyed to the colours required at any given time.

Dyed wool yarns are used for three main applications: certain worsted fabrics, some knitted fabrics, and carpets. For worsted yarns, some are dyed by hand in a

Fig. 7.8 Hank dyeing: on the right a 400 lb load of undyed yarn being lowered into the dyebath by crane. On the left, carriers of dyed yarn ready for removal.

manner similar to that described for the dyeing of cotton yarns in hank form, but machines which maintain the yarn in a stretched condition are mostly used. Such machines consist of two wheels fitted on each end of a horizontal spindle. The yarn is stretched between two wooden rods, one of which is attached along this spindle and the other across the periphery of the wheels. The yarns are thus fixed radially and, when the machine is fully loaded, produce the appearance of a drum of wool. The 'drum' is immersed in the dyeing solution up to the spindle level and made to revolve. There is also a slow rotation of the rods supporting the yarns, and the whole of the equipment is covered in to preserve heat. Using such an arrangement, each portion of the yarns receives an equal dyeing treatment. Worsted yarn can also be dyed as a warp, and also in cheese or cop package form, in the manner used for cotton dyeing and already described. Much of the worsted yarn processed is intended for hard-wearing fabrics, such as men's suits, and very fast dyes are required. Fast chrome colours are often used.

For yarns intended for hosiery applications, the types of yarns selected will be determined by price and the type of knitted garment to be made. Men's socks and jerseys are washed frequently, and the so-called 'acid' dyes having a good wash fastness are mostly used. On the other hand, fastness to washing is of less importance than uniform colouring and light fastness for the pale shades most frequently used for women's coats and costumes, and here the easily levelling acid dyes are mainly used.

The dyeing of carpet yarns, usually carried out by the carpet manufacturer himself, is normally performed by hanging the hanks from rods in a tank and circulating the dyeing solution through the yarns by means of propellers. The rods are usually supported in a frame so that all can be raised or lowered together, and finally transferred to other tanks for washing and other treatments. For carpet yarns, dyes which show good properties of penetration and fastness to light are used, and they should also be fast to rubbing and water. Colour matching is done by viewing the cut ends of the yarns. Some of the sulphuric acid used in the dye bath is often retained in the yarn by omitting rinsing after dyeing, since the presence of the acid stiffens the material to give a better pile and also improves the rubbing and water fastness of the dye.

The winch machine, already described in the case of cotton, is mostly used for the dyeing of wool fabric lengths. Almost every class of dye which can be applied to wool is used for some application. Fabrics intended for heavy wear are usually dyed with fast mordant colours: the salts of chromium, particularly sodium bichromate or chromate, are mainly used for mordanting wool. The so-called 'acid' dyes are used in considerable amounts for dyeing wool fabrics for a range of applications. These dyes are simple to apply, it being necessary only to immerse the wool for the requisite time in a boiling, acidified solution of the dye. A slower rate of dyeing, and improved levelness, can be obtained by adding Glauber's salt. Sulphuric acid is used mainly to acidify the solution, and an excess of acid is often added to prevent too great a degree of softening of the fibres, which would lead to permanent creasing. The material to be dyed is immersed in the dye solution, and the latter is then taken to the boil in about 45 min, after which the fabric is moved about in the boiling solution for a further 30 to 45 min. Usually, very little dyeing takes place until the boiling temperature has been reached.

Direct cotton dyestuffs can be applied to wool, but they tend to give unlevel shades which are not particularly fast to rubbing. Since wool is sensitive to alkalis, such dyes have to be applied in a neutral or near-neutral state and there is a risk of obtaining too soft a fibre and developing creasing.

The basic dyestuffs go on to wool readily without the help of dyeing assistants, but they are sensitive to hard water and give colours which are not very resistant to light and rubbing.

The vat range of dyestuffs is used where very good fastness properties are required; uniforms, bathing wear, babies' wear, and coloured blankets are examples. For such applications, the necessary degree of fastness is difficult to obtain with acid or chrome colours. Vat dyes are usually applied to loose wool or wool in yarn form. If fabrics are dyed, machines must be used in which the material is kept completely submerged in order to avoid oxidation.

Silk yarns can be dyed in hank form by hand, or in hank dyeing machines fitted with rotating arms which hold the silk. Silk fabrics are usually dyed in winch machines. Nearly every type of dyestuff used for cotton and wool can be applied to silk, and similar methods of application are used. Basic and acid dyestuffs are in common use but where exceptional fastness characteristics are required, as is the case with silk embroidery yarns, mordant and vat dyestuffs are employed. The boiling-off liquor obtained during degumming plays an important part in the dyeing of silk. The addition of this gum solution to the dyebath appears to assist quite markedly in obtaining uniform dyeing results.

7.2.7 Raising

The 'raising' process consists in passing a rough surface over the face of the fabric to pluck some of the fibres and raise them so that they stand out from the body

Fig. 7.9 Woven woollen blanket cloth being 'raised' on a metallic teazle raising machine.

of the material. The process is applied to wool, cotton and synthetic fibre fabrics in order to achieve a number of objectives. Thus, raising is carried out in order to develop a pile on the surface of the fabric so as to change its appearance and give it a warmer handle, to soften the feel of the material, to obscure the fabric structure, and to soften the outlines of a design.

Two types of raising machines are used; one makes use of teazles obtained from thistle plants as the plucking means and the other uses rollers covered with card wire. In the first type of machine, a large diameter cylinder carries frames filled with the teazles. It is usual to wet the teazles with hot water to soften them before packing them into the frames. Care must be taken during this packing to ensure that a level surface is obtained in order to avoid unraised portions of the fabric surface. A roller above the teazle cylinder guides the fabrics and determines what arc of wrap is obtained on the cylinder; this roller can be moved to allow any desired wrap. The cylinder rotates at quite a high speed, and the fabric travels through the machine in contact with the teazle points at a slower speed.

The teazles wear away quickly and attempts have been made to replace them by small rollers covered with specially prepared wire clothing. The wire clothing is in the form of one inch wide strips carefully wrapped round the rollers to given an even raising surface. Steel pins, preferably flat or oval in cross-section, are fixed to the rollers with the edges in the direction of rotation. The small wire-covered rollers are mounted on the outside of a cylinder and, as the latter revolves, a belt also causes each roller to turn about its own axis. The speeds of rotation can be varied. The fabric is fed under ten-

Fig. 7.10 Raising: fabric in open width being passed between rotating rollers. The rollers are covered with fine wires which pull the surface fibres out of the fabric to give a soft, fluffy appearance.

sion over the cylinder. It can pass in contact with the points once, or its ends can be joined together to form an endless belt and each portion passed over the points a number of times. In one type of machine, the rollers rotate and the fabric moves in a direction opposite to that of the cylinder surface, the amount of raising produced depending upon the relative speeds of the fabric and rollers. In another type of machine, used principally in the manufacture of blankets, the fabric passes through the machine in the same direction as the cylinder turns but the small rollers rotate in the opposite direction. In yet another type of machine, a rotating roller is covered with card wire and is lowered to make contact with the fabric supported on a flat bed so that it lifts up the pile.

The looser the fabric construction, the more easily will it respond to raising. High twist yarns resist raising more than low twist yarns, and fabrics made from doubled yarns are less easily raised than those constructed from single yarns. After raising, fabrics made from merino wools have a short, dense pile, while the ones made from coarser and longer wools have a more shaggy appearance after being treated.

Wool fabrics can be raised in the wet or dry condition. Wet raising, using either acid or soap solutions, is usually done on the teazle type of machine, while dry raising, usually employed in the manufacture of blankets and velours, is normally carried out on the card wire machine.

Raised wool fabrics are often cut level using a cropping machine. The blade cylinder must be accurately parallel to the surface of the fabric and, as the pile fibre ends are cut, they are removed by suction.

7.2.8 Weighting of silk

Silk has always been an expensive material to produce and, to add to the cost of the final material, there is a loss in weight of something like 25% during degumming. It is believed that, originally, it was a desire to make up this degumming loss that led to the practice of 'weighting' the silk. Weighting is a process whereby certain chemical substances are deposited inside the silk fibres to produce an increase in weight and thickness of each fibre. Since that time, weighting has been extended far beyond the cancelling out of the degumming loss in order to cheapen the material as much as possible. If weighting is kept within reasonable limits, then it can add a fullness and richness of handle to silk, the latter being largely due to the swelling of the fibres which takes place at the same time. Excessive weighting reduces the ability of the silk to resist wear and can produce a rapid deterioration in the material.

The compound normally used for silk weighting is stannic chloride, and the method used consists in soaking the silk in solutions of stannic chloride and sodium phosphate alternately, with a thorough rinse in water after each immersion in the stannic chloride, and a rinse and treatment with acid after each immersion in the sodium phosphate bath. This process is repeated until the desired gain in weight has been obtained. The treated material is then given a final treatment in a bath containing sodium silicate.

7.2.9 Anti-shrink treatments applied to wool

Finished wool fabrics can shrink during wear and washing for two reasons; they can suffer from both 'relaxation' shrinkage and 'felting' shrinkage.

Relaxation shrinkage arises from the fact that, in the finishing processes, the fabric is left under strain by being pulled and extended in length. Further if a piece of fabric contracts too much during scouring or dyeing, it is usually pulled back to its former width during drying on a stenter. It is also necessary to impart a small amount of stretch in width to remove creases from the fabric. When the fabric which is strained in this way, or a garment made from it, is immersed in water, relaxation takes place and shrinkage occurs until the strains have been relieved. In order to eliminate relaxation shrinkage, a pre-shrinking process, usually referred to as 'London shrinking', can be carried out in which the fabric is wetted and then allowed to dry free from strain. The moisture can be applied either by folding the fabric along with a damp cotton cover and allowing to stand for a time, by spraying the fabric with water, or by passing the fabric between rollers, the lower one of which is picking up water the whole time. After being suspended until it dries, the fabric is then re-pressed in a strain-free condition.

The felting properties of wool were referred to earlier, and if felting occurs during laundering, the fabric will contract in area and gain in thickness. Once felting has taken place there is no treatment which can be applied to restore the fabric to its original condition. As explained earlier, felting takes place as a result of the migration of the wool fibres due to the scales with which their surfaces are covered, and the only way in which wool can be made less prone to felting shrinkage is to modify, or even remove entirely, the scale structure. To be successful the treatment should make the wool non-felting without materially affecting its other properties. The earliest method of reducing the felting tendency was to treat the material for a short time with an acidified solution of sodium hypochlorite, when active chlorine was released and attacked the scales on the surfaces of the fibres. The process was usually carried out in a winch machine or a dolly. The cloth was first wetted out and the correct amount of acid added. The sodium hypochlorite was then added gradually to enable the fabric to absorb the chlorine evenly as the acid and hypochlorite reacted together to produce chlorine. With such a process close control is necessary to make certain that the mixture remains acid throughout otherwise the maximum stability towards felting will not be obtained and, in addition, there will be a risk of yellowing of the fabric. After wet chlorination in this way, the fabric is rinsed in a solution of sodium sulphite or sodium bisulphite to remove any unreacted chlorine which may remain. Wet chlorination is not easy to apply and uneven treatment can result. It has been suggested that this unevenness can result from the great affinity which wool has for chlorine and which results in the fibres on the outside of a yarn becoming overtreated. For this reason, various modifications to the process have been made to provide a slower liberation of the chlorine and a more uniform absorption by the wool material. Wet chlorination is relatively cheap to carry out and is still used quite frequently when careful control can be exercised.

More recently, a method of treatment using dry chlorine has been developed by the Wool Industries Research Association. The wool is first dried to a moisture content of about 8% and then placed in a suitably lined autoclave. The air is pumped out of the autoclave and then chlorine gas admitted. When the gas has been absorbed, the material is removed when, in some cases, it is treated with sodium bisulphite.

There are other methods of chemical treatment which are applied to reduce the felting shrinkage of wool and which do not rely on chlorination. In some of these methods, chemicals are applied in organic solvents; when a wool fibre is wetted it swells and is more easily penetrated, whereas a suitable organic solvent will not produce swelling and tends to confine the chemical action to the scale layer of the fibre. An important shrink-resist treatment now in wide-scale use employs a combination of oxidation followed by reduction. The fabric is first treated with permonosulphuric acid, and then with sodium bisulphite, the latter providing the necessary reducing action.

These chemical treatments degrade the surface of the wool fibre and make its scales less effective in promoting felting. They do, also, affect the general properties of the material to some extent, and methods of coating the fibre surfaces to make the scales less effective, and yet not interfering with the material forming the body of the fibre, have been investigated. A number of polymers have been used as coating materials. In one method, the wool is treated with a solution of anhydrocarboxyglycine in an ethyl acetate/water mixture for several hours, when a thin film is deposited on the surface of each fibre to cover the scales. Other polymer substances have been used as coatings and applied in a similar manner. In some treatments, the material is impregnated with the chemical substance in a non-polymerized form, and the polymers are developed inside the fibres by baking. Melamine resins have been introduced into the fibres in this way. In a still later process, a very thin layer of resin is actually formed on the surface of the fibre by immersing the wool successively in solutions of two materials which, on making contact, will deposit the resin film. No heat curing is required in this case and high speeds of treatment can be attained, and the only additional steps required are rinsing and scouring.

7.2.10 Moth-proofing of wool

Wool garments and fabrics are liable to be damaged by the common clothes moth. Actually, it is not the moth itself which does the damage but the larvae or grubs which hatch out from the eggs she lays and which attack the wool fibres as a source of food. The grubs prefer the finer wools, which means that they are liable to do the greatest amount of damage to the more expensive merino garments.

Attempts have long been made to protect wool garments from moth damage. The earlier methods were to make the vicinity of the garments as unpleasant as possible to the moth and so discourage her from laying her eggs there. Moths do not like naphthalene, and the presence of this substance in moth balls was intended to drive the moths away rather than destroy them. The use of moth balls is probably the least effective way available for protecting garments from moth attack. There are other chemicals which can be used in a similar way and it is claimed that these can actually kill the clothes moth. This may be true, provided a sufficiently high concentration of vapour can be provided, but it is unlikely that the moth will ever experience concentrations of these magnitudes normal measures.

More effective protection is given by substances which actually kill the insect. Contact insecticides can be applied to the wool garments or fabrics. These kill by contact and will destroy not only the moth but any grubs which hatch out. An insecticide, 'Dieldrin', is widely used for this purpose; it is usually applied in the form of an emulsion added to the processing solutions applied to the fabrics, and it is effective in action. Unfortunately the effects of the insecticide treatment are not really permanent since they tend to be removed in subsequent dry-cleaning operations.

The most effective method of protecting wool materials against moth damage is probably by the use of chemicals which, while making the wool unpalatable or even poisonous to the moth grubs, are also capable of being attached permanently to the wool fibre rather in the manner of a colourless dye. Such chemicals are usually applied by the dyer and finisher, and are virtually unaffected by washing and dry-cleaning.

7.2.11 Decatizing

This is a treatment given to a wool fabric towards the end of the finishing processes to give a flat surface and a soft, lustrous finish. In decatizing, a length of cotton cloth is first wrapped round a perforated hollow metal cylinder in order to prevent staining of the wool fabric, and then the wool fabric, along with a length of cotton 'wrapper', is wound on to the cylinder. A few extra yards of cotton wrapper are wound on to form an outside cover. The cylinder and its covering fabrics are then placed inside a large cylindrical pressure vessel which is sealed. The outer shell of the vessel is heated to prevent water condensing on it and falling on to the wool fabric to mark it. Steam under pressure is now introduced into the vessel and the pressure held for the necessary time, after which the steam is shut off, the pressure reduced, and the cylinder with the fabric round it moved to a cooling frame.

The primary object of decatizing is, as indicated, to obtain a smooth, soft, lustrous wool fabric, but the process can be used to produce novel effects on the surface of the fabric. Thus, a raised pattern in string can be formed on the surface of the cotton wrapper and this wrapper then wound in contact with the wool fabric during decatizing; a longer steam-blowing time than normal will probably be used. When the process has been completed, the wool fabric will then have the string pattern permanently impressed on its surface.

Care is needed if faults are to be avoided during decatizing. The main faults arise from water-marking and unlevel treatment. Metal stains can also be introduced. The tension in the wound fabric is also important. If it is too low, then the setting and lustre will be unsatisfactory and the fabric can have a wavy appearance. If the tension is too high, the fabric will be thin and lack 'body'.

7.2.12 Permanent pleating of wool

In the wool fibre, the chain molecules are joined, one to the other, by lateral bonds, referred to as the bisulphide bonds. When the fibre is distorted, say by bending or stretching, the bonds extend somewhat in the manner of pieces of elastic and, when the distorting force is removed, the bonds recover and pull the fibre back into its former shape and dimensions. Treatments have been found which break these lateral bonds in the wool fibre, thus eliminating the elastic recovery, and, while the fibre is still in its distorted condition, reform them so that the distorted state now becomes the stable form of the fibre. The elasticity of the bonds is now utilized to pull the fibres back into the distorted shape if an attempt is made to straighten them. Such treatments form a basis for a permanent pleating process for wool.

To effect permanent pleating, the wool fabric is first treated with a reducing agent to break the lateral bonds; sodium bisulphite was the first used. The fabric is then folded into its pleated or creased form and pressure and hot steaming are applied, while it is in its folded form, for something like 3 min. The combined effect of the heat and moisture reforms the bonds while the pleats are being held in place. Upon releasing the fabric, it will have become permanently pleated, and the pleats will be retained even when the fabric has been washed a considerable number of times.

The Si-Ro-Set process for permanently pleating wool used a 2% solution of ammonium thioglycollate as the reducing agent, and this was sprayed on to the fabric along the crease and pressure and steam were applied. This compound and sodium bisulphite have largely been replaced by another substance, monoethanolamine bisulphite, used as a 3% solution, and this is now in common use as a permanent pleating agent.

If a wool fabric has been treated in this way, the garment maker-up runs into difficulties. He now finds it impossible to use a hot pressing to insert the simple creases he requires as a normal method of constructing the garment. The difficulty has been overcome by the development of 'delayed-cure' treatments. The reducing agent is now applied to the entire fabric and then dried at as low a temperature as possible. The wool is now in the intermediate stage, and it only requires the garment maker to apply a water spray before hot-pressing the creases or pleats in position, when curing is completed and a permanent effect is obtained.

7.3 MAN-MADE FIBRES

7.3.1 Scouring and bleaching

Man-made fibres are produced as a result of chemical processes and do not contain as much impurity as natural fibres. The impurities they are likely to contain are the small amounts which they have picked up during the processes to which they have been subjected. It follows that scouring and bleaching are not such important processes as is the case with natural fibres, scouring simply being required to remove the processing impurities, while bleaching is rarely necessary since the fibres or yarns have usually been bleached during the early stages of manufacture.

The cleaning of the man-made fibres is usually a relatively simple operation which involves a treatment with a mild soap or synthetic detergent solution. It should be remembered, however, that the term 'man-made fibres' covers a whole range of materials having very different resistances to chemical attack, and the cleaning agents which can be used must be selected accordingly if safe processing is to be obtained.

Viscose rayon is formed from cellulose which has a less compact structure than that of cotton, and so it is more sensitive than cotton to attack by acids and alkalis. Viscose

Fig. 7.11 A viscose rayon wash machine.

materials require as mild a scouring treatment as possible. They can be boiled in a dilute alkaline solution but, if too high a concentration is used, they can lose weight during the processing and finish with a harsh handle. Polynosic rayon, having a more compact structure than normal viscose rayon, is less affected by chemical treatment and, in fact, behaves very much like cotton in respect to its treatment with alkaline solutions.

Cellulose acetate should not be treated with solutions of caustic soda, or even sodium carbonate. Such alkaline treatment causes the acetate fibre material to 'saponify' and produce regenerated cellulose, when the properties valued in the acetate fibre will be lost. It is usual for ammonia to be used in the scouring bath when cleaning cellulose acetate material, and any tendency to saponification is avoided.

The regenerated protein fibres also have an open, loose structure, and are more readily attacked by chemicals than the natural animal fibres such as wool. In their cases also, a mild scouring treatment should be used.

In general, the synthetic fibres are very resistant to chemical attack, some of them remarkably so, and more severe scouring treatments can be used. Usually, however, it is not necessary to go beyond a scour in a warm solution of trisodium phosphate to which a little detergent has been added. Synthetic fibre fabrics can change in dimensions during hot scouring, and any creases introduced at the same time will be difficult to remove. For these reasons, the fabric is sometimes first subjected to a 'partial setting' treatment before hot scouring. This partial setting may be accomplished by treating the material with boiling water, or dry heat above the temperature of boiling water. Another treatment which can be used is to wind the fabric in an open condition on to a perforated metal tube and blow steam through it. When a synthetic fibre fabric has to be scoured before a high temperature treatment is applied, cleaning can be carried out in a machine in which the material rests on a moving belt and the hot scouring solution is sprayed down on to it from above. Care is taken to avoid a tight fabric build up which would lead to creases being introduced. A winch type of machine is often used for the scouring and bleaching of rayon fabrics.

Some synthetic fibres are not a perfect white as they are received from the fibre producer. Polyacrylonitrile fibres tend to be slightly yellow, and the high temperature setting treatments given to most synthetic fibre fabrics to obtain stabilization can produce a slight discoloration. Synthetic materials respond less well to the standard bleaching treatments used for natural fibres. Sodium hypochlorite and hydrogen peroxide are only moderately effective; in any case such materials have a harmful effect on nylon. Sodium chlorite and peracetic acid, applied under acid conditions, give more satisfactory results, but quite severe conditions of treatment may be necessary to obtain good whites.

7.3.2 Dyeing

a. Rayons. Viscose rayon is dyed in yarn and fabric forms. Hank dyeing is often used for yarns but many processors now dye 'cakes' of the yarns since it eliminates the winding operations needed to prepare the hanks and later to convert back into the useable package form, and the fibres remain undisturbed in the yarns. In dyeing hanks of viscose yarn, great care is needed in handling to avoid broken filaments. Much hank dyeing is done by hand with the hanks hanging from sticks into open dye vessels heated by open steam pipes. Special hank dyeing machines are also used; the hanks pass round rollers which rotate and carry the yarn through the dye solution, reversing automatically after a specified number of rotations.

Fig. 7.12 A conical pan dyeing machine.

In the hot dye solution, viscose rayon swells quite appreciably and is weaker when in this condition. Undue stress during dyeing should, therefore, be avoided. Knitted fabrics are usually dyed on winch machines operated in such a way as to impose as little strain as possible on the material. Woven fabric can be dyed using either the winch or jigger machines. Staple fibre viscose fabrics can also be dyed using the jigger machine, but they then tend to become somewhat hard and 'boardy' due to the tension applied during the processing; an improved handle is obtained if winch dyeing is used.

Direct cotton dyestuffs are used mainly for normal viscose materials, since they give a wide range of colours of good fastness characteristics, and are easy to apply. They are applied in much the same way as for cotton. The dyeing is usually carried out at a temperature below the boil, say at 90°C (194°F), and at these temperatures level dyeing results are obtained.

When brilliant shades are required on viscose materials the basic dyestuffs are used. The material is entered with the dye bath cold and containing a small amount of acetic acid mixed with the dye solution. The temperature of the bath is raised slowly to 65°C (149°F) and held at that value until dyeing is completed. For deep shades, and shades of maximum fastness to light and washing, a mordant is used.

When exceptional wash fastness is required, the sulphur dyestuffs can be applied to viscose materials. These are employed in a manner very similar to that used for cotton, but with less salt introduced into the bath to assist take-up of dye. Sulphur dyestuffs do not cover variations in the viscose very well. A high degree of fastness to light and washing can be obtained on viscose yarns by the use of the vat dyestuffs. These are applied in a manner similar to that used for cotton, but it is necessary to retard the rate of dyeing since the viscose yarns take up the dye rapidly.

The dyeing of high-tenacity viscose is not easy by reason of its poor affinity for dyestuffs normally used on rayon. The direct cotton dyestuffs are mostly used and high concentration dyebaths have to be employed. Fortunately, the material is mostly used in the undyed state.

Cuprammonium rayon is dyed in much the same way as viscose but its greater affinity for dyestuffs usually means that the salt added to the dyebath is eliminated or used in very much reduced amounts.

Acetate rayon has very little affinity for the usual dyestuffs. Acetate cannot be boiled during dyeing as this destroys its lustre and, as already mentioned when discussing scouring, caustic alkali solutions cannot be used. There was an obvious need for a special dyestuff for acetate rayon and finally, after much experimental work, the 'dispersed' dyes were developed. These are very simply applied. The dyestuff is incorporated in the dyebath, with soap or soluble oil to keep the dye particles in suspension, and the temperature raised to about 85°C (185°F) when dyeing takes place without any other additions being made to the dyebath.

b. Synthetics. The majority of the synthetic fibre materials have a low, or even negligible water absorption and, it follows, they also show no inclination to admit dyes readily. Many of them are difficult to dye in the normal way. Nylon has some affinity for water and can be dyed relatively easily using the dispersed acetate dyestuffs applied in much the same way as for acetate rayon. Very level dyeing is obtained but the light fastness on nylon is not as good as on acetate rayon while the washing fastness is similar for both materials. The water-soluble acid wool dyestuffs, and the direct cotton dyes, can be applied to nylon from solutions in water at temperatures just below the boil and a good wet fastness is obtained. The basic dyes can also be applied to nylon but the light and wet fastnesses are not particularly good. Nylon shows very little affinity for the vat dyestuffs.

The polyacrylonitrile fibres are difficult to dye and, originally, the only satisfactory way of colouring them was to incorporate coloured pigments during manufacture. It is possible, however, to incorporate substances in them having dye-attractive characteristics, and a number of such modified polyacrylonitrile fibres are on the market.

It is not easy to dye polyester fibres using normal methods of dyeing. Pale shades of a reasonably satisfactory degree of fastness can be obtained using the dispersed acetate dyestuffs. Deeper shades cannot be produced in this way. Dyeing of the natural and regenerated fibres is made relatively easy by the fact that they are swelled by being immersed in the water of the dyebath. This loosens up the internal structure of the material so that the dyestuffs readily penetrate into the fibres. Contact with water produces no swelling of polyester fibres, or indeed the majority of the synthetic fibre materials, and swelling and loosening of the structure to give easier dyeing must be accomplished in other ways. This is accomplished by chemicals which are available and do produce swelling in synthetic fibre materials. If one of these is added to the dyebath, loosening of the fibre structure takes place and much easier dye penetration is obtained. These chemical compounds which help dye penetration are referred to as 'carriers'. Carrier dyeing is carried out very successfully on a number of synthetic fibre materials using normal cotton and wool dyes, and once the carrier has been subsequently washed away thoroughly, good standards of fastness are obtained.

Fig. 7.13 A dyeing range for the continuous dyeing of polyester fibres.

An alternative method of loosening the fibre structure so that it will accept dye more readily, is to dye at temperatures higher than the boiling point. In order to obtain these elevated temperatures it is necessary to dye under pressure; a pressure of 15 p.s.i (10460 kg/m^2), for instance, will allow a temperature of 121°C (250°F) to be obtained. Special high-pressure dyeing machines have been devised and proved to give very satisfactory results on materials which require a long time for the dye to penetrate into the individual fibres. Apart from the synthetic fibres, high temperature dyeing can be aplied to the majority of fibres. Under pressure, wool, for instance, can be dyed in something like 2 min. Cellulose acetate fibres cannot be dyed in this way, since the use of such high temperatures produces severe tendering of the material.

7.3.3 Heat setting

All thermoplastic man-made fibres tend to contract towards their original unstretched length when subjected to moist or dry heat. Reference has already been made to a partial heat-setting treatment which can be given to the materials to eliminate permanent creasing and, at the same time, this tendency to contract, before hot scouring and dyeing are undertaken. This heat setting is only effective provided the setting temperature used is not exceeded in any treatment the material may receive subsequently; if it is exceeded, the heat setting which was given will be no protection against the creasing and dimensional instability which will be obtained under these new conditions. It follows that, while the partial setting conditions used may be such as to enable satisfactory processing to be carried out, they may not necessarily have been sufficiently severe to ensure dimensional stability during laundering, when boiling in water followed by hot ironing may be used. It is usual therefore, to give fabrics made from thermoplastic materials a final setting treatment as part of the finishing process. This final setting ensures that laundering can be carried out without any fear of shrinkage taking place at the temperatures used, and no more than a light ironing using a warm iron is required to smooth out any wrinkles which may have been introduced into the material during the cleaning processes.

Various heat-setting methods have been devised, and special machines designed for the purpose. The setting conditions the finisher will select will be in accordance with the application for which the material is intended, and the type of thermoplastic material used, but the general procedure is the same in each case. While the material is being heat-set it must be held to the required finished dimensions; if not so restrained, the materials will shrink under the heat applied to them. The methods of applying this restraint to fabrics are indicated below, but if garments are being treated, they are stretched over frames of a suitable shape and size. Moist or dry heat can be used in setting, it usually being necessary to go to a somewhat higher temperature with dry heat to obtain an equivalent degree of setting. Actually dry heat gives a rather different setting effect when compared with moist heat, and it is often claimed to improve the handle and drape of nylon fabrics.

Setting using dry heat can be obtained by taking the fabric under tension over heated cylinders. Alternatively, the fabric can be held out in open width on an enclosed stenter frame when the required temperature is reached by the use of hot air or from radiant heaters. Only a short time of exposure to dry heat is necessary. Setting using moist heat can be effected by winding the fabric on to a perforated metal tube covered with a few protective layers of cotton material, and placing the wound tube inside an air-tight chamber when it can be subjected to pressurized steam. In this case, the fabric must be wound evenly and uniformly to avoid creasing or wrinkling which would become set in the material, and under sufficient tension to hold the fabric in place against the contraction which the application of heat would otherwise produce.

Continuous setting machines capable of giving a controlled treatment up to a temperature of 260°C (500°F) are available. Calendering is sometimes used as a final operation to smooth out the fabric and make it more attractive to the purchaser.

In the case of regenerated cellulose fibre materials, dimensional stability is improved by methods similar to those applied to cotton. It should be remembered, however, that such fibres have a less compact structure than cotton and greater care in carrying out the various treatments is necessary if damage to the fabrics is to be avoided.

The same treatments which are applied to thermoplastic man-made fibre fabrics to set them in an open flat condition to which they will return after wrinkling, and thus avoid creases being formed, can also be used to insert permanent creases or pleats in the material. Thus, instead of holding the fabric in a flat condition during heat-treatment, it is folded and held in a pleated formation while the heat is applied. The setting temperature used for fabrics made from nylon, polyester, or polyacrylic fibres will probably be of the order of 150°C (302°F) so that, when hot water is used to launder such permanently pleated fabrics, the pleats will remain sharp and distinct.

Regenerated cellulose fibres are not thermoplastic and they do not have the inherent property of being able to be set into permanent pleats. They can be made capable of taking pleats, however, by a treatment similar to that used to make cotton crease-resisting. A partially-polymerized resin is introduced into the fibres and dried. The im-

Fig. 7.14 Machine pleating: the fabric is protected by tissue paper, folded and hot pressed at about sixty pleats per minute.

pregnated fabric is then folded and held in a pleated form and the resin it contains 'cured' at a high temperature to complete its polymerization. The pleats produced in this way will be fast to washing. Cotton fabrics can, of course, be permanently pleated in the same way.

7.3.4 Rayon crêpe fabrics

There is a substantial market for rayon crêpe fabrics. Crêpe fabrics are formed from specially twisted yarns which, when subjected to wet finishing, become distorted to give a cockled or 'pebbled' appearance to the fabric surface. Only yarns which are capable of absorbing moisture to an appreciable extent are suitable for use by the normal crêping methods. Of the rayons, viscose rayon and cuprammonium rayon are suitable materials while acetate rayon responds less well to the processing. A typical rayon crêpe fabric will have, therefore, a warp of viscose threads and a weft of acetate threads, or vice versa, depending on whether or not the fabric is to be a warp or weft crêpe fabric. The fabric which is mostly prepared is the one having acetate threads in the warp direction and viscose threads in the weft. In that case the viscose weft threads are prepared with a high degree of twist, something like 30 or more turns per inch being applied. The twisted threads are steamed to set the twist in the yarn and this twist will remain without any tendency to untwist unless the threads are wetted. Two types of weft thread are prepared, one having twist in one direction and the other twist in the opposite direction so that, by using these in sequence in the fabric, first one and then the other, the final material is 'balanced' and remains free from overall distortion. It is usual to apply a fugitive dye to the twisted yarns, which will be removed easily in the final finishing processes, to distinguish a yarn of one twist direction from one of the opposite twist.

In weaving crêpe fabrics, probably two weft threads of one particular twist direction will be inserted, and these will be followed by two with the opposite twist, and so on until the length of fabric has been completed. At this stage, an apparently normal, flat fabric will be produced with two sets of coloured lines running across it. The crêpe effect is then developed during finishing.

If the fabric is boiled, the weft threads will untwist and kink mainly in the spaces between the warp threads, some kinking in one direction and the others in the opposite direction, to form a cockled effect over the surface of the fabric. The boiling water-bath used normally contains soap, since this enables the threads to move more easily. Various methods of introducing the fabric into the boiling soap solution are used. In one, the fabric is suspended in a loose form from a pole which is lowered into the boiling solution; from time to time a number of such loaded poles will be moved along the vat to draw the fabric through the solution and help to provide uniform crêping over the whole surface of the material. In another method, the fabric is drawn along a sloping surface and boiling soap solution is sprayed down on to it. Whatever method of developing the crêpe effect is used, care must be taken to keep the fabric as free as possible from tension, and from creases which would be extremely difficult to remove subsequently.

It is difficult to obtain a uniform crêpe effect over the whole surface of a piece of fabric and, unless uniformity is obtained, the appearance will be spoilt. It has been found that more certain results are obtained if the fabric, previous to being boiled to develop the crêpe effect, is run through an embossing calender when a pattern of finely spaced lines, similar to that obtained from the actual creping treatment, is impressed into the surface of the fabric. This embossing appears to induce the kinks to form in something like the same pattern and a more even effect is obtained.

During the crêpe development, the weft threads shrink quite appreciably and the fabric is reduced in width. At the same time, the warp threads have to accommodate themselves to the changes in the weft and their interlacing waviness becomes more pronounced, thus leading to a contraction in the warp direction too. In order to finish the fabric to its final dimensions and appearance, the material has to be pulled out in both directions and set in this condition. Since some of this stretch in a subsequent laundering treatment would be recovered and some shrinkage take place, the makers of the fabric usually recommend that dry-cleaning should be substituted for washing.

7.3.5 Modification of handle and appearance

In general, the methods of modifying the handle and appearance of rayon fabrics which are used are similar to those employed during the finishing of cotton. It must be remembered, however, that viscose rayon fabrics are more easily extended and damaged when they are in a wet condition and care is needed in handling them during the finishing treatments.

When rayon first made its appearance, it was valued for its bright lustre and sheen. Later on the fashion changed and there was a greater demand for fabrics with a more subdued lustre. Attempts were made to achieve this by applying delustring agents, such as China clay, to the fabrics, and binders such as gums and starches were incorporated with the delustring agent to obtain a better adhesion of the particles on to the fabric. It was difficult to obtain uniform results in this way and, in spite of the binders used, the delustring powder left the surface of the fabric during wear and was lost completely after a few washes. Attempts were made to improve on the results of these early treatments. These followed the lines of trying to deposit the dulling agents within the material of the fabric itself. In one, the fabric was treated with a solution of a suitable substance, such as barium chloride, and then transferred to a second bath containing, say, sodium sulphate. The two chemicals reacted inside the material itself and a dense, white precipitate of barium sulphate was formed. The delustring obtained in this way was much faster to washing and similar methods are still in use.

Probably the most dense and opaque material which is readily available is titanium oxide, and finely ground titanium oxide is now produced in large quantities and incorporated in paints. Unfortunately, it is not possible to form titanium oxide inside the rayon fibres as was done in the case of barium sulphate, and the material can only be applied externally. Very finely ground titanium oxide penetrates more thoroughly into the fabric and, the better the penetration, the greater the resistance to removal during washing. However, continued washing will eventually remove all the titanium oxide, no matter how fine its particle size may be, and methods of anchoring the material in position were sought. No dulling agent which will be attracted by the rayon material after the manner of a dye is known, but investigations showed that certain pigments, including titanium oxide, could be compounded with an organic substance which had an affinity for rayon and these composite delustring agents were readily absorbed by the rayon material. Such composite dulling agents can be applied as a separate treatment, or they can be added to the dye bath during dyeing or to the last rinse after dyeing.

The most satisfactory method of producing a dulling effect in both the rayons and the synthetics is to incorporate titanium dioxide in the spinning solutions when the fibres are being extruded through the spinneret. In that case, the light diffusing particles are suspended inside the filaments themselves and are absolutely permanently in position and completely free from any adverse effects produced by laundering or any other treatments. Synthetic fibres, such as the polyamides and the polyesters, are offered for sale containing various amounts of titanium dioxide pigment and thus have varying degrees of dullness.

Cellulose acetate material can be given a dull appearance by prolonged boiling in a soap solution. The effect of the boiling is to give striated and pitted surfaces to the fibres and thus destroy the lustre. The method is often used to delustre cellulose acetate fabric, the amount of delustring applied being a function of the temperature and time of application. The fabric must be free from tension for a satisfactory delustring by this method to be obtained, even a moderate amount of tension preventing sufficient dulling from being produced. For this reason, the treatment cannot be given using a jigger type of machine which does pull on the fabric and winch machines are normally used. There is some difficulty with winch machines in keeping the soap solution at the boiling point and modifications to the normal winch machine have to be made. Cellulose triacetate fibres do not respond to a similar delustring treatment.

Delustred cellulose acetate material can have its lustre restored by swelling the fibres while tension is applied. Thus, delustred cellulose acetate can be treated with a solution of acetic acid in water and then dried while it is in a stretched condition, when most of its original lustre will be regained. However, material which has recovered its lustre in this way is more easily dulled in hot soap solution than it was originally, and it is not even necessary to go to the boiling point to obtain delustring.

The man-made fibres are, in general, made under such conditions that a relatively soft handle is obtained. If required, the degree of softness can be improved still further and any harshness which may be introduced as a result of further processing can be removed by various means. Thus, when vat dyes are applied to rayons, a strongly alkaline dyebath is used and this can produce harshness in the materials. Again all textile materials dried under tension develop a harshness of handle, and rayons dried at elevated temperatures tend to lose softness. As was the case for cotton, mechanical means can be used to produce some softening. Thus calendering will flex the fibres and soften them and also, by flattening the surfaces of the fabrics, the handle of the fabric can be altered by this means. Steaming or 'decatizing' are particularly effective with rayon fabrics and there is less risk of damaging the materials than with calendering. The raising or brushing of man-made fibre fabrics to obtain a raised pile can produce a soft, pleasant handle and, due to the amount of air held in the pile, the handle is made warmer at the same time. Raising processes have been dealt with earlier in the chapter.

It is quite common to treat man-made fibre materials with so-called 'softening agents' in an attempt to give them a handle nearer to those made from the natural fibres. These softening agents are 'oily' compounds which have the effect of reducing the friction between the individual fibres, and a large number of such compounds are now available. Many of the compounds have a strong affinity for the textile materials and become attached after the manner of colourless dyes and are resistant to washing. They can be applied as a warm solution in water and are taken up readily by the fabric. It is not wise to apply too much softening agent since it is possible for the individual fibres to become slippery and slide easily one over the other when tension is applied, thus producing an apparent loss in strength.

With cellulose acetate fabrics a 'moiré' finish is sometimes applied. In considering the finishing of the vegetable fibre fabrics earlier in the chapter (section 7.1.5) reference was made to the 'schreinering' of cotton fabrics in which a pattern of closely spaced fine lines was impressed into their surfaces. The moiré finish is a type of schreiner treatment but, instead of passing the fabric between the engraved rollers of a calender, the cellulose acetate fabric is in contact with another fabric as it goes through the roller nip so that the thread formation of the latter fabric is impressed upon the cellulose acetate. The increased lustre produced is not distributed uniformly over the surface of the acetate fabric but has a more wavy appearance which is the moiré finish.

7.3.6 Anti-static finishes

In chapter 4 it was explained how electrostatic charges, usually referred to as 'static', can build up on the newer synthetic fibres, and on cellulose acetate fibres, and cause considerable trouble during processing. Much of the difficulty arises from the fact that such fibres have a low water absorption so that the provision of a humidified atmosphere only goes a certain way towards discharging the very considerable amounts of static which do accumulate. Natural fibres like cotton and wool are just able to avoid a static build-up in the dry state, while in a humid atmosphere their substantial moisture absorptions ensure that any charge which may develop will leak away quickly.

During the various stages of preparing fabrics from synthetic fibres, steps can be taken to make static problems less troublesome. The provision of a moist atmosphere does help, by providing a film of moisture on each fibre to make the latter a little more conducting. The fibres can be treated with a conducting oil to achieve a similar effect. Finally, the air surrounding the machine, or at least the parts in proximity to the synthetic material, can itself be made electrically conducting by the use of 'ionizing' equipment. These remedies are only effective during manufacture and there still remains the possibility that the fabric will pick up quite large static charges during wear, causing the material to cling to the body of the wearer, and cause discomfort. It is quite common, therefore, for synthetic materials to receive a so-called anti-static finish in an attempt to eliminate the trouble during wear and use.

For a few applications, such as curtains and drapings, it may not be necessary to have an anti-static finish which is completely fast to washing, but for the majority of fabrics and garments a 'permanent' finish is obviously highly desirable. Much research has been carried out on the subject and a whole range of treatments, claimed to give durable anti-static properties, have been covered by patents. In many of these, the compounds have been applied in a soluble form on to the surfaces of the fibres and then 'cured' in position by the use of heat in the presence of a catalyst. One method which is of interest is to coat the fibres with a partially-cured resin which can subsequently be treated at an elevated temperature to produce cross-linking and so enclose each fibre in an electrically-conducting skin which is unaffected by washing.

Various methods of providing relatively durable anti-static finishes are now available, and to these must be added the possibility of anti-static materials being incorporated in the fibres themselves at the spinning stage.

7.3.7 Flame-proofing

Amongst man-made fibre yarns and fabrics there exists a whole range of burning characteristics and degrees of fire hazard. Viscose and cuprammonium rayons are flammable and burn rather like cotton, as one would expect from their compositions. Cellulose acetate is less flammable than the rayons but it melts and burns quite readily. In general, the regenerated protein fibres burn rather like wool and have a high ignition temperature and slow rate of flame propagation; they would not therefore be considered to constitute a fire hazard. The thermoplastic fibre materials are difficult to classify since they tend to melt and shrink back from the flame. They are, however, less flammable than the cellulose fibres and, when nylon and polyester burn, the molten polymer falls away carrying the flame with it and there is usually no propagation of the flame along the fabric. The hot, molten material falling on to other materials could in such cases be considered to constitute an additional fire hazard: Materials such as Saran and modacrylics do not burn but shrink away from the flame.

As far as all textile fibres are concerned, only certain types of fabrics can be considered to constitute really serious hazards. Before a flame can spread, the material must receive an ample supply of oxygen. It follows that light-weight material and particularly light-weight nets, and brushed and raised fabrics which hold a large amount of air in between the fibres, are in a condition to support rapid flame propagation. When these materials of open construction are made from fibres which burn readily, such as the rayons and cellulose acetate, then some type of flame proofing treatment is essential if the possibility of serious accidents taking place is to be avoided. Flame-proofing research has been concentrated mainly on these easily-burning materials.

The rayons can be treated using the compounds and methods employed in the fire-proofing of the natural vegetable fibres such as cotton. It should be remembered that the fabrics made from rayon yarns are already less strong than fabrics of similar construction made from cotton, and since a substantial amount of chemical must be deposited on the fabric to make the treatment effective, the method of treatment should produce little or no tendering.

A number of processes for the flame-proofing of cellulose acetate are in existence, but they depend on the use of substances, such as phosphoric esters, which act as plasticizers for the material and cannot be applied, therefore, in sufficient quantities to ensure complete proofing. Some reduction in the flammability of cellulose acetate can be obtained by the application of a 10% solution of ammonium thiocyanate, but the treatment swells the fibres in the fabric and reduces the overall strength. Many of the later flame-proofing processes which are applied to cellulose acetate make use of expensive chemicals such as ethylenediamine dihydrobromide.

Much work has been done on developing flame-proofing treatments for the synthetic fibres where this is required. It is claimed that the application of a thiourea-formaldehyde resin to nylon is effective but it is doubtful if such a treatment will resist laundering sufficiently well. Similar treatments have been tried on the polyester and acrylic materials but with no greater success than with nylon. In all these cases the most effective treatment is to incorporate the flame-proofing agent in the fibres themselves at the spinning stage. This can of course, produce changes in the properties of the fibres and research is required in order to find compounds which are effective and yet produce only the minimum of change in the properties. The Monsanto Company has produced a flame retardant acrylic fibre for use in the manufacture of carpets, while other manufacturers have blended their acrylic fibres with about 30% of modacrylic fibre to obtain flame retardant carpets.

Very considerable difficulties are experienced in applying flame-proofing treatments to blends of, say, cotton and nylon. If the nylon content is greater than about 30%, the charred treated cotton holds the molten nylon globules on to the fabric and prevents them from falling away so that the whole fabric is burned. The same is true of blends of cotton with many other of the synthetic fibres, and the only solution would appear to be to use fibres which have been made inherently flame retardant, as mentioned above, in blends with cotton.

7.4 LITERATURE

Perkin Centenary, London: 100 Years of Synthetic Dyestuffs, Oxford, Pergamon Press, 1958.

J. C. CAIN and J. F. THORPE. *Synthetic Dyestuffs and the Intermediate Products from which they are Derived*, London, Griffin, 1946.

H. C. SPIEL. *Textile Chemicals and Auxiliaries*, New York, Reinhold, and Chapman & Hall, 1957.

S. R. COCKETT and K. A. HILTON. *Basic Chemistry of Textile Colouring and Finishing*, London, National Trade Press, 1955.

S. R. COCKETT and K. A. HILTON. *Basic Chemistry of Textile Preparation*, London, National Trade Press, 1955.

E. R. TROTMAN. *Dyeing and Chemical Technology of Textile Fibres*, London, Griffin, 1964.

S. R. COCKETT. *Dyeing and Printing*, London, Pitman, 1964.

S. R. COCKETT and K. A. HILTON. *Dyeing of Cellulosic Fibres and Related Processes*, London, Leonard Hill, 1961.

R. S. HORSFALL and L. C. LAWRIE. *Dyeing of Textile Fibres*, London. Chapman & Hall, 1946.

R. W. LITTLE. *Flameproofing*, London, Chapman & Hall, 1949.

A. J. HALL. *Handbook of Textile Finishing*, London, National Trade Press, 1957.

J. T. MARSH. *Introduction to Textile Bleaching*, London, Chapman & Hall, 1956.

J. T. MARSH. *Introduction to Textile Finishing*, London, Chapman & Hall, 1957.

Introduction to Textile Printing. Butterworths in association with ICI Ltd, 1964.

J. T. MARSH. *Mercerizing*, London, Chapman & Hall, 1951.

R. W. MONCRIEFF. *Mothproofing*, London, Leonard Hill, 1950.

H. U. SCHMIDLIN. *Preparation and Dyeing of Synthetic Fibres*, (trans. N. Meitner and A. F. Kertess) London, Chapman & Hall, 1963.

E. KNECHT and J. B. FOTHERGILL. *Principles and Practice of Textile Printing*, London, Griffin, 1952.

W. TAUSSIG. *Screen Printing*, London, Clayton Aniline, 1950.

J. T. MARSH. *Self-Smoothing Fabrics*, London, Chapman & Hall, 1962.

C. L. BIRD. *Theory and Practice of Wool Dyeing*, Society of Dyers and Colourists, 1963.

J. L. MOILLIAT (Ed.). *Waterproofing and Water-Repellency*, Amsterdam, Elsevier, 1963.

R. W. MONCRIEFF. *Wool Shrinkage and its Prevention*, London, National Trade Press, 1953.

CHAPTER 8

Synthetic Resins and Plastics

8.1 INTRODUCTION

'Plastics' are materials based on organic compounds which, by the application of heat and pressure, applied separately or at the same time, can be made to flow and take up pre-determined shapes, after which they can be hardened and set so as to retain their shape.

As we shall see in this chapter, the plastics now available are many and varied and differ widely in their physical and chemical characteristics and in their final usable form. They have all, however, assumed the mouldable, plastic state at some stage of manufacture.

8.2 HISTORICAL DEVELOPMENT

Natural resins and plastic materials have been well known and used for many hundreds of years; lac, for instance, in the form of shellac coatings and cast mouldings. Gutta percha was known in the Western world in the early 1600s and later was applied as a cable insulating material and for moulding purposes. Natural rubber had been used for a very long time before the investigational work carried out between the years 1820 to 1850 by Hancock and Goodyear resulted in easily mouldable rubber products and in the development of the vulcanization treatment to make the material of much wider practical application. Hancock and Goodyear, acting independently of each other, heated rubber with an excess of sulphur, and obtained a hard product which became known as 'ebonite' or 'vulcanite', and this was the very first 'thermosetting' plastic to be manufactured.

In 1856 cellulose nitrate was used by Parkes as a dressing for woven fabrics in an attempt to make them waterproof, and he later developed it as a moulding material, when it became known as 'Parkesine'. Parkes was, therefore, the first man to apply a chemically-modified polymeric material as a 'thermoplastic' on a commercial scale.

In 1870, Hyatt in the USA took out a patent for a tough material using cellulose nitrate and camphor. This material become known as 'celluloid'. Some twenty years later, casein plastics prepared by reacting together milk protein and formaldehyde were developed in Germany. They became known under the trade name of 'erinoid'.

In 1892 was laid down the foundation of the plastics industry of today. In that year, Bayer reported that phenols and aldehydes could be reacted together to produce resinous products and a few years later Smith took out patents in Britain covering the preparation of phenol-aldehyde resins. It was left to Baekeland in the USA, however, to discover how to control and modify the reactions between the two types of chemicals so as to produce useful, practical products. The Bakelite Co. was formed in the USA to exploit his findings. The phenolic resins perfected by Baekeland were the world's first commercially successful truly synthetic resins.

Others were stimulated to work along similar lines, and in 1918 John reacted together urea and formaldehyde to produce a urea-formaldehyde resin, pale in colour, which could be pigmented to produce brightly-coloured moulded articles.

Cellulose acetate was used as an aircraft 'dope' during the 1914–18 war and, after the war, a means of preparing textile fibres from it was perfected. In 1927, suitable plasticizers for cellulose acetate were discovered and it was used in preference to celluloid as a moulding material. Cellulose acetate continued to be used on a large scale in the preparation of injection mouldings until early in the 1950s.

A period covering a relatively few years, between 1930 and 1940 saw the development of the important thermoplastic materials we know today. The German firm of I.G. Farben produced polystyrene in 1930 and the Dow Chemical Co. in America also began commercial exploitation of the material. About the same time, polyvinyl chloride was discovered and developed on a commercial scale. In 1931 a means of producing polyethylene (polythene) was discovered in the laboratories of ICI Ltd. in Great Britain, in a study of the effects of very high pressures on certain organic systems, a white solid was more or less accidentally produced and proved to be a polymer of ethylene. A plant was constructed to manufacture the polymer. Also, just prior to the 1939–45 war, ICI produced polymethyl-methacrylate, which they marketed as the transparent plastic, 'perspex'; this material was used in large quantities during the war for the glazing of aircraft.

As a result of the research of Carothers, nylon was developed in the USA by the Du Pont Co. The material was first applied as a textile fibre material in the mid-1930s and later as a general moulding material. The Du Pont Co. also developed polytetrafluoroethylene in 1941 and established a production plant in 1943.

When the 1939–45 war ended, attempts were made to find applications for the newer synthetic materials which were now available, but many of the applications were quite unsuitable and indifferent results were obtained. The period up to about 1955 was spent in improving the quality and reproducibility of the existing plastics. Since that time, a number of new thermoplastic materials have made their appearance. These include high-density polythene and the lightweight polypropylene. In the USA the Du Pont Co. has developed the acetal resins, and the polycarbonates have been produced in both the USA and Germany. Further research, directed towards the development of high-impact polystyrenes, has led to the perfecting of the so-called ABS (acrylonitrile-butadiene-styrene) plastics. The would-be user now has a wide range of materials and fund of expertise to draw upon in making his selection for any particular application.

8.3 MOLECULAR ARRANGEMENT

8.3.1 Formation of large molecules

Plastic materials all have one characteristic in common; they are built up from very large molecules, which Staudinger named 'macromolecules'. These so-called macromolecules may be many hundreds, if not thousands, of times larger than the original unit molecules used in their construction. The substances which consist of the individual unit molecules destined to be potential 'links' in the long chains, are referred to as 'monomers'. Those which result when the unit molecules join up to form the macromolecules are called 'polymers', and the chemical reaction by which the change from monomer to polymer takes place is called polymerization. For the unit molecules to be able to link up to form macromolecules, they must be capable of forming two valency bonds, one with a molecule in front and another with a molecule behind, and they are usually stimulated to join up in this way by the presence of a suitable catalyst and the manipulation of the reaction conditions, such as changes in temperature and pressure.

These are two main ways in which polymer chains can be built up; one is known as an 'addition' reaction, the other as a 'condensation' reaction. In an addition polymerization, the unit molecules simply add together, one on the end of another, to build up a large molecule, without the formation of another substance as an intermediate stage. Thus, a molecule of ethylene, $CH_2{=}CH_2$, is potentially reactive and, under suitable conditions, one will link up with another to form a polymer molecule, polyethylene or polythene:

$$—CH_2—CH_2—CH_2—CH_2—$$

Polyethylene is thus formed by an addition reaction. Another example of addition polymerisation is the linking up of unit vinyl chloride, $CH_2 = CHCl$, molecules to form polyvinyl chloride:

$$—CH_2—CH{\cdot}Cl—CH_2—CH{\cdot}Cl—CH_2—CH{\cdot}Cl—$$

In the case of addition reactions, once started the build-up into large molecules is extremely rapid.

In condensation polymerization, more than one chemical entity, usually two, are involved. Before a reactive unit, capable of joining up with others to form a large molecule, can be formed, an intermediate step must take place. This is usually a chemical reaction in which a small molecule, usually water, is eliminated or expelled. Thus, in the preparation of polyhexamethylene adipamide or nylon 6.6, two organic compounds, hexamethylene diamine and adipic acid, are reacted together as follows:

$$\underset{\text{adipic acid}}{HOOC(CH_2)_4COOH} + \underset{\text{hexamethylene diamine}}{NH_2(CH_2)_6NH_2} \longrightarrow \underset{\text{hexamethylene adipamide}}{—HOOC(CH_2)_4CONH(CH_2)_6NH_2—} + \underset{\text{water}}{H_2O}$$

The hexamethylene adipamide unit thus formed has reactive ends and is capable of being made to join up with other similar units to form the polymer, polyhexamethylene adipamide. A condensation polymerization is a step-like, relatively slow, reaction.

Materials produced by joining units end-to-end to form long chains are referred to as 'chain polymers'. Many well-known plastics have this chain construction. As the chains build up, the monomer changes from a gas or mobile liquid to a solid, by reason of the loss of freedom of movement of the units as they are anchored at both ends. When such a material is heated, parts of the chains begin to vibrate, and when sufficient amplitude of vibration has been reached the chains slide easily one over the other to take up new positions. The material is thus capable of flowing into the shape of a mould containing it. When the heat is removed, and the material cools down, the vibrations of the parts of the chains are reduced, any tendency to flow is eliminated, and the mass of material becomes 'set' in this new shape. Plastics which become mouldable simply by the application of heat, and are then set by cooling, are known as 'thermoplastics', and their shapes can be changed any number of times by re-heating, re-moulding, and then cooling to obtain setting.

In addition to forming linear chains, chain molecules may also develop side chains or 'branches'. These branches, which are usually short compared with the length of the main chain, may vary in length and distribution along the chain; they can have a pronounced effect upon the physical properties of the material.

There is also a possibility that an occasional short link may be thrown across from one polymer chain to another, serving to 'tie' the material together at these points. This is referred to as 'cross-linking'. Cross-linking can occur accidentally while polymerization is taking place, or it can sometimes be brought about deliberately by chemical means when it is desired to give additional stability to the polymer structure. Polyethylene, for instance, has been made more suitable for cable coverings operating at temperatures up to 90°C by inducing a certain amount of cross-linking using chemical treatments with peroxides.

If at least one of the reacting unit molecules is capable of forming more than two bonds, there is a possibility that a three-dimensional structure, a kind of organic 'scaffolding', will be built up. Once this complete cross-linking has been built up, no amount of vibration of the molecular portions can loosen up the structure and cause the macromolecules to slide. However much they are heated, such materials remain hard and infusible and cannot be remoulded in any way. This type of material is referred to as being 'thermosetting'.

It has been found possible to construct macromolecules from two, or even more, different kinds of units, and such units can link up in several different ways. The products resulting from the joining up of different kinds of units are known as 'copolymers', as distinct from the 'homopolymers' which are constructed from units of

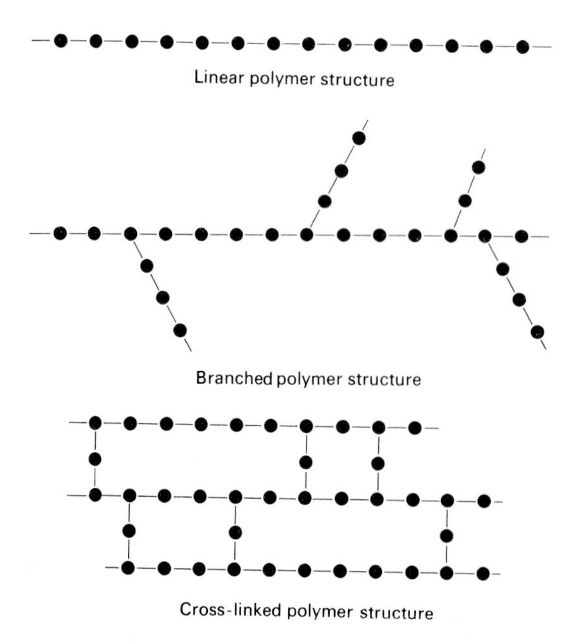

Fig. 8.1 Diagram illustrating the various types of polymer constructions.

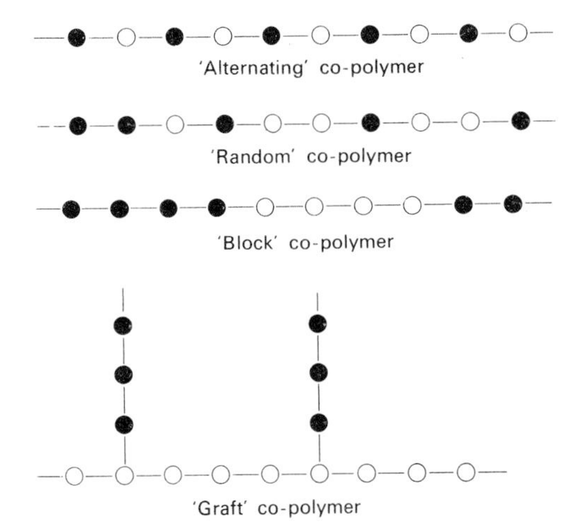

Fig. 8.2 Diagram illustrating the main linking-up possibilities for two types of unit molecule.

the same kind. Copolymerization is more in evidence with addition polymerization. There are several ways in which the two sets of units can arrange themselves in the polymer chain. Thus, if the units alternate along the chain, a unit of one being followed by a unit of the other, they will produce an 'alternating' copolymer. A random arrangement of the units within the chain, to produce a 'random' copolymer, is usually more important from a commercial point of view. Another arrangement has a number of units of one type together in a block, followed by a block of the other type, and so on all along the molecular chain, forming a 'block' polymer. A so-called 'graft' copolymer is produced when the main chain is built up of one type of unit only, and the other type of unit is confined to side-chains. No single polymer in a group may have all the properties desirable for a particular application, and yet the group may contain most of them; by copolymerizing together some members of the group it may be possible to provide a range of properties nearer to the ideal for that application.

8.3.2 Polymerization techniques

Theoretically, the preparation of a polymer is carried out by using a straightforward chemical reaction. The monomer, either as a gas or a liquid, is brought into contact with a suitable catalyst, or initiator, under specific conditions of temperature and pressure. The polymerization reaction may be carried out in bulk, in a solution in a suitable solvent, in suspension, or as an emulsion.

Bulk polymerization represents the direct approach, and gives very satisfactory products. It is used to produce some of the well-known plastic materials in the forms in which they are required. There are, however, difficulties associated with its use. The chemical reaction in which the polymer is produced from the monomer is accompanied by the evolution of a considerable amount of heat, together with an increase in viscosity and decrease in thermal conductivity. In addition, the polymerization is accompanied by a substantial contraction in volume of the mass and this, again, can present practical difficulties. Nevertheless, polymethyl methacrylate sheets are made by carrying out the polymerization between two sheets of plate glass. The shrinkage, and the amount of heat developed, are reduced by partially polymerizing the methyl methacrylate to form a syrupy solution of polymer in monomer, and this is fed into the space between two glass plates, separated round the edges by a flexible tube which can contract in response to the shrinkage. An alternative method of preparing the 'syrup' is to dissolve a given weight of polymethyl methacrylate in the monomer. The syrup, containing a catalyst, is now completely polymerized by heating followed by carefully controlled cooling as the reaction develops. Another well-known thermoplastic, polystyrene, is made by first partially polymerizing the monomer, and then introducing it at the top of a tall tower which is maintained at a higher temperature at the bottom than at the top; this ensures a better conversion ratio and also boils off any styrene monomer which may be present in the polymer. The styrene is polymerized as it travels down the tower, the heat of reaction being controlled carefully by heating and cooling coils, and it leaves the tower at the bottom in the form of molten filaments which are hardened by cooling, disintegrated mechanically, and then passed forward as a moulding powder.

As already indicated, there are quite severe practical difficulties associated with the exothermic reaction during bulk polymerization, and various other polymerization techniques have been devised. One of these is solution polymerization, in which a suitable solvent is used to dilute the reacting materials. In addition, the heat of reaction can to some extent be dispelled by stirring the solution. In the polymerization of styrene in this way, for instance, the monomer and the solvent are thoroughly mixed together, and the solution is forced firstly through a reactor which heats it up to start the reaction, then through others to cool it and reduce the heat developed during the polymerization, and then other reactors to keep the reaction in progress but under accurate control. The material is then heated to drive off the solvent and unchanged monomer, which are recovered and re-used. The molten polymer emerges from the heated vessel as thin filaments when it is solidified by cooling, broken up, and packed ready for use. The difficulties associated with solution polymerization are the complete removal of the solvent, solvent recovery problems, and fire and toxicity hazards.

Another method, widely employed for the polymerization of styrene, is 'suspension polymerization'. The monomer is stirred vigorously with water in which it becomes suspended as tiny droplets. The reaction is initiated by the introduction of a water-soluble catalyst. Since the droplets are of such small dimensions and are suspended in a low viscosity fluid, the removal of the heat developed presents little problem. As the formation of the polymer proceeds, the mixture of monomer and polymer makes the surfaces of the droplets sticky, with a tendency to adhere to each other. This adhesion is prevented by adding a suspension agent, such as talc or polyvinyl alcohol, which forms a coating round each droplet. The polymerization takes place within the droplets themselves and the material is produced in the form of small beads. After the process has been completed, any unreacted monomer which may be present can be removed by steam distillation, after which the polymer can be washed and dried. There are disadvantages associated with suspension polymerization. Unless the conditions are accurately maintained, the mass of droplets can aggregate and settle down on the bottom of the reaction vessel. Very large reaction vessels are required, since much of the volume is occupied by water. The polymer beads have to be washed and dried, and drying can cause some discolouration. Finally, suspension polymerization is more suited to batch operation and it is difficult to convert it into a continuous process.

The course of the chemical reaction can also be modified by using an emulsion polymerization technique. In this case, the reaction mixture contains a certain amount of an emulsifying agent, such as soap, and a catalyst which is water-soluble. The mixture is stirred in a reaction vessel and the monomer becomes suspended as an emulsion of tiny droplets. Heat may be applied and polymerization started within the droplets.

Once the polymer particles have been formed, steam can be used to drive off the small amount of unchanged monomer. The particles can then be coagulated to give a material which is more easily handled. Filtering, washing, and compressing into pellets follow, after which the dry pellets can be ground to moulding powder.

Continuous processing of polyvinyl chloride by the emulsion method is in use. The vinyl chloride monomer, catalyst, emulsifying agent, and water, enter at the top of a reaction vessel. The emulsification takes place in the top few feet of the vessel, and the temperature of the vessel is accurately controlled along its entire length by means of a water jacket. After the reaction is completed, this taking from 3 to 5 hours depending on the initiator used, the emulsion is pumped from the bottom of the vessel into a vacuum evaporator, from which any residual monomer is taken off for re-use. The emulsion particles are either spray-dried, or coagulated and dried. The resulting powder is washed, filtered and compressed. It is then dried, ground to a fine powder, and packed ready for use.

Acrylic ester polymers are largely prepared by emulsion polymerization, by a method very similar to that already described.

8.4 PROCESSING TECHNIQUES FOR PLASTICS

One of the main practical advantages of plastics is that they can be formed or moulded into shape, and then hardened or 'set' to retain that particular shape. Once an accurate mould has been prepared, therefore, even quite complicated plastic replicas can be produced rapidly, cheaply and with a high degree of accuracy.

Various methods of producing shaped plastic products are now known. The one selected in a particular case will be determined by a number of factors. Thus, the nature of the plastic itself, the accuracy of reproduction required, the use to be made of the plastic moulding, cost, and so on will have a big influence on the method to be employed.

The forms in which the various plastic materials are offered for shaping and moulding, and the shaping and moulding techniques which have been devised, are discussed below.

8.4.1 Moulding powders

Phenol-formaldehyde resins, and the thermosetting plastics in general, are not suitable in their purely resin form for the majority of commercial applications since they lack toughness and tend to be brittle. They are usually compounded with 'fillers' to improve the final mechanical properties, with catalysts or 'accelerators' and lubricants to give satisfactory moulding characteristics, and with pigments to make the final moulded product more attractive. It is usual first to make the resin, grind it up finely, and then mix it with the other ingredients. The choice of filling materials, which serves to anchor the plastic, will be determined by the mechanical properties required in the end product. Wood flour is often used, although materials such as cotton, paper and other fibrous fillers can be employed. The lubricant included to prevent sticking to the polished surfaces of the mould is normally a compound such as stearic acid. When heat and pressure are applied to the powder which has been fed into the mould the material must first soften sufficiently to flow under the applied pressure into even the tiniest cavities and then it must set hard rapidly so that it can be ejected from the mould ready

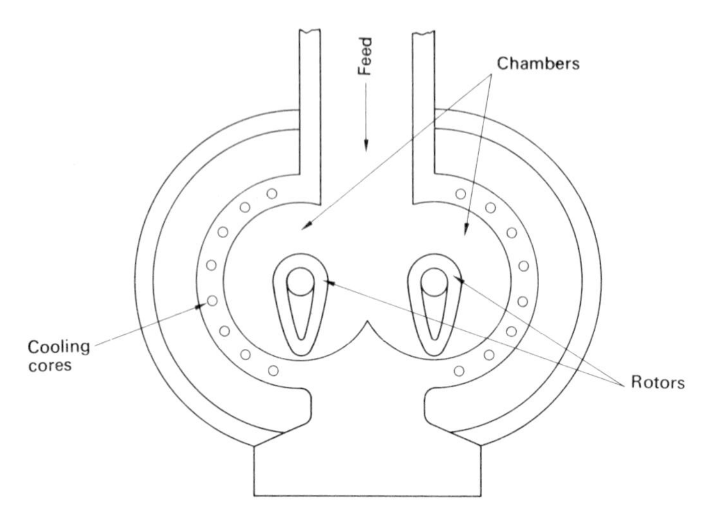

Fig. 8.3 The chambers in a Banbury mixer.

for the next cycle. This rapid setting is helped by the presence of the accelerator which speeds up the chemical action; lime or magnesium salts are often used for this purpose. The pigments incorporated in the moulding powder are similar to those employed in the manufacture of rubber and paints. The main white pigment used is titanium dioxide, but zinc oxide is also suitable, while carbon is the principal black pigment.

The original resin must be only partially 'cured' in order that it will soften and become plastic enough to flow into the mould when heat and pressure are first applied, after which further heat will complete the curing and setting. Some carefully limited extra curing takes place during blending to form the moulding powder. This blending is usually started in a Banbury mixer, but mixing can also take place on open rollers such as are used in the rubber industry. In the Banbury mixer, blades or rotors revolve inside mixing chambers; the rotors and the walls of the mixing chambers may be heated or cooled. The treated powder is removed at the base of the machine, from where it is conveyed to a set of adjustable hollow steel rollers (which may also be heated or cooled)

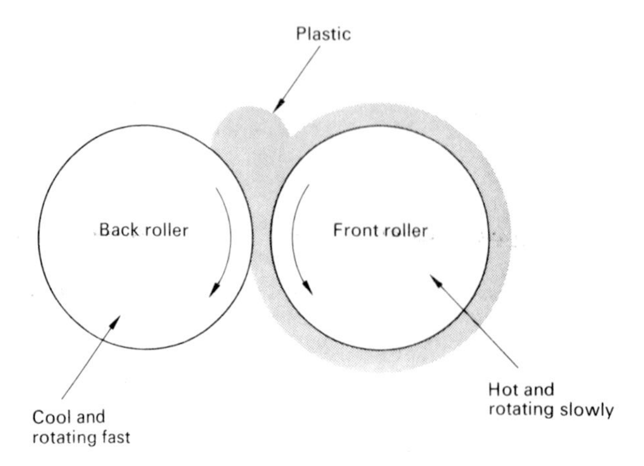

Fig. 8.4 Mixing rollers.

where the final intimate mixing is completed. The rollers rotate at different speeds to shear the mixture and thoroughly blend it. At the commencement of the final stage of mixing, the rollers are steam heated. They rotate inwards towards each other, and the material is fed into the rollers and scraped off from time to time and folded back on to the top of the rollers to ensure thorough mixing (this is called calendering). It may then be fed on to a similar pair of cold rollers where any further curing is arrested and the material is formed into a thin, hard sheet. This is crushed into fine powder, when a number of batches may be blended together to obtain a uniform product.

In general, the thermoplastic materials do not require fillers to improve their mechanical properties. The polymer is produced from a starting monomer, with the aid of a suitable catalyst and processing conditions, and the solid polymer is then powdered. An alternative method, for certain polymers, is to prepare the moulding powder by emulsion polymerization as already described.

8.4.2 Casting

Casting was one of the early methods of shaping articles from plastics. The material, in liquid form, is poured into a mould, where it is made to harden and set. When the setting is completed, the cast article can be removed from the mould.

In casting thermoplastic materials, the liquid monomer to which a catalyst has been added can sometimes be used. In such cases, the large amount of heat evolved during polymerization, and the high contraction in volume which takes place during the change from liquid to solid, pose substantial problems. Partial polymerization to convert the monomer from a mobile liquid into a more viscous syrup is of benefit. There is, however, still substantial heat generation and shrinkage, which must be allowed for.

Thermoplastic materials can also be cast from a solution of a polymer in a solvent which is subsequently evaporated away; this method is used in the production of plastic films. Another method is to use a mixture of finely divided polymer particles suspended in a plasticizer suitable for that particular polymer. An example of this method is the use of polyvinyl chloride pastes which are made by stirring the polymer powder with a plasticizer. Such pastes will maintain their paste condition for long periods if kept cool, but they can be made to set by heating at a temperature of about 170°C.

With certain thermoplastics, nylon for instance, it is possible to make castings direct from the molten polymer. For example, nylon 6 can be cast directly from the monomer caprolactam. Other methods sometimes used for the casting of thermoplastics are rotary casting and centrifugal casting.

For the casting of thermosetting plastics, the starting material is, in nearly all cases, a partially-cured liquid, which is heated in the mould to complete the curing under the influence of the catalyst still present. The viscosity of the resin is important–it can influence the properties of the end product. Again, heat development and shrinkage during setting can be troublesome but the addition of substantial amounts of filler material helps reduce their effects. Castings are made, in particular, from unsaturated polyester resins in the embedding of biological specimens and the production of buttons, for example, and from epoxy resins for electrical applications. Polyurethanes and silicone resins are also used in the same way. Shaped castings are usually produced from split moulds which can be made from, for instance, aluminium or reinforced plastics. Inserts to be embedded in the castings can be held in the moulds. Large quantities of resins mixed with fillers, accelerators, and colouring agents, have been cast to make floors in industrial premises.

8.4.3 Spreading and spread coating

Certain polymers, e.g. polyvinyl chloride, can be produced in the form of pastes. Polymer pastes enable fabrics to be readily coated with plastic materials, as in the manufacture of 'leathercloth' used for upholstery. As already mentioned, a polyvinyl chloride paste is prepared by dispersing the polymer in a suitable plasticizer. This can be mixed with fillers and pigments in blade-type mixers, a stabilizer such as white lead or lead silicate being added during the mixing to improve the resistance to weathering. Pastes containing fillers tend to have a short shelf life and should be used immediately after mixing. In practice, the best results are obtained by grinding the fillers, pigments, etc. thoroughly with some of the plasticizer before they are added to the paste.

The equipment used for spreading the pastes on to the fabric surface is similar to that which is used for a similar purpose in the rubber industry. A knife is supported at an adjustable height above the surface which supports the fabric. The paste is fed down in front of the knife and is carried forward under the knife blade by the moving fabric. Thus the plastic material is spread over the surface of the fabric, and the thickness of the coating applied depends on the height of the blade above the fabric, the type of fabric used, and the speed with which the fabric is moving. The coating on the fabric can then be set, either by being passed over steam-heated drums, or by directing infrared heat down on to it. A temperature of 150°C is the minimum used for polyvinyl chloride paste. After heating, the material is ready for use, or it can be given a pattern by being passed between embossing rollers.

8.4.4 Dip coating and paste moulding

Dip coating is normally used to give a protective or decorative finish to metal articles. Such finishes are particularly useful to act as a flexible cushion and prevent scratching of the surfaces in their contact with other articles. Thus, metal wire baskets and crates can be dip coated to reduce noise and protect the contents.

The metal construction to be coated is first heated and then dipped into a polymer paste. Any surplus paste is allowed to drain away, and then the dipped article is carried through a heated oven to cure the coating. The thickness of coating applied is determined by the viscosity of the paste used, and the initial temperature to which the metal article was heated.

A variation of the above technique is often referred to as 'fluidized bed' coating. The solid polymer, in the form of a powder (polythene is mostly used for this purpose), is placed in a tank equipped with a porous ceramic base and air is driven up through this base and through the powder. The air keeps the plastic particles in a state of agitation, when the particle mass acts rather like a liquid. The article to be coated is heated, dipped into the fluidized bath, and allowed to drain. A second heating of the article then melts the plastic particles adhering to its surface and produces a smooth, even coating.

A technique known as 'dip moulding' is also largely carried out using polyvinyl chloride paste. The method was originally developed in order to fabricate hollow articles. A one-piece or split hollow metal mould is heated and a specific amount of paste is poured into it. The mould is rotated to distribute the polymer paste evenly over the inside surface. Under the influence of the heat, the paste first sets and then melts to produce a smooth, even coating. The mould and its casting are then cooled and the casting is removed, either by opening the split mould or, since the casting is flexible, by stripping it away from the inner surface. In an automatic version of the process a series

plates, and a pressure of about 150 kg/cm^2 is then applied to them, until the resin is set, so the sheets are bonded together by the resin to form a composite plate. The best known product made in this way is hard plastic-impregnated paper or indurated cloth, known under such trade names as Pertinax and Formica for hard paper, and Novotex or Ferrozall for indurated cloth. These semi-fabricated products are mainly used for industrial purposes; but if a decorative effect is desired, e.g. for use in the building industry, the top layer of such a product can be pigmented. A similar system of compression moulding flat heated plates is used for polyester fibre-glass panels, both flat and corrugated. Here, too, the glass fibre or glass cloth is first treated with the resin binder, and is then known as 'pre-impregnated glass fibre'. Alternatively, the necessary resin is applied in liquid form to the glass layer placed in the press. The processing pressure can be comparatively low, usually between 10 and 50 kg/cm^2.

8.4.7 Injection moulding

The first injection moulding machines were constructed in the middle of the nineteenth century and were intended for the working of metals (Sturgiss, 1849). For making products from the first practicable thermoplastic material, cellulose nitrate, special-purpose moulding machines were built (Hyatt, 1872). The process is now fully developed for mass production and has grown to become one of the most important moulding methods for articles in plastics.

The fabrication cycle is as follows: The starting material, in powder or granular form, is moved through a heated container with the aid of a plunger or (in modern machines) a screw. During this process the material is compressed and melted. The molten mass is then injected through relatively narrow orifices into a cold die of the required shape. After a cooling time, which depends on its wall thickness, the product can be removed.

The screw machine gives better mixing and can melt a larger quantity of the material in a given time. Some of the work done by the rotating screw on the molten mass is also converted into heat. To inject the molten plastic into the mould cavity, the screw is moved axially. The temperature of the molten mass is between 200° and 400°C, depending on the plastic used, and the injection pressure can be as much as 2000 kg/cm^2.

It is clear that the force necessary to keep the mould locked must be greater than this pressure. Both mechanical and hydraulic locking systems are in use. For large products,

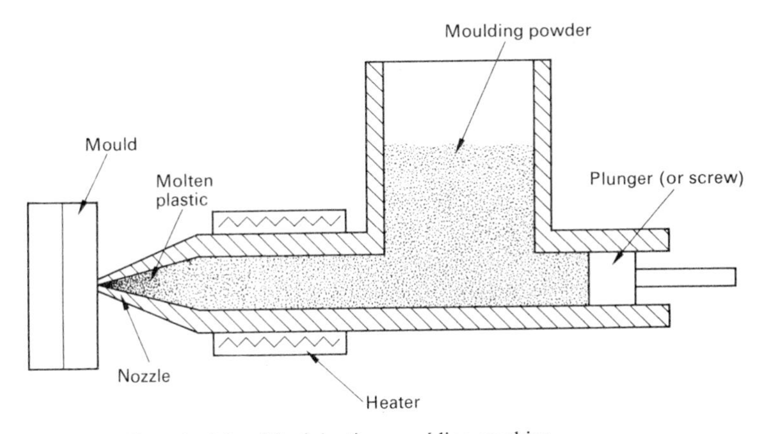

Fig. 8.6 The principle of the injection moulding machine.

Fig. 8.7 A large injection moulding machine capable of applying 'shot weights' at injection pressures of up to 1 750 kg/cm^2 (25 000 p.s.i.).

locking forces up to 3 000 tons may be necessary. The production cycle may vary from a few seconds for small thin-walled articles to some minutes for a large thick-walled product.

The procedure here described is for a thermoplastic material. However, in principle, the process is suitable for both thermoplastic and thermosetting plastics. In the fabrication of thermosetting plastics, setting must take place in the heated die and not before. The procedure is therefore more critical. The die is kept at a temperature of 150° to 200°C, depending on the type of thermosetting plastic in use.

Die costs are very high and are only justified if the number of articles made from them is large enough – 100 000 or more. A die can have one, or more than one, cavity depending on the size of the press and on the type of product. In view of the high pressures and the large output, a die must satisfy exacting demands and must be made of high-grade steel.

In order to reach the required dimensional accuracy it is necessary, when designing the product, to take into account the shrinkage arising when the molten plastic cools in the die. The cooling time (or setting time) frequently determines the rate of production. For these reasons, injection moulding machines are constructed with several dies (each of which may have more than one cavity) and one injection unit. The dies are then filled in turn, and a higher output can thus be achieved.

Articles produced by injection moulding comprise, among others, engineering and electronic products, household articles, toys, packaging, and motor car components. Metal inserts can be cast-in, and this may simplify subsequent assembly.

Fig. 8.8 A prototype 1500-ton lock injection moulding machine designed to produce large mouldings.

8.4.8 Transfer moulding

Transfer moulding is a special type of compression moulding. Here the moulding compound is not introduced directly into the opened die cavity, but into a container, which is placed next to the cavity and forms part of the die assembly. This container thus has about the same temperature as the cavity. When the die is fully closed an auxiliary plunger forces the compound through narrow channels into the cavity or, more commonly, into several cavities. The moulding powder is preferably heated and can readily take up heat in the narrow channels from the heated die so that it flows readily. The pressure on the plunger, known as the transfer plunger, is between 500 and 800 kg/cm^2.

The transfer moulding press must be of the double-acting type and capable of a transfer operation in addition to the die opening and closing operations; this makes it expensive. On the other hand, the setting time can be considerably shortened, partly by preheating the powder, but mainly by the heat taken up while the material passes through the injection channel. When it arrives in the cavity, the material is already completely liquid, with a resulting saving in setting time. The reduction in the operating cycle may amount to about 50%. A second important advantage of the method is that, if the die is appropriately constructed, no excess material need be allowed, since the filling system ensures that the cavity is properly filled. This means that flash is absent; only the usually very small 'gating' must be removed and finished off.

The technique lends itself to multiple transfer moulding using several cavities grouped round a central transfer plunger, giving high output. Moreover, the quality of the product is, as a rule, very good, because the resin is thoroughly heated before it reaches the die cavity. The process can be used successfully even if the wall thickness of the product shows marked variation or has unusually large dimensions.

8.4.9 Extrusion

In the extrusion of plastics the fused polymer is forced through an orifice the profile of which determines the cross-section of the product.

The basic component of an extruder is the screw, which rotates in a cylinder which is usually heated. Screw diameters vary between 10 and 300 mm. Powdery or granular material is compressed by the rotating screw and made to fuse while it passes through the cylinder. To ensure that the material can be properly worked, a melting temperature of 200° to 400°C may be necessary. A pressure as high as 500 kg/cm² can be built up in the orifice, depending on the form of the screw, the viscosity of the fused material, and the resistance of the orifice.

The output of an extrusion press, or extruder, depends on the speed at which the screw rotates and on the die resistance. The dwell or cycle time, and therefore the quality of the fused material, can be influenced in this way. The material is subjected to thorough mixing while it passes through the extrusion cylinder. For these reasons extruders, usually of large capacities, are also used in the processing industry for mixing-in of pigments, fillers and other additives.

Extruders are also used for continuous mixing in the production of, for instance, film by calendering.

Various types of extruder are known with, among other features, one or several screws, with screw designs and operating principles adapted to special processes, and with axially moving screws. Many specialized types have been designed, in particular to ensure intensive mixing. The heating of the extrusion stock is obtained not only by heat transfer from the heated cylinder, but also by conversion of much of the work done by the rotating screw into heat. Extruders based on this last principle have been designed with a high-speed screw as the only source of heat. It may be necessary to incorporate a vent half-way down the cylinder, to allow gaseous products to escape.

Apart from being devices for mixing and granulation, extruders can be used to make the following products: pipes, hoses, sections, rod, wire, wire and cable sheathing, flat film and blown film, and sheeting. For the sheathing of wire and cable the wire to be sheathed is passed through a specially designed orifice and covered with a layer of plastic.

Mesh-type structures can also be extruded, using rotating components in the orifice assembly. Products with different colours or made from different plastics can also be obtained in this way; several extruders are here used and the material from each is brought together, as required, in the orifice. This method is used, for instance, for producing the identification colours on cable covering, for rigid sections with elastic ribs used for sealing purposes, and for film built up from several layers with combined properties, used for packaging purposes.

For many products it is necessary to install 'calibration' equipment, directly after the orifice, to support products which are not yet dimensionally stable. Frequently water cooling is used at the same time. With the aid of pulling devices, the cooled product is taken at an even speed away from the orifice and passed through the calibration equipment and through the cooling bath. In the production of hollow articles such as pipes, sections, hoses and blown film, either internal pressure or an external vacuum

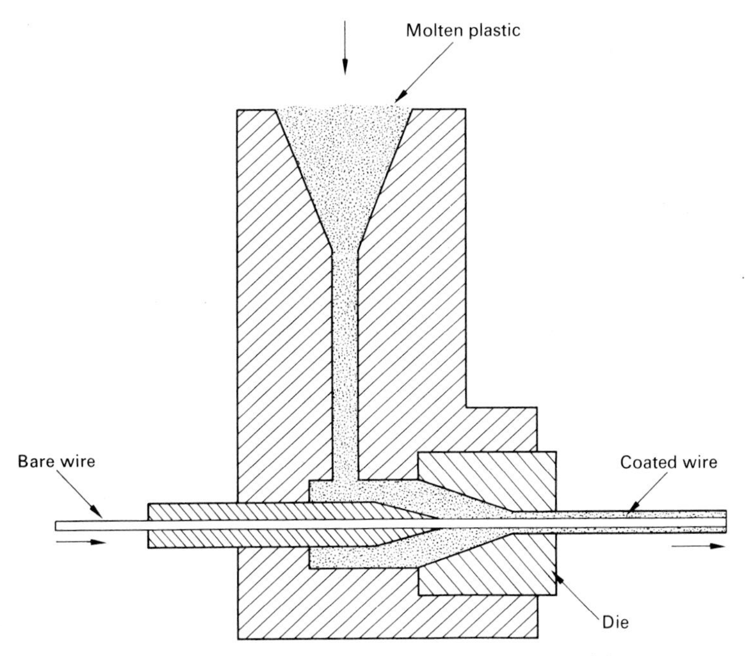

Fig. 8.9 Method of coating a wire with plastic insulating material.

is used to hold the products in the required shape. The finished product can be cut to length or coiled on drums, depending on the rigidity and the end use.

8.4.10 Drawing

The forming of hollow articles from thermoplastic film and plate, known as 'drawing' was initially carried out by applying a vacuum to one side of the sheet of plastic, laid over a hollow mould. To permit forming, the sheet is pre-heated or heated *in situ* so that the material is drawn into the cavity by the application of the vacuum. It is naturally greatly stretched during this process, very much so with a deep cavity. As a result, at certain points, in particular in the corners and edges of the bottom parts of the mould, the material is made thinner. The aim is, of course, to obtain as uniform a wall thickness as possible.

Modifications in the equipment were soon made, all aimed at a better distribution of the material. For instance, an auxiliary plunger can be used to force the heated sheet part of the way into the die cavity, the final shape being obtained by vacuum action. Another, more complex, method is the system known as 'drape-mould'. Here the film is stretched over a square or rectangular moulding box below which a die with the external shape of the final product is placed. The free-hanging film is heated and then blown up by applying pressure in the moulding box, as a result of which it assumes the form of a bubble above the moulding box. The mould is then moved upwards into the bubble-shaped space. Only then is the remaining space between the plastic film and the mould evacuated. Satisfactory uniform wall thicknesses are thus obtained, but the process is fairly complex.

For the production of beakers, cups and similar containers for the packaging industry, a simpler system is now normally used. The film is stretched over a ring or a hollow

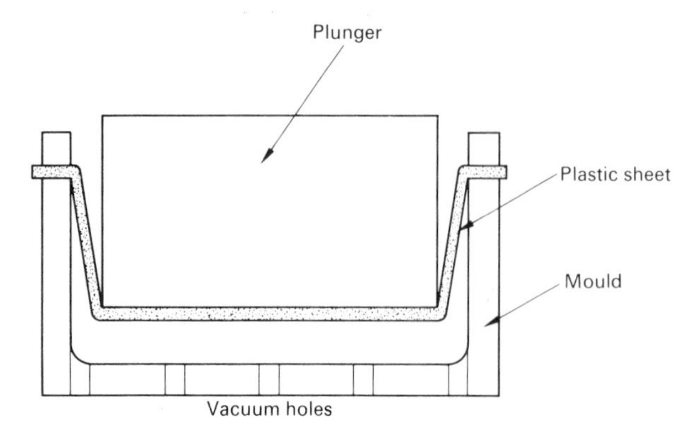

Fig. 8.10 Moulding an article from plastic sheet.

chamber, and a ram in the shape of the final product forces the plastic film into the mould, in much the same way as when forming metal sheet. With the film qualities which are now available, this system can give very satisfactory results as regards the wall thickness distribution and, it follows, the strength of the product. Moreover, this is a speedy method because evacuation becomes unnecessary.

The quality of the products thus obtained depends on the temperature at which the deformation takes place, as well as the uniformity of wall thickness. If the forming or drawing takes place at a low temperature–and this may seem reasonable in order to obtain rapid cooling and thus a quick production cycle–the drawing operation may produce so much internal stress that the product tends to return to its initial condition of a flat film or sheet, especially as the temperature is increased. It is even possible, by sufficient heating, to allow the initial flat shape to be assumed again completely. On the other hand, if the drawing temperature is too high, tears and cracks may occur during the deformation process, as the liquid condition is approached. To obtain the optimum quality it is therefore necessary to find the highest temperature at which the material will still remain intact during deformation. Some materials are more suitable for drawing than others. Polystyrene, which has been referred to above, is particularly popular because it has a large deformation range. Other materials which are also suitable, although they are less frequently used, are cellulose acetate, cellulose acetobutyrate and methyl methacrylate. The polyolefins are generally difficult to work, if it can be done at all, because they have a particularly short deformation range. Polythene of the low-density type cannot in practice be shaped into products from the plate form, but high-density polythene can, although to a limited extent. Polyvinyl chloride has a good deformation range but particular care must be taken to ensure uniform heating, especially for non-plasticised PVC, to prevent sudden degradation (depolymerization) in hot spots.

One of the attractive points of drawing as a method of fabrication is that for short runs simple dies or moulds of wood, plaster, or aluminium can be used. This applies less in the case of mass production where, to ensure trouble-free operation, the use of more expensive metal dies is justified.

8.4.11 Blow moulding

Blow moulding is a method of forming hollow articles by inflating a hot pre-moulded tubular shape (or 'parison') so that it expands to make contact with the walls of a mould, after which it is set by cooling. The blow-mouldability of thermoplastic materials is influenced by the temperature and melt history of the shape at the moment of blowing. The method is similar in principle to that used to make blown glass articles.

A type of blow moulding was first used to manufacture toys from celluloid. Two sheets of celluloid were clamped together between the two halves of a mould. These sheets were heated, and air or steam pressure was introduced between them. Thus they were blown out to make contact with the mould surfaces, when they were cooled and set and could be removed from the mould. After the 1914–18 war, celluloid was replaced by the less flammable cellulose acetate.

Towards the end of the 1930s, the Plax Corporation and the Continental Can Company both had patents covering the manufacture of plastic bottles by first extruding a tube-shaped parison, and expanding this under air pressure to take up the shape of a bottle mould. It was not until the early 1950s, however, that a really satisfactory blowing material – polythene – became available. When the two patents referred to above expired in 1958, other manufacturers began to take an interest in the process, and both polythene and rigid polyvinyl chloride bottles were produced in large numbers, and production has continued to increase up to the present time.

For blowing a bottle, a parison, rather like a thick-walled test-tube in shape, is first extruded and, while still hot, hung between the opened halves of a metal mould. The mould is closed and air is then forced into the hot parison, which is blown out until it makes contact with the mould walls where it becomes cooled and set. The ratio of the diameter of the parison to that of the mould cavity is called the 'blow ratio'.

The cycle time needed to make one bottle depends, among other things, on the total of the times required to open, to close, and to move, the mould. Roughly 60 to 80% of the cycle time is taken up in cooling the blown article sufficiently to enable it to be handled. The cooling efficiency can be improved by a suitable choice of mould materials and the surface finish given to them. Attempts to increase the output of the blow moulding process have included the use of more automatic operations, providing blowing moulds on either side of the injection mould, and the use of multiple-cavity injection and blowing moulds.

A later development in this method of making hollow articles is often referred to as 'extrusion blowing'. This makes use of a continuously extruded tube of the plastic material. The tube can be trapped at suitable intervals between the two halves of a split mould and, as the mould closes, it seals off both ends of this portion of tube. Compressed air is introduced into the sealed-off portion by the insertion of a blowing-pin,

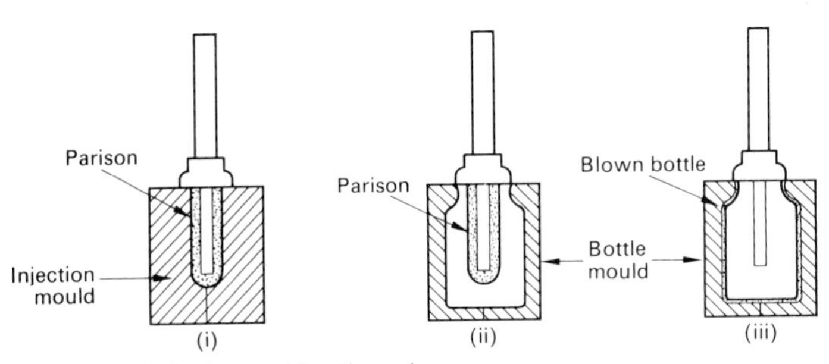

Fig. 8.11 Three stages in the blowing of a plastic bottle.

Fig. 8.12 Production of bottles by blow-moulding.

and the material is inflated. On touching the cold walls of the mould, the material cools and sets, when the blown bottle can be removed, and the cycle repeated. In a variation of this method, often referred to as the 'cold tube' process, the tube is extruded in advance and allowed to cool and set. When it is required to be used, the tube is cut into pieces of a pre-determined length and these are heated and blow-moulded. An advantage of this two-stage process is that the extruded tube can easily be transported and stored.

A hollow container can be blow-moulded from flat sheet. After heating and clamping, the now flexible sheet is drawn with a rounded-off plunger over the neck part of the blow mould. The plunger is then pulled back and the sheet is blown to form the article. The method produces a substantial amount of waste material, but this can be re-used.

Yet another variation of the method is provided by machines in which two rolls of plastic sheet are fed so that they are held, one above the other and with a pre-determined spacing between them. The sheets are heated, a blow-pin is entered at a central point between them, and the upper and lower halves of a mould are closed on to the sheets to seal off the entire perimeter. Air under pressure is applied between the plastic sheets and a vacuum is also applied at each of the mould halves. The sheets expand to take up the shapes of the mould halves, the material is then cooled, the mould is opened and two new sheets come into position to allow the cycle to be repeated.

In principle, all thermoplastic materials can be blow moulded. The selection of a suitable material, apart from its blowability and design aspects, will depend on the final use of the end product. Packaging offers a big outlet for blow-moulded articles, but here cost is of prime importance. Blow moulding materials in substantial use are polythene, polypropylene, rigid polyvinyl-chloride, polystyrene, polycarbonate and polyamide. There would appear to be a future for foamed material containers; polystyrene foam with a density of about 40 kg/m^3 is suitable for this purpose. Good results have been obtained with the direct blowing of a foam parison, and also by using a pre-shape of foamed sheet material.

8.4.12 Machining plastics

In general, plastics can be machined using the existing tools and methods without any particular difficulty, many with standard wood-working machines; others in the ways which would normally be applied to soft metals.

During the machining of plastics, it is necessary to remember the particular characteristics of the material being worked. Thus, if too heavy cuts are taken during the machining of thermoplastics, a considerable amount of heat may be developed during the cutting action, and the plastic may become soft and sticky. Cutting tools should be kept sharp, light cuts are recommended, and precautions should be taken to allow as much of the heat generated as possible to leak away.

8.4.13 Laminated plastics and resins

Lamination is the process of combining two or more layers of sheet or film to form a single, composite sheet. There are two distinct methods of laminating; high pressure and low pressure. In the production of laminates such as Formica, Warerite and Tufnol high pressures are used because thermosetting resins, such as phenol-formaldehyde, urea-formaldehyde, or melamine-formaldehyde resins are employed and these produce water in the form of steam during the curing process; high pressures are needed to prevent this steam from separating the layers during curing.

The usual type of laminate consists of layers of paper or fabric cemented together with set resin. The layer material is passed through a bath containing, say, a solution of phenolic resin and dried in a heated chamber. The impregnated material is then cut into equal-sized sheets which are stacked up one on the other and then squeezed under high pressure (about 1000 p.s.i. (70 kg/cm^2)) between the heated plates of a press until the resin is fully cured. The plates of the press are then cooled, when the laminated sheet or block can be removed. Because of the high pressure used, the equipment required to produce large sheets is expensive.

Relatively cheap, decorative laminated sheets can be produced by making the bulk of the sheet from, say, a relatively cheap phenol-formaldehyde resin, and the top surface from a decorated paper finished with a layer of transparent urea-formaldehyde or melamine-formaldehyde.

If resins such as polyesters or epoxides are used, there is no need for high pressures since the resins set without the formation of water. No steam is thus generated during the heat setting, and the only pressure needed is one which will keep the resin-impregnated sheets in contact. Very large sheets can be produced without much difficulty.

8.4.14 Reinforced plastics and resins

An important industry has been built up round the production of articles made by the shaping of so-called 'reinforced plastics'. The resins generally used in the preparation of such articles are the thermosetting resins, and the reinforcing materials are tough, fibrous products.

Hand-operated forming methods are popular, particularly for polyester and epoxy resins, by reason of their ease of operation, their flexibility, and the low investment costs. Setting is achieved without raising the temperature by making use of appropriate catalysts, accelerators, or setting agents.

The reinforcement material used is mostly glass fibre since it has a high strength, a favourable modulus of elasticity, excellent chemical resistance, and moderate cost. The fibre can be used as a matting built up from glass fibre cut to length, or in the form of a piece of woven fabric. Other reinforcement materials are used, such as cotton or sisal, selected synthetic fibres, asbestos fibres, metal filaments, or, more recently, carbon fibres, but all to a comparatively small extent.

As a rule, manual forming is adopted for individual articles and small production runs. Since neither pressure nor heat is involved, it is possible to use cheap and easily-workable mould materials, but no great dimensional accuracy or consistency in quality of production can be achieved. The mould is a replica of the external or internal shape of the article, depending on which face of the final article is to be smooth and polished. The mould must first be coated with a suitable 'parting' agent, which is a 'gelled' coat of non-reinforced resin. The article is now built up on the mould, by applying successive layers of reinforced resin material. During the building-up, local additional reinforcement such as stiffening ribs, inserts, etc., can be incorporated at points likely to experience high stress. The layers are then cured at room temperature without pressure. After setting, the assembly can be removed from the mould. The method allows the production of shaped articles having one smooth face which has been in contact with the polished mould; the other face shows the structure of the last layer of reinforcement which was added.

An alternative manual method is by spraying the glass fibres and resin together on to the surface of the mould. The resin and accelerators are supplied from tanks in the blowing machine, and blades cut up the glass fibres from yarn wound on to bobbins, so that both the resin and filler are blown forward together. Spraying is continued until the desired thickness of moulding has been built up. The method is often used to construct very large articles indeed – ships' hulls, for instance. It can also be used to cover large flat areas, such as floors, tank walls and so on.

The use of vacuum or air pressure techniques has improved the manual methods in a number of ways. The reinforced resin is deposited on the mould surface as already described, and then a flexible sheet is placed over it and an airtight seal made at its edges. The space under the sheet is then evacuated so that it compresses the resin while it is setting. An alternative way of producing the same effect is to fix a flexible bag above the mould and inflate this under air pressure. These two methods can also be used when the material is contained in the two halves of a mould, the exhaustion or inflation of a flexible bag forcing the two parts of the mould together while setting is taking place and thus giving two smooth surfaces on the final moulded article. A further variation is to place the dry reinforcement material inside a closed mould and then force the impregnating resin under pressure into the mould.

When large production runs and consistent accuracy of form of the final article are required, compression moulding methods are used. Again, glass-fibre reinforced polyester and epoxy resins are used in large quantities but the various techniques employed differ as regards mould design, pre-treatment and filling of the material, pressure used, and curing temperature. For runs of a few hundred articles cold compression methods are sometimes used, employing moderate pressures and relatively inexpensive moulds. The required quantity of glass-fibre reinforcement, cut to size, together with the resin, are introduced into the bottom part of the mould. The mould is closed, thereby exerting pressure on the contents and distributing the resin evenly through the reinforcing material. The mould reaches a temperature of about 60°C, as a result of the heat generated during the setting action.

Hot compression methods are also used, generally for large moulding runs since short cycle times are thereby obtained. The moulds are usually of steel or cast iron. The resin and the reinforcement, cut to size, are introduced separately into the mould. The mould temperature for polyester resins is of the order of 100° to 120°C; for epoxy resins it is much higher, about 180°C. For complicated shapes the reinforcement material may be first preformed in a special machine using a temporary binding agent. The preform is then placed in the mould, the resin added, and heat and pressure applied. Alternatively, the reinforcement can be impregnated with the resin and a semi-curing carried out; this is then filled into the mould and the curing is completed.

Winding techniques can be used to make hollow tubes. A reinforcement 'bandage' of glass fibre or other material is impregnated with resin and wound round a mandrel, and the whole is then hardened. Products of excellent quality and high strength can be made by this method.

A method similar to the one used to make extruded sections in thermoplastics can be employed for reinforced thermosetting plastics. The machine has sections which progressively deal with the material, supply the impregnation, the shaping through one or more dies, hardening, and cutting to standard length. The most critical parts of the machine are the dies which give the correct contours to the material which is fed to them in a condition intermediate between a liquid and a solid.

Centrifugal casting of reinforced plastics is sometimes used to make tubes for such applications as drain-pipes and low-pressure conduits. The mixture of resin and reinforcement is introduced into a hollow tubular mould which then rotates at high speed while the resin hardens under the influence of heat. Fairly long cycle times are required and the method is discontinuous.

Thermoplastics are sometimes reinforced with a fibrous filler to improve the mechanical properties. The reinforcing material is usually glass fibre but asbestos and some synthetic fibres can be incorporated. The reinforced polymer is fabricated by injection moulding or extrusion in exactly the same way as for the unreinforced material. The moulding material is fed to the machine in one of these forms:-

(i) Ready-prepared compound of thermoplastic and glass fibre in granular form.

(ii) 'Concentrate', which has a very high ratio of fibre to resin. This is blended with the straight thermoplastic to give the final plastic/fibre mixture required in the article.

(iii) Plastic powder and cut fibres in the required proportions.

8.4.15 Foaming

Plastic foam is made from two components; the plastic and a gas distributed throughout it. If the distribution is discontinuous and the voids in the material are separated, then the foam is said to be of the 'closed-cell' form; if continuous, an 'open-cell' foam is obtained. Foams can be based on thermoplastic and on thermosetting resins.

The gas is developed within the still-plastic mass by the use of a blowing agent. The gas can be developed from the agent either by physical or chemical means. Physical agents may be liquids or gases. Liquids are evaporated using either externally-applied heat or heat developed by the curing action. If a gas is used, a reduction in pressure is employed to allow the gas to expand and bubble its way through the plastic mass. Chemical blowing agents can be solids or liquids; the gas is usually developed by decomposing the blowing agent by heat, or by a chemical reaction between two components during which a gas is released.

Polystyrene can be foamed by four different methods. The oldest, and still widely used, method is by polymerization in suspension. The blowing agent can be added at the same time so that the polymer particles which are formed contain a certain amount of the agent. On heating, softening of the polymer particles takes place and, at the same time, the blowing agent evaporates to produce the foam. A similar method is used to prepare foams from copolymers of styrene with acrylonitrile, and with methacrylate. Another method of polystyrene foam production which is becoming popular, is to feed granules of the polymer into an extruding machine and, at the same time, introduce a liquid or gaseous blowing agent into the plasticizing section of the machine. Foaming takes place as the material leaves the extruder; a mixture of citric acid and sodium bicarbonate are sometimes added to give additional gas and produce an exceptionally well-foamed material. Extruded foam can also be produced by feeding into the extruding machine a mixture of styrene monomer and azoisobutyrodinitrile; the latter acts as a polymerization catalyst and also as a blowing agent. In the fourth method of foaming polystyrene, a mixture of polystyrene and a blowing agent is injected through nozzles into a mould, when foaming takes place. This last method produces foamed polystyrene of a higher density than the other techniques, the skin round the foam being particularly dense.

The production of foamed polyurethane is based on the reaction between isocyanate and polyol. An added fluid blowing agent, such as Freon 11, is made to evaporate by the heat of the reaction. It is also possible to add a low-boiling liquid, such as Freon 12, under pressure; foaming takes place at room temperature when the pressure is released. The products are known, in both cases, as 'Freon blown' foams. Alternatively, the reaction between isocyanate and water can be used to release carbon dioxide gas, and the product is then known as 'water blown' or 'CO_2 blown' foam. In the above processes, the addition of the isocyanate and the polyol separately is referred to as a 'two-shot' method. A 'one-shot' method can be used in which the two substances are introduced together, when the plastic polymer and the foaming action are produced at the same stage of the operation. In yet a further case, a pre-polymer is used; the isocyanate is allowed to react with part of the polyol, the reaction with the remainder taking place at a second stage.

Polyvinyl chloride can be foamed by introducing air or an inert gas into the polymer, either mechanically or by compounding with chemical blowing agents which decompose under the action of heat to produce a gas. A variety of vinyl foams with different cell structures and degrees of softness can be prepared by appropriate choice of the manufacturing technique. One process uses a gas under pressure dissolved in the plasticizer portion of a polyvinyl chloride paste, when a release of the pressure causes foaming; heat fusing follows to give a final product of medium to low density. In another process, air is whipped into a specially formulated paste to give a higher density product containing smaller cells.

Polythene foams are of two types: high-density foam of 20 to 30 lb/ft^3 (300–500 kg/m^3), and low-density foam of 2 lb/ft^3 (30 kg/m^3). The low-density material is made by mixing a foaming agent with the hot molten polymer under pressure and then releas-

Fig. 8.13 Rigid expanded polystyrene for use in packaging.

ing the pressure followed by cooling to obtain setting. High-density polythene foam is produced by heating a mixture of polymer and blowing agent until the polymer softens and the blowing agent releases gas throughout the mass.

The practical applications for foamed plastics are many and varied. Polyvinyl chloride foam has been used in ship construction. Flexible (soft) polyurethane foam is used as a buoyancy material , in packaging, in mattresses, and in covering for furniture. Polythene and polypropylene foams are able to absorb heavy impact loads and for this reason are very satisfactory for the protection of fragile equipment in transit. Foamed polythene is also very suitable for covering electrical wire by reason of its outstanding dielectric properties. Phenolic foams are used primarily for roofing panels and in sandwich constructions; they have the merit of being self-extinguishing. Epoxy and silicone foams have only a limited field of application, mainly for encapsulating electrical components. Silicone foams are useful in applications requiring relatively high temperature stability. Urea-formaldehyde foam has a specialized use as thermal insulation in cavity walls.

When selecting the type of plastic foam for a particular application, cost must be balanced against the importance of the physical and chemical properties for that specific use. It may also be necessary to sacrifice some buoyancy in order to obtain an adequate mechanical strength. A substantial range of plastic foams is now available to enable the would-be user to make his selection for a particular application while bearing all these factors in mind.

8.5 POLYOLEFINS

These are polymerization products of the unsaturated aliphatic hydrocarbons known as olefins. The best known member of the class is polythene.

8.5.1 Polythene

a. History. In 1933 a polymer of ethylene was produced for the first time by ICI Ltd. in the UK. From 1933 onward, work proceeded on a commercial process, and particularly during the Second World War the invention of radar offered a great stimulus for further development because of the excellent high-frequency electrical properties of the material. Many other applications were also devised, but it was only some time after the war that polythene became recognized as a general-purpose plastic. The output of 8 grammes in 1935 grew to roughly 3 000 000 tons in 1966.

b. Structure and classification. Polythene has the simplest chemical structure of all plastics, consisting only of carbon and hydrogen in a straight chain:

$$\begin{array}{ccccccccccccccccccc} & H & & H & & H & & H & & H & & H & & H & & H & & H & \\ & | & & | & & | & & | & & | & & | & & | & & | & & | & \\ - & C & - & C & - & C & - & C & - & C & - & C & - & C & - & C & - & C & - \\ & | & & | & & | & & | & & | & & | & & | & & | & & | & \\ & H & & H & & H & & H & & H & & H & & H & & H & & H & \end{array}$$

A varying number of side branches can be linked to the main chain. These branches, which are mainly methyl groups, influence the physical structure. Polythene is a semi-crystalline polymer; that is, some of the molecules are present in an ordered crystalline arrangement, known as crystallites. As the number of side branches increases, the ability to pack the molecules closely decreases, and with it the amount of crystallization. This is reflected in the density of the polymer: it varies from 0.92 g/cm^3 for soft polythene (also called low-density polythene or high-pressure polythene) to 0.96 g/cm^3 for hard polythene (known alternatively as high-density or low-pressure polythene). The degree of branching can be determined by infrared analysis. The mechanical and physical properties, and therefore the suitability for certain applications, vary with the density, which is therefore an important quality.

c. Properties. The properties of polythene are influenced by density and molecular weight. The latter is usually expressed as the melt index (see section 8.20. 1*c*), important as regards processing. Table 8.1 gives the physical and mechanical properties. Polythene is a fairly inert material; it is resistant to dilute acids and bases, but can be attacked by oxidizing chemicals (chlorine, iodine, nitric acid). Chlorinated solvents cause polythene to swell or dissolve, especially at elevated temperatures; alcohols, esters, aldehydes and ketones hardly affect polythene, if at all, at room temperature. Polythene tends to corrode under the influence of surface-tension lowering agents. Thus a soap solution may cause small fissures to arise under stress, and these can determine the life of a polythene product. However, a good deal is known about this, and appropriate allowance can be made in designing articles. Polythene is also sensitive to oxygen, especially when exposed to ultra-violet light, and it is therefore unsuitable for outdoor applications in an unprotected form. To overcome this drawback, special stabilizers are used and also protective pigments such as carbon black or titanium dioxide.

Table 8.1 Properties of polythene

Property	*Units*	*Soft polythene*	*Hard polythene*
Density	g/cm³	0·92	0·96
Refractive index	nD	1·51–1·52	1·53–1·54
Melting point, cryst.	°C	105–110	125–130
Specific heat	kcal/kg°C	0·6	0·55
Tensile strength	kg/cm²	100–150	250–350
Elongation at break	%	20–250	20–200
Elastic modulus	kg/cm²	1 500–2 000	4000–6000
Notch impact strength (Charpy)	kg-cm/cm²	> 100	15–20
Electrical breakdown strength	kV/mm	> 20	> 20
Resistivity	ohm/cm	> 10^{17}	> 10^{17}
Dielectric constant	E(10^2–10^{10}Hz)	2·3	2·3
Dielectric loss factor	tan d	< 3×10^{-4}	< 3×10^{-4}

d. Commercial products. There are now many manufacturers of polythene, which is marketed under trade names such as Alkathene (UK), Bakelite, Alathon (USA), Lupolen (Germany) and so on. The material is always supplied in granular form. Practically every brand is marketed with a range of melting points, the choice depending on the particular application and the method of fabrication.

e. Production. Polythene is produced by the polymerization of ethylene gas (C_2H_4), obtained in the preparation of petroleum fractions or from natural gas. The following processes are used most commonly for the production of polythene:

(i) *High-pressure process.* Developed by I.C.I. Ltd. in the 1930s, this process has been refined and improved over the years. The polymerization is based on a free-radical reaction with a free-radical initiator, such as a peroxide, added to the purified ethylene. Polymerization takes place at a pressure of from 1 500 to 2 500 kg/cm² and at a temperature of about 100° to 200°C. The fused polymer is provided, if necessary, with ultra-violet stabilizers, colouring agents, and the like in an extruder, and this is followed by granulation and packaging.

(ii) *Low-pressure process (Ziegler).* Polythene can be obtained by the polymerization of ethylene at low pressures at a temperature of about 80°C under the influence of an alkyl metal halide (e.g. $TiCl + Al(C_2H_5)_3$). The catalyst is dissolved in an organic solvent (e.g. pentane or cyclohexane) and the ethylene is then introduced into this solution. The polymer is precipitated in the form of a white powder. The washed and dried polythene can be compounded with auxiliary substances, such as are referred to under (i) above.

(iii) *Phillips process.* In the Phillips process a solid catalyst is used containing a metal oxide (e.g. chromium oxide) as the active substance. The catalyst is dispersed in a liquid, and the resulting slurry is fed into a reaction vessel together with ethylene and a solvent. Polymerization then takes place at a temperature of 140°C and a pressure of 35 atm. Changes in the reaction conditions give different molecular weights. The polymer is separated from the solvent after leaving the reaction vessel and further processing stages are as described above. A modification of the Phillips process has been developed by the Standard Oil Co. of India, a different metal oxide, e.g. molybdenum oxide, and also higher temperatures and pressures being used.

f. Some important copolymers. There are a number of copolymers of polythene with quite different properties. The most important of these are as follows. (i) *Ethylene-propylene copolymer.* This is a copolymer which, by reason of the side branches that develop, does not crystallise but remains amorphous and rubbery. It is known as E-P rubber. (ii) *Ethylene-butylene copolymer.* This is used in the Phillips process to obtain a polymer with a lower density and a somewhat greater resistance to stress corrosion cracking. (iii) *Ethylene-vinyl acetate copolymer (EVA).* This is an amorphous plastic which can be worked in the same way as polythene to produce tubing, sections, film and bottles. (iv) *Ethylene-ethyl acrylate copolymer.* This is used for injection moulding, blow moulding, and the extrusion of articles with special properties, and as a modifier for other copolymers. A special product is obtained by the subsequent chloridization of polythene to give it a better resistance to ultra-violet light and thus enable it to be used outdoors. This product is also used to improve the impact strength of polyvinyl chloride.

g. Uses. The versatility of polythene is shown by the number of applications. Nearly all the usual methods of fabrication are possible but an appropriate grade of polythene must be chosen. Higher density gives greater rigidity. Material with a high melt index flows easily but always has poorer mechanical properties. For injection moulding, therefore, material with a fairly high melt index is used, because good fluidity is here important. For extrusion processing the melt index is lower, and for high-grade film or tubes the melt indices used are less than 1. Melt index values are somewhat lower for similar applications in high-pressure polythene. (See section 8.20, page 607.)

Film and sheet material represent important uses of low-pressure polythene. Film is mainly produced by blowing. Tubes are made in both low-pressure and high-pressure polythene. Many products are made by injection moulding, from cheap toys to high-grade engineering products. In the design of injection moulds, allowance must be made for the high shrinkage of the material arising from crystallization.

Other applications of polythene are bottles (especially high-pressure polythene, for instance, for non-rigid squeeze bottles), cable sheathing (for high-frequency electric conductors), coating of paper for packaging purposes (mainly low-pressure polythene) and filament (high-pressure polythene). Strip made from film and stretched in one direction is very strong lengthways and is used for making industrial fabric, to replace jute, for instance.

8.5.2 Polypropylene

a. History. The polymerization of polypropylene was started on a commercial basis in the United States (Hercules Powder Co.) and carried out with stereo-specific catalysts developed by Natta (Italy).

b. Structure and classification. Like polythene, polypropylene is built up of carbon and hydrogen atoms but it differs from that material by the presence of CH_3 groups as side branches:

$$\begin{array}{cccccccccccccccc} & H & & H & & H & & H & & H & & H & & H & & H \\ & | & & | & & | & & | & & | & & | & & | & & | \\ - & C & - & C & - & C & - & C & - & C & - & C & - & C & - & C - \\ & | & & | & & | & & | & & | & & | & & | & & | \\ & CH_3 & & H & & CH_3 & & H & & CH_3 & & H & & CH_3 & & H \end{array} \quad \text{isotactic polypropylene}$$

Because of the asymmetrical carbon atom, to which the methyl group is attached, stereo-regularity plays a role. Theoretically there are three possible polypropylenes–atactic, isotactic and syndiotactic molecular arrangements. In the atactic polymer the side branches on either side of the main chain are arranged at random, in the isotactic product all are either on one side or the other (i.e. it is stereo-regular), and in syndiotactic polypropylene they alternate from one side to the other. Only isotactic polypropylene is important in practice. It is a semi-crystalline polymer of low density (approx. 0.90). Polypropylene is fairly sensitive to oxidation owing to the presence of the many tertiary carbon atoms. The polymer is produced in different molecular weights and, as for polythene, the melt index is used for grading the product.

c. Properties. Polypropylene is one of the lightest known plastics. Owing to its strength, which for a thermoplastic is fairly high, polypropylene offers a favourable strength/weight ratio. Table 8.2 gives the mechanical and physical properties. The chemical resistance is practically the same as for polythene but the oxidation sensitivity of polypropylene is a little greater. Antioxidants or ultra-violet absorbers are likewise added to this product, in particular when it is intended for outdoor applications. The tendency to stress corrosion is substantially lower than for polythene. Allowance must be made for a drop in impact strength at sub-zero temperatures.

d. Commercial products. Polypropylene is marketed by a number of manufacturers under various trade names: Propathene (UK); Profax (USA); Carlona P (Netherlands) and so on.

The material is supplied in the form of granules with different melt index values. Special stabilized grades are produced for outdoor use.

e. Production. The basic material is propylene (C_3H_6) obtained from the cracking of petroleum. This is polymerized with an organometallic catalyst at 70° to 80°C and 6 to 8 atm pressure. Catalyst residue is removed from the product, which is then compounded with auxiliary substances.

Table 8.2 Properties of polypropylene

Property	*Units*	*Value*
Density	g/cm^3	0.89–0.91 (depending on crystallisation)
Melting point, cryst.	°C	165–170
Specific heat	kcal/kg°C	0.55
Tensile strength	kg/cm^2	300–400
Elongation at break	%	200–700
Elastic modulus	kg/cm^2	10000–12000
Notch impact strength (Charpy)	kg-cm/cm^2	5–10
Electrical breakdown strength	kV/mm	> 20
Resistivity	ohm/cm	$> 10^{16}$
Dielectric constant		2.0–2.1
Dielectric loss factor	tan d	$< 5 \times 10^{-4}$

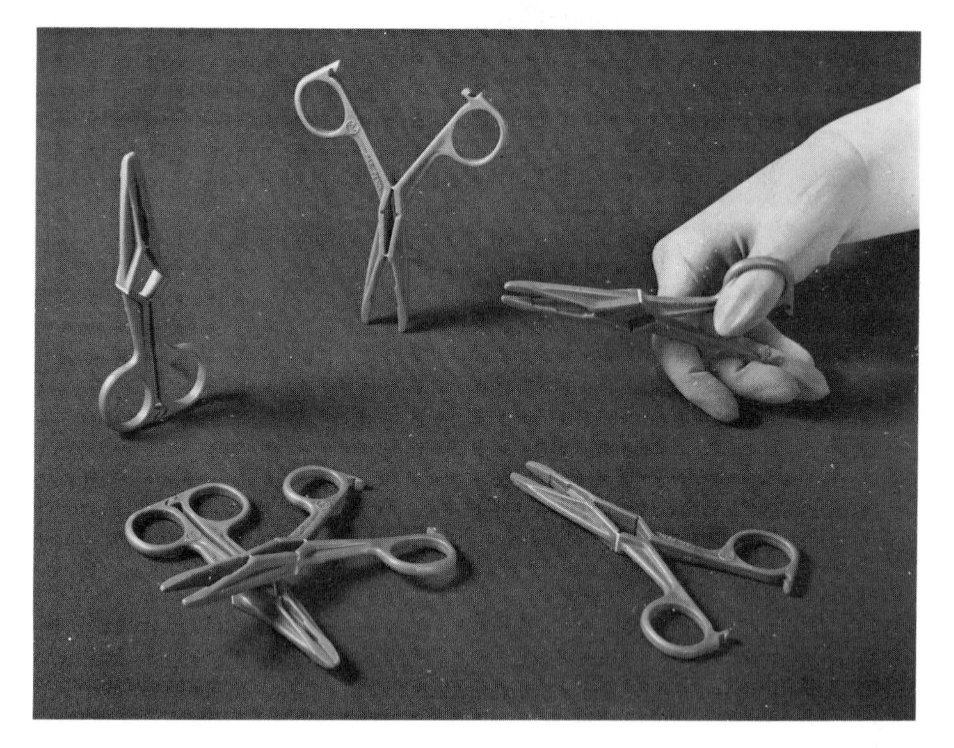

Fig. 8.14 Disposable medical forceps made by injection-moulding polypropylene.

f. Some important copolymers. It forms copolymers with ethylene (see under polythene) and vinyl chloride (see under polyvinyl chloride).

g. Uses. Nearly all the usual thermoplastic fabrication methods are used for polypropylene. Special polypropylene products are biaxially expanded film and monofilament. Polypropylene is an eminently suitable material from which to prepare thin films which readily lend themselves to the production of split-film fibres.

8.5.3 Other polyolefins

a. Isotactic poly-α-butylene. Like propylene, α-butylene can be polymerized stereo-specifically with Ziegler-Natta catalysts (see volume 4), to give a fairly lightweight plastic. It is slightly less rigid and less tough than polypropylene. It can be extruded, injection-moulded or calendered (i.e., passed through rolls).

b. Polyisobutylene. Polymerization of isobutylene gives a rubber-like product, used in the adhesives industry and elsewhere.

c. Polymethylpentene. Poly-4-methylpentene-1 is the lightest plastic known at present, with a density of 0.83 g/cm^3. It is a transparent material with a tensile strength of 280 kg/cm^2, elongation at break of 15%, elastic modulus of 15 000 kg/cm^2 and softening point at 240°C, and has good chemical resistance and electrical properties.

8.6 POLYSTYRENE

This section deals with those thermoplastics which are synthesized with sytrene as the most important constituent. The following are discussed:

Homopolymers of styrene.
Mixtures of homopolymers with rubber-like polymers.
Copolymers of styrene with other monomers.
ABS plastics.

8.6.1 History

A reference to a solid to be produced from 'styrol' (styrene) appeared as early as about 1930 (Bonastre and Simon). An explanation of this phenomenon was given by two English chemists, Blythe and Hofmann, who called the solid 'metastyrol'. The first patent was taken out by Kronstein in 1900. Staudinger recognized the true molecular structure and proposed the present name polystyrene.

Commercial production was started in Germany and America about 1930. However, it was not before 1937 that the material could be so controlled as to make large-scale production possible.

The demand for rubber during the Second World War was responsible for the enormous rise in the production capacity for the monomer, styrene. When, after the war, much of this became generally available, polystyrene was used for a variety of purposes, some of which were unsuitable. The poor reputation which polystyrene acquired at the time probably checked its growth, but at the same time encouraged the development of 'modified' polystyrene, with better properties. The impact strength was improved mainly by the incorporation of synthetic rubbers; copolymers also played a role. Further development led to the acrylonitrile-butadiene-styrene (ABS) plastics. These are tailor-made materials which have well-balanced properties so as to be useful over a wide range of applications. Further developments are taking place; they include transparent MBS plastics (methylmethacrylate-butadiene-styrene), cold forming ABS, fibreglass-reinforced styrenes, and galvanizable (metallizable) ABS.

8.6.2 Structure and classification

The polystyrenes are commonly synthesized from one or more of the following monomers: styrene, $C_6H_5CH = CH_2$, a colourless fluid with a density of $d_4^{20} = 0.91$ g/cm^3, a boiling point of 145.2°C, and with a structural formula:

H
|
C=CH$_2$
|
(benzene ring)

styrene

Acrylonitrile, a colourless fluid with a density of $d_4^{20} = 0.80$ g/cm^3, boiling point of 77.3°C, and structural formula:

H
|
C=CH$_2$
|
C≡N

acrylonitrile

Butadiene, a gas at room temperature with a density of $d_4^{-6} = 0.65$ g/cm³, boiling point of −4.4°C, and structural formula:

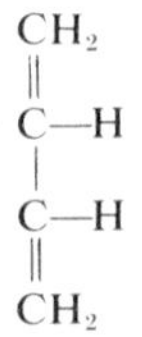

$$\begin{array}{c} CH_2 \\ \| \\ C\text{—}H \\ | \\ C\text{—}H \\ \| \\ CH_2 \end{array}$$

1,3 butadiene

Styrene is readily polymerized exothermally to the straight polystyrene (homopolymer):

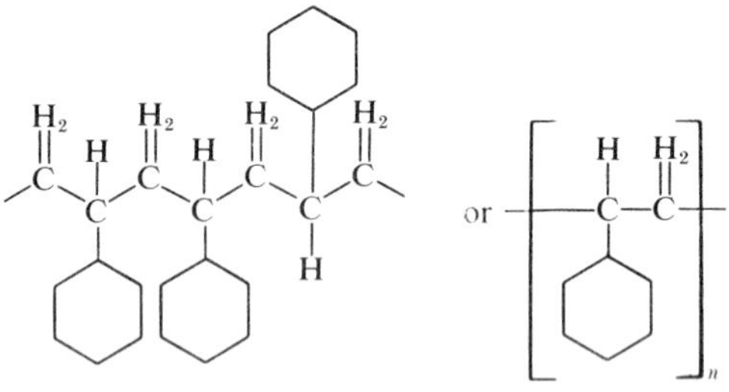

polystyrene

The commercial polymers are mainly atactic and the crystallite content is negligible. Crystalline isotactic polystyrene is known but has not yet been used commercially. The various types of homopolymers differ in molecular weight distribution, in molecular weight (n–2000), in purity, and minor addition quantities. Modifications based on styrene can be grouped as follows: (i) Mixtures of polystyrenes with synthetic rubbers. The rubbers are mostly based on butadiene-styrene copolymers. The rubber content amounts to 4–6% as a rule; sometimes it is higher. (ii) Copolymers of styrene with one or two (or possibly three) monomers. The copolymer with acrylonitrile (SAN), usually in the proportion of 76 to 24, is well known. Less well known are copolymers with methyl styrene, divinylbenzene and maleic acid anhydride. The copolymer of styrene and acrylonitrile (SAN) has the following molecular structure:

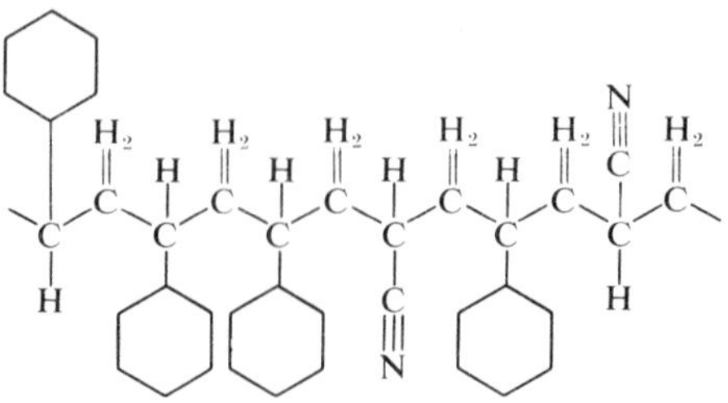

(iii) ABS plastics are two-phase systems. One phase consists of the continuous matrix of the hard plastic. The other is a more or less rubbery mass, which is dispersed in the first phase. The best properties are obtained if the rubber is finely distributed, without one phase being dissolved in the other, but so that the rubber particles and hard plastic are held together as firmly as possible. The rubbery phase is obtained by the addition of emulsions of styrene and acrylonitrile to a poly butadiene emulsion. This mixture is then polymerized. A 'graft' copolymer is obtained, as the styrene-acronitrile copolymer is united on the polybutadiene. In order to lessen the solubility of this rubbery phase in styrene, the polybutadiene is usually partially cross-linked. The rubbery phase as a copolymer of butadiene and styrene, or as a terpolymer of butadiene, styrene, and acrylonitrile, has turned out to be less effective and is being used less frequently. The plastic phase can be obtained by the copolymerization of styrene and acrylonitrile, as described under (ii). ABS plastic is now produced by the mechanical mixing of the two components in the melt, in such a way that the rubber phase is dispersed in the plastic phase.

Many different compositions are possible in such systems, but most ABS plastics keep within the following ranges: 15–25% acrylonitrile, 5–30% butadiene and 50–75% styrene. A typical composition is 22 : 10 : 68. The compositions are formulated to obtain an optimum compromise between price, processing characteristics, shaping characteristics and quality.

8.6.3 Properties

a. Straight polystyrene. This is a clear colourless substance, easy to process. The electrical properties are very good. However, the material has a low impact strength and high susceptibility to stress-corrosion; even low internal or external stresses can easily lead to fracture. The maximum temperature at which it can be used is relatively low. Table 8.3 gives some properties of such a general-purpose material, together with those of the improved polymers described below.

b. High impact polystyrene. If rubbers are added the strength, modulus, hardness, and heat resistance generally decrease while the impact strength and bending strength increase. The material becomes opaque.

c. SAN polymer. If the second monomer is acrylonitrile all the properties are improved. The material remains clear and is usually light in colour. The electrical properties are only slightly inferior to those of straight polystyrene.

d. Acrylonitrile-butadiene-styrene (ABS) polymers. One example is given in table 8.3. The data for such compounds suggest that the styrene chains ensure rigidity, hardness and processing capacity, that the butadiene chains are responsible for improved impact strength and resistance to stress corrosion, and the acrylonitrile chains for heat resistance and an increase in general stability. The ABS polymers are opaque. The electrical properties are only slightly inferior to those for the straight polymers of styrene. ABS plastics have recently come into use for engineering applications, for which glass fibre reinforcement is employed because of the improved dimensional stability achieved, better heat resistance, higher modulus, and lower coefficient of expansion. An attractive transparent material is said to be obtained if the acrylonitrile is replaced by methylmethacrylate (MBS plastics).

8.6.4 Commercial products

The polystyrenes are commercially available in different forms, but chiefly as granules, suitable for extrusion and injection moulding. Special grades are frequently offered for the extrusion of tubes, films, and sections and by judicious modification of the basic polymer a wide range of properties can be offered. Apart from granules, the polystyrenes are also marketed as powders, 'pearls', and emulsions. The number of

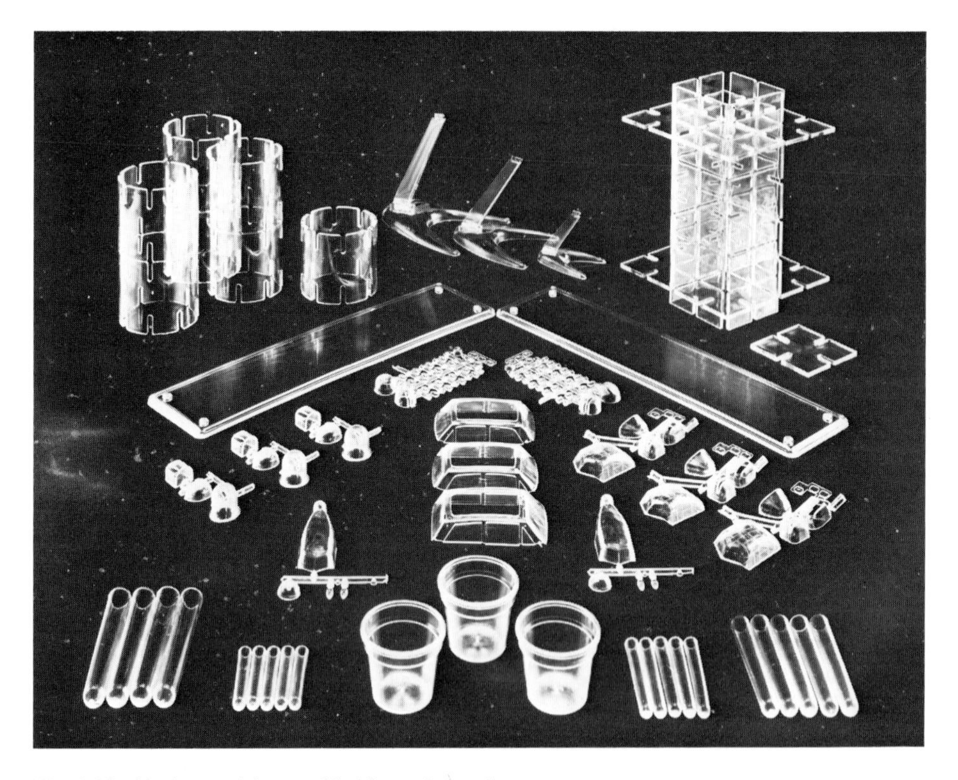

Fig. 8.15 Various articles moulded from clear polystyrene.

Table 8.3 Properties of styrene plastics

Properties	*General purpose*	*High impact*	*SAN*	*ABS*
Tensile strength, p.s.i.	5500–8000	3500–4500	10000–12000	6500–8500
Elongation, %	1.5–2.5	20–30	3–4	10–40
Modulus (in tension), p.s.i.	$4–5 \times 10^5$	$3.3–4.5 \times 10^5$	$5.0–5.5 \times 10^5$	$3.5–4.5 \times 10^5$
Hardness, Rockwell	70–80	45–60	85–90	80–90
Flexural strength, p.s.i.	8000–19000	no failure	15000–19000	9000–13000
Impact strength (notched Izod $\frac{1}{2} \times \frac{1}{2}$ in.)	0.25–0.35	0.65–0.90	0.40–0.50	1.5–4.5
Coefficient of thermal expansion, °C	$6–8 \times 10^{-5}$	$7–8 \times 10^{-5}$	$7–8 \times 10^{-5}$	$7–8 \times 10^{-5}$
Deflection temp. under load (Comp. molded Spec.) °F @ 264 p.s.i. fibre stress	165–190	160–180	190–210	185–195
Colour possibilities	unlimited	translucent-opaque	unlimited	opaque
Specific gravity (natural only)	1.05	1.03–1.06	1.07	1.04–1.72
Moisture absorption, %	0.03–0.04	0.07–0.08	0.25	0.10–0.25
Loss factor, 10^6 cycles/sec	0.0001–0.0005	0.0004–0.0010	0.007	0.001–0.008
Dielectric constant, 10^6 cycles/sec	2.50–2.65	2.55–2.70	2.8	3.0–4.8
Dielectric strength, v/mil; $\frac{1}{8}$ in. thick	> 500	> 450	> 440	> 450
Volume resistivity, ohm/cm	$10^{18}–10^{19}$	$10^{16}–10^{17}$	2.6×10^{16}	$> 10^{15}$

commercial products is very large indeed. They are marketed by a number of companies under a variety of trade names; e.g. Polystyrol, Novodur (Germany); Bexan SAN, Bexan ABS (UK); Lustrae (France); Urtal (Italy); Stynom, Lustrex (USA).

8.6.5 Production

Polystyrenes are manufactured in two stages. In the first stage the monomer styrene, possibly together with other monomers, is converted into the polymer, and impurities are removed. In the second stage the polymer is granulated, other substances frequently being added at the same time; rubbers, other polymers, and also oxidizers, pigments, and the like, are such additions.

The first stage of the production process can be carried out in different ways. Mass, solution, emulsion and suspension polymerization are all used, the two latter processes being the most common. Some suspension polymers ('pearl polymers') and also certain emulsion polymers, can be used by themselves; an example is the rubbery component of ABS plastics, which is generally obtained by emulsion polymerization. However, most of it is compounded. For simple formulations, e.g. straight polystyrene, it will be sufficient to use preliminary mixing, followed by extrusion to rod and disintegrating to granules. For complex formulations, e.g. ABS plastics, more vigorous mixing is necessary.

The compounding of the polystyrene can have a great influence on the properties of the product. This applies especially to complex two-phase, or multi-component systems such as the ABS plastics.

8.6.6 Uses

Because they comprise a large group of materials with widely different properties, polystyrenes have a corresponding range of uses. Their foremost application is for articles or components made by injection moulding; refrigerator parts, radios, toys, and cars are a few examples. Extruded products include tubes, sections, and sheet material. The tubes are finding a ready market in the United States where about 10% of all ABS products are extruded to tubing. Sheet material is made into a variety of articles by deep-drawing. At one end of the range, thin film is deep-drawn to make containers and beakers. At the other end, relatively large parts deep-drawn from thick sheet, such as components for scooters, boats, and cars, can be made.

The widening of the scope for styrene plastics has been stimulated by specialized processes. The foaming process has been developed for the straight polystyrenes, because of its relatively easy foamability, the good properties (due to the high modulus) of this hard closed-cell foam, and especially the fact that the expansion gives the chain molecules an orientation, with the result that the normal brittleness of styrene disappears. Densities range from under 1 lb/ft^3 to 16 lb/ft^3 (16–250 kg/cm^3). The properties of these foams include strength, rigidity (for sheets, tiles, cores for sandwich construction), particularly good thermal insulation, and shock resistance and energy absorption (valuable for packaging material). Its low weight in conjunction with the closed-cell structure gives the material great buoyancy.

The electroplating of ABS plastics, usually with chromium, is attractive because of the high adhesion between the plastic and metal. Applications are chiefly in the decorative field; the potential scope is believed to be very great indeed. Metal-coated ABS plastics are used, for instance, by many large motor car manufacturers, and electrical applications are developing strongly.

A specialized application–polystyrene latex–is to be found in styrene polymer dispersions containing 40 to 50% solids in water. They are used primarily in the dye-stuff industry in formulations of modern water-based dyes.

8.7 POLYVINYL CHLORIDE (PVC)

8.7.1 History

The monomer was first described by Regnault in 1835. In 1912 a patent was granted for a commercial production method, based on a catalysed reaction with hydrochloric acid. Only after 1930 was a start made with the industrial production of PVC and its copolymers, following extensive development work, especially in the United States and Germany. In Germany a copolymer of vinyl chloride and methyl acrylate ('Troluloid') came on the market in 1931, and the production of the homopolymer was started in 1935. In the United States the possibility of plasticizing PVC was discovered in 1932. That country was particularly active in developing the copolymers in the 1930s. Stimulated by the Second World War, PVC has become a very important plastic, because of a combination of favourable properties, the numerous possibilities of compound formation and fabrication techniques, and its low price (PVC accounts for about one quarter of the world production of plastics.)

8.7.2 Processing methods and structure

a. Monomer. Vinyl chloride has the following structure:

$$\begin{array}{cc} H & H \\ | & | \\ C & = C \\ | & | \\ H & Cl \end{array}$$

Its boiling point is $-13.4°C$ (760 mm Hg). It is only very slightly soluble in water (0.11%). With the aid of catalysts, normally based on peroxides or persulphates, it can be activated and polymerized. The classical production method (from acetylene) is as follows:

$$HC\equiv CH + HCl \xrightarrow[HgCl_2]{180°C} CH_2{=}CHCl$$

The petrochemical, ethylene, is the basis of commercial manufacture, the following process being carried out:

$$H_2C = CH_2 + Cl_2 \longrightarrow \underset{\text{dichloroethane}}{CH_2Cl - CH_2Cl}$$

and then

$$CH_2Cl - CH_2Cl \xrightarrow{400°C} CH_2 = CHCl + HCl$$

A modern process is represented by the following equation, with cycling of hydrochloric acid:

$$\begin{aligned} 2HCl + \tfrac{1}{2}O_2 + Cl_2 + 2C_2H_4 &\longrightarrow 2C_2H_4Cl_2 + H_2O \\ 2C_2H_4Cl_2 &\longrightarrow 2C_2H_3Cl + 2HCl \end{aligned}$$

b. Polymer. PVC has the following structure:

```
        H    H    H    H    H    H    H    H    H
        |    |    |    |    |    |    |    |    |
 - - - -C----C----C----C----C----C----C----C----C- - -
        |    |    |    |    |    |    |    |    |
        H    Cl   H    Cl   H    Cl   H    Cl   H
```

In practice, the molecule is not so regular but somewhat branched and the chlorine atoms are for the most part irregularly distributed about the carbon chains (atactic distribution). Hence PVC is mainly an amorphous plastic which can have a low proportion of crystalline material (approx. 10%). Using new polymerization techniques, it is possible in principle to produce a stereo-regular polymer. With the greater regularity of molecular structure, this type of PVC is largely crystalline, with a melting point above 200°C. Because of the higher price and the more difficult processing, this type is not yet used in practice.

The following polymerization techniques are commonly used for the industrial production of PVC: (i) *Emulsion polymerization.* Vinyl chloride is emulsified in water under pressure, and afterwards polymerized. The procedure used is either discontinuous or continuous. The polymerizate can be further processed by coagulation with an electrolyte; this is followed by washing, jet drying, or roller drying. The emulsion can also be used by itself for certain purposes, but for many applications the coagulated type, from which the emulsifiers have largely been removed, is used; otherwise the clear colour, electrical properties and resistance to moisture would be impaired. (ii) *Suspension (pearl) polymerization.* The monomer is mechanically suspended in water and the suspension is stabilized with very small quantities of protective colloids. The end product is much purer than the emulsion polymer and also has a coarser grain. This product is often preferred for clear-colour products and for electrical applications. (iii) *Mass polymerization.* This method of polymerization has long been used on a small scale. As no auxiliary agents are used, apart from the catalyst, the polymer is very pure. The resulting polymer precipitates, as it is insoluble in the monomer. After filtration the monomer is removed from it.

Apart from the difference in the polymerization technique used, the various types of PVC differ also in their mean molecular weights. Mean molecular weights are often indicated by a 'K' value, which is derived from the viscosity of a dilute solution of the polymer.

8.7.3 Copolymers and modifications

Copolymerization is normally used to improve the ease of processing or solubility (in organic solvents) and/or to improve certain mechanical properties. Vinyl chloride is copolymerized with vinyl acetate, acrylonitrile, butylacrylate, vinylidene chloride and propylene. Vinyl acetate is used most frequently at concentrations of from 2 to 15%. The higher the concentration, the better the fluidity during processing and the greater the solubility; applications are gramophone records, flooring compounds, and coatings. The copolymerizate with acrylonitrile (35–40%) is chiefly made into fibre. Propylene as copolymer improves the impact strength. The copolymers based on vinyl chloride and vinylidene chloride have long been known. For instance, crystalline and not easily soluble copolymers with about 15% vinyl chloride are on the market under the trade name 'Saran' for use as grease-resistant and gas-tight coatings. Fairly soluble copolymers with about 20% polyvinylidene chloride are made by Goodrich (Geon 200 series).

A much used chemical post-treatment of PVC is chlorination (up to 62–65% chloride). According to the procedure employed, this gives either products with better solubility (fibres, adhesives) or a product with a higher softening point (hot-water pipes).

8.7.4 Properties

In the discussion of the properties of PVC, a distinction must be made between rigid PVC and the plasticized type: see table 8.4.

a. Rigid PVC. Rigid PVC is, at room temperature, a hard, horn-like and colourless material. At lower temperatures it becomes harder and fairly brittle, while its hardness gradually decreases as the temperature rises. At about 80°C it becomes soft and elastic. In the elastic region, PVC can be readily shaped at between 120°C and 150°C by vacuum forming or bending. Above 150°C, the elasticity decreases gradually and plasticity begins to set in. At about 170°C, rigid PVC becomes nearly completely plastic and can be shaped above this temperature by extrusion, injection moulding or calendering. At about 180°C it begins to decompose fairly rapidly; this shows by hydrochloric acid separation and discoloration. Decomposition can be slowed down by adding a small quantity of stabiliser. Ultra-violet absorbers are employed where necessary, to improve the resistance to light-induced changes; anti-oxidants are likewise used. As a result of these additions, PVC can be worked at elevated temperatures and have a long service life. For applications such as food packaging and water pipes, the choice of stabilizer is restricted in view of the fact that many stabilizers (e.g. lead, cadmium and barium compounds) are toxic. However, compounds considered non-toxic can be made up on the basis of calcium-zinc epoxy and certain dioctyl tin compounds. Compared with many other thermoplastics, the chemical resistance of PVC is very good, and it is therefore much used in the chemical industry. In particular, it is resistant to weak acids, strong bases, alcohols, and aliphatic hydrocarbons; it can be slightly affected by strong

Table 8.4 Room temperature properties of rigid and plasticized PVC (typical values)

Property	*Unit*	*Rigid PVC*	*Plasticized PVC*[2]	*Notes*
Density	g/cm³	1.38–1.40	1.20–1.36	
Tensile strength	kg/cm²	500–600	150–200	30% DOP[1]
Elongation at rupture	%	15	200–300	30% DOP
Modulus of elasticity	kg/cm²	30 000	13 000	25% DOP
	–	–	500	50% DOP
Impact strength (notched)	kg-cm/cm²	2	–	
Brittle point	°C	–	−16	20% DOP
Heat distortion (Vicat)	°C	78–82	50–60	20% DOP
Linear coefficient of expansion	per°C	8×10^{-5}	$7\text{–}10 \times 10^{-5}$	
Dielectric strength	kV/cm	180	200–300	
Dielectric constant[3]	–	3.2–3.4	3–9	
Power factor[3]	–	0.015–0.020	0.015–0.020 and higher	
Gas permeability (oxygen)	ml/min/cm²/sec/cm Hg $\times 10^{-10}$	0.5–1.2	6	23% DOP
Water absorption	mg	20	20–100	

[1]Dioctylphthalate.
[2]Largely dependent on amount and type of plasticizer.
[3]Dependent on temperature, plasticizer concentration, and frequency.

acids. PVC swells or dissolves in ketones, esters, chlorinated hydrocarbons, and aromatic hydrocarbons.

Rigid PVC becomes brittle at about 0°C. This restricts its use in applications where impact strength is important. To overcome this drawback, while retaining its other properties, auxiliary substances are added in the form of synthetic rubbers (butadiene acrylonitrile), ABS, chlorinated polyethylene or certain acrylates; or copolymerization is used. Many of these additions improve the workability at the same time, and, not infrequently, this is the main reason for using them. The maximum operating temperature under load for rigid PVC is generally put at 60°C. Above this temperature creep becomes too great.

Because of its high chlorine content (57%), PVC is self-extinguishing; on burning, a large amount of hydrochloric acid forms. In common with most other plastics, PVC is notch-sensitive; that is, a notch, scratch, inhomogeneity, or sharp fold may initiate tearing under load.

b. Plasticized PVC. If plasticizers are added to PVC, products ranging from semi-rigid to very flexible are obtained, depending on the amount of plasticizer added. Mixed plasticizers are frequently used; particularly well known are dioctylphthalate and tricresyl phosphate. Plasticized PVC is a physical mixture; this means that the plasticizer can be removed by extraction or evaporation. It may also exude as a result of segregation, due either to incorrect composition or ageing. Some plasticizers are liable to become mouldy under certain conditions. Specific fungicides are used as a remedy.

Plasticizers have a large influence on the properties of PVC. Tensile strength decreases, and elongation increases, with the quantity used. Also, hardness and brittleness at low temperatures depend directly on the amount of plasticizer present, the brittleness being determined, in addition, by the type of plasticizer used. Resistance to chemicals is also determined by the plasticizer. Loss of plasticizer, followed by hardening, may occur as a result of extraction by organic solvents or from saponification by alkalis.

Colouring material incorporated in the plasticizer must be insoluble, to inhibit migration. Both organic and inorganic pigments are used. Any such auxiliary substances should not impair the heat resistance and light fastness. Greases or lubricants are frequently added to prevent sticking during compounding or processing, but the presence of lubricants can lead to exudation, making the surface dull and greasy.

8.7.5 Fabrication

Both rigid and plasticized PVC can be worked to give semi-finished or end products on conventional equipment such as extruders, injection moulding machines or calenders. Normally, after preliminary mixing, the mixtures are compounded in machines such as friction rolls and/or internal mixers (Banbury, Ko-Kneter, etc.) and mixer-extruders. At this stage, stabilizers, possibly plasticizers, internal and external lubricants (respectively, to improve fluidity and prevent sticking during the processing), modifiers (to influence properties and workability), and pigments and fillers are mixed thoroughly to a homogeneous mass. This can be followed directly by calendering (section 8.4.1) or by granulation for subsequent fabrication in extruders or injection moulding machines. The practice of feeding 'dry blends' directly into fabricating machines is gaining ground steadily. Dry blends are free-flowing mixed powders coagulated to varying degrees and containing all the necessary or desired components.

A method frequently employed for making PVC products starts from a paste. PVC powders (also copolymers) can be prepared in such a way that, when stirred together with plasticizer, they give a paste which is stable at room temperature. This paste, also called a 'plastisol', can be worked by spreading, casting, dipping, slush moulding, wheel casting and spraying. If necessary, the viscosity is lowered by adding organic thinners. Such products are then known as 'organosols'. After the forming operation the paste is subjected to a heat treatment (gelation), to allow the PVC and plasticizer to dissolve one in the other. This is normally carried out at a temperature above 160°C to achieve maximum mechanical properties.

Both rigid and plasticized PVC, the latter mainly as the plastisol, are foamed by means of blowing agents. Chemical blowing agents are mainly used for plasticized PVC and physical blowing agents for rigid PVC.

8.7.6 Final shaping

Excellent welded joints can be produced in rigid PVC, using a hot wire or hot air jet. This method is used for making tank linings or whole tanks, for producing large-diameter tubes from sheet, and for the jointing of tubes. Friction welding may be used in some special cases. Bonded joints in rigid PVC are easily produced. Chlorinated PVC dissolved in methylene chloride is frequently used as an adhesive. For plasticized PVC an adhesive on a nitrile rubber base is commonly used.

As PVC has a fairly high dielectric loss value, it is very easy to weld by high-frequency heating. High frequency welding is used for nearly all constructions based on plasticized PVC film.

Forming with the aid of a vacuum or compressed air is used a great deal for rigid PVC. This material is also very suitable for machining operations (milling, turning, drilling), provided that appropriate tools are used and care is taken to ensure that the material is not overheated. Rigid PVC tubes and sheet can be readily bent, provided they are pre-heated throughout their thickness. Owing to the polar properties of PVC printing and lacquering present few problems.

8.7.7 Uses

a. Rigid PVC. In the chemical industry, rigid PVC has long been used on a large scale in the form of calendered and/or moulded sheet. Some typical examples are tank liners, hoods, exhausters and pump housings. Welded or extruded tubing is finding a growing market for the transport of gases and liquids. Other well-known applications are rainwater goods, golf clubs, drainage pipes and electric cable ducts. Rigid PVC film, produced by calendering or with the aid of the flat film or blow process, is used in the packaging industry; for instance, margarine containers are directly or vacuum-formed from calendered film. For a number of applications PVC is stretched biaxially, which greatly improves its mechanical properties in both directions. Magnetic tape is made from film stretched in only one direction. A rapidly growing application in the packaging industry is the non-returnable bottle for beer, wine and edible oils. Rigid PVC foam of different densities is used in sandwich-type constructions (aircraft, boats, houses), in floats (boat construction), insulation, and for battery separators. Long-playing gramophone records are nearly all made from PVC polymer on a vinyl acetate basis. A similar copolymer is also much used for hard, filled flooring tiles.

Fig. 8.16 High frequency welding of clear PVC sheet.

Chlorinated PVC finds a use as an adhesive for PVC tubes and as a corrosion-resistant material in chemical plants. Its use in hot-water pipes is gaining ground. Copolymers with 85 to 90% vinylidene chloride are used as monofilament and coatings for packaging materials.

b. Plasticized PVC. Because of the many applications these are listed systematically under the method of their fabrication.

Calendering (possibly extrusion)

Film: Curtains (also for showers)
Rainwear
Inflatable toys, mattresses, boats
Cases (for spoons etc.)
Portfolio covers
Tablecloths
Top-layer floor covering
Roofing insulation
Water tank (liner)
Adhesive tape and film
Covering for sheet metal (as lacquered metal)
Packaging

Extrusion

Sheet: Flooring (possibly laminated)
Table covering

Tubing: Garden hoses and the like
Slip-on tubing
Pipe covering
Insulation (low and high tension)
Cable sheathing
Wire covering (e.g. for fences)
Fine meshing
Packaging
Sacks

Sections: Sealing strips
Beading
Buffer protecting strip

Injection Moulding:

Footwear
Toys
Components

Spreading, Dipping Coating (of paste):

Leathercloth, usually on fabric support, foamed for upholstery
Handbags
Coated metal sheet (as lacquered metal)
Coated wire netting (fencing)
Coated tubing and rod assemblies (bed frames, chairs, plastic racks, etc.)
Gloves
Door knobs
Floor covering (applied e.g. to jute)
Printing rollers

Pressing:

Floor tiles

8.7.8 Commercial products

PVC products and materials are marketed in most industrial countries, under a variety of trade names; e.g. Breon, Corvic (UK); Geon, Vinylite (USA); Vinoflex (Germany) and so on.

8.8 OTHER VINYL POLYMERS

8.8.1 Polyvinylidene chloride

a. History. As early as 1836 Regnault described a white deposit obtained trom 1,1-dichloroethane (vinylidene chloride). He considered this deposit to be an isomer of the dichloroethane. It is now known that peroxide is readily formed by atmospheric air acting on impure vinylidene chloride. The polymerization is initiated by the peroxide, and it is therefore certain that Regnault had produced polyvinylidene chloride. Staudinger and Feisst published in 1930 the result of an investigation on the polymerization of vinylidene chloride. The development of this new plastic was pursued in the following years. It soon became clear that the homopolymer is not entirely suitable for commercial applications owing to its marked tendency to crystallization. Because of this, the softening point is too close to the decomposition point, so that the usual processing methods cannot be used. The copolymers, especially with vinyl chloride and acrylonitrile, have lower softening points and in 1938 these copolymers came on the marked under the name 'Saran' (Dow Chemical Corp.). In Germany also the production stage was reached before the Second World War: 'Diurit' of I.G. Farben contained about 85% vinylidene chloride, 13% vinylchloride and 2% acrylonitrile. It was mainly used for the manufacture of brushes. A vinylidene chloride/ethylacrylate copolymer was spun to make yarn for weaving into parachute fabric. After the war, the output of vinylidene chloride copolymers rose rapidly. Its strength, its chemical resistance and, above all, its low permeability to gas, ensured that this type of plastic became one of the most important packaging materials.

b. Structure. Vinylidene chloride ($Cl_2C{=}CH_2$) undergoes head-to-tail polymerization. The structural formula of the polymer is:

$$
\begin{array}{ccccccccc}
 & H & & Cl & & H & & Cl & \\
 & | & & | & & | & & | & \\
-- & C & - & C & - & C & - & C & -- \\
 & | & & | & & | & & | & \\
 & H & & Cl & & H & & Cl &
\end{array}
$$

It was found from X-ray diagrams that the bond angle between three successive carbon atoms is approximately 120°. This is an appreciable deviation from the tetrahedral valency angle (109° 28'). The reason for this is the dense occupation of the polymer chain by the large chlorine atoms. The rotation in the chain about the C—C bond is greatly impeded, so that the polymer has a very rigid chain. The dense occupation by polar chlorine atoms and the regular structure, with the chlorine atoms placed in the form of a spiral around the chains, favour crystallization. The polymer is easily decomposed at the temperatures which are used in processing, HCl being released. The insolubility of polyvinylidene chloride is due to the pronounced tendency to crystallization. The usual plasticizers are not tolerated by polyvinylidene chloride; they are rejected. Flexible segments are incorporated in the chains by copolymerization, and the regular structure is thus disturbed, so that the softening point decreases and the solubility increases.

c. Properties and applications. Polyvinylidene chloride is produced in the form of white flakes or white powder. It is obtained in the amorphous condition by rapidly cooling the molten material. This condition is stable below 0°C, but crystallization takes place at room temperature. At temperatures between 60 and 150°C, the rate of crystallization is high. The softening point is between 180 and 200°C. At this temperature, HCl is found to separate and at 220°C dissociation proper occurs. Traces of iron salts or copper salts catalyse the dissociation so that special materials which do not promote decomposition of the polymer must be used for the machine tools with which polyvinylidene chloride comes into contact while it is hot. Polyvinylidene chloride is insoluble at room temperature and is not then attacked by the majority of chemicals. At higher temperatures, e.g. at 100°C, trichloroethane, dioxane, tetralin and chlorinated aromatic substances can act as solvents. Because polyvinylidene chloride cannot be mixed with most plasticizers copolymers have gained much ground. The especially good chemical stability, the low gas permeability, and the good mechanical properties of the homopolymer are exploited in the copolymers. The softening point is modified so that most fabrication techniques can be used. Normal heat stabilizers, which are very effective for other materials such as polyvinyl chloride, are not suitable for vinylidene chloride copolymers. Because of their crystallization properties, copolymers with high contents of vinylidene chloride demand careful processing. Thus, a copolymer after extrusion and cooling in a water bath will be soft. During storage at room temperature, partial crystallization occurs and the product becomes hard and dimensionally stable. This crystallization can be adjusted by a controlled heat treatment. Compression moulding and injection moulding methods of shaping are equally suitable. To ensure a rapid cycle of operations, it is important to use not cooled, but heated moulds; this stimulates crystallization in the mould. Directional crystallization is produced in film and fibres and the products thus gain high strength and form a dense barrier against gases and vapours (and also aromatic substances). Fibres become abrasion-resistant by stretching and can therefore be woven at high speeds.

The most commonly used copolymerization monomers are vinyl chloride and acrylonitrile, but vinyl acetate, styrene, acryl and methacryl esters and butadiene can also be used. Copolymers with vinyl chloride can be prepared using all ratios of monomers. Those with more than 70% vinylidene chloride are largely crystalline, and below this percentage they are amorphous. The softening point passes through a minimum of about 5°C at approximately 60% vinylidene chloride. With increasing contents of vinylidene chloride or vinyl chloride the softening point rises close to the values of the respective homopolymers. For 15% vinylidene chloride with vinyl chloride, the working temperature, which for polyvinyl chloride amounts to 170° to 180°C, drops to about 135°C. The properties of these products are similar to those of polyvinyl chloride. Depending on the relative monomer proportions, the copolymers of vinylidene chloride are either hard and tough, or rubbery and elastic. They are impact-resistant, resistant to most chemicals, and resist decay under any atmospheric conditions. Owing to the high chlorine content (between 57 and 73%) they become self-extinguishing directly after being removed from a flame. They are highly suitable for packaging purposes because of their extremely low gas permeability. A large part of the output is therefore made into film, mostly by blowing up an extruded tube, super-cooling, and orienting it. The film is easily welded and is readily printed. Cellophane is coated on both sides with polyvinylidene chloride copolymer for packaging purposes. Polyvinylidene chloride copolymers are also used for upholstery and car seat covering, carpets, brushes, filter cloth, fishing nets, mosquito netting, sun screens, tubing for the chemical industry, conveyor belts and impregnated paper.

d. Commercial products. Some trade names for polyvinylidene chloride copolymers include Darlan, Saran (USA); Geon (UK); Diofan, Vylen (Germany); Sarkel (France); there are many others.

e. Production. Vinylidene chloride is prepared from acetylene. In the first stage, 1, 1, 2-trichloroethane is formed:

$$HC\equiv CH + HCl + Cl_2 \longrightarrow ClH_2C{-}CHCl_2$$

In the second stage, HCl is split off with the aid of sodium or calcium hydroxide:

$$ClH_2C{-}CHCl_2 \xrightarrow{-HCl} CH_2{=}CCl_2$$

Vinylidene chloride is a colourless liquid, which boils at 31.7°C. In the absence of oxygen and in the dark, the monomer can be stored safely. In contact with air, peroxides are formed which decompose to produce phosgene, formaldehyde and hydrogen chloride. The peroxides give rise to polymerization which, for instance on heating, can be explosive. Prior to use the peroxides must be removed by washing. Peroxide formation is prevented by using phenols or amines as inhibitors. It is recommended, when the monomer is to be kept for a long time, to cover it with a soda solution. Before polymerization, the inhibitors and impurities are removed by washing in solutions of sodium hydroxide and ferrous sulphate. The vinylidene chloride is then distilled under nitrogen. Volatile methanol compounds, which would lower the heat resistance of the polyvinylidene chloride, can be removed by treatment with ethylene oxide in the presence of a little water. The polymerization to give polyvinylidene chloride is, on the whole, similar to that of vinyl chloride. The polymer is insoluble in the monomer, so that with mass polymerization a solid mass forms even for a low conversion; the product obtained in this way is therefore of non-uniform composition. For this reason, polyvinylidene chloride and the copolymers are prepared, on an industrial scale, by the emulsion or suspension techniques, the procedures being similar to those employed for the polymerization of vinyl chloride. Using ferrosalts as components of redox catalysis systems is not recommended, as this affects the heat stability of the polymer; preference should be given to organic catalyst systems. Polymerization with peroxide catalysts should be carried out at a temperature below 50°C in order to prevent oxidation of the vinylidene chloride. The molecular weights lie between 10000 and 100000 according to composition. Crystallized polyvinylidene chloride is a heavy plastic; its specific gravity is about 1.85.

8.8.2 Fluorine plastics

a. History. Five different plastics are grouped together under the collective name fluorine plastics. These are built up of two or more of the elements carbon, fluorine, chlorine and hydrogen; carbon and fluorine are always present. The five polymers are given below.

Polytetrafluoroethylene (PTFE)

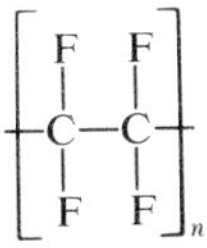

$$\left[\begin{array}{ccc} & F & F \\ & | & | \\ - & C - & C - \\ & | & | \\ & F & F \end{array}\right]_n$$

Polytrifluorochloroethylene (PTFCE)

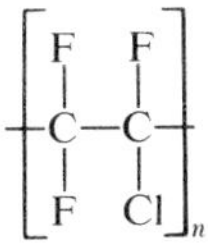

$$\left[\begin{array}{ccc} & F & F \\ & | & | \\ - & C - & C - \\ & | & | \\ & F & Cl \end{array}\right]_n$$

Fluorinated ethylene propene copolymer (PFEP)

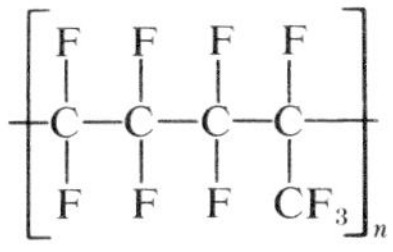

$$\left[\begin{array}{ccccc} & F & F & F & F \\ & | & | & | & | \\ - & C - & C - & C - & C - \\ & | & | & | & | \\ & F & F & F & CF_3 \end{array}\right]_n$$

Polyvinyl fluoride (PVF)

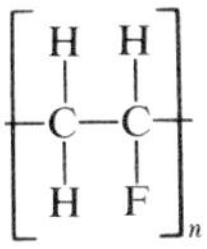

$$\left[\begin{array}{ccc} & H & H \\ & | & | \\ - & C - & C - \\ & | & | \\ & H & F \end{array}\right]_n$$

Polyvinylidene fluoride (PVF_2)

$$\left[-\overset{\displaystyle H}{\underset{\displaystyle H}{\overset{|}{\underset{|}{C}}}} - \overset{\displaystyle F}{\underset{\displaystyle F}{\overset{|}{\underset{|}{C}}}} - \right]_n$$

Polytrifluorochloroethylene (PTFCE) was made in Germany as early as 1934. However, the molecular weight of this product was too low for commercial use. The further development of fluorine plastics took place in America, where it was stimulated by research work on new materials for the isotope separation of uranium hexafluoride. The fluorinated hydrocarbons (e.g. Freon, Genetron) and other refrigerants also contributed. Du Pont de Nemours for a long time maintained a monopoly of non-stick coatings marketed as 'Teflon'. When the basic patent expired, other manufacturers became active in this field, e.g. Allied Chemical and Dye Corp., Farbwerke Hoechst, Montecatini.

The fluorine polymers are expensive and generally very difficult to process, yet owing to their properties, which cannot be equalled by any other plastic, they have an assured market, by far the greatest share of which is taken by polytetrafluoroethylene. New developments, particularly in the field of elastomers, can be expected in the next few years.

b. Structure. The polymers are usually built up linearly and have a head-to-tail structure. Of great importance for the properties of this group of polymers, in particular of the highly fluorinated types, is the extraordinarily strong C—F bond. Because of its small radius, the fluorine atom imparts to the polymers very high resistance to chemicals and very high heat stability. Highly fluorinated plastics show practically no affinity for other materials, so that the coefficient of friction is extremely low. They are, therefore, excellent self-lubricating bearing materials. The drawbacks arising from these properties are the insolubility of the plastic and its high melting point, which in some cases lies above the dissociation temperature so that shaping is difficult.

c. Processing, polymerization, properties and uses

(i) *Polytetrafluorethylene (PTFE).* The monomer tetrafluoroethylene is obtained by pyrolysis from CHF_2Cl (monochlorodifluoromethane; that is, the refrigerants Freon-22 and Genetron-141). It is highly reactive and the presence of traces of oxygen can cause an explosion; the oxygen content must therefore be kept below 0.002%. Moreover, small amounts of impurities are detrimental to the properties of the polymer. Hence the monomer must be thoroughly purified. Strictly enforced precautionary measures are used in the preparation of the monomer and in the polymerization. The monomer can be stabilized by the addition of such readily oxidizable substances as terpenes, aliphatic amines and mercaptans. Polymerization takes place by a radical mechanism, the initiators used being peroxides or azodinitriles. Generally, the polymerization is carried out in an emulsion or in suspension at a temperature of about 60°C and at a pressure of about 60 kg/cm^2, the polymer being obtained as a powder by coagulation. The dispersions can also be concentrated and processed as such.

PTFE is a white, highly crystalline plastic. It is fairly flexible and feels greasy. The molecular weight is estimated at about 500000; it cannot be determined exactly because of the insolubility of the polymer. Above the first-stage transition point (327°C) it becomes translucent and can be shaped. Above 400°C it disintegrates into the monomer and other (toxic) decomposition products. Smoking in premises where PTFE is stored or being processed is very dangerous because it is dissociated at the temperature of burning tobacco. Even small quantities of these decomposition products, when inhaled directly, are extremely injurious.

The good mechanical properties are retained even at very low temperatures, so that PTFE can be used in the range from −100° to 250°C. The polymer is only attacked by fused alkali metals. The electrical properties are very good. The coefficient of friction with metals (other than those just referred to) lies at about 0.02–0.04. The only drawbacks are the high thermal expansion coefficient and the flow of material when cold under load, but these effects can be reduced by the use of fillers.

The processing of PTFE is difficult. It is shaped by compression sintering techniques. Extrusion is possible but very high pressures and temperatures are necessary; even then, the extrusion rate is very low, at about 1 m/h. This is probably the reason why PTFE is extruded in the form of a paste based on a hydrocarbon carrier. The hydrocarbon is removed afterwards. Machining operations are possible without difficulty. Dispersions of the polymer can be applied to other materials and sintered after drying. To make a bonded joint, the PTFE is first etched with a reagent such as sodium dissolved in liquid ammonia to give a surface which an epoxy adhesive will grip.

PTFE is used in applications for which its resistance to chemicals and its temperaure stability are the controlling factors; for instance, in pipe lines, fittings, seals, gaskets, valves, etc. The dispersions are used for the impregnation of sealing material. The low coefficient of friction of the material makes it particularly suitable for the lining of bearings, sealing of axles and the like.

Trade names of PTFE plastics include: Teflon (USA); Fluon (UK); and Hostaflon TF (Germany).

(ii) *Polytrifluorochloroethylene (PTFCE).* The monomer, trifluorochloroethylene, is produced by the separation of chlorine from 1,1,2-trifluoro-1,2,2-trichloroethane (that is, the refrigerant F-113, Genetron-226). The monomer is less reactive and easier to handle than tetrafluoroethylene. However, it must be highly purified. Polymerization is carried out either as an emulsion or suspension, or by mass polymerization, at a temperature under 0°C. The last of these methods involves a polymerization time of some weeks but gives a colourless product with very good properties.

PTFCE has a molecular weight of 100000 to 500000. It is harder than PTFE but can be shaped by the usual methods. On the other hand, resistance to chemicals and temperature stability is a little inferior to that of PTFE. Decomposition of PTFCE may occur during the processing unless it is stabilized by phenols or oxidizing salts. PTFCE can be used in the temperature range from −200° to +200°C; it softens at 240°C and decomposes at approximately 300°C. Like PTFE, its decomposition products are toxic. PTFCE swells in chlorinated hydrocarbons, e.g. carbon-tetrachloride and chloroform. At elevated temperatures it is soluble in dichlorobenzotrifluoride, enabling the molecular weight to be determined. Cold flow is less pronounced than in PTFE.

Certain copolymers with PTFCE have gained importance; for instance, those with various concentrations of vinylidene fluoride. This copolymer becomes brittle less quickly than the homopolymer of TFCE, and it is also possible to ensure solubility in certain solvents such as esters and ketones.

The telomers (low molecular weight polymers) of TFCE are waxes and find specialized uses, on the basis of their high resistance to chemicals and their temperature stability. PTFCE can be press-worked, extruded and injection-moulded, but the machine components which come into contact with the polymer must be resistant to the corrosive effect. PTFCE is used for the same applications as PTFE, but is preferred where a low cold flow is important. PTFCE is also used in large quantities for cable insulation. Film is used for packaging which can be sterilized and the dispersions for the coating of metals by spraying or dipping; the coatings are then treated at about 250°C. Permeability to water vapour is extremely low. Well-known trade names include: 'Kel-F' (USA); Halon (USA); Hostaflon-C (Germany).

(iii) *Fluorinated ethylene-propylene copolymer* (*PFEP*). The production of tetrafluoroethylene is described in connection with that for PTFE. The co-monomer hexafluoropropylene (also perfluropropylene) is prepared by chlorine separation from fluorochloropropane, or by the pyrolysis of tetrafluoroethylene. Copolymerization of these two monomers is very sensitive to the reaction conditions. It takes place in aqueous environment at a pressure of 50 kg/cm^2 and a temperature of about 90°C, the proportion of the co-monomers in the reaction medium being 90–50% perfluoropropylene and 10–50% tetrafluoroethylene. However, in the copolymer the perfluoropropylene content is substantially lower.

Although PFEP melts at a lower temperature than PTFE (290°C) it is very stable at lower temperatures, so that it can be processed by conventional methods. The viscosity of the melt is much lower than that of PTFE.

The temperature range in which PFEP can be used extends from approximately −100° to +200°C. Its chemical stability is somewhat lower than that of PTFE and the thermal decomposition products are equally toxic.

Apart from the described limitations, the applications lie in the same fields as for the fluoropolymers discussed above. PFEP is supplied by du Pont de Nemours under the trade name Teflon-100 X or Teflon-FEP.

(iv) *Polyvinyl fluoride* (*PVF*). PVF has three hydrogen atoms for every fluorine atom. Certain physical and chemical properties are therefore different from those of fluoropolymers containing no hydrogen.

The monomer vinylfluoride is prepared by the addition of hydrogen fluoride to acetylene, with the aid of a catalyst,

$$CH{\equiv}CH + HF \xrightarrow[\text{or Hg compounds}]{\text{CuCN}} CH_2{=}CHF$$

or by the pyrolysis of 1,1-difluoroethane or 1-fluoro-1-chloroethane:

$$CH_3{-}CHF_2 \xrightarrow{-HF} CH_2{=}CHF$$

The monomer was already known in the nineteenth century but not before 1941 was commercial polymerization practicable (by Farbenfabrik Bayer). Polymerization takes longer than for vinylchloride. Impurities have a disturbing effect, so that oxygen and acetylene must be thoroughly removed. The polymerization takes place in the presence of water at a temperature of 50° to 150°C and a pressure of 100 to 500 kg/cm^2, a peroxide or an azodinitrile being used as an initiator. The polymer is obtained in the form of a powder or dispersion. Temperature and pressure variations have a great influence on the molecular weight and the structure. The molecular weight of the commercial product lies in the range 25000 to 250000. Above 100°C the polymer is soluble in dioxane, tricresyl phosphate, nitroparaffin, chlorohydrocarbons, dimethylformamide, and certain ketones and nitriles. It has a melting point around 200°C and can be used up to about 150°C. Polyvinyl fluoride can be plastically deformed above

200°C, but injection moulding can be used only with grades of relatively low molecular weight. Film is usually cast from solution.

The important properties of PVF are its high tensile strength and elongation, its good elastic properties, high stability to weathering, and low gas permeability; hence it is mostly used in the form of film for packaging and for protection against atmospheric exposure. Brushes, tubing and rod are also made from this plastic and it is used for the impregnation of textiles and leather. PVF is marketed by Du Pont Corp. under the trade mark 'Tedlar'.

(v) *Polyvinylidene fluoride* (*PVF_2*). Polyvinylidene fluoride came on the market in 1961. It is obtainable in the form of a white powder, tablets, or dispersions in dimethylacetamide, dimethylphthalate and diisobutylketone. The monomer, vinylidene fluoride, is prepared by fluorination of methyl chloroform (1, 1, 1-trichloroethane) followed by the removal of hydrogen chloride:

$$CH_3{-}CCl_3 \xrightarrow{HF} CH_3{-}CClF_2 \xrightarrow{-HCl} CH_2{=}CF_2$$

Polymerization takes place at high pressure in water. The polyvinylidene fluoride contains more hydrogen than the other fluorine plastics and therefore is a little less inert than them. Even so, it is resistant to the effect of most acids, alkalis and solvents. It has a high tensile strength, good electrical properties, good abrasion resistance, and a low cold flow. It can be used over a temperature range from about −60° to +150°C. Fabrication is possible by extrusion, injection moulding and compression moulding. The fabrication temperatures lie within 200° to 250°C.

Because of its resistance to chemicals, polyvinylidene fluoride is used for tubes, valves, and pumps in the chemical industry. Its good electrical properties make it suitable for cable insulation. Mixed with pigment, this plastic is sprayed on metal. These coatings are practically non-porous and offer protection against corrosion over a very long period. By copolymerization with ethylene and hexafluoropropylene, an elastomer is obtained.

Polyvinylidene fluoride is marketed under the name 'Kynar' by the Pennsale Chemical Corp.

8.8.3 Polyvinyl alcohol (PVAL)

a. History. Polyvinyl alcohol was first produced by Hermann and Haehnel in 1925. The method used by them consists in the interchange of the ester radicals of polyvinyl acetate and alcohol:

$$\left[\begin{array}{c} -CH_2-\underset{\displaystyle O-\overset{\displaystyle |}{\underset{\displaystyle O=C-CH_3}{}}}{\underset{|}{CH}}- \end{array}\right]_n + nROH \longrightarrow \left[-CH_2-\underset{\displaystyle OH}{\underset{|}{CH}}-\right]_n + n\left(CH_3-C\begin{matrix}{\nearrow O}\\{\searrow OR}\end{matrix}\right)$$

Direct synthesis is not possible because the monomer, vinyl alcohol, is not stable; it will convert directly into acetaldehyde.

In Germany, polyvinyl alcohol was manufactured some time before 1945 by I.G.-Farben and Wacker-Chemie (product trade names, 'Mowility' and 'Polyviol' respectively). A large part of the output was used as an emulsion for textile treatment and adhesives. Important applications were highly durable antistatic hammers (used in the metal-working industry) and antistatic drive belts ('Drawerit'). Polyvinyl alcohol was placed on the market by Du Pont Corp. in the United States in 1939.

b. Structure. The usual method of preparing polyvinyl alcohol is by the hydrolysis of a polyvinyl ester, generally polyvinyl acetate. The molecular weight and structure of the starting material thus contribute to the development of the properties of the polyvinyl alcohol produced. Variations in the types of catalysts and amounts used, and the reaction conditions, can influence the degree of hydrolysis obtained and, as a result, a wide range of PVAL grades is available. The following three grades are distinguished on the basis of molecular weight and the concentration of solution in water in which they are offered: 'low viscous' (mol. wt. approx. 30000 and concentration 10–15%); 'medium viscous' (mol. wt. approx. 130000 and concentration 15–20%); and 'high viscous' (mol. wt. approx. 200000 and concentration 20–30%). The commercial grades are usually about 88% or 99–100% hydrolysed. The fully hydrolysed polymer will only dissolve in hot water; polymers with about 88% hydrolysis will dissolve in cold water. This behaviour is due to the increase in the number of hydrogen bridges as the degree of hydrolysis increases, as a result of which hydration of the polymer is progressively inhibited. A similar behaviour is observed in cellulose which, in spite of the great number of hydroxyl groups, is insoluble in water.

c. Properties and applications. Polyvinyl alcohol is a white powder, soluble in warm water; partially hydrolysed PVAL is soluble in cold water. The solution concentration which can be obtained depends on the molecular weight and the degree of hydrolysis. PVAL does not dissolve in simple alcohols, esters, hydrocarbons, ketones, oils and fats. Gels are obtained with glycol, glycerol and amides. Owing to its solubility and structure, PVAL is an excellent emulsifier and dispersion agent. As such it is finding applications in two-phase systems of various types–for instance in the emulsion and suspension polymerization of many different monomers, as protective colloids, in paints, pharmaceuticals and cosmetics, and in textiles.

The polar hydroxyl groups in PVAL allow complex-forming and cross-linking reactions to be carried out, so that fibres and films can be made water-insoluble. The products have excellent resistance to ultra-violet light radiation and show outstanding resistance to ageing. Being hydrophilic, PVAL fibre is very suitable for making underwear, blankets, clothing etc. It is also used for surgical goods (Synthofil).

Chemicals can be packaged in water-soluble PVAL sacks (herbicides etc.). The film-forming character of PVAL, together with its adhesive properties and high tensile strength, have opened up extensive applications in the field of adhesives and binders. Water-insoluble and water-sensitive products can be made by modifications of the polymer, a valuable feature for coatings on paper, fabrics and metals.

The polymer is not attacked by hydrocarbons, oils and greases, whilst most gases (except water vapour and ammonia) diffuse only very slowly through it. This has led to the production of PVAL petrol tubes and petrol pump diaphragms. PVAL is also used for impregnating seals resistant to oil, grease and petrol. Leather-like materials are produced by the addition of plasticizers.

d. Production. PVAL is produced by the interchange of ester radicals, by means of hydrolysis or saponification from a polyvinyl ester, usually polyvinyl acetate (PVAC). The polyvinyl acetate used is prepared by emulsion or suspension polymerization, generally with polyvinyl alcohol as an emulsifier or dispersion agent, in basic or acid media; the PVAL produced differs accordingly. In the basic medium, the product consists of a mixture of PVAL and PVAC; if an acid is used, a PVAL/PVAC copolymer is obtained. A drawback of the acid hydrolysis is the presence of acid residue, for instance sulphate, which is undesirable when PVAL is modified to polyvinyl acetals.

The interchange of ester radicals from PVAC to PVAL usually takes place in methanol under the influence of an acid or basic catalyst. The PVAL is insoluble in the medium and is precipitated. It is then filtered off, washed and dried.

PVAL is often characterized by two figures placed after the commercial designation: one of these gives the K value and is thus an indication of the molecular weight, while the other is the hydrolysis percentage.

e. Copolymers. PVAL is basically a copolymer of vinyl alcohol. Only completely hydrolysed PVAL can be regarded as a homopolymer. If copolymers of vinyl acetate are copolymerized with other monomers (vinyl chloride, acrylates, etc.) and then hydrolysed, copolymers of vinyl alcohol with these co-monomers are formed. Treatment of polyvinyl alcohol with aldehydes gives polyvinyl acetals, which can also be regarded as copolymers if partially hydrolysed PVAL has been used as a basis, or if partial acetalising has taken place. Of commercial importance are primarily the polyvinyl acetals, in particular polyvinyl formal and polyvinyl butyral.

f. Commercial products. Like many plastics, PVAL is marketed in many countries and under a variety of trade names, e.g.: Elvanol (USA); Alcotex, Pevalon (UK); Polyviol (Germany).

8.8.4 Polyvinyl esters

a. History. In 1912 vinyl acetate was obtained by Klatte as a byproduct during the preparation of ethylidene acetate from ethyn (acetylene) and acetic acid:

$$CH{\equiv}CH \xrightarrow{CH_3COOH} \underset{\text{vinyl acetate}}{CH_2{=}CH{-}OCOCH_3} + \underset{\text{ethylidene acetate}}{CH_3{-}CH(OCOCH_3)_2}$$

By taking steps to prevent the addition of the second molecule of acetic acid, it was possible to increase the output of vinyl acetate. In 1916 polyvinyl acetate (PVAC) was put on the market. About 1930 substantial production of polyvinyl acetate and its copolymers began roughly at the same time in America and Germany.

The introduction of the ester groups has a softening effect, and in the case of polyvinyl acetate this is so pronounced that the polymer is unsuitable for the fabrication of articles. Higher ester groups enhance the softening. The only lower ester, polyvinyl formate, cannot be prepared in a simple way. The following monomers have become important: vinyl acetate, vinyl chloroacetate (fairly hard film), vinyl propionate and vinyl stearate.

b. Structure. PVAC has a normal head-to-tail structure:

$$\text{--}CH_2\text{—}\underset{\underset{O=C-CH_3}{|}}{\underset{|}{\overset{}{CH}}\atop O}\text{—}CH_2\text{—}\underset{\underset{O=C-CH_3}{|}}{\underset{|}{CH}\atop O}\text{—}CH_2\text{—}\underset{\underset{O=C-CH_3}{|}}{\underset{|}{CH}\atop O}\text{--}$$

The polyvinyl acetate radical is very reactive, so that linkage transfer takes place during the polymerization. The branches thus introduced can arise at two points: at the carbon atom bonded to the oxygen in the chain, and at the acetate group.

For certain applications, e.g. enamel varnishes, a branched product is necessary; in particular, vinyl chloroacetate and vinyl benzoate give strongly branched products.

c. Properties and applications. PVAC is too elastic and too soft to be used for the fabrication of articles. This applies even more to the higher esters. However, the polyvinyl esters form film easily and adhere remarkably to many materials. PVAC is by far the most important among the group of polyvinyl esters.

PVAC is a colourless, odourless plastic with good resistance to light. Its specific gravity is 1.18. Its softening point depends on the molecular weight, being about 80°C for low-molecular and about 150°C for high-molecular products. PVAC is soluble in methanol, acetone, esters, benzene, toluene and methyl chloride. It is insoluble in water, petroleum spirit and ether, cyclohexanol and glycol.

The molecular weight is indicated in commercial products by the K value. The properties of PVAC are adjusted by the addition of plasticizers (up to 25%). An internally plasticized product is obtained by copolymerization with vinyl butyrate or vinyl stearate.

Their properties make polyvinyl esters, especially PVAC, outstanding basic materials for the preparation of lacquers, paints, adhesives and textile auxiliaries. Dissolved in volatile solvents, PVAC is the film-forming constituent of dipping and spraying lacquers. Another important application is in emulsion paints. On drying, PVAC is converted to a strongly homogeneous and water-resistant film which, however, is able to bind water and then to expel it.

Solutions of PVAC in volatile solvents are used as quick-drying adhesives. Dispersions of PVAC are used as industrial adhesives, for instance for joining together wood, leather, artificial leather and paper. The latex obtained by an emulsion polymerization of vinyl acetate with the aid of polyvinyl alcohol as the emulsifier, is here particularly useful. The textile industry uses PVAC in finishing operations.

d. Production. The most important method of producing vinyl esters consists in the addition of carboxylic acids to acetylene. Thus, vinyl acetate is obtained by heating acetylene with acetic acid in the presence of mercury, cadmium or zinc salts, a temperature above 70°C being chosen so as to prevent the formation of ethylidene acetate:

$$CH{\equiv}CH + CH_3COOH \longrightarrow CH_2{=}CH{-}OOCCH_3$$

A reaction in the vapour phase is much used. The catalyst is finely divided zinc acetate applied to a support of active carbon. The reaction temperature is 170° to 200°C.

Another, less commonly used method is by the addition of acetic acid to ethylene oxide, followed by the removal of water:

$$\underset{\diagdown O \diagup}{H_2C{-}CH_2} + CH_3{-}COOH \longrightarrow HOCH_2CH_2OOCCH_3 \xrightarrow{-H_2O} CH_2\,CH{-}OOCCH_3$$

Certain other vinyl esters can be obtained, by the interchange of ester radicals, from acetate:

$$CH_2{=}CHOOCCH_3 + HOOCR \longrightarrow CH_2{=}CHOOCR + HOOCCH_3$$

In this way it is possible to prepare vinyl formate, which cannot be obtained directly.

Vinyl acetate has a strongly irritating smell. Its boiling point is 72°C (760 mm). In order to prevent premature polymerization, a small amount of inhibitor (e.g. hydroquinone) is added to stabilize it.

Vinyl acetate can be polymerized by mass, solution, emulsion, or suspension techniques. In mass polymerization the process is interrupted at an early stage with low conversion, in order to avoid obtaining strongly cross-linked products. Using solution polymerization, it is possible to obtain products with low K values, depending on the tendency to chain transfer by the chosen solvent. The polyvinyl acetate thus obtained normally has a lower mean molecular weight than the product derived from emulsion or suspension polymerization. It can be isolated for working up into commercial forms or, by an appropriate choice of solvent and concentration, it is possible to process the solution further directly. Suspension or pearl polymerization is used a great deal for vinyl acetate. Suspension stabilizers include neutral electrolytes, kaolin, aluminium oxide, barium sulphate, polyvinyl alcohol, starch, water-soluble cellulose derivatives, polyvinyl pyrrolidone and other substances. Catalysts soluble in the organic phase are used as initiators; for instance, organic peroxides, azobisisobutyronitrile. In emulsion polymerization, the initiator is soluble in the water phase and normally is potassium persulphate. PVAC obtained by emulsion or pearl polymerization has a high K value and very regular structure. If desired, the emulsions can be processed directly into further products.

e. Some important copolymers. The copolymers of polyvinyl acetate, as well as the monomer, have long been known. In 1916, F. Klatte was granted a patent for copolymers of vinyl acetate. The possibility of copolymerization, and hence the preparation of products with specific properties, became apparent. The pronounced reactivity of the vinyl ester radical makes it very suitable for the production of copolymers with many other monomers, but the structures of styrene and of vinyl acetate are such that copolymerization of these two monomers is impossible.

Copolymers from among the different vinyl esters themselves can be prepared without difficulty. This is important with respect to the internal plasticizing of polyvinyl acetate by incorporating varying amounts of vinyl butyrate or vinyl stearate. Vinyl formate/vinyl stearate copolymers can be partially hydrolysed, because the formate group hydrolyses very readily, but the stearate group not so readily. PVC can be plasticized by copolymerization with vinyl acetate. The copolymer is soluble in solvents in which PVC homopolymer is not dissolved. Saponification of the vinyl acetate groups yields adhesives and cements which, through their hydrophilic groups, have good adhesion to glass and wood. Continued hydrolysis makes the copolymers insoluble in petroleum spirit, so that they become suitable for pipe connections and seals. Gramophone discs are also made of PVC/PVAC copolymers.

Copolymerization of vinyl acetate with unsaturated acids, e.g. maleic acid and their esters, or hemi-esters, leads to modified products which have improved properties for special applications, e.g. in the textile industry. Saponification of these products introduces hydroxyl and carboxyl groups into one and the same chain. Intermolecular and intramolecular esters, and lactonic linkage, impart greater chemical stability and also greater hardness and brightness to surface layers and films of this material. Used as adhesives, they have a higher strength.

Copolymers of vinyl esters with olefins, e.g. ethylene, are soft. This type of plastic is suitable for making flexible products, such as hoses, packaging, and films. Rubbers can be made by cross-linking these copolymers. Saponification of the copolymers gives a water-insoluble polyvinyl alcohol/ethylene copolymer, even if the ethylene percentage is quite small.

Acrylic esters are copolymerized with vinyl acetate to give internally plasticized PVAC; the higher acrylic esters give the best results. A small quantity of vinyl acetate in an acrylonitrile polymer considerably improves the dyeability of acrylonitrile fibres. Certain copolymers of vinyl acetate, e.g. those with vinylidene cyanide, and a terpolymer of vinyl acetate with acrylonitrile and vinylpyridene, are promising synthetic fibres.

f. Commercial products. Polyvinyl esters are marketed in many industrial countries under a variety of names. These include: Mowilith, Vinnapas (Germany); Elvacet, Polyco (USA); Synresyt (Holland). PVC/PVAC copolymers include: Vinoflex (Germany); Vinyon (USA); Viplavil (Italy).

A PVC/polyvinyl propionate polymer 'Propiofan' is made in Germany, and adhesives based on polyvinyl esters are available in Germany and USA.

8.8.5 Polyvinyl acetals

a. History. The first patent for the production of polyvinyl acetal dates back to the period 1924-26. In Canada these patents were taken out in the name of Shawinigan Products Corp., and in Germany in the name of Consortium für Electro-Chemische Industrie and I.G.-Farben. Commercial production was developed in the following few years and after 1930 the work tended to concentrate on applications.

The use of polyvinyl butyral for making safety glass was patented in 1938 and this has since become one of the most important applications, certainly in America. The next important consumer of polyvinyl acetals is the lacquer and paint industry (for wash-primers). In other fields the polyvinyl acetals have made little headway because of their relatively high price.

b. Structure. Acetals are produced from aldehydes and alcohols under the influence of an acid catalyst, e.g. hydrochloric acid. An aldehyde can react either with one (hemi-acetal) or two alcohol groups (acetal). Both hydroxyl groups can form part of one molecule, so that polyvinyl alcohol can be modified with aldehydes to give polyvinyl acetals:

$$\left[-CH_2-\underset{\displaystyle OH}{\underset{|}{CH}}-CH_2-\underset{\displaystyle OH}{\underset{|}{CH}}-\right]_n + RCHO \overset{H^+}{\rightleftharpoons} \left[-CH_2-\underset{\displaystyle O}{\underset{|}{CH}}-CH_2-\underset{\displaystyle O}{\underset{|}{CH}}-\right]_n \text{(O–CH(R)–O bridge)} + H_2O$$

The reaction tends to take place with two neighbouring hydroxyl groups, so that intermolecular acetalizing (that; is the reaction of the aldehydes with two different molecule chains), and therefore cross-linking and gelling, are unlikely. Only at a higher degree of acetalizing (over 80%) must this be taken into account. The acetalizing is reversible, so that polyvinyl acetals can be decomposed by acids. They show better resistance to alkalis.

c. Properties and applications. The polyvinyl acetals, like the polyvinyl esters, are derivatives of polyvinyl alcohol. They generally have a higher softening point than the polyvinyl esters, so that they show better the true characteristics of a plastic. The properties can be varied within wide limits because the character of the polyvinyl acetal is determined, not only by the average molecular weight, but also by the aldehyde used and the ratio between the hydroxyl, ester, and acetal groups. Polyvinyl esters with different degrees of saponification can be used as raw material. Copolymers of vinyl esters with other, non-saponifiable vinyl monomers, such as vinyl chlorides and ethylene, can be used; and it is possible to acetalize with various aldehyde mixtures or with ketones. The formulation of the polyvinyl acetal can be influenced by the choice of reaction conditions and reaction time. From this it follows that a wide range of polyvinyl acetals with quite different properties can be produced.

The polyvinyl acetals can be mixed with a great number of plasticizers but the addition of plasticizers is in most cases unnecessary because plasticizing groups can be incorporated by the choice of the aldehyde component or the aldehyde mixture (internal plasticizer). Polyvinyl acetals are themselves sometimes added by way of plasticizers or softening components, for instance to phenol formaldehyde and urea formaldehyde resins. However, for certain applications a separate plasticizer must be added to polyvinyl acetals.

The commercial polyvinyl acetals are white or light yellow powders. For the polyvinyl butyrals the Vicat softening point (see section 8.20.2 *f.*) lies between 65 and 130°C, varying with the substituent ratio. The higher the K value and the lower the degree of acetalization, the higher the softening point. The softening point drops whenever the hydrocarbon residue from the aldehyde used for the acetalizing rises. Polyvinyl acetals have outstanding light stability and adhere particularly well to most materials, properties which make polyvinyl butyral highly suitable for safety glass.

The composition of polyvinyl acetal has a great influence on its solubility. The normal commercial polyvinyl acetals are soluble in alcohols, glycol, ethers, ketones, esters and chlorinated hydrocarbons. Most polyvinyl acetals are not attacked by petrol, oils and greases. They are therefore used for oil and petrol pipes and for seals which must be resistant to hydrocarbons. Polyvinyl formal, in particular, is used in petrol-resistant lacquers for petrol tanks.

The principal use of polyvinyl formal (PVFM) is for plasticizing the phenol formaldehyde resins used as an insulation varnish for electric wire, so that the wire can be wound without damage to the insulation. Acetaldehyde and propionaldehyde are used infrequently for acetalizing. The next aldehyde in the series, butyraldehyde, gives polyvinyl butyral (PVB), which is by far the most important representative of the class of polyvinyl acetals. It is used for lacquers, wash primers, safety glass, adhesives, textile processing, and moulding compounds.

The adhesion and light fastness of polyvinyl acetals make these plastics a valuable lacquer base. In solution they can be applied by brushing, spraying or dipping. Without solvent they can be used in the flame spraying process. With phenol formaldehyde, or urea formaldehyde, they give flexible wire insulation, and thermosetting lacquers which are resistant to weak alkalis. Owing to their film-forming capacity, polyvinyl acetals are suitable for making protective films which, after service, can be readily removed from the substrate.

Especially in America, safety glass is made from glass panels cemented together with a film of polyvinyl butyral between them. Before 1936 cellulose derivatives were used for this purpose, but afterwards were completely supplanted by PVB. The formulation of polyvinyl butyral film has been the subject of many research projects, as a result of which a product was found which does not become yellow, has outstanding adhesion to glass, is impact-resistant, and withstands both high and low temperatures without any significant change in its properties. Polyvinyl butyral can be cast on to the glass or be applied to it in the form of a film. The cementing takes place under heat at a high pressure. The films can naturally be used for many other applications, possibly mixed with other plastics.

A small proportion of the total production of polyvinyl acetal is used for special metal adhesives, adhesive tape, household sponges, peel-off films for water proofing of textiles, brake lining adhesives, and articles which must be resistant to petrol. However, these applications are unimportant compared with those described above. The high price of polyvinyl acetals is partly responsible for this.

d. Production. The starting material for the production of polyvinyl acetals is polyvinyl alcohol or polyvinyl ester. In the first case there is direct acetalizing to the hydroxyl groups. The polyvinyl esters are saponified and at the same time acetalized. This process can be readily controlled because the acetalizing takes place at a higher rate than the hydrolysis of the ester groups. There are therefore practically no free hydroxyl groups present during the reaction. Also, partially saponified PVAL is used, so that the degree of saponification of the starting product, and the degree of acetalization of the polyvinyl acetal, can be accurately controlled. The acetalizing takes place in solution or emulsion. Normally, mineral acids are used as catalysts, but organic acids and acid salts are also reported in the literature.

Where a polyvinyl acetal with non-acetalized hydroxyl groups is to be prepared, a solvent (e.g. water) is chosen in which PVAL is dissolved readily and polyvinyl acetal not at all. After a certain time, during which the reaction takes place in a homogeneous medium, the product contains so many acetal groups that it dissolves no longer and precipitates. Complete acetalizing is thus prevented. The process can also be started the other way round, in an organic medium in which PVAL is insoluble. When a certain degree of conversion is reached, the product goes into solution and can then be completely acetalized in a homogeneous phase. The condensation product, water, which forms during acetalizing, can be distilled off azeotropically. The end product is precipitated from the solution with the aid of a non-solvent for polyvinyl acetal, e.g. water.

e. Commercial products. Polyvinyl formal is offered under various trade names: Formadue (Germany); Formex, Formvar (USA); as is polyvinyl butyral: Alvar, Butvar (USA); Mowital, Trosifol (Germany) etc.

8.8.6 Polymethylvinylketone and related polymers

a. History. Methylvinylketone has been known since 1906. It is a specially reactive monomer and polymerizes rapidly. The homopolymer has a softening point of about 60°C and is of no practical importance. The copolymerization parameters suggest that methylvinylketone can be copolymerized with many other monomers. By this means methylvinylketone has found some use.

The toxicity of methylvinylketone is one reason for its limited use. Phenylvinylketone is somewhat less toxic owing to the lower vapour tension and the presence of the aromatic nucleus. Methylisopropenylketone offers the same physiological hazard but gives a homopolymer with a softening point at about 150°C. This is due to the additional methyl groups, which supply molecular rigidity.

b. Structure. In common with most of the other vinyl polymerizates, polymethylvinylketone has a head-to-tail structure:

$$-CH_2-\underset{\underset{CH_3}{|}}{\underset{C=O}{|}}{CH}-CH_2-\underset{\underset{CH_3}{|}}{\underset{C=O}{|}}{CH}-$$

c. Properties and applications. The low softening point and the poor stability of polymethylvinylketone make this polymer unsuitable for use as a true plastic. However, the charge distribution and resonance potential in the molecule make it a good copolymerization partner for many other vinyl monomers, such as butadiene, styrene, acryl monomers, vinyl ethers and vinyl esters; but it has been found that the properties of such copolymers are not different from those obtainable by other co-monomers which are less toxic than methylvinylketone.

Methylvinylketone copolymers have found a limited field of application in lacquers, insulation material and synthetic rubber ('Buna-K'). The dyeing capacity of acrylic fibres has been improved by incorporation of methylvinylketone.

Polyisopropenylvinylketone has a softening point of about 150°C. Polyphenylvinylketone also has a relatively high softening point and its properties may be modified by introducing substituents at the aromatic nucleus. Transparent plastics (glasses) are made from these polymers.

d. Production. The monomer, methylvinylketone, can be prepared by the addition of water to the acetylene dimer, vinylacetylene:

$$\underset{\text{vinylacetylene}}{CH_2{=}CH-C{\equiv}CH} + H_2O \rightarrow \underset{\text{methylvinylketone}}{CH_2{=}CH-\underset{\underset{O}{\|}}{C}-CH_3}$$

Another method, by which isopropenylmethylketone can also be produced, consists in the removal of water from the appropriate ketone:

$$CH_3-\underset{\underset{O}{\|}}{C}-CHR-CH_2OH \xrightarrow{-H_2O} CH_3-\underset{\underset{O}{\|}}{C}-CR{=}CH_2 \quad (R = H, CH_3)$$

Similarly, vinylketones can be obtained by the removal of halogen hydrogen on halogenized ketones.

Methylvinylketone is a clear liquid (b.p. 81°C) with a penetrating odour and toxic action: it causes eye-watering and attacks the mucous membranes, seriously damages the skin and affects the kidneys, even at low concentrations. Methylvinylketone is soluble in most organic solvents. The presence of water has no influence on the monomer. It is highly reactive, and at room temperature can be kept only for a few weeks, when it must be protected against light. Injudicious storage gives rise to formation of dimers. Cooling to a low temperature increases its shelf life considerably. Hydroquinone and glacial acetic acid are used for stabilizing the monomer.

Methylisopropenylketone has the same properties. Its boiling point is a little higher, at 98°C.

Phenylvinylketone, b.p. 118°C, is somewhat less volatile. This, together with the presence of an aromatic nucleus, lessens the toxicity.

Mass, suspension or emulsion polymerization is possible for vinylketones. Initiation takes place by inorganic or organic peroxides, depending on the polymerization process used. Polymerization is also induced by light, heat and oxygen.

8.9.7 Polyvinyl ethers

a. History. Methylvinyl ether was obtained in 1891 by Favorski by addition of methanol to acetylene and afterwards developed by Reppe to an industrial process. The interest centred on polymers with similar properties to those of the vinyl esters but with more difficultly saponifiable substituents in the chain. As was to be expected, the choice fell on the polyvinyl ethers, which appeared in 1928.

Apart from some special cases, the vinyl esters can only be polymerized by cationic initiation and propagation mechanisms, because of the high negative polarity of the double bond for the different vinylalkyl ethers. With certain other vinyl monomers, copolymerization is possible by the radical mechanism, for instance with vinyl chloride, acrylates and methacrylates.

The search for suitable initiators brought to light a further characteristic of polyvinyl ethers. Depending on the initiator system used, two structural modifications are possible. In 1947, Schildknecht and others discovered that the use of BF_3-etherates leads to a stereospecific, and possibly isotactic polymer, while BF_3, supplied as gas, initiates the formation of an atactic polymer. The properties of the two modifications differ, the isotactic polyvinyl ether being capable of crystallization. In the Second World War, polyvinyl ethers were, in fact, produced in Germany with the aid of different initiator systems, but the difference in the properties of the products (oil, wax, or solid) is explained by the difference in molecular weight or by the presence of chain branching.

b. Structure. Polyvinyl ethers can be produced in two stereospecific modifications, depending on the polymerization conditions:

$$-CH_2-\underset{\displaystyle OR}{\overset{\displaystyle H}{\overset{|}{\underset{|}{C}}}}-CH_2-\underset{\displaystyle OR}{\overset{\displaystyle H}{\overset{|}{\underset{|}{C}}}}- \quad (\text{I}) \qquad \text{and} \qquad -CH_2-\underset{\displaystyle OR}{\overset{\displaystyle H}{\overset{|}{\underset{|}{C}}}}-CH_2-\underset{\displaystyle H}{\overset{\displaystyle OR}{\overset{|}{\underset{|}{C}}}}- \quad (\text{II})$$

The isotactic polymer (I) arises during slow polymerization at −60°C initiated by a BF_3-ether complex. This polymer is capable of crystallization; it is then harder and has a lower specific gravity than polymer (II). Modification (II) is produced with aid of gaseous BF_3 as the initiator, also at −60°C. Under identical conditions the molecular weight of polymer (I) is lower than that of polymer (II): 200000 compared with 500000. It has been shown in a number of ways that the steric configuration, rather than branching or different molecular weight, is the reason for the difference in the properties of the two types.

c. Properties and applications. Polyvinyl ethers are oils, waxes or rubbers. Owing to the plasticizing effect of the ether substituents, they are not suitable for the fabrication of shapes, so that they cannot be counted as true plastics.

Polyvinyl ethers are not saponifiable or resistant to dilute acids, salt solutions and dilute or concentrated alkalis. Polyvinylmethyl ether is soluble in cold water, though if the temperature is increased above 30°C the polymer is precipitated from the solution; but the higher polyvinylalkyl ethers are insoluble.

The polyvinyl ethers normally dissolve readily in organic solvents, the solubility increasing with the number of carbon atoms of the alkyl residue. An important property for industrial applications is the compatibility of the polyvinylmethyl ether with other plastics. The high-molecular weight prevents exudation when used as a plasticizer, so that the properties of the mixture do not change with time. Polyvinyl ethers have good adhesion to many materials. They are therefore used, either alone, or in combination with other polymers, for adhesives in the metal and leather industries and for making adhesive plaster and tape. The adhesion of lacquers is improved by the addition of polyvinyl ethers; at the same time they act as plasticizers, so that the lacquer film becomes less brittle. Emulsion paints frequently contain polyvinyl ethers as a component.

The higher polyvinyl ethers, such as polyvinyloctadecyl ether, are of a waxy consistency, and are used in the manufacture of furniture and floor polishes. Polyvinylethyl ether and polyvinylisobutyl ether form the basis of chewing gum. Dentists' moulding materials are made from polyvinyl ethers. These two applications require a very pure product known as 'type K'.

Polyvinylmethyl ether is used for the impregnation of high-tension cable insulation.

Polyvinyl ethers are used, mainly in Britain and Germany, for strippable coatings. Heated articles are dipped into the solution of polyvinyl ether so that a film forms on the surface, giving it temporary protection. Subsequently this film can easily be stripped off. Polyvinylmethyl ether is used as gelling agent in the application of neoprene foam as a carpet backing.

d. Commercial products. Some well-known trade names of polyvinyl ethers include: Garbel, Gedovyl (France); Lutonal (Germany) and Synresine (Netherlands).

The type is normally indicated by a figure giving the K-value, followed by a letter and possibly M for polyvinylmethyl ether, A for the ethyl homologue, and I for the isobutyl homologue.

e. Production. The best-known processes are the removal of hydrogen chloride from chlorinated dialkyl ethers, the reaction between vinyl halides and alkali alcoholates, and the removal of alcohol in acetals. The last method has been in use for some time.

With acetylene and the appropriate alcohol as starting materials the reaction is:

$$HC{\equiv}CH + 2ROH \longrightarrow CH_3{-}CH(OR)_2 \longrightarrow CH_2{=}CH{-}OR + ROH$$

The removal of alcohol takes place at a temperature between 200° and 400°C and under the influence of a metal catalyst, normally finely divided platinum or palladium.

Alcohol vinylization, following Reppe, is frequently used at present. This is a versatile method, which can be used for many hydroxyl compounds. Acetylene, diluted with nitrogen, is brought in contact with the alcohol at a pressure of 2 to 20 atm, depending on the volatility of the end product. The reaction is catalysed by caustic potash, but many other catalysts are equally practicable. The reaction can be carried out continuously in a vinylization tower. It is described by the following equations:

$$HC{\equiv}CH + ROK \rightarrow K{-}CH = CH{-}OR$$
$$K{-}CH{=}CH{-}OR + ROH \rightarrow CH_2 = CH{-}OR + ROK.$$

The vinyl ethers are soluble in many solvents, e.g. propane, dichloroethane. They are not readily soluble in water. Vinylmethyl ether is a gas at room temperature (b.p. 6°C). Vinylethyl ether boils at 36°C. The boiling point of the higher homologues rises steadily and for vinyl-*n*-butyl ether it is 94°C. They are colourless liquids. The odour is not unpleasant and they are not harmless to the mucous membranes.

Polymerization of the vinyl ethers takes place by the cationic mechanism. The excess charge at the bouble bond causes the monomer to be highly reactive to electrophilic reagents. Friedel-Crafts catalysts, such as BF_3, $AlCl_3$ are therefore very appropriate initiators. It is likely that co-catalysts, e.g. traces of water, here play a part by forming complexes such as $(BF_3OH)^-H^+$. The proton becomes attached to the double bond and forms a CH_3-CH(OR) cation, which initiates the polymerization. Iodine, sulphur dioxide and acidic substances such as active carbon impregnated with phosphoric acid can act as initiators.

Vinyl ethers are polymerized by mass or solution polymerization. Mass polymerization is carried out in an autoclave, a small quantity of the reactants being introduced at first. After the polymerization has started the remainder of the monomer and the catalyst are added continually. Polymerization conditions determine the type of product.

In the solution method two separate solutions of monomer and initiator in propane at −60°C are fed into an airtight chamber where they combine to give a single stream of polymerization solution. The propane evaporates by the heat of polymerization and is abstracted. The polyvinyl ether is removed, purified and dried. Continuous production is thus possible.

f. Some important copolymers. Polyvinyl ethers can easily be made to copolymerize with the aid of cationic initiation.

Radical polymerization, in the mass, solution, or emulsion, of vinyl ethers with unsaturated acids, esters or anhydrides, gives products which are commercially important as protective colloids or thickeners, for the textile and leather industries. A particularly important example is the copolymer of vinylisobutyl ether and various acrylates. Emulsion copolymers and terpolymers with styrene and acrylic acid can be used directly as colloids and thickeners, also for paper coating.

The radical polymerization of vinyl ethers with dibasic α, β unsaturated carboxyl acids, such as maleic acid, or with the ester and anhydrides derived from them, has some theoretical interest, since while neither the vinyl ethers nor the specified acids and their derivatives can separately be made to polymerize with radical initiation, no difficulty arises with copolymerization.

8.8.8 Acrylics

a. History. The polymerizates of acrylic acid, $CH_2{=}CHCOOH$, methacrylic acid, $CH_2{=}C(CH_3)COOH$, and their derivatives such as acrylonitrile, $CH_2{=}CHCN$, are known collectively as acrylics. The most important derivatives are the esters, nitriles, and amides.

Acrylic acid has been known for over a hundred years and its polymerization was discovered and studied in the last century. However, the history of acrylics begins properly in 1901; in that year Röhm in Tübingen made known the results of an investigation of the preparation and properties of acrylic acid esters. Röhm, together with Haas, built a plant for the production of acrylates. Apart from the laboratories of Röhm and Haas in Germany and America, important work in the development in this field was carried out by du Pont, ICI and I.G. Farbenindustrie.

This group of plastics has become widely known in the manufacture of transparent organic 'glass'. Large quantities of acrylics and its copolymers are also used in many other applications. Polyacrylonitrile is chiefly known as a synthetic fibre.

b. Structure. The polymethacrylates generally show the head-to-tail structure, such as is found in most vinyl polymers.

$$\begin{array}{cccccc} & CH_3 & & CH_3 & & CH_3 \\ & | & & | & & | \\ -CH_2- & C & -CH_2- & C & -CH_2- & C- \\ & | & & | & & | \\ & COOR & & COOR & & COOR \end{array}$$

Some acrylic acid derivatives show also the head-to-tail and tail-to-tail type of polymerization.

Because of the methyl groups, the chains of methacryl polymers are more rigid than those of acrylates. The softening point of polymethylmethacrylate lies accordingly about 90 degrees higher than that of polymethylacrylate (90°C and 0°C, respectively). The lengthening of the alkyl component of the ester group has a softening effect, which is reflected more clearly in the methacryl polymers compared with the acryl polymers: the softening points of the octyl esters are, respectively, −20°C and −60°C. On further lengthening of the alkyl groups, the hydrocarbon character dominates. The softening point of the polyacrylates rises rapidly and that of the polymethacrylates more slowly.

The physical properties, e.g. melting point and solubility, are largely controlled by the secondary valency forces present between the polar chain substituents. In polyacrylic acid, polymethacrylic acid and the amide derivatives, these forces are so great that the polymers decompose on heating as the softening point is reached. They cannot therefore be deformed thermoplastically. Polyacrylonitrile also shows a special behaviour. The α hydrogen atom is so activated by the CN group that very strong hydrogen bridges arise between them. Polyacrylonitrile, therefore, has a very high softening point (250° to 330°C) and is only soluble in certain solvents (e.g. dimethylformamide). In polymethacrylonitrile the α hydrogen atom is lacking, so that hydrogen bridges cannot occur. The presence of an active hydrogen atom can easily lead to chain transfer, resulting in branched polymers. This does not arise in the polymerization of methacryl compounds unless polymerization is carried too far, e.g. with a molecular weight of over 100 000. In mass polymerization this may even lead to cross-linking and gelling of the whole contents of the polymerization reaction vessel.

An undesirable behaviour of the highly reactive acryl monomers is 'popcorn' polymerization: the monomer, methylacrylate is particularly susceptible. Pipelines and taps can thus become stopped up and even damaged. The emergence of popcorn is a result of the high polymerization activity of methyl acrylate. In the methacrylates and the higher acryl esters it is encountered less frequently. Popcorn is a porous grain of polymer. It usually has an extremely high molecular weight and occurs under conditions which lead to an extremely low rate of polymerization. Popcorn does not arise in a normal polymerization process. It acts as a nucleus for further polymerization and then rapidly increases in size. The reasons for popcorn polymerization are usually complete absence of oxygen, excessive storage temperature, or contamination with traces of iron or water.

The presence of the two substituents, $-CH_3$ and $-COOCH_3$, at one carbon atom of methylmethacrylate leads to stereoregularity in the polymethylmethacrylate. If radical initiation is used, a predominantly syndiotactic polymer is obtained at room temperature or above, but with many irregularities. This type is therefore called atactic. Polymerization at lower temperatures (e.g. −40°C, initiated by ultra-violet radiation) yields syndiotactic polymethylmethacrylate. Anionic initiation at a low temperature (−70°C) in polar solvents leads also to a syndiotactic product, but in an apolar solution and at about 0°C isotactic structure is obtained. The different structures have different associated properties. The derivatives of acrylic acid which have a voluminous substituent, e.g. N-substituted acrylamides and acrylates of branched alcohols, show such behaviour. By the hydrolysis of a stereospecific polymethacrylate it is possible to obtain the corresponding stereospecific polymethacrylic acid.

Using stereospecific catalysts it is possible to polymerize α-ethylacrylonitrile and higher homologues, which with the conventional initiators show no tendency to homopolymerization, to give rapidly crystallizing products. They are dissolved by organic solvents, are resistant to strong acids, and melt at 200° to 300°C. They can be deformed plastically.

Cross-linked acrylics can be obtained in several ways; by copolymerization with bifunctional monomers, or by cross-linking reactions at the chain substituents.

c. Properties and applications. The chemical composition of polyacrylic acid, as the basic polymer of the whole class of acrylics, permits the production of many derivatives. The presence of the carboxyl group (—COOH) in polyacrylic acid enables it to be readily modified. The esters, amides, and nitriles have proved to offer the best combination of properties for the plastics industry. Acryl monomers polymerize rapidly, and copolymerization with most other vinyl monomers takes place very easily. Depolymerization (breakdown of the polymer molecules) on other polymers yields products with specific properties. The method used for the polymerization, whether in solution, suspension or emulsion, likewise influences the properties and performance of the product, so that acrylic polymers and copolymers have a wide range of applications.

Polymethacrylic acid is a hard brittle substance. Plastic deformation is not possible, because when the temperature is increased, the formation of hydride, and therefore cross-linking, occurs. Owing to the presence of many carboxyl groups, polymethacrylic acid adheres very strongly to many surfaces. The carboxyl groups impart to it the character of a polyelectrolyte: if an aqueous solution of the weakly acid polymethacrylic acid is neutralized, the viscosity of the solution increases greatly, owing to ionization of the polymethacrylate formed. The negatively charged chain substituents repel each other, so that the polymer molecule expands and provides a greater contribution to the viscosity. Because of the more flexible chains, polyacrylic acid shows this effect more strongly than polymethacrylic acid. By the addition of a strong acid, dissociation is inhibited and the viscosity of the solution is lowered. Polymethacrylic acid is accordingly used as a thickener. Owing to their insensitivity to earth alkalis, the methacrylic acid copolymers are important in the oil industry. The viscosity and the stability of drilling solutions are adjusted and improved by the addition of polymethacrylic acid copolymers with low molecular weight. In washing powder manufacture, good compatibility with hard water is imparted to the products by copolymerization with methacrylic acid. The salts are used as thickeners in toothpaste, creams, brilliantine, and shampoo. Some few tenths of one per cent of an ammonium salt of polyacrylic acid is added to emulsion paints, to prevent coagulation and ensure uniform distribution of the pigment. Wall paints frequently contain copolymers based on acrylic acid, neutralized with ammonium hydroxide to prevent coagulation. Enamel varnishes prepared with methacrylic acid copolymers or their derivatives show excellent adhesion to metal. The acid groups can be used at the same time as hardening accelerators. An important application of polymethacrylic acid polymers and copolymers is in the field of textile dressing. Acid-type finishing agents are used for polyamide fibres, because they have good adhesion to the amide groups. For other synthetic fibres, e.g. polyacrylonitrile and polyester, salts of polymethacrylic acid or their copolymers with methacrylates are preferred. In the leather industry the pores in the leather are filled with copolymers of methacrylic acid to prevent loss of dye and escape of the softening agents from the leather. More recent uses include the clarification of cloudy liquids, waste water purification, processing of mineral products, ion exchange, and soil improvement. The pharmaceutical industry uses copolymers of methacrylic acid to coat certain drugs. Such coated preparations are insensitive to saliva and stomach acids, but in the alkaline environment of the intestines the coating dissolves and the drugs are released. If a basic methacrylate is used instead of methacrylic acid, the preparation will dissolve prematurely in the stomach.

In 1964 a new class of plastics, the 'ionomers', composed of ethylene and small quantities of co-monomers containing carboxyl groups, e.g. methacrylic acid, was marketed by the Du Pont Co. The chains are cross-linked by electrostatic forces via ions of sodium, potassium, magnesium or zinc. At elevated temperatures these cross-links are temporarily broken so that the ionomer can be plastically deformed by normal techniques. The films made from these plastics are very clear and strong and are not attacked by solvents, oils and greases. The film remains flexible even at very low temperatures (down to about $-100°C$). The ionomers have already found uses as non-porous, very strong, and transparent packaging films, blown bottles, tubes, hoses, and metal-coated articles.

Polymethylmethacrylate is, in the form of 'acrylic glass', the most important and best known representative of the polymethacrylic acid esters. The acrylic glass is made in the form of rod, tubing and sheet by polymerization *in situ*: this means that the polymerization takes place with the simultaneous final shaping of the product. The monomer is partly pre-polymerized, so that a solution of polymer in monomer is obtained, which has the desired viscosity for the casting into the mould. Methylmethacrylate polymerizes under the influence of radiation, heat, or radical-forming substances such as peroxides or azodi-isobutyronitrile. In certain cases, when polymers with a stereo-regular structure are to be produced, anionic polymerization is used; methacrylates do not polymerize by the cationic mechanism. For the fabrication of rod, the compound is poured into aluminium pipes; tubing is made by centrifugal casting. The large polymerization shrinkage of acrylic glass (20 to 25%) facilitates the withdrawal of the product from the mould.

Polymethylmethacrylate sheets are cast between two plates of plate glass or polished metal. Polymethacrylate is itself a poor thermal conductor so that heat accumulation will occur unless the polymerization is interrupted. Isothermal (constant temperature) polymerization is obtained by the addition of regulators, usually mercaptanes, and by controlled cooling of the plates with air or water. Provision must be made for shrinkage of the sheet in the direction of its thickness during polymerization, and for the production of optically perfect acrylic glasses impurities must be absent. To prevent crazing (which would reduce the clarity) and the occurrence of stress corrosion (to which acrylic polymers are susceptible) the product should be annealed. Acrylic glass can be machined readily. Joints can be made by riveting, bolting, welding, or cementing. Cementing is carried out with solvents, polymethacrylate solution, or polymerization bonding agents consisting of monomeric methacrylate solutions to which an initiator has been added. Stresses must be relieved by heat treatment both before and after the cementing. Acrylic glasses can be deformed plastically by heating. The types with a high molecular weight (10^4 to 3×10^6) do not fuse but they become soft and depolymerize at still higher temperatures. The products with lower molecular weights fuse normally and can be processed in extruders and injection moulding machines. The injection moulding and extrusion compounds in polymethacrylate contain a small amount of polyacrylate to improve fluidity. Material flow is less pronounced than in most other plastics – the softening points lie between 180° and 220°C. Fountain pens, rear lights, components of electrical equipment, and many other products can be made in this way. No finishing treatment is necessary.

Acrylic glass is used in applications which can make profitable use of the material's special optical properties, high impact strength even at low temperatures, resistance to weathering, and low density. Articles made from acrylic glass include advertising panels, illuminated sign posts and traffic signs, window panes for buses, aircraft and trains, illumination decorations, car rear lights and reflectors, transparent pipe lines, safety covers for machines, transparent demonstration models, tableware items, etc. For the windscreens of motor vehicles acrylic glass is too soft and easily scratched.

Fig. 8.17 A large dome, 70 ft (21 m) diameter and 35 ft (10 m) high, built from moulded 'Perspex' acrylic sheet.

Dental prostheses are made from pre-polymerized methylmethacrylates. Just as in contact lenses, the smooth surface and physiological harmlessness are important factors for these applications. The high light transmission of acrylic glass is exploited in optical fibres. If a number of filaments are bunched together and illuminated from one end the light is totally reflected from the fibre walls and illuminates any cavities into which it is introduced. The polymers of the higher methacrylic acid esters, e.g. the ethylhexyl and lauryl methacrylates, are being used increasingly as lubricant additives, for the purpose of adjusting the viscosity of oil for particular requirements. Polyacrylonitrile is an important synthetic fibre. In addition, large quantities of acrylonitrile are copolymerized to elastomers and ABS plastics. Synthetic elastomers are mainly copolymers with butadiene (Buna-N, GRA, and nitrile rubber). A small quantity of acrylonitrile, copolymerized with acrylic and methacrylic acid esters, has a favourable influence on their resistance to solvents. Larger contents of acrylonitrile (over 50%) improve the impact strength of the product. The disadvantages are the light yellow colour, the lower softening and vitrification points, and the fact that these copolymers cannot be worked by extrusion and injection moulding techniques.

The ABS plastics (acrylonitrile-butadiene-styrene) have been developed in recent years. As three monomers are present, many different combinations are possible. ABS plastics can be produced by mixing an acrylonitrile/butadiene copolymer with an acrylonitrile/styrene copolymer, but also by grafting acrylonitrile and styrene onto polybutadiene. ABS plastics are rigid and yet impact-resistant, have a hard surface, are resistant to many chemicals, and easy to work by conventional shaping techniques, such as extrusion and injection moulding. They are excellent for metallizing. The above properties make this class of plastic particularly suitable for refrigerator components, tubing for electrical equipment, typewriters, bottles, motor car components etc.

Polymethacrylamides and their copolymers, e.g. with methacrylic acid or other acid monomers, are used as sedimentation agents for the clarification of cloudy fluids, for the preparation of mineral products (e.g. kaolin) and purification of waste water. In paper production they are used for improving the wet strength. This application is based on linking the cellulose chains to the acrylamide molecules via aluminium ions, which are derived from the alum which has also been added. Polyacrylamides are used as dispersion agents in the pearl polymerization of other vinyl monomers. In emulsion polymerization they act as protective colloids.

d. Commercial products. Some well-known trade names of acrylics are: Plexiglas, Acrifix, Degulan (Germany); Diakon, Perspex (UK); Lucite, Creslan, Crofton (USA); Edimet, Crilat (Italy); Carboulon (Japan).

Trade marks for ABS polymers include: Abstrene (UK); Abson, Lustran (USA); Afcolene (France); Novodur, Terluran (Germany); Lastilac (Italy).

e. Production. Acrylic acid and its monomer derivatives are synthesized from ethylene, acetylene and propylene by a number of methods. With the increase in propylene production, its conversion to acrylic compounds is becoming more economic. The following methods are commercially important:

(i) Ethylene oxide is prepared from ethylene either by direct oxidation or by way of chlorohydrin. The ethylene oxide is reacted with hydrocyanic acid to ethylene cyanohydrin. By hydrolysis, acrylic acid is obtained which, if desired, is converted in the same stage to acrylester:

$$\underset{\diagdown O \diagup}{H_2C-CH_2} + HCN \longrightarrow \underset{OH\quad CN}{CH_2-CH_2} \xrightarrow[-H_2O\cdot ROH]{H_2SO_4} CH_2=CH-COOR$$

A direct route to the acrylic acid is possible by the addition of carbon monoxide to the ethylene oxide, catalysed by nickel tetracarbonyl:

$$\underset{\diagdown O \diagup}{H_2C-CH_2} + CO \xrightarrow{Ni(CO)_4} CH_2=CH-COOH$$

(ii) Following Reppe, acetylene is converted into acrylic acid by treatment with nickel tetracarbonyl, water, and acid:

$$4CH{\equiv}CH + Ni(CO)_4 + 4H_2O + 2HCl \longrightarrow 4CH_2=CH-COOH + NiCl_2 + H_2$$

Replacing water with alcohol yields the corresponding acrylate ester. A variation of this method is the addition of carbon monoxide to acetylene, catalysed by nickel halogenides, nickel sulphide, nickel tetracarbonyl or other nickel compounds:

$$CH{\equiv}CH + CO + ROH \longrightarrow CH_2=CH-COOR$$

(iii) From keten and formaldehyde β-propiolactone is obtained:

$$CH_2=C=O + HCOH \longrightarrow \begin{matrix} CH_2 & - & CH_2 \\ | & & | \\ O & - & C=O \end{matrix}$$

This polymerizes very vigorously to a polyester

$$-O-CH_2-CH_2-COO-CH_2-CH_2-CO-$$

From this a very pure acrylic acid is obtained by pyrolysis. The acryl esters are obtained by heating the β-propiolactone with alcohols in the presence of alkyl sulphates as catalysts:

$$\begin{matrix} CH_2 & - & CH_2 \\ | & & | \\ O & - & C=O \end{matrix} + ROH \longrightarrow CH_2=CH-COOR + H_2O$$

Treatment of β-propiolactone with ammonia yields acrylamide.

(iv) A recent method of preparing acrylic acid starts from propylene. This is oxidized catalytically to acrylic acid:

$$CH_2=CH-CH_3 \xrightarrow{O_2} CH_2-CH-COOH$$

(v) Some methods for the preparation of acrylates proceed via the acrylonitrile, which is important in its own right for the polymerization to polyacrylonitrile:

$$\underset{\diagdown O \diagup}{H_2C-CH_2} + HCN \longrightarrow HO-CH_2-CH_2-CN \xrightarrow{-H_2O} CH_2=CH-CN$$

$$CH{\equiv}CH + HCN \xrightarrow[NH_4Cl]{Cu} CH_2=CH-CN$$

$$CH_2=CH_2 \xrightarrow{O_2} CH_3-CHO \xrightarrow{HCN} CH_3-CHOH-CN \xrightarrow{-H_2O} CH_2=CH-CN$$

$$CH_2=CH-CH_3 \xrightarrow{O_2} CH_2=CH-CHO \xrightarrow{NH_3}$$

$$CH_2=CH-CH=NH \xrightarrow{O_2} CH_2=CH-CN$$

The preparation of acrylic acid or acrylester from acrylonitrile takes place by hydrolysis or alcoholysis, e.g.:

$$CH_2{=}CH{-}CN \xrightarrow[H_2O]{H_2SO_4} CH_2{=}CH{-}CONH_2 \,.\, H_2SO_4 \xrightarrow{R.OH} CH_2{=}CH{-}COOR$$

The methacryl monomers are obtained commercially by the following methods:

(vi) Cyanohydrin is formed from acetone and hydrocyanic acid, in a basic medium. By hydrolysis or alcoholysis, by way of the amide, this is converted to methacrylic acid or a methacrylic acid ester:

$$CH_3{-}\overset{\overset{\displaystyle O}{\|}}{C}{-}CH_3 + HCN \xrightarrow{OH^-} CH_3{-}\underset{\underset{\displaystyle CH_3}{|}}{C}(OH){-}CN \xrightarrow{H_2SO_4} CH_2{=}\underset{\underset{\displaystyle CH_3}{|}}{C}{-}CONH_2 \,.\, H_2SO_4$$

$$\xrightarrow{CH_3OH} CH_2{=}\underset{\underset{\displaystyle CH_3}{|}}{C}{-}COOCH_3 + (NH_4)HSO_4$$

(vii) Catalytic oxidation of isobutene yields α-methyl lactic acid which, by removal of water and esterification, is converted to methacrylic acid esters:

$$CH_3{-}\underset{\underset{\displaystyle CH_3}{|}}{C}{=}CH_2 \xrightarrow{O_2} CH_3{-}\overset{\overset{\displaystyle OH}{|}}{\underset{\underset{\displaystyle CH_3}{|}}{C}}{-}COOH \xrightarrow[H_2SO_4]{ROH} CH_2{=}\underset{\underset{\displaystyle CH_3}{|}}{C}{-}COOR$$

The methacryl monomers are highly reactive compounds, which can polymerize in an explosive manner. Precautions must therefore be taken to prevent polymerization during preparation and storage. An effective method when manufacturing it is the addition of phenol to the vessel and a counter current of phenol passing through the distillation column. Large quantities of methacryl monomers must not be kept for longer than a few hours unless an inhibitor is present. Air present in the storage vessel has an inhibiting effect, but the vapour mixtures are highly explosive, so that care must be taken to avoid sparking. The storage vessels can be made in stainless steel or in iron protected by galvanizing or a phenol resin coating. Safety valves must be incorporated.

During storage, hydroquinone, hydroquinone methylether or pyrogallol are added as inhibitors. The methacryl monomers are somewhat less reactive than the acryl monomers, so that less inhibitor is necessary for stabilization. The inhibitor is removed by washing, or the monomer is distilled off. The temperature in the upper part of the cooler must then be under 5°C. As copper has an inhibiting effect, it is recommended as a cooler, and copper salts may be added to the acrylic acid monomer for stabilization. Acrylonitrile is stabilized with amines.

The esters have an irritating effect on eyes and nose. Methacrylic and acrylic acid are fairly strong acids owing to the electron-attracting effect of the vinyl group. Most methacryl monomers are fluid at normal temperatures; acrylic acid is solid below 12.5°C, the methacrylic acid below 15°C. The boiling points of the most common methacryl monomers lie around 100°C (acrylic acid, 141°C; methyl acrylate, 80°C; ethyl acrylate, 99.5°C; methacrylic acid, 106.5°C; methylmethacrylate, 100.3°C; ethylmethacrylate, 116.5°C; acrylonitrile, 77.3°C).

f. Polymerization. The polymerization of methacryl monomers takes place, on a commercial scale, nearly always by the radical mechanism. In special cases, where a particular stereospecific structure is required, the polymerization is anionic. The radicals are supplied by peroxides or azobisisobutyronitrile. In many cases the presence of regulators which determine the molecular weight, e.g. mercaptans, is desirable. Without these chain-transfer additions, the molecular weight would be so high that the product could no longer be worked thermosplastically. Mass polymerization is not generally used in the industry – because the heat of polymerization cannot be conducted away satisfactorily – except for the production of acrylic glass. The parts to be made in this way must be neither too large nor too thick; tubular and flat shapes can be produced satisfactorily. Techniques using solution, emulsion, and suspension are easier to manage. For acryl polymers used as lacquer base, textile auxiliaries, lubricant additives, and the like, the products made by solution polymerization are particularly suitable, as separation for drying is unnecessary. If an aqueous dispersion is required, the polymerization is carried out in an emulsion or suspension. The molecular weights attained are high in this case and the polymerization proceeds rapidly. The polymers can also be obtained as dry powder, or in the form of pearls. Suspension polymerization gives a purer product than emulsion polymerization.

8.8.9 Polyvinyl pyrrolidone (PVP)

a. History. The production of polyvinyl pyrrolidone on an industrial scale was made possible by the acetylene research of J. W. Reppe. The monomer, vinyl pyrrolidone, was prepared in about 1937 from acetylene, formaldehyde, hydrogen, and ammonia and in 1942 it was produced in commercial quantities by I. G.-Farbenindustrie. The most interesting application at the time was the use of 2.5% aqueous solutions ('Periston' and 'Kallidon') as a substitute for blood serum. The stability of the solutions, and their compatibility with all blood groups are valuable features. The polymer, even with a molecular weight up to 500000, has an excellent solubility both in water and in many organic solvents, and this is a cogent reason for the present wide range of polyvinyl pyrrolidone applications. Frequently the polymer plays only the part of an auxiliary, and the various formulations require only small percentages of it. The wide range of applications and the large consumption of products containing PVP make the production of polyvinyl pyrrolidone economically attractive.

b. Production. The monomer, vinylpyrrolidone, is prepared from acetylene, formaldehyde, hydrogen and ammonia. First, butanediol is produced:

$$HC{\equiv}CH + 2H_2CO \longrightarrow HOCH_2{-}C{\equiv}C{-}CH_2OH \xrightarrow{H_2} HO{-}CH_2{-}CH_2{-}CH_2{-}CH_2{-}OH$$

This is converted with the aid of a copper catalyst into the lactone, from which the pyrrolidone is formed with ammonia:

$$HOCH_2{-}CH_2{-}CH_2{-}CH_2OH \xrightarrow[200°C]{Cu} \begin{array}{l} H_2C{-}CH_2 \\ \;|\qquad\;\; O \\ H_2C{-}C{=}O \end{array} \xrightarrow{NH_3} \begin{array}{l} H_2C{-}CH_2 \\ \;|\qquad\;\; NH \\ H_2C{-}C{=}O \end{array}$$

By the addition of acetylene, vinylpyrrolidone is obtained:

$$\begin{array}{l} H_2C{-}CH_2 \\ \;|\qquad\;\; NH \\ H_2C{-}C{=}O \end{array} + HC{\equiv}CH \longrightarrow \begin{array}{l} H_2C{-}CH_2 \\ \;|\qquad\;\; N{-}CH{=}CH_2 \\ H_2C{-}C{=}O \end{array}$$

Vinylpyrrolidone (b.p. 95°C, m.p. 13.9°C) is colourless and soluble in water and many organic solvents. It is resistant to alkalis but not to acids, by which it is decomposed giving off acetoaldehyde, and polymerizes slowly under the influence of light and heat.

Both mass and solution polymerization are possible. The solubility of both monomer and polymer in water excludes the use of emulsion and suspension polymerizations. Mass polymerization has the drawback that the rise in temperature causes the product to become yellow or brown: yet it still contains up to 10% monomer, which must be extracted because of its toxic effect. When polymerized in aqueous solution temperature control is better and the polymer is white and has a narrower molecular weight range. Monomer is practically absent, so that the viscous PVP solution can be further processed as such. PVP can also be obtained by jet drying. Polymerization in aqueous solution is mostly used in industry.

The polymerization is initiated by the redox system, $H_2O_2 + NH_3$. Polymers with molecular weights of about 200000 are thus obtained. Even higher molecular weights can be obtained using azodi-isobutyronitrile as an initiator. The polymer has the following structure:

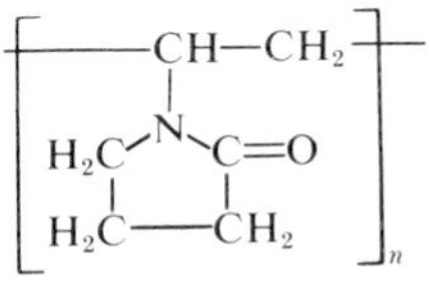

$$\left[\begin{array}{c} -CH-CH_2- \\ | \\ H_2C\diagup^{N}\diagdown C=O \\ | \qquad | \\ H_2C\text{——}CH_2 \end{array} \right]_n$$

Cationic initiation is also possible, for instance with a BF_3/di-ethylether complex. Vinylpyrrolidone is not suitable for anionic initiation. The properties of polyvinyl pyrrolidone can be varied within wide limits by copolymerization with other vinyl monomers. Many such copolymers with styrene, vinyl acetate, acrylates etc., are being investigated.

c. Applications. Polyvinylpyrrolidone is physiologically harmless, has a high solubility in many solvents, and readily forms complexes. In many cases it fulfils the function of a protective colloid. It can form films and shows a very good adhesion on many materials. These properties assure it a wide field of applications.

In the pharmaceutical industry PVP is used as a synthetic blood serum, as a binding agent for tablets, as a complexor with powerful drugs (which are thus gradually liberated, so that the prescribed doses can be greatly increased) and as a substance to be administered in cases of poisoning. The cosmetics industry uses PVP in hair sprays, shampoos, hair rinses and dyes, shaving soaps and lotions, creams and toothpastes.

PVP is of interest to the textile industry chiefly because of its complex-forming properties. The polymer is insensitive to the condition of the water and therefore is used in the stripping of dyes from cellulose fibres. PVP is incorporated in synthetic fibres to improve the dye affinity and the hydrophilic character of the fibres. Other uses are for the clarification and stabilization of beer, wine, and whisky, decolorization of paper pulp, and the manufacture of soap, detergents, polishing agents, water-soluble adhesives, and ink. In pearl polymerizates PVP is used as a protective colloid.

8.9 POLYESTERS

8.9.1 Definition and history

This group is more limited than the title would suggest because, properly speaking, a polyester consists of macro-molecules which have ester groups as their most prominent functional groups. For instance, 'plexiglass' is a polyester and also a polycarbonate, but is not counted among the polyesters proper.

The reaction between polyhydroxy compounds and polybasic acids leads to the formation of proper polyesters. Condensation polymerization by the interchange of ester radicals was already known in the nineteenth century. The first commercial esters, prepared from glycerol and phthalic acid, date back to 1915. In 1929 Carothers published '*Recurring condensation of monomers*', in which techniques are described which have remained essentially unchanged to the present day. Between 1930 and 1940 the rapid hardening of unsaturated polyester with the aid of unsaturated monomers was discovered, and this method was used commercially on a limited scale between 1940 and 1950. Rapid growth occurred only after 1950. In 1966 the world production was 400 m. lb (nearly 200000 tonnes). Polyesters are condensation products of dibasic acids with glycols, hence mainly linear polymers (saturated or unsaturated) are produced. The saturated polyesters, such as polyethyleneterephthalate, are used in films and fibres, while unsaturated polyesters, after they are hardened, have a wide variety of uses.

8.9.2 Structure and classification

Polyesters are produced by a reaction between an alcohol and an acid, during which water is set free. An unsaturated polyester is constructed in this way:

—P—G—O—G—P—G—O—G—

where P is phthalic acid, G the glycol, and O the unsaturated dibasic acid component. In such a linear polyester the double bonds are used by the unsaturated acids to obtain cross-linking with a second unsaturated compound, usually styrene (M), and a structure of the following type is obtained:

```
         |               |
—P—G—O—G—P—G—O—G—
         |               |
         M               M
         |               |
—P—G—O—G—P—G—O—G—
         |               |
```

The polyesters are usually marketed dissolved in the cross-linking monomer in the form of a syrup, to which inhibitors are added. The choice of unsaturated acids is limited: maleic acid or fumaric acid are normally used. The polyesters made from fumaric acid show greater reactivity, and a higher heat distortion temperature and rigidity, but they are also more expensive. The mechanical properties, such as flexibility and rigidity, can be modified by using isophthalic acid and the aliphatic dibasic acids, adipic, azelaic and sebacic acids, and the flame extinction capacity by incorporating halogenated acids.

8.9.3 Production

The polyesters are produced by an interchange of ester radicals in the glycols and dibasic acids, in a stainless-steel vessel provided with stirrers, means of heating and cooling, and a distillation column. The reaction takes place under an inert gas and starts at about 80° to 90°C. When the reaction becomes less vigorous, the temperature is raised to about 190°C and then slowly up to 200°C. The end point is determined by a particular acid value. The resin is afterwards cooled and, in the case of unsaturated polyesters, diluted with monomer.

8.9.4 Properties and applications of linear polyesters

One of the best-known applications of polyesters is in the preparation of synthetic textile fibres. Carothers, in his search for suitable fibre materials, studied a range of fibre-forming polyesters before abandoning them in favour of the polyamides, and it was Whinfield and Dickson of the Calico Printers Association who, at a later date, produced the only material of this class to become really successful as a fibre material: polyethylene terephthalate. This compound, best known under the trade names 'Terylene' (ICI) and 'Dacron' (Du Pont), gives fibres having a density of 1.38 g/cc., a melting temperature of 250°C, an excellent tensile strength, a negligible moisture absorption, and a good chemical resistance.

Polyethylene terephthalate has been produced in the form of moulding and extrusion material. The principal characteristics are a high gloss, a good scratch resistance, and a high rigidity.

Polyethylene terephthalate film is produced in an amorphous form by quenching the extruded film, reheating it to a temperature of between 80° and 100°C and stretching to three times its linear dimensions in all directions to develop some crystallinity and increase the strength. The film is then subjected to an annealing treatment at between 180° and 200°C while it is held under restraint, when further crystallinity develops and any tendency to shrink is reduced. Such films are used mainly for electrical purposes and particularly as a separating material in condensers. They are also used for recording tapes and, when metallized, as a decorative material.

8.9.5 Crosslinked polyesters

Crosslinking takes place by the reaction of the monomer with the linear unsaturated polyester under the influence of an initiator.

a. Properties. The properties of the hardened products naturally depend on the polyester and monomer used. By varying these factors it is possible to ensure a product with chemical resistance, heat stability, slow burning, special flexibility, and other desirable properties.

b. Specific applications. The general-purpose type, prepared from phthalic anhydride, maleic anhydride and propylene glycol, is used for large shaped parts consisting of laminated material, and for the production of mouldings and castings. The types stabilized against the influence of light are used in the fabrication of transparent and translucent sheet for glazing roofs etc. and in the opaque form, for bathroom equipment and the like. A special 'lay-up' type is used in boat construction and for other large structures.

The type with good chemical resistance is used particularly for tanks, pipework, ducting and certain utensils. The heat-resistant type is used in the production of dies, for mouldings produced at low pressure, for laminated material produced under pressure, for heat-resistant castings and coatings, and various aircraft components. Flame-resisting polyesters are used for similar purposes and also for impregnation. The flexible type is used for impregnation, and for modifying hard resins to obtain better viscosity, for touching up damaged motor car bodywork, and for the sealing of ceramic pipes.

Fig. 8.18 A motor launch constructed from glass fibre reinforced polyester laminate.

Finally, special-purpose types exist for making buttons and imitation marble, for patching up damaged motor car bodywork and for impregnated products.

8.10 POLYAMIDES

8.10.1 History and definition

The synthetic thermoplastic polymers described as polyamides are better known by the name 'nylons'. They are linear polymers with the characteristic acid amide group —CO.NH_2— linked by shorter and/or longer methylene groups —CH_2.

The first nylons, produced in 1928 by Carothers in the United States, were condensation products of diamines and diacids. The polymers had particularly high strength and elasticity, but a low melting point. Not before 1935 did Carothers and his colleagues obtain a high-melting polymer, known as a super-polyamide, which was produced by the condensation of hexamethylene-diamine and adipic acid. In 1938 the first yarns were spun from this material, and in the following year nylon stockings had largely replaced those made from other materials in America.

The polyamides produced by condensation from diamines and diacids are designated by the number of carbon atoms in the diamine, followed by the number of carbon atoms in the diacid. Thus, the polyamide from hexamethylene-diamine (6 carbon atoms) and adipic acid (6 carbon atoms) is called nylon 6/6 or 6.6.

Scientists of I.G.-Farbenindustrie in Germany succeeded, at about the same time, in condensing ϵ-amino-caprolactam, in a manner similar to nylon 6/6, to produce a super-polyamide. This nylon, obtained from amino acid or one of its derivatives, is designated by the number of carbon atoms in the unit; six in the case of caprolactam. Nylon 6 was first produced in Europe in 1943 on a commercial basis.

Although the polyamides were originally developed as fibre-forming polymers (see chapter 4), they have steadily gained importance in other fields, e.g. for injection moulding and extrusion, and to a smaller extent for the blowing of bottles.

8.10.2 Structure

The chain of polyamides has the following structure:

$$\diagup CH_2 \diagdown CH_2 \diagup CH_2 \diagdown CH_2 \diagup CH_2 \diagdown NH \diagup CO \diagdown CH_2$$

The presence of polar amino leads to a strong attraction between the molecules giving rise to the high melting point of the polyamide. The longer the carbon chain, the less the attraction possible between the different chains, and this is reflected in a lowering of the softening point of the polymer. Thus, the softening point of nylon 6/10 (125°C) is lower than that of nylon 6/6 (265°C). The longer carbon chains result also in a reduction of the water absorption capacity. The substitution of an alkyl group for a hydrogen molecule decreases the melting point and increases the solubility. Thus, N-methylated nylon 6/6 is a viscous syrup.

8.10.3 Properties

Polyamides are generally strong and relatively rigid materials, of a creamy colour and somewhat translucent. They have a high impact strength and are particularly tough. Nylons show excellent chemical resistance, especially to aromatic and aliphatic hydrocarbons, ketones and esters. Phenols and formic acid are specific solvents, and these are used during the determination of the molecular weight of polyamides. Nylons are not attacked by alkalis and the majority of salt solutions, but are dissolved by saturated solutions of potassium thiocyanate and lithium bromide. Nylons are attacked by strong acids and oxidizing agents. Water, alcohol, and similar compounds are absorbed by nylons; they have thus a plasticizing effect and cause the polymer to swell.

Some properties of the most important types of nylon are given in table 8.5.

8.10.4 Nylon 6 (polycaprolactam)

a. Chemical structure and classification. Nylon 6, or polycaprolactam, can be formed by the condensation of ϵ-amino caproic acid, but this method is of no commer-

Table 8.5 Some properties of the most important types of nylon

	Nylon type			
Properties	6/6	6/10	6	11
Density, g/cc	1.14	1.09	1.13	1.04
Melting point, °C	265	220	215	185
Molecular weight	14–18000	16–22000	16–22000	20–25000
Water absorption, %	3.4	2.6	3.6	2.3
Extractable with H_2O, %	0.5–0.9	0.5–0.8	5–10	0.5–1.2
Fibre tensile strength, p.s.i.	10000	9500	10000	8500
Fibre elongation at break %	22	30	35	38

cial importance. The ϵ-amino caprolactam is mostly used as monomer. The opening of the lactam ring by heat in the presence of a catalyst is the basis for the formation of polycaprolactam:

$$\underset{\epsilon\text{-caprolactam}}{n[NH(CH_2)_5CO]} \xrightarrow{\text{catalyst}} \underset{\text{nylon 6}}{[NH(CH_2)_5CO]_n}$$

An important factor, which influences the equilibrium, is the rate of the reaction by which the lactam is formed. If this rate is high, the formation of lactam takes precedence and hardly any polymerization occurs. There is a wide choice of catalysts; both acid and alkaline materials are used. The most important acid catalysts are adipic acid, formic acid, phosphoric acid and acetic acid; the most important alkaline catalysts are sodium hydroxide, sodium carbonate and bicarbonate, potassium carbonate and metallic sodium. Water also catalyses the polymerization, but at high concentrations it has an inhibiting effect.

In 1955 Wichterle and his co-workers achieved the polymerization of ϵ-caprolactam with the rejection of water in the presence of alkaline catalysts. This polymerization takes place very rapidly and uses an ionic mechanism, by which a metal-organic lactam is formed as an intermediate product. The average degree of polymerization is high but it cannot be controlled, and the melt cannot therefore be spun. A variation of this method is the *in situ* polymerization in a non-aqueous medium in the presence of an alkaline catalyst and an accelerator at a temperature which is lower than the melting point of the polymer formed. Castings in nylon 6 can be made by this technique.

The interest in the production of articles in nylon 6 by the casting of caprolactam monomer and *in situ* polymerization, is due to the following three factors: (i) The low melting point of caprolactam and its low heat of polymerization simplify the heat extraction problem in the production of massive parts: (ii) Because of the low viscosity of the fused monomer and the relatively low shrinkage no excess pressure is needed, and the moulds are therefore inexpensive; (iii) The process gives the tough nylon casting in one stage, starting from the monomer.

However, a limitation of the process is the requirement that the caprolactam monomer necessary for castings must be very pure.

b. Production. The caprolactam monomer can be obtained in a number of ways. The primary requirement is a very high purity. The conventional process is based on cyclohexanone, which is obtained from cyclohexanol by catalytic dehydrogenation; the cyclohexanol is in turn formed by hydrogenization from phenol. The cyclohexanone is reacted with hydroxylamine to give cyclohexanonoxime:

$$\underset{\text{cyclohexanone}}{C_6H_{10}O} + \underset{\text{hydroxylamine}}{H_2NOH} \longrightarrow \underset{\text{cyclohexanonoxime}}{C_6H_{10}NOH + H_2O}$$

This oxime is then heated with oleum and, after neutralization with ammonia, it is converted into ϵ-amino caprolactam:

$$\underset{\text{oxime}}{C_6H_{10}NOH} \xrightarrow{\text{oleum}} \underset{\epsilon\text{-amino-caprolactam}}{\underline{HN(CH_2)_5\,C{=}O}}$$

Another method for the production of ϵ-caprolactam is known as the Snia Viscosa process. The ϵ-caprolactam is here obtained from hexahydrobenzoate by interaction with nitrosyl sulphuric acid:

$$\underset{\text{hexahydro benzoate}}{C_6H_{11}COOMe} + \underset{\text{nitrosyl sulphuric acid}}{NOHSO_4} \xrightarrow[SO_3]{H_2SO_4} \underset{\epsilon\text{-caprolactam}}{HN(CH_2)_5\,C{=}O} + MeHSO_4 + CO_2$$

In the process of Toyo Rayon Ltd., nitrosyl chloride acts on cyclohexane under the influence of ultra-violet radiation to form cyclohexanonoxime hydrochloride:

$$\underset{\text{cyclohexane}}{C_6H_{12}} + \underset{\text{nitrosyl chloride}}{NOCl} \xrightarrow[HCl]{U.V.} \underset{\text{oxime hydrochloride}}{C_6H_{10}NOH.HCl}$$

This oxime hydrochloride is then converted into ϵ-caprolactam in the presence of oleum.

The hydrolytic polymerization of ϵ-caprolactam can be carried out either in batches or continuously. In the batch process caprolactam, water (as catalyst) and a stabilizer (to fix the molecular weight) are heated for 12 to 15 hours under nitrogen at 255° to 260°C and 12 to 15 atm pressure. The equilibrium mixture formed consists of approximately 90% polymer. Polymer granules can be obtained from this by extrusion; the granules are then extracted with water, to remove monomer and low-molecular constituents, dried, and packaged so that they are airtight.

In the continouus process, which can be carried out either under pressure or without pressure, a mixture of caprolactam, water and stabilizer is passed through a reaction column heated to 250° to 260°C. A continuous stream of polymer leaves the column; after being cooled it is further processed to granules, in the same way as in the batch process.

In the technique used for *in situ* polymerization of caprolactam, the caprolactam is fused in two equal parts. Catalyst is added to one half and accelerator to the other. Both these highly fluid melts are thoroughly mixed in the mould at 100° to 160°C, and after about 30 seconds polymerization sets in. This is seen by the fact that the initially clear melt becomes cloudy. After a short time the shaped product becomes detached from the mould owing to the shrinkage that occurs. Apart from finished articles, these block polymers are also marketed as semi-fabricated products.

c. Commercial products. Nylon 6 is marketed by many of the established manufacturers of plastics (e.g. Du Pont, ICI, Basf) under various names.

8.10.5 Nylon 6/6

a. Chemical structure and classification. Nylon 6/6 (or 6.6) is the polycondensation product obtained from adipic acid and hexamethylenediamine. The polymerization process is carried out in two stages. The first reaction is the neutralization of adipic acid with the base, hexamethylenediamine, to obtain the nylon salt, hexamethylenediammonium adipate:

$$\underset{\text{hexamethylene diamine}}{H_2N(CH_2)_6NH_2} + \underset{\text{adipic acid}}{HOOC(CH_2)_4COOH} \longrightarrow \underset{\text{nylon salt}}{H_2N(CH_2)_6NH{\cdot}CO(CH_2)_4COOH} + H_2O$$

The second stage is the dehydration of the salt and the polycondensation to the polymer, polyhexamethylene adipamide:

$$n[H_2N(CH_2)_6NH{\cdot}CO(CH_2)_4COOH] \longrightarrow [HN(CH_2)_6NH{\cdot}CO(CH_2)_4CO]_n + nH_2O$$

nylon salt — nylon 6/6

b. Production. The monomers can be obtained in a number of ways and from very different raw materials. The processes based on phenol, furfural and butadiene are of commercial importance.

Phenol is first hydrogenated to cyclohexanol and this is catalytically dehydrogenated to cyclohexanone. Cyclohexanone is then oxidized directly with nitric acid in the presence of a catalyst, to adipic acid:

$$\underset{\text{cyclohexanone}}{C_6H_{10}O} \xrightarrow{HNO_3} \underset{\text{adipic acid}}{(CH_2)_4(COOH)_2}$$

The adipic acid is converted to adiponitrile with ammonia and a catalyst, at 300° to 350°C. From adiponitrile is obtained hexamethylenediamine by hydration with a Cu/Co catalyst at 125°C and 600 atm:

$$\underset{\text{adipic acid}}{(CH_2)_4(COOH)_2} \xrightarrow{NH_3} \underset{\text{adiponitrile}}{(CH_2)_4(CN)_2} \xrightarrow{H_2} \underset{\text{hexamethylene-diamine}}{(CH_2)_6(NH_2)_2}$$

For a long time adiponitrile was obtained from furfural which was reacted by decarbonylation and then hydration to tetrahydrofurane. Then 1.4-dichlorobutane was obtained by a treatment with HCl at 180°C and this, with sodium cyanide, gives the adiponitrile:

$$\underset{\text{furfural}}{CH(CH)_2COCHO} \longrightarrow \underset{\text{furane}}{CH(CH)_2CHO} + CO$$

$$\underset{\text{furane}}{CH(CH)_2CHO} \xrightarrow{H_2} \underset{\text{tetrahydrofurane}}{CH_2(CH_2)_2CH_2O} \xrightarrow{HCl} \underset{\text{1.4-dichlorobutane}}{(CH_2)_4Cl}$$

$$\underset{\text{1.4-dichloro-butane}}{(CH_2)_4Cl} \xrightarrow{NaCN} \underset{\text{adiponitrile}}{(CH_2)_4(CN)_2} \xrightarrow{H_2} \underset{\text{hexamethylene-diamine}}{(CH_2)_6(NH_2)_2}$$

The use of butadiene as a raw material for the starting substance in nylon producion is of more recent date. Butadiene is chlorinated to dichlorobutene and this is reacted with hydrogen cyanide to produce 1.4-dicyanobutene which, by hydrogenation, is converted to adiponitrile and then to hexamethylenediamine:

$$\underset{\text{butadiene}}{CH_2(CH)_2CH_2} \xrightarrow{Cl_2} \underset{\text{dichlorobutene}}{ClCH_2(CH)_2CH_2Cl} \xrightarrow{HCN} \underset{\text{dicyanobutene}}{CNCH_2(CH)_2CH_2CH}$$

$$\underset{\text{dicyanobutene}}{CHCH_2(CH)_2CH_2CH} \xrightarrow{H_2} \underset{\text{hexamethylenediamine}}{(CH_2)_6(NH_2)_2}$$

The polycondensation of hexamethylenediamine and adipic acid is carried out in two stages. The first stage is the formation of the nylon salt. For this, a 60 to 80% diamine solution in methanol is added to a hot 20% solution of the dicarboxylic acid in methanol. The salt, which dissolves with difficulty in methanol, is precipitated. The hexamethylene adipamide is recrystallized from water; the melting point is 202° to 205°C. The second stage is the polycondensation of the hexamethylene adipamide and this can be carried out either in batches or continuously. In the batch process a concentrated solution of the nylon salt is heated with a polymerization regulator in water under nitrogen in a reaction vessel at 280° to 290°C for 1 to 2 hours, when the pre-condensate, which remains dissolved in water, can form. The pressure and temperature are reduced slightly and the reaction continued for about 6 hours to obtain a polymer with high molecular weight. The fused polymer can then be extruded or spun. The melt must not come into contact with oxygen, as this would result in a discoloration of the polymer. In continuous polymerization the reaction mass is slowly moved from one end of the reaction vessel to the other. During this process both the monomer and the polymer are continuously added or removed. The same three stages take place as in batch polymerization. First, polymerization is started in the presence of water; next, the polymerization is continued while the water is being removed, and finally a melt, free from water, is obtained with a high molecular weight.

In the production of nylon 6/6 it is unnecessary to remove unreacted monomer from the solid polymer, as is the case for nylon 6; the nylon salt is completely converted.

The processing of the polymer to granules is obtained in the same way as for nylon 6. The polymer is cast on to a slowly rotating drum cooled internally by water. The polymer is afterwards cut into small pieces. The granules thus obtained are finally packaged so that they are airtight to prevent absorption of moisture from the air. Moisture causes 'blistering' during extrusion.

c. Commercial products. Nylon 6/6 is offered under a variety of names and by a number of companies.

8.10.6 Other polyamides

It is clear that a great variety of nylon compositions can be obtained with attractive properties. In choosing, the properties and production costs are balanced against each other; the costs of the intermediate products amount to a considerable proportion of the total cost.

a. Nylon 6/10. Nylon 6/10 is produced from hexamethylenediamine and sebacic acid. The production process is the same as for nylon 6/6, except for a somewhat lower polymerization temperature owing to the lower melting point of the 6/10 polymer. Because of its lower melting point it can also be worked more easily than the 6/6 polymer. It is marketed by many of the makers of other nylons.

b. Nylon 11. The polymerization of 10-amino-undecaic acid to nylon 11 is a relatively simple process, which consists only of heating the acid to, or above, 200°C, when the water which is formed is distilled off. The last stage of the reaction proceeds at reduced pressure, to obtain a high molecular weight.

The fused polymer can be spun directly to fibre, or extruded and broken down to granules.

The advantage of nylon 11 over other polyamides is its low water absorption.

Castor oil, an inexpensive material, is the starting material; however, the many operations necessary to obtain the amino-undecaic acid make it a relatively expensive monomer.

c. Nylon 12. Nylon 12 is obtained by polymerization from laurinlactam. Laurinlactam has a low ring tension and as a result ring opening under the influence of water takes place quite slowly. Polymerization is substantially accelerated by the addition of acids, which are used as end group stabilizers. The polymer is thermally more stable than the homologues with shorter chains, so that polymerization can be carried out at 300°C. The properties of nylon 12 are similar to those of nylon 11, but production of laurinlactam is easier than that of the monomer for nylon 11, while the raw material for laurinlactam, cyclododecatriene, can be obtained by trimerization from butadiene, which is an inexpensive starting material. Nylon 12 melts at 175°C and its density is 1.03.

d. A number of other types, among them nylons 7, 8 and 9, are either in a development stage or are being produced on a small scale, especially in Europe.

8.10.7 Copolyamides

The various amide-forming components can be copolymerized to obtain lower melting points and better transparency, flexibility and/or solubility and, in some cases, greater toughness. For instance, a copolymer of 6/6 and 6/10 nylon (50:50) has a better transparency and melts at a lower temperature (205°C) than either homopolymer. Lower melting points of the polymers give, in turn, less trouble during production and cause smaller shrinkage cavities. However, the improvements obtained by copolymerization are generally accompanied by a drop in some important properties of the homopolymers. For this reason, the consumption of homopolymers far exceeds that of the copolymers.

In some cases more than two amide-forming components are used to prepare terpolymers, which have a higher flexibility, and a better solubility in alcohol-water systems. Thus, tough films can be obtained which are resistant to chemicals and are used for a variety of coatings.

8.10.8 Polyamide resins

Polyamide resins are soluble condensation products from ethylenediamine or polyethyleneamine with higher branched unsaturated dicarboxylic acids.

The raw materials for the branched dicarboxylic acids are vegetable fatty acids with multiple unsaturated acids, such as linoleic and ricinoleic acids, which are converted by heating into dimerized and trimerized systems.

The components are mixed at a low temperature in an autoclave. The temperature is then slowly raised to 100°C to remove free water. Finally, the reaction is continued and completed under vacuum. The following two groups of polyamides must be distinguished, depending on the degree of polymerization: a low-molecular tough resin product which melts below 55°C, and a high-molecular amorphous polyamide resin of thermoplastic character. These strongly cross-linked polyamide resins have amino end groups which can react with epoxy and phenol resins to give an adhesive.

8.10.9 Applications

The first commercial use of polyamides was for tooth-brush bristles. This was in 1938 and was followed in 1939 by the appearance on the market of stockings made from nylon yarn. Other applications of the fibres followed in quick succession. They were used, for instance, for parachute cord, safety belts, fishing lines, brushes, and various types of cord. The advantages of using nylon in these applications lie in its strength, elasticity and durability – it can be stretched up to 40% in length. Apart from women's clothes, where it replaces silk, it is used for night-wear and underwear, gloves, shirts, synthetic fur, etc.

The types of nylon used as textile fibre are 6, 11 and 6/6 and, on a smaller scale, 6/10. These are also used for nylon thread and bristles, but especially 6/10 and 11 in applications where the minimum water absorption is required. For textile fibres it is important that the polymer has the highest possible softening point, and this is the reason why nylons 6/6 and 6 are preferred.

Nylon mouldings are used in applications requiring high mechanical strength and toughness and relatively high melting points. Suitable types are therefore 6, 6/6, 6/10 and 11; these have low melt viscosities, so that they flow readily and fill the mould completely.

Nylon has a very low coefficient of friction and is therefore an excellent material for gear wheels, bearings, and the like, in particular in the food and textile industries, where contact with oil and other lubricants must be avoided. It is widely used for small mechanical components in cameras, domestic appliances, medical apparatus etc.

When used in place of light metals, nylon is often reinforced with glass fibre. As a result, the modulus of elasticity, the tensile strength, and the heat distortion temperature of the material, increase.

The largest use of the extrusion technique for nylon lies in the production of films and coatings of nylon on other materials for the packaging industry. The food industry uses nylon film extensively for packaging because of its high resistance to tearing, low gas permeability, and good formability.

Because it is chemically inert, extruded nylon tubes and bottles are widely used in the chemical and pharmaceutical industries.

Nylon 6, 6/6, 6/10 and 11 are used extensively as a covering for wire and cable, and also as nylon tubing, where rough usage is experienced, as in motor vehicles. There are various grades, some with a high melt viscosity in order to facilitate the extrusion process, and others consisting of plasticized copolymers, to ensure a flexible end product.

Castings in nylon 6, obtained by *in situ* polymerization of caprolactam, are used for large shaped parts, which would be difficult to obtain by other methods, for instance ships screws and the like. At the other end of the scale, since the cost of the mould for the casting process is relatively low, cast components in small quantities are economic.

The resins normally used for adhesives are the polyamides known as Versamide and Versalon, and copolymers of 6, 6/6 and 6/10, or modified types such as the N-alkoxy nylons. All these polyamides are fairly low-melting and readily soluble in alcohol and alcohol/water mixtures and can be applied by heating or as a solution. Polyamide adhesives are used widely for small packages, and in book binding.

Thermosetting adhesives on a polyamide base are used for the bonding of wood, in honeycomb structures for aircraft, etc. Epoxy and phenol resins are used as setting agents. Phenol resins, in this application, require higher temperatures and pressures, compared with epoxy resins, because water is set free.

Polyamide resins in alcohol and other solvents have found a specialized use as lacquers. Polyamide lacquers are of low viscosity and have good adhesion on metal surfaces; they form a thin, tough, tear-resistant film. Some types of lacquer form a film at room temperature, and others under heat, among them the copolyamides.

8.11 POLYETHERS

8.11.1 Polyacetal

Although polymerization products of methanol (formaldehyde) had long been known, it was a very long time before it was found possible to obtain a dimensionally and thermally stable thermoplastic material. The work began in about 1950, but it was not until 1960 that such a polymer was marketed. The development work was carried out by chemists of the Du Pont Co. at Wilmington; a parallel development took place at nearly the same time in the laboratories of the American Celanese Corp.

The polymerization technique of Du Pont is carried out in a hydrocarbon medium with amine catalysts. The chain structure of the polymer is very simple:

$$
\begin{array}{ccccccccccccccc}
 & & H & & & & H & & & & H & & & & H & & \\
 & & | & & & & | & & & & | & & & & | & & \\
-- & - & C & - & O & - & C & - & O & - & C & - & O & - & C & - & -- \\
 & & | & & & & | & & & & | & & & & | & & \\
 & & H & & & & H & & & & H & & & & H & &
\end{array}
$$

The polymer of the American Celanese Corp. is also a polymer of methanal, but a small quantity of ethylene oxide contributes to the polymerization; it is thus a copolymer.

Polyacetal is largely crystalline and not transparent. It resists weathering well and, in contrast with many other thermoplastics, shows little cold flow. (It has been called the aluminium among the plastics.) It is sensitive to notch effects. Its water absorption is very low and resistance to organic solvents, oils, greases, and alkalis very good. Acids attack it, especially at elevated temperatures. The polymer can be used over a temperature range of from −40° to 90°C.

Because of its good mechanical properties, polyacetal is used mainly for parts of machines and equipment, and also in the motor industry for gears, bearings, clutches, small fuel pumps, and the like. It has replaced metal in many of these applications. Polyacetal is shaped chiefly by injection moulding but tubing and rod are extruded on a limited scale. A minor application in America is aerosol packages produced by blowing, e.g. for cosmetics.

8.11.2 Polyethers from glycols and alkylene oxides

Under certain conditions it is possible to obtain linear chain polymers–the polyethylene oxides–from either ethylene glycol or oxide. The polymerization of ethylene oxide under strictly controlled conditions yields a range of greases and waxes (according to molecular weight) known under the trade name of 'Carbowax'.

In 1958, high molecular weight crystalline ethylene oxide polymers were produced, under the trade name 'Polyox', by the Union Carbide Corp. They are similar in appearance to polythene, but are miscible with water in all proportions at room temperature. The products are strong and extensible, but the tensile strength falls markedly under very humid conditions. The polyethylene oxides can be injection moulded, extruded, and calendered. The average medium-high molecular weight product of this type is shaped by the various techniques over a temperature range of 90° to 130°C, its melting point being of the order of 66°C. The polymers are of interest for making water-soluble packaging films. They are also of value as the basis of textile sizes and thickening agents.

Polymers of polypropylene oxide have also been prepared. Like polyethylene oxide, they are relatively cheap to produce. The materials are prepared by polymerizing the oxide in the presence of propylene glycol as an initiator and a caustic catalyst at a temperature of the order of 160°C.

Polymers of tetrahydrofuran were introduced, under the trade name 'Teracol' by the Du Pont Co. in 1955.

8.11.3 Epoxy resins (ethoxyline resins)

The production of epoxy resins from phenol polyalcohols was first discussed in 1930 in the German literature by Blumen. Only in 1939 was the first patent for epoxy resins granted in Germany. A modest production took place in 1943, when the production patent rights were granted in Switzerland to Castan.

Epoxy resins are produced by a reaction between epichlorohydrin and diphenylolpropane (also called bisphenol-A) in an alkaline medium. First a linear chain molecule is produced; this changes into a cross-linked structure with thermo-setting properties. Both liquid and solid products can be formed in the reaction.

The setting reaction (that is, the transition from a thermoplastic to a thermosetting resin) is determined by the reactivity of the remaining epoxy groups and the hydroxyl groups.

Setting is promoted by hardeners, which can be divided into two groups:

(i) Polyvalent amines, polyamides and polysulphides. These promote setting at room temperature. Some commonly used amine hardeners are ethylenediamine, diethylenetetramine and triethylenetetramine.

(ii) Acids, acid anhydrides, urea, and melamine resins, which react only at elevated temperatures.

Epoxy resins are remarkable for their good adhesion on many materials, particularly non-porous ones. The electrical properties are outstanding and the resistance to alkalis and non-oxidizing acids good to moderate. Aromatic and chlorinated hydrocarbons, as well as most esters and ketones, produce swelling. The water absorption is very low. The stable temperature range lies for most types between about −40°C to about 90°C. The mechanical properties are good. Although fairly hard, epoxy resins have an adequate impact strength. The hot-setting types have, as a rule, somewhat better mechanical properties than the cold-setting types.

Production was first carried out at Ciba A.G. in Basle, the resin being marketed as 'Araldite'. There are now a number of other manufacturers and trade names.

The range of applications for epoxy resins has grown over the years. Initially they were used only as adhesives, especially for materials such as glass, porcelain, and metals. It was then applied to the sealing of insulators for transformers, insulators cast completely in epoxy resin, and encapsulation of components. The paint industry has been making use for some years of the good adhesion and chemical resistance of epoxy resins, in particular for the protection of metals. Applications range from the coating of preserving cans to the treatment of steel plates in ship building.

Epoxy resins are also used with glass reinforcement, in sheet form for electrical insulation and printed circuits, and as shaped parts where polyester resins are not satisfactory, e.g. where they are not strong enough or their chemical resistance is unsatisfactory, since the price of epoxy is considerably higher than that of polyester resins. Other uses include epoxy-sand mixtures for bridge roadways, for the cambering and surfacing of roads, for floors of industrial buildings, and for the outside finish of facades of buildings.

8.12 POLYARYL ETHERS

8.12.1 Polysulphone

This polymer was developed in the laboratories of the Union Carbide Corp. and marketed in 1965.

The structure consists of chains of benzene rings connected by sulphone, isopropylidene, and other links.

The material can be used over a temperature range from − 100° to 140°C. The mechanical and electrical properties are very good, and it is therefore suitable for structural applications. Shaping, which hitherto has been mainly confined to injection moulding, is difficult and requires very accurate temperature control. Its resistance to acids and alkalis is excellent. Aromatic hydrocarbons are solvents for polysulphone.

The price of the material is still high; this, together with the fact that the material is difficult to shape, has limited the applications to special components in electronic and electrical engineering. A particular example is alkaline accumulator cases.

8.12.2 Polyphenylene oxide (PPO)

This fairly recent polymer was first described in 1964 and it came on the market in 1965. The development work was carried out by the General Electric laboratory at Schenectady (USA). The research laboratory of AKU at Arnhem has also worked in the same field.

PPO is a linear non-crystalline polyether. The raw material is 2,6-dimethyl phenol. A new technique was developed for polymerization; oxidative coupling with the aid of oxygen and catalysts. The structure is as follows:

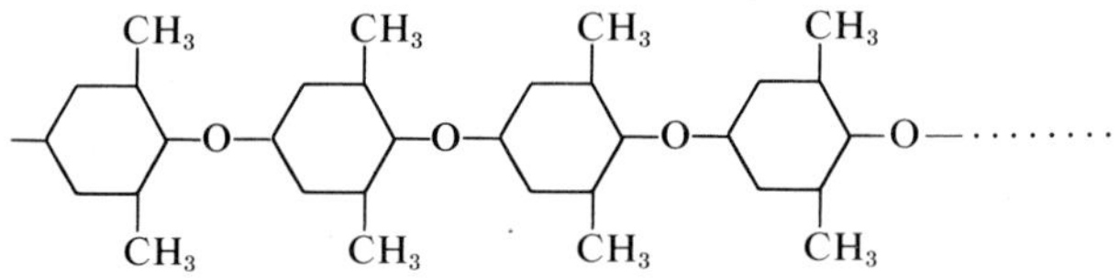

CH_3 CH_3 CH_3 CH_3

—⬡—O—⬡—O—⬡—O—⬡—O— · · · · · · · ·

CH_3 CH_3 CH_3 CH_3

PPO is stable up to 175°C. Its mechanical properties are outstanding, and stability of shape is better than that of most other thermoplastics. The electrical properties are good and are retained even at high temperatures. The linear coefficient of thermal expansion, 2.5×10^{-5} cm/cm °C, is much lower than that of most other thermoplastics (usually $7-12 \times 10^{-5}$ cm/cm °C). Creep is low even at elevated temperatures. Its density is 1.06 g/cc. Water absorption is low (0.1% in seven days) while the resistance to acids and alkalis is good. Chlorinated and aromatic hydrocarbons are solvents for PPO.

The international abbreviation, PPO, has been registered as a trademark in both America and Europe. PPO can be worked by most of the techniques used for plastics, such as injection moulding, extrusion, blow moulding, or vacuum forming. The applications are in those fields in which high heat resistance and good mechanical properties are important, e.g. parts for electrical appliances such as electric irons, electric cookers, fans, surgical instruments, components of switching systems, computers and the like.

A mixture of PPO and polystyrene, known as Noryl, was placed on the market some time after PPO and quickly gained a foothold. The mechanical properties are similar to those of PPO but the maximum operating temperature is much lower–about 85°C. At this temperature, Noryl has about the same dimensional stability and also shows very little creep. Noryl is used mainly in the construction of instruments and apparatus, particularly where great dimensional stability and also little deformation under load are important requirements. The material is supplied in various colours, also with a glass fibre filler.

8.13 POLYURETHANES

8.13.1 History

The chemistry of polyurethanes started with the organic chemistry of the isocyanates as prepared by Wurtz in 1848. In 1884 Hentschel developed the most convenient method of preparing isocyanates; that of phosgenation of primary amines. Until the 1930s no real commercial application was found. The present line of investigation of the polyurethanes began in 1937, for it was then that Dr Otto Bayer decided to experiment with addition products of diisocyanates as a means of producing fibres equal or superior to nylon, which would not be covered by the Du Pont patents on nylon.

8.13.2 Structure and preparation

All urethane polymers have one feature in common: they contain urethane linkages, formed by the reaction of an isocyanate with a compound containing hydroxyl groups. To form a polymer, polyfunctional types must be used. The urethane polymer, however, is a complex structure which may contain, in addition, urea groups derived from the reaction of an isocyanate group with water or an amine. Allophanates and biurets from the reaction of isocyanates with urethane and urea linkages, respectively, may also be present. The basic reaction between the isocyanate group and hydroxyl group can be represented by the following:

$$\underset{\text{isocyanate}}{\text{R—NCO}} + \underset{\text{hydroxyl}}{\text{R}'\text{—OH}} \longrightarrow \underset{\text{urethane}}{\text{R}\,\vdots\,\underset{|}{\overset{\text{H}}{\text{N}}}\text{—}\overset{\text{O}}{\overset{\|}{\text{C}}}\text{—O}\,\vdots\,\text{R}'}$$

In the presence of water the isocyanate reacts as follows:

$$\text{R—NCO} + \text{H}_2\text{O} \longrightarrow \underset{\text{unstable carbamine acid}}{\text{R—}\overset{\text{H}}{\overset{|}{\text{N}}}\text{—}\overset{\text{O}}{\overset{\|}{\text{C}}}\text{—OH}} \rightarrow \underset{\text{amine}}{\text{R—NH}_2} + \text{CO}_2^-$$

The amine in its turn may react with isocyanate, thus:

$$\text{R—NCO} + \text{R—NH}_2 \longrightarrow \underset{\text{substituted urea}}{\text{R—}\underset{\text{H}}{\underset{|}{\text{N}}}\text{—}\underset{\text{O}}{\underset{\|}{\text{C}}}\text{—}\underset{\text{H}}{\underset{|}{\text{N}}}\text{—R}}$$

In general, isocyanates react with organic substances containing active hydrogen. The products formed (urethanes and substituted urea) can again react with isocyanate, thus:

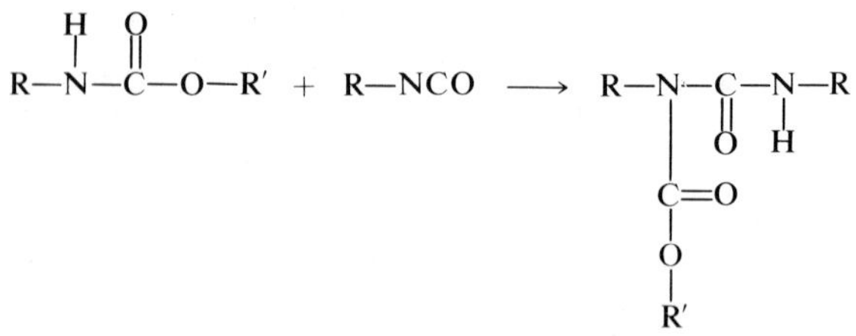

$$\mathrm{R{-}\overset{\overset{H}{|}}{N}{-}\overset{\overset{O}{\|}}{C}{-}O{-}R' + R{-}NCO \longrightarrow R{-}\underset{\substack{|\\ C{=}O\\ |\\ O\\ |\\ R'}}{N}{-}\underset{\substack{\|\\ O}}{C}{-}\underset{\substack{|\\ H}}{N}{-}R}$$

allophanate

$$\mathrm{R{-}\underset{\substack{|\\ H}}{N}{-}\underset{\substack{\|\\ O}}{C}{-}\underset{\substack{|\\ H}}{N}{-}R + RNCO \longrightarrow R{-}\underset{\substack{|\\ C{=}O\\ |\\ N{-}H\\ |\\ R}}{N}{-}\underset{\substack{\|\\ O}}{C}{-}\underset{\substack{|\\ H}}{N}{-}R}$$

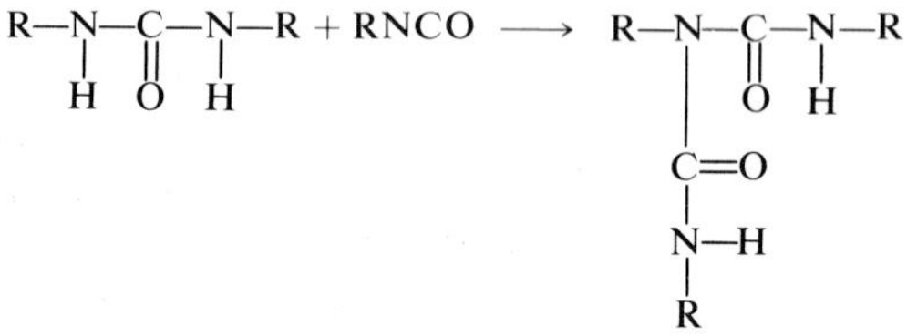

biuret

One of the two main components of the urethane polymer is the di-isocyanate. Commercially the di-isocyanates are made by the reaction of di-amines with phosgene. The chief one is the toluene di-isocyanate (TDI), an 80:20 mixture of 2.4- and 2.6-toluene di-isocyanate isomers. Much smaller quantities of the 65:35 isomer mixture are used. Small amounts of 100% 2.4-toluene di-isocyanate are sometimes used for special formulations.

Diphenylmethane-isocyanate (MDI) is next in importance. Its structural configuration is such that a more symmetrical polymer is produced. MDI finds its greatest use in elastomers, coatings, fibres, adhesives and castings. More recently, crude types of MDI have been used for one-shot rigid foams. Another isocyanate which has been useful for rigid urethane foams is polymethylene polyphenylisocyanate (PAPI).

The compound which furnishes the hydroxyl group for the reaction has a great influence on the properties of the final urethane polymer. The sources of hydroxyl groups for almost all commercial uses of urethane polymers are polyethers, polyesters, and naturally-occurring hydroxyl-bearing oils such as castor oil. Of these, the polyethers (mainly polyoxypropylene) are the most important. Although polyoxypropylene diols have some application, the polyethers primarily used are triols or polyols of higher functionality and are based on glycerine, trimethylolpropane, sorbitol, methylglucoside and sucrose. Polyesters, which are based on di-basic acids (such as adipic) and polyols (such as di-ethyleneglycol) have been largely displaced by polyethers, particularly for foams but are still important for some coating foams, and elastomers. Castor oil goes mainly into coatings, elastomers and adhesives.

8.13.3 Foams

The major application for urethanes lies in the field of flexible and rigid foams, chiefly the former, though the potential market for rigid types is increasing. Rigid foams differ from flexible foams in that they are made from polyols having greater functionality and lower molecular weight to provide for a higher cross-linking. Both foam types are prepared with the aid of a blowing agent.

Flexible urethane foams are made chiefly from polyethers. The polyether foams cost less than polyester types and have more desirable properties for seating applications. They are based on polyoxypropylene diol of approximately 2000 molecular weight and polyoxypropylene triols with molecular weights up to 4000. These triols generally use glycerine as a basic material. Rigid urethane foams are also formulated largely from polyethers, although minor amounts of polyesters are still used in special applications. With the development of di-isocyanates which permit one-shot moulding with polyethers, the use of polyols in rigid foam is increasing. These polyethers are based on compounds such as sorbitol methylglucoside, sucrose, and certain aromatic derivatives. Polyethers for rigid foams have hydroxyl numbers in the range of 350 to 600, compared to 40 to 75 for polyethers used in flexible foam. Other ingredients in urethane foam formulations include various types of catalysts, silicone surfactants and/or emulsifiers. Tin salts, such as stannous octoate and dibutyltin dilaurate, along with amine catalysts such as triethylenediamine, are used for one-shot polyether foams. The surfactant is used to promote fine cell size and stabilize the rising foam until gelation occurs. In one-shot foams, the foam is produced as part of the polymer-forming process. Two-shot foams are made from a pre-polymer, with the aid of water, catalysts etc.

8.13.4 Coatings

The main characteristics of polyurethane coatings are toughness, hardness and wear resistance, combined with outstanding flexibility, and abrasion and chemical resistance: but these properties depend on the class of urethane and the particular formulation. The chief disadvantages of the urethane coatings appear to be a yellowing tendency after application, and higher cost than some competitive systems. The urethane coating in common use, based on TDI, invariably yellows on exposure, and improvement has been sought by the use of stabilizers. The second approach to improved colour retention is in the new types of isocyanates, foremost among which is hexamethylenedi-isocyanate. Very substantial reductions in cost have been achieved by shortening purification processes, and the use of isocyanates based on MDI in which a final distillation step is eliminated. This results in a stable liquid isocyanate of low viscosity containing no volatile solvents so that irritant vapours are completely absent. Pigmented enamels of brilliant colours are readily obtainable with standard pigments even though the isocyanate itself has a very dark colour. Only pure white and some pale pastel shades cannot be obtained. The solvent-free, low-viscosity isocyanates now available may be combined with low-viscosity polyols, such as castor oil and numerous polyethers, to produce coatings with a high solids content, especially when adequately pigmented. A small proportion of volatile solvent is sometimes included to secure properties not otherwise obtainable, such as compatibility of components, viscosity adjustment, activity control etc. The advantages of solvent-free coatings are numerous – greater economy by minimizing loss upon application, lower hazards from flammability and toxicity, improved adhesion, minimized porosity of film caused by passage of solvent, and greater thickness of film deposited. Since protection and durability are directly proportional to film thickness, the desired coating thickness can be obtained with fewer coats and hence lower application costs. Their highest potential is as high-build primers and undercoats or as one-coat systems.

8.13.5 Elastomers

Applications of urethane elastomers in the metal-forming field are of major importance and are increasing. They are currently being used as forming pads, replacing expensive precision metal dies. Reported advantages include reduced set-up time and the elimination of drawmarks and of the expensive machining required with matched steel dies etc.

The use of urethane elastomers in the more typical and better known mechanical rubber products (such as shock absorbent pads, bearings, gears, solid tyres, drive belts etc.) continues to grow at a steady rate. Urethane elastomers are now being used for monolithic flooring such as terrazo floors. The urethane chemicals used are generally pre-polymers prepared from inexpensive polyols and poly-isocyanates, such as toluene di-isocyanate and the newer polymeric isocyanates. These pre-polymers are mixed with fillers such as clay or coloured, ground chips of thermoplastic resins. They are supplied either as 100% reactive systems or as solvent-containing systems. Curing of the flooring compound is accomplished at room temperature. Products of the same general nature as are used in flooring, and in a range of hardnesses, are being used as sealants etc.

8.14 POLYCARBONATES

8.14.1 History

Polyesters of carbonic acid with di-valent phenols have been known since 1900; but only in about 1956 was it found, in the laboratories of Bayer in Germany and of General Electric in the United States, independently of each other, that in particular the ester carbonates of 2.2-bis (4-hydroxyphenyl)propane were polymers with very attractive properties.

8.14.2 Structure

The polyarylcarbonates produced on a commercial basis are all carbonic acid esters of 2.2-bis(4-hydroxyphenyl)propane. This material is often called bisphenol A. The polyaryl carbonates are polymeric condensation products with the following structure:

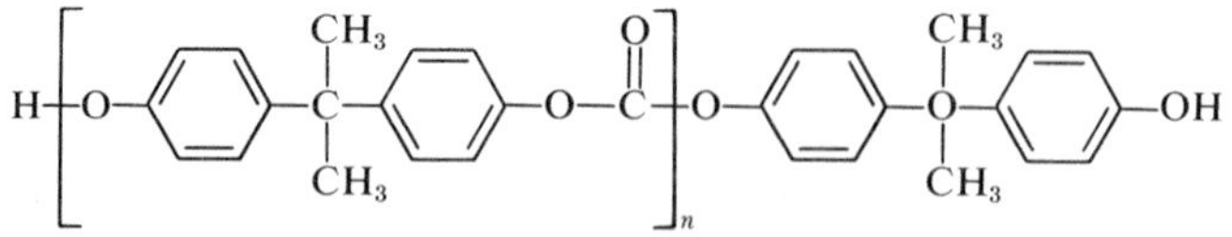

$$H\left[O-C_6H_4-C(CH_3)_2-C_6H_4-O-\overset{O}{\overset{\|}{C}}\right]_n O-C_6H_4-O(CH_3)_2-C_6H_4-OH$$

The value of n is between 100 and 400.

Instead of bisphenol A many other aromatic dihydroxy compounds can be used in principle. Examples are hydroquinol, dihydroxydiphenylsulphoxide and the ring-halogenized derivatives of bisphenol A. Copolymers can also be prepared from them.

8.14.3 Properties

Polycarbonates are tough, impact-resistant, transparent and slow burning. The dimensional stability, thermal resistance, and mechanical and electrical properties are all good. The creep strength of polycarbonates, both at room temperature and elevated temperatures, is outstanding among thermoplastics. Dimensional tolerances of about 0.001 mm (0.025 in) are possible; also, the electrical properties are fairly constant over a wide range of temperatures, humidities, and frequencies. Ultra-violet stabilizers can be used to give grades appropriate for optical work. The mechanical properties of polycarbonates can be substantially improved by reinforcement with short glass fibres. At room temperature polycarbonates are resistant to water, dilute organic and inorganic acids, oxidizing and reducing agents, and neutral and acid salts. Polycarbonate is attacked by alkalis, amines, ketones, esters and aromatic hydrocarbons. Methylene chloride, ethylenedichloride, cresol and dioxane are good solvents.

A limitation of the polycarbonates is a tendency to craze or crack under tensile stress. This effect can be significant in the presence of various solvents and vapours such as dioxane and carbon tetrachloride. Therefore, it is necessary to test samples or prototype parts under actual conditions of use for applications subject to unusual exposure.

All normal processing techniques for thermoplastic materials such as injection moulding, extrusion, compression moulding, and drawing are used for polycarbonates. During these processes the moisture content of the polycarbonate must be below 0.03%, otherwise the melt viscosity will drop rapidly, because of hydrolytic degradation of the polymer, and poor quality parts will result. Films and coatings may be cast from solution, and coatings applied by dipping or spraying. The unique combination of properties of polycarbonates makes possible the cold-forming of the material at room temperature, during which process stretching and orientation of the polymer chains take place. At elevated temperatures the binding forces weaken and the chains tend to return to their original position. In the case of polycarbonate the high heat distortion temperature (270° to 280°F) under stress, and the negligible creep, allow the shape to be maintained up to temperatures not far from the heat-distortion temperature. Products made of polycarbonate can be machined and finished by all the conventional techniques.

8.14.4 Commercial products

In West Germany polycarbonate is sold under the trade mark 'Makrolon', and in the United States 'Merlon' and 'Lexan'. Polycarbonate is commercially available in the form of pellets, powder, film, sheet, rod, and tubing, depending on the intended application.

8.14.5 Production

Three processes for the manufacture of polycarbonates are of commercial importance. In the process using the interaction of ester radicals, the polycarbonate is produced from diphenyl carbonate with bisphenol A as follows:

$$n \cdot C_6H_5{-}O{-}\overset{O}{\overset{\|}{C}}{-}O{-}C_6H_5 + n \cdot HO{-}C_6H_4{-}\underset{CH_3}{\underset{|}{\overset{CH_3}{\overset{|}{C}}}}{-}C_6H_4{-}OH \longrightarrow$$

diphenyl carbonate — bisphenol A

$$\left[{-}C_6H_4{-}\underset{CH_3}{\underset{|}{\overset{CH_3}{\overset{|}{C}}}}{-}C_6H_4{-}O{-}\overset{O}{\overset{\|}{C}}{-}O{-}\right]_n + 2n \cdot C_6H_5OH$$

The reaction takes place at an elevated temperature in a vacuum, and the phenol formed is distilled off. A drawback of this method is that it is very difficult to obtain high molecular weights.

A second process is the direct reaction of phosgene with bisphenol A in a solvent which dissolves the product formed and chemically binds the hydrochloric acid set free. Pyridine is very suitable for this purpose.

$$n\,.\,HO-C_6H_4-C(CH_3)_2-C_6H_4-OH + n\,.\,COCl_2 \xrightarrow{\text{pyridine}} \left[-O-C_6H_4-C(CH_3)_2-C_6H_4-O-\overset{\overset{O}{\|}}{C}-\right]_n + 2n\,.\,HCl$$

In the third process the polymer is formed by means of an interfacial reaction. The monomer, bisphenol A, is dissolved in a solution of caustic soda, and this is then stirred into an organic solvent for the polymer which will be formed. The caustic soda solution and the solvent remain as two phase. Phosgene is then introduced into the mixture, and the reaction takes place at the liquid-liquid interface. The ionic ends of the growing polymer molecule are soluble in the caustic solution, whilst the rest of the molecule dissolves in the organic solvent. A small quantity of a quaternary compound must be present as a catalyst to enable the reaction to take place. The polymer which is produced by the reaction is recovered by washing the organic solvent with water, neutralizing the caustic soda and either evaporating off the solvent or precipitatating out the polymer by the addition of a non-solvent reagent.

8.14.6 Applications

Polycarbonates are valuable for making transparent products with good electrical and mechanical properties which are maintained at elevated temperatures (up to 130°C). As they combine resistance to staining and transparency with toughness and good heat resistance, polycarbonates are gaining ground in domestic and commercial equipment which involves the handling of food. By virtue of the thermal resistance and the low dimensional tolerances that can be obtained they are being used increasingly as constructional materials to replace metals. This applies to parts of machinery, insulating handles, and motor car components. Polycarbonates are also finding a use in safety helmets.

8.15 PHENOLIC PLASTICS

8.15.1 History

Phenolic resins are thermosetting condensation products which arise in the reaction of phenols and aldehydes under the catalytic influence of either bases or acids.

The first investigation in this field was carried out by A. von Bayer in 1872. Some decades later a great demand arose for a good substitute for shellac, and as a result interest focussed again on the resinous products based on phenol and formaldehyde as described by Bayer. In America, Baekeland laid the foundation, during 1905–9, for a method of making a relatively strong and durable plastic (Bakelite) with the aid of high pressures and temperatures. The use by Aylsworth in 1911 of hexamethylenetetramine as hardener in 'Novolak' moulding powders was a major step forward towards ensuring high-speed production of consistent products. Since about 1926 a number of companies have engaged in the production of phenolic resins, and although subject to competition from newer plastics the phenolics continue to maintain their important position.

8.15.2 Chemical structure and classification

The manufacture of phenolic resins from phenol and formaldehyde in an aqueous medium, with acid or basic catalysts, rests in the first instance on a primary addition reaction of formaldehyde with the active ortho- and para-points of the phenol molecule. Reactive methylol phenols are thus formed. If phenol is added in excess monomethylol phenol (reaction I) is mainly formed and this, in a secondary condensation or resinification reaction, is converted into polynuclear methylene phenols (e.g. reaction II).

(I) OH (phenol, * * *) + CH_2O ⟶ OH, CH_2OH (* *)

(*active points)

(II) OH, CH_2OH (* *) + OH, CH_2OH (* *) $\xrightarrow[\text{or } OH^-]{H^+}$ OH … CH_2 … OH, CH_2OH + H_2O

A methylol group is here left over which can again react with a further monomethylol phenol molecule to give a polynuclear product. It is possible for reaction II to take place indirectly between ortho-type methylol groups, especially if a base is used as catalyst. In this case methylene ether bridges are formed, which by the elimination of formaldehyde can be converted into methylene bridges (reactions IIa and IIb):

(IIa) OH, CH_2OH (* *) + HOH_2C, OH (* *) $\xrightarrow{OH^-}$ OH, $CH_2{-}O{-}CH_2$, OH (* * * *) + H_2O

(IIb) OH, $CH_2{-}O{-}CH_2$, OH (* * * *) ⟶ OH, CH_2, OH (* * * *) + CH_2O

The split-off formaldehyde forms a methylol group again at one of the nuclei of the product obtained, and further reactions can take place with methylol phenols. Phenol being present in excess, the condensation stops after some time because a reaction takes place with a phenol molecule which has no methylol group for further reactions (e.g. reaction III).

(III)

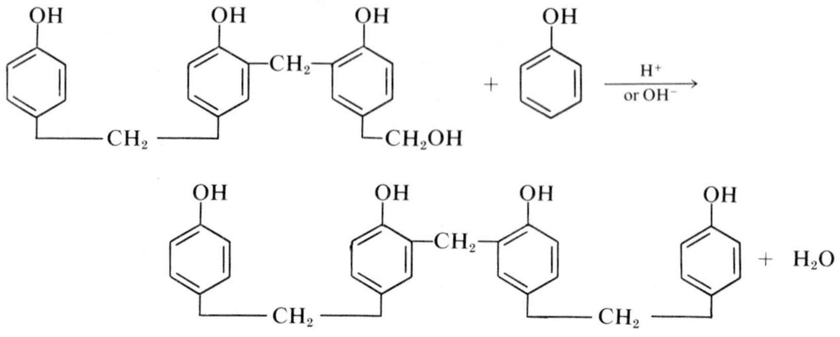

OH OH OH OH

CH_2 CH_2 CH_2OH + $\xrightarrow[\text{or } OH^-]{H^+}$

OH OH OH OH

CH_2 CH_2 CH_2 + H_2O

Further condensation reactions can then take place only by supplementary addition of formaldehyde.

If formaldehyde is present in excess, significant quantities of polymethylol phenols are formed in the primary addition reaction (e.g. reaction IV)

(IV)

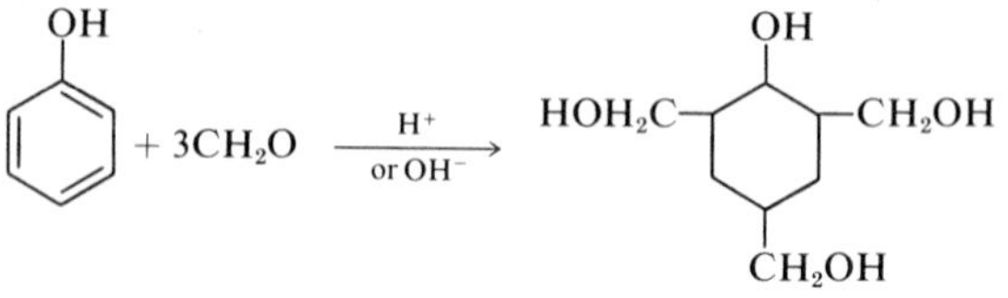

OH + $3CH_2O$ $\xrightarrow[\text{or } OH^-]{H^+}$ HOH_2C OH CH_2OH CH_2OH

The polymethylol phenols (similar to the monomethylol phenols) are converted in secondary condensation reactions into polynuclear methylene ether phenols or methylene phenols. Since much less non-substituted phenol is present, and frequently several methylol groups occur at one nucleus, further condensation to macromolecular products takes place.

Depending on the reaction conditions (degree of acidity, temperature, time, etc.) there arises in the first instance a conglomerate of multinuclear methylene phenols and methylene ether phenols. With increasing condensation the mean molecular weight increases; this is reflected in a marked increase in viscosity. The ratio of phenol to formaldehyde (P/F) determines whether, and to what extent, gelation and setting of the resinous mass occurs at a certain degree of reaction. For gelation and setting to occur, the molecular P/F ratio must be equal to or less than 1.

Chemically, hardening proceeds by cross-linking of the phenol nuclei, by continued reaction of methylol groups with free ortho and para points. In the final condition there arises a macromolecular three-dimensional network of phenol atoms linked together by methylene groups. As the condensation generally takes place at elevated temperatures, most ether bonds are converted to methylene groups by setting free formaldehyde.

Phenol resins evolve in three stages, which are connected with the degree of condensation:

1. Resol or A-stage (beginning of condensation): the resin is fluid, soluble, and still contains much water.

2. Resitol or B-stage (continued condensation, slight cross-linking): insoluble, rubbery.

3. Resite or C-stage (final condition of the cured product): infusible and insoluble.

There are no sharp boundaries between these stages, which merge one into the other. The mechanical and physical properties of the resite, or end stage, depend greatly on the degree of hardening and the manner in which this hardening has taken place. The chemical structure of the resite stage can be described as a compact network of phenol nuclei, which are linked together by CH_2 and CH_2—O—CH_2 groups (isogel structure). The properties of the resite depend greatly on its microstructure. Hardening to the resite stage should, naturally, take place only after the final shape of the end product has been obtained.

A useful intermediate product (pre-condensate) can be obtained by one of the following two methods:

1. By using phenol in excess, which will stop the reaction.
2. By using formaldehyde in excess in an alkaline medium. In the presence of a base the reaction takes place reasonably quickly only at an elevated temperature and more slowly than in the presence of an acid. The reaction can therefore be discontinued temporarily by lowering the temperature.

The two methods of pre-condensation impart different working properties and demand different production techniques. Phenol resins are therefore divided into two major types:

1. Two-stage or 'novolak' resins. (Here the condensation takes place in two stages – first with excess phenol and then with excess formaldehyde, which is added subsequently.)
2. Single-stage resins or 'resols' (the condensation takes place in one process, with excess formaldehyde, and is only temporarily interrupted).

8.15.3 Production and properties of the precondensates

The starting materials for the manufacture of phenolic plastics are mainly phenols and formaldehyde. Other aldehydes (e.g. furfural) are of secondary importance. Initially the phenols were obtained almost exclusively by the fractional distillation of coal tar; both phenol and mixtures of cresols and xylenols were used in varying concentrations and degrees of purity. At the present time, pure synthetically prepared phenol is generally used, which is produced on a large scale as a starting material for various plastics (nylon, epoxy resin, polycarbonates and others). The formaldehyde, in the form of a 37% aqueous solution (formaline), is prepared by a catalytic oxidation from methyl alcohol in the gas phase. Formaldehyde polymer (para-type) and a compound with ammonia and formaldehyde, hexamethylenetetramine ('hexa') are also frequently used as substances yielding formaldehyde.

a. Novolak resins These are prepared in large tanks provided with heating, cooling and stirring facilities. The phenol and the formaline are mixed in a ratio of $P/F \geqslant 1.1$. Usual catalysts are hydrochloric, oxalic or phosphoric acids. If the formaldehyde content is too high the condensation will take place at a very high rate and give insoluble and infusible products which are useless for further processing.

After boiling for some time the layer of resin is separated from the aqueous solution, and after further evaporation and post-condensation it is dried and poured off hot. When cool, the brittle resinous mass is broken down and milled.

Novolak resins are stable compounds which can be mixed with fillers, accelerators, plasticizers and other substances as desired. They are readily soluble in various alcohols, ketones, esters, and ethers and can be fused without any further condensation taking place. Novolak consists of polymeric methylene phenols with about 5 to 8 phenol units per molecule. Approximately one active point per phenol nucleus is still present. Only by further addition of formaldehyde or substances yielding formaldehyde such as 'hexa' does a three-dimensional conglomerate with a resite structure arise. Novolak resins have an indefinitely long life.

b. Resol resins. In the presence of alkalis the condensation to polynuclear compounds takes place much more slowly compared with the acid state. A higher formaldehyde concentration can therefore be used: e.g. P/F = 1/1.1 – 1/1.5 (for moulding and laminating); P/F = 1/1.5 – 1/2.5 (for casting resins and adhesives). The reaction is carried out in tanks, in aqueous or alcoholic solution, with small quantities of catalysts (NH_4OH, NaOH, $Ba(OH)_2$, etc.). The condensation must be carefully controlled to prevent premature gelation. The resols are normally used in the viscous condition for various further processing. Solid resols can be obtained by careful evaporation. On prolonged heating the resols convert to hardened resite; much water is here set free as a result of the chemical reaction of the reactive methylol phenols, which for the most part make up the resols. Subsequent addition of strong inorganic acids gives rise to a very rapid and vigorous reaction to resite. This process is used in the manufacture of foamed phenolic plastics. Resols used as starting materials have a very limited storage life owing to the presence of reactive groups.

8.15.4 Properties of hardened phenol resins

Fillers or reinforcement materials are nearly always added to phenol resins because in the hardened condition they are fairly brittle, the impact strength is low, while the modulus of elasticity as well as the hardness are usually very high. The properties of the end products naturally depend greatly on the type and amount of these additions. For instance, the inadequate mechanical strength of phenol resins can be greatly improved by fibrous fillers.

In the hardened resite stage, the material has a good resistance to water and chemicals but a low resistance to strong alkalis. The material is completely infusible and thus retains its shape until decomposition takes place. Below 150°C its initial mechanical properties are completely maintained and it can even be used at temperatures of 150° to 200°C for long periods. The flammability of nearly all phenolic plastics is very low; they can be regarded as self-extinguishing.

A drawback of phenolic plastics is their dark brown colour, which restricts their use to industrial applications.

8.15.5 Production and uses of novolak resins

The great importance of novolak resins lies primarily in the field of moulding powders. A typical characteristic of these resins is that the final shaping of the products requires fairly high pressures to prevent escape of the volatile constituents formed by the progressive condensation reaction. The prime use of moulding powders based on novolak resins is for the production of articles and components in the electrical engineering field. Typical applications are switches, components of radio and television sets and of motor vehicles, telephones, etc. Some trade marks are: Bakelite, Durex, Durite, Philite.

Novolak moulding powders are made by milling the resin and then mixing with fillers and with 'hexa' as hardener at about 110° to 120°C. Wood flour is often used as filler.

During the mixing process a further condensation of the novolak resin takes place by reaction with formaldehyde set free from the hexa; the viscosity of the resin increases, so that after some time the mixture becomes detached from the mixing rolls. The mixing time is so adjusted and controlled as to ensure a convenient compromise between adequate fluidity of the moulding compounds and a short setting time in the final moulding cycle. Large mouldings naturally require a greater fluidity than small ones.

Auxiliary substances (solvents, colouring agents, catalysts etc.) are added at the mixing stage. The mixed material is milled down to the required particle size and used as moulding powder in the fabrication of the end products.

8.15.6 Manufacture and uses of resol resins

Initially a great deal of resol was made into articles by casting the resin in the form of rod or plate, followed by hardening; the castings are then machined to shape. There is still a good demand for cast resin in the button and jewellery industry.

The main uses of resols are as binding agents in the production of laminates based on cotton, paper, asbestos, glass-fibre and the like, and as adhesives for making plate, hardboard, safety glass, etc. The properties of laminated materials depend greatly on the quality and type of the reinforcement used. The most important applications of resol resin laminates are for electrical insulation and as structural and decorative sheet. Trade marks are Fiberite, Micarta, Farlite, Durestos and Phenolite.

Glass-fibre reinforced resol resins are used in large quantities in aircraft and rocket construction. Since high temperatures and high pressures are necessary to ensure satisfactory hardening, the choice of shape is limited to simple types (plate, rod and tubing).

Other important uses of resols are, for instance, in the production of friction clutches and belt pulleys for motor cars, where they are used as binding agents for the granular friction material. Novolak with hexa is also used for this application, usually in combination with resols.

In metal foundries resins are being used very successfully as binders for sand moulds, and resol resins form a durable adhesive in plywood production.

Phenol resins are widely used in the furniture and woodworking industries. They are not suitable for metal bonding because of their brittleness and pronounced shrinkage. In the aircraft industry modified adhesives are used, which consist of combinations of resols with thermoplastics (e.g. polyvinyl formal) in order to improve the elasticity.

8.16 AMINO PLASTICS

8.16.1 History

By amino plastics are generally understood the condensation products of formaldehyde (occasionally also other aldehydes) with organic compounds containing amino or amide groups; primarily, however, they are thermosetting resins which are produced by the action of formaldehyde (CH_2O) on aqueous solutions of urea ($NH_2 \cdot CO \cdot NH_2$) and melamine ($C_3H_6N_6$). In properties and processing these resins are in certain respects comparable to the phenol resol resins; they also cover the same areas of application. Other basic materials such as thio-urea, dicyano diamide, guanidine, hydantoin and the like are in fact used, but the resins made from them have found no wide application and play a subsidiary role as auxiliary chemicals to obtain specialized effects in the preparation of urea-melamine resins.

The commercial development of urea resins took place mainly between 1925 and 1935; the development was stimulated by the favourable results obtained for phenol resins by the high pressure moulding technique. British Cyanide Co. Ltd. was one of the first companies to succeed, in about 1926, in making moulding powders, from which end products could be produced in any desired colour. A major improvement in the properties, in particular heat and water resistance, was achieved when melamine resins were introduced by Henkel in 1935. A further advance was the introduction of alcohol-modified melamine and urea resins. These are easily soluble in organic solvents and were therefore incorporated in lacquers in combination with the alkyd resins.

Benzoguanamine and ethylene urea resins became very important for the treatment of textiles, as they imparted crease resistance to cotton fibres, and improved the water-repellent and ironing properties.

8.16.2 Chemistry of the urea and melamine resins

The primary stage in the production of amino plastics is the formation of methylol groups by the reaction of formaldehyde with amide and amino groups:

$$RNH_2 + CH_2O \longrightarrow R{-}NH{-}CH_2OH$$

For instance, for urea, carried out in aqueous solution:

$$NH_2{-}\underset{\underset{\displaystyle O}{\|}}{C}{-}NH_2 + CH_2O \longrightarrow NH_2{-}\underset{\underset{\displaystyle O}{\|}}{C}{-}NH{-}CH_2OH$$

The reaction product is monomethylol urea. The second amino group can also be converted. With large excess of formaldehyde even trimethylol urea is indicated. These methylol compounds generally have a very good solubility in water. Aqueous solutions in which the hydrogen of the amino groups in the urea is replaced to varying degrees by CH_2OH groups are thus obtained, depending on concentration ratio, acidity, and temperature. Even in weakly acid solutions the methylol compounds continue to react with any free amino groups present, giving rise to methylene ureas, while water is set free:

$$H_2N{-}\underset{\underset{\displaystyle O}{\|}}{C}{-}NH{-}CH_2OH + H_2N{-} \xrightarrow{H^+} H_2N{-}\underset{\underset{\displaystyle O}{\|}}{C}{-}NH{-}CH_2{-}NH{-} + H_2O$$

As the condensation proceeds, the molecules grow together to form larger units. Water solubility decreases steadily as the methylol groups and amide configurations disappear. Finally a point is reached where the increasingly more viscous syrup begins to gel. If heating is continued a hard, colourless and infusible mass is obtained (the resin is then said to be 'cured') while water and formaldehyde are set free.

The condensation can be stopped at any stage by neutralizing the solution with a little alkali. Mixed resins with different degrees of viscosity can be prepared in this way. In view of the fact that the condensation reaction depends greatly on the acidity, the latter must be closely controlled to avoid premature gelation.

In the final curing, part of the excess formaldehyde is frequently again set free by the hydrolytic separation of methylene ureas; this demonstrates the reversible character of the condensation. The relatively low water resistance of the cured resin points in the same direction. This is in contrast with phenol formaldehyde resins, in which the resin formation is irreversible and the methylene bridges cannot be separated off hydrolytically.

The production of melamine formaldehyde resins is along much the same lines. The melamine is reacted with formaldehyde:

$$\underset{\text{melamine}}{H_2N\text{—}C_3N_3(NH_2)\text{—}NH_2} + \underset{\text{formaldehyde}}{CH_2O} \longrightarrow H_2N\text{—}C_3N_3(NH_2)\text{—}NHCH_2OH$$

In neutral or slightly basic aqueous formaldehyde solutions, methylol melamines are formed first. A large molar excess of formaldehyde is commonly used (3 to 6 mol per mol melamine); as a result, a very high functionality (e.g. hexamethylolmelamines) is obtained and, at the same time, the insolubility of the primary and secondary condensation products is enhanced.

8.16.3 Modified urea and melamine resins

The methylol groups of amino resins make it possible to obtain products with specified desirable properties such as the non-aqueous resins which are widely used in the lacquer industry. The primary condensation products can, in the presence of a large excess of alcohol, be converted to the corresponding ether compounds:

$$R\text{—}NH\text{—}CH_2OH + R'OH \xrightarrow{H_+} R\text{—}NH\text{—}CH_2OR' + H_2O$$

The solubility of the modified resin in organic solvents (e.g. turpentine, xylene, etc.) increases with higher alcohols. By reason of the properties and cost, preference is generally given to butylated resin ($R'=C_4H_9$). Setting is obtained by further heating at 120–150°C. There again arise CH_2 bridges, by reaction of other groups with amino or imino groups, e.g.

$$\text{—}N\text{—}CH_2\text{—}O\text{—}C_4H_9 + H_2N\text{—} \rightarrow \text{—}N\text{—}CH_2\text{—}NH\text{—} + C_4H_9OH.$$

On the other hand, a reaction of methylol groups with glycerol, glycols etc. increases the water solubility of the resin by introducing some free hydroxyl groups on further condensation. This effect can also be obtained by a reaction with sodium bisulphite, when a readily soluble sulphonate resin is formed.

$$R\text{—}NH\text{—}CH_2OH + NaHSO_3 \rightarrow R\text{—}NHCH_2\text{—}SO_3\text{—}Na + H_2O$$

8.16.4 Industrial uses

a. Production and processing of moulding powder. Amino plastics are nearly always processed in combination with fillers. A resin solution is prepared first by reaction of about 40% formaline solution with urea or melamine. For urea-formaldehyde the molar ratios normally lie at 1 : 1.2 to 1.5 and for melamine somewhat higher, at 1 : 1.5 to 3. By controlling the reaction time, temperature, and acidity, a viscosity and stability which will ensure good mixing with the filler is imparted to the precondensates. Wood flour and bleached cellulose pulp are the preferred fillers for urea resins. The fibre length of these materials has a major influence on the mechanical properties; long fibres give the finished product a good impact strength. In order to improve the electrical properties, particularly of melamine plastics, mineral fillers such as mica, asbestos or crushed quartz are incorporated.

Impregnation takes place in a kneading machine. This is a critical process and a correct procedure is necessary to ensure a good result. At this stage of mixing the various auxiliary substances are normally added, such as lubricants, plasticizers, pigments, catalysts and other materials. Next the crumbly mass is dried with care in ovens and then milled with or without other auxiliary materials. Important factors such as moisture content, fluidity, bulk density and curing rate are measured and controlled at regular intervals.

Powerful presses are used for making the moulding powders into finished products. The powder is first compacted by cold pressing; the press-work cycle can be shortened by preheating, using, for instance, a high-frequency technique. The press temperature lies between 130° and 180°C; for melamine it is higher than for urea resins. In view of the gas evolution the pressure is generally lifted for short periods to allow the gas to escape so as to prevent cracking, blisters etc.

b. Adhesives and laminates. Owing to their hydrophilic character, the amino plastics make excellent adhesives for wood and other cellulose-containing material, paper, textiles, etc. For this purpose, urea or melamine is condensed with formaldehyde to a resin of the required viscosity and solids content. Cold-setting can be obtained with the aid of acid catalysts such as ammonium chloride, which reacts with the formalde-

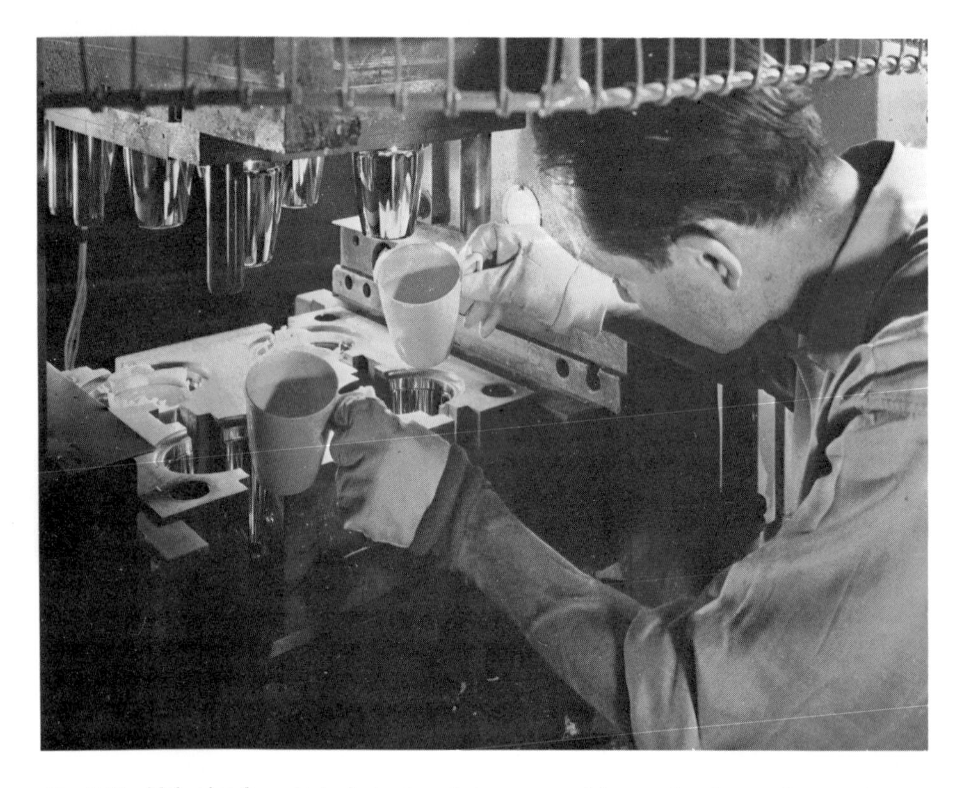

Fig. 8.19 Melamine-formaldehyde beakers being removed from a four-impression mould on a compression press.

hyde present in excess to give hydrochloric acid, but if a catalyst is used the shelf life of the adhesive is limited.

Adhesive layers based on urea formaldehyde are not resistant to boiling and are sometimes attacked by micro-organisms. An improvement in this respect can be obtained by using melamine resins or a combination with phenol resins, wood flour frequently being used as a filler. The adhesives are also marketed dry and sometimes mixed with 'latent' catalysts (i.e. catalysts which become effective only under certain conditions). However, the powders are very hygroscopic, and attention must therefore be given to storage conditions. They are mixed with water prior to use. The adhesives are used in carpentry work, and the manufacture of plywood, chipboard etc.

c. Auxiliaries in the textile industry. Amino plastics play an important part in making cotton and other textile fibres crease-resistant and non-shrinking. The fibres are impregnated for this purpose with dilute aqueous solutions of condensates of low molecular weight. A certain quantity of resin, normally between 5 and 10%, is dissolved, while salts or weak acids are added at the same time as catalysts to the resin solution, to promote the curing in the dried textiles. This takes place at a temperature of about 140° to 160°C. Afterwards the fabric is washed to remove unconverted resin and catalyst.

The methylol melamines or resins with ether groups have a better resistance to washing compared with the cheaper urea resins. To improve the mechanical strength and softness of the fabric, various auxiliary substances are frequently added (polyoxyethylene esters, quaternary ammonium salts, etc.). See also chapter 7.

d. Lacquers. Amino plastics play an important part in the lacquer industry as a component of stoving enamels. By reacting the methylol groups with alcohols, the condensates become soluble in organic solvents while retaining their hardening capacity. Lacquer base coats with different properties can be obtained by varying the formaldehyde concentration, reaction time, type of alcohol used and other factors.

The cheaper urea lacquer resins have largely been replaced by melamine resins, which have an excellent resistance to heat and water. A precondensate is first obtained from melamine and formaldehyde; alcohol is added (usually butanol) and heating is continued under the influence of an acid catalyst while water, and finally excess alcohol, are removed by azeotropic distillation. It is important that the degree of etherization is balanced with the degree of condensation. Xylene and other aromatic hydrocarbons are frequently added to promote the separation of alcohol and salts. The butylated amino resins are clear, colourless, viscous liquids having good compatibility with cheap aliphatic petroleum fractions.

The film-forming properties are excellent, but the hardened layers are brittle and adhesion to metal is equally unsatisfactory. It is therefore common practice to mix the resins with plasticizers, or other resins, to increase both flexibility and adhesion; non-drying, oil-modified alkyd resins are suitable. Owing to the presence of active hydroxyl and carboxylic acid groups in the alkyd resin, chemical bonding presumably takes place between the components during the curing process, thus leading to homogeneous and very stable coatings. A variety of pigments can, of course, be added to increase the hardness and obtain the required colour. Stoving takes place at temperatures of 120° to 150°C in 30 to 60 minutes. The curing time can be shortened by the addition of weak acids (e.g. acid alkyd phosphate).

Coatings of particularly good quality have recently been produced by co-condensation with epoxy resins. Another interesting development is the use of methyl ethers of amino resins as a component in water-soluble resin systems.

e. Some other applications. Urea and melamine resins are used in the paper industry to improve the wet strength of certain kinds of paper. To increase the adsorption on the cellular fibres, the resins are reacted in a colloidal solution of positively charged particles. Other applications include foamed materials which, however, have the drawbacks of being highly sensitive to water and not resistant to micro-organisms. Further uses are as binding agents for the cores used in sand casting, ion exchangers on a methylol melamine base, and as leather tanning agents.

8.17 CELLULOSE DERIVATIVES

8.17.1 Introduction

Cellulose has been very aptly described as 'the chemical that grows'. Cellulose, which is the raw material for a whole family of chemical products, is indeed the main constituent of all plant life. Plants as such cannot be used indiscriminately in the production of 'chemical' cellulose, and selected types of cellulose are produced for specific applications; and furthermore these have to be specially purified and freed from the non-cellulosic contaminants which vary from plant to plant. Some plants are useless either for technical or economic reasons, whilst others are reasonably pure in the natural state, and it is with these, such as cotton linters and wood pulp, that the cellulose 'conversion' industry is primarily concerned.

As long ago as 3500 B.C. the Egyptians made papyrus from the pith of aquatic reeds, followed in the early Christian era by the production of paper, and concurrently the use of cotton cellulose in the production of fibres. It is not surprising that when the production of synthetic fibres was attempted, this was approached along the lines of the chemical conversion of cellulose. Again, it was quite logical to produce a fibre in the form of a continuous filament as does the spider and silk-worm. This was in fact attempted and accomplished, and useful fibres were produced by the extrusion of various solutions of cellulose in the latter part of the nineteenth century. Industry was thus provided with a new group of raw materials, the cellulose derivatives, which in addition to their use as fibres were eventually applied as films and formable masses, and gave birth just over a century ago to the plastics industry. Strangely, these pioneer cellulose derivatives (notably cellulose nitrate and cellulose acetate) were not recognized as in any way related to the ill-defined phenolic condensation products of Baekeland for example, until Staudinger eventually established his theory that organic polymers are built up of long-chain macro molecules, of which cellulose was perhaps the outstanding example.

Chemical analysis had early established that the long chains of cellulose consisted of cellulose units each containing three hydroxyl groups, and as alcohols they were thus capable both of etherification and esterification. Over the past century many esters have been prepared based on both inorganic and organic acids, but mainly the latter. Of these a mere half-dozen remain commercial today. This is not so much due to their intrinsic short-comings, but to the fact that the synthetic fibres have supplanted those derived from cellulose (the chief outlet). Nevertheless, cellulose is still the finest chain molecule available, and the advent of an annual quick-growing plant of high cellulose content could restore the situation.

8.17.2 The cellulose esters

Although a large number of cellulose esters and mixed esters have been reported in the literature during the past 50 years, industrial production today is limited to cellulose nitrate, cellulose acetate and its mixed esters, and to a small extent cellulose propionate.

a. Cellulose nitrate. The conversion of cellulose into the nitrate is frequently erroneously referred to as nitration, whereas in actual fact it is esterification. Although theoretically it is possible to esterify progressively the three hydroxyl groups in cellulose, those esters approximating to the di-nitrate are the only ones of commercial interest. These give a corresponding nitrogen content of 10.7 to 11.1% (the plastics (celluloid) and film grades) 11.2 to 11.3% (mainly used for lacquers) and 12.4 to 13.5% (the explosives (cordite) grades). All these materials present an explosive or fire hazard, the potential danger increasing with the nitrogen content. In consequence they are used in industry suitably 'damped', rarely if ever in the dry fibrous state.

In the preparation of cellulose nitrate suitably purified cotton linters or wood pulp are treated batchwise with a controlled mixture of nitric acid, sulphuric acid and water, the mixture being such that the cellulose does not lose its fibrous structure at any stage of the process. Control of the acid proportions and of the temperature of the reaction mass throughout the process are of great importance. Other essential factors are good and uniform contact between the cotton and the acid mix, and also the recovery of the spent acid, on which much of the overall economy of the process depends. When esterification has been carried to the required nitrogen content and product solubility (which is influenced by the molecular weight), according to the end-use for which it is destined, the residual acid is removed by centrifuge, and a number of batches are then bulked for washing. When this is complete the mixed sulphuric-nitric esters which are the secondary products of the reaction are removed, a step vital to the safe storage and ultimate use of the product. Where the cellulose nitrate is to be used for a special purpose such as the production of high quality film or sheet, near water-white colour is usually specified. This is accomplished by bleaching in the cold by processes very similar to that applied to the original cellulose. The final step, usually referred to as dehydration, is in fact the substitution of denatured ethyl alcohol for the bulk of the water present. Since cellulose nitrate is not completely insoluble in the alcohol, this step usually occasions a loss of up to 1% of the bulk, but since the extract is essentially unwanted low-nitrogen content esters, the process is in effect a final purification.

In its final alcohol-damped form cellulose nitrate closely resembles the original cotton, but the characteristic cotton convolutions have been removed, with the result that the mass has lost the soft handle of the original material. Cellulose nitrate is stored alcohol-damped, and very stringent regulations are applied to control the storage conditions until it is finally processed into film, sheet or lacquer. Although the manufacture of cellulose nitrate is quite obviously accompanied by considerable risks, these are now well understood and the necessary safeguards taken, so that serious accident is relatively rare.

Excellent though cellulose nitrate is in many ways, it has the serious drawback that it is highly flammable, indeed explosive. Some degree of improved safety was obtained by mixing of cellulose nitrate and non-flammable cellulose acetate, but this involved further process steps not without hazard themselves. Mixed esters were also produced, either by nitrating cellulose acetate or by acetylating cellulose nitrate. Here again the results were by no means entirely satisfactory, and were abandoned when better and safer alternative raw materials such as cellulose triacetate, cellulose acetate butyrate, or more recently polyester, became available.

b. Cellulose acetate. The success of cellulose nitrate, particularly for the manufacture of photographic film, had inspired many attempts towards the end of the last century to produce similar products, but without the fire risks attendant on the nitrate. Of these, cellulose acetate was by far the most successful. It was initially produced by Schützenberger in 1865. Subsequent development and production received a strong fillip from the First World War, especially when acetone-soluble acetate was used as a 'dope' for aircraft. When the war ended in 1918 so did this urgent need, and other uses had to be found for cellulose acetate plants in Europe and USA. The natural alternative was fibre production, an outlet which became commercially feasible when René Clavelle solved the dyeing problem with his 'Ionamine' dyestuffs. Once acetate fibre (rayon) was established the economic production of cellulose acetate was assured, and with it the foundation also for acetate plastics.

Though there are many variations in the acetylation as commercially operated, it consists essentially of two stages, the acetylation proper to the near tri-ester degree, followed by a 'ripening' to reduce this to the di-ester with corresponding solubility in common non-toxic solvents such as acetone (the tri-ester is soluble only in some organic acids and chlorinated hydrocarbons). Although numerous acetylation mixtures have been documented, the most usual consists of a mixture of acetic acid and acetic anhydride with concentrated sulphuric acid as catalyst (direct or heterogeneous route). A more recent variant is the use of methylene chloride as a solvent diluent (homogeneous route): this has the advantage that it makes possible the removal of excessive heat by the controlled evaporation of the methylene chloride.

Traditionally the production of cellulose acetate was a batchwise operation. More recently a process was developed by the I.G.-Farbenindustrie at Dormagen which has the advantage that it is continuous at the acetylation stage, and can be used to produce esters up to the triacetate, and by solution (homogeneous) or fibrous (heterogeneous) alternative routes.

Where the homogeneous solvent method is applied, when required degree of solubility has been attained by ripening, the cellulose acetate is separated in flake form by the controlled precipitation with water. The preparation of the solid acetate having been accomplished either by precipitation from solution, or direct in the fibrous form, it is washed free from acetylating mixture. Finally, it is treated with sulphuric acid to remove the sulpho-acetic residues which have low thermal stability.

Although the acetylation of cellulose appears relatively simple and straight-forward: there are many difficulties, not the least of which is the variability of the cellulosic material which necessitates treatment specific to the type and batch, coupled with the fact that prior to the introduction of continuous acetylation it was only possible to handle batches of 100 to 150 lb (45 to 75 kg) of cellulose. Batch-to-batch quality variation is minimized by bulk-blending of the finished material. Rigorous control of the temperature at the acetylation and ripening stages is necessary to give products of the required solubility (acetyl value), viscosity and molecular weight distribution. At every stage, although instrumentation is used wherever possible, much depends on the judgement of the operative. All this to some extent accounts for the rapidity with which the production of cellulose acetate has lost ground compared with the truly synthetic polymers, such as polyethylene and polyvinyl chloride.

c. Cellulose propionate. A number of organic esters of cellulose are described in the literature, but apart from the acetate only cellulose propionate (marketed as 'Forticel') has attained commercial status. The preparation of this ester is very similar to that of the triacetate, using propionic anhydride in place of acetic anhydride. Both these may be used together, giving the mixed ester, acetate propionate ('Tenite Propionate'). These materials vary in their solubility and their compatibility with plasticizers, and in consequence their physical properties can be varied to meet specialist applications, particularly where water-resistance, toughness and surface finish are of importance.

d. Cellulose mixed esters: cellulose acetate butyrate (CAB). Many efforts have made to produce mixed cellulose esters of organic acids, to give improved water resistance and general dimensional stability. One of the earliest materials to attain any degree of commercial success was cellulose acetate butyrate, and it is still the only one which can be rated fully commercial today. Production of this mixed ester has been confined to Germany and the United States: in Britain the process workers refused to handle butyric acid under the conditions of esterification because of its unpleasant smell. The process follows broadly that of cellulose acetate. It starts with purified cotton linters which are pretreated and then esterified with a mixture of butyric acid, acetic anhydride, and concentrated sulphuric acid catalyst. Stringent temperature control at all stages is essential, as with cellulose acetate, which it resembles closely as a white flaky granular material. CAB (as it is usually designated) shows a wide range of butyryl and acetyl content according to the end use for which it is destined, e.g. film or injection moulding compound. For the latter purpose the mixed ester requires less plasticizer and has better flow properties, giving products which are tougher and of improved water resistance. Not surprisingly, similarly improved properties are found in cellulose acetate propionate, but the difference is not usually such as to justify the increased cost which is consequent on the lower production bulk. Superior weathering properties and increased dimensional stability have enabled CAB to regain some of the ground lost earlier to the synthetic polymers.

8.17.3 The cellulose ethers

Although cellulose ethers have potentially better stability than esters, only ethyl cellulose of the truly thermoplastic ethers has survived commercial trials. Other, non-plastic cellulose ethers – methyl cellulose, hydroxylethyl cellulose and sodium carboxymethyl cellulose – have attained commercial significance, but mainly as textile-treating agents, and as thickening additives for food-stuffs.

a. Ethyl cellulose. Cellulose in the form of bleached linters or wood-pulp is first converted to 'alkali-cellulose' by treatment with concentrated aqueous sodium hydroxide. This is separated from excess alkali in the form of a disintegrated 'crumb' of controlled moisture content varying from 10 to 15%, an important factor in determining the final properties of the product. The alkali-cellulose 'crumb' is treated with the required amount of ethyl chloride in a nickel autoclave, alkalinity being preserved throughout the reaction, which is maintained at a pressure of 80° to 150°C for 6 to 24 hours. Excess reactants are then removed by distillation, the ethyl cellulose being precipitated by the addition of hot water and finally thoroughly washed and dried.

The properties of the product depend on the reaction conditions. The ethoxy content (degree of substitution) is controlled by the proportions of the reactants and by the temperature and duration of the reaction. Ethyl cellulose containing 48 to 49% ethoxyl shows the widest range of solubility characteristics, whilst hardness and moisture absorption decrease progressively with increasing ethoxyl content. Ethyl cellulose is very tough, a property which is retained at temperatures as low as $-40°C$. It is soluble in a wide range of solvents, and is more compatible with a wider range of plasticizers, resins and waxes than any other cellulose derivative. Its many specialist applications range from strip-coating compositions to strong mouldings.

8.17.4 Applications

Although cellulose can be regarded as the pioneer plastic material, it is seldom now able to compete with the newer polymers such as nylon, PVC, polystyrene etc. except for certain special purposes (e.g. filter tips for cigarettes) and where the gain in properties (such as dimensional stability) outweighs the extra cost.

8.18 SILICONES

8.18.1 History

The silicones, which are chemically better described as polyorganosiloxanes, have been known for some time. The chemistry of these compounds was developed mainly by F. S. Kipping who, between 1899 and 1944, published over 50 papers on this subject. Modern developments, based on Kipping's work, date from the Second World War, and resulted in important applications for both liquid and solid polyorganosiloxanes. Dow Corning Corporation, in the United States, was the first to produce silicones on an industrial scale and to investigate their potential commercial applications. Development centred initially on heat-resisting resins with good dielectric properties.

8.18.2 Structure

Silicones can be represented chemically by the general formula

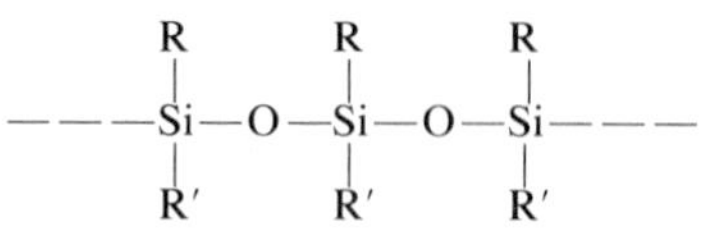

$$---\underset{R'}{\overset{R}{\underset{|}{\overset{|}{Si}}}}-O-\underset{R'}{\overset{R}{\underset{|}{\overset{|}{Si}}}}-O-\underset{R'}{\overset{R}{\underset{|}{\overset{|}{Si}}}}---$$

These products can sometimes have a cyclic structure apart from the linear structure. The R and R′ groups are usually methyl and/or phenol groups. Part of these groups may consist of hydrogen, halogen, or other substituents, to impart special properties. All these groups may also be present in the molecule as end groups. Products with low molecular weight are liquids, with viscosities increasing directly with molecular weight. At high molecular weights the products form resinous substances. The products with very high molecular weights are frequently rubbery, and, after being vulcanized give the well-known silicone rubbers.

8.18.3 Properties

Silicones have a number of important specific properties, which are constant over a wide temperature range up to, and exceeding, 200°C. Further, the products are water-repellent and have very low surface tensions. They are also chemically, physically, and physiologically inert. At the same time, most silicones have attractive electrical properties: the fluids show good resistivity, dielectric strength, dielectric constant, and power factor, whilst the resins and elastomers have a good resistance to corona discharge, and can operate at high voltages and temperature.

8.18.4 Commercial products

Silicones are offered on the market in various types by a great number of manufacturers in several industrial countries under trade names like: Siloset (UK); Viscasil (USA); Silastene (France). Some of these products are supplied in the form of solutions or emulsions for specialized applications.

8.18.5 Production

The important monomers for making silicones are dimethyldichlorosilane and diphenlydichlorosilane. Silicones are formed by reaction with water. Trimethylchlorosilane, which can be added at the same time, forms the end groups of the stable polymer.

The chlorosilane is dissolved in a suitable solvent and then mixed with water to which has been added the substances used to control the reaction. Since a considerable amount of heat may be developed, a means of controlled cooling is provided.

At the end of the reaction, the polymer/solvent layer is separated off and neutralized, and a certain amount of solvent is distilled off until the desired solids content has been achieved. At this stage the solution contains a mixture of the various forms of polymer (linear, branched and cross-linked) and these are all of low molecular weight. An increase in molecular weight and hence in the viscosity of the solution is obtained by heating in the presence of a catalyst, usually zinc octoate, at 100°C. The resin thus produced is cooled and stored in suitable containers.

Cross-linked silicones can be obtained by the addition of monoalkyltrichlorosilanes.

Dimethyldichlorosilane and diphenyldichlorosilane are produced by the reaction of silicon with methylchloride and chlorobenzene, respectively, at elevated temperatures in the presence of a catalyst, for which copper is suitable.

$$\underset{\text{chlorobenzene}}{2C_6H_5Cl} + \underset{\text{silicon}}{Si} \xrightarrow[400°C]{\text{Cu catalyst}} \underset{\text{diphenyldichlorosilane}}{(C_6H_5)_2SiCl_2}$$

The trialkylchlorosilanes and monoalkyltrichlorosilanes referred to above are obtained in minor quantities as by-products. The various monomers are obtained from the reaction mixture by difficult fractional distillation. This synthesis is called the direct process. The other synthesis, which is used on an industrial scale, is known as the Grignard process, and was discovered in 1904 by Kipping. Magnesium chips suspended in ether are first allowed to react with methyl chloride to methylmagnesium chloride.

$$\underset{\text{methyl chloride}}{CH_3} + \underset{\text{magnesium}}{Mg} \longrightarrow \underset{\text{methylmagnesium chloride}}{CH_3MgCl}$$

This is followed by a reaction with silicon tetrachloride, giving a mixture of methylchlorosilanes:

$$SiCl_4 + CH_3MgCl \longrightarrow H_3CSiCl_3 + MgCl_2$$
$$H_3CSiCl_3 + CH_3MgCl \longrightarrow (H_3C)_2SiCl_2 + MgCl_2$$
$$(H_3C)_2SiCl_2 + CH_3MgCl \longrightarrow (H_3C)_3SiCl + MgCl_2.$$

The monomers formed are again separated by fractional distillation. The process can be so controlled that mainly dimethyldichlorosilane is produced. The method is also suitable for other alkyl and aryl chlorides, and therefore is more versatile than the direct process. Mixed monomers, such as methylphenyldichlorosilane, can also be produced. Various other reactive groups can be incorporated at the same time. As large amounts of ether and other flammable solvents are used, great care must be taken in the manufacturing plant to avoid the risk of fire and explosion.

8.18.6 Uses

Liquid silicones as such, or in solution and emulsion, have a wide variety of applications. They are used as mould release or parting agents in the plastics and rubber industries, as de-foamers and heat transfer fluids, and in polishes. Silicone greases are applied as dielectric water-repellent coatings. Silicone resins, being thermally stable, are used as impregnating varnishes, moulding compounds, encapsulants, and in laminates. These laminates can be reinforced with glass fibre and asbestos, and fillers (silica) can be added. Uses include fire walls, coil forms, motor slot wedges, rocket cases, and radomes. Silicone elastomers and resins are used for embedding electrical and electronic components, also as sealants. Silicone rubber can also be used as a self-releasing mould.

Special silane monomers, e.g. vinyltrichlorosilane, are used as coupling agents between glass fibres and thermoplastic or thermosetting resins (unsaturated polyester resins) in glass-fibre reinforced plastic products. The mechanical and electrical properties, especially under wet conditions, are very much improved as a result of better bonding. Masonry and garments can be made water-repellent by treatment with suitable silane monomers or polymers.

8.19 PLASTICIZERS AND PLASTICIZING

Plasticizers are essentially non-volatile or low-volatile solvents, in which the plastic is dissolved at an elevated temperature which depends on the type of plastic. They are generally high-boiling liquids or low-melting solids which, when incorporated in a plastic, reduce the brittleness and improve the flexibility whilst reducing the elastic modulus. These changes in properties are all due to the fact that the plastic-plasticizer aggregate lowers the vitrification point.

The first use of a plasticizer goes back to 1868 when Hyatt discovered that camphor can make nitrocellulose flexible; this started an eager search for other reasonably priced materials with similar properties. Another important stage in the development of plasticizers was reached in 1912 with the introduction of triphenyl phosphate, soon followed by tricresyl phosphate.

In the period between the two world wars the phthalate plasticizers made their appearance, and the rapid growth of the vinyl industry after the Second World War resulted in the plasticizer industry adapting itself in quantity and quality to such developments.

8.19.1 Mechanism of plasticizing

A non-plasticized plastic, e.g. rigid polyvinyl chloride, consists of a number of long polymer chains which are restricted in mobility owing to the presence of intermolecular forces of attraction between the chains, and by mechanical entanglement. A rigid, inflexible material is the result. If heat is applied to such a product, the forces of attraction are diminished, the molecules become more mobile, and the plastic assumes a certain plasticity. The greater freedom of movement of the molecules allows the plastic to be worked and shaped. A plasticizer, which normally is a material with a relatively low molecular weight (300 to 500), acts in a similar way to heat. It can penetrate between the polymer chains, thus reducing their mutual attraction and allowing the polymer greater flexibility.

8.19.2 Miscibility with the plastic

To serve as a plasticizer, the material must be able to mix readily with the plastic and give a stable solution. Stability is achieved by the formation of polar bonds between the plastic and plasticizer. (A polar bond arises by attraction between two opposite electric charges.) In the case of polyvinyl chloride $(-CH_2-CHCl-)_n$ the chlorine atom, being an electron acceptor, renders the neighbouring carbon atom somewhat positive. Dipoles are thus formed at right angles to the axis of the PVC chain. (A dipole consists of a positive and a negative charge, which are held at a distance from each other.) Because of these dipoles the chains are subjected to forces of attraction, represented schematically as follows:

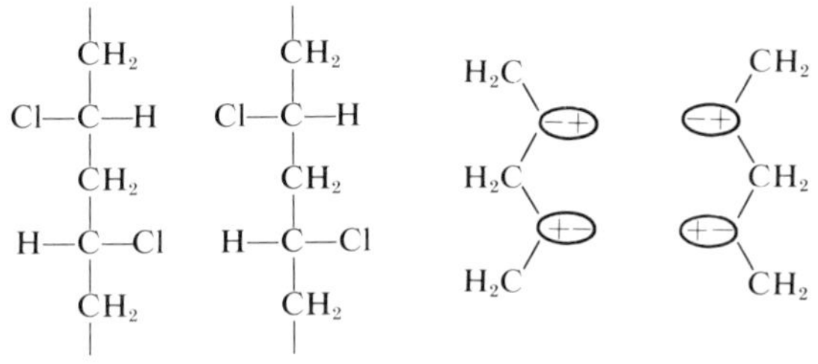

```
      |            |
     CH2          CH2           H2C                 CH2
      |            |               \               /
 Cl—C—H      Cl—C—H             (−+)         (−+)
      |            |               /               \
     CH2          CH2           H2C                 CH2
      |            |               \               /
 H—C—Cl      H—C—Cl             (+−)         (+−)
      |            |               /               \
     CH2          CH2           H2C                 CH2
      |            |
```

Owing to the great number of dipoles, the intermolecular forces between two PVC chains are so strong that they can be overcome only at about 90°C (the vitrification point of PVC) by the greater energy of heat motion.

In order to effect polar bonding between polymer chains and plasticizer, one, or more than one, polar groups must be available to the plasticizer in addition to a number of apolar (i.e. carrying no charge) groups. These apolar groups act as a lubricant between the polymer chains, while kept in position by the polar bond between the plastic and plasticizer.

Polar groups of plasticizers are normally ester groups but can also comprise an epoxy ring, or ether oxygen, nitrogen, sulphur, or chlorine. Aromatic rings close to a polar group, which are polarized by induction, and thus contribute to the influence of the polar group, are also important.

Polar groups in the plasticizer are therefore, necessary, but good mixing is possible only if the polar character of the plasticizer matches that of the plastic. This adaptation of the plasticizer polarity is obtained by the length of the alkyd groups, depending on the alcohol used for esterification. It is found, for instance, for polyvinyl chloride, that dimethyl phthalate and diethyl phthalate mix poorly because the polarity is too high. Miscibility is excellent with the longer alkyl groups, e.g. dibutyl phthalate and diethylhexyl phthalate. As the length of the alkyl group is increased still further, the polarity becomes weaker and miscibility is again less good. For instance, ditridecyl phthalate can still be used, but mixing is unsatisfactory for even longer alkyl chains, e.g. one with 17 carbon atoms.

It can generally be predicted from the chemical formula of a substance whether a plasticizing effect is likely to arise. Table 8.6 indicates the possibilities as regards polarity and polarisability, from which the plasticizing characteristics are derived.

To determine, in a practical way, whether a plasticizer mixes well with polyvinyl chloride, a rolling test (rolling the mixture into a thin sheet) is generally carried out at 160° to 170°C using a mixture consisting of 100 parts by weight PVC, 50 parts by weight plasticizer, and 2 parts by weight heat stabilizer. If the plasticizer mixes well, a 'hide' showing good transparency and flexibility is thus obtained, which can be regarded as a homogeneous solution of PVC in the plasticizer. The solubility can vary with the grade and type of plasticizer and is indicated by the time necessary to produce a hide which can be handled. This entry into solution is known as 'gelation' and the time necessary for it to occur as 'gelation time'. It can be stated, generally, that aromatic phosphates (tricresyl phosphate) gel faster than aliphatic phosphates (trioctyl phosphate) and phthalates (diethylhexyl phthalate), and phthalates in turn faster than adipates, sebacates, and azelates. It is also found that a difference in the rate of gelation corresponds to the difference in the length of the alkyl group. For instance, a PVC mixture with dibutyl phthalate gels more rapidly than one with dioctyl phthalate, and one with dihexyl azelate more rapidly than with dioctyl azelate.

8.19.3 Permanence

A good plasticizer must have a number of properties which ensure that the plasticized PVC retains its flexibility over a long period and preferably also above the room temperature range.

Table 8.6 Plasticizing possibilities

Chemical structure symbol			*Performance as plasticizer*	*Example*
Polar	*Polarizable*	*Apolar*		
1+	–	–	Not usable	Dimethyl oxalate
2+	+	–	Usable	Aromatic phosphate
3+	–	+	Usable	Dihexyl sebacate Dihexyl adipate Dioctyl azelate
4+	+	+	Usable	Dioctyl phthalate
5–	+	–	Usable only in combination with 2, 3 or 4 (secondary plasticizer)	Benzyl naphthalene
6–	+	+	Usable only in combination with 2, 3 or 4 (secondary plasticizer)	Triphenyl propane
7–	–	+	Not usable	Paraffin

The additional properties required of a plasticizer are these:

1. Good compatibility with the polymer at operating temperatures (unmixing or segregation results in material transfer to the surface of the product; this is known as sweating or blebbing).
2. Low volatility (the vapour pressure of the plasticizer must be low at the temperatures at which the material is processed and used).
3. Good resistance to migration towards other solids with which it comes into contact.
4. Good resistance to extraction by fluids such as motor spirit, oils, greases, water and detergents.
5. Good chemical stability to light and heat.

8.19.4 Efficiency of the plasticizer

The smaller the amount necessary to obtain the desired flexibility, the more efficient is the plasticizer. Generally, the efficiency of a plasticizer is measured by the amount necessary to obtain a standard modulus, or standard hardness. The efficiency is then given by the parts in weight of plasticizer for 100 parts in weight of plastic necessary to obtain, e.g., an elongation of 100% under a load of 90 kg/cm^2. If 50 parts by weight are necessary for plasticizer type A and only 45 parts for plasticizer type B, the efficiency of type B will be greater than that of type A.

The figures in table 8.7 show the influence of the alkyl group for a number of phthalate esters.

8.19.5 Types of plasticizer

With respect to solvent power, plasticizers are divided into two clearly distinguished groups – primary plasticizers with generally good dissolving power, and secondary plasticizers which usually have little or no affinity for the plastic and are used because they are inexpensive or in order to improve very specialized properties. Primary plasticizers, normally esters, are therefore the true plasticizers, which are further divided into monomer types (the most important group), polymer, and epoxy types.

a. Monomer plasticizers. A typical monomer plasticizer is a diester such as dioctyl phthalate, obtained by the reaction of one molecule of phthalic acid anhydride with two molecules of octyl alcohol. Other important plasticizers of this type, apart from the phthalates, are the multiple esters of phosphoric acid, and aliphatic carboxylic acids, e.g. adipic acids.

Table 8.7 Influence of the alkyl group on plasticizing efficiency

Plasticizer	*Efficiency (grammes of phthalate necessary for 100% elongation at 90 kg/cm^2)*
Dibutyl phthalate	47
Dihexyl phthalate	49
Dioctyl phthalate	54
Ditridecyl phthalate	69

By far the most commonly used are the phthalate esters, as they combine very good mixing power with low volatility and good low-temperature stability. They are also relatively inexpensive and have a good chemical stability. DOP (dioctyl phthalate) and DIOP (di-isooctyl phthalate) are the most popular plasticizers of the phthalate group for PVC, and DOP is considered the standard plasticizer. If extremely low volatility and low hygroscopicity are required (for electrical applications) esters with longer groups are used, e.g. decyl, di-isodecyl and tridecyl phthalate. Phthalate esters obtained by combination esterification with an aliphatic and an aromatic alcohol, such as butylbenzyl phthalate, are used if rapid gelation with the PVC is desired. For the plasticizing of the more polar cellulose nitrate the low-molecular phthalates are preferred, e.g. dibutyl, diethyl or dimethyl phthalate. These plasticizers tolerate a high degree of dilution with thinners and are therefore used in the cellulose lacquer industry. Dimethyl, diethyl and dibutyl phthalate are good plasticizers also for polyvinyl acetate and cellulose acetate, but aromatic phthalates are suitable too.

Phosphate esters, e.g. tricresyl phosphate, were once the most popular plasticizers, but because of their somewhat toxic nature they lost ground to the phthalate esters. Nowadays they are mainly used for their flame-resistant and fugus-resistant properties. Apart from tricresyl phosphate, other important types for PVC are trioctyl phosphate, with its good flame-resistance combined with good low-temperature flexibility, and certain combination esters, e.g. cresyldiphenyl phosphate with octyldiphenyl phosphate. These phosphates are notably less toxic than tricresyl phosphate. For the plasticization of the more polar cellulose derivatives, use is also made of the more polar phosphates, such as triphenyl, tributyl, and tributoxyethyl phosphates.

Plasticizers of the aliphatic carboxylic acid ester type are produced by the esterification of a divalent aliphatic acid with an alcohol. The most important plasticizers of this group are the adipates, sebacates, and azelates; they are used primarily because of their superior low-temperature properties compared with the phthalates. Dioctyl adipate (DOA, esterified with 2-hexylhexanol) is the most important adipate. In addition to its good low-temperature properties it shows high efficiency (compared with DOP, less DOA is necessary to obtain the same modulus). Against this, DOA has the major drawback of being highly volatile. The volatility is substantially reduced by using the unbranched octyl alcohol instead of diethyl hexanol. Among the azelates, the most frequently used are di-2-ethyl azelate (DOZ), di-isooctyl azelate (DIOZ) and dihexyl azelate (DMZ). The azelates give somewhat better low-temperature flexibility compared with the adipates, combined with lower volatility and slightly better miscibility with PVC. The least volatile of the aliphatic esters are the sebacates, but their use is limited because of their high price. Aliphatic carboxylic esters are also used for the plasticization of polyvinyl acetate and cellulose derivatives. Esters of glycollic acid are used in the plasticization of cellulose acetate, cellulose nitrate, and polyvinyl acetate.

Citrate monomers are employed for non-toxic applications. The most important for PVC is acetyltributyl citrate, while triethyl citrate and acetyltriethyl citrate are used for polyvinyl acetate and the cellulose derivatives. Other monomer plasticizer types are the esters of pentaerythritol, and trioctyl trimellitate. These substances have excellent electrical properties at elevated temperatures and very low volatility.

b. Polymer plasticizers. Polymer plasticizers are linear polyesters obtained by the reactions of dibasic acids, e.g. adipic acid or sebacic acid, with a bifunctional alcohol, e.g. propylene glycol. The reactions are frequently stopped by the addition of a monovalent acid, such as octanoic acid, or lauric acid. A great variation in molecular weight is thus possible, but polyesters with a molecular weight of 2 to 20 times that of DOP are normally used. Polymer plasticizers generally have a higher viscosity compared with monomers, and they are used in applications which require very low volatility and/or good resistance to extraction and migration. A drawback of these polymer plasticizers is their relatively low flexibility at low temperatures. Where this property is required, part of the polymer is replaced by monomer.

c. Epoxy plasticizers. Epoxy plasticizers are normally obtained by the epoxidation of unsaturated oils and fats. The epoxy ring formed enables compounds thus produced to be mixed with PVC, although to a limited extent, and the epoxy group imparts to the PVC mixture additional stability to the effects of heat and light. The two most important types are epoxidized unsaturated triglycerides, e.g. soya bean oil, and epoxidized esters of oleic acid and tall-oil fatty acid. Epoxy plasticizers are generally used in combination with others, e.g. phthalates. In view of the high price and the limited miscibility with PVC, about 10% of the total plasticizer should consist of epoxy compounds; such an amount gives a satisfactory improvement in the stability of the PVC resin.

d. Secondary plasticizers. Secondary plasticizers for PVC are, by themselves, only in a very loose sense true plasticizers, because most of the materials on the market have a low dissolving power (if any) for PVC. They are used in the vinyl industry for two reasons: to lower the price of the product, and/or to obtain one or more specific properties. The cheap materials are usually dark in colour and in general have poor miscibility. Typical examples are certain petroleum fractions and both chlorinated and unchlorinated polythene. They can only be used in combination with a good primary plasticizer such as DOP, at concentrations of up to about 20% of the total plasticizer. The more expensive types are lighter in colour and frequently also have better, or even much better, miscibility. This group comprises the chlorinated paraffins which, moreover, impart a drier surface, lower flammability, and lower volatility.

Finally, there is a group which normally costs more than, say, DOP but in return confers improved properties such as permanence and fluidity, and greater flexibility at low temperatures. To this expensive group belong a number of diesters of adipic acid, azelaic acid and sebacic acid; although these do not mix well with PVC on their own, miscibility presents no problems when they are used in combination with DOP.

8.20 TESTING OF PLASTICS

When plastics first came into use, attempts were made to assess their properties using test methods which had been employed for such materials as metals and ceramics. As the fields of application for the new materials widened, it became clear that they had characteristics which differed markedly from those of the longer established materials. The mechanical properties of plastics were observed to be dependent on time, temperature, and load, in contrast to those of the conventional materials for which dependence on such factors usually only became noticeable at very high temperatures.

It was obviously necessary to be in a position to obtain reliable test data in order to exercise production control, carry out standardization, meet the demands of customers, identify applications, and enable research and investigation to be carried out. Over the years, therefore, special tests more suited to the characteristics of the materials have been devised and used, and conventional tests modified for application to the new materials. The material manufacturer and the fabricator are interested primarily in working properties like the melt viscosity and stability during processing, while the user is more concerned with such properties as impact strength, bursting pressure, stability of shape, and compressive strength of the final article. With such fields of interest in mind, the tests can be divided into three main groups:

1. Tests carried out on the basic, unshaped material.
2. Determination of the properties of the shaped material.
3. Quality assessment of the end product.

Many of the tests under (1) and (2) have been standarized and enable test values to be compared with a reasonable degree of certainty, while those under (3) are so dependent on the particular applications that it has only been possible to standardize in a few large-user applications, such as mass-produced pipes and crates.

8.20.1 Testing of the basic materials in granular or powder form

a. Chemical composition. For many plastics, the exact composition of the basic materials has a pronounced influence on the properties of the final product. The composition is determined by general quantitative and qualitative chemical analysis. Among the various determinations which should be made are: the nature and amounts of the components of the resin, the plasticizer content, the amount of stabilizer present, and the filler content.

b. Water content. A poor moulding may result from too high a water content of the material. The free water content can be determined analytically, but a satisfactory estimation may also be obtained by weighing a quantity of the powder and then drying it to a constant final weight.

c. Flow properties. In processing plastics, the flow characteristics of the molten material are of obvious interest.

An early test devised to examine the flow behaviour of thermosetting moulding powders was the ‘cup flow test’. A test cup or beaker of standard size is moulded in a mould of specified dimensions under fixed conditions. The time of flow is measured in seconds from the instant that pressure has been applied, to the instant the ‘flash’ or overspill at the top of the moulding ceases to move. Moulded cups with bases of varying thicknesses are obtained. The results are expressed as the number of seconds of flow observed.

Another test method, which was devised in the USA, used the so-called ‘Rossi-Peakes’ flow tester. This allows an estimate to be made, not only of comparative plastic flows, but also of ‘curing’ times, and can be applied to both thermosetting and thermoplastic materials. A preformed pellet of the material under test is placed on top of a plunger, carried up into a cavity, and then pressed upward under a standard pressure. The cavity is maintained at a constant standardized temperature throughout and, as the pellet softens, the pressure forces it to flow through a narrow channel. Plastic flow ceases when the material is cured, and the distance of flow along the channel is

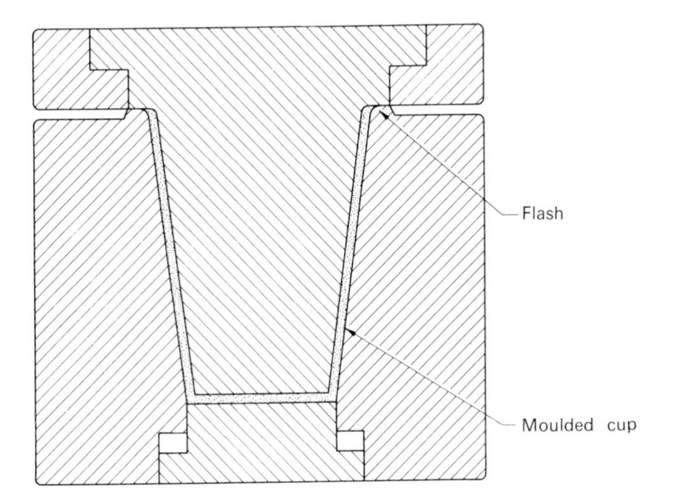

Fig. 8.20 Cup flow test for determining the flow characteristics of a plastic.

recorded on a chart. When a thermoplastic material is tested on the equipment, a moulded or machined test piece is used and, since now no curing takes place, the distance of flow along the channel in a given time, say two minutes, is recorded.

Instruments operating along similar lines to that of the Rossi-Peakes tester continue to be used. The flow through a narrow channel of specified dimensions at different temperatures, pressures, and flow rates, is examined and because of the complex behaviour of polymers in the molten state the conditions are often matched, as closely as possible, to those which will be experienced during actual production moulding.

The flow properties of a thermoplastic material are closely related to its average molecular weight and its molecular weight distribution, i.e. the average chain length and the distribution of the chain lengths. The lower the molecular weight, the more easily the material will flow. Early on, the 'melt index' was taken as a measure of the mouldability of the plastic material. The molten polymer was forced through a specified orifice under standard conditions of temperature and pressure, and the weight of material emerging in a fixed period, say ten minutes, was taken to be a measure of the melt index. The method is no longer used for this purpose; it still serves, however, as an indirect determination of the average molecular weight of the material.

d. Density. The density of a plastic material can sometimes be an indication of its properties. In polyethylene, for instance, a higher density indicates a higher degree of crystallinity and a greater rigidity.

Density can be determined by the usual immersion method, using pieces of a moulding. Another method, often used, is to introduce a particle of the material into a 'density column'. The column is formed in a tall glass cylinder by introducing two liquids of different densities (e.g. xylol at the top, and the heavier carbon tetrachloride at the bottom). Successive layers of mixtures of the two liquids, of progressively differing density, are added. The column is lightly stirred and left to stand. The density is usually calibrated using small glass floats of a known range of densities. The particle of plastic material is dropped into the cylinder, when it will sink to a depth corresponding to its density, which can be read off.

8.20.2 Testing the formed material

These tests are, on the whole, standardized methods of determining such properties as mechanical characteristics, temperature resistance, and ageing. The tests are usually carried out on test pieces which have been moulded, or cut by machining or by punching from sheet. Test results on moulded test pieces can depend on the conditions of moulding. If a test piece is made by machining, then the finish given to it is important.

a. Tensile tests. Tensile strength is the load per unit area of cross section required to break the material under test. Plastics have a wide range of properties. They can, however, be divided into two types: materials which are relatively rigid and have quite a small extensibility, and others which are soft and highly extensible.

In carrying out the test, a moulded sample of the material is clamped between the jaw grips of a tensile machine, and the jaws are then separated. Most modern tensile machines are equipped with recording equipment which measures the applied force and the extension this produces in the test piece. The process is continued until the specimen breaks, when the tensile strength, elongation at break, the yield stress, and the modulus of elasticity, can be obtained from the recordings on the chart. In Great Britain a test piece of dumb-bell shape is used for rigid materials, while for the more extensible materials the test piece is parallel-waisted. In both cases, the test pieces must conform to the dimensions laid down in the standard specifications.

Elongation at break is only of importance in studying the mechanical behaviour of rigid and brittle materials. For the more extensible materials this elongation can be very high, and is of some practical importance in the quality assessment of raw materials and in the investigation of ageing processes. As regards the latter, a decrease in the elongation at break in an old material compared with freshly produced material is an indication of degradation.

In general, the tensile test results are dependent on the temperature at which the test was carried out and the rate of application of the load. The important factor is not the absolute rate of testing but the relative change in length in unit time.

b. Compressive strength tests. In this test, a suitable test piece is compressed between two parallel metal plates. The test piece can be cylindrical or rectangular in cross-section; hollow tubes have also been used. The ratio of height to diameter of the test piece is specified to avoid buckling under compression. A continuous record of the load and contraction is obtained.

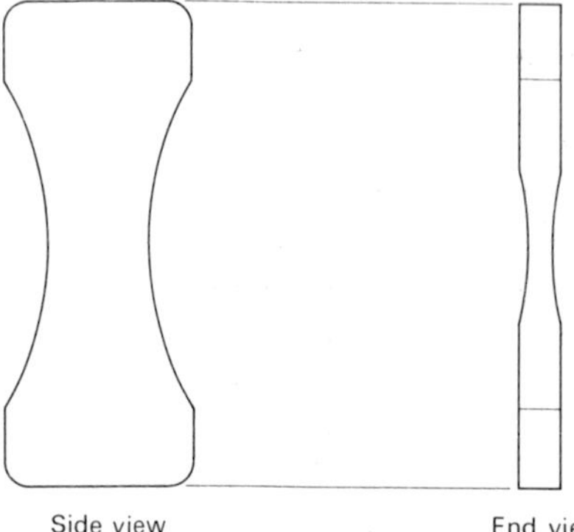

Fig. 8.21 Type of test piece used for tensile tests on rigid plastics.

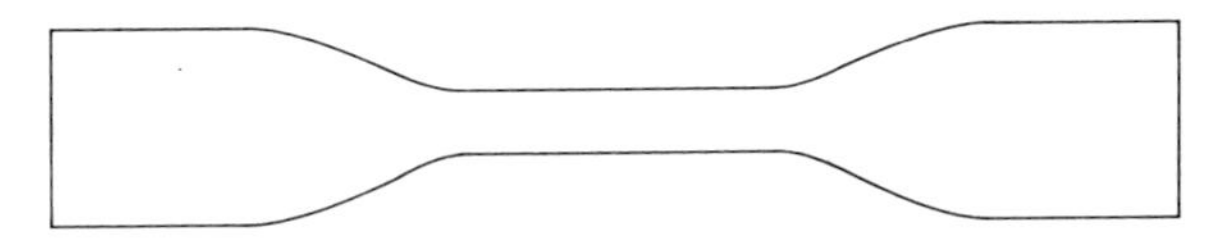

Fig. 8.22 Type of test piece used for tensile tests on extensible plastics.

For brittle materials, the breaking strength can be determined and is of some limited value. With more flexible materials, it is only possible to determine a yield point.

c. Determination of bending strength. In this test, a test piece of rectangular cross-section is supported near its ends and a central load is applied. The recording equipment indicates the deflection of the bar at various values of the load. For brittle, rigid materials, the bending strength and the deflection of the specimen at break can be found. For more flexible materials, the force required to produce a specified deflection is noted. Test bars of standard shape and size are used, but the dimensions differ from one country to another. The rate of bending can affect the results significantly and is specified in the various test procedures.

d. Impact strength determination. This test provides information regarding the ability of a plastic material to withstand sudden or impact loads under the conditions of the test. The test can be carried out using an excess of striking energy, in which case the amount of energy absorbed to produce fracture is determined, or it can be performed to determine the average amount of energy to break a certain number of the test pieces used, say half.

In one type of machine the test specimen is rigidly fixed as a cantilever in an upright position so that it will be struck at the bottom part of the swing of a loaded pendulum arm, the length of the pendulum arm, and the angle at which it is released, being specified. The height the pendulum arm reaches after fracturing the test piece, in comparison with its starting position, allows the energy absorbed to be calculated.

Another type of test depends on impact at different energy levels, either by varying the dropping height of a constant weight, or by using a constant height and varying the size of the weight. In the first of these cases, account must be taken of the fact that the

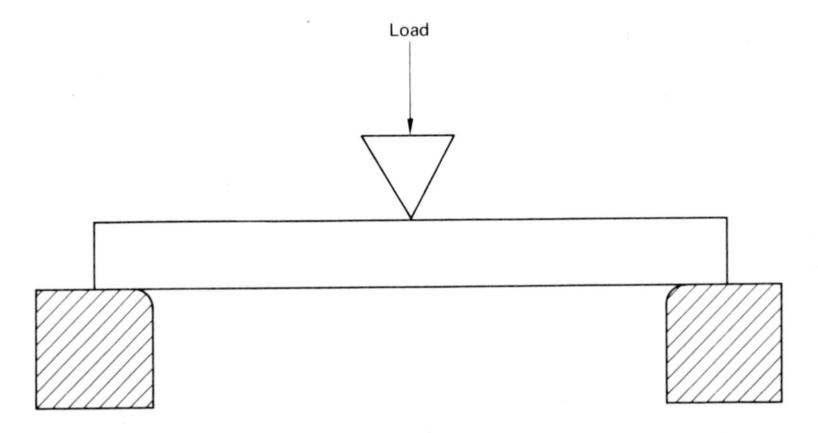

Fig. 8.23 Testing the flexural properties using a standard moulded plastic bar.

speed at impact is changing in addition to the energy of impact. This means that results obtained by the two methods are not necessarily comparable. Either method allows the establishment of an average impact value from several specimens or a figure for the energy absorbed by fracture when say, half the number of specimens are broken.

Impact tests can be carried out on notched or unnotched test pieces; the shape and size of the notch is specified. The difference between the two sets of results indicates to a designer how far he can go in using re-entrant angles in his design if a satisfactory impact strength is to be retained.

e. Hardness tests. For certain types of plastic, PVC for instance, which is widely used as a floor-covering material, it is of value to know how it resists indentation under load. The specimens to be tested for 'hardness' are of a specified thickness and greater than a specified area. The test piece is supported on a rigid, horizontal surface and a plunger, terminating at its lower end in a ball of a stated diameter, is pressed on to the specimen under load for a period of, say, 5 seconds. At the end of the specified time the load is removed, and a measurement of the residual indentation is made after a standard recovery period has elapsed. Alternatively, the depth of penetration of a ball of specified dimensions, measured a standard time after a fixed load has been applied, is sometimes used as a method of test.

f. Temperature stability tests. It is of obvious practical importance to determine permissible minimum and maximum temperatures of use. The only satisfactory methods are to carry out short-term and long-term loading and impact tests at different temperatures, in order to be able to indicate the limiting temperature at which both the short and long-term properties of the material are still acceptable for a particular application. It is also necessary to determine whether any unacceptable ageing of the material, which will result in embrittlement, for instance, will take place at this temperature.

For high-temperature testing, the following methods have been laid down:

1. In one method, a three-point bending load is applied to a test bar immersed in an oil bath. The temperature of the bath is then raised at a rate of 120°C per hour, and the temperature at which a certain deflection of the bar is obtained is noted.
2. In the Martens test, a test bar is clamped at one end and a downward load applied at its free end. The temperature of the environment is then raised at a rate of 50°C per hour. The temperature at which a specified deflection of the loaded end of the bar occurs is noted.
3. In the Vicat temperature test, a needle point is placed on a plate of the material to be tested with a specified load on the needle. The whole is then heated up in an oil or air bath at a rate of 50°C per hour. The temperature is noted at which the needle penetrates a depth of 1 mm into the test specimen.

Such tests are useful for production control. They are of particular value when comparing the softening characteristics of materials of the same type, e.g. different grades of PVC or ABS plastics.

In order to study the characteristics of plastic materials at low temperatures the torsional modulus is measured as the temperature is progressively lowered. This enables the temperature to be determined at which a specified degree of rigidity is obtained.

A further method of test, often used in the USA, is to carry out a special impact test in a cooling bath and determine the temperature at which half a specified number of test specimens are fractured. The test is of use in comparing plastics of the same type with each other. A special version of the test procedure exists to enable it to be carried out on plastic films.

g. Wear tests. A wear test is available in which a test piece, under a specified load, is held against a rotating abrasive disc. The number of rotations required to produce a definite reduction in thickness of the specimen, or a specified loss in weight, is taken as a measure of the material's resistance to wear.

h. Ageing tests. A number of techniques have been used for the accelerated ageing of plastics. The purpose of such tests is to obtain an acceleration of the ageing such as would occur in use, by the use of higher temperatures, higher concentrations of the surrounding medium, and continuous and high intensity illumination. While an exact correlation with the results obtained in normal use is not possible, the methods are of value for indicating differences between the members of a group of plastics. The types of tests carried out are: ageing for fairly long periods at an elevated temperature and noting how properties such as tensile strength, impact strength, and elongation at break are affected by the temperature used; the exposure of test pieces to intense sources of artificial light, often accompanied by simulated rain, and determining the effect on the visual appearance and the mechanical test results; determining the changes in the mechanical test results brought about by boiling the specimens for a given time: determining how a particular material will withstand contact with various chemicals and solvents, the criterion of change being an alteration in appearance or in the results of mechanical tests.

i. Other tests. Tests for other properties, such as the thermal expansion coefficient, electrical properties, water permeability of films, thermal conductivity and specific heat, give results of value for particular applications. These are carried out by conventional methods and the details need not be discussed here.

8.20.3 Testing the end product

Tests can be carried out on the product as a whole, or on specimens cut from it. The latter may not give completely reliable results; internal stresses which may have existed in the article can be relieved when the specimens are cut, and the article may exhibit an orientation with, for instance, a greater strength in one direction than in another. Wherever possible, it is desirable to carry out tests on the complete article.

a. Deformation behaviour. If a product is to be used under a constant load, it is important to test it for its resistance to such a condition. Usually, the tests are made at different loads and temperatures, the deformation obtained with time being observed. From the results obtained, the behaviour under the practical conditions which will be met can be determined by extrapolation. Bottle crates and plastic pipes have been examined in this way.

b. Long-term strength. Pipes have been tested for their ability to withstand internal pressures. The pipe must support a specified internal pressure for a given length of time according to the type of plastic used.

It is possible for a product to develop cracks in the course of time as a result of internal stresses which are present, without any external load being applied. Accelerated tests have been carried out by immersing the product in a bath containing a solution of a surface-active agent at an elevated temperature.

c. Determination of impact strength. An impact load can be applied to certain shapes of product–pipes for instance. In one test, a large number of plastic pipes are built up into a V-shaped heap and a specified blow is given to it; the number of pipes which are broken is determined. For large-diameter pipes, the article is tested for impact strength at various points round its circumference.

The whole plastic article can be subjected to a drop test. The energy applied will depend on the weight of the article and the height of drop. Bottles are often tested in this way by being filled and then allowed to drop from varying heights on to a concrete floor or metal plate; crates are also often subjected to a drop test.

d. Determining changes in shape. Internal stresses may be present in the product, and these could cause changes in shape, and distortion in the course of time. The test carried out consists in holding the product at a certain temperature, depending on the type of material used, and noting its dimensions before and after treatment. In addition, after being heated, the product is examined for the development of cracks, holes, and blisters.

e. Flammability. All plastics are organic substances and will burn under the right conditions, but the ease with which they ignite and combust depends on the type of plastic and its physical form. The intrinsic flammability of the commonly-used plastics has already been mentioned in the appropriate sections on properties and uses. In general, the thermosetting materials have good fire resistance. Thus, the phenolics blister at 200° to 300°C when exposed to heat, then decompose and char without melting. The ignition point is high–between 600° and 900°C. Thermoplastic materials melt when heated; they may also decompose and ignite. The flammability of some thermoplastics can be reduced by chemical modification. For example, polyesters may be rendered slow-burning by cross-linking, and the flame extinction improved by incorporating halogenated acids. PVC is self-extinguishing because hydrochloric acid is set free when it is decomposed by heat. Cellulosic plastics may ignite readily or with difficulty. Cellulose nitrate, for example, has an ignition point of 300° to 375°C and must be regarded as a highly flammable and dangerous substance, but fortunately other cellulose derivatives, e.g. esters and ethers, have a better fire resistance.

Flammability can be assessed in terms of a burning rate (in inches or centimetres per minute, or square inches or centimetres per second) if the conditions (such as the thickness in the case of sheet material) are specified. Such tests are useful for comparing the fire resistance of material in the same form.

In practice the behaviour of plastics depends not only on the intrinsic flammability but also on the physical form, notably the ratio of surface area to volume. Liquids may constitute a fire hazard, especially if volatile solvents are present. The risk may be minimized by good house-keeping. Compact solids, with a low area/volume ratio, do not easily ignite, and they burn only slowly; end-products in thermosetting plastics are often of this type. Extended solids (e.g. film, fibres, fabrics and foam) have a large surface area and some are highly flammable (e.g. cellulose nitrate film, polystyrene foam) if there is ready access to air. On the other hand nylon fabric does not burn easily because the plastic is self-extinguishing. Where possible, extended solids should be made of plastics of high intrinsic fire resistance.

f. Biodegradability. The capacity of certain plastics to break down under the influence of sunlight, moisture, air and micro-organisms – their biodegradability – is becoming increasingly important as the use of plastics for disposable end-products continues to grow. Millions of tons of containers and wrappings are discarded every year, and some of this gives rise to unsightly accumulations of waste material and problems in disposal as garbage.

Plastics are in general durable and highly resistant to weathering. These qualities, so advantageous in service, can be a drawback when the product has served its purpose and is thrown away. The problem is acute with single-use material such as food cartons, packaging, non-returnable bottles and similar consumer products. Although at present these plastics form only about 2 to 3% by weight (but about double this in volume) of all refuse, the proportion is increasing. In the USA the usage of throw-away plastics is greater, and so is the disposal problem. Articles made of natural cellulose, e.g. paper cartons, wrapping and boxes, disintegrate gradually by weathering and the action of micro-organisms, and if collected are easily compressed and finally burnt. Plastic containers are bulky and not easily decomposed. If incinerated they may give off unpleasant fumes; left on open tips they will not rot down.

To deal with this problem biodegradable plastics are being developed. These are designed to disintegrate on exposure to the weather, ultra-violet light, micro-organisms, or a combination of these, so that they function normally in service but decompose when discarded. A difficulty here is that the products are likely to have limited shelf life and it would not be easy to ensure that they retained their integrity until disposed of. Thorough testing would be necessary. An alternative approach is to substitute repeated use for single use where appropriate. Thus, milk is packaged in disposable waxed cartons or in glass bottles with a life of some 50 journeys. Non-returnable bottles would present a garbage problem, but a re-usable plastic bottle might be an attractive substitute for glass. Another partial solution to the pollution problem is to collect and then 'recycle' the plastic waste by converting it into a usable, if lower-grade, material or by incorporating it in concrete, building board etc. The success of such operations depends on the availability of the waste, the cost of collecting and processing it, and the value of the product.

8.21 STATISTICS

8.21.1 Polyolefins

The tables below provide details of consumption in the major areas, covering a five-year period, and also provide details of the divisions between different applications for 1968 and 1971.

Table 8.8 Polyolefin consumption ('000 tonnes)

	1967	1968	1969	1970	1971
Polypropylene					
USA	260	340	455	415	490
Japan	165	275	350	450	460
W. Europe	151	195	245	310	345

Table 8.8 (continued)

	1967	1968	1969	1970	1971
Low density polyethylene					
USA	1 100	1 310	1 530	1 725	1 890
Japan	450	485	600	635	706
W. Europe	1 080	1 300	1 665	1 940	2 210
High density polyethylene					
USA	445	490	640	675	745
Japan	135	165	235	245	289
W. Europe	370	400	555	625	685

Source: Shell International Chemical Company.

Table 8.9 Applicational patterns 1968 (%)

	UK	*Japan*	*USA*
Low density polyethylene			
Film	52	60	47
Extrusion coating	5	13	14
Wire/cable coating	9	7	13
Injection moulding	17	10	16
Pipe/tube	3	1	3
Blow moulding	11	9 (Blow moulding and Others combined)	2
Others	3		5
Total	100	100	100
High density polyethylene			
Injection moulding	32	37	25
Fibres*	4	28	—
Blow moulding	48	21	48
Pipe	4	2	6
Wire/cable coating	4	—	4
Film/sheet	4	9	6
Others	4	3	11
Total	100	100	100
Polypropylene			
Injection moulding	52	39	52
Fibres*	30	26	28
Sheet/pipe	3	—	2
Film	14	23	10
Others	1	12	8
Total	100	100	100

*Includes film tape and monofilament.
Source: Shell International Chemical Company.

Table 8.10 Applicational patterns 1971 (%)

	UK	*Japan*	*USA*
Low density polyethylene			
Film	57.5	59.0	57.0
Extrusion coating	4.5	12.4	9.4
Wire/cable coating	7.9	7.5	10.8
Injection moulding	14.2	6.7	14.8
Pipe/tube	2.0	0.7	1.8
Blow moulding	9.7	3.1	1.4
Others	4.0	10.6	4.8
Total	100	100	100
High density polyethylene			
Injection moulding	29.8	37.1	26.2
Fibres†	3.5	26.6	0.6
Blow moulding	49.3	22.0	51.0
Pipe	4.8	1.6	7.1
Wire/cable coating	1.0	—	2.2
Film/sheet	6.3	8.6	5.5
Others	4.3	4.1	7.4
Total	100	100	100
Polypropylene			
Injection moulding*	51.3	43.7	45.0
Fibres†	29.4	21.9	32.0
Sheet/pipe	2.2	2.0	1.1‡
Film	14.8	23.6	9.7
Others	2.3	8.8	12.2
Total	100	100	100

*Includes minor quantities for blow moulding.
†Includes film tape and monofilament.
‡Pipe only.
Source: Shell International Chemical Company.

8.21.2 Polystyrene

Table 8.11 gives the production figures in 1.000 metric tons for some of the major producing countries.

Table 8.11 Production of polystyrene

	1966	1967	1968	1969	1970	1971
USA	1081	1084	1313	1516	1609	1700
France	88	87	98	117	131	149
Italy	97	117	139	158	184	181
UK	106	113	129	148	162	161

Source: Growth of World Industry, United Nations, 1972.

There is a tendency for the total styrene plastics consumption to level off, the production of straight polystyrene to fall, and the production of ABS plastics to increase. The increase is at a higher rate than predicted. The USA production figures for 1970, as estimated in 1964, were actually reached in 1967. This must be attributed to the good properties of the ABS plastics. On the other hand the latter are much more expensive and therefore less used. The situation in mid-1968 emerges from the following American prices in cents/lb.

Acrylonitrile	14.50
Butadiene	11.75
Styrene	8.50
Polystyrene	15.50
Impact polystyrene	18–22
Styrene-acronitrile copolymer (SAN)	26
ABS plastics	32–36

Source: Modern Plastics Encyclopedia and *European Chemical News.*

The figures show clearly that the margin between the raw material price and the plastics price is currently much smaller for polystyrene than it is for ABS plastics. It can be expected that in the future the prices, in particular for ABS plastics, will drop considerably while consumption of the latter will gradually increase, partly at the expense of straight polystyrene and impact polystyrene.

8.21.3 Polyvinylchloride

Table 8.12 gives the production figures in 1.000 metric tons for some of the major producing countries.

Table 8.12 Production of polyvinylchloride

	1966	1967	1968	1969	1970	1971
USA	981	971	1 195	1 375	1412	1 556
France	235	255	280	369	412	457
Germany	409	487	622	732	777	846
UK	202	228	271	283	315	314

Source: Growth of World Industry, United Nations, 1972.

Of the total production capacity in Western Europe, upwards of 98% is nylon 6 and 6/6. The production of nylon 11 is only 1% of the total; nylon 6/10 together with all the other polyamides make up the remainder.

8.21.4 Polyamides

Table 8.13 gives in round figures (in 1.000 metric tons) the world production of polyamides.

Table 8.13 Production of polyamides

1963	750
1965	1090
1967	1610
1969	1790
1971	2225

8.21.5 Urea and melamine

Output has risen sharply over the past 20 years, especially in the fields of binding agents and paper treatment. The drop in the prices of the raw materials, in particular melamine, has substantially contributed to the steadily rising sales. It should also be borne in mind that both urea and melamine are by-products of fertilizer manufacture based on nitrogen compounds.

World distribution over the major fields of consumption for 1962–1965 was as follows:

Adhesives – for wood	20%
Other materials	15%
Moulding powders	25%
Coatings	8%
Textile treatment	12%
Paper treatment	20%

In the United States, the total production rose from about 100 000 tons in 1950 to over 200 000 tons in 1962, amounting to 6–7% of the overall plastics production in the United States. Other important producers are Japan, Britain, West Germany, and France.

8.21.6 Urethanes

Distribution of the market for methane is given in table 8.14.

Table 8.14 Distribution of the urethane market (in '000 tons)

	1963	*1968*
Flexible foam	72	125
Rigid foam	22	63
Elastomers	7	20
Fibres	3	10
Adhesives	5	9
Coatings	5	15
	114	242

8.21.7 Cellulose

Statistics for the United States may be representative of world trends. These show that output of cellulose nitrate fell from nearly 400 m. lb in 1946 to a mere 20 m. in 1961. By contrast the collective cellulose esters remained fairly constant around 150 m. lb from 1946–66, then increased to nearly 200 m. lb in 1969. It is forecast that production will gradually increase, reaching a possible 300 m. lb by 1980 The more sophisticated cellulosics have established for themselves areas of special application which they are likely to hold against the synthetic resins despite a price disadvantage.

8.22 LITERATURE

A. E. LEVER (Ed.). *The plastics manual.* London, Scientific Press, 1966.

J. A. BRYDSON, *Plastics Materials.* London, Iliffe, 1966.

J. G. COOK, *Your Guide to Plastics*, Watford, Merrow, 1968.

J. H. BRISTON and C. C. GOSSELIN. *Introduction to Plastics*, London, Newnes, 1968.
W. M. SMITH. *Vinyl Resins*. New York, Reinhold, 1958.
T. H. FERRIGNO. *Rigid Plastic Foams*. New York, Reinhold, 1967.
T. T. HEALY. *Polyurethane Foams*. London, Iliffe, 1964.
G. OTT, H. M. SPURLEN, and M. W. GRIFFIN. *Cellulose and its Derivatives*. New York, Interscience, 1954.
V. E. TARSLEY, W. FLAVELL, P. S. ADAMSON, and N. G. PERKINS. Cellulose Plastics. London, Iliffe, 1964.

CHAPTER 9

Photographic Materials and Processes

9.1 HISTORY

9.1.1 Early experiments

The first step to present-day photography was reported by Aristotle who, about 350 B.C., described the principle of the *camera obscura*, as illustrated. Light passing through a hole in one wall of a room forms an inverted image, of the scene outside, on the opposite wall of the room. The larger the hole the brighter the image, but the less sharp it is. Later, in the sixteenth century, a lens was used instead of a simple hole to give a brighter image without loss of sharpness. Both in the earlier and later forms, various designs of camera obscura were used by artists to guide them in drawing and painting—and ultimately the form with lens became the essential form of the camera as we know it today.

Although the basic principle of the photographic camera was already in existence, nearly two hundred years elapsed before serious attempts were made to convert the optical image, produced by the camera, into a permanent image by means of light-sensitive materials, i.e. without tracing with a pencil or paint. The light-sensitivity of silver salts had already been discovered by the German physician J. H. Schulze in 1727

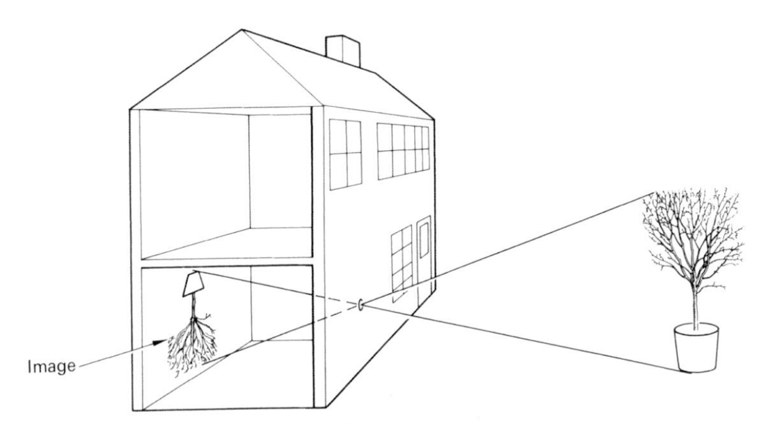

Fig. 9.1 Early form of *camera obscura*.

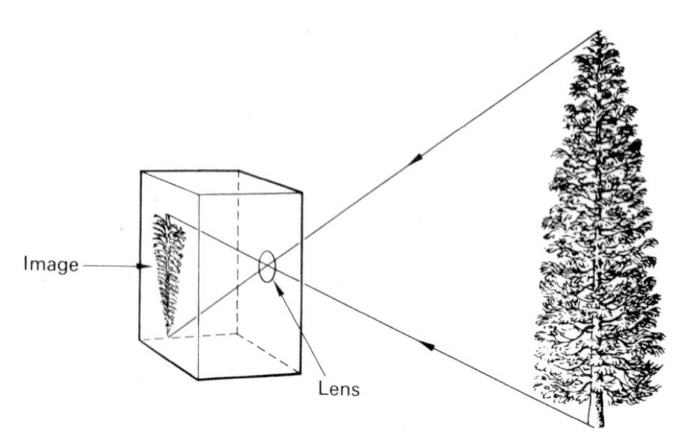

Fig. 9.2 *Camera obscura* with lens forming the basis of the modern camera.

who recorded the formation of shadow images in a mixture of chalk and silver nitrate, but he did not propose their use in photography. The first attempts to use light-sensitive materials for the production of permanent images with a camera were made by the Englishmen Sir Humphry Davy and Tom Wedgwood in 1802. They produced an image on a paper coated with silver chloride but the paper soon blackened completely due to the presence of unexposed silver chloride.

The first person to make a stable photograph with a camera was the Frenchman Josephe Nicéphore Niépce in 1816. He used a metal plate covered with a thin layer of bitumen which hardened on exposure to light, so that, by dissolving in oil of lavender the portions not exposed Niépce obtained an image in relief in the bitumen.

Meanwhile Louis Daguerre carried out experiments, but without success, with a simple wooden camera provided with a silvered copper plate, on which a layer of silver iodide had been formed by means of iodine vapour. Daguerre continued these experiments in association with Niépce and, after the death of the latter in 1833, in collaboration with Niépce's son Isidore. By chance, Daguerre found in 1837 that with the aid of mercury vapour an image could be formed on an exposed silver iodide plate, the mercury adhering to the exposed areas. He dissolved the unexposed silver iodide with a concentrated salt solution and the result was a positive image of light mercury amalgam and relatively dark silver. Later he replaced the salt solution by a solution of sodium thiosulphate to dissolve the unexposed silver iodide.

9.1.2 Development of modern silver halide processes

In 1841, following on experiments over eight years, the Englishman William Henry Fox Talbot made photographs on transparent paper which had been made light-sensitive with silver iodide, and which after exposure was treated with a solution of gallic acid and silver nitrate. This solution had the property of precipitating silver in those places where small grains of silver (produced by the decomposition of silver iodide by the action of light) are present. After dissolving the unexposed silver iodide by means of sodium thiosulphate (hypo) and drying, positive prints were made by exposing another piece of sensitized paper in contact with the negative; and then developing the positive in gallic acid, fixing in hypo and washing. The technique invented by Talbot constitutes the basis of present-day photography, in spite of the fact that Talbot's photographs were of much poorer quality than those of Daguerre. The poor quality of Talbot's photographs was caused by the fact that the paper did not form a uniformly transparent medium.

In 1847, Niépce de Saint Victor, a nephew of Josephe Niépce, improved Talbot's process by using a glass plate as the support. Since the light-sensitive silver iodide does not adhere to the glass plate he mixed the iodide with albumen in order to ensure good adhesion to the surface.

After the discovery of cellulose nitrate (gun cotton) a process was developed by Le Gray and Scott-Archer independently in which silver iodide in a collodion layer formed the sensitive surface.

In this process potassium iodide (or preferably an iodide-bromide mixture) is added to cellulose nitrate dissolved in alcohol and ether which together form collodion. A glass plate is coated with the collodion solution and, when set, is immersed in a solution of silver nitrate, producing a sensitive layer of silver iodide-bromide in the pores of the collodion. The plate must be exposed while wet and the latent image is developed in a solution of pyrogallol and silver nitrate.

The necessity of using a wet plate was an obvious disadvantage in field work. At a later stage it was found possible to produce a plate that could be dried, and even stored, before exposure. This so-called 'collodion emulsion' plate, like the earlier 'wet plate' was, however, much less sensitive than the gelatine emulsion plate described below.

The major innovation was the invention of a light-sensitive material, composed of silver bromide and gelatin, by Richard Leach Maddox in 1871. This material, applied on a glass plate, could be used like the collodion emulsion but in a dry form and due to this property it superseded the wet process.

In 1873, the German Herman W. Vogel discovered the phenomenon of optical sensitization. Since silver halides are sensitive only to light of short wave length (ultra-violet and blue light) his discovery was of great importance for the improvement of light-sensitive materials. Vogel found, in fact, that silver iodide could be made sensitive to green and yellow light by the addition of certain dyes such as cyanine and quinoline red to the light-sensitive material.

One year later, Becquerel successfully prepared the first silver halide emulsion to be sensitive to red light as well, by using chlorophyll for this purpose.

In 1887 H. Goodwin developed a method for the production of thin transparent and flexible celluloid film from a mixture of cellulose trinitrate, camphor and ethanol, and soon this material was used as a support for silver bromide-gelatin layers. The way was then clear for the development of roll films, miniature films, cine films and the like, as they exist at the present time. Since celluloid films are highly flammable, they have now been replaced by safety bases made of cellulose acetate or other plastics.

9.1.3 Development of colour photography

In 1856, the British physicist J. C. Maxwell demonstrated that all colours in nature can be synthesized by mixing red, green and blue light. Various experiments were performed in the second half of the nineteenth century by producing a picture of the same object on three light-sensitive layers by exposing one of the layers through a red filter, the second through a green, and the third through a blue filter. After development the negative pictures were copied to produce positive pictures and then projected so that their images were superimposed by means of three projectors each provided with the filter of the corresponding negative. In this way a projected picture in true colours is obtained. Since the colours are added to each other on the projection screen the process is named an 'additive' process.

However the development of a good process for the production of colour photographs was not possible before the discovery of optical sensitization and the light-sensitive material based on gelatin and silver bromide in the years 1871–1873.

In 1868 Ducos du Hauron indicated that the three colour records needed for the additive process could all be made on one plate by covering it with a mosaic of microscopic blue, green and red filters. This prompted work in which the mosaic was made up of coloured spots or squares and the resulting image on a panchromatic emulsion was developed in a reversal process to give a positive image. However, it was not until 1903 that the first successful process was marketed. This was the Lumière *Autochrome* plate in which the plate consisted of glass covered with a random mosaic of starch grains dyed red, green and blue. This starch grain layer was covered with a varnish and then over-coated with a panchromatic silver-halide emulsion. The emulsion was exposed *through* the grains and processed to give a *positive* image, which showed the original scene in colour. Later systems used regular geometrical mosaics. Materials of this type were the most important starting points for the production of colour photographs. In the years between 1935 and 1940 the additive processes were superseded by the present-day subtractive colour materials which have three different light-sensitive layers, one on the top of the other.

The process using three different filters is still used in photomechanical production of printing plates. The colour separation images are used to produce the different printing plates by which different colours can be printed, one after the other.

9.1.4 Non-silver halide materials

The light-sensitive property of bitumen (asphalt) has already been mentioned in the reference to Niépce's work.

The second light-sensitive non-silver halide material, which has been of commercial importance, is based on ferric salts. In 1842 Sir John Herschel discovered that various ferric salts are reduced to ferrous salts when exposed to light and that the ferrous salts can be converted into a blue dye (Prussian blue) on treating with potassium ferricyanide. This is the basis of blue-print paper.

In 1852 the Englishman William Fox Talbot discovered that a mixture of gelatin and potassium dichromate hardens and becomes more resistant to water on exposure. In 1852 he filed a British patent application for a process for the production of printing plates based on steel plates covered with this mixture. The hardening of gelatin and various other products by a dichromate salt formed the basis for many photographic processes for the production of printing plates.

The fourth light-sensitive non-silver halide material was based on the discovery of aromatic diazo compounds by Griess in 1858. Green, Cross and Bevan produced the first diazotype copy in 1890 when they discovered that the exposed areas of a diazo-compound coated on cloth lose their property of forming dyes. In 1923 diazotype paper (two-component diazotype paper) was marketed for the first time by Kalle in Germany under the trade name 'Ozalid' and soon (in 1927) a modified version (one-component diazotype paper) was marketed in the Netherlands by Van der Grinten under the trade-name 'Océ'.

More recently many other light-sensitive materials have been developed. Of these, polycondensates of diazonium salts, the naphthoquinone diazides, photopolymers and the photoconductive materials used in electrophotographic reproduction are some of the most important.

9.2 SILVER HALIDE MATERIALS AND PROCESSES

9.2.1 Composition of the light-sensitive material

The light-sensitive material is composed basically of a suspension of silver halide grains in a viscous colloid. The silver halides are silver chloride, silver bromide and silver iodide. The colloid is usually gelatin. Originally the gelatin was only intended to prevent settling of the silver halide and, at the same time, to allow the penetration of treatment solutions. Nowadays it is known that gelatin plays an active role in the processes which result in the formation of an image.

The suspension of silver halide crystals in gelatin is usually described as an 'emulsion'. In fact this is an unfortunate name since the term emulsion is commonly used for a dispersion of a liquid in another, non-miscible, liquid. The term *suspension* would be strictly correct, but it is never used in practice.

As a gelatin layer has a very low strength, layers of a photographic emulsion are always supported by a base material such as paper, plastic, foil, glass or metal.

9.2.2 Mechanism of latent image formation and development

When a silver halide grain is exposed to light, the light is absorbed. Light consists of quanta of energy and when such a quantum is absorbed by a halogen atom of the silver halide crystal, one electron may become excited so intensively that it is released from a halogen ion. This electron is taken up by a silver ion in the grain to form a metallic silver atom. Thus if bromine is the halogen atom:

$$Br^- + \text{light energy} \longrightarrow Br + e^-;$$
$$e^- + Ag^+ \longrightarrow Ag \downarrow$$

When a number of quanta are absorbed by a layer of silver halide crystals, a corresponding appropriate number of silver nuclei are formed in the crystal layer. These nuclei are too small to produce a visible image, but form a latent image which can be made visible by treatment with a developer solution which increases the amount of silver at the places where the nuclei are present.

Two types of different developer are available; namely the physical and the chemical developer.

The physical developer comprises basically a silver nitrate solution which contains in addition a suitable reducing agent. This solution is able to deposit silver from the silver nitrate on to the nuclei of the silver halide layer.

The chemical developer is made up of an aqueous solution of a reducing agent which is able to reduce silver halide in the light-sensitized grains to metallic silver. The reducing agents are selected from a group of organic compounds which can only attack exposed silver halide grains. Nearly all modern developers are chemical developers, and for these the development process can be represented by the following equation:

$$\underset{\substack{\text{developing}\\ \text{agent}}}{D} + \underset{\substack{\text{silver}\\ \text{bromide}\\ \text{grain}}}{2AgBr} + \underset{\substack{\text{essential alkali}\\ \text{for developer}\\ \text{to act}}}{2NaOH} = 2NaBr + \underset{\substack{\text{silver}\\ \text{image}}}{2Ag \downarrow} + \underset{\substack{\text{oxidized}\\ \text{developer}}}{DO} + H_2O$$

9.2.3 Production of the light-sensitive emulsion

a. Special precautions. Although the operations in the production of a light-sensitive emulsion are basically simple, the problems are enormous, since the quality and properties of the resulting product are affected by various impurities, temperature, humidity and light.

The presence of impurities such as hydrogen sulphide and sulphur dioxide in the surrounding air may be disastrous. Also care must be taken that the atmosphere is free from dust particles which become visible when an image on a photographic film is enlarged many times. Dust has been a special problem in periods of relative high radio-activity since radio-active dust particles which are far too small to become visible may effect the light-sensitive emulsion during storage.

For all these reasons the production of photographic materials must be performed in air-conditioned rooms. In addition to purification, air conditioning must also include rigid temperature and humidity control, to ensure that the light-sensitive emulsion gets the desired properties.

Silver halides without dye additions are sensitive to blue and ultra-violet light. For this reason all the processes with these materials must be carried out in red light. When sensitized with dyes their spectral sensitivity is extended and consequently they can be handled only in low intensity light of a colour to which they are insensitive. With orthochromatic materials, sensitive up to about 540 mμ, red light may be used but materials sensitive to the whole visible spectrum (known as panchromatic) can only be made in the dark.

b. Starting materials. The starting materials must be very pure. Even traces of impurities can be responsible for a completely valueless product. The purity must often be tested by making test emulsions since quantities which are too small for chemical analysis may affect the properties of the light-sensitive emulsion.

The major starting materials are water-soluble halogen salts such as potassium bromide silver nitrate, water and gelatin.

The silver nitrate must be free from heavy metal salts, such as copper and lead salts, as even very small traces of these can decrease the light sensitivity of the emulsion.

The water must not only be free from dust and dissolved salts, but also organic matter and microorganisms must be removed completely.

Gelatin is a substance the properties of which vary greatly depending on the source and the production method. Gelatin contains varying amounts of various products which promote or inhibit the ripening process (see below) and as a result of this the light-sensitivity of the emulsion may increase or decrease accordingly. In addition the properties of the resulting emulsion are affected by physical properties such as the viscosity, solidification point and acidity of the gelatin. The relation between the composition and properties of the gelatin on the one hand and the properties of the light-sensitive emulsion on the other hand is not clear and is usually determined empirically by preparing test emulsions (see also 9.2.3d).

c. Production of the emulsion. Gelatin and one or more halogen salts are carefully mixed with pure water at a temperature of about 50°C to form a collidal solution. An ammoniacal silver nitrate solution is then slowly added with stirring at a carefully controlled temperature which is usually not higher than 50°C. Very fine crystals of silver halide are then precipitated in the gelatin emulsion which contains an excess of alkali halide. In an alternative precipitation process a silver nitrate solution, without ammonia, is slowly added at a temperature selected between 60 and 75°C.

The resulting product, which is still not highly light-sensitive, is allowed to remain for a certain time at a constant temperature which is usually chosen between 40 and 50°C. During this first ripening process small silver halide crystals dissolve in the excess of alkali halide solution and the dissolved silver then crystallizes again on to the surface of the larger silver halide crystals, as the result of which the light-sensitivity is increased and the silver halide grains attain approximately their final size and form. The ripened emulsion is allowed to solidify by cooling and is forced through a sieve as a result of which thin shreds (noodles) are formed, which are washed with cold clean water so that the soluble nitrates and the excess of alkali halides diffuse out.

The purified product is heated again to a temperature which is usually between 45 and 55°C and subjected to the after-ripening or chemical sensitization process, which is sometimes called the second ripening stage. In this process the silver halide crystals do not grow further but are modified by the various substances contained in the gelatin as the result of which the light sensitivity increases again.

During, or at the end of the after-ripening, various substances are introduced into the emulsion, depending on its type. These include, among others, optical sensitizers to obtain correct colour sensitivity, potassium bromide or special organic compounds to prevent fogging, saponin and alcohol which affect the surface tension of the emulsion, tannic acid, formalin or alum to harden the gelatin, and starch to reduce shininess. A flow sheet of the manufacture of light-sensitive emulsion is shown on the following page.

d. Control of the properties of the emulsion. The size of the silver halide crystals increases when the temperature or the treatment time in the precipitation or first ripening process is increased and the result of this is a higher sensitivity and in general a lower image contrast. The increasing light-sensitivity with increasing crystal size can be explained by assuming that, once a crystal contains a single silver nucleus, as a result of exposure to light, it is blackened completely on development so that with large crystals a relatively small number of nuclei can cause an extensive blackening effect.

The light sensitivity can also be modified by using different halides. Silver chloride has a lower light sensitivity than silver bromide and the sensitivity of silver bromide can be increased by precipitating a silver bromide-iodide from a bromide solution containing 2–8% by weight of an iodide. The addition of silver iodide also results in a lower imagé contrast.

The light sensitivity of an emulsion can be markedly increased by mixing a complex gold compound such as potassium aurothiocyanate with the gelatin.

Some constituents of gelatin such as thiourea and decomposition products of gelatin are active as sensitizers in the second ripening process and can increase the light sensitivity considerably. It is believed that the sulphur compounds present in the gelatin produce small spots of silver compounds on the surface of the silver halide grains and thus affect their sensitivity.

An increased understanding of the effect of various compounds and the second ripening process has made it possible to prepare films of extreme sensitivity without increasing the crystal size to an excessive degree.

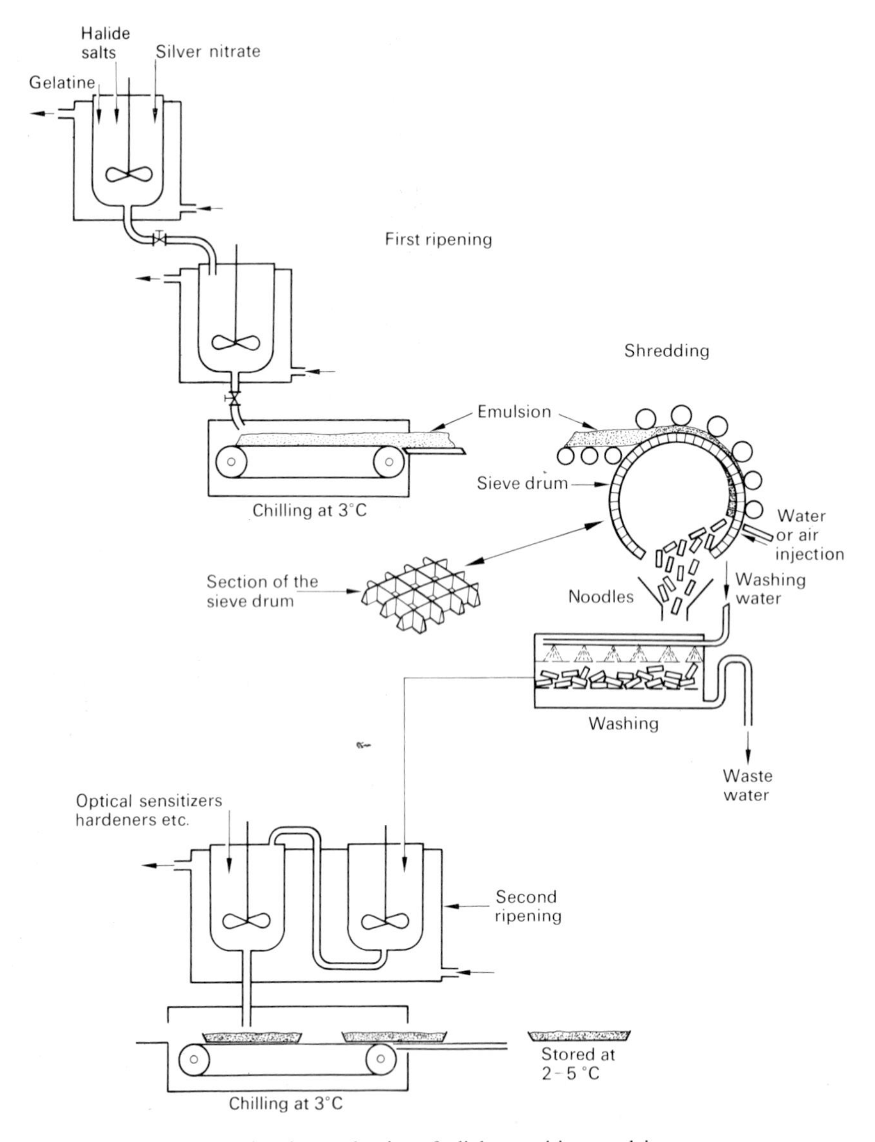

Fig. 9.3 Diagram showing the production of a light-sensitive emulsion.

9.2.4 Production of photographic films and papers

A light-sensitive emulsion based on gelatin must be provided with a supporting base material. All base materials must be free from traces of substances which can attack the emulsion (see precautions in section 9.2.3). To avoid contamination film and paper bases must be manufactured in air conditioned rooms and water used in the production processes must be free from dangerous chemicals (especially sulphur compounds), dust and radioactive particles. The elimination of dust is particularly important in processes for the production of transparent films.

a. Film bases. Films used as an emulsion base must satisfy certain special requirements as to flexibility, strength, transparency, dimensional stability, resistance to various chemicals, etc.

The dimensional stability is very important and particularly so in various copying processes such as the reproduction of technical drawings and graphic arts. Accordingly the base should have high dimensional stability when subject to heat and moisture. Celluloid satisfied such requirements but because of its high flammability has been replaced by materials such as cellulose acetate-butyrate, styrene, polyethylene terephthalate and bisphenol polycarbonate.

The surfaces of all these materials must be specially prepared, e.g. by an intermediate coating, to ensure that the gelatin emulsion adheres.

b. Paper bases. Paper used for contact printing and enlarging paper must also satisfy several requirements. It must have a high wet strength and resist the various chemicals, including alkalis and acids present in developing and fixing solutions. In addition the paper must not contain substances which can diffuse into the light-sensitive layer or cause stains; its colour must not change under the conditions of treatment and its dimensions must not alter relatively to those of the gelatin layer when the gelatin coated paper becomes moist or even wet.

Usually the paper is made from high purity sulphite or sulphate pulp with a mixture of natural resin, fatty acid and an aminoplast as adhesive, titanium dioxide or pure kaolin as pigment and gelatin or starch as surface coating. The paper is made on a common paper-making machine, but, to avoid contamination with heavy metal compounds, the parts in contact with the paper are made from stainless steel or non-metallic materials. The paper weight varies from about 45 g/m^2 for thin document paper to 240 g/m^2 for double weight enlarging paper.

Some papers are coated only with a thin layer of pure gelatin but usually the paper is coated with a mixture of gelatin and barium sulphate (baryti) together with dyes for tinting purposes and formalin or chrome alum to harden the gelatin. The mixture of gelatin solution and baryti is usually applied to the paper by means of rollers and evenly distributed over the surface by reciprocating brushes.

c. Glass as a base material. Glass plates coated with a light-sensitive emulsion are still used but the consumption is steadily declining. Before applying the emulsion the glass is coated with an intermediate layer of waterglass or of gelatin which has been hardened with chrome alum to ensure that the emulsion adheres to the glass.

d. Coating of films and paper with emulsion. The coating processes are basically the same for film and paper. Films and paper to be coated are available in the form of rolls containing up to around 1 200 m (4 000 ft) of material with a width of about 1.25 m (about 4 ft).

In the trough coating process the base material is dipped into the molten light-sensitive emulsion by passing it beneath a roller which is partially immersed in the emulsion contained in a long shallow trough (see Fig 9.4).

In the roller coating process the base material is also passed beneath a roller but the latter is not immersed in the molten emulsion. The emulsion is applied on to the base material by means of a second roller which is immersed in the emulsion, and transfers the emulsion to the base.

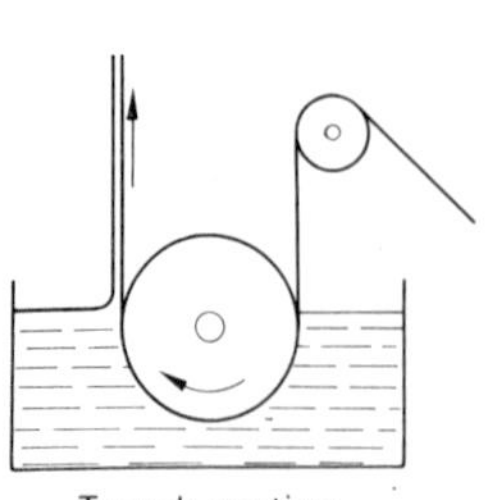

Fig. 9.4 Diagram illustrating trough coating.

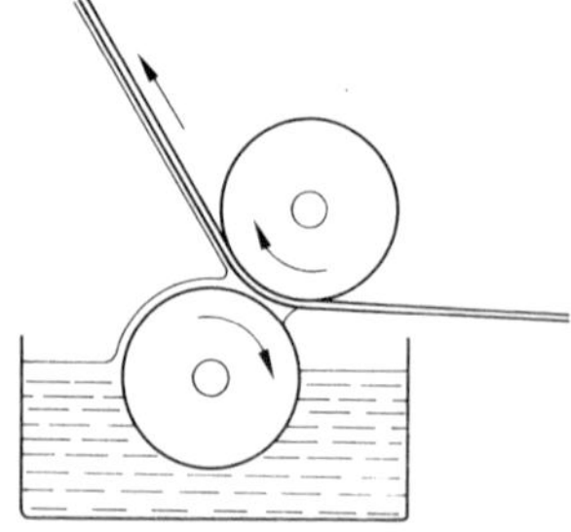

Fig. 9.5 Diagram illustrating roller coating.

The major problem in applying emulsion layers to a base material is the formation of a layer of uniform thickness. An uneven layer results in a picture with darker or lighter areas, and in colour material, areas of different colour would be produced after exposure and development.

To promote the formation of uniform layers the surface tension of the emulsion is lowered, usually by using saponin. In addition the coating must be cooled rapidly. The latter is performed by passing the coated material through a chill box which is kept at 3°C.

If the viscosity and temperature of the emulsion, and the speed of coating are carefully selected, coatings of a thickness down to 2 μ can be obtained. More coatings can be applied on one side of the base material if the emulsion is solidified by cooling after each coating stage. For example, to produce colour films, up to 10 coatings with a total thickness of 25 μ can be applied in this way.

Before entering the drying chamber the coated material is looped by means of rods which are transported by endless conveying chains and the resulting festoons are slowly passed through the drying chambers by means of the same rods (see Fig. 9.6).

Great care must be taken in drying the coated film or paper and, for this reason the drying chambers for coated films and papers are provided with air conditioning equipment (see special precautions in section 9.2.3). Drying is gentle initially after which the temperature is raised, without, however, melting the gelatin. After drying, a short re-humidifying stage is provided to adjust the moisture content in the finished material. The moisture content prevents the formation of electrostatic charges and thus harmful electrical discharges; in addition, the formation of cracks when handled is inhibited.

Most films are also provided with a layer of non-sensitized gelatin on the reverse side. This is necessary because a swollen gelatin layer shrinks markedly on drying, so that the film would curl. In many types of films the gelatin layer on the underside contains the so-called anti-halation dyes, which absorb all the light passing through the film. Without these dyes, this light would be reflected from the rear surface of the film and would pass through it a second time, thus producing a poor quality image.

9.2.5 Properties of photographic films and papers

a. Sensitometry and the characteristic curve. A photographic film or paper which has been exposed to light and developed shows varying degrees of blackening depending on the 'exposure' which it has received. To a first approximation the 'exposure' is given by the product (light intensity × time of exposure), $E = I.t$. In other words, doub-

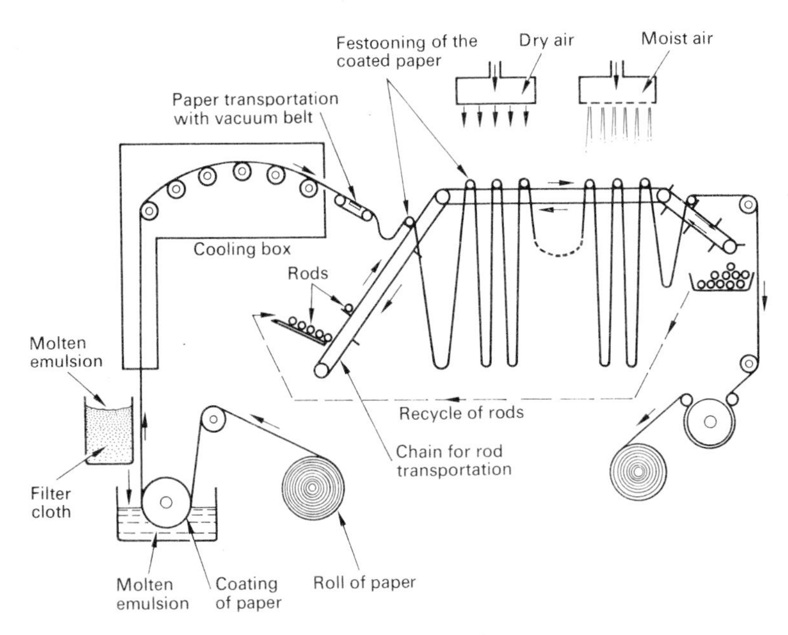

Fig. 9.6 Simplified diagram of a paper-coating plant.

ling the light intensity has the same photographic effect as doubling the time of exposure. This 'reciprocity law', for exposures to light, fails for very short times and high intensities and also breaks down at the other end of the scale with long times and very low intensities, but for practical purposes this 'reciprocity failure' need not be further considered here.

The examination of the effect of exposure and development on photographic materials is known as 'sensitometry'. Even in the days of the daguerreotype plate attempts at speed measurement had been made, but simple measurements of the smallest amount of light necessary to produce the first visible effect on a plate failed to give all the information required. The first classical experiments to give a scientific basis to sensitometry were made in England by Hurter and Driffield and published in 1890.

This work forms a convenient starting point for discussing the properties of photographic materials. When light of intensity I_0 is passed through a developed photographic film a fraction of the incident light is absorbed and the remainder, I_t is passed on. The ratio I_0/I_t is known as the opacity of the film. In practice it is convenient to work with the density, which is the logarithm of the opacity, $D = \log_{10} I_0/I_t$, since the density is also a measure of the amount of silver in the film.

The first step in sensitometry is to subject a strip of film to a graded series of exposures. This may be done either by a constant light intensity through a graded series of densities (step wedge) or by an intermittent series of exposures produced by a disc having cut-out sectors of varying angles rotated in front of a constant light source (sector wheel method). The resulting densities after development are measured by means of a visual or a photoelectric densitometer, the latter being the more usual.

After development the resulting densities, D, are plotted against log E to give a typical 'characteristic curve' as shown in Fig. 9.7. It will be seen that, at the low-exposure end the curve is almost horizontal and that even for zero exposure a certain

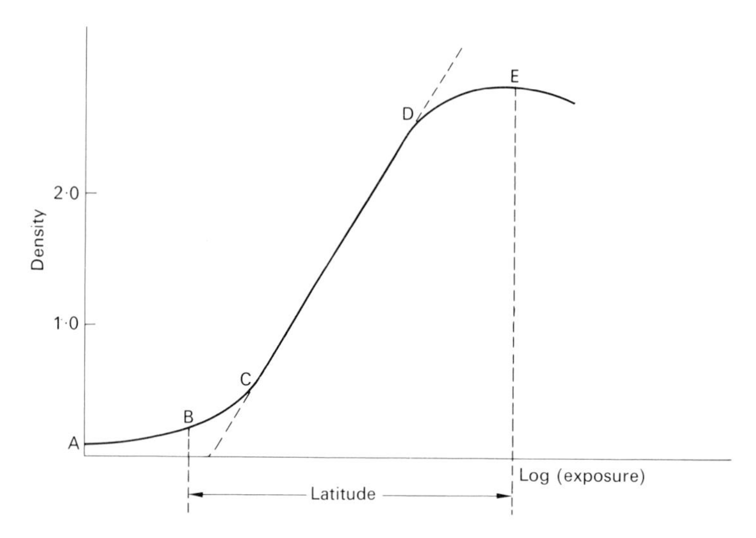

Fig. 9.7 A typical characteristic curve.

small density (the 'fog level') is produced. After a certain threshold exposure the curve begins to rise from B to C, until it reaches the more or less straight portion CD. After this, from D to E, the curve rises less steeply. On the straight portion density is proportional to log E, i.e., $D = \gamma \log E$, where γ is a constant, known as the *gamma*, though occasionally the term *development factor* is used. The slope of the curve (gamma, contrast), γ, increases as development time is increased until it reaches a maximum $\gamma\infty$ (gamma infinity) after which no further change occurs except, perhaps, increase of fog.

In the past a film was considered to be 'under exposed' if the densities produced fell on the proportion BC and 'over exposed' if they fell on DE. Only exposures falling on the straight portion could be considered as satisfactory, since a necessary condition for a true rendering of the subject was that densities should be proportional to log E. For this reason Hurter and Driffield called the straight portion of the curve the 'period of correct representation'. The characteristic curves of modern photographic emulsions do not always show the straight line portion CD. For these materials the *average* slope over a given density range is a measure of the *contrast*, this term being used in preference to gamma which is more usually associated with the slope of the straight-line portion of the curve. Where the two reference points correspond to the approximate minimum and maximum densities used in practice, the slope is called the *contrast index* (Fig. 9.9).

b. Latitude, development factor and gradation. The range of intensities over which a light-sensitive material can be used is the *latitude* of the material. As shown by the characteristic curve the latitude is limited so that a light-sensitive material is able only within certain limits to reproduce both the lightest and darkest portions of the object photographed. The degree of contrast (i.e. the ratio between the lightest and darkest portion) of the object to be photographed is often greater than the maximum attainable ratio between the lightest and darkest parts of a light-sensitive material.

To overcome this difficulty, light-sensitive films are made in such a way that, with suitable development, the contrast of the developed image may be varied. Generally, as illustrated previously, increase of development time increases the contrast and

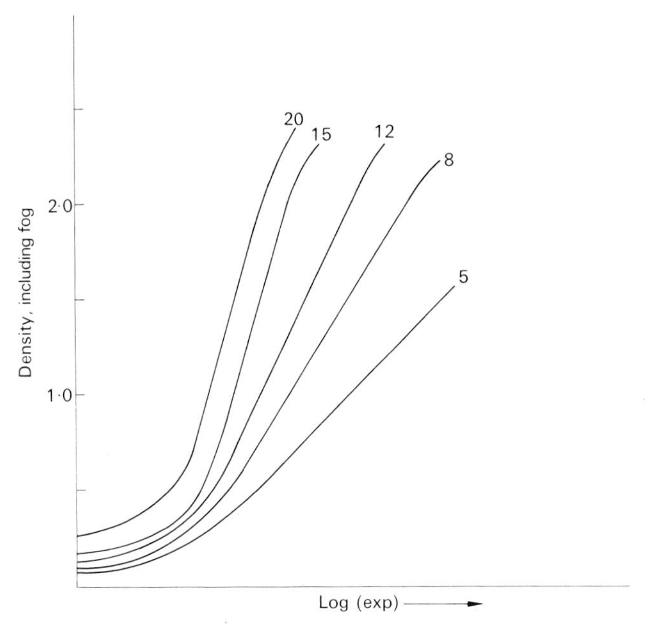

Fig. 9.8 The slope (gamma, contrast) increases at first with increase of development time (5, 8, 12 . . . minutes) and finally reaches a constant value of gamma infinity.

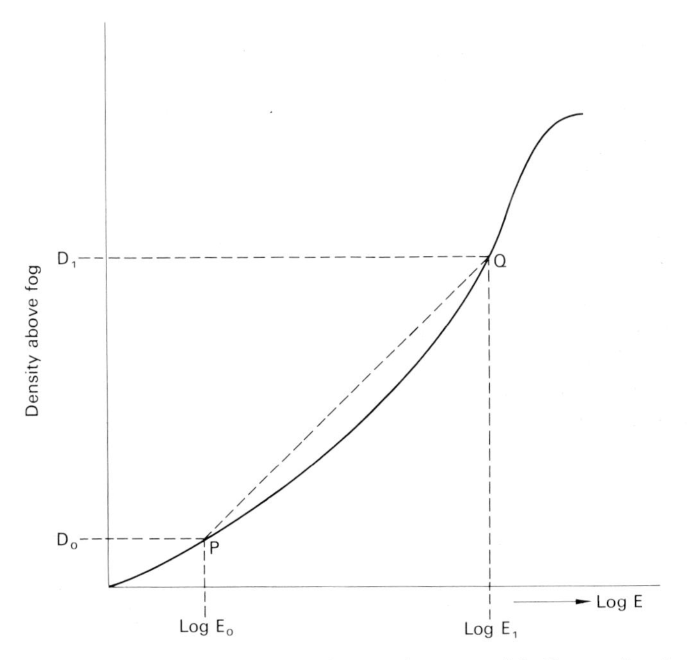

Fig. 9.9 Where the characteristic curve has no straight line portion the average slope of contrast over a given density range (here D_0 to D_1) is given by:

$$\text{slope} = \frac{D_1 - D_0}{\log E_1 - \log E_0}$$

Note: for clarity the fog density has been eliminated.

reduces the latitude. Figure 9.10 shows the wider latitude associated with low image contrast.

In practice it is generally preferable to use a photographic film which has the best contrast for the subject, rather than to modify the contrast by development, which results in an effective change in film speed.

For photographs of subjects such as landscapes, buildings, and people a value of γ of 0.55–0.65 for the negative is usual when the positive images are formed by projection, as in making enlargements. For contrast prints a slightly higher gamma, say up to 0.7 or 0.8, is generally preferred. The corresponding *contrast index* figures are somewhat lower.

The gamma and contrast index are not exclusive properties of the light-sensitive material but depend also on the development time and the type of developer chosen.

For light-sensitive emulsions on paper, the situation is essentially the same as for films. Enlargement papers are generally developed to $\gamma\infty$ whereas for films, development is stopped before this point is reached. The development factor of photographic papers has been adjusted to this practice. The maximum attainable density of photographic paper is much smaller than for films, since the image is seen by means of light reflected from its surface.

Gamma values are not quoted for photographic papers. Usually, papers with different contrasts are described by the terms: extra hard, hard, normal, special, soft and extra soft or with the numbers 0 to 5, but notations of different manufacturers are usually not comparable. A *hard* paper is used when printing a low contrast negative, and vice versa.

c. The speed. A further property of light-sensitive materials can be obtained from the characteristic curve: namely the speed. There have been a large number of methods for measuring the speed of photographic films, for example Hurter and Driffield prolonged the straight portion of the curve to meet the base line and deduced

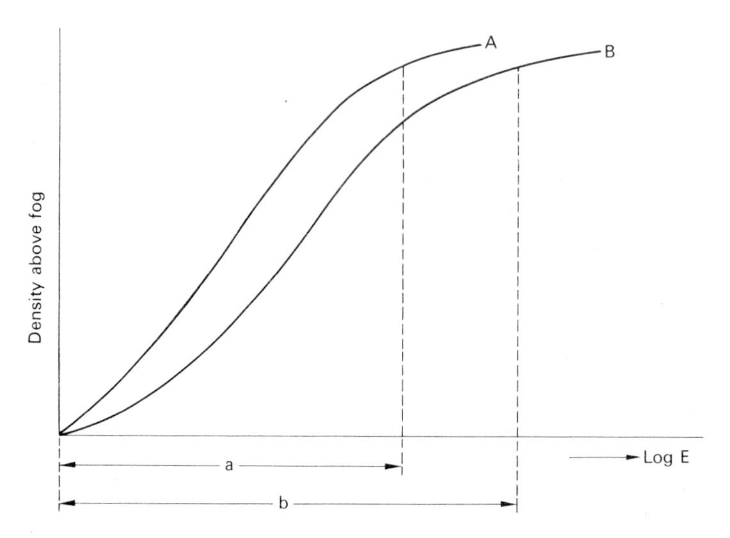

Fig. 9.10 The exposure latitude, a, of the higher contrast curve A is less than that, b, for the lower contrast curve B.

a speed number for the exposure value at the point of intersection. Later methods were based on the minimum exposure required to give a defined density on development. Today the major methods in use are the ASA (American) and DIN (German) speeds, which are both included in the International Standards Organization (ISO) film speed Standard. These speeds apply only to continuous tone exposures, and not to exposures with, say, infra-red light or x-rays. Essentially the film is exposed and processed under well-defined critical conditions, after which the characteristic curve is drawn and the speed calculated as defined in the ISO, DIN, and ASA Standards. The major difference between the DIN and ASA speeds is that the first is based on a logarithmic scale, with a difference of three corresponding to a doubling (or halving) of the speed, whereas the ASA scale is arithmetic, as shown in table 9.1. The higher the speed value the shorter is the exposure.

These speed values are based on the use of a standard developer. Other developers may give other speeds as judged by the exposure necessary to give an effective image. For example, a film intended for use with a standard type of developer will lose effective speed if a fine grain developer is used instead. On the other hand, the speed may be effectively doubled if a more energetic developer is used.

The ASA and DIN speed values do not apply to the papers used for making prints from the camera negative. In fact, the actual light sensitivity of photographic printing papers is very much less than that of the film used in the camera, but this low speed is usually unimportant in the printing process.

d. Grain size, graininess and granularity. In emulsions for ordinary photography the silver halide grains are not uniform in size. For this reason it is usual to quote the *average area* of grains by dividing the total projection area by the number of grains involved. Typical figures for the average area of grains have been given as $0.49\mu^2$ and $0.93\ \mu^2$ for fine-grain and high-speed roll films respectively.

Because of this inherent granular structure the developed image also shows a structure, which is sometimes described simply as 'grain' but nowadays is more usually called *graininess*. This structure may not always be evident to the unaided eye, particularly with the fine-grain materials and some magnification may be necessary to make the *graininess* evident.

The faster the emulsions, the more evident the structure, because such emulsions are based on larger silver halide grains than the slower emulsions.

Table 9.1 DIN and ASA speed values

Relative speed	DIN	ASA
1	6	3
2	9	6
4	12	12
8	15	25
16	18	50
32	21	100
64	24	200
128	27	400
256	30	800

It must be noted, however, that it is not the grains themselves which are revealed, but the image structure of the grains subsequent to development. The situation is complicated because:
1. the silver halide grains may themselves coalesce;
2. the halide grains are not spread in a single plane in the emulsion layer but also distributed throughout its depth. Consequently when the image is viewed by transmitted light some grains may overlap those in other planes in the emulsion layer;
3. in development, the silver formed by one grain may coalesce with the silver in an adjacent grain. In addition, there are other development effects, such the adjacency effects as described below, which complicate the image structure.

Although the graininess of the image is largely dependent on the initial grain size in the emulsion, the formulation of the developer and the development conditions can have important modifying effects.

A simple way of comparing the graininess in two images is to determine what linear magnification is possible before the graininess becomes evident. This is only a *rough* way because the images themselves may have a different type of grain structure, e.g. one may have a diffused 'woolly' appearance whereas the other may seemingly be formed of discrete grains. Even assuming that the grains are of the same form, great care is necessary in making this sort of comparison. Areas of *uniform* and *similar* density in the two images must be used for the comparison to be valid. It is essential to ensure:
1. that the positive images at the varying magnifications are processed identically;
2. that both images are examined at the same distance;
3. that the same printing or projection system is used for both negatives.

The results of such a test will show, for instance, that for a fast emulsion the image structure is revealed at around a linear magnification of 4x whereas a fine-grain image may allow a 12x linear or more magnification before the graininess starts to show. The test is very subjective and the results at the best are only very approximate. The method has been described to indicate the effects of the graininess rather than to commend its use as a comparison method.

Granularity. Whereas graininess is a visual impression of the structure of the image, granularity is a measure of the *local density variations* which give rise to what we see as graininess. Briefly the image is scanned by a microdensitometer so that the local density variations are measured. The average deviation in density from the mean density of the scan can then be measured. This figure will vary with the aperture of the microdensitometer so it is multiplied by the square root of the area of the aperture to give granularity figures. Detailed specifications of some photographic films now include root-mean-square (RMS) granularity figure. Typical figures are given in table 9.2.

Table 9.2 Typical RMS granularity values for negative films of different ASA speeds

Film speed	Granularity
25 to 32	9
64 to 100	9.5 to 10
250 to 300	15 to 17
400 to 600	22 to 25

e. Acutance. Many attempts have been made to find some way of measuring image sharpness and of giving it a value; and various terms have arisen in this search. The terms *acutance* and *contour sharpness* were introduced some years ago in the USA and Germany respectively. This factor is a measure of the sharpness at a boundary corresponding to white light on the one side and complete darkness on the other, usually formed by making an exposure with a knife edge in contact with the emulsion layer. Ideally the image should be as shown in Fig. 9.11 which represents the density variations measured along a line at right angles to the boundary. In fact, due to irradiation, which is the sideways spread of light in the emulsion, the edge is less well defined as in Fig. 9.12 which applies to (i) an image of low acutance, and (ii) an image of high acutance.

It should be noted that the shape of this curve can also be affected by adjacency effects during development. (see later section on adjacency effects)

f. Resolving power. For many years it has been usual to determine the resolving power of film/development systems by photographing test charts of fine lines of diminishing widths which have a separation equal to their width. As the lines get narrower and closer together the recording system fails to separate the lines. The resolving power is defined as the highest number of lines per millimetre which are resolved *in the image*. The figure obtained depends on the contrast of the original test chart. Consequently two figures are often quoted, the larger applying to a high contrast test and a lower one applying to a low contrast test. For a test chart with a constant ratio of 30 to 1 the resolving powers of high-speed panchromatic emulsions and of process materials used in printing are 40 to 60 and 140 to 220 lines per millimetre respectively.

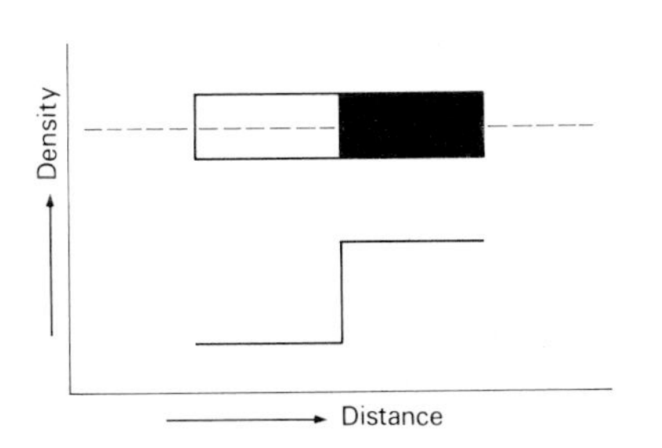

Fig. 9.11 Ideal density trace across a sharp density curve.

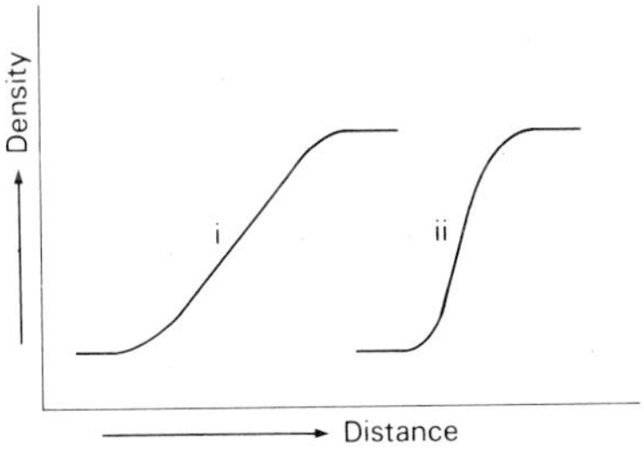

Fig. 9.12 These traces across the density bound ary show (i) low acutance, (ii) high acutance.

g. Modulation transfer function. The validity of the resolving power has been challenged on the grounds that real subjects do not show the sudden variations of density found in the test charts described. As a consequence a new method has emerged which has been described in the literature by all of the following terms: modulation transfer function, sine-wave response, optical transfer function, contrast transmission factor, transfer function and modulation transfer. When determining this function the test object is not in the form of lines having sharp edges. Instead it consists of a series of density variations between a constant maximum and a constant minimum with the high density peaks coming closer as the frequency increases.

Detailed specifications for films now include modulation transfer data. In practice the information is given in the form of a graph showing the percentage contrast of the lines relative to the optimum instant contrast.

h. Adjacency effects. These are effects arising in development which cause deviations in density which are not in the image to which the film was exposed. If films are developed in a vertical plane and there is insufficient agitation of the developer, light streaks may occur below dark image areas and dark streaks below light image areas. The light streaks occur because the spent developer flowing down contains an excess of bromide from the development process and this retards development. The dark streaks are due to relatively unused developer flowing down over the denser areas and locally accelerating development.

Where a very light and very dark area are adjacent, an 'overshoot' effect can give rise to *Mackie lines*, which appear as a dark line just within the dark area and a light line just within the light area (see Fig. 9.13).

When the images are dark circular spots, the image density may be slightly greater just within the boundary for a similar reason. This is known as the *Eberhard effect* and is illustrated in Fig. 9.14.

i. Colour sensitivity. The last property of films to be considered is the colour sensitivity. It should be mentioned here that a black-and-white film does not, of course, reproduce colours, but has to convert the colours into shades of grey in accordance with their brightness.

Silver halides, as already stated, are sensitive only to ultraviolet, blue and blue-green light and are not affected by other colours. The result is that a photographic film based on silver halides as such remains unexposed on the places illuminated with green, yellow or red light. When the negative image on such a film is used to produce a positive image on paper all colours with the exception of blue and blue-green are reproduced as black.

To obtain true colour sensitivity, colour sensitizers must be added to the light-sensitive emulsion. To be effective, the dyes must possess the property of being adsorbed on to the surface of the silver halide crystals. They take up energy from light of long wavelength and then give it up again in a form which is able to affect the silver halides on to which they have been adsorbed.

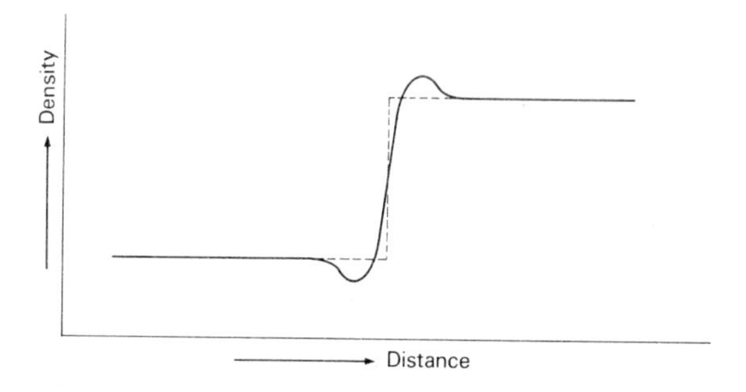

Fig. 9.13 Micro-densitometer trace showing Mackie lines.

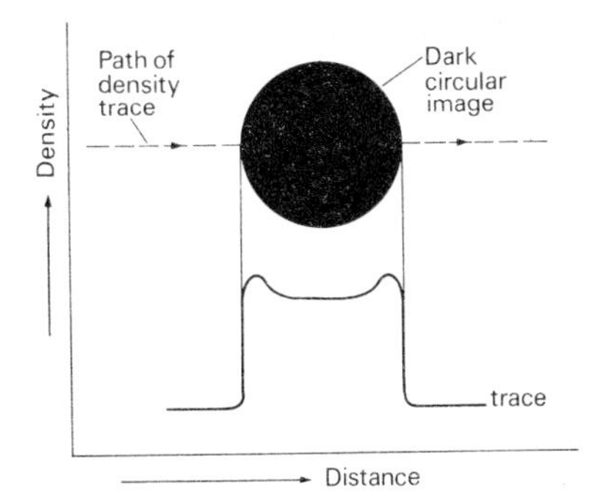

Fig. 9.14 Micro-densitometer trace for Eberhard effect.

Usually a combination of dyes is used to obtain true colour sensitivity in an emulsion. It has been possible, in photographic films, to simulate the sensitivity of the eye to the various colours. Such films are named panchromatic films in contradistinction to orthochromatic films which are sensitive to blue, green and yellow but not to red light. Almost all the films used for general photographic purposes are now panchromatic, but orthochromatic films and films sensitive only to the blue-green and blue end of the spectrum are still used for certain purposes.

Light-sensitive emulsions on paper are, in general, sensitive only to blue and blue-green light. Nothing more than this is necessary since the negative, from which black-and-white prints or enlargements are to be made is not usually coloured. One type of light-sensitive paper, however, marketed under the name 'Panalure paper' is panchromatic so that it can be used for the production of black-and-white prints and enlargements from colour negatives.

9.2.6 Developing solutions

Development of photographic emulsions on films and papers is effected by aqueous solutions of organic reducing agents capable of converting the exposed silver halide into silver without affecting the unexposed silver halide, as already mentioned in section 9.2.2. Various other chemicals are added to the developer solution to control the reduction process.

a. Developing agents. A large number of organic compounds can be used as developing agent. Among them may be mentioned: polyhydroxybenzenes, para-aminophenols and their derivatives, and para-phenylenediamine and its derivatives.

Developing agents of the polyhydroxybenzene group are hydroquinone, catechol and pyrogallol:

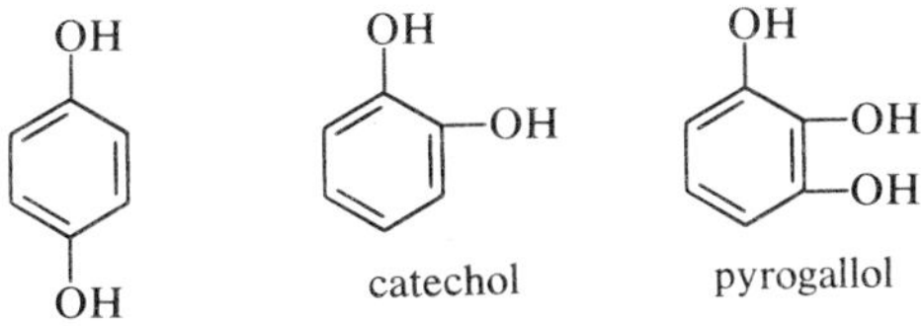

hydroquinone   catechol   pyrogallol

These developing agents are active only in strongly alkaline solutions, they are typically slow working and produce a rather coarse grain structure. The most important substance of this group is hydroquinone, which produces a hard gradation. Hydroquinone is mainly used in combination with Metol or Phenidone (see below). Pyrocatechin and pyrogallol tan gelatin and form a colour image from oxidized developing agent in addition to the silver image. The gradation produced by these developers can be modified from soft to hard by modifying the alkalinity.

The most important aminophenols are para-aminophenol, para-methylaminophenol (Metol, Elon) and 2,4-diaminophenol (Amidol). Unlike the polyhydroxybenzenes, *p*-aminophenol and *p*-methylaminophenol produce a rather fine grain, a soft gradation and are rapid working in alkaline solutions. In weakly alkaline or neutral solutions these developing agents are slow working and produce a still finer grain. 2,4-Diaminophenol, the third important member of this group fogs photographic emulsions in alkaline solutions. It is a rapid working developing agent in neutral solutions: it can be used as a rather slow working developing agent in slightly acidic solutions without fogging the emulsion.

p-aminophenol 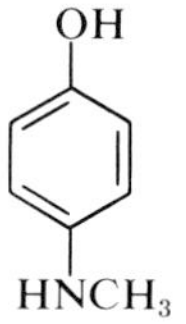 | *p*-methylaminophenol 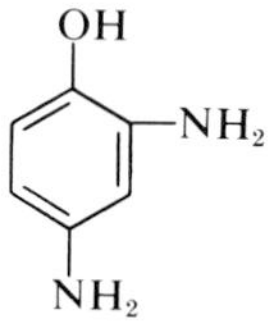 | 2,4-diaminophenol

A useful derivative of aminophenols is para-hydroxyphenylglycine (Glycin). Like para-phenylenediamine, this developing agent produces a very fine grain and is slow working in neutral or slightly acidic solutions. Derivatives of para-phenylenediamine are used in colour developers (see section 9.3.2).

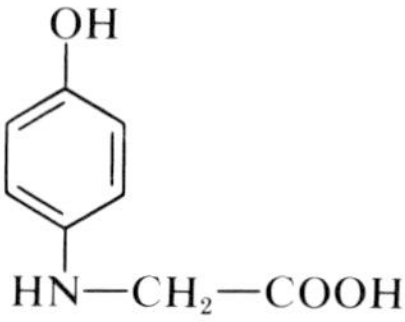

p-hydroxyphenylglycine  | *p*-phenylenediamine 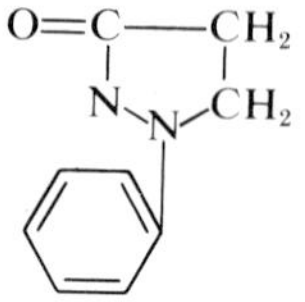 | Phenidone

An important developing agent with a quite different structure is 1-phenyl-3-pyrazolidone (Phenidone) which somewhat resembles Metol in its developing properties. Phenidone is much more soluble in alkaline solutions than Metol and it activates hydroquinone. For these reasons Phenidone can be used in smaller quantities than Metol when it replaces Metol in Metol-hydroquinone combinations. In addition developers based on the combination of Phenidone and hydroquinone can be marketed as concentrated solutions, unlike developers based on the Metol-hydroquinone combination.

Due to the typical differences in developing properties of Metol and Phenidone on one hand and hydroquinone on the other hand, by far the largest number of modern developer formulae are based on the use of the combination Metol-hydroquinone in which, sometimes, Metol is replaced by Phenidone.

b. Composition of developing solutions. Developing solutions usually contain about 3 to 10 g of one or more developing agents per litre of water. Various other chemicals are added, of which alkaline substances, sulphites and restrainers are the most important.

Alkaline substances such as sodium carbonate, alkaline hydroxides and borax are added to control the activity of the developing solution and for this reason the alkaline substances are often named accelerators. The activity of the developer rises with increasing alkalinity.

Sodium sulphite and potassium metabisulphite are used as preservatives. These substances inhibit the oxidation of the developing agents by oxygen from air. In addition they control the pH of nearly neutral developing solutions.

The fourth important component is the restrainer, which prevents fogging of the photographic emulsion. The most important restrainer is potassium bromide. Other examples of organic restrainers are para-nitrobenzimidazole and dinitrobenzimidazole.

Other substances are also used in small amounts in order to achieve special properties in the developing solution; many of these are trade secrets.

It is possible to a large extent to vary the speed of development and the gradation, grain structure and speed of the light-sensitive emulsion by varying the constituents of the developing solution. In this context, however, it must be noted that the fundamental properties of the emulsion, as made, will primarily control the results obtained. It is always best to choose the most appropriate film for the work in hand, rather than to try to achieve similar image characteristics by modified development of a less suitable film.

9.2.7 The chemical processing of photographic materials

a. Negative and positive processing. The sequence of operations in the processing of black and white negative or positive images is:

developing
rinsing (or use of stop bath)
fixing
washing
drying

In a processed photographic emulsion, the quantity of silver is related to the intensity of the light which has fallen on it, and consequently, the blackest parts on a developer emulsion correspond with the lightest parts of the object photographed. Thus the densities in the image on an emulsion are complementary to the brightnesses in the subject to which the emulsion has been exposed. A positive image is obtained by exposing a second emulsion (on film or paper) via the negative image.

Assuming that the film is developed in a dish or tank, the development time is usually in the range of 5 to 10 minutes at 20°C (68°F) with solution agitation: the development time for paper is generally in the range of 90 to 120 seconds. The rate of development increases with the degree of agitation of the *solution*, and with increase in development temperature. For uniform results both should be standardized. To obtain the best results the manufacturer's instructions should be followed closely. With some modern developers and some current automatic processing equipment the times mentioned for developing (and for other parts of the process) may be materially reduced.

After development, the image must be fixed by removing the unexposed silver halide. Usually, fixing is performed by treatment, for about 5–10 minutes, in a 20% aqueous solution of sodium thiosulphate ('hypo').

To prevent the reduction of thiosulphate ions to sulphide ions by the developer (which would cause brown stains of silver sulphide) the sodium thiosulphate solution is acidified to stop the activity of the developer. Potassium metabisulphite, which also stabilizes the sodium thiosulphate solution, is mainly used for this purpose. A good fixing solution generally contains a hardener, such as potassium alum, which hardens the gelatin and so prevents it softening in the washing process or when drying by heat. More rapid fixing is achieved by using ammonium thiosulphate instead of the sodium salt.

Sometimes the film is placed in a 'stop bath' of about 1% acetic acid immediately after development to stop the developing process and to reduce the effect of developer alkali on the fixing solution. If a stop bath is not used the film must be rinsed before being transferred to the fixing solution.

After all the unexposed silver halide has been removed by the fixing solution, the emulsion is still saturated with the chemicals and byproducts of the fixing process. If these are not removed the image may become discoloured in time. Accordingly the film or paper should be thoroughly washed after fixation. The time required is shortest when the emulsion is exposed to a rapid flow of fresh water. More time is necessary if the process is carried out in a tank or dish as the chemical-laden water must also be washed from the vessel.

Drying has a very important bearing on the image quality. It should be done carefully so as to avoid any mechanical damage to the emulsion or marks due to uneven drying, and without exposing the wet emulsion to dust or lint which, once attached, is almost impossible to remove.

b. Reversal processing. In the reversal process a positive image is produced in a photographic material without the intermediate production of a separate negative image. The negative image is first obtained by development, and this image is then dissolved in an acid solution of potassium permanganate. The remaining unexposed silver halide is then exposed to light and developed into a positive black image by means of a second development followed by rinsing, fixing, washing and drying. For satisfactory results films specially manufactured for reversal processing must be used. The exposure for reversal films is much more critical than that for negative films as errors in exposing the latter can generally be corrected when printing the positive image.

9.3 VARIOUS SILVER HALIDE MATERIALS AND THEIR USE

9.3.1 Black and white materials for general still photography

a. Negative and positive materials in still photography. Films and papers for general still photography are available in various sizes. Besides the sheet films with dimensions between 6.5 × 9 cm (about $2\frac{1}{2} \times 3\frac{1}{2}$ inch) and 50 × 60 cm (about 20 × 24 inch), 120, 127 and 620 roll films are available. In addition 35 mm films are used with the smaller format cameras: some of these are sold in ready-to-load magazines, as for 126 films. A recently introduced camera uses 16 mm film, specially perforated in a light-tight ready-to-load cartridge.

Papers for the production of prints from negatives are usually sold as cut sheets with sizes ranging from about 9 cm (3.5 in) square to 50 × 60 cm (19.6 × 23.6 in). Papers in rolls for photofinishing (about 9 cm wide) and for enlarging (20.3 (8 in) and 25.4 cm (10 in) wide) are also made for commercial work. They are often available with different contrasts to enable them to be matched to the image contrast of the negative. For some types there is a choice of image colour, base tint and surface texture.

b. Cinephotography. When a motion picture is taken, a long, perforated film is moved intermittently past the place where an image is projected by the lens of the cine camera. The film is moved by means of claws which fit into perforations in the film to advance the film a 'frame' at a time. A series of pictures is obtained by opening the shutter of the lens at the moments that the film is stationary; this takes place 24 times per second, for images which are intended for use as 'sound' films and 16 times per second for 'silent' films. Projection on to a screen is achieved in a similar way by covering the lens during the periods in which the film is moving for a brief interval after the light stimulus has ceased. In practice, if the pictures are projected at the rate of sixteen or more pictures per second or more, a second picture has already appeared before the first has disappeared from memory, so that the periods during which the lens is covered are not noticeable.

For amateur use, 8 mm wide film is available both for the older standard 8 mm frame and for the newer Super-8 frame which gives a larger picture. Unfortunately the equipment for the two systems is not interchangeable. Both systems use reversal type emulsions as only one copy is usually required and the use of a single film reduces the costs. Some amateur work is done with 16 mm wide films but this size is also used by the professional. For larger formats films 35 mm, 65 mm or 70 mm wide are used. All films for cinephotography are perforated to make use of the claw mechanism for advancing the film during exposure and projection. Colour reversal films for amateur use, and colour negative stock are steadily replacing black-and-white films for motion picture work.

9.3.2 Colour negative films and papers with couplers in the emulsion

a. Composition. Both negative colour films and positive colour papers consist of a base coated with six or more gelatin layers of which three are in the form of light-sensitive silver bromide emulsions.

In negative colour films, the under layer is sensitive only to red light, the intermediate layer to green light and the top layer to blue light, so that together they are sensitive to the whole range of visible light. The selective sensitivity is obtained by the addition of different optical sensitizers. Thin separation layers of gelatin are applied between the light sensitive layers and a protecting layer of hard gelatin is finally applied. The separating layer between the blue sensitive layer and the green sensitive layer comprises a yellow filter to protect the under layers from blue and ultraviolet light to which they are more or less sensitive. For simplification, at this stage, the 'colour masking' system incorporated in source negative systems will be omitted, and will be discussed later.

A similar 'integral tri-pack' of emulsions and filter may be used for the positive printing paper. Not all such papers use this system, however, and the sequence of the light-sensitive layers may be changed, in which case the yellow filter layer is omitted.

Each light-sensitive layer contains a colour coupler which produces a colour with the oxidation products of the development process. The colour in each layer is complementary to its sensitivity; namely yellow, cyan (blue-green) and magenta in the blue-sensitive, red-sensitive and green-sensitive layers respectively. Repetition of the process during printing onto a tri-pack colour paper or film gives the colour rendering and tone graduations in the original subject.

Many difficulties have had to be overcome in the choice of the colour couplers and sensitizers. The couplers must form a colour with a predetermined absorption spectrum to reproduce a colour correctly and in addition the couplers should not migrate from one layer into another, i.e. they must be incapable of diffusing. Against this, however, is the requirement that, during development or fixing, the processing chemicals must be able to diffuse rapidly in and out of the light-sensitive layers.

b. Exposure and development. On exposure in a camera, the various layers of a colour film are affected by the light at various points, depending on the colour of the image which has been projected on to it. Thus three different latent images are formed in the light-sensitive layers. These are developed by means of a developer which must satisfy two requirements. It must, firstly, reduce the exposed silver halide to silver, and secondly, it must be converted into an oxidation product which, at the same point as that at which the reduction took place, forms a transparent dye of a complementary colour with the coupler present in each light-sensitive layer.

The developer solution has much the same composition as a black-and-white developer, with the exception of the preservative (sodium sulphite), which is partially replaced by hydroxylamine hydrochloride. This is necessary since large quantities of sulphite inhibit the dye formation. The development time is about 5 to 10 minutes. To keep the characteristic curves of the three different light-sensitive layers coinciding or at least parallel, it is necessary to follow the manufacturers' instructions closely. A difference of 1 deg. C in the development temperature, for example, can decrease the colour quality considerably.

Subsequent to development the negative material must be washed or treated in a stop bath. This is followed by bleaching, often in a ferrocyanide bath, which removes the *silver* of the image, leaving only the dye image. Treatment in a hypo bath is next, to remove the unexposed silver halide, and this is followed by washing and drying. A hardening and stabilizing process is often included prior to the final wash. Sometimes the two processes of bleaching and fixing are done in one solution known as a 'bleach-fix'.

c. Colour masking. A colour negative would result in an ideal print only if the transmission of the three dyes were ideal. Yellow dyes usually meet this requirement since they absorb the blue and transmit all the green and red light. However the magenta dye does not only absorb the green light but also a portion of the blue light and the result is that it behaves partially as a yellow dye. Similarly the blue-green dye does not only absorb red light but also a portion of the blue and green light, as shown in the figure. To compensate the unwanted absorption a low density positive yellow image must be produced together with the negative magenta image and a low density red image must be produced together with the negative cyan image.

Usually the couplers in the various layers are used for this purpose. In some films a coloured coupler is applied. (The colour disappears at the places where the oxidized developing agent forms a new dye with the coupler.) In other films the colourless coupler remaining at the unexposed areas is coloured by suitable agents. Colour masking accounts for the orange colour of most colour negatives.

d. Colour reversal processes. Colour transparencies are obtained by reversal processing of an integral tri-pack film. The colour coupler is incorporated in the emulsion except for special materials which are processed by the manufacturer or his agent. For this reason, the system incorporating its own colour couplers will be described. The emulsion coatings on the film are colour sensitized similarly to those used in the negative-positive system. The first stage in processing is basically black-and-white development which acts on the exposed silver halide grains. After washing, or treatment in a stop bath, the film is hardened. The film is then well exposed to light so rendering the residual silver halide developable: sometimes this exposure is achieved by chemical means. This is followed by colour development in which the residual halide is developed and the dye image is formed. After washing, the silver images are bleached out. Finally after rinsing, the film is fixed, washed and dried.

In this sequence of processing the exposed silver halide grains corresponding to the image are developed and then bleached away. This leaves the unexposed grains to form the *positive* image when they are exposed and colour-developed. Consider a portion of the image which was a *pure* red: silver halide in this position in the red-sensitive layer will be exposed and used up in the first exposure. Accordingly, after development and bleaching, there will be no silver halide in this area to form a cyan image after fogging and colour development. As the colour was *pure* red, the superposed blue- and green-sensitive layers will have been unexposed in the first exposure, so that after fogging and colour development these will form dye images of yellow and magenta

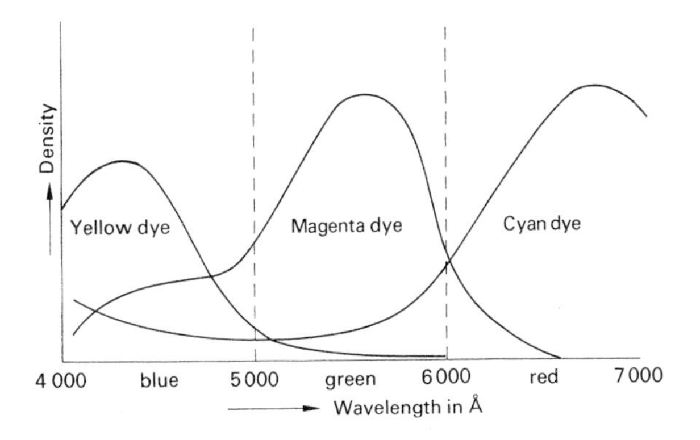

Fig. 9.15 Absorption curves for the dyes used in a colour negative film.

respectively. White light incident on these filter layers will appear to be red when viewed from the other side: in other words the filters together transmit red light. In the image areas corresponding to green the silver halide will be left for development only in the red- and blue-sensitive layers, so that this portion of the image will be rendered by a cyan and a yellow dye image which together transmit green light.

As with other reversal processes, colour reversal systems have less exposure latitude than the negative-positive systems.

e. Available forms. Colour negative films, papers and films for the production of positive images from negatives, and reversal films are available in the same dimensions as black-and-white materials. The reversal film, which is uncommon in black-and-white still photography, is extensively used in colour photography.

Many manufacturers produce at least two films with different emulsion speeds: 8 and 16 mm reversal cinematographic films for amateur use are also available.

Owing to the different spectral composition of daylight and artificial light, special colour films are made for the two forms of illumination, since with reversal films compensation cannot be made during development: it is true, of course, that films intended for exposure to daylight can be used with artificial light and vice versa with the aid of colour filters during exposure, but this has disadvantages (e.g., loss of speed) and is better avoided by the use of the correct material for the job.

In the positive-negative process corrections can be made during printing, by exposing through colour filters to offset any colour distortion which could arise from reciprocity failure in very long or short exposures. This method is generally satisfactory for amateur work but to minimize the problems arising from reciprocity failure special negative films are available for the professional who must use very short or very long exposures.

9.3.3 Dye destruction colour print materials

These materials are restricted to print making and use a paper or white plastic foil coated with three light-sensitive layers. The under layer is sensitive to red, the intermediate layer to green and the top layer to blue light, as in the colour coupler materials. Unlike the latter, however, the dyes are present in the sensitized layers before development: a cyan (blue-green) dye in the red sensitive layer, a magenta (purple) dye in the green sensitive layer and a yellow dye in the top layer. These dyes differ from those formed in the colour coupler materials in that they are very lightfast azo-dyes, which are usually more stable to light. Typical films are the Cilchrome and Ciba materials.

After exposure, the Cilchrome film is developed, stopped and fixed in solutions of the type used for black-and-white materials. Then the material is treated in a dye bleaching solution which destroys the dyes selectively at the places where the silver of the negative black-and-white image is present. The quantity of dye bleached is proportional to the quantity of silver present. The subsequent processing stages involve the bleaching of the silver image, and fixing out of the residual silver halides, with appropriate intermediate rinsing. A positive transparent image is thus produced from a positive transparency.

The Ciba colour print material, for the production of positives from negatives, is basically the same as the Cilchrome material, but like the black-and-white reversal film it is processed in a first and second black-and-white developer with intermediate dissolving of the first negative image and re-exposure. The next stages are the same as with the Cilchrome process.

9.3.4 Materials for reprography

Reprography (photocopying) is the photographic production of a duplicate from two-dimensional objects such as documents, book pages, letters, and other images on paper, textiles or other flat materials.

If contact-size reproduction of an original is required, many types of light-sensitive materials are available. These include not only the silver halide materials discussed in this section and in section 9.3.5 relating to the stabilization process, but also the non-silver-halide materials such as blue prints, diazotype materials, photosensitive semi-conductors used in electrophotography and the heat sensitive materials used in the 'Dual-Spectrum' process (see section 9.5). Reduced or enlarged size reproduction is achieved by means of a camera.

In order to reduce the space required for storing documents, greatly reduced copies may be made on 16 mm or 35 mm films, or on microcards or microfiche. Usually these are negative copies though positive ones are possible. The recording medium must have a very fine grain structure, develop to high contrast and give excellent definition, so that the copy contains all the information which is in the original. Linear reductions of 12, 20 or even 40 × are used. The greater the reduction the more critical the exposing and reading equipment and the greater the demand on the sensitized film. Usually the documents are read in the form of the projection image on a so-called 'reader;' some of these are designed to make, at will, a copy which can be read without magnification.

9.3.5 Rapid photographic processes

a. High speed development. In the high speed development process, a length of exposed film is allowed to move intermittently through an apparatus into which three liquids are squirted, one after the other, on to the sensitive layer for development, fixing and washing.

The developer is activated by replacing the commonly used alkali carbonates by alkali hydroxides. The high-speed fixing solution is a 40 to 50% solution of ammonium thiocyanate: as this tends to soften gelatin, up to 5% formalin may be added to ensure that the gelatin is adequately hardened. If used at fairly high temperature each chemical treatment is no longer than 2 seconds, and is followed by drying with compressed air for 6 seconds. By combining the development apparatus with a camera which automatically takes a picture every 15 seconds, the aim of the process can be achieved with a minimum of operations, and a picture is available about 12 seconds after exposure. This process has been used for producing photographs from a radar screen.

b. Stabilization process. In the stabilization process a photographic paper is exposed and developed in the usual way, but the unexposed silver halide is not removed by dissolving in a solution of sodium thiosulphate: instead the exposed and developed paper is treated in a stabilizer bath. In the latter the unexposed silver halide is converted into a complex which is insensitive to light. Aqueous solutions of alkali thiocyanates or thiourea are most frequently used as stabilizers.

This process has the advantage in that the time consuming processes of fixing with sodium thiosulphate and washing with water are not required. This method, however, has the disadvantage that the resulting prints are not as stable as normally fixed and washed prints.

The stabilization process has been used mostly in office copying.

c. The diffusion transfer process. By the use of this method a semi-dry copy of a document can be made in less than 1 minute.

The negative material is a paper coated with a high contrast light-sensitive silver halide emulsion, and the positive material, which is not light-sensitive, consists of paper with a coating containing very fine particles in the form of 'colloidal silver nuclei'.

As the negative paper has high contrast it can be exposed by the reflex method. In this, the document to be copied is placed against the light-sensitive side of the negative paper and light is passed through the paper. Image formation in the light-sensitive layer depends on the fact that the intensity of the light reflected by the document depends on the brightness of its various parts.

The exposed material is passed through a 'developer' and then, while still wet, squeezed together with a sheet of positive paper through a set of rubber rollers so that both coated layers are adjacent and in close contact. As a result, the unexposed silver halide in the negative layer is dissolved in a halide solvent, such as 'hypo', in the 'developer', and diffuses into the coating on the positive paper. Here the silver nuclei cause precipitation of silver in the areas corresponding to the unexposed areas in the negative. The result is a positive image. As the silver halide from the unexposed areas of the negative paper is consumed, it is not possible to make more than a few positives from one negative. The reproduction of half-tones is poor but not quite impossible. The images may, in time, show some discolouration due to the residual chemicals but they are entirely adequate for normal commercial papers. They can, however, be given greater permanency by washing, but this is rarely, if ever, done.

In a modified diffusion transfer process the developing agent is incorporated in the negative paper and after exposure the negative paper is treated in an alkaline solution (activator).

d. The dye transfer or gelatin transfer process. In this process the handling of materials is the same as in the diffusion transfer process, but the composition of the materials is different. Again, the reflex method of exposure is used as shown below.

The negative paper is coated with hardened gelatin in which a silver halide and a dye-forming compound are incorporated. When the exposed layer is developed with a tanning developer, the gelatin becomes hardened on the exposed areas. Simultaneously a portion of the dye-forming compound is oxidized and forms a dye with the unoxidized portion of the dye-forming compound. When the exposed and developed negative material is squeezed against a piece of ordinary untreated paper, a thin layer of unhardened dyed gelatin (from the unexposed areas) is transferred to the paper and thus a positive copy is obtained.

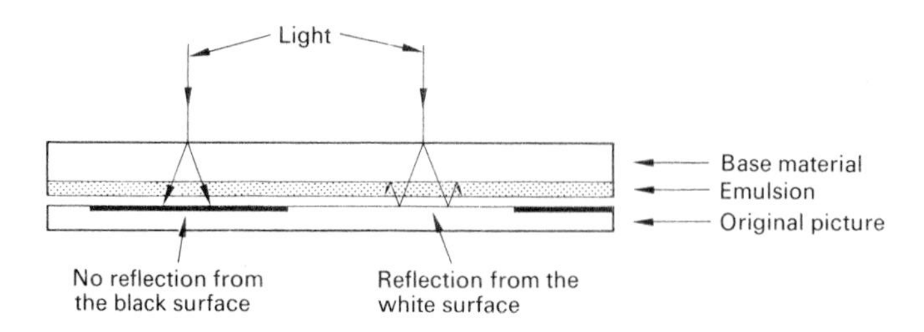

Fig. 9.16 Exposure of a light-sensitive material by the reflex method.

The dye is not completely stable in the presence of light, but its tendency to fade is eliminated if the positive paper is treated with thiourea before use. On thiourea-treated paper, the unexposed silver halide, transferred with the gelatin, is slowly converted into metallic silver and silver sulphide and thus the fading of the dye is counterbalanced.

About 5 copies can be made from one sheet of negative paper (named matrix); if a paler picture can be accepted in the last few copies as many as 8 can be made. Like the diffusion transfer process, the dye transfer process is used in office copying. The gelatin transfer process does not produce acceptable half-tones and can therefore only be used for line pictures and text.

e. The Polaroid-Land process. This process is a refinement of the diffusion transfer process. By carefully controlling the amount of silver halide in the negative material and reducing the contrast, the material is suitable for the reproduction of half-tones and consequently it is suitable for general photography by means of a camera.

In the Polaroid-Land process, a roll of negative paper and one of positive paper are placed in a specially designed camera or in a slide designed to fit to cameras with interchangeable backs. Small capsules filled with developing solution are fixed at a certain distance from each other on the positive paper. After exposure, the exposed piece of negative, together with a piece of positive paper of the same size, is passed through two rollers, so that one of the capsules is broken, and the developer is spread out between the two layers. Development of exposed silver halide takes place and also diffusion of unexposed silver halide from the negative paper into the positive paper. Since small silver nuclei are present in the positive coating, the unexposed silver halide is developed by the same developer. After development, which requires about 30 seconds, the negative is removed from the positive paper.

In addition to negative paper which has to be thrown away, stabilized negative film is also available. If these negatives are given an after-treatment within a few days, they can be kept as long as a common negative film. An unlimited number of copies can then be made in the dark-room, using normal contact and enlargement papers.

f. The Polacolor process. The Polacolor process is a diffusion transfer process

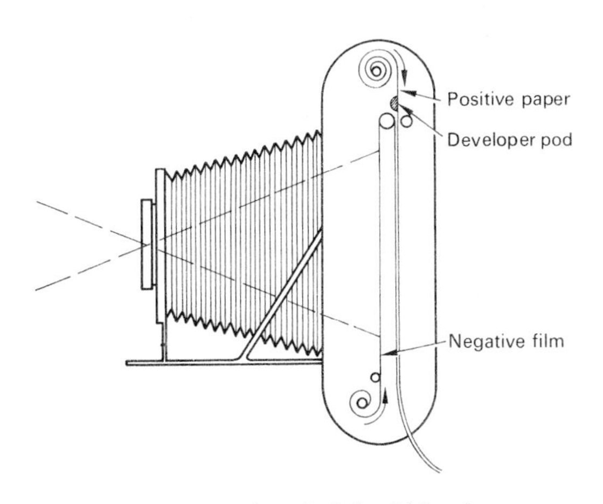

Fig. 9.17 Cross-section of a Polaroid-Land camera.

for the production of colour pictures. Like the common colour negative film, the Polacolor negative film contains three light-sensitive emulsion layers of which the upper layer is sensitive to blue light, the intermediate layer to green light and the under layer to red light. In contradistinction with the common colour negative film, the Polacolor negative film contains a dye-developer layer under each light-sensitive emulsion layer. The dye-developer comprises a dye and a developing agent which are chemically combined. The blue-green dye occurs in the red sensitive emulsion layer, the purple dye in the green sensitive layer and the yellow dye in the blue sensitive layer.

The camera or slide used for exposure and development is basically the same as that for black-and-white Polaroid-Land films and papers. When, after exposure, the negative film and a positive paper are squeezed together, a capsule containing an alkaline activator paste is broken open and the alkaline activator is distributed uniformly between the film and paper. The exposed areas of each light-sensitive layer are developed by the dye-developer substance present in that layer and from these areas no dye (which is fixed to the developer molecules) can diffuse out of the negative film. The dye-developer substances under the unexposed areas can pass through all the layers and are transferred to the positive paper.

The positive paper is not light-sensitive, but is coated with three layers. The under layer contains acidic substances which neutralize the alkaline substances from the activator paste, the intermediate layer inhibits the diffusion of acidic substances to the negative film and the top layer contains the mordant which insolubilizes the dye-developer which has diffused from the unexposed areas in the negative, thus giving a positive colour image. After one minute the negative film can be separated from the positive paper. No after treatment is required.

9.3.6 Some special scientific techniques and materials

a. High-speed photography. In ordinary photography, the exposure time is limited by the maximum shutter speed which does not permit exposure times shorter than about 1/2000 of a second. Shorter exposure times are, however, often required, e.g. for the study of explosions, in the investigation of the phenomena which occur in front of and behind a rapidly moving projectile, when studying the operation of an atomizer, or in the successive phases which occur in the breaking of materials and the flight of insects and birds.

If one picture is required, use is made of electronic flash for which the discharge has a duration of as little as one-millionth of a second. Such a short exposure time 'freezes' the most rapid motion.

If a series of pictures is required, a special high-speed cine camera is used. These cameras do not operate on an intermittent system like common cinematographic cameras since, at the high speeds used, the film would not be strong enough to be moved by claws fitted into perforations at the edge. In these special high speed cine camera the images are caught in the correct position by means of moving prisms, mirrors or lens systems. The most common high speed cine cameras can take up to 3000 pictures per second resulting in a picture showing about 200 times slower action when the film is projected at a speed of 16 pictures per second. High-speed cine cameras have been made, which can take pictures at the *rate* of 10 million pictures a second, for a very short time. The photographs obtained cannot be projected continuously to show the phenomena recorded, and have to be analyzed frame by frame.

b. Infrared photography. In everyday language the word 'light' is taken to mean radiation visible to the human eye. In photography it has a wider significance and includes all radiation, of wavelengths longer and shorter than those of the visible spectrum, capable of affecting a photographic emulsion or a photoelectric cell. Beyond visible light of long wave-length (red) lies the infrared and beyond visible light of short wave-length (blue-violet), the ultraviolet. X-rays and gamma-rays have still shorter wave-lengths.

It is possible to make infrared perceptible in a number of ways. Objects which emit or reflect infrared radiation with a wave-length up to about 1300 mμ can be photographed with silver halide emulsions sensitized to this region of the spectrum. A filter which allows only infrared radiation to pass through is placed in the front of the lens. As infrared radiation is less easily scattered than visible light, it will penetrate haze and so may be used to advantage in long-distance landscape photography and for photography from the air. The reflection and transmission by an object of infrared radiation may differ materially from that of visible light. Accordingly, infrared photographs may show differences in objects which are not visible in ordinary lighting. As a result, infrared photography may help in showing the alterations in forged documents, and may reveal the text in charred documents. Infrared photography is also used in spectrography.

An infrared colour film somewhat similar to the common colour film is also available: it has three light-sensitive layers but these are sensitive to green, red and infrared light and contain yellow, purple (magenta) and blue-green (cyan) colour couplers respectively. All the layers are also sensitive to blue light, but the latter is eliminated by means of a yellow filter. Objects photographed with this material are usually reproduced in colours quite different from those seen visually. The value of this type of film is that, in the processed image, it shows differences in infrared reflectance in the various parts of the scene. It is therefore particularly valuable for differentiating between healthy deciduous foliage which reflects infrared light strongly and diseased foliage which does not. Water, on the other hand absorbs infrared light so that drainage channels and water courses can be readily revealed by air photography with this type of film. It has also been used in military work for distinguishing between real foliage and imitation foliage use in camouflage.

c. Ultraviolet photography. As silver halide emulsions are naturally sensitive to this form of radiation, the conversion of ultraviolet radiation into a visible radiation is usually not required. Ultraviolet radiation with wavelengths down to about 330 mμ can be recorded on normal emulsions by means of the usual optical systems of a camera. However for photography with ultraviolet light of shorter wavelength special emulsions with a low gelatin content and optical systems made from quartz must be used because gelatin and glass absorb ultraviolet radiation with wavelength below 330 mμ. For this type of work, a filter absorbing visible light but transmitting ultraviolet light must be placed over the camera lens. In this way, ultraviolet photography may be used to distinguish between materials which appear the same in visible light but have different ultraviolet absorption, e.g. stains on clothing.

As ultraviolet light can excite fluorescence in some materials, these may be detected by illuminating the specimen with ultraviolet light and photographing them through a filter which transmits visible light only. Such images may be recorded on black-and-white or on colour films.

An important application of ultraviolet photography is ultraviolet photomicrography. The resolving power of a microscope is increased when using ultraviolet owing to its short wave length. Ultraviolet radiation down to about 200 mμ can be used in photomicrography.

d. Radiography. Radiography is photography by means of röntgen rays (X-rays), which, like light, are part of the electron magnetic spectrum and have very short wavelength ranging from 0.01 to 10 mμ.

Light-sensitive emulsions based on silver halides are sensitive to X-rays but as most of the rays pass through the emulsion without being absorbed, emulsions for radiography have a much higher silver halide content than emulsions intended for exposures to light. The exposure time may be substantially reduced by sandwiching the film between intensifying screens, held in uniform contact by pressure in a cassette. The screens are made of a stiff backing coated with a chemical, such as calcium tungstate, which emits blue-violet light when excited by X-rays. Images made in this way are therefore formed both by the X-rays absorbed by the emulsion and the light emitted from the screens. Some loss in definition always occurs due to the grain size of the excited chemical.

X-rays are capable of traversing matter, and are therefore used for the study of the interior of materials and living beings, e.g. for the detection of internal cracks in metallic materials and fractures or diseased areas of the human body. In general a shadow picture is taken by passing a beam of X-rays through the object in question. Depending on the absorption, varying amounts of the radiation are transmitted and act on the sensitive material placed behind the object in question. Non-bony portions of the human body also often show differences in opacity to X-rays. For this reason, diseased conditions of the lungs can be detected. Other organs such as kidneys and blood vessels, which are not differentiated by X-rays, can be made opaque to them by injecting or ingesting a radiation-absorbing substance into the body cavities: for example, a barium compound drunk by the patient will reveal the digestive tract. In industrial radiography the images may be formed by X-rays or by gamma-rays.

e. Autoradiography. Autoradiography is the photography of details of tissues which have absorbed radio-active isotopes. For this purpose, stripping films are used. These are films in which a thin layer of collodion has been inserted between a fine grain emulsion and the base. The emulsion layer can be stripped off, and the flexible and very thin emulsion layer thus obtained can be carefully placed in contact with the surface of the object being studied. In due course, the radio-activity affects the halide grains and the image can be developed *in situ* to reveal the distribution of the absorbed radio-active element in the specimen.

This method is used, for example, with very thin tissue sections on microscope slides, so that the photographic image produced can be examined under a microscope, to determine the distribution of the radio-active element.

9.4 DICHROMATE-COLLOID MATERIALS AND PROCESSES

The light-sensitivity of dichromated colloids arises from the fact that certain colloids, such as albumen, fish glue, gelatin, gum arabic and polyvinyl alcohol are tanned by the decomposition products of the chromate or dichromate ion. When an image is projected onto a layer of dichromated colloid, the latter is tanned at the exposed area.

The extent to which this takes place increases with the light intensity, and the colloid is thereby made correspondingly less 'sensitive' to water. By treatment with water, the unexposed colloid can be caused to swell or even dissolved away.

Swollen portions have the property of repelling greasy inks but absorbing aqueous dyes; the tanned portions, on the other hand accept greasy inks but repel aqueous substances.

When the unexposed portions are removed with water, it is possible to etch the corresponding area of the base material and the result is a relief image. Such a relief image can, for example, be prepared by using a mixture of a dichromate and saponified shellac; in the presence of light the latter is converted into a substance which is less soluble in ethyl alcohol.

Dichromated-colloid layers must be prepared by the user as the colloid is also slowly tanned in the dark. Despite their short keeping life, dichromated colloids are widely used in various photomechanical printing processes (i.e. mechanical printing of pictures by means of a photographically prepared printing plate). Some examples of the photographic manufacture of printing plates are given below.

9.4.1 Intaglio printing processes

In the intaglio process the image in the printing plate or printing cylinder consists of recesses in the surface. To print from such a plate, this is first inked uniformly, and the surface is then wiped clean with a knife. Paper is then pressed onto the surface of the plate when some, or most, of the ink transfers to the paper.

Most modern intaglio ('gravure') printing is done from cylinders ('roto gravure'). The starting materials are a paper coated with a layer of pigmented gelatin, and a coppered steel cylinder or a polished copper plate which may be wrapped around a cylinder. The paper is made light-sensitive by means of a solution of potassium dichromate, dried and exposed to light through a transparent positive copy of the picture to be printed. The exposed paper is moistened and the exposed coating is simultaneously laminated to the polished copper surface. The gelatin sticks to the copper surface and, by treatment with warm water the paper can be removed. Depending on the intensity of the light during exposure, varying amounts of gelatin are removed by washing with warm water, so that a picture in relief is obtained. (Fig. 9.18)

The sequence of operations: exposure on paper, transfer to the copper surface, removal of paper and development of the gelatin layer on the metal surface, is logical if it is borne in mind that the gelatin is tanned (imagewise) from the surface downwards. It is thus from the other side of the layer alone that the untanned parts can be dissolved. If the exposed side of the gelatin layer were developed, the tanned layer would be detached and the half-tones lost.

After development, the relief picture is dehydrated with alcohol and dried. Etching is then carried out with a ferric chloride solution. This slowly penetrates through the gelatin layer and dissolves copper when it reaches the copper surface. Depending on the thickness of the gelatin layer, the etching liquid takes a shorter or longer time to reach the copper and etches for a correspondingly shorter or longer time. The result is a pattern of recesses with different depth, corresponding to the gelatin relief image. After removal of the gelatin the cylinder or plate is 'proofed' and if necessary chromium plated for wear resistance.

Ink is applied to the plate on the cylinder by means of a roller (plate) or by rotating the cylinder in an ink trough. At the same time a wiping device (doctor blade) removes excess ink. The ink consists of a pigment and a resin binder in a volatile liquid. Paper or textile brought into contact with the printing plate soaks up the ink out of the etched recesses and a copy of the original picture results. Since an ink with a rather low pigment concentration is used, various shades of grey or colour are reproduced by differences in the thickness of the ink layer. This thickness depends on the depth of the recesses in the printing plate.

9.4.2 Screens in intaglio printing plate making

Without particular precautions, the intaglio printing process, mentioned above, can only be used for the printing of line pictures. If a printing plate is made from a continuous tone picture, the result is a plate provided with both very small and very large recesses of varying depths. These recesses cannot be filled to the same level with ink and in addition the ink from one area will spread to neighbouring areas of the etched surface and the wrong shades will be produced. To avoid this, the recessed image in the copper surface must be broken up into small compartments ('cells') of near constant area, but varying depth.

The oldest photogravure method comprises the use of a flat copper plate which is dusted with bitumen powder. When this plate is etched, after applying and developing the exposed light-sensitive layer, the copper surface is not affected under the dust particles. The result is a relief of varying depth which is interrupted by a plurality of thin pillars. This method is still used occasionally for the production of short run plates for flat bed printing.

In modern rotogravure processes use is made of a printing cylinder and the relief in the cylinder is broken up in another way. In these processes the light-sensitive paper is pre-exposed to light through a screen, which usually consists of a black glass plate or black film provided with a pattern of evenly spaced transparent lines in two directions at an angle of 90° to each other (crossline screen). The pattern usually contains 60 to 80 lines per cm. The width of the transparent parts of the plate or film is about one-third of that of the opaque parts.

The pre-exposed light-sensitive paper is exposed image-wise and is then wet-laminated to the copper plate, as described in 9.4.1. After development and etching an image consisting of a regular array of small square cells of varying depth from about 3 μm to about 50 μm results in the copper surface of the plate or cylinder.

9.4.3 Letterpress printing plates

In letterpress printing (or relief printing) the ink image is formed from the un-

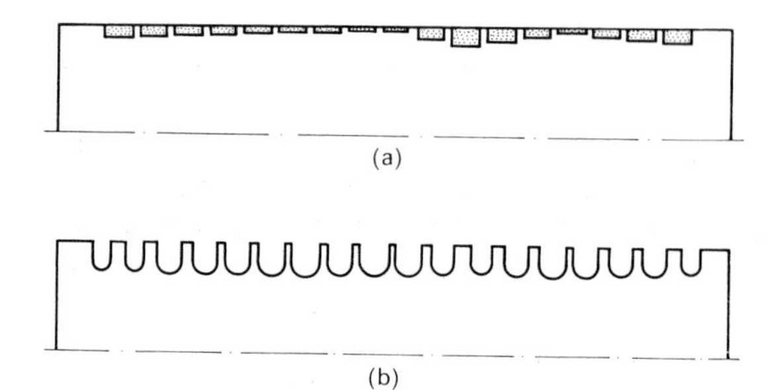

Fig. 9.18 Cross-section of (a) inked intaglio and (b) un-inked letterpress printing plates.

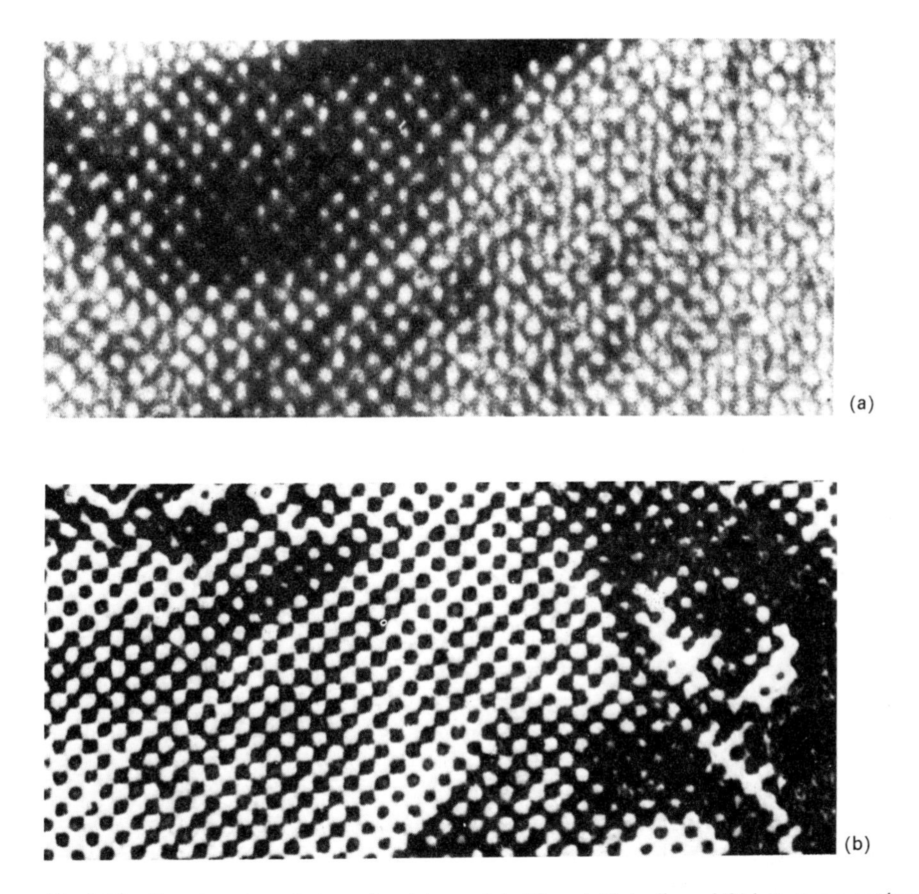

Fig. 9.19 Greatly enlarged view of a picture printed from (a) intaglio and (b) letterpress or planographic printing plates.

etched parts of the plate, which are all at the same level. The exposure of a letterpress printing plate must therefore be made through a transparent negative, and not through the positive of the original as in intaglio printing.

The starting material is usually a copper, zinc or magnesium plate, covered with a light-sensitive layer of fish glue or saponified shellac with ammonium dichromate. If fish glue is used, the exposed plate is developed with a powerful jet of water which removes the unexposed fish glue. After drying, the developed image is heated to about 250°C. The exposed fish glue is thereby converted into a glassy layer which is resistant to etching liquids.

Until recently etching of zinc or magnesium was carried out with dilute nitric acid. To avoid undercutting the side walls of the portions standing in relief, etching with this acid must be performed in a number of stages. Between the stages, a layer of grease (or ink) is applied on the top and sides of the relief image, by means of a roller. The greased plate is sprinkled with a powdered resin or asphalt which only adheres to the grease-covered areas. By heating, the powder is melted to cover the tops and sides of the relief image with a homogeneous resin layer which prevents etching of the tops and sides.

This multi-stage etching process has now been superseded by a one-stage 'powderless' etching process in which etching of zinc or magnesium plates is performed with dilute nitric acid to which have been added certain organic substances consisting of a mixture of a water-soluble detergent and a liquid which is immiscible with water e.g. a mixture of sulphonated hydrocarbons and diethyl benzene. This etching liquid, if applied by means of suitably designed equipment etches downward only and not sideways. It is believed that the organic substances present in the etching liquid are selectively deposited on the sides of the etched relief image.

When etching is complete the fish glue is removed and the plate is ready for use (see Fig. 9.18). It is mounted on a block to conform to standard thickness, and is inked up by means of a roller which applies a stiff ink based on pigmented polymerized linseed oil. Care needs to be taken that only the flat plateaus of the relief image are inked. Paper is then pressed on to the inked surface (by means of a platen) and the paper is removed. The ink film splits between the paper and the block. The ink on the paper is allowed to dry by oxidation in the air. For a description of the formation of 'half-tones' in letterpress see 9.4.5.

9.4.4 Planographic printing plates

The planographic (or lithographic) printing plate is essentially flat and the printing image is formed by water-repellent areas, while the non-image areas are the remaining water-attracting surface which is wetted before and during printing. The printing ink which is basically composed of a concentrated dispersion of a pigment in a viscous oil, adheres well to the water-repellent areas but is repelled by the wet, water-attractive background. The ink is applied by means of rollers and can be transferred to paper by contacting with the latter ('direct lithography'). However, the ink is usually not transferred directly to the paper, but is 'offset' onto a cylinder covered with a rubber 'blanket', which, in turn, transfers the ink image to the paper (offset lithography). Various types of planographic printing plates can be made with bichromated colloids.

a. Surface plates sensitized with dichromated albumin. The dichromated albumin plate is prepared by applying a light-sensitive layer of a dichromate and albumin to a fine-grain zinc or aluminium plate. After exposure through a negative original the unexposed parts are removed and the clean metallic areas are made water-accepting by treating with solutions of water-attractive substances (e.g. solutions of phosphates and gum arabic). Ink applied to the plate only adheres to the exposed albumin parts. The dichromated albumin plate has never been succesful since only a relative small number of prints (up to 1 000) can be made with this plate due to the low wear resistance of the exposed albumin. Nowadays diazocompounds and photopolymers are used for the production of surface plates (see pages 659 and 664).

b. The deep-etch plate. The deep-etch plate is a very important planographic printing plate which is made by a reversal process in which, after development, the unexposed areas are made ink-accepting by after treatment. The result is a positive working plate, i.e. a plate for positive prints can be made after exposure through a positive original (as opposed to the dichromated albumin plate which must be exposed under a negative). The deep-etch plate is prepared by coating a fine grained zinc or aluminium plate with a light-sensitive layer of a dichromate and a water-attractive (hydrophilic) colloid such as gum arabic or polyvinyl alcohol. After exposure through a positive original the plate is developed to remove the unexposed parts of the light-sensitive colloid and the shiny parts of the metal are weakly etched until the fine grains are just removed. Development is usually performed by means of a solution of lactic acid and calcium chloride. Etching is effected by treating with a weakly acidic solution of common salt. Development and etching may also be performed in a single stage. After washing and drying, the plate is treated with a lacquer (e.g. a solution of a phenol-formaldehyde resin) which adheres well to the shiny, smooth parts of the metal, but does not adhere to the tanned colloid. To improve the adhesion of the ink, used in the last stage of the process, the lacquer is treated with a greasy substance (the grease base). The tanned colloid is then softened with dilute sulphuric acid and removed by means of a brush. The metal thus exposed is made water-attracting (hydrophilic) by treating with a solution of acid salts (such as diammonium phosphate) followed by treating with a solution of gum arabic. Ink applied on to the plate only adheres to the grease base on the lacquer but is repelled by the wet metallic portions. Since the image is formed by the wear resistant lacquered parts of the plate, long runs (up to about 50000 prints) can be made with this deep-etch plate.

If anodized aluminium is used for the production of the deep-etch plate, about 100000 prints can be made, and if the plate is not lacquered after etching but chemically treated with a solution containing copper salts to deposit metallic copper on the image areas, about 200 000 prints are possible. The coppered deep-etch plate is mainly used in the USA. In Europe the multimetallic plate, described below, is used for very long runs.

c. The multimetallic plate. Like the deep-etch plate the multimetallic plate is a very important planographic printing plate. The most common multimetallic plates have a copper layer which has been coated with a chromium top layer. The copper layer may be a self-supporting copper plate (bimetallic plate) or may be a thin copper layer supported by a steel plate (trimetallic plate). The multimetallic plate is coated with a light-sensitive layer of a dichromate and a colloid such as gelatin or polyvinylalcohol. The light-sensitive layer is exposed under a positive original and developed to remove the unexposed parts. The resulting plate is treated with a weakly acidic solution of a salt to remove the chromium layer from the bare metal parts. (The acid etches the chromium and the salt prevents attack of the image formed by the exposed parts of the colloid). When the chromium is completely removed from the parts which are not covered by the colloid, the latter is removed too and after carefully cleaning the result is a planographic printing plate provided with a negative image of chromium on a background of copper. Since the chromium parts selectively accept water and the copper parts accept printing ink, positive prints are obtained on printing.

Like intaglio and letterpress printing plates, multimetallic plates are used if long runs (50000 prints or more) are required. Up to about one million prints (and even more) can be produced with the multimetallic plate.

9.4.5 Screens in letterpress and planographic printing

The letterpress and planographic plates imaged as described above can be used only to print text and line drawings. The various shades of grey of a continuous tone picture cannot be reproduced as such as the image-forming parts of the plate all lie in the same plane and consequently can transfer only as ink layers of constant thickness. Various shades of grey can only be produced by breaking up the image into dots in such a way that the number of dots per unit area, the 'screen ruling', remains constant but the diameter of the dots varies with the shade of grey. As shown in the previous figures large dots form a darker grey than small dots with the same screen ruling.

The image is broken up by means of a screen which consists not of sharply defined elements, as in intaglio printing, but of unsharp (usually chess-board shaped) dots which steadily become fainter toward their edges. The screens usually have 24 to 80 lines (grains per cm) depending on the type of paper on to which printing is to take place. Such a screen (contact-screen) is used in contact with a 'lith film' (a photographic film with a very hard gradation) when the latter is exposed to or through a continuous-tone image, which may be a reflection original or a transparent positive or negative. The result from the combination of the fuzzy screen and a sharp continuous tone image, is an image composed of dots of which the diameter depends on the brightness of the original picture. (Such an image is named a half-tone). Alternatively, the continuous-tone image can be photographed with a camera in which a sharply drawn screen is placed at a carefully calculated distance between the lens and film. Together with the sharp image of the original picture, a fuzzy shadow of the screen is projected on to the film. The processed dot image is then contact-exposed on to the light-sensitive coating of the printing plate (Fig. 9.20).

9.4.6 Photogelatin printing or collotype printing plates

Photogelatin printing is a special form of planographic printing in which half-tones can be printed without a screen, because a screen is formed by the surface structure of the gelatin.

In this process, a light-sensitive layer of gelatin and potassium or ammonium dichromate is poured on to a glass or synthetic resin plate, dried at relatively high temperatures, and then cooled under carefully controlled conditions. The result is a grainy surface layer. The grains stick out slightly above the surface of the plate, and are all harder at the centre than at the edges. On exposure to light through a continuous-tone negative, and after washing in cold water and drying, the plate is treated with aqueous glycerol so that the unexposed parts are made to swell. Unswollen grains of varying diameter remain on the exposed areas. The varying diameters of the grains are caused by the fact that the grains are to some extent affected at their edges by the glycerol solution, depending on the intensity of the light to which they have been exposed. Since, as in planographic printing, a greasy ink is used, which adheres only to the unswollen dots, the picture obtained has very fine grain structure.

The half-tones are thus formed by the varying diameter of the grains in the gelatin

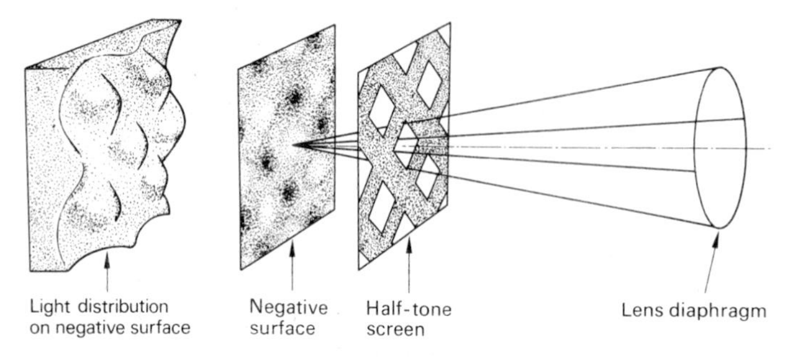

Fig. 9.20 Principle of half-tone dot reproduction.

layer. Since the grains are very small, pictures of very good quality are obtained, which reproduce accurately the various shades of grey. The difficulty with this process is the fragility of the image on the plate, so that only a limited number of prints can be made (about 1 000).

9.4.7 Photomechanically prepared stencils for screen printing

A light-sensitive layer prepared from a dichromate and gelatin or glue can also be applied to a tissue of fine yarns or fine copper wire. After exposing to light under a drawing the gelatin or glue is removed from the unexposed parts by washing with water and the result is a pattern of small holes. Ink is pressed through the small holes and a corresponding pattern of ink is produced on a sheet when it is contacted with the reverse side of the tissue. This printing process is named silk screen printing, screen printing or stencil printing.

The silk screen process is also used in the large-scale production of printed circuits metal, textiles, plastics and paper. If a thick ink or paint is used, relief pictures can be obtained on various materials. This relief picture process is used for letter-heads, greeting cards, art prints etc.

The silk screen process is also used in the large scale production of printed circuits for electronic devices. In one process, a light-sensitive stencil is exposed through a negative of a drawing and after development it is used to print a plastic pattern on to a very thin copper layer on a temporary support. The resulting positive picture of unprotected copper is thickened with copper by electroplating. The picture side of the plate is then stuck on to an electrically insulating plate and the temporary support is removed. Finally the very thin copper layer is removed (Fig. 9.21).

9.5 OTHER NON-SILVER HALIDE MATERIALS AND PROCESSES

9.5.1 Blueprints

The production of blueprints of engineering drawings is based on the fact that, in the presence of organic substances, ferric salts are converted into ferrous salts when exposed to light.

In the blueprint process, use is made of paper which has been coated with ferric ammonium citrate and potassium ferricyanide. On exposure under a drawing on translucent paper the ferric ammonium citrate is converted into a ferrous compound and the latter forms a blue dye (Prussian blue) with the ferricyanide. Under the black lines of the drawing the ferric ammonium citrate is not converted and can be removed by washing with water. The resulting picture consists of white lines on a blue background.

The blue print process is still used occasionally for copying drawings but it has largely been superseded by processes based on diazonium salts.

9.5.2 Diazotype materials and processes

Diazotype materials and processes are based on two properties of diazonium salts. Firstly, diazonium salts decompose under the influence of light and, secondly, they form dyestuffs by reacting (coupling) with various aromatic compounds. Thus, if a diazotype material is exposed to light through a translucent line diagram, the diaz-

Fig. 9.21 Stage in the production of printed circuits. Plates with electrolytically thickened drawings are being withdrawn from an electroplating bath.

onium salt under the image parts is not decomposed and is able to react with aromatic amines and phenols.

Diazonium salts can only be used in diazotype materials if they satisfy the following conditions. They must be stable at normal temperatures, they must have a high light-sensitivity, and their coupling products must be light-fast dark dyes. In addition the diazonium salts must form light-decomposition products which are colourless and remain colourless under atmospheric conditions. These diazonium salts are sensitive to blue and ultraviolet light (from about 300–450 mμ).

The coupling compounds are usually used in the presence of a buffer to ensure the desired activity and to take up the acid formed in the coupling reaction.

a. Diazotype papers and films. Papers, specially prepared textiles and film bases such as cellulose acetate films are used as the support for the diazonium salts. Papers are usually coated with an aqueous solution of the diazonium salt without the addition of a binding agent. However non-porous plastic films must be coated with a solution containing a binding agent (usually a polymer) in addition to the diazonium salt, otherwise the salt would not adhere to the surface.

The coating equipment used for coating diazotype material is similar to that used in the coating of silver halide emulsions (see page 629).

Basically there are two types of diazotype materials, the two-component material and the one-component material.

In the *two-component* diazotype system both the diazonium salt and the coupling compound are present in the light-sensitive material. The layer is kept slightly acidic with a non-volatile acid such as tartaric acid or boric acid to prevent the reaction between the diazonium salt and the coupling compound. The acid medium also stabilizes the diazonium salt. The latter is of a slow working type which does not couple under acid conditions. After exposure, the unexposed areas are developed with ammonia. The layer then becomes alkaline and at the area where the diazonium salt has not been decomposed by light a dye is formed.

In the *one-component* diazotype system, the diazonium salt alone, with a weak acid as a stabilizer, is present in the light-sensitive layer and the developer solution contains the coupling compound in an alkaline medium (e.g. a borax buffer solution). In this process, rapid working diazonium salts, which couple even in weakly acid media, can be used. This makes it possible to prepare acid developer solutions which remain stable for longer periods than the alkaline solutions. The developer is applied to the exposed paper or film by means of grooved rollers, in such a way that some developer is also applied to the back, so that the material does not curl up. The grooved rollers make it possible to minimize the application of developer solution, in order to keep the light-sensitive material as dry as possible. Although the dimensional stability in this 'moist' process is not as good as in the dry two-component process, it is much better than in the wet blueprint process.

b. Screened diazotype foil. Common diazotype papers and foils must be used with translucent negatives and therefore cannot be used for copying of pictures printed on heavy paper or paper which contains a picture on both sides. In these cases exposure by the reflex method or by means of a lens would be the alternative methods. Unfortunately the reflex method could not be applied to normal diazotype materials since the light passing through the light-sensitive layer would decompose all of the diazonium salt. Exposure by means of a lens forming a projected image would also be unpractical because of the very long exposure time required.

Shortly after the Second World War the problem was solved by means of a transparent diazotype material provided with a very fine crossline screen on the light-sensitive layer (the screened diazotype foil or reflex screen foil). The screen was made up of black intersecting lines (about 120 per cm) on a transparent base and covered about 75% of the screen area. Copying involved contacting the support of the diazotype foil with the original picture and exposing through the screen. The light passed through the screen and the light-sensitive layer and was reflected by the picture. At the places where the screen allowed light to pass through, the diazonium salt was decomposed. At the places where the direct light was kept back by the black lines of the screen, the diazonium salt could only be decomposed by the light reflected from the original picture. At these places, therefore, the amount of light and thus the amount of diazonium

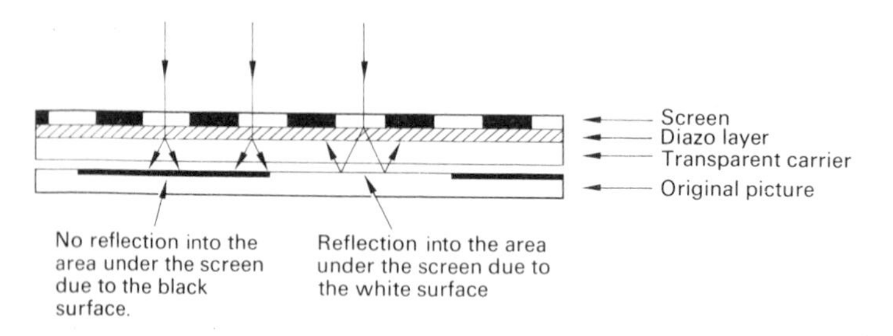

Fig. 9.22 The principle of printing on to screened diazo-type foil.

salt decomposed depended only on the brightness of the original picture. After removal of the screen and development a copy of the original was obtained, from which, by print-through exposure, copies could be made on diazotype paper.

The reflex-screen foil was used for copying continuous tone originals e.g. (pages from magazines etc.), but, for this purpose it has now been replaced by electrophotographic processes since the reflex-screen foil was expensive and needed long exposure times.

c. Diazotype printing foil. Diazotype printing foil is a modified diazotype paper which is suitable for the manufacture of a planographic printing plate. The diazonium salt is applied to a paper of good wet-strength which has been coated with a composition containing clay and a hydrophilic binding agent (e.g. a water-insoluble polyvinylalcohol). The diazonium salts selected for this foil form water-repelling and ink-accepting dyes with coupling agents such as phloroglucinol. Like the one-component diazotype paper the printing foil is exposed through a positive transparent original and developed with a solution of a coupling compound to form an ink-accepting dye on the unexposed parts of the foil. The exposed parts, consisting of the decomposition product of the diazonium salt and the hydrophilic clay-binding agent composition accept water on wetting. Up to about 200 prints can be made with such printing plates. Since the printing plate is very cheap and can be exposed under a cheap original (e.g. a typed image on a translucent paper) the small number of prints obtainable is not a disadvantage.

If the diazotype printing foil is made from paper which has been treated to smooth its surface and to prevent penetration of water runs of up to about 1 500 may be made.

9.5.3 Surface plates based on diazo compounds

a. Negative working surface plates. The oldest diazo printing plate of the surface type is composed of a water-soluble polycondensation product of formaldehyde and a 4-N-phenyl *p*-aminobenzenediazonium salt applied to a cellulose ester foil, the surface of which has been saponified. The solubility in water of the polycondensation product is caused by the ionic diazonium salt groups. On exposure, nitrogen is evolved and, as a result of this, the ionic diazonium salt groups are destroyed and the condensation product becomes non-ionic and thus insoluble in water. The saponified cellulose ester surface is hydrophilic and can be wetted with water but the exposed water-insoluble polycondensation product repels water and accepts printing ink. As the ink-accepting area corresponds to the exposed parts and thus to the bright parts of the original, a negative original must be used for exposure to produce positive prints.

In modified surface plates, of the type described above, the saponified cellulose ester film is supported by paper or plastic foils, or the saponified cellulose ester film is replaced by paper coated with a polyvinylalcohol-clay layer. Depending on the composition of the carrier or hydrophilic surface about 1 500 to 5 000 prints may be made with these surface plates.

A very succesful modification of the negative working surface plate consists basically of an aluminium plate coated with a water-soluble polycondensate of a diazonium salt. As diazonium salts are not stable in the presence of aluminium the latter is generally anodized or coated with sodium silicate, potassium zirconium fluoride, titanium compound, acrylic polymer or another suitable product before applying the diazonium salt. Like the planographic printing plates sensitized with bichromated colloids the aluminium must be fine grained in order to improve the wettability of the water accepting area. Without further treatment about 5 000 prints may be made with such a plate but runs up to about 50 000 may be made if the developed plate is treated with an oleophilic lacquer solution or emulsion which selectively adheres to the image parts. Solutions or emulsions containing phenol formaldehyde resins or epoxy resins may be used for this purpose.

b. Positive working surface plates. In addition to the short-run surface plate described in the preceding section (the diazotype printing foil), long-run surface plates, which can be exposed through a positive original, are also available. Basically these plates consist of an aluminium foil coated with a light-sensitive mixture of an ortho-diazo-oxonaphthalene derivative and a novolak. The ortho-diazo-oxonaphthalene derivatives and also their mixtures with novolak are insoluble in alkaline solutions, but on exposure they are converted into products which are soluble in such solutions. Runs up to about 50 000 can be made if the carrier consists of anodized aluminium.

Surface plates coated with positive or negative working diazo compounds have been very successful on account of their stability which makes it possible to produce pre-sensitized plates, i.e. plates coated with the light-sensitive material on an industrial scale, a relatively long time before use.

c. Surface plates by transfer process. All printing plates described up to now must be exposed through transparent or translucent original pictures. If the original picture is opaque, it is usually photographed on a transparent film to get a transparent image. A transparent image is not required if the transfer foil is used for the production of a planographic printing plate.

Transfer foil is composed of a screened cellulose ester film which has been coated with a light-sensitive mixture of a polycondensate of a diazonium salt and a colloid which has the property of becoming sticky on wetting with water but loses this property after exposure (e.g. gelatin or polyvinyl alcohol). The screen is applied between the light-sensitive layer and the cellulose ester film by embossing the film and filling the resulting grooves with a mixture of carbon black and a suitable binding agent.

The light-sensitive layer is contacted with the opaque positive picture and exposed through the back of the foil (compare exposure of the screened diazotype foil). After exposure, the foil and a transfer sheet are wetted, pressed together and pulled apart. The sticky unexposed portions together with the carbon black on the corresponding area is transferred to the transfer sheet to give a positive image on the latter. A negative image of exposed material remains on the foil.

The transfer sheet may be an aluminium foil or a paper with a hydrophilic surface.

About 1000 prints can be made when the paper is used and up to about 10000 prints with the aluminium foil if the image is fixed on to the foil by heating.

9.5.4 Photopolymers

Photopolymers are polymeric products or mixtures of monomeric and polymeric products which are cross-linked or polymerized to less soluble products by the action of light. Many photopolymer compositions have been patented but only three types are of commercial importance now.

a. Cinnamic ester resins. The cinnamic ester resins are cinnamic esters of polyvinylalcohol, cellulose or epoxy resins. When exposed to ultraviolet light the cinnamic ester groups of one molecule react with those of other molecules and thus the polymer is cross-linked. The unexposed portions of a layer of the photopolymer, which has been exposed through a transparent original, can be selectively dissolved in an organic solvent mixture and a negative image of ink-accepting polymer results. Because of their good resistance to acids these photopolymers are used in the production of bimetallic printing plates and printed circuits. Surface plates on anodized aluminium are also made with this product. Runs up to about 100 000 can be obtained with these plates due to the high wear resistance of the cinnamic ester resins.

b. Acrylic and methacrylic esters. Acrylic and methacrylic esters polymerize when exposed to light in the presence of a light-sensitive catalyst such as benzoin, acetophenone or anthraquinone. Since the monomeric products are liquid or soft solids they are usually mixed with polymers. Polymers containing free carboxylic acid groups, e.g. copolymers of acrylic acid or methacrylic acid and styrene or another vinylcompound, may be used for this purpose. After exposure, the unexposed portions can be dissolved in aqueous alkaline solutions containing an organic solvent which is miscible with water. Acrylic and methacrylic ester photopolymer compositions are used for the production of letterpress printing plates and negative working surface plates. Etching of the letterpress printing plate and removal of the exposed product is not required since a very strong relief image is formed by the polymerized areas.

c. Acrylamides and methacrylamides. Basically the light-sensitive composition consists of a mixture of a polyamide (nylon) and an acrylamide or methacrylamide such as N,N-methylene bisacrylamide. On exposure in the presence of a light-sensitive carbonyl compound such as benzoin the mixture polymerizes and cross-links to a product which is insoluble in alcohol solutions. The photopolymerisable amide-compositions are used for the production of letterpress printing plates. As with letterpress printing plates based on acrylic or methacrylic esters, printing is performed by the tops of the relief image formed by the exposed photopolymer. The exposed photopolymer is strong enough to enable runs in excess of 200000 impressions.

9.5.5 Thermographic materials and processes

According to a limited definition, a thermographic material is a material which is exposed to heat instead of light, but usually any copying material in which use is made of heat-sensitive products is named a thermographic material. The development of thermographic materials is based on the increasing need for a material which does not require a further treatment after exposure or can be treated without the use of liquid (especially for copying of letters or other documents in the office).

In one process use is made of a paper coated with a compound which, on heating, decomposes to form coloured decomposition products. Infrared radiation of short wavelength is used to expose the material by the reflex method. The image on the original picture, if it contains metal or carbon black, absorbs the infrared radiation and emits heat radiation of long wavelength which decomposes the heat-sensitive product.

In another thermographic process, the reflex method is applied to an oil-impregnated paper and oil is evaporated at the places where heat is generated. This oil is absorbed in a second sheet of paper and can be coloured by means of a pigment which only adheres to the oil-containing areas.

Since no heat is generated if the original picture does not contain carbon or metal, the thermographic processes mentioned above cannot be used for copying images which do not contain these products. Owing to this disadvantage, these processes were superseded by the electrophotographic processes.

A more successful thermographic process, which avoided this difficulty, is the 'Dual Spectrum' process. In this process an intermediate sheet is first exposed to light (in contact with the original) by the reflex method. This intermediate sheet has been impregnated with organic reducing agents which decompose at the areas corresponding with the white portions of the original picture. In the next step, the exposed intermediate sheet is placed in contact with a paper sensitized with an organic silver salt and irradiated with infrared. As a result of this, the reducing agents in the unexposed areas of the intermediate sheet distil across to the sensitized paper and convert the silver compound into metallic silver and a positive picture is obtained.

Another successful thermographic process, which is used for copying microfilms, is the vesicular process. The material used in the vesicular process comprises a transparent plastic foil in which a diazonium salt is incorporated. On exposure the diazonium salt releases nitrogen gas which coalesces to microscopic bubbles on heating the plastic foil. The result is an opaque image in the transparent foil.

9.5.6 *Electrophotography*

Electrophotography is based on the photoconductivity of certain materials, of which selenium and zinc oxide are the most important. These materials show a high electrical resistance in the dark and a low electrical resistance when exposed to light.

a. Indirect electrophotography. In the indirect electrophotography process (Xerography) use is made of an earthed aluminium plate which has been coated with photoconductive selenium. In the dark, the selenium coating is an electrical insulator and its surface can be given a positive electrostatic charge up to about 700 volts by moving the plate under a series of metal wires connected to an electric potential of about 7000 volts. The electrostatic charge is the result of ionization of air of which the positive ions move to the selenium-coating and the negative to the wires (corona discharge). The picture to be copied is then projected on to the charged surface by means of a lens, as in a camera. At the exposed areas the selenium coating becomes electrically conductive and the charge leaks away to the earthed metal plate. The unexposed areas remain electrostatically charged.

A powdered mixture of a pigment such as carbon black and a thermoplastic resin, negatively charged by friction against another material (e.g., resin coated beads or fur brushes), is then applied to the selenium surface and adheres only to the charged (unexposed) areas.

The negative powder image on the selenium surface is then transferred to a sheet of common writing paper by charging the latter by means of positive corona-discharge wires and placing the resulting positively charged paper over the powder image. The process is completed by heating the paper to a temperature at which the thermoplastic resin melts as a result of which the powder is fixed on the paper (see illustration below). Fully automated equipment for the process is available.

More than one copy can be obtained from one latent image on the selenium surface. The selenium surface can be re-charged and exposed many times and thus light-fast copies can be quickly obtained at low cost.

The indirect electrophotographic process is also used for the manufacture of planographic printing plates. In this case the sheet of transfer paper is replaced by a paper which has been coated with a hydrophilic layer which may be, for example, a polyvinyl alcohol-clay layer. The image produced on such a paper accepts printing ink while the hydrophilic layer repels the ink in the wet state.

b. The direct electrophotographic process. The direct electrophotographic process (Electrofox process) differs from the indirect electrophotographic process in that the photoconductive layer is applied not to a metallic plate but on the copy paper itself. In this process a paper coated with a layer of zinc oxide in a plastic binder is negatively charged by a corona discharge unit in the dark. On exposure the negative charge flows away from the exposed parts and the unexposed parts remain negatively charged. The electrostatic image is made visible by applying a positively charged powder consisting of a mixture of a thermoplastic resin and a pigment which adheres only to the charged areas. The powder image is fixed by heating.

The direct electrophotographic process is of growing importance in office copying of opaque originals and in planographic printing.

For planographic printing the copy on the zinc oxide paper may be converted into a

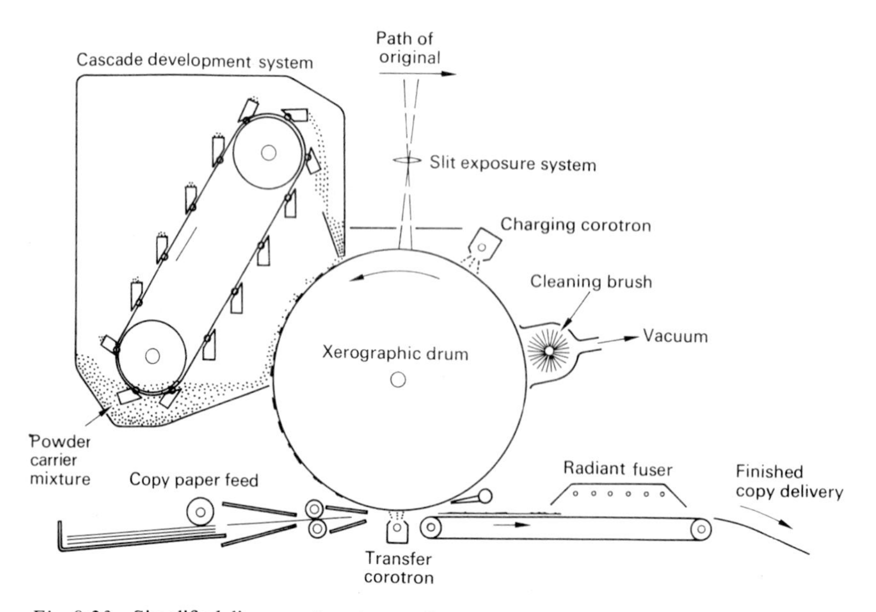

Fig. 9.23 Simplified diagram of equipment for xerographic copying of documents.

planographic printing plate by treating with an acidic solution of potassium ferrocyanide which covers the zinc oxide surface with hydrophilic zinc ferrocyanide. The powdered parts of the surface remain water-repellent and ink-accepting.

The direct electrophotographic process is also used for the production of metallic planographic printing plates. In this case a powder picture is produced on a metallic plate which has been coated with an organic photoconductor. The powder picture is fixed by heating and the remaining (unexposed) photoconductor is removed by means of a solvent. The result is a plate of which the powdered areas are ink-attractive and the bare metal areas are ink-repellent in the wet state.

9.6 LITERATURE

History

DR. W. BAIER. *Geschichte der Fotografie*, Halle, VEB Fotokino Verlag, 1964.

B. NEWHALL. *The history of photography*, New York, The Museum of Modern Art, 1964.

Photography in general

H. BAINES and E. S. BOMBACK. *The science of photography*, London, Fountain Press, 1967.

R. CROOME and F. G. CLEGG. *Photographic gelatin*, London, Focal Press, 1965.

H. FRIESER, G. HAASE and E. KLEIN. *Die Grundlagen der Photographischen Prozesse mit Silber-halogeniden* (3 Vol), Frankfurt am Main, Akademische Verlagsgesellschaft, 1968.

P. GLAFKIDES. *Photographic Chemistry*, London, Fountain Press, 1958–1960.

C. J. JACOBSON. *Developing*, London, Focal Press, 1966.

C. J. JACOBSON and L. A. MANNHEIM. *Enlarging*, London, Focal Press, 1967.

L. F. A. MASON. *Photographic processing chemistry*, London, Focal Press, 1966.

C. E. K. MEES and F. H. JAMES. *The theory of the photographic process*, New York, Macmillan, 1966.

E. MUTTER. *Kompendium der Photographie* (3 Vol), Berlin-Borsigwalde, Verlag für Radio-foto-kinotechnik GmbH., 1963.

C. B. NEBLETTE. *Photography, its materials and processes*. Princeton N.Y., Van Nostrand, 1966.

F. PURVES et al. *The Focal encylopedia of photography* (2 Vol), London, Focal Press, 1965.

ULLMANNS *Encyclopädie der technische Chemie* (Vol 13) p. 603–696: Photography. München, Urban and Schwarzenberg.

V. L. ZELIKMAN and S. M. LEVI. *Making and coating photographic emulsions*, London, Focal Press 1964.

T. H. JAMES and G. C. HIGGINS. *Fundamentals of Photographic Theory*, New York, Morgan and Morgan, 1960.

Colour photography

H. BERGER. *Agfacolor*, Wuppertal, W. Girardet, 1968.

J. H. COOTE. *Colour prints*, London, Focal Press, 1968.

E. S. BOMBACK. *Manual of colour photography*, London, Fountain Press, 1964.

R. W. G. HUNT. *The reproduction of colour*, London, Fountain Press, 1967.

E. MUTTER. *Farbphotographie, Theorie und Praxis.* (Handbuch der wissenschaftlichen und angewandten Photographie. Vol. 4) Wien, Springer-Verlag, 1967.

L. A. MANNHEIM and VISCOUNT HANWORTH, *D. A. Spencer's Colour Photography in Practice*, Focal Press, London and New York, 1966.

Infrared and ultraviolet photography

G. SPITZIG. *Grenzbereiche der Fotografie*, Seebruck am Chiemsee, Heering Verlag, 1968.

W. CLARK, *Photography by Infrared*, Chapman and Hall, 2nd Edition, 1946.

Reproduction and printing processes

O. R. CROY. *Camera copying and reproduction*, London, Focal Press, 1964.

H. BAUM. *Grundsätzliches und Wissenswertes vom Tiefdruck*, Leipzig, Verlag für Buch und Bibliothekswesen, 1960.

H. M. CARTWIGHT, Photo Engraving, *Ilford Graphic Arts Manual.* Vol. 1. Essex, Ilford Ltd, 1961.

J. H. DESSAUR and H. E. CLARK. *Xerography and related processes*, London, Focal Press, 1965.

M. S. DINABERG. *Photosensitive diazocompounds*, London, Focal Press, 1964.

W. R. HAWKEN. *Copying methods manual*, Chicago, American Library Association, 1966.

R. E. KIRK and D. F. OTHMER. *Encyclopedia of Chemical Technology*, Vol 16 (1968) p. 494–546 Printing processes, Vol 17 (1968) p. 328–378 Reprography, New York, Interscience.

J. KOSAR. *Light-sensitive systems*, New York, John Wiley, 1965.

A. H. SMITH. 'The use and application of synthetic coatings in Photo-Lithography', *Printing Technology* II (April 1967) pages 19–39.

R. M. SCHAFFERT. *Electrophotography*, London, Focal Press, 1965.

K. STOTZER. *Handbuch der Reproductionstechnik*, (4 Vol.). Frankfurt am Main, Polygraph Verlag, 1962–1967.

'The revolution in office copying'. *Chemical and Engineering News*. July 13, 1964 pages 114–131, July 20, 1964 pages 84–96.

H. R. VERRY, *Microcopying Methods*, Focal Press, London and New York, 1963.

Index